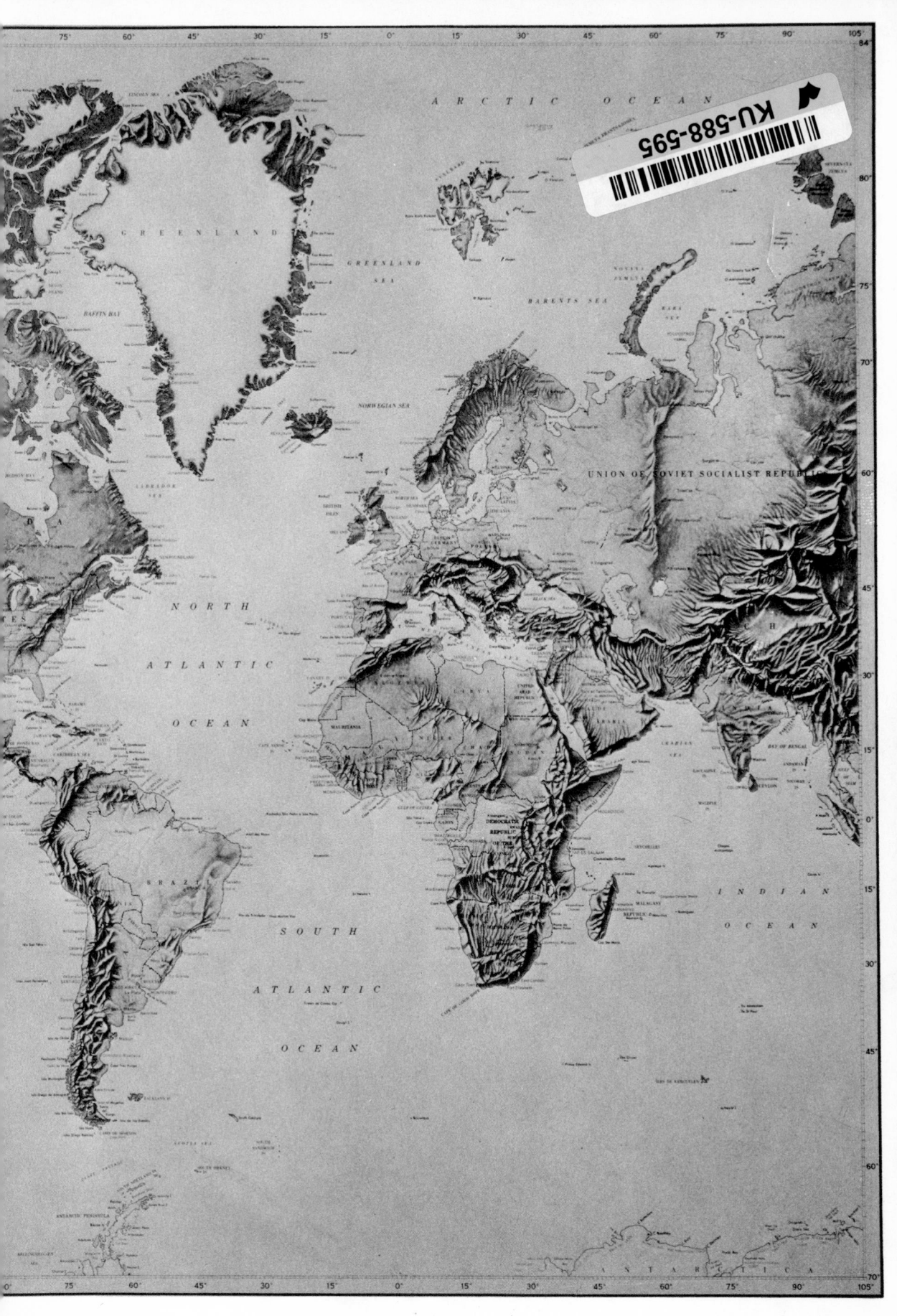

WORLD FACTS AND FIGURES

A Unique, Authoritative Collection of
Comparative Information About Cities,
Countries, and Geographic Features of
the World

VICTOR SHOWERS

A WILEY-INTERSCIENCE PUBLICATION

JOHN WILEY & SONS, New York • Chichester • Brisbane • Toronto

Library of Congress Cataloging in Publication Data:

Showers, Victor, 1910-
    World facts and figures.

    "A Wiley-Interscience publication."
    Rev. and enl. ed. of the author's The world
in figures, published 1973.
    Bibliography: p.
    Includes index.
    1. Geography—Tables, etc. 2. Cities and
towns—Statistics. I. Title.
G109.S52 1979   910'.21'2   78-14041
ISBN 0-471-04941-7

Printed in the United States of America

10 9 8 7 6 5 4 3 2 1

To THELMA

**whose encouragement and help
cannot be measured in English or metric units**

ACKNOWLEDGMENTS

A descriptive brochure recently issued by the Library of Congress notes that two-thirds of its 18 million books are printed in 470 different foreign languages; also that its cartographic collection alone numbers $3\frac{1}{2}$ million maps and atlases. The availability of this mass of material a few miles from my home—and in particular the innumerable local and national histories, guidebooks, and statistical reports included therein—made it not only possible but pleasurable for me to do the continual cross-checking required to verify most of the information contained in this book. In this endeavor I was ably assisted by many specialists in the Library, who freely gave of their time and expertise to help in locating all the important sources and to interpret difficult foreign-language passages.

To say that the Library of Congress was indispensable to the research that resulted in *World Facts and Figures* is not, however, to say that it was sufficient. I am, in fact, indebted to many other agencies and individuals both here and abroad. Much of the most recent population data, for example, was obtained either by writing to foreign census bureaus or by interviewing demographers of the US Bureau of the Census, and I want to thank both for their splendid cooperation in this regard. I also received valuable suggestions in the acquisition of worldwide climatic data from meteorologists of the US National Climatic Center at Asheville, North Carolina, and newly published statistics on Canadian geography from scientists of Environment Canada at Ottawa, Ontario. Both Asheville and Ottawa, incidentally, are delightful cities to visit.

For the most current information on foreign countries, I repeatedly called at a number of the embassies in Washington, the personnel of which were invariably cooperative. In this connection, it is only fair to note that some governments, such as Denmark, East and West Germany, the Netherlands, Norway, Sweden, and the UK, publish so much information that it is scarcely necessary to visit their embassies for that purpose. This said, I want to convey special thanks for their assistance to the Embassies of Australia, Brazil, Cameroon, Canada, Chile, Colombia, Egypt, Finland, France, India, Indonesia, Iran, Italy, Japan, Madagascar, Mexico, New Zealand, Pakistan, the Philippines, Portugal, Romania, South Korea, Spain, Switzerland, Turkey, the USSR, and Venezuela.

Among the many specialized collections in Washington area libraries other than the Library of Congress, I made the most profitable use of those in the Atmospheric Sciences and Bureau of the Census Libraries of the US Department of Commerce; the main libraries of the Departments of Interior and Transportation; the National Geographic Society Library; the Columbus Memorial Library of the Organization of American States; the United Nations Information Centre; and the public libraries of the District of Columbia, Fairfax County, and Alexandria, Virginia. In all of them, I found the librarians to be unstintingly helpful.

Finally, I want to extend my thanks to the many correspondents in this country and abroad who provided answers to specific questions that could not be found in published materials or who located and either gave or lent photographs for my book. I am especially grateful to the Central Bureau of Statistics of Indonesia, which provided from unpublished sources the name and latest census population of the largest settlement on each of 40 Indonesian islands and thereby made it possible to publish this important information for the first time.

VICTOR SHOWERS

1217 Shenandoah Road
Alexandria, Virginia 22308
January 1979

CONTENTS

ILLUSTRATIONS

World Facts and Figures

CURRENT TRENDS IN WORLD FACTS AND FIGURES

W*orld Facts and Figures* records the existence of 222 individual countries, as compared with 224 set forth in my earlier book, *The World in Figures*, which was published in 1973. This reflects the addition of four countries and the deletion of six. In addition, 13 countries are now listed under new names, seven of them in Africa alone[1]:

New Name of Country	Former Name
Belize	British Honduras
Benin	Dahomey
Central African Empire	Central African Republic
Djibouti	French Territory of the Afars and the Issas
Guinea-Bissau	Portuguese Guinea
Madagascar	Malagasy Republic
Namibia	South-West Africa
Papua New Guinea	Papua *and* New Guinea (Australia)
Samoa	Western Samoa
Solomon Islands	British Solomon Islands
Sri Lanka	Ceylon
Western Sahara	Spanish Sahara

Apart from new countries and new names, there have been many significant changes in world facts and figures during the past six years. The purpose of this summary is to highlight some of these changes and thereby to delineate the more important trends in the various fields covered herein that are subject to rapid change.

COUNTRIES, CENSUSES, AND WORLD POPULATION

Of the 222 countries cited above, only two—China and India—account for more than a third of the world's total population. Sixty-two percent of that total live in just 10 countries. Although one of these 10 countries—the USSR—occupies 15% of all land area, the combined areas of all 10 represent only 38% of the world's total land area. Excluding the

---

[1] Still another African country, Rhodesia, may become Zimbabwe before this book is published.

1

USSR, the 10 leading countries in population hold 57% of all the people on earth but only 25% of the land area.

It is not alone the more densely settled countries, however, that are concerned with problems of population. Increasingly, and in both industrialized and underdeveloped nations, there is concern that overpopulation may be calamitous. The poorer countries feel threatened with widespread malnutrition if not mass starvation. The richer ones foresee a deterioration in the quality of life that, given the relatively high educational level and aspirations of their peoples, might produce social and political disturbances of equal intensity.

One result of this growing concern has been a pronounced effort to ascertain the facts. Countries large and small are refining their birth and death records and evaluating the numbers of their inhabitants with increasing frequency, either by compulsory registration or national censuses. Whereas formerly even the most advanced countries generally conducted population censuses at intervals of 10 years, many now reduce the interval to five or even fewer years.

Table 10a of the Country Gazetteer gives striking evidence of the trend. The new censuses recorded in this gazetteer are those taken after 1971, the earlier ones having been listed in *The World in Figures*. No fewer than 87 such censuses are tabulated therein. By year and continent, they break down as follows:

National Censuses Recorded since 1971

Continent	1972	1973	1974	1975	1976	1977	Total
Africa	2	4	7	5	7	1	26
Asia	3	3	4	6	3	1	20
Europe				4	1	1	6
North America		2	4		1		7
Oceania	1	3	3	1	10	1	19
South America	3	1	2	2	1		9
Total	9	13	20	18	23	4	87

POPULATION GROWTH RATES

A digital clock, mounted above a busy intersection in Washington, D.C., and monitored by the Environmental Fund, purports to show present world population; the figures change 172 times/minute. Whether the excess of births over deaths is actually that high is questionable,[1] since demographers have recently detected a noticeable slowing down of the "world population clock," but the display is at least a dramatic demonstration of the population trend.

Another way of evaluating this trend is to consider annual growth rates. By mid-1976 the population of the world was approximately 4,025,000,000, as compared with about 3,600,000,000 (as estimated in *The World in Figures*) six years earlier. At an annual

---

[1] The biggest question mark is China, which has published no national census results since 1953.

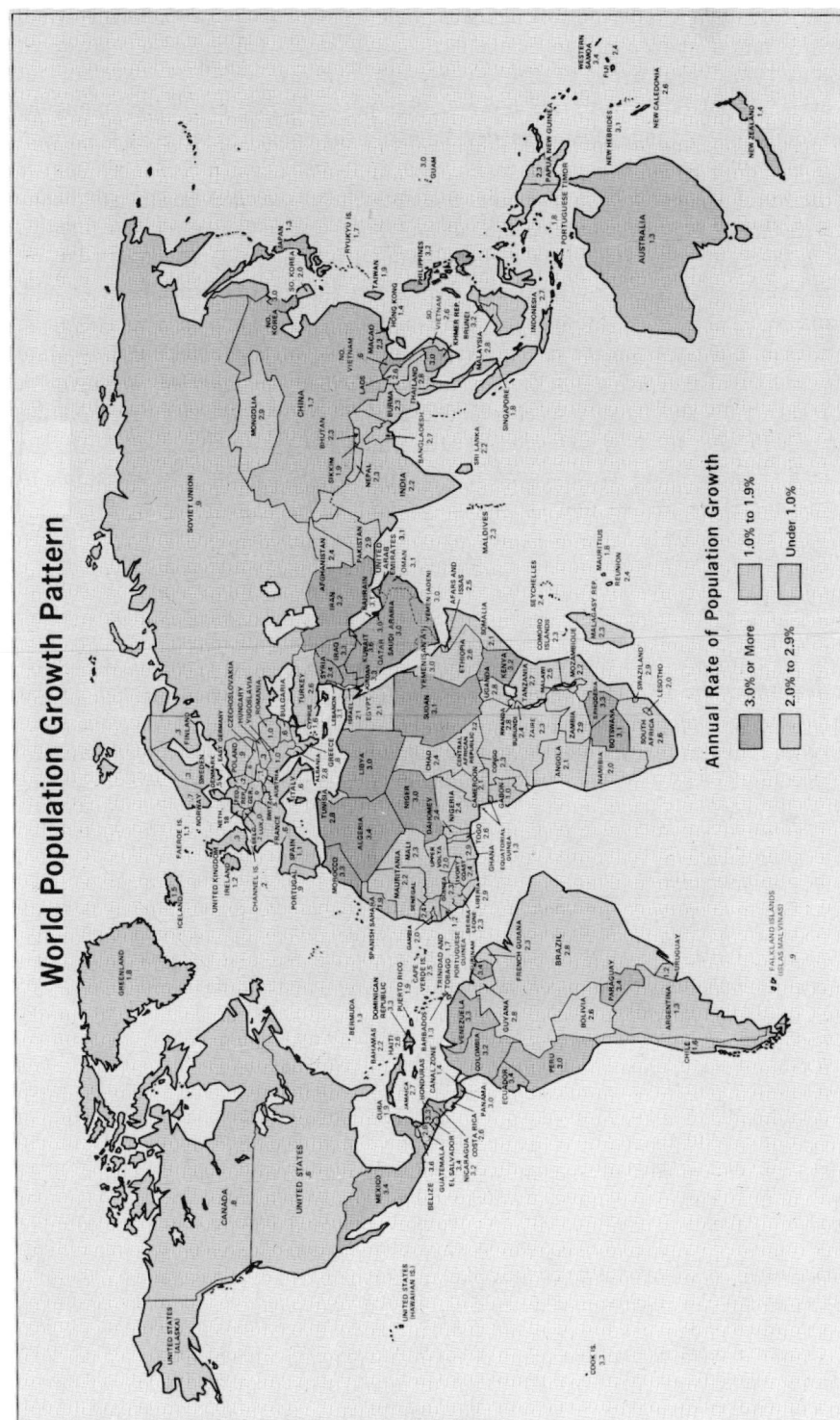

Figure 2. World population growth pattern. This map was published in 1974, but growth rates have not changed markedly since that time. (Credit: International Statistical Programs Center, US Bureau of the Census.)

3

growth rate of 1.9%, it is presumably increasing now by more than 75,000,000 a year.[1] If this growth rate were to continue for the next decade, world population would exceed 5 billion by 1988. Yet as late as 1950 it was only half that large.

The overall population growth rate, however, conceals greatly different rates among the various continents and individual countries. In Africa the annual growth rate from mid-1970 to mid-1976 was 3.3%, and in South America it was 2.2%. In Asia it was a little above the world average at 2.1%. The other three continents recorded growth rates below the overall figure. In North America the rate was 1.5%, and in Oceania it was 1.7%. In Europe, however, the annual increase amounted to only 0.7% between 1970 and 1976.

Of the seven largest countries, Brazil had the highest annual growth rate between mid-1970 and mid-1976—2.7%. Indonesia at 2.5% and India at 2.1% were also above the world average. But the estimated rate of increase for China was slightly below the average at 1.6%, and that for Japan was 1.3%. The USSR and USA were far below, with growth rates of only 1.0% and 0.8%, respectively. Of all large countries, Mexico may have the highest growth rate, for it is estimated to have grown at a yearly rate of 4.3%. On the other hand, East Germany is the only large country to have sustained an actual loss in population during the past six years.

BIRTH RATES

As might be expected from the data cited above, 19 of the 25 countries with the lowest birth rates during the 1971–1975 period were in Europe. The average rate for all 25 was 14.4/thousand of population, as compared with an average rate of 16.1 for the same 25 countries five years earlier. West Germany had the lowest rate of all—10.8/thousand—down from a rate of 15.7 in 1966–70. Many other countries with low birth rates exhibited a downward trend, as shown below:

Country	1966-70 Rate	1971-75 Rate
East Germany	14.5	11.5
Finland	15.4	13.1
Belgium	15.1	13.2
Austria	16.8	13.4
Sweden	14.5	13.5
Switzerland	17.1	13.7
UK	17.1	14.2
Denmark	15.9	14.6
Netherlands	18.9	14.9
Norway	17.4	15.5
USA	17.9	15.5
Canada	18.0	15.9
Italy	17.8	15.9
France	16.9	16.0
Australia	19.9	19.3

[1] By comparison, the 172/minute excess of births over deaths cited above would yield an annual increase of 90,463,000.

In a few countries, however, the birth rate increased between 1966–70 and 1971–75:

Country	1966-70 Rate	1971-75 Rate
Hungary	14.6	16.1
Bulgaria	16.0	16.2
USSR	17.4	17.9
Poland	16.4	18.0
Czechoslovakia	15.4	18.4
Japan	17.8	18.7

Very few of the countries with high birth rates keep accurate records, and it is impossible to establish a trend.

DEATH RATES

Of countries with a population of at least 5 million, Taiwan had the lowest death rate during the most recently reported five-year period, followed in order by Cuba, Japan, Malaysia, Venezuela, and Canada. These six countries showed a decline of 10% from the previously reported five-year period, as indicated below:

Country	1966-70 Rate	1971-75 Rate
Taiwan	5.3	4.7
Cuba	6.5	5.6
Japan	6.8	6.5
Malaysia	8.3e (1966–68)	6.6 (1970–74)
Venezuela	7.8e (1965–70)	7.0e (1970–75)
Canada	7.4	7.4
Average	7.0	6.3

A review of the 10 largest countries with comparatively low death rates shows a decline of 7% in the rate from the late 1960s to the early 1970s. Excluding China, however, the decline amounted to only 2%.

Country	1966-70 Rate	1971-75 Rate
China	15.3e (1965–70)	10.3e (1970–75)
USSR	7.8	8.7
USA	9.5	9.2
Japan	6.8	6.5
Brazil	9.5e (1965–70)	8.8e (1970–75)
Mexico	9.4	8.0
West Germany	11.9	11.9
Italy	9.8	9.8
UK	11.7	11.9
France	11.0	10.6
Average	10.27	9.57

As with birth rates, so few of the countries with high death rates keep accurate statistics that it is impossible to establish a trend.

LIFE EXPECTANCY

In the so-called developed world at least, life expectancy continues to inch upward, and women are increasing their already commanding lead over men, which now averages about six years. Norway and Sweden lead all other countries in this respect with an average expectancy of 74.9 years.

The latest statistics for the 25 countries with highest life expectancies show an average expectancy of 72.5 years for both sexes, 75.5 years for women and 69.4 for men. These 25 countries had an average expectancy of 71.0 years (73.7 for women, 68.2 for men), according to statistics published in *The World in Figures* six years ago. In the USA women now live, on the average, 76.5 years, or 7.8 years longer than men, and in the USSR, 74 years, or 10 years longer.

In the underdeveloped world, the data on life expectancy are less current and reliable, but estimates of from 35 to 45 years for both sexes are common. That improvement is taking place, however, is indicated by the documented increase in India, from 41.2 years in 1951–60 to 46.3 years in 1961–70. In both periods, incidentally, men lived, on the average, 1.5 years longer than women in India.

ENERGY PRODUCTION AND CONSUMPTION

When the various sources of energy, such as coal, petroleum, natural gas, and electricity, are converted to a common caloric value, it appears (from Table 7k) that the USA alone produces nearly a quarter of the world total and that three countries out of 222—the USA, USSR, and Saudi Arabia—produce approximately half. These three countries, which together produce 49.3% of the world's energy, have only 11.9% of its population. That is one measure of the energy problem.

Another measure lies in the rates of consumption. The USA and USSR alone, with less than 12% of the world's population, consume 47% of its total energy. On the other hand, China, India, Indonesia, Brazil, Nigeria, Bangladesh, and Pakistan together account for more than 47% of the world's population but consume only 10% of its energy.

Although it is the second largest consumer of energy, the USSR still produces more than it uses. But the USA, with the greatest consumption per capita of any large country, produces only 86.7% as much energy as it uses. And many other industrialized nations are far more energy deficient, as shown in the table below, where they are listed in inverse order of the deficiency:

Country	Ratio of Energy Production to Energy Consumption	Country	Ratio of Energy Production to Energy Consumption
Argentina	93.4%	Hungary	62.1%
Czechoslovakia	77.4%	West Germany	50.2%
East Germany	70.4%	South Korea	49.5%
UK	62.3%	Austria	39.2%

Country	Ratio of Energy Production to Energy Consumption	Country	Ratio of Energy Production to Energy Consumption
Bulgaria	36.0%	Sweden	16.9%
Greece	32.6%	Italy	16.1%
Taiwan	31.9%	Belgium	15.4%
Spain	26.7%	Japan	9.3%
France	22.7%	Finland	6.8%
Switzerland	21.3%	Denmark	0.9%

PRODUCTIVITY AND PERSONAL INCOME

In recent years, great emphasis has been placed on the gross national product (GNP) of various countries as a rough measure of their productivity or economic activity in monetary terms. When converted into a common currency, such as the US dollar, the GNP data can be used to compare one country's productivity with another's. In conjunction with population statistics, they also serve as an inexact measure of personal income.

For the latest year for which national account statistics are available, 1976, Table 7n indicates that the USA had a GNP more than twice that of any other country and almost equal to the combined GNP of the USSR, Japan, and West Germany, which ranked second, third, and fourth, respectively. All four had GNPs exceeding 400 billion US dollars. Below this level, however, productivity was sharply reduced. Only seven other nations—France, China, the UK, Canada, Italy, Brazil, and Spain—had a GNP of more than 100 billion dollars, and only eleven nations in all had one of between 50 and 100 billion dollars.

On a per capita basis, the USA ranked sixth, trailing three oil-producing nations in the Middle East and two industrial nations in Europe. The per capita ranking of the other members of the "big four" shows even more clearly the poor correlation between national productivity and personal income. Among 222 countries, the USSR ranked 44th in GNP per capita in 1976, while Japan ranked 27th and West Germany 9th. Conversely, the countries that ranked first, second, and third in per capita GNP—Kuwait, the United Arab Emirates, and Qatar—were not even among the 50 leaders in total GNP, and Switzerland and Sweden, which ranked fourth and fifth on a per capita basis, were 22nd and 16th, respectively, in the total grouping.

MEGALOPOLISES

Including their adjacent suburban areas, large cities, with but few exceptions, continued to grow during the period between 1970–71 and 1975–76. New York remained in first place, but Mexico, in eleventh place five years earlier, replaced Tokyo as the second largest. London, which was the most populous city for many decades and ranked eighth five years ago, dropped to eleventh place. The 10 largest urban areas, with 1970–71 rankings in parentheses, are now given, in order, as follows: New York (1), Mexico (11), Tokyo (2), Shanghai (3), Buenos Aires (5), Paris (7), Los Angeles (6), Moscow (9), Peking (4), and Sao Paulo (13). The listing is questionable, however, since there is no recent estimate for Calcutta, which ranked tenth in 1970–71 and is certainly growing faster than some of the cities now ranked above it; it was probably larger by 1976 than Moscow, Peking, or Sao Paulo. And two other cities, Seoul and Cairo, are also increasing rapidly in population; though not yet among the first 10, they may well join that select group of megalopolises by 1980. Altogether, 163 cities are now estimated to have a million or more inhabitants, as compared with 140 five years ago.

Considering only those populations within city limits—a less realistic approach—Shanghai remained the world's largest city, but Tokyo, even though it lost population between 1970 and 1975, replaced Peking as second largest. Again, but in this case partly through annexation, Mexico rose phenomenally, from 26th place in 1970-71 to third place in 1975-76. The first ten in this category, with 1970-71 rankings in parentheses, are, in order: Shanghai (1), Tokyo (3), Mexico (26), Peking (2), Moscow (6), New York (4), Sao Paulo (9), Seoul (8), Tientsin (13), and Bombay (7).

SKYSCRAPERS

Of the 20 highest buildings standing 40 years ago—buildings of 600 ft (183 m) or more—17 were in New York City and all but seven were completed between 1928 and 1931, including the two tallest, the Empire State and the Chrysler. The Empire State, at 1250 ft (381 m), retained its rank for 41 years until it was overtaken by the World Trade Center in 1972. But the Trade Center was only a symbol of a phenomenal skyscraper-building boom that began in the mid-1960s and spread throughout much of the world, transforming the skylines of cities as far apart as Chicago, Toronto, Houston, San Francisco, London, and Tokyo as radically as the 1928-31 boom altered the skyline of New York. The boom seems now to be slowly abating, but by 1977 it had produced at least 75 buildings that rose to a height of 600 ft or more. Of the 20 highest buildings standing today, exactly half are in New York City, and 11 of the 20 were completed between 1967 and 1977. The World Trade Center itself was superseded in 1974 by the Sears Tower of Chicago as the world's tallest skyscraper; it was topped off at 1454 ft (443 m). Incidentally, that marvel of 1913, New York's Woolworth Building, which rose to a height of 792 ft (241 m), is now in twentieth place.

BRIDGES

Although several new types of long-span bridges have been developed in recent years, all bridges with spans greater than 1800 ft (549 m) are still of the cable suspension type. Of these, the Verrazano-Narrows Bridge in New York has held the record at 4260 ft (1298 m) since 1964. It will soon be surpassed, however, by a bridge over the Humber River that is now nearing completion near Hull, England, with a designed span of 4626 ft (1410 m). An even longer span of 5840 ft (1780 m) is projected to cross the Akashi Strait as part of an ambitious plan to link the two Japanese islands of Honshu and Shikoku by means of 18 main bridges to intervening islands along three separate routes, the first of which is scheduled for completion by 1987. In addition, the fourth and tenth longest span bridges were both completed in 1973, the former across the Bosporus Strait at Istanbul, Turkey, joining the continents of Europe and Asia, and the latter across the Kammon Strait between the islands of Honshu and Kyushu in Japan.

Except for suspension bridges, those with the longest spans continue to be either cantilever or steel arch designs. Of the former, the third and fourth longest were completed in 1974 and 1973 at Osaka, Japan, and Chester, Pennsylvania, USA. The longest of the cantilevers is still the famous Quebec Bridge across the Saint Lawrence River, which was opened in 1917 after long delays and two fatal mishaps. Until quite recently, the two longest steel arch bridges were the Bayonne Bridge across Kill van Kull between New York and Bayonne, New Jersey, and the Sydney Harbour Bridge in New South Wales, Australia. Completed in 1931 and 1932, respectively, these two bridges have main spans of almost identical lengths. In 1977 they were finally surpassed by a bridge over the New River near Fayetteville, West Virginia, USA.

TUNNELS

The Japanese now seem to be monopolizing the construction of long railroad tunnels. The Seikan, which is being built to connect Hokkaido and Honshu Islands by a 33.5-mi (54-km) underwater crossing of Tsugaru Strait, will upon completion be nearly three times longer than any other such tunnel in the world. The original target date was 1979, but difficulties have been encountered that may delay its opening by as much as 4 years. The second longest railroad tunnel under construction, Shimizu III, will also surpass in length the famous Simplon Tunnels between Italy and Switzerland, which are the longest such tubes now open. And, ranking just below the Simplons, is the New Kammon, longest completed underwater tunnel, which was opened between Honshu and Kyushu Islands in 1974.

The two Simplon tubes are each 12.3 mi (19.8 km) in length. By comparison, the longest highway tunnel in use is the 7.25-mi (11.7-km) Mont-Blanc, which pierced that celebrated mountain between France and Italy in 1965. Now, however, two new transalpine road tunnels, each longer than the Mont-Blanc, are nearing completion. These are the 7.9-mi (12.7-km) Mont-Cenis Tunnel, also known as Frejus, which is directly south of Mont-Blanc and also on the Franco-Italian border, and the 10.1-mi (16.3-km) Gotthard, or Saint-Gotthard, Tunnel, which lies farther east in south-central Switzerland. It is interesting to note that these two record-setting highway tunnels, both scheduled to open in 1978, parallel the two oldest transalpine railroad tunnels, which bear the same names and were completed, respectively, in 1871 and 1882.

DAMS

Although in recent years dams have been much criticized on environmental grounds, there is no indication that the rate at which they are being built has been slowed. Only two of the 10 highest ones in place or under construction, and only eight of the 20 highest, were finished before 1974. In that year Switzerland's Grande Dixence Dam, the record holder since 1962, was surpassed in height by the Nurek, on the Vakhsh River in Tadzhikistan, USSR, and no sooner was the Nurek in place than a still higher dam was projected for the same river.

Fewer of the largest dams (in volume of structure) are of such recent origin, but even in this category seven of the 20 leaders, including those under construction, date from 1974 or later. Of these seven, two are in the USSR and one each in the Netherlands, Pakistan,

Syria, the USA, and Venezuela. The Fort Peck Dam on the Missouri River in Montana, USA, was the largest of all for 35 years, but in 1975 it finally yielded precedence to the Tarbela Dam on the Indus River in Pakistan.

UNIVERSITIES

The latest available data on university student enrollments indicate that the State University of New York is still the world's largest in this respect, with some 350,000 registered at all campuses in 1975-76 as compared with 314,000 five years earlier. On the other hand, the City University of New York jumped from seventh to second place in the list, with a total of 251,000 students, even though all but one of the institutions that outranked it five years earlier also augmented their enrollments. The Université de Paris, with 233,000 students in 1975-76 and 166,000 five years before, remained in third place, while the University of Calcutta dropped from second to fourth place despite a rise of 31,000 in enrollment.[1] The University of California, which ranked fifth just as it did in 1970-71, reported an increase of 35,000 and, at the latest count, had approximately 182,000 students, of whom 35,000 were at Berkeley and 60,000 at Los Angeles. The number of students registered at the presently 10 largest universities was 1,954,000 in 1975-76, as compared with 1,452,000 in 1970-71, a gain of 35%. Five years ago the hundred largest universities all had enrollments of 30,000 or more; now the same number of universities have at least 35,000 students.

LIBRARIES

Six years ago *The World in Figures* listed three of the five largest libraries as being in the USSR, but noted that the size of library collections in that country was often exaggerated because it was customary to report the number of items rather than the number of volumes, as was the general rule elsewhere. In 1974, however, comprehensive statistics became available on both the number of items and the number of volumes in libraries of the Soviet Union; so it is now possible to make valid comparisons of the size of library collections throughout the world. As indicated in Table 9g, the Library of Congress in Washington, with nearly 18,000,000 volumes at the latest count, has without question the largest collection of all. The British Library in London, which incorporates the British Museum Library but includes other collections, ranks second, with about 12,150,000 volumes, and the V. I. Lenin State Library in Moscow ranks third with 11,750,000. Next in order come the Harvard University Library at Cambridge, Massachusetts, and the New York Public Library in New York City. The M. E. Saltykov-Shchedrin State Public Library in Leningrad, which was listed in second place six years ago, is in fact the sixth largest library, followed by the Bibliothèque Nationale in Paris. The Biblioteca Academiei Republicii Socialiste Romania in Bucharest, the National Diet Library in Tokyo, and the Yale University Library at New Haven, Connecticut, USA, rank eighth, ninth, and tenth, respectively.

---

[1]University enrollments in India are not strictly comparable with other enrollments. See note 1, p. 292.

INTRODUCTION

World Facts and Figures is a digest of useful and up-to-date information about each of the 222 countries into which the world is now divided, over 2000 of its most important cities, and more than 2500 other geographic and cultural features. It is the only known work in any language that provides this information on a worldwide and strictly comparable basis.

As indicated in the table of contents, the book is divided into six main sections. The first of these covers physical features—seas, islands, rivers, mountains, lakes, and waterfalls. The second and third sections contain a number of valuable comparisons of countries and cities, respectively. Section four is devoted to outstanding cultural features, such as buildings, bridges, tunnels, dams, universities, and libraries.

The last two main sections of the book are unique tabular gazetteers, of countries and cities, respectively. In these, the countries are listed alphabetically by continent, and the cities are entered alphabetically by country. Following the six main sections are a selected bibliography, arranged by subject, and a cross-reference index of approximately 9000 names.

Each list or table in World Facts and Figures has been constructed to give as much practical information as is consonant with clarity. As an example, the principal table of rivers, listing 670 of the longest and most important, gives the length of the stream, the area of its drainage basin, its average discharge rate, its outflow (sea, lake, river, etc.) and the subdivision and country in which it is located, the latitude and longitude at its mouth, and its alternate and former names, if any. The basic table of islands, listing 414 of the largest and best-known, gives geographic coordinates, alternate and former names, principal body of water, subdivision and country in which located, area and highest elevation, and name and population of the largest settlement.

In addition to hundreds of the highest mountains and waterfalls, the geographic statistics cover every river on earth that is at least 500 mi (800 km) long, every natural lake of more than 300 mi^2 (780 km^2), and every island of at least 1000 mi^2 (2590 km^2), plus many smaller but well-known examples of each category.

In compiling the tables in this book, the latest official data have been used whenever they were available. When of necessity unofficial figures are given, they have been verified for authenticity to the greatest possible extent, but they are still identified as unofficial. Wherever appropriate, the date of the information supplied is given, and all information is revised to August 1978. Although each table is reasonably self-explanatory, its exact contents, arrangement, and coverage are described at the beginning of the table.

Because Arabic numerals are now universally employed, World Facts and Figures can be understood without difficulty by people in all countries. To increase its usefulness in countries that have adopted the metric system and to obviate the necessity of conversion, measurement data are given in both English and metric units.

Another important feature of the book is the provision of latitude and longitude (to the nearest minute) for every city and almost every physiographic feature listed. This has been done to facilitate their location on maps and to make possible a rapid comparison of geographic positions. Many of the features cited, and even a few of the cities, cannot be found in current world atlases, either because the maps in these atlases have too small a scale or because they are not sufficiently up-to-date, but supplied with latitude and longitude, the reader can still locate them with respect to better-known cities and features that are shown in these atlases.

No effort has been spared to make this digest as accurate and complete as possible. The Selected Bibliography cites some of the more important sources from which information has been derived. Arranged as it is by subject and including only the most useful publications in each field, this bibliography should prove most valuable to readers who seek more detailed information on any subject covered by *World Facts and Figures*. Yet, many of the figures supplied in this book are not found in any of the works cited in the Selected Bibliography. A considerable number of the dates of foundation for libraries here provided, for example, are not given in any of the standard library directories listed in the bibliography.

It is for this reason—to uncover more information—as well as for purposes of verification that innumerable books and periodicals of a more specialized nature—population census reports, guidebooks and histories of particular localities, meteorological and water supply bulletins, and civil engineering and architectural journals—have also been consulted in the preparation of this work. In addition, correspondence with governmental agencies here and abroad has yielded some basic data published neither in the official statistical yearbooks nor elsewhere.

World Facts and Figures gives striking evidence of a shrinking earth. Only a few years ago the compilation of an authoritative digest of comparative statistics of this scope would have been impossible. Now, with jet aircraft flying travelers to every corner of the globe and television flashing instant pictures of important events by satellite, there is no longer a *terra incognita* anywhere. Census-taking has become almost universal, meteorological data are collected in nearly every important community, and some of the most underdeveloped nations vie with the most "advanced" in the production of official statistical yearbooks.

Nor is it only that we know more about the remote regions of the world. We have simultaneously acquired far greater knowledge of our own and neighboring countries. Forty years ago the longest watercourse of the Mississippi River (from the source of the Red Rock to the Gulf of Mexico) was universally regarded as the longest river in the world; its length was officially stated to be 4221 mi, or 6793 km. By 1939, however, the US Geological Survey had reduced that figure to 3988 mi (6418 km), and 10 years later it was again reduced to 3872 mi (6231 km). Further revisions have decreased the length to 3741 mi (6021 km), and the Mississippi is now known to rank well behind the Nile and the Amazon as the third longest of the world's great rivers.

It will be noted that this work contains no statistics on agriculture, mining, manufacturing, or commerce. Since such statistics are subject to rapid change and would be misleading after a short lapse of time, their presentation here would be of questionable value. Moreover, official and adequate data in these fields are already published annually in the *Statistical Yearbook* of the United Nations (reference 84).

ABBREVIATIONS USED IN THIS WORK

AD	anno Domini (i.e., since the birth of Christ)
alt	alternate name (or names)
Apr	April
Aug	August
avg	average
B	bay
BC	before Christ (other dates are AD)
bef	before
bldg	building
c	century
C	Celsius (or centigrade)
ca	circa
cfs	cubic feet per second
ctry	country
Dec	December
e	estimated (other population data are from census returns)
E	east
ed	edition
Ed	editor
elev	elevation (i.e., altitude above sea level)
exc	excluding
F	Fahrenheit
Feb	February
for	former name (or names)
ft	feet
ft^2	square feet
G	gulf
GNP	gross national product
I	island
in	inches
inc	including
Jan	January
Jul	July
Jun	June
km	kilometers
km^2	square kilometers
L	lake
lag	lagoon
lib	library
m	meters
m^3	cubic meters
m^3/sec	cubic meters per second
Mar	March

mi	miles
mi^2	square miles
mm	millimeters
mt	mountain (peak or massif)
mts	mountains (range or system)
N	north
ND	no date
Nov	November
NP	no place
nr	near (i.e., within approximately 10 mi, or 16 km)
NWT	Northwest Territories
O	ocean
Oct	October
off	official name (or names)
p	page
pp	pages
R	river
s	with adjacent suburban areas
S	south
sec	seconds
Sep	September
UC	under construction
UK	United Kingdom of Great Britain and Northern Ireland
univ	university
unp	unpaged
US, USA	United States of America
USSR	Union of Soviet Socialist Republics
v	volume (or volumes)
volc	active volcano
W	west
wf	waterfall
yd^3	cubic yards
yr	years
°	degrees (of temperature)
●	unofficial data
*	capital (of a country or subdivision)
***	(used to divide tables into two parts, the portion above the division representing all-inclusive coverage and that below the division, selective coverage)

GENERAL EXPLANATORY NOTES

An explanation of the contents, arrangement, and coverage of each statistical table and tabular gazetteer in this work precedes the table or gazetteer in question. The notes in this section are of a more general nature or pertain to more than one table or gazetteer.

Spelling

The spelling of geographic names accords generally with that in the various gazetteers prepared by the US Office of Geography (see reference 129). Accent and other diacritical marks are omitted for simplicity.[1] Exception to this spelling rule is, of course, made when the name has been changed since publication of the relevant gazetteer and when, as evidenced by a consensus of later authorities, the spelling is manifestly outdated. The spelling of nongeographic names is that adopted officially or that approved by recognized authorities in the appropriate field. The names of dams, for example, are either given their official spelling or spelled in conformity with the *World Register of Dams* (reference 225).

Conventional and Other Names

Many cities, countries, and physiographic features have conventional names in English that differ from their official (sometimes called vernacular) names. The difference may be marked (as Albania for Shqiperi) or minimal (as Lyons for Lyon). In either case, the conventional English name is always given in this work, and the initial citation of the city, country, or physiographic feature is by this name if it is in general use. The official name or names, marked "off," are also given, however, in appropriate places in the tables and gazetteers, and if no "off" name appears in these places, it may be assumed that the conventional name is also an official one.

Two other sorts of names appear in the tables: alternate ("alt") names and former ("for") names. If no "off" name is listed, the "alt" name or names are other official names or (rarely) English names infrequently used. Often these other official names are names used in another country. The Danube River, flowing through eight countries in Europe, has no fewer than six official names, none of which is Danube, the conventional English name. Sometimes, however, there are two or three official names within a single country. Helsinki and Helsingfors are both official names for the capital of Finland, since both Finnish and Swedish are official languages in that country.

Former ("for") names are, as the term implies, names that were once in general use but have become obsolete. Ancient (i.e., classical) names are not given in this book, but names dating from more recent historical periods are entered unless the name in question was used for a very brief time.

The conventional and other names given for countries are their so-called short names, for example, Mexico (official spelling identical), not United Mexican States (off Estados

---

[1] Readers versed in foreign languages will inevitably notice the omission of these diacritical marks. Since their primary function, however, is to indicate proper pronunciation, with which this work is not concerned, their inclusion would be of limited value at best. On the other hand, the marks are so numerous and so frequent in occurrence that their inclusion would serve to clutter the book, confuse the English-speaking reader, and raise serious problems in alphabetizing.

Unidos Mexicanos). To save space and because the abbreviations are in common use, three large countries are identified by abbreviations, as follows:

UK, for United Kingdom of Great Britain and Northern Ireland
USA, for United States of America
USSR, for Union of Soviet Socialist Republics

In the citation of foreign physiographic features, that portion of the name signifying river, mountain (or mount), lake, or the like is usually deleted, and an abbreviation for the English-language equivalent is substituted for it, unless, of course, this equivalent is a tabular heading. Thus the river known in Brazil as Rio Negro is listed simply as Negro in Tables 3, 3b, and 3c and as Negro R (for River) in Table 11c. Similarly, the Chinese lake Poyang Hu is cited as Poyang or Poyang L.[1] When, however, the foreign designation for the physiographic feature is little known and especially when the particular feature is almost never cited in English-language works without its full name, even though this makes the citation redundant, that portion of the name signifying river, and so forth, is retained but is parenthesized; for example, (Tonle) Sap L, Kizil (Irmak) R. Portions of conventional geographic names that are redundant or misleading but are invariably used are also retained and parenthesized, for example, Dead (Sea). The Dead Sea is actually a saltwater lake.

It goes without saying that a cross-reference entry appears in the index for each unconventional name, whether it is an alternate, former, or official one.

Alphabetization

Some of the individual tables, as well as the index, are arranged alphabetically. Alphabetizing in this work is letter-by-letter, in accordance with the English alphabet, without regard for spaces, punctuation marks, or foreign combination letters. Thus Newark precedes New Britain, and Chile precedes Cienfuegos, even though the latter words are Spanish and in Spanish "ch" is a separate letter that follows the letter "c." Abbreviations are alphabetized as they are spelled: for example, USA falls among the "us's," not among the "un's." When, however, names begin with words like "Saint" that are often abbreviated in other books, these words are spelled out. And names beginning with Arabic numerals (e.g., 345 Park Avenue) are alphabetized as if the numerals were spelled out.

Translation and Transliteration

Since the spelling of geographic names is in accord with the pattern set by the gazetteers prepared by the US Office of Geography, it follows that the systems of transliteration adopted by that office have been employed in this work. This accounts, for instance, for the spelling "Gorkiy" in preference to "Gorki," "Gorkii," or "Gorky" in the names of dams, universities, and libraries. Personal names aside, however, the official designations of universities and libraries in countries that do not use the Roman alphabet are not transliterated. In line with the rule followed by many authorities in the field of education, these designations are translated into English, whereas the names of like institutions

---

[1] But note that foreign-language designations for river, and so on, used in countries where the language in question does not prevail, are not deleted. For example, Rio Grande R, not Grande R, designates the river in the southwestern USA.

in countries where the Roman alphabet prevails are left unchanged.[1] Thus the national library of Thailand is listed under Bangkok (in Table 11c) as "National Library," but the national library of France is entered under Paris as "Bibliotheque Nationale." When universities and libraries are commonly known by two names, as sometimes happens in countries with two official languages, both names are given. For example, under Bloemfontein, South Africa, there is listed, "University of the Orange Free State (alt Universiteit van die Oranje-Vrystaat)."

"University" Defined

For purposes of inclusion in and exclusion from this work a university is defined as an educational institution that, either of itself or in conjunction with constituent or affiliated institutions, offers undergraduate work leading to a bachelor's degree and graduate work leading to at least a master's degree. This broad definition of "university" includes some institutions that call themselves colleges, but it is felt that a more restrictive definition would be too exclusive. In 1970 there were in the United States only 298 universities that offered graduate work leading to a doctor's degree (universities in the narrow sense), whereas there were 826 that offered graduate work leading to at least a master's or a second professional degree (universities in the broad sense employed in this book). Because of the different standards that prevail in various countries, it is not always possible to apply this definition to the letter, but every effort has been made to adhere to it as closely as possible.

So-called colleges that meet the definition, as well as other institutions of university level that do not have the word "university" or its foreign-language equivalent in their names, are therefore listed in Table 11c, and also in Table 9f if they qualify by size. These institutions, however, are identified by the designation "univ" placed in parentheses after their names. On the other hand, a number of so-called universities offer no graduate work. These are also included in the listings if they are independent institutions, but they are identified by the designation "college" placed in parentheses after their names. Branch campuses of a university centered elsewhere are not listed, unless they are of university level in the sense already defined.

Rounding

Many of the statistical figures in this book are rounded. This means, for example, that a population of 151,743 is given as 152,000 (151,499 would, of course, become 151,000). Population data are invariably rounded to the nearest 1000. However, when the original figure is less than 1000, the rounding is done to the nearest 100. Other figures in the book may be rounded to the nearest 10 ft (m) (mountain elevations) or to the nearest three significant digits[2] (island areas). The extent of the rounding is described in the explanatory notes preceding each table and tabular gazetteer.

---

[1] This is not the case with capitalization, which for consistency and to avoid confusion follows English-language rules. In some foreign languages, proper adjectives are not capitalized, and there are many other differences.

[2] This means, for example, that the area of Madagascar, which is greater than 100,000 mi^2 and 100,000 km^2, is given as 227,000 mi^2, or 587,000 km^2; and that of Reunion, which is less than 1000 mi^2 and less than 10,000 km^2, is given as 969 mi^2, or 2510 km^2.

Although the rounding of all population figures to the nearest 1000 obviously saves print and space, since the word "thousands" can be inserted into the tabular heading and the figure printed as 152 instead of 152,000, the main purpose of rounding is to promote understanding—to make the statistical tables easier to read and the data they contain easier to comprehend. To one untrained in statistics, it might appear that rounding sacrifices accuracy, but this is not necessarily so. If done with discrimination, it may actually enhance statistical significance. To say that Mount Lucania is 17,147 ft high is really misleading, for mountain elevations cannot be determined with such accuracy. As for populations, they tend to grow so fast and the census techniques by which they are ascertained are so inexact that expressions of their size to the last digit are meaningless.

Divisions of Countries

No statistical data are given for political or regional divisions of countries, nor are they listed in the index. Nevertheless, such divisions are cited throughout the book for the larger countries, to facilitate the location of cities and physiographic features and for general interest. Only "great" divisions are listed (e.g., regions rather than departments for France), and these are cited by conventional or official[1] names without alternative designations. The following kinds of divisions are listed:

Argentina — provinces
Australia — states and territories
Brazil — states and territories
Canada — provinces and territories
China — provinces and regions
Czechoslovakia — regions
France — regions
India — states
Indonesia — islands
Italy — regions
Japan — islands
Libya — provinces
Malaysia — divisions
Mexico — states and territories
New Zealand — islands
Philippines — islands
Saudi Arabia — regions
South Africa — provinces
Spain — regions
Tanzania — divisions
UK — divisions
USA — states
USSR — republics
West Germany — states
Yugoslavia — republics

[1] Official names are used for the divisions of France, Italy, Spain, and West Germany because only a few of these have different conventional English names.

Latitude and Longitude

Further to facilitate geographic location, latitude and longitude to the nearest minute are given for all cities and almost all physiographic features. Almost exclusively, the latitudes and longitudes are derived from the gazetteers prepared by the US Office of Geography; these are based on official large-scale maps of the relevant countries and regions. They can therefore be depended upon for accuracy. For reasons of simplicity, they are given in this form: 50.46N, 6.06E (this defines the geographic position of Aachen, West Germany, and means 50 degrees 46 minutes north of the equator and 6 degrees 6 minutes east of Greenwich, a section of London).

Special Symbols

Every effort has been made to obtain official data for the various statistical tables. When such data are unavailable, the most reliable unofficial figures are given. In Tables 2 through 5c, unofficial data are identified by the symbol ● placed before the name of the physiographic feature in question.

All population figures are from census returns except those for which an "e" is placed after the year of record; the "e," of course, indicates that the figure has been estimated, usually by official sources. For many cities, two sets of population figures are given. Figures without a symbol are for the city proper; those followed by an "s" are for the city and adjacent suburban areas. Preferably, the latter figures are for the urbanized area or conurbation only, but some officially reported suburban areas include rural districts.

The asterisk (*) is another special symbol used in this work. This symbol indicates that the city cited is the capital of a country or subdivision. If the asterisk is placed before the name of the city, it is the capital of the country under which it is listed. If the asterisk is placed before the name of a subdivision, the city cited is the capital of that subdivision. It should be noted that some countries have more than one capital.

Finally, most of the tables that are arranged by some measurement, rather than alphabetically, are divided into two parts separated by three centered asterisks (***). When this division is made, the portion of the table that is above the asterisks represents all-inclusive coverage, and the portion that is below, selective coverage.

AREA AND POPULATION
OF CONTINENTS

Contents

Conventional name of continent.
Area, including inland waters and adjacent islands, in thousands of square miles and square kilometers, and percentage of world total.
Latest population, in thousands, and percentage of world total.
Density of population, per square mile and per square kilometer.
World totals.

Arrangement

By continent, alphabetically.

Coverage

All continents.

Rounding

Areas are rounded to the nearest 1000 mi^2 and km^2; populations, to the nearest 1000; densities of population, to the nearest two significant digits.

Special Feature

World totals are equivalent to world land area, population, and density of population.

Entries

7.

AREA AND POPULATION OF CONTINENTS

Continent	Area (1000 mi²)	(1000 km²)	(%)	Population (thousands)	(%)	Density of Population (per mi²)	(per km²)
Africa	11,696	30,293	20.2	431,209	10.7	37	14
Antarctica	5,396	13,975	9.3	0	0	0	0
Asia	17,179	44,493	29.6	2,349,048	58.4	137	53
Europe	3,956	10,245	6.8	660,476	16.4	167	64
North America	9,442	24,454	16.3	346,418	8.6	37	14
Oceania	3,454	8,945	6.0	23,446	0.6	6.8	2.6
South America	6,887	17,838	11.9	214,684	5.3	31	12
World	58,009[1]	150,243	100[1]	4,025,281	100	69	27

[1] Not equal to total because of rounding.

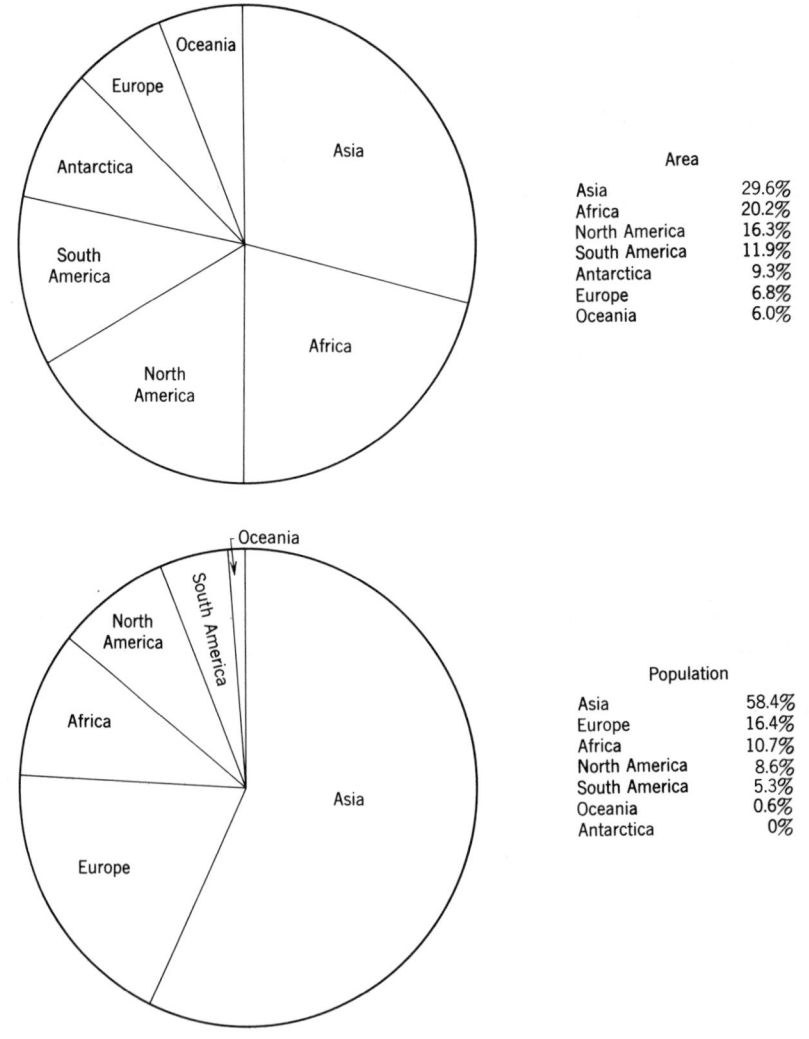

Area	
Asia	29.6%
Africa	20.2%
North America	16.3%
South America	11.9%
Antarctica	9.3%
Europe	6.8%
Oceania	6.0%

Population	
Asia	58.4%
Europe	16.4%
Africa	10.7%
North America	8.6%
South America	5.3%
Oceania	0.6%
Antarctica	0%

Figure 3. Area and population of continents.

1
LARGEST SEAS

Contents

Conventional and alternate names of sea.
Area in thousands of square miles and square kilometers.

Arrangement

By area.

Coverage

All large seas, including those commonly designated as oceans, gulfs, bays, straits, and the like, generally recognized by oceanographic authorities.

Rounding

Areas are rounded to the nearest 1000 mi^2 and km^2.

Entries

65.

TABLE 1. LARGEST SEAS

Sea	Area (1000 mi²)	Area (1000 km²)	Sea	Area (1000 mi²)	Area (1000 km²)
Pacific O			Timor Sea	237	615
with adjacent seas	70,017	181,344	Andaman (alt Burma) Sea	232	602
without adjacent seas	64,186	166,241	North Sea	232	600
Atlantic O			Chukchi (alt Chuckchee) Sea	225	582
with adjacent seas	36,415	94,314	Great Australian Bight	187	484
without adjacent seas	33,420	86,557	Beaufort Sea	184	476
Indian O			Celebes (alt Sulawesi) Sea	182	472
with adjacent seas	28,617	74,118	Black Sea	178	461
without adjacent seas	28,350	73,427	Red Sea	174	450
Arctic O			Java (alt Jawa) Sea	167	433
with adjacent seas	4,732	12,257	Sulu Sea	162	420
without adjacent seas	3,662	9,485	Yellow (alt Huang, Hwang) Sea	161	417
Coral Sea	1,850	4,791	Baltic Sea	149	386
Arabian Sea	1,492	3,863	G of Carpentaria	120	311
South China (alt Nan) Sea	1,423	3,685	Molucca (alt Maluku) Sea	119	307
Caribbean Sea	1,063	2,754	Persian G (alt G of Iran)	93	241
Mediterranean Sea	971	2,516	G of Siam (alt G of Thailand)	92	239
Bering Sea	890	2,304	G of Saint Lawrence	92	238
B of Bengal	839	2,172	G of Aden	85	220
Sea of Okhotsk	614	1,590	Makassar (alt Macassar, Makasar)		
G of Mexico	596	1,543	Strait	75	194
G of Guinea	592	1,533	Ceram (alt Seram) Sea	72	187
Barents Sea	542	1,405	B of Biscay	71	184
Norwegian Sea	534	1,383	G of Oman	70	181
G of Alaska	512	1,327	Aegean Sea	69	179
Hudson B	476	1,232	G of California	68	177
Greenland Sea	465	1,205	Adriatic Sea	51	132
Arafura Sea	400	1,037	Flores Sea	47	121
Philippine Sea	400	1,036	Bali Sea	46	119
Sea of Japan	378	978	G of Bothnia	45	117
East Siberian Sea	348	901	G of Tonkin	45	117
Kara Sea	341	883	Savu Sea	41	105
East China (alt Tung) Sea	290	752	Irish Sea	40	103
Solomon Sea	278	720	White Sea	35	90
Bandá Sea	268	695	Bass Strait	29	75
Baffin B	266	689	English Channel (alt La Manche)	29	75
Laptev Sea	251	650	Sea of Azov	15	38

2
LARGEST ISLANDS, BY CONTINENT

Contents

Latitude and longitude, in degrees and minutes, at center of island.

Conventional and other (alternate, former, and official) names of island.

Principal body of water, and subdivision and country in which located.

Area in thousands of square miles and square kilometers.

Highest elevation in feet and meters.

Name of largest city and its latest population, in thousands, with year of census or estimate (e).[1]

Arrangement

By area, under continent.

Coverage

Above ∗ ∗ ∗ , all islands with an area of at least 1000 mi^2 (2590 km^2).

Below ∗ ∗ ∗, other well-known islands, including the largest in certain countries.

Rounding

Areas are rounded to the nearest three significant digits.

Elevations are rounded to the nearest 10 ft (m).

Entries

414.

---

[1]On some islands the "largest city" is actually a small town or even a village. No attempt has been made to differentiate in the index, where each is identified simply as a city.

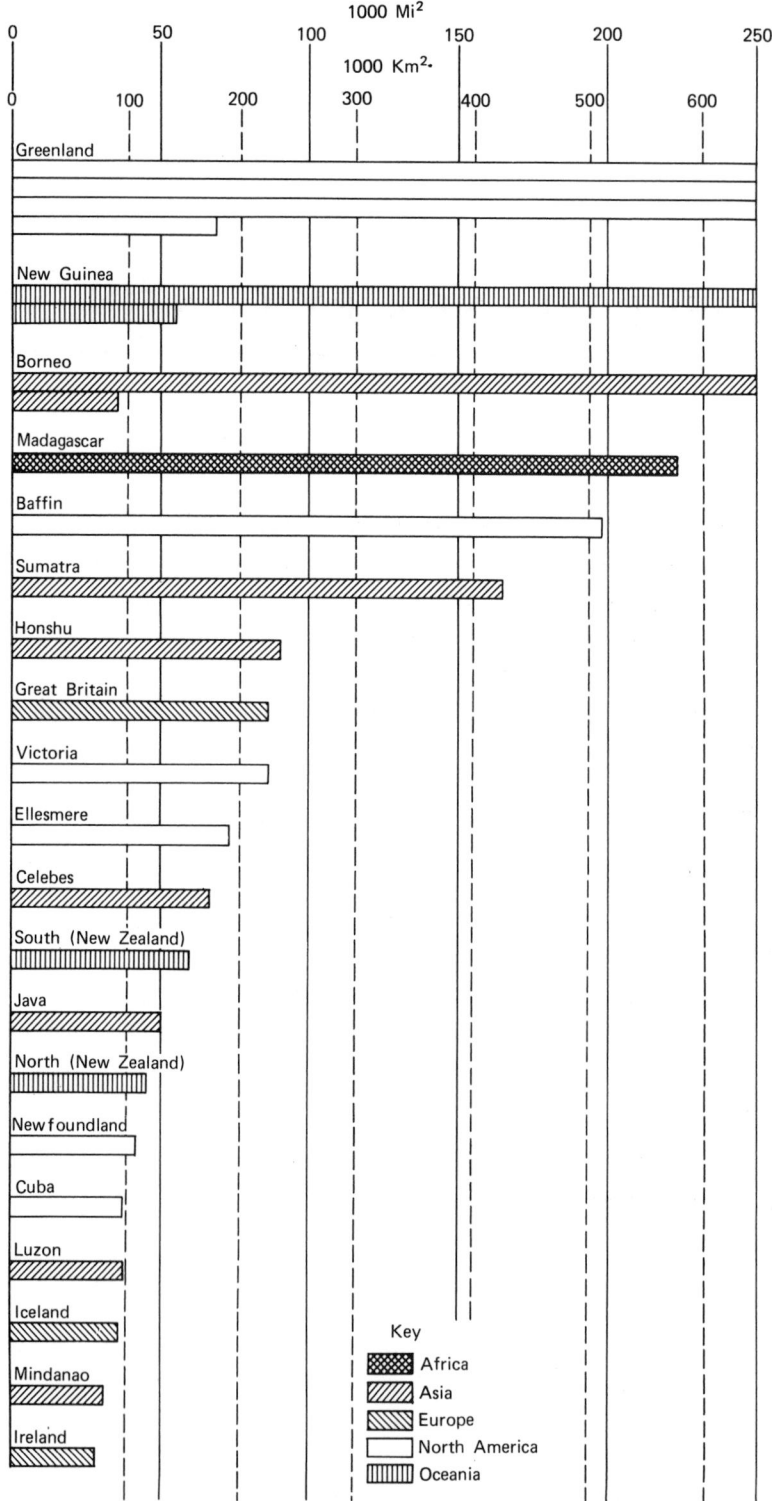

Figure 4. Twenty largest islands.

TABLE 2. LARGEST ISLANDS, BY CONTINENT
AFRICA

Latitude and Longitude	Island	Principal Body of Water	Location	Area (1000 mi²)	Area (1000 km²)	Highest Elev (ft)	Highest Elev (m)	Largest City and Its Latest Population (thousands), with Date
20.00S, 47.00E	Madagascar (alt Madagaskara)	Indian O	Madagascar	227	587	9,470	2890	Tananarive: 452 (1975)
49.30S, 69.30E	• Kerguelen	Indian O	French Southern and Antarctic Lands	2.32	6.00	6,430	1960	Port-aux-Francais: 0.08 (1973e)
			* * *					
21.06S, 55.36E	Reunion (for Bourbon)	Indian O	Reunion	0.969	2.51	10,070	3070	Saint-Denis: 66 (1967), 104s (1974)
3.30N, 8.42E	Macias Nguema (for Fernando Po)	Bight of Bonny	Equatorial Guinea	0.779	2.02	9,840	3000	Malabo (for Santa Isabel): 19 (1970e)
28.19N, 16.34W	Tenerife (for Tenerife)	Atlantic O	(Canary Islands), Spain	0.745	1.93	12,200	3720	Santa Cruz de Tenerife: 160 (1975e)
20.18S, 57.35E	Mauritius [for (Ile de) France]	Indian O	Mauritius	0.720	1.86	2,710	830	Port Louis: 141 (1976e)
28.20N, 14.00W	Fuerteventura	Atlantic O	(Canary Islands), Spain	0.642	1.66	2,650	810	Puerto del Rosario (for Puerto de Cabras): 6 (1970)
6.10S, 39.20E	Zanzibar (alt Unguja)	Indian O	Zanzibar, Tanzania	0.640	1.66	430	130	Zanzibar: 68 (1967)
28.00N, 15.36W	Gran Canaria (alt Grand Canary)	Atlantic O	(Canary Islands), Spain	0.592	1.53	6,400	1950	Las Palmas: 341 (1975e)
11.35S, 43.20E	• Grande Comore	Mozambique Channel	Comoros	0.443	1.15	8,120	2470	Moroni: 12 (1974e)
15.05N, 23.40W	Sao Tiago (alt Santiago)	Atlantic O	Cape Verde	0.384	0.994	4,570	1390	Praia: 21 (1970)

AFRICA

Latitude and Longitude	Island	Principal Body of Water	Location	Area (1000 mi²)	Area (1000 km²)	Highest Elev (ft)	Highest Elev (m)	Largest City and Its Latest Population (thousands), with Date
5.10S, 39.48E	• Pemba	Indian O	Zanzibar, Tanzania	0.380	0.984	300	91	Wete: 8 (1967)
15.42N, 40.08E	• Dahlac (alt Dahlak, Grand Dahlac)	Red Sea	Ethiopia	0.347	0.900	low elev		
0.12N, 6.39E	Sao Tome (for Sao Thome)	G of Guinea	Sao Tome and Principe	0.330	0.854	6,640	2020	Sao Tome (for Sao Thome): 17 (1970)
29.00N, 13.40W	Lanzarote	Atlantic O	(Canary Islands), Spain	0.302	0.782	2,200	670	Arrecife: 22 (1970)
17.05N, 25.10W	Santo Antao	Atlantic O	Cape Verde	0.302	0.782	6,490	1980	Ribeira Grande: 2 (1950)
32.44N, 17.00W	Madeira	Atlantic O	(Madeira Islands), Portugal	0.286	0.741	6,110	1860	Funchal: 38 (1970)
28.40N, 17.52W	La Palma (alt Palma)	Atlantic O	(Canary Islands), Spain	0.256	0.662	7,960	2430	Santa Cruz de la Palma: 13 (1970)
16.05N, 22.50W	Boa Vista	Atlantic O	Cape Verde	0.239	0.620	1,270	390	Sal Rei: ? [island population 4 (1970)]
33.48N, 10.54E	Djerba (off Jarbah)	G of Gabes	Tunisia	0.197	0.510	160	50	Houmt Souk (off Hawmat as Suq): 9 (1966)
14.55N, 24.25W	Fogo	Atlantic O	Cape Verde	0.184	0.476	9,280	2830	Sao Filipe: 2 (1960)
12.15S, 44.25E	Anjouan	Mozambique Channel	Comoros	0.164	0.424	5,170	1570	Mutsamudu: 8 (1966)
12.50S, 45.10E	Mayotte	Mozambique Channel	Mayotte	0.144	0.374	1,970	600	Sada: 2 (1966)
28.06N, 17.08W	Gomera	Atlantic O	(Canary Islands), Spain	0.136	0.353	4,880	1490	San Sebastian de la Gomera: 5 (1970)
16.35N, 24.15W	Sao Nicolau	Atlantic O	Cape Verde	0.132	0.343	4,280	1300	Ribeira Brava: 8 (1950)
15.15N, 23.10W	Maio (for Mayo)	Atlantic O	Cape Verde	0.104	0.269	1,430	440	Porto Ingles: ? [island population 3 (1970)]
27.45N, 18.00W	Ferro (off Hierro)	Atlantic O	(Canary Islands), Spain	0.102	0.264	4,920	1500	Valverde: 3 (1970)

AFRICA

Latitude and Longitude	Island	Principal Body of Water	Location	Area (1000 mi²)	Area (1000 km²)	Highest Elev (ft)	Highest Elev (m)	Largest City and Its Latest Population (thousands), with Date
16.50N, 25.00W	Sao Vicente	Atlantic O	Cape Verde	0.088	0.227	2,540	770	Mindelo: 29 (1970)
4.40S, 55.30E	Mahe	Indian O	Seychelles	0.055	0.142	2,990	910	Victoria (alt Port Victoria): 23s (1977)
15.57S, 5.42W	Saint Helena	Atlantic O	Saint Helena	0.047	0.122	2,700	820	Jamestown: 1 (1975)
37.05S, 12.17W	Tristan da Cunha	Atlantic O	Saint Helena	0.038	0.098	6,760	2060	Edinburgh: ? [island population 0.3 (1975)]
7.57S, 14.22W	Ascension	Atlantic O	Saint Helena	0.034	0.088	2,870	870	Georgetown: ? [island population 1 (1975)]

ANTARCTICA

Latitude and Longitude	Island	Principal Body of Water	Location	Area (1000 mi²)	Area (1000 km²)	Highest Elev (ft)	Highest Elev (m)	Largest City and Its Latest Population (thousands), with Date
71.00S, 70.00W	● Alexander (for Alexander I)	Bellingshausen Sea	Antarctica	16.7	43.2	10,300	3140	uninhabited
79.30S, 49.30W	● Berkner	Weddell Sea	Antarctica	1.50	3.88	3,200	980	uninhabited
67.15S, 68.30W	● Adelaide	Bellingshausen Sea	Antarctica	1.27	3.30	10,500	3200	uninhabited

* * *

| 69.45S, 75.15W | ● Charcot | Bellingshausen Sea | Antarctica | 0.772 | 2.00 | 2,000 | 610 | uninhabited |

ASIA

Latitude and Longitude	Island	Principal Body of Water	Location	Area (1000 mi²)	Area (1000 km²)	Highest Elev (ft)	Highest Elev (m)	Largest City and Its Latest Population (thousands), with Date
1.00N, 114.00E	● Borneo (alt Kalimantan)	South China Sea	Brunei-Indonesia-Malaysia	285	737	13,450	4100	Banjarmasin: 283 (1971)
0.00, 102.00E	● Sumatra (off Sumatera)	Andaman Sea	Indonesia	164	425	12,470	3800	Medan: 636 (1971)
36.00N, 138.00E	● Honshu (alt Hondo)	Pacific O	Japan	88.0	228	12,390	3780	Tokyo: 8647, 11,540s (1975)
2.00S, 121.00E	● Celebes (off Sulawesi)	Celebes Sea	Indonesia	67.4	174	11,340	3450	Ujung Pandang: 434 (1971)
7.30S, 110.00E	● Java (off Jawa)	Indian O	Indonesia	50.0	129	12,060	3680	Jakarta: 5490 (1975e)
15.00N, 121.00E	Luzon	Pacific O	Philippines	40.4	105	9,610	2930	Manila: 1454, 4863s (1975)
8.00N, 125.00E	Mindanao	Pacific O	Philippines	36.5	94.6	9,690	2950	Davao: 482 (1975)
44.00N, 143.00E	● Hokkaido (for Ezo, Yezo)	Pacific O	Japan	30.1	78.0	7,510	2290	Sapporo: 1241 (1975)
51.00N, 143.00E	Sakhalin (for Karafuto, Saghalien)	Sea of Okhotsk	Russia, USSR	29.5	76.4	5,280	1610	Yuzhno-Sakhalinsk: 131 (1976e)
7.30N, 80.30E	Sri Lanka (alt Ceylon; for Serendib)	Indian O	Sri Lanka	25.2	65.3	8,280	2520	Colombo: 592, 850s (1974e)
23.30N, 121.00E	Taiwan (alt Formosa)	Pacific O	Taiwan	13.8	35.8	13,110	4000	Taipei: 2089 (1976e)
33.00N, 131.00E	● Kyushu (alt Kiushu)	Pacific O	Japan	13.8	35.7	5,870	1790	Kitakyushu: 1058 (1975)
19.00N, 109.30E	Hainan	South China Sea	Kwangtung, China	13.1	34.0	6,160	1880	Hoihow: 402 (1958e)
8.50S, 126.00E	● Timor	Timor Sea	Indonesia	10.2	26.3	9,720	2960	Kupang: 49 (1971)
1.00N, 128.00E	● Halmahera (for Djailolo; off Jailolo)	Molucca Sea	Indonesia	6.95	18.0	6,260	1910	Jailolo (for Djailolo): 3 (1971)

ASIA

Latitude and Longitude	Island	Principal Body of Water	Location	Area (1000 mi²)	Area (1000 km²)	Highest Elev (ft)	Highest Elev (m)	Largest City and Its Latest Population (thousands), with Date
33.45N, 133.30E	• Shikoku	Pacific O	Japan	6.86	17.8	6,500	1980	Matsuyama: 367 (1975)
3.00S, 129.00E	• Ceram (off Seram)	Banda Sea	Indonesia	6.62	17.2	10,020	3050	Masohi (alt Amahai): 2 (1971)
8.30S, 121.00E	• Flores	Flores Sea	Indonesia	5.50	14.2	7,820	2380	Ende: 27 (1971)
79.30N, 97.00E	Oktyabrskoy Revolyutsii (alt October Revolution)	Arctic O	Russia, USSR	5.47	14.2	3,170	960	uninhabited
8.40S, 118.00E	• Sumbawa (for Soembawa)	Indian O	Indonesia	5.16	13.4	9,350	2850	Raba: 41 (1971)
12.00N, 125.00E	Samar	Pacific O	Philippines	5.05	13.1	2,790	850	Catbalogan: 18 (1970)
10.00N, 123.00E	Negros	Sulu Sea	Philippines	4.90	12.7	8,090	2470	Bacolod: 223 (1975)
10.30N, 118.30E	Palawan (for Paragua)	South China Sea	Philippines	4.55	11.8	6,840	2080	Puerto Princesa: 12 (1970)
75.45N, 138.44E	Kotelnyy	Arctic O	Russia, USSR	4.50	11.7	1,230	370	Iloilo: 227 (1975)
10.42N, 122.33E	Panay	Sulu Sea	Philippines	4.45	11.5	6,730	2050	Pangkalpinang: 75 (1971)
2.15S, 106.00E	• Bangka (alt Banka)	Java Sea	Indonesia	4.37	11.3	2,310	700	uninhabited
78.40N, 102.30E	Bolshevik	Arctic O	Russia, USSR	4.35	11.3	3,070	930	Waingapu (for Waingapoe): 16 (1971)
10.00S, 120.00E	• Sumba (for Sandalwood, Soemba)	Indian O	Indonesia	4.31	11.2	4,020	1220	Calapan: 11 (1970)
12.50N, 121.10E	Mindoro	South China Sea	Philippines	3.76	9.73	8,480	2590	Nicosia: 117s (1974e)
35.05N, 33.15E	Cyprus (off Kibris, Kypros)	Mediterranean Sea	Cyprus	3.57	9.25	6,410	1950	uninhabited
80.30N, 95.00E	Komsomolets	Arctic O	Russia, USSR	3.48	9.01	2,560	780	Namlea: 4 (1971)
3.24S, 126.40E	• Buru (for Boeroe)	Banda Sea	Indonesia	3.47	9.00	7,970	2430	

ASIA

Latitude and Longitude	Island	Principal Body of Water	Location	Area (1000 mi²)	Area (1000 km²)	Highest Elev (ft)	Highest Elev (m)	Largest City and Its Latest Population (thousands), with Date
71.00N, 179.30W	Wrangel (off Vrangelya)	Chukchi Sea	Russia, USSR	2.82	7.30	3,600	1100	Ushakovskiy: ?
10.50N, 124.52E	Leyte	Visayan Sea	Philippines	2.78	7.21	4,430	1350	Tacloban: 50 (1970)
45.00N, 148.00E	Iturup (for Etorofu)	Pacific O	Russia, USSR	2.60	6.72	5,360	1630	Kurilsk: 2 (1940e)
75.00N, 149.00E	Novaya Sibir (alt New Siberia)	East Siberian Sea	Russia, USSR	2.39	6.20	250	76	Bolshoy Zimovye: ?
8.30S, 115.00E	• Bali	Indian O	Indonesia	2.17	5.62	10,310	3140	Denpasar: 98 (1971)
8.45S, 116.30E	• Lombok	Indian O	Indonesia	2.10	5.43	12,220	3730	Mataram: 35 (1971)
7.00S, 113.30E	• Madura (for Madoera)	Java Sea	Indonesia	2.04	5.29	1,550	470	Pamekasan: 56 (1971)
75.30N, 144.00E	Faddeyevskiy	Arctic O	Russia, USSR	1.93	5.00	200	61	uninhabited
2.50S, 108.00E	• Billiton (off Belitung)	Java Sea	Indonesia	1.85	4.80	1,670	510	Tanjungpandan (for Tandjoengpandan): 37 (1971)
1.05N, 97.35E	• Nias	Indian O	Indonesia	1.84	4.77	2,910	890	Gunungsitoli (for Goenoengsitoli): 9 (1971)
73.35N, 142.00E	Bolshoy Lyakhovskiy	Arctic O	Russia, USSR	1.78	4.60	890	270	
10.23N, 123.50E	Cebu (for Zebu)	Visayan Sea	Philippines	1.71	4.42	3,320	1010	Cebu: 408 (1975)
5.00S, 122.55E	• Butung (alt Buton; for Boetoeng)	Banda Sea	Indonesia	1.62	4.20	3,900	1190	Baubau: 18 (1971)
9.50N, 124.10E	Bohol	Mindanao Sea	Philippines	1.49	3.86	2,630	800	Tagbilaran: 33 (1970)
12.30N, 54.00E	• Socotra (alt Sokotra; off Suqutra)	Indian O	South Yemen	1.40	3.63	4,900	1490	Tamrida (alt Hadibu): 3 (1960?e)
7.48S, 126.18E	• Wetar	Banda Sea	Indonesia	1.40	3.62	4,630	1410	Ilwaki: 0.2 (1971)

ASIA

Latitude and Longitude	Island	Principal Body of Water	Location	Area (1000 mi²)	Area (1000 km²)	Highest Elev (ft)	Highest Elev (m)	Largest City and Its Latest Population (thousands), with Date
12.15N, 123.30E	Masbate	Visayan Sea	Philippines	1.26	3.27	2,280	700	Masbate: 18 (1970)
1.20S, 98.55E	• Siberut (for Siberoet)	Indian O	Indonesia	1.22	3.17	1,330	410	Muarasiberut (for Moearasiberoet): 8 (1971)
7.36S, 131.25E	• Yamdena (for Jamdena)	Arafura Sea	Indonesia	1.10	2.86	low elev		Saumlaki: 2 (1971)
1.48S, 124.48E	• Taliabu (for Taliaboe)	Molucca Sea	Indonesia	1.10	2.85	4,330	1320	Todeli: 0.5 (1971)
6.35S, 134.20E	• Trangan	Arafura Sea	Indonesia	1.02	2.65	290	90	Rebi: 0.5 (1971)

* * *

Latitude and Longitude	Island	Principal Body of Water	Location	Area (1000 mi²)	Area (1000 km²)	Highest Elev (ft)	Highest Elev (m)	Largest City and Its Latest Population (thousands), with Date
1.30S, 127.45E	• Obi (alt Obira)	Ceram Sea	Indonesia	0.951	2.46	5,290	1610	Wayaloar: 1 (1971)
1.20S, 123.10E	• Peleng	Molucca Sea	Indonesia	0.929	2.40	3,470	1060	Tataba: 0.5 (1971)
0.35S, 127.30E	• Bacan (for Bachan, Batjan)	Molucca Sea	Indonesia	0.913	2.36	6,930	2110	Labuha (for Laboeha): 2 (1971)
8.15S, 124.45E	• Alor (alt Ombai)	Banda Sea	Indonesia	0.900	2.33	5,790	1760	Kalabahi: 10 (1971)
19.06N, 93.48E	• Ramree (off Yanbye)	B of Bengal	Burma	0.888	2.30	3,000	910	Kyaukpyu: 7 (1953)
50.25N, 155.50E	Paramushir (for Paramushiro)	Pacific O	Russia, USSR	0.788	2.04	5,960	1820	Severo-Kurilsk: 2 (1947e)
58.50N, 164.00E	Karagin (off Karaginskiy)	Bering Sea	Russia, USSR	0.772	2.00	2,990	910	Ostrovnoy: ?
2.33N, 95.55E	• Simeulue (for Simeuloe)	Indian O	Indonesia	0.712	1.84	2,180	580	Sinabang: 7 (1971)
33.20N, 126.30E	• Cheju (for Quelpart, Saishu)	East China Sea	South Korea	0.710	1.84	6,400	1950	Cheju (for Chyei Chyu, Saishu): 58, 106s (1970)

ASIA

Latitude and Longitude	Island	Principal Body of Water	Location	Area (1000 mi²)	Area (1000 km²)	Highest Elev (ft)	Highest Elev (m)	Largest City and Its Latest Population (thousands), with Date
2.20N, 128.25E	● Morotai	Molucca Sea	Indonesia	0.695	1.80	4,100	1250	Berebere: 0.8 (1971)
55.00N, 137.42E	Bolshoy Shantar	Sea of Okhotsk	Russia, USSR	0.691	1.79	2,300	700	Shantar: ?
5.00S, 122.30E	● Muna (for Moena)	Banda Sea	Indonesia	0.658	1.70	1,460	440	Raha: 12 (1971)
55.00N, 166.15E	Bering (off Beringa)	Pacific O	Russia, USSR	0.641	1.66	2,460	750	Nikolskoye: 0.5 (1948e)
22.30N, 90.45E	● Dakhin Shahbazpur	B of Bengal	Bangladesh	0.612	1.59	low elev		Bhola: 9 (1961)
44.10N, 146.00E	Kunashir (for Kunashiri)	Pacific O	Russia, USSR	0.598	1.55	5,970	1820	Yuzhno-Kurilsk (for Furukamappu): ?
12.30N, 92.50E	Middle Andaman	B of Bengal	Andaman and Nicobar, India	0.593	1.54	1,680	510	none [island population 11 (1971)]
13.45N, 124.15E	Catanduanes	Philippine Sea	Philippines	0.552	1.43	2,510	760	Virac: 10 (1970)
13.15N, 92.55E	North Andaman	B of Bengal	Andaman and Nicobar, India	0.531	1.38	2,400	730	none [island population 8 (1971)]
11.45N, 92.10E	South Andaman	B of Bengal	Andaman and Nicobar, India	0.520	1.35	1,060	320	Port Blair: 26 (1971)
26.45N, 55.45E	● Qeshm (alt Qishm; for Kishm, Tawila)	Strait of Hormuz	Iran	0.515	1.33	1,310	400	Qeshm (alt Qishm; for Kishm): 7 (1976)
6.34N, 122.03E	Basilan	Celebes Sea	Philippines	0.494	1.28	3,320	1010	Basilan: 126 (1975)
26.42N, 128.11E	Okinawa	Pacific O	Japan	0.471	1.22	1,600	490	Naha: 295 (1975)
1.05N, 104.30E	● Bintan (alt Bintang)	South China Sea	Indonesia	0.415	1.07	1,140	350	Tanjungpinang (for Tandjoengpinang): 44 (1971)
7.00N, 93.50E	Great Nicobar	Indian O	Andaman and Nicobar, India	0.403	1.04	2,100	640	none [island population 0.2 (1961)]
29.47N, 48.10E	Bubiyan (alt Bobian)	Persian G	Kuwait	0.333	0.863	low elev		uninhabited
38.00N, 138.25E	Sado	Sea of Japan	Japan	0.331	0.857	3,850	1170	Ryotsu: 23 (1970)
3.40S, 128.10E	● Ambon (alt Amboina)	Banda Sea	Indonesia	0.294	0.761	3,410	1040	Ambon: 80 (1971)

ASIA

Latitude and Longitude	Island	Principal Body of Water	Location	Area (1000 mi²)	Area (1000 km²)	Highest Elev (ft)	Highest Elev (m)	Largest City and Its Latest Population (thousands), with Date
28.15N, 129.20E	Amami(-Oshima)	Pacific O	Japan	0.274	0.709	2,280	690	Naze (alt Nase): 44 (1970)
19.12N, 72.54E	•Salsette	Arabian Sea	Maharashtra, India	0.246	0.637	1,530	470	Bombay (in part): 2584 (1971)
26.00N, 50.30E	Bahrain (alt Bahrein; off Bahrayn)	Persian G	Bahrain	0.217	0.563	440	140	Manama: 110 (1975e)
1.22N, 103.48E	Singapore (off Singapura)	South China Sea	Singapore	0.210	0.543	580	180	Singapore: 2278 (1976e)
5.24N, 100.14E	Penang (off Pinang)	Strait of Malacca	Peninsular Malaysia, Malaysia	0.108	0.280	2,720	830	George Town: 269, 332s (1970)
0.41N, 127.24E	•Tidore	Molucca Sea	Indonesia	0.045	0.116	5,680	1730	Soasiu (alt Tidore; for Soasioe): 1 (1971)
22.15N, 114.11E	Hong Kong (alt Hsiangchiang, Hsiangkang)	South China Sea	Hong Kong	0.029	0.075	1,800	550	Hong Kong (in part): 996 (1971)
0.48N, 127.20E	•Ternate	Molucca Sea	Indonesia	0.025	0.065	5,630	1710	Ternate: 35 (1971)
18.59N, 72.50E	Bombay	Arabian Sea	Maharashtra, India	0.024	0.062	180	55	Bombay (in part): 3387 (1971)

EUROPE

Latitude and Longitude	Island	Principal Body of Water	Location	Area (1000 mi²)	Area (1000 km²)	Highest Elev (ft)	Highest Elev (m)	Largest City and Its Latest Population (thousands), with Date
53.30N, 2.30W	•Great Britain (alt Britain)	North Sea	UK	84.4	219	4,410	1340	London: 2553, 7111s (1975e)
65.00N, 18.00W	Iceland (off Island)	Atlantic O	Iceland	39.7	103	6,950	2120	Reykjavik: 85, 99s (1974e)

EUROPE

Latitude and Longitude	Island	Principal Body of Water	Location	Area (1000 mi²)	Area (1000 km²)	Highest Elev (ft)	Highest Elev (m)	Largest City and Its Latest Population (thousands), with Date
53.25N, 8.00W	• Ireland (alt Eire)	Atlantic O	Ireland-UK	32.5	84.1	3,410	1040	Dublin: 568, 680s (1971)
75.00N, 61.30E	Novaya Zemlya (north island)	Kara Sea	Russia, USSR	18.9	48.9	5,080	1550	Russkaya Gavan: ?
78.45N, 16.00E	West Spitsbergen (off Vestspitsbergen)	Arctic O	Svalbard	15.3	39.5	5,630	1720	Longyearbyen: 0.9 (1970?e)
72.00N, 54.00E	Novaya Zemlya (south island)	Barents Sea	Russia, USSR	12.8	33.3	4,400	1340	Krasino: ?
37.30N, 14.00E	Sicily (off Sicilia)	Mediterranean Sea	Sicilia, Italy	9.81	25.4	10,760	3280	Palermo: 673 (1976e)
40.00N, 9.00E	Sardinia (off Sardegna)	Mediterranean Sea	Sardegna, Italy	9.19	23.8	6,020	1830	Cagliari: 240 (1976e)
79.48N, 22.24E	North East Land (off Nordaustlandet)	Barents Sea	Svalbard	5.79	15.0	2,510	760	none
42.00N, 9.00E	Corsica (off Corse)	Mediterranean Sea	Corse, France	3.37	8.72	8,890	2710	Ajaccio: 51 (1975); Bastia: 51 (1975)
35.29N, 24.42E	Crete (for Candia; off Kriti)	Mediterranean Sea	Greece	3.19	8.26	8,060	2460	Iraklion: 78, 85s (1971)
55.30N, 11.45E	Zealand (off Sjaelland)	Baltic Sea	Denmark	2.71	7.02	410	130	Copenhagen: 562, 1287s (1975e)
77.45N, 22.30E	Edge	Barents Sea	Svalbard	1.94	5.03	1,900	580	none
57.15N, 9.50E	Vendsyssel-Thy[1]	Skagerrak Strait	Denmark	1.81	4.68	450	140	Frederikshavn: 25 (1970), 35s (1975e)

[1] Usually regarded as part of the Danish mainland, i.e., Jutland (off Jylland) Peninsula, but actually an island since 1825, when Lim Fjord of Kattegat Strait reached the North Sea, cutting the peninsula in two.

EUROPE

Latitude and Longitude	Island	Principal Body of Water	Location	Area (1000 mi²)	Area (1000 km²)	Highest Elev (ft)	Highest Elev (m)	Largest City and Its Latest Population (thousands), with Date
38.34N, 24.24E	Euboea (off Evvoia)	Aegean Sea	Greece	1.41	3.65	5,720	1740	Chalcis (off Khalkis): 36 (1971)
39.30N, 3.00E	Majorca (off Mallorca)	Mediterranean Sea	Baleares, Spain	1.40	3.63	4,740	1440	Palma: 277 (1975e)
70.00N, 59.30E	Vaygach	Kara Sea	Russia, USSR	1.31	3.38	560	170	
69.05N, 49.15E	Kolguyev	Barents Sea	Russia, USSR	1.24	3.20	540	170	
57.30N, 18.33E	Gotland (for Gottland)	Baltic Sea	Sweden	1.16	3.00	270	83	Visby: 25 (1975e)
55.20N, 10.30E	Fyn (for Funen)	Baltic Sea	Denmark	1.15	2.98	430	130	Odense: 137 (1970), 169s (1975e)
80.30N, 49.00E	Zemlya Georga	Barents Sea	Russia, USSR	1.12	2.90	1,360	420	uninhabited
80.45N, 46.00E	Zemlya Aleksandry	Barents Sea	Russia, USSR	1.08	2.80	1,250	380	Nagurskoye: ?
58.25N, 22.30E	Sarema (alt Saaremaa; for Osel)	Baltic Sea	Estonia, USSR	1.03	2.68	180	54	Kingisepp (for Arensburg, Kuressaare): 12 (1970)

* * *

Latitude and Longitude	Island	Principal Body of Water	Location	Area (1000 mi²)	Area (1000 km²)	Highest Elev (ft)	Highest Elev (m)	Largest City and Its Latest Population (thousands), with Date
68.30N, 16.00E	Hinn(oya)	Norwegian Sea	Norway	0.849	2.20	4,150	1270	Harstad: 16 (1970)
58.05N, 6.40W	• Lewis with Harris	Atlantic O	Scotland, UK	0.770	1.99	2,620	800	Stornoway: 5 (1971)
57.20N, 6.15W	• Skye	Sea of the Hebrides	Scotland, UK	0.643	1.67	3,310	1010	Portree: 2 (1971)
39.10N, 25.50E	Lesbos (alt Mytilene; off Lesvos)	Aegean Sea	Greece	0.629	1.63	3,180	970	Mytilene (for Kastro; off Mitilini): 23 (1971)
69.20N, 17.30E	Senja	Norwegian Sea	Norway	0.614	1.59	3,310	1010	Gryllefjord: 0.9 (1970)
36.10N, 28.00E	Rhodes (off Rodhos)	Mediterranean Sea	Greece	0.541	1.40	3,990	1210	Rhodes (off Rodhos): 32 (1971)

EUROPE

Latitude and Longitude	Island	Principal Body of Water	Location	Area (1000 mi²)	Area (1000 km²)	Highest Elev (ft)	Highest Elev (m)	Largest City and Its Latest Population (thousands), with Date
56.45N, 16.38E	Oland	Baltic Sea	Sweden	0.519	1.34	170	51	Borgholm: 11 (1976e)
54.46N, 11.30E	Lolland (for Laaland)	Baltic Sea	Denmark	0.480	1.24	98	30	Nakskov: 16 (1970), 17s (1975e)
58.50N, 22.40E	Khiuma (alt Hiiumaa; for Dago)	Baltic Sea	Estonia, USSR	0.382	0.989	180	54	Kardla: 1 (1934)
60.16N, 1.16W	Mainland (Shetland Islands)	Atlantic O	Scotland, UK	0.378	0.979	1,470	450	Lerwick: 6 (1971)
56.25N, 5.54W	● Mull	Atlantic O	Scotland, UK	0.367	0.951	3,180	970	Tobermory: 0.6 (1971)
54.25N, 13.24E	Rugen	Baltic Sea	East Germany	0.358	0.926	400	120	Sassnitz; 13 (1971)
38.22N, 26.00E	Chios (off Khios)	Aegean Sea	Greece	0.325	0.842	4,260	1300	Chios (off Khios): 24, 30s (1971)
38.15N, 20.35E	Cephalonia (off Kefallinia)	Ionian Sea	Greece	0.302	0.781	5,340	1630	Argostolion: 7 (1971)
37.47N, 25.30W	Sao Miguel	Atlantic O	(Azores), Portugal	0.288	0.747	3,630	1100	Ponta Delgada: 20 (1970)
60.15N, 20.00E	Ahvenanmaa (alt Aaland, Aland)	G of Bothnia	Finland	0.285	0.738	420	130	Maarianhamina (alt Mariehamn): 9 (1970)
40.00N, 4.00E	Minorca (off Menorca)	Mediterranean Sea	Baleares, Spain	0.266	0.689	1,170	360	Mahon: 19 (1970)
55.48N, 6.12W	● Islay	Atlantic O	Scotland, UK	0.235	0.609	1,540	470	Bowmore: ? [island population 4 (1971)]
39.40N, 19.42E	Corfu (off Kerkira)	Ionian Sea	Greece	0.229	0.592	2,970	910	Corfu (off Kerkira): 29 (1971)
55.10N, 15.00E	Bornholm	Baltic Sea	Denmark	0.227	0.588	530	160	Ronne: 15 (1970), 15s (1975e)
54.14N, 4.33W	(Isle of) Man	Irish Sea	Isle of Man	0.227	0.588	2,030	620	Douglas: 20 (1976)

EUROPE

Latitude and Longitude	Island	Principal Body of Water	Location	Area (1000 mi²)	Area (1000 km²)	Highest Elev (ft)	Highest Elev (m)	Largest City and Its Latest Population (thousands), with Date
39.00N, 1.25E	Ibiza (alt Iviza)	Mediterranean Sea	Baleares, Spain	0.219	0.568	1,560	470	Ibiza (alt Iviza): 17 (1970)
59.00N, 3.15W	Pomona [alt Mainland (Orkney Islands)]	Atlantic O	Scotland, UK	0.207	0.536	880	270	Kirkwall: 5 (1971)
54.48N, 11.58E	Falster	Baltic Sea	Denmark	0.198	0.514	140	44	Nykobing Falster: 20 (1970), 26s (1975e)
39.54N, 25.21E	Lemnos (off Limnos)	Aegean Sea	Greece	0.184	0.476	1,410	430	Mirina: 4 (1971)
37.48N, 26.44E	Samos	Aegean Sea	Greece	0.184	0.476	4,700	1430	Samos (for Limin Vatheos): 5 (1971)
54.00N, 14.00E	Usedom (alt Uznam)	Baltic Sea	East Germany–Poland	0.172	0.445	200	60	Swinoujscie (for Swinemunde): 27 (1970)
38.28N, 28.20W	Pico	Atlantic O	(Azores), Portugal	0.167	0.433	7,710	2350	Lajes do Pico: 2 (1970)
37.02N, 25.35E	Naxos	Aegean Sea	Greece	0.165	0.428	3,300	1010	Naxos: 3 (1971)
55.36N, 5.15W	• Arran	Firth of Clyde	Scotland, UK	0.165	0.427	2,860	870	Corrie and Brodick: 1 (1971)
45.05N, 14.35E	Krk (for Veglia)	Adriatic Sea	Croatia, Yugoslavia	0.158	0.408	1,870	570	Krk (for Veglia): 2 (1971)
44.40N, 14.25E	Cres (for Cherso)	Adriatic Sea	Croatia, Yugoslavia	0.156	0.404	2,130	650	Cres (for Cherso): 2 (1961)
37.52N, 20.44E	Zante (alt Zacynthus; off Zakinthos)	Ionian Sea	Greece	0.155	0.402	2,480	760	Zante (off Zakinthos): 9 (1971)
38.43N, 27.13W	Terceira	Atlantic O	(Azores), Portugal	0.153	0.397	3,350	1020	Angra do Heroismo (alt Angra): 14 (1970)
43.19N, 16.40E	Brac (for Brazza)	Adriatic Sea	Croatia, Yugoslavia	0.153	0.395	2,550	780	Supetar: 1 (1961)
50.40N, 1.17W	• (Isle of) Wight	English Channel	England, UK	0.147	0.381	790	240	Newport: 22 (1971)

EUROPE

Latitude and Longitude	Island	Principal Body of Water	Location	Area (1000 mi²)	Area (1000 km²)	Highest Elev (ft)	Highest Elev (m)	Largest City and Its Latest Population (thousands), with Date
37.50N, 24.50E	Andros	Aegean Sea	Greece	0.147	0.380	3,310	1010	Andros: 2 (1971)
40.40N, 24.40E	Thasos	Aegean Sea	Greece	0.146	0.379	3,950	1200	Thasos (for Limin): 2 (1971)
62.08N, 7.00W	Streymoy (off Stromo)	Atlantic O	Faeroe Islands	0.144	0.373	2,590	790	Torshavn: 11 (1975e)
71.00N, 8.20W	Jan Mayen	Greenland Sea	Norway	0.144	0.372	7,470	2280	uninhabited
56.00N, 5.54W	•Jura	Sound of Jura	Scotland, UK	0.140	0.363	2,570	780	Craighouse: ? [island population 0.2 (1971)]
56.50N, 8.45E	•Mors	Lim Fjord	Denmark	0.140	0.363	290	89	Nykobing Mors: 9 (1970)
38.43N, 20.38E	Leucas (for Santa Maura; off Levkas)	Ionian Sea	Greece	0.117	0.302	3,800	1160	Leucas (for Santa Maura; off Levkas): 7 (1971)
35.40N, 27.10E	Karpathos (for Scarpanto)	Mediterranean Sea	Greece	0.116	0.301	4,000	1220	Karpathos (for Pigadhia): 1 (1971)
43.07N, 16.45E	Hvar (for Lesina)	Adriatic Sea	Croatia, Yugoslavia	0.116	0.300	2,050	630	Hvar (for Lesina): 2 (1961)
36.50N, 27.10E	Kos (alt Cos; for Coo)	Aegean Sea	Greece	0.112	0.290	2,780	850	Kos (for Coo): 8 (1971)
44.30N, 15.00E	Pag (for Pago)	Adriatic Sea	Croatia, Yugoslavia	0.110	0.285	1,140	350	Pag (for Pago): 2 (1961)
55.00N, 10.50E	•Langeland	Baltic Sea	Denmark	0.110	0.284	85	26	Rudkobing: 4 (1970), 7s (1975e)
40.10N, 25.50E	Gokceada (for Imbros, Imroz)	Aegean Sea	Turkey	0.108	0.280	1,960	600	Gokce (for Imbros, Imroz, Panagia): 3 (1970)

EUROPE

Latitude and Longitude	Island	Principal Body of Water	Location	Area (1000 mi²)	Area (1000 km²)	Highest Elev (ft)	(m)	Largest City and Its Latest Population (thousands), with Date
36.15N, 23.00E	Cythera (for Cerigo; off Kithira)	Mediterranean Sea	Greece	0.107	0.278	1,660	510	Kithira (for Kapsali): 0.3 (1971)
42.57N, 16.55E	Korcula (for Curzola)	Adriatic Sea	Croatia, Yugoslavia	0.107	0.276	1,860	570	Blato: 5 (1961)
35.53N, 14.27E	Malta	Mediterranean Sea	Malta	0.095	0.246	830	250	Sliema: 20 (1975e)
38.38N, 28.03W	Sao Jorge	Atlantic O	(Azores), Portugal	0.092	0.238	3,500	1070	Velas: 2 (1970)
42.46N, 10.17E	●Elba	Mediterranean Sea	Toscana, Italy	0.086	0.224	3,340	1020	Portoferraio: 11 (1976e)
54.28N, 11.08E	Fehmarn	Baltic Sea	Schleswig-Holstein, West Germany	0.071	0.185	89	27	Burg: 6 (1971)
40.27N, 25.35E	Samothrace (off Samothraki)	Aegean Sea	Greece	0.069	0.178	5,250	1600	Samothrace (off Samothraki): 1 (1971)
38.34N, 28.42W	Faial (for Fayal)	Atlantic O	(Azores), Portugal	0.065	0.168	3,420	1040	Horta: 6 (1970)
49.13N, 2.07W	Jersey	English Channel	Channel Islands	0.045	0.116	450	140	Saint Helier: 28 (1971)
49.27N, 2.36W	Guernsey	English Channel	Channel Islands	0.024	0.063	350	110	Saint Peter Port: 16 (1971)

NORTH AMERICA

Latitude and Longitude	Island	Principal Body of Water	Location	Area (1000 mi²)	Area (1000 km²)	Highest Elev (ft)	(m)	Largest City and Its Latest Population (thousands), with Date
73.00N, 42.00W	●Greenland (off Gronland)	Atlantic O	Greenland	823	2131	12,140	3700	Godthab: 9 (1976e)
68.00N, 70.00W	Baffin (for Baffin Land)	Baffin B	NWT, Canada	196	507	6,750	2060	Frobisher Bay: 2 (1976)

Figure 5. Century-old houses and modern apartment blocks at Godthab, the capital of Greenland, earth's largest island. (Credit: Ministry of Foreign Affairs of Denmark.)

NORTH AMERICA

Latitude and Longitude	Island	Principal Body of Water	Location	Area (1000 mi²)	Area (1000 km²)	Highest Elev (ft)	Highest Elev (m)	Largest City and Its Latest Population (thousands), with Date
71.00N, 114.00W	Victoria	Viscount Melville Sound	NWT, Canada	83.9	217	2,150	660	Cambridge Bay: 0.7 (1971)
81.00N, 80.00W	Ellesmere	Arctic O	NWT, Canada	75.8	196	8,580	2620	Grise Fiord: 0.1 (1971)
49.00N, 56.00W	Newfoundland	Atlantic O	Newfoundland, Canada	42.0	109	2,670	810	Saint John's: 87, 143s (1976)
21.30N, 80.00W	Cuba	Caribbean Sea	Cuba	40.5	105	6,480	1970	Havana: 1838 (1974e)
19.00N, 71.00W	Hispaniola (for Hayti; off Haiti, Santo Domingo)	Atlantic O	Dominican Republic–Haiti	29.2	75.6	10,420	3170	Santo Domingo: 923 (1975e)
73.15N, 121.30W	Banks	Arctic O	NWT, Canada	27.0	70.0	2,400	730	Sachs Harbour: 0.1 (1971)
87.00N, 75.00W	Devon	Baffin B	NWT, Canada	21.3	55.2	6,300	1920	uninhabited
80.30N, 92.00W	Axel Heiberg	Arctic O	NWT, Canada	16.7	43.2	8,400	2560	uninhabited
75.15N, 110.00W	Melville	Viscount Melville Sound	NWT, Canada	16.3	42.1	3,500	1070	uninhabited
64.20N, 84.40W	Southampton	Hudson B	NWT, Canada	15.9	41.2	1,750	530	Coral Harbour: 0.4 (1976)
72.40N, 99.00W	Prince of Wales	Viscount Melville Sound	NWT, Canada	12.9	33.3	830	250	uninhabited
49.00N, 125.00W	Vancouver	Pacific O	British Columbia, Canada	12.1	31.3	7,220	2200	Victoria: 63, 218s (1976)
73.15N, 93.30W	Somerset	Lancaster Sound	NWT, Canada	9.57	24.8	1,600	490	uninhabited
76.00N, 100.30W	Bathurst	Viscount Melville Sound	NWT, Canada	6.19	16.0	1,480	450	uninhabited
76.45N, 119.30W	Prince Patrick	Arctic O	NWT, Canada	6.12	15.9	600	180	Mould Bay (weather station)
69.00N, 97.30W	King William	Queen Maud G	NWT, Canada	5.06	13.1	450	140	Gjoa Haven: 0.3 (1971)

NORTH AMERICA

Latitude and Longitude	Island	Principal Body of Water	Location	Area (1000 mi²)	Area (1000 km²)	Highest Elev (ft)	Highest Elev (m)	Largest City and Its Latest Population (thousands), with Date
78.30N, 104.00W	Ellef Ringnes	Arctic O	NWT, Canada	4.36	11.3	2,000	610	Isachsen (weather station)
73.13N, 78.34W	Bylot	Baffin B	NWT, Canada	4.27	11.1	6,600	2010	uninhabited
18.15N, 77.30W	Jamaica	Caribbean Sea	Jamaica	4.24	11.0	7,400	2260	Kingston: 170 (1973e), 626s (1975e)
46.00N, 60.30W	Cape Breton	Atlantic O	Nova Scotia, Canada	3.98	10.3	1,750	530	Sydney: 31, 89s (1976)
67.50N, 76.00W	Prince Charles	Foxe Basin	NWT, Canada	3.68	9.53	50	15	uninhabited
57.30N, 153.30W	Kodiak	Pacific O	Alaska, USA	3.67	9.51	3,400	1040	Kodiak (for Pavlovsk Gavan): 4 (1975e)
18.15N, 66.45W	Puerto Rico (for Porto Rico)	Atlantic O	Puerto Rico	3.35	8.67	4,390	1340	San Juan: 472 (1972e) 820s (1970)
69.45N, 53.20W	Disco (alt Disko)	Davis Strait	Greenland	3.31	8.58	6,300	1920	Godhavn: 1 (1970)
49.30N, 63.00W	Anticosti	G of Saint Lawrence	Quebec, Canada	3.07	7.95	620	190	Port-Menier: 0.4 (1971)
75.15N, 94.30W	Cornwallis	Barrow Strait	NWT, Canada	2.70	6.99	1,350	410	Resolute Bay: 0.2 (1971)
55.30N, 132.45W	Prince of Wales	Pacific O	Alaska, USA	2.59	6.70	3,160	960	Craig: 0.4 (1975e)
53.00N, 132.00W	Graham	Pacific O	British Columbia, Canada	2.46	6.37	3,940	1200	Masset: 2 (1976)
24.26N, 77.57W	Andros[1]	Great Bahama Bank	Bahamas	2.30	5.96	100	30	Nicolls Town: 4 (1970)
46.30N, 63.00W	Prince Edward	G of Saint Lawrence	Prince Edward Island, Canada	2.18	5.66	460	140	Charlottetown: 17, 25s (1976)
62.30N, 83.00W	Coats	Hudson B	NWT, Canada	2.12	5.49			uninhabited
57.50N, 135.40W	Chichagof	G of Alaska	Alaska, USA	2.08	5.40	3,500	1070	Hoonah: 0.8 (1975e)

[1] Actually several (unnamed) islands.

NORTH AMERICA

Latitude and Longitude	Island	Principal Body of Water	Location	Area		Highest Elev		Largest City and Its Latest Population (thousands), with Date
				(1000 mi²)	(1000 km²)	(ft)	(m)	
78.00N, 97.00W	Amund Ringnes	Peary Channel	NWT, Canada	2.03	5.26	1,000	300	uninhabited
77.45N, 111.00W	Mackenzie King	Hazen Strait	NWT, Canada	1.95	5.05			uninhabited
10.30N, 61.15W	Trinidad	Atlantic O	Trinidad and Tobago	1.86	4.83	3,080	940	Port of Spain: 60 (1973e)
73.17N, 106.45W	Stefansson	Viscount Melville Sound	NWT, Canada	1.72	4.45	1,120	340	uninhabited
63.30N, 170.30W	Saint Lawrence	Bering Sea	Alaska, USA	1.71	4.43	2,070	630	Gambell: 0.4 (1975e)
57.45N, 134.25W	Admiralty	Chatham Strait	Alaska, USA	1.65	4.27	4,640	1410	Angoon: 0.7 (1975e)
60.06N, 166.20W	Nunivak	Bering Sea	Alaska, USA	1.62	4.21	1,670	510	Mekoryuk: 0.2 (1975e)
70.45N, 26.00W	Milne Land	Greenland Sea	Greenland	1.62	4.20	6,230	1900	
54.45N, 164.00W	Unimak	Pacific O	Alaska, USA	1.61	4.16	9,370	2860	False Pass: 0.06 (1970)
57.00N, 135.00W	Baranof	G of Alaska	Alaska, USA	1.60	4.14	4,530	1380	Sitka: 6 (1975e)
72.30N, 24.00W	•Traill	Greenland Sea	Greenland	1.60	4.14	4,520	1380	
40.50N, 73.00W	Long	Atlantic O	New York, USA	1.40	3.63	390	120	New York (in part): 4372 (1975e)
62.00N, 79.50W	Mansel	Hudson B	NWT, Canada	1.23	3.19	300	90	uninhabited
53.00N, 81.20W	Akimiski	James B	NWT, Canada	1.16	3.00			uninhabited
55.35N, 131.20W	Revillagigedo	Behm Canal	Alaska, USA	1.14	2.97	4,590	1400	Ketchikan: 8 (1975e)
56.46N, 133.25W	Kupreanof	Frederick Sound	Alaska, USA	1.09	2.82	3,980	1210	Kake: 0.6 (1975e)
78.30N, 110.30W	Borden	Arctic O	NWT, Canada	1.08	2.80	500	150	uninhabited
45.50N, 82.20W	Manitoulin	Huron L	Ontario, Canada	1.07	2.77	1,120	340	Little Current: 1 (1976)
53.45N, 167.00W	Unalaska	Pacific O	Alaska, USA	1.06	2.76	6,680	2040	Unalaska: 0.4 (1975e)
52.45N, 131.50W	Moresby	Pacific O	British Columbia, Canada	1.01	2.62	3,440	1050	Sandspit: 0.5 (1971)

* * *

NORTH AMERICA

Latitude and Longitude	Island	Principal Body of Water	Location	Area (1000 mi²)	Area (1000 km²)	Highest Elev (ft)	Highest Elev (m)	Largest City and Its Latest Population (thousands), with Date
52.55N, 128.50W	Princess Royal	Princess Royal Channel	British Columbia, Canada	0.869	2.25	5,500	1680	Butedale: 0.01 (1971)
21.40N, 82.50W	(Isle of) Pines (off Pinos)	Caribbean Sea	Cuba	0.849	2.20	1,020	310	Nueva Gerona: 17 (1970)
60.40N, 164.50W	Nelson	Etolin Strait	Alaska, USA	0.843	2.18	1,500	460	Tanunak: 0.3 (1970)
56.30N, 134.05W	Kuiu	Chatham Strait	Alaska, USA	0.750	1.94	3,000	910	uninhabited
58.15N, 152.35W	Afognak	Pacific O	Alaska, USA	0.721	1.87	2,550	780	uninhabited
53.15N, 168.20W	Umnak	Pacific O	Alaska, USA	0.687	1.78	7,050	2150	Nikolski: 0.06 (1970)
21.05N, 73.18W	Inagua (alt Great Inagua)	Windward Passage	Bahamas	0.596	1.54	110	33	none [island population 1 (1970)]
53.40N, 129.50W	Pitt	Principe Channel	British Columbia, Canada	0.531	1.38	3,150	960	uninhabited
26.38N, 78.25W	Grand Bahama	Northwest Providence Channel	Bahamas	0.530	1.37	34	10	Freeport: 15 (1970)
14.40N, 61.00W	Martinique	Atlantic O	Martinique	0.425	1.10	4,580	1400	Fort-de-France: 94 (1967), 99s (1974)
53.25N, 130.12W	Banks	Hecate Strait	British Columbia, Canada	0.382	0.989	1,760	540	uninhabited
26.28N, 77.05W	Abaco (alt Great Abaco)	Atlantic O	Bahamas	0.372	0.963	100	30	Marsh Harbour: 3 (1970)
16.10N, 61.40W	Basse-Terre	Caribbean Sea	Guadeloupe	0.364	0.943	4,870	1480	Basse-Terre: 15 (1974)
29.20N, 113.25W	Angel de la Guarda	G of California	Baja California, Mexico	0.330	0.855	4,310	1320	uninhabited
52.15N, 127.40W	King	Burke Channel	British Columbia, Canada	0.312	0.808	5,500	1680	uninhabited
15.25N, 61.20W	Dominica	Atlantic O	Dominica	0.290	0.751	4,750	1450	Roseau: 18 (1976e)
29.00N, 112.25W	Tiburon	G of California	Sonora, Mexico	0.290	0.751	4,000	1220	none

NORTH AMERICA

Latitude and Longitude	Island	Principal Body of Water	Location	Area (1000 mi²)	Area (1000 km²)	Highest Elev (ft)	Highest Elev (m)	Largest City and Its Latest Population (thousands), with Date
18.51N, 73.03W	Gonave	G of Gonave	Haiti	0.270	0.700	2,480	750	Anse-a-Galet: 0.5 (1950)
13.55N, 60.59W	Saint Lucia	Atlantic O	Saint Lucia	0.238	0.616	3,140	960	Castries: 4, 39s (1970)
16.20N, 61.25W	Grande-Terre	Atlantic O	Guadeloupe	0.219	0.566	910	280	Pointe-a-Pitre: 24 (1974)
48.00N, 88.50W	(Isle) Royale	Superior L	Michigan, USA	0.210	0.544	1,310	400	none
45.30N, 73.35W	•Montreal	Saint Lawrence R	Quebec, Canada	0.201	0.521	760	230	Montreal: 1081 (1976)
7.27N, 81.45W	•Coiba	Pacific O	Panama	0.200	0.518	1,390	420	none [island population 1 (1970)]
25.10N, 76.14W	Eleuthera	Atlantic O	Bahamas	0.200	0.518	170	51	Governor's Harbour: 3 (1970)
20.25N, 86.55W	Cozumel	Caribbean Sea	Quintana Roo, Mexico	0.189	0.490	low elev		Cozumel (off San Miguel de Cozumel): 6 (1970)
23.15N, 75.07W	Long	Atlantic O	Bahamas	0.173	0.448	180	54	none [island population 4 (1970)]
48.10N, 122.33W	Whidbey	Puget Sound	Washington, USA	0.172	0.445	320	98	Oak Harbor: 11 (1975e)
12.10N, 69.00W	Curacao	Caribbean Sea	Netherlands Antilles	0.171	0.444	1,220	370	Willemstad: 50 (1970e), 94s (1960)
13.10N, 59.33W	Barbados	Atlantic O	Barbados	0.166	0.431	1,110	340	Bridgetown: 9, 115s (1970)
24.27N, 75.30W	Cat	Atlantic O	Bahamas	0.150	0.388	210	63	The Bight: 2 (1970)
46.13N, 83.57W	Saint Joseph	Huron L	Ontario, Canada	0.141	0.365	1,400	430	Hilton Beach: 0.2 (1971)
46.02N, 83.43W	Drummond	Huron L	Michigan, USA	0.136	0.352	880	270	Drummond Island: 0.2 (1970e)

NORTH AMERICA

Latitude and Longitude	Island	Principal Body of Water	Location	Area (1000 mi²)	Area (1000 km²)	Highest Elev (ft)	Highest Elev (m)	Largest City and Its Latest Population (thousands), with Date
13.15N, 61.12W	Saint Vincent	Atlantic O	Saint Vincent	0.133	0.344	4,050	1230	Kingstown: 22 (1973e)
12.06N, 61.42W	Grenada	Atlantic O	Grenada	0.120	0.311	2,760	840	Saint George's: 7 (1970), 30s (1975e)
29.34N, 91.52W	Marsh	G of Mexico	Louisiana, USA	0.117	0.303	5	2	uninhabited
11.15N, 60.40W	Tobago	Atlantic O	Trinidad and Tobago	0.116	0.301	1,900	580	Scarborough (for Port Louis): 1 (1960)
12.15N, 68.27W	Bonaire	Caribbean Sea	Netherlands Antilles	0.111	0.288	780	240	Kralendijk: 0.8 (1960)
17.09N, 61.49W	Antigua	Atlantic O	Antigua	0.108	0.280	1,320	400	Saint John's: 24 (1975e)
44.20N, 68.18W	Mount Desert	Atlantic O	Maine, USA	0.108	0.280	1,530	470	Bar Harbor: 2 (1970)
26.50N, 97.13W	Padre	G of Mexico	Texas, USA	0.099	0.256	40	12	none
45.37N, 73.43W	• Jesus	Ottawa R	Quebec, Canada	0.095	0.246	160	50	Laval: 246 (1976)
41.24N, 70.32W	Martha's Vineyard	Atlantic O	Massachusetts, USA	0.093	0.241	310	94	Vineyard Haven: 2 (1970)
28.28N, 80.40W	Merritt	Banana River Lag	Florida, USA	0.093	0.241			Merritt Island: 29 (1970)
17.45N, 64.45W	Saint Croix	Caribbean Sea	Virgin Islands (USA)	0.084	0.218	1,160	360	Christiansted: 3 (1970)
25.02N, 77.24W	New Providence	Northeast Providence Channel	Bahamas	0.080	0.207	120	37	Nassau: 102s (1970)
19.20N, 81.15W	Grand Cayman	Caribbean Sea	Cayman Islands	0.076	0.197	50	15	George Town: 4 (1970)
33.24N, 118.25W	Santa Catalina	Pacific O	California, USA	0.075	0.194	2,110	640	Avalon: 2 (1975e)
12.30N, 70.00W	Aruba	Caribbean Sea	Netherlands Antilles	0.075	0.193	620	190	Oranjestad: 16 (1968e)
21.47N, 71.43W	Grand Caicos (alt Middle Caicos)	Atlantic O	Turks and Caicos Islands	0.073	0.189			none [island population 0.4 (1970)]
46.55N, 71.00W	(Ile d')Orleans	Saint Lawrence R	Quebec, Canada	0.072	0.186	290	89	Beaulieu: 0.8 (1976)

NORTH AMERICA

Latitude and Longitude	Island	Principal Body of Water	Location	Area (1000 mi²)	Area (1000 km²)	Highest Elev (ft)	Highest Elev (m)	Largest City and Its Latest Population (thousands), with Date
17.20N, 62.45W	Saint Kitts (off Saint Christopher)	Caribbean Sea	Saint Kitts-Nevis	0.068	0.176	3,710	1130	Basseterre: 16 (1970)
17.38N, 61.48W	Barbuda	Atlantic O	Antigua	0.062	0.161	140	44	Codrington: ? [island population 1 (1960)]
24.02N, 74.28W	San Salvador (for Watling)	Atlantic O	Bahamas	0.059	0.153	120	37	none [island population 0.8 (1970)]
15.56N, 61.16W	Marie-Galante	Atlantic O	Guadeloupe	0.058	0.149	670	200	Grand-Bourg: 3, 7s (1967)
40.35N, 74.09W	Staten	New York B	New York, USA	0.057	0.148	410	120	New York (in part): 325 (1975e)
18.08N, 65.25W	Vieques (alt Crab)	Caribbean Sea	Puerto Rico	0.051	0.132	660	200	Vieques: 2 (1970)
17.10N, 62.34W	• Nevis	Caribbean Sea	Saint Kitts-Nevis	0.050	0.129	3,230	990	Charlestown: 2 (1970)
41.16N, 70.05W	Nantucket	Atlantic O	Massachusetts, USA	0.046	0.119	100	31	Nantucket: 2 (1970)
16.44N, 62.11W	Montserrat	Caribbean Sea	Montserrat	0.039	0.102	3,000	920	Plymouth: 1 (1970)
18.04N, 63.04W	Saint-Martin (alt Sint-Maarten)	Caribbean Sea	Guadeloupe–Netherlands Antilles	0.033	0.086	1,390	420	Marigot: 3 (1967); Philipsburg: 2 (1970?)
18.21N, 64.55W	Saint Thomas	Atlantic O	Virgin Islands (USA)	0.028	0.073	1,550	470	Charlotte Amalie: 12 (1970)
40.47N, 73.57W	Manhattan	Hudson R	New York, USA	0.022	0.057	270	82	New York (in part): 1429 (1975e)
18.27N, 64.36W	• Tortola	Atlantic O	Virgin Islands (UK)	0.021	0.054	1,780	540	Road Town: 3 (1973e)
18.20N, 64.45W	Saint John	Caribbean Sea	Virgin Islands (USA)	0.020	0.052	1,280	390	Cruz Bay: 1 (1970)
32.18N, 64.45W	Bermuda	Atlantic O	Bermuda	0.014	0.036	260	79	Hamilton: 3, 14s (1970)

OCEANIA

Latitude and Longitude	Island	Principal Body of Water	Location	Area (1000 mi²)	Area (1000 km²)	Highest Elev (ft)	Highest Elev (m)	Largest City and Its Latest Population (thousands), with Date
5.00S, 140.00E	● New Guinea (alt Irian, Papua)	Pacific O	Indonesia–Papua New Guinea	305	790	16,500	5030	Port Moresby: 113 (1976)
44.00S, 170.00E	South (New Zealand)	Pacific O	New Zealand	58.2	151	12,350	3760	Christchurch: 172, 295s (1976)
38.00S, 175.40E	North (New Zealand)	Pacific O	New Zealand	44.2	114	9,180	2800	Auckland: 151, 743s (1976)
42.00S, 147.00E	Tasmania (for Van Diemen's Land)	Indian O	Tasmania, Australia	24.9	64.4	5,310	1620	Hobart: 132s (1976)
5.40S, 151.00E	New Britain (for Neupommern)	Bismarck Sea	Papua New Guinea	14.6	37.8	8,000	2440	Rabaul: 27 (1971)
21.30S, 165.30E	New Caledonia (off Nouvelle-Caledonie)	Coral Sea	New Caledonia	6.47	16.7	5,380	1640	Noumea: 42 (1969), 56s (1976)
7.53S, 138.23E	● Dolak (alt Kolepon; for Frederik Hendrik)	Arafura Sea	Irian Jaya, Indonesia	4.16	10.8	low elev		Kimaam: 8 (1971)
19.30N, 155.30W	Hawaii	Pacific O	Hawaii, USA	4.04	10.5	13,800	4210	Hilo: 26 (1970)
18.00S, 178.00E	Viti Levu	Pacific O	Fiji	4.01	10.4	4,340	1320	Suva: 64 (1976), 80s (1966)
6.00S, 155.00E	Bougainville	Pacific O	Papua New Guinea	3.88	10.0	10,210	3110	Kieta: 2 (1971)
3.20S, 152.00E	New Ireland (for Neumecklenburg)	Pacific O	Papua New Guinea	3.34	8.65	7,500	2290	Kavieng: 3 (1971)
11.40S, 131.00E	Melville	Timor Sea	Northern Territory, Australia	2.40	6.22	low elev		Snake Bay: ?
9.32S, 160.12E	Guadalcanal	Solomon Sea	Solomon Islands	2.17	5.63	8,030	2450	Honiara: 15 (1976)
16.33S, 179.15E	Vanua Levu	Pacific O	Fiji	2.14	5.53	3,500	1070	Lambasa (alt Labasa): 2 (1966)

OCEANIA

Island	Latitude and Longitude	Principal Body of Water	Location	Area (1000 mi²)	Area (1000 km²)	Highest Elev (ft)	Highest Elev (m)	Largest City and Its Latest Population (thousands), with Date
Espiritu Santo (alt Santo; for Marina)	15.15S, 166.50E	Coral Sea	New Hebrides	1.93	5.00	6,190	1890	Santo (alt Luganville): 4 (1972e), 5s (1967)
Malaita	9.00S, 161.00E	Pacific O	Solomon Islands	1.75	4.53	4,700	1430	Auki Station: 0.9 (1970)
Kangaroo	35.50S, 137.06E	Indian O	South Australia, Australia	1.68	4.35	900	270	Kingscote: 1 (1971)
Santa Isabel (alt Ysabel)	8.00S, 159.00E	Pacific O	Solomon Islands	1.55	4.01	4,000	1220	Kia: 0.6 (1970)
San Cristobal (alt Makira, San Cristoval)	10.36S, 161.45E	Pacific O	Solomon Islands	1.35	3.50	4,100	1250	Kirakira: 0.4 (1970)
•Waigeo (for Waigeoe)	0.14S, 130.45E	Pacific O	Irian Jaya, Indonesia	1.25	3.25	3,280	1000	Tapokreng: 8 (1971)
Choiseul	7.00S, 157.00E	Pacific O	Solomon Islands	1.20	3.10	3,500	1070	Sasamunga: 0.5 (1970)

* * *

Island	Latitude and Longitude	Principal Body of Water	Location	Area (1000 mi²)	Area (1000 km²)	Highest Elev (ft)	Highest Elev (m)	Largest City and Its Latest Population (thousands), with Date
Malekula (alt Mallicolo)	16.15S, 167.30E	Coral Sea	New Hebrides	0.965	2.50	2,920	890	Kolivu: 0.5 (1967)
•Biak	1.00S, 136.00E	Pacific O	Irian Jaya, Indonesia	0.950	2.46	3,390	1030	Biak:19 (1971)
Groote Eylandt	14.00S, 136.40E	G of Carpentaria	Northern Territory, Australia	0.950	2.46	520	160	none
•Yapen (for Japen)	1.45S, 136.15E	Sarera B	Irian Jaya, Indonesia	0.940	2.43	4,910	1500	Serui (for Seroei): 14 (1971)
New Georgia	8.15S, 157.30E	New Georgia Sound	Solomon Islands	0.900	2.33	3,300	1010	Paradise (alt Menakasapa) 0.4 (1970)

OCEANIA

Latitude and Longitude	Island	Principal Body of Water	Location	Area (1000 mi²)	Area (1000 km²)	Highest Elev (ft)	Highest Elev (m)	Largest City and Its Latest Population (thousands), with Date
40.00S, 148.00E	Flinders	Bass Strait	Tasmania, Australia	0.802	2.08	2,700	820	Whitemark: 0.2 (1966)
11.37S, 130.23E	Bathurst	Timor Sea	Northern Territory, Australia	0.786	2.04	low elev		Bathurst Island Mission: ?
20.48N, 156.20W	Maui	Pacific O	Hawaii, USA	0.729	1.89	10,030	3060	Kahului: 8 (1970)
13.36S, 172.22W	Savaii	Pacific O	Samoa	0.703	1.82	6,090	1860	Safotu: 1 (1966)
1.52S, 130.10E	•Misool	Ceram Sea	Irian Jaya, Indonesia	0.676	1.75	3,250	990	Misool: 4 (1971)
47.00S, 167.40E	Stewart	Pacific O	New Zealand	0.674	1.75	3,210	980	Halfmoon Bay (alt Oban): 0.3 (1966)
2.05S, 147.00E	•Manus (alt Admiralty)	Pacific O	Papua New Guinea	0.633	1.64	2,360	720	Lorengau: 4 (1971)
21.30N, 158.00W	Oahu	Pacific O	Hawaii, USA	0.608	1.57	4,020	1230	Honolulu: 344 (1975e), 442s (1970)
22.05N, 159.32W	Kauai	Pacific O	Hawaii, USA	0.553	1.43	5,170	1580	Kapaa: 4 (1970)
9.30S, 150.40E	•Fergusson	Solomon Sea	Papua New Guinea	0.518	1.34	6,000	1830	Salamo: ? [island population 13 (1971)]
20.53S, 167.13E	Lifou (alt Lifu)	Coral Sea	New Caledonia	0.462	1.20	200	60	none [island population 7 (1974)]
2.30S, 150.15E	•New Hanover (alt Lavongai; for Neuhannover)	Pacific O	Papua New Guinea	0.460	1.19	2,870	870	Taskul: ? [island population 7 (1971?)]
13.55S, 171.45W	Upolu	Pacific O	Samoa	0.430	1.11	3,610	1100	Apia: 32s (1976)
9.06S, 152.50E	•Woodlark (alt Murua)	Solomon Sea	Papua New Guinea	0.430	1.11	1,200	370	Kulumadau: ? [island population 2 (1971)]
39.50S, 144.00E	King	Bass Strait	Tasmania, Australia	0.425	1.10	700	210	Currie: 0.7 (1971)
17.37S, 149.27W	Tahiti	Pacific O	French Polynesia	0.402	1.04	7,340	2240	Papeete: 25, 65s (1971)

OCEANIA

Latitude and Longitude	Island	Principal Body of Water	Location	Area (1000 mi²)	Area (1000 km²)	Highest Elev (ft)	Highest Elev (m)	Largest City and Its Latest Population (thousands), with Date
10.00S, 151.00E	• Normanby	Solomon Sea	Papua New Guinea	0.400	1.04	3,600	1100	Esaala: ? [island population 10 (1971)]
18.48S, 169.05E	• Erromango (alt Eromanga)	Pacific O	New Hebrides	0.376	0.975	2,950	900	Dillon's Bay: 0.1 (1967)
17.40S, 168.25E	• Efate (alt Vate)	Pacific O	New Hebrides	0.353	0.914	2,140	650	Vila (alt Port-Vila): 3 (1967), 13s (1973e)
5.15S, 154.35E	• Buka	Pacific O	Papua New Guinea	0.320	0.829	1,650	500	Gagan: ?; Hanahan: ? [island population 32 (1971)]
11.30S, 153.30E	• Tagula (alt Sudest)	Coral Sea	Papua New Guinea	0.310	0.803	3,000	910	none [island population 2 (1971)]
9.22S, 150.16E	• Goodenough (for Morata)	Solomon Sea	Papua New Guinea	0.290	0.751	8,500	2590	Bolubolu: ? [island population 11 (1971)]
21.08N, 157.00W	Molokai	Pacific O	Hawaii, USA	0.259	0.671	4,970	1510	Kaunakakai: 1 (1970)
21.30S, 168.00E	Mare	Coral Sea	New Caledonia	0.248	0.642	490	150	none [island population 4 (1974)]
13.26N, 144.43E	Guam	Pacific O	Guam	0.212	0.549	1,330	410	Tamuning: 8 (1970)
7.30N, 134.36E	Babelthuap	Pacific O	Pacific Islands (USA)	0.153	0.396	700	210	Airai: 0.7 (1973)
20.50N, 156.55W	Lanai	Pacific O	Hawaii, USA	0.141	0.365	3,370	1030	Lanai City: 2 (1970)
6.55N, 158.15E	Ponape	Pacific O	Pacific Islands (USA)	0.129	0.334	2,580	790	Kolonia (alt Ponape): 5 (1973)
19.02S, 169.52W	Niue (for Savage)	Pacific O	Niue Island	0.100	0.259	230	69	Alofi: 1 (1971), 4s (1976e)
21.10S, 175.10W	Tongatapu	Pacific O	Tonga	0.099	0.257	270	82	Nukualofa: 18 (1976)
16.50S, 151.25W	Raiatea	Pacific O	French Polynesia	0.092	0.238	3,390	1030	Uturoa: 3 (1971)
21.55N, 160.10W	Niihau	Pacific O	Hawaii, USA	0.072	0.186	1,280	390	none [island population 0.2 (1970)]

OCEANIA

Latitude and Longitude	Island	Principal Body of Water	Location	Area (1000 mi²)	Area (1000 km²)	Highest Elev (ft)	Highest Elev (m)	Largest City and Its Latest Population (thousands), with Date
9.45S, 139.00W	Hiva-Oa	Pacific O	French Polynesia	0.058	0.150	4,130	1260	Atuona: 0.7s (1962)
14.18S, 170.42W	Tutuila	Pacific O	American Samoa	0.052	0.135	2,140	650	Pago Pago: 3 (1974)
17.32S, 149.50W	Moorea (for Eimeo)	Pacific O	French Polynesia	0.051	0.132	3,960	1210	Afareaitu: 0.9s (1962)
8.54S, 140.06W	Nuku-Hiva	Pacific O	French Polynesia	0.046	0.120	3,860	1180	Taiohae: 0.5s (1962)
15.12N, 145.45E	Saipan	Pacific O	Northern Mariana Islands	0.046	0.119	1,550	470	Saipan: 13 (1973)

SOUTH AMERICA

Latitude and Longitude	Island	Principal Body of Water	Location	Area (1000 mi²)	Area (1000 km²)	Highest Elev (ft)	Highest Elev (m)	Largest City and Its Latest Population (thousands), with Date
54.00S, 69.00W	• Grande de Tierra del Fuego	Atlantic O	Argentina–Chile	18.7	48.4	8,090	2470	Ushuaia (Argentina): 3 (1970); Porvenir (Chile) 2 (1960)
1.00S, 49.30W	• Marajo	Atlantic O	Para, Brazil	18.5	48.0	66	20	Soure: 9 (1970)
11.30S, 50.15W	• Bananal	Araguaia R	Goias, Brazil	7.72	20.0	590	180	Wari-Wari: ?
42.30S, 73.55W	• Chiloe	Pacific O	Chile	3.24	8.39	2,650	810	Ancud: 12 (1970)
49.20S, 74.40W	• Wellington	Trinidad G	Chile	2.61	6.75	3,540	1080	
51.55S, 58.45W	• East Falkland (alt Malvina del Este)	Atlantic O	Falkland Islands	2.44	6.31	2,310	700	Stanley (alt Port Stanley): 1 (1974e)
53.45S, 72.45W	• Santa Ines	Pacific O	Chile	2.12	5.50	4,390	1340	
0.10N, 50.10W	• Caviana	Atlantic O	Para, Brazil	1.92	4.97	low elev	20	none
1.00S, 51.30W	• Grande de Gurupa	Amazon R	Para, Brazil	1.88	4.86	66	20	none
51.50S, 60.00W	• West Falkland (alt Malvina del Oeste)	Atlantic O	Falkland Islands	1.68	4.35	2,300	700	none [island population 0.5 (1962)]

SOUTH AMERICA

Latitude and Longitude	Island	Principal Body of Water	Location	Area (1000 mi²)	Area (1000 km²)	Highest Elev (ft)	Highest Elev (m)	Largest City and Its Latest Population (thousands), with Date
55.15S, 69.00W	● Hoste	Pacific O	Chile	1.59	4.11	4,300	1310	
0.30S, 91.06W	● Isabela (alt Albemarle)	Pacific O	Ecuador[1]	1.45	3.76	5,600	1710	Villamil: ? [island population 0.3 (1962)]
54.15S, 36.45W	● South Georgia	Atlantic O	Falkland Islands	1.45	3.76	9,620	2930	Grytviken Harbour: ? [island population 0.02 (1973e)]
53.00S, 72.30W	● Riesco	Strait of Magellan	Chile	1.20	3.11	5,460	1660	
			* * *					
44.40S, 73.10W	● Magdalena	Moraleda Channel	Chile	0.998	2.58	5,450	1660	
55.05S, 67.40W	● Navarino	Nassau B	Chile	0.955	2.47	3,900	1190	
54.10S, 71.50W	● Clarence	Strait of Magellan	Chile	0.863	2.24	3,000	910	
0.02S, 49.35W	● Mexiana	Atlantic O	Para, Brazil	0.592	1.53	low elev	low elev	none
2.36S, 44.14W	● Sao Luis (for Maranhao)	Atlantic O	Maranhao, Brazil	0.465	1.20	low elev	low elev	Sao Luis: 168 (1970), 330s (1975e)
0.38S, 90.23W	● Santa Cruz (alt Indefatigable)	Pacific O	Ecuador[1]	0.358	0.927	2,550	780	Puerto Ayora: ? [island population 0.6 (1962)]
11.00N, 64.00W	Margarita	Caribbean Sea	Venezuela	0.355	0.920	4,800	1460	Porlamar: 32 (1971)
0.25S, 91.30W	● Fernandina (alt Narborough)	Pacific O	Ecuador[1]	0.253	0.655	4,500	1370	uninhabited
0.14S, 90.45W	● San Salvador (alt James, Santiago)	Pacific O	Ecuador[1]	0.199	0.515	2,970	910	uninhabited

[1] Galapagos (off Colon) Islands.

SOUTH AMERICA

Latitude and Longitude	Island	Principal Body of Water	Location	Area (1000 mi²)	Area (1000 km²)	Highest Elev (ft)	Highest Elev (m)	Largest City and Its Latest Population (thousands), with Date
0.50S, 89.26W	• San Cristobal (alt Chatham)	Pacific O	Ecuador[1]	0.185	0.480	2,490	760	Baquerizo Moreno: ? [island population 1 (1962)]
27.07S, 109.22W	• Easter (alt Rapa Nui; off Pascua)	Pacific O	Chile	0.045	0.117	1,970	600	Hanga Roa: ? [island population 1 (1960)]

[1] Galapagos (off Colon) Islands.

2a
LARGEST ISLAND IN EACH COUNTRY

Contents

Conventional name of country.
Conventional and other (alternate, former, and official) names of island.
Principal body of water in which located.
Country (if any) with which island is shared.
Area in thousands of square miles and square kilometers.

Arrangement

By country, alphabetically.

Coverage

All countries with an area of at least 5000 mi^2 (12,950 km^2) having or sharing an island of significant size.

Rounding

Areas are rounded to the nearest three significant digits.

Entries

85.

TABLE 2a.   LARGEST ISLAND IN EACH COUNTRY

Country	Island	Principal Body of Water	Country with which Shared	Area (1000 mi^2)	(1000 km^2)
Argentina	● Grande de Tierra del Fuego	Atlantic O	Chile	18.7	48.4
Australia	Tasmania (for Van Diemen's Land)	Indian O		24.9	64.4
Bangladesh	● Dakhin Shahbazpur	B of Bengal		0.612	1.59
Belize	● Ambergris (Cay)	Caribbean Sea		0.060?	0.155?
Brazil	● Marajo	Amazon R		18.5	48.0
Brunei	● Borneo (alt Kalimantan)	South China Sea	Indonesia, Malaysia	285	737
Burma	● Ramree (off Yanbye)	B of Bengal		0.888	2.30
Canada	Baffin (for Baffin Land)	Baffin B		196	507
Chile	● Grande de Tierra del Fuego	Atlantic O	Argentina	18.7	48.4
China	Hainan	South China Sea		13.1	34.0
Costa Rica	Chira	G of Nicoya of Pacific O		0.020	0.052
Cuba	Cuba	Caribbean Sea		40.5	105
Cyprus	Cyprus (off Kibris, Kypros)	Mediterranean Sea		3.57	9.25
Denmark	Zealand (off Sjaelland)	Baltic Sea		2.71	7.02
Dominican Republic	Hispaniola (for Hayti; off in Dominican Republic Santo Domingo)	Atlantic O	Haiti	29.2	75.6
Ecuador	● Isabela (alt Albemarle)	Pacific O		1.45	3.76
Egypt	● Tiran	Red Sea		0.030	0.070
Equatorial Guinea	Macias Nguema (for Fernando Po)	Bight of Bonny		0.779	2.02
Ethiopia	● Dahlac (alt Dahlak, Grand Dahlac)	Red Sea		0.347	0.900
Falkland Islands	● East Falkland (alt Malvina del Este)	Atlantic O		2.44	6.31
Fiji	Viti Levu	Pacific O		4.01	10.4
Finland	Ahvenanmaa (alt Aaland, Aland)	G of Bothnia		0.285	0.738
France	Corsica (off Corse)	Mediterranean Sea		3.37	8.72
French Guiana	● Cayenne	Atlantic O		0.060	0.155
Germany					
East	Rugen	Baltic Sea		0.358	0.926
West	Fehmarn	Baltic Sea		0.071	0.185
Greece	Crete (for Candia; off Kriti)	Mediterranean Sea		3.19	8.26
Greenland	● Greenland (off Gronland)	Atlantic O		823	2131
Guinea-Bissau	● Orango	Atlantic O		0.120	0.300

Country	Island	Principal Body of Water	Country with which Shared	Area (1000 mi²)	(1000 km²)
Haiti	Hispaniola (for Hayti; off in Haiti Haiti)	Atlantic O	Dominican Republic	29.2	75.6
Honduras	• Roatan	Caribbean Sea		0.077	0.200
Iceland	Iceland (off Island)	Atlantic O		39.7	103
India	Middle Andaman	B of Bengal		0.593	1.54
Indonesia					
in Asia	• Borneo (alt Kalimantan)	South China Sea	Brunei, Malaysia	285	737
in Oceania (Irian Jaya)	• New Guinea (alt Irian, Papua)	Pacific O	Papua New Guinea	305	790
Iran	• Qeshm (alt Qishm; for Kishm, Tawila)	Strait of Hormuz		0.515	1.33
Ireland	• Ireland (alt Eire)	Atlantic O	UK	32.5	84.1
Italy	Sicily (off Sicilia)	Mediterranean Sea		9.81	25.4
Jamaica	Jamaica	Caribbean Sea		4.24	11.0
Japan	• Honshu (alt Hondo)	Pacific O		88.0	228
Korea					
North	• Sinmi (for Shimmi; off Shinmi)	Korea B		0.020	0.053
South	• Cheju (for Quelpart, Saishu)	East China Sea		0.710	1.84
Kuwait	Bubiyan (alt Bobian)	Persian G		0.333	0.863
Madagascar	Madagascar (alt Madagaskara)	Indian O		227	587
Malaysia	• Borneo (alt Kalimantan)	South China Sea	Brunei, Indonesia	285	737
Mexico	Angel de la Guarda	G of California		0.330	0.855
Netherlands	• Schouwen-Duiveland	North Sea		0.086	0.222
New Caledonia	New Caledonia (off Nouvelle-Caledonie)	Coral Sea		6.47	16.7
New Hebrides	Espiritu Santo (alt Santo; for Marina)	Coral Sea		1.93	5.00
New Zealand	South (New Zealand)	Pacific O		58.2	151
Nicaragua	• Ometepe	Nicaragua L		0.106	0.275
Norway	Hinn(oya)	Norwegian Sea		0.849	2.20
Oman	• Masirah	Arabian Sea		0.250	0.650
Panama	• Coiba	Pacific O		0.200	0.518
Papua New Guinea	• New Guinea (alt Irian, Papua)	Pacific O	Indonesia	305	790
Philippines	Luzon	Pacific O		40.4	105
Portugal					
in Africa (Madeira Islands)	Madeira	Atlantic O		0.286	0.741
in Europe	Sao Miguel	Atlantic O		0.288	0.747
Puerto Rico	Puerto Rico (for Porto Rico)	Atlantic O		3.35	8.67

Country	Island	Principal Body of Water	Country with which Shared	Area (1000 mi²)	Area (1000 km²)
Saudi Arabia	• Farasan al Kabir	Red Sea		0.120	0.300
Sierra Leone	• Sherbro	Atlantic O		0.260	0.670
Solomon Islands	Guadalcanal	Solomon Sea		2.17	5.63
South Africa	• Marion	Indian O		0.140	0.350
Spain					
in Africa (Canary Islands)	Tenerife (for Teneriffe)	Atlantic O		0.745	1.93
in Europe	Majorca (off Mallorca)	Mediterranean Sea		1.40	3.63
Sri Lanka	Sri Lanka (alt Ceylon; for Serendib)	Indian O		25.2	65.3
Svalbard	West Spitsbergen (off Vestspitsbergen)	Arctic O		15.3	39.5
Sweden	Gotland (for Gottland)	Baltic Sea		1.16	3.00
Taiwan	Taiwan (alt Formosa)	Pacific O		13.8	35.8
Tanzania	Zanzibar (alt Unguja)	Indian O		0.640	1.66
Thailand	• Phuket (for Salang)	Andaman Sea		0.206	0.534
Trinidad and Tobago	Trinidad	Atlantic O		1.86	4.83
Tunisia	Djerba (off Jarbah)	G of Gabes		0.197	0.510
Turkey					
in Asia	Marmara	Sea of Marmara		0.045	0.117
in Europe	Gokceada (for Imbros, Imroz)	Aegean Sea		0.108	0.280
UK	• Great Britain (alt Britain)	North Sea		84.4	219
USA					
in North America	Kodiak	Pacific O		3.67	9.51
in Oceania (Hawaii)	Hawaii	Pacific O		4.04	10.5
USSR					
in Asia	Sakhalin (for Karafuto, Saghalien)	Sea of Okhotsk		29.5	76.4
in Europe	Novaya Zemlya (north island)	Kara Sea		18.9	48.9
Venezuela	Margarita	Caribbean Sea		0.355	0.920
Vietnam	• Phu Quoc	G of Siam		0.230	0.596
Yemen					
North	• Zuqar	Red Sea		0.070	0.180
South	• Socotra (alt Sokotra; off Suqutra)	Indian O		1.40	3.63
Yugoslavia	Krk (for Veglia)	Adriatic Sea		0.158	0.408

3
LONGEST RIVERS, BY CONTINENT

Contents

Latitude and longitude, in degrees and minutes, at mouth of river.

Conventional and other (alternate, former, and official) names of river and of its tributaries (if any) constituting the longest watercourse.

Outflow (sea, lake, river, etc.), and subdivision and country in which located.

Total length of watercourse, in miles and kilometers.

Area of drainage basin of river and all its tributaries, in thousands of square miles and square kilometers.

Average discharge rate, in thousands of cubic feet per second and in cubic meters per second. The discharge rate given is that of the main stream, preferably averaged over several recent years and measured at the gauging station registering the greatest discharge (or nearest the mouth of the river).

Arrangement

By length of watercourse, under continent.

Coverage

Above * * *, all rivers with continuous watercourses of at least the following lengths:

Africa—500 mi (800 km)
Asia—500 mi (800 km)
Europe—300 mi (480 km)
North America—300 mi (480 km)
Oceania—300 mi (480 km)
South America—500 mi (800 km)

Below * * *, other well-known rivers, including the longest in certain countries and certain border rivers of strategic importance. Also included are all nontributary rivers in the USA at least 250 mi (400 km) long.

Rounding

Lengths are rounded to the nearest 10 mi (km); drainage basin areas, to the nearest 1000 mi² (km²); average discharge rates, to the nearest 1000 cfs (100 when less than 10,000) and to the nearest 10 m³/sec (one when less than 1000).

Entries

670.

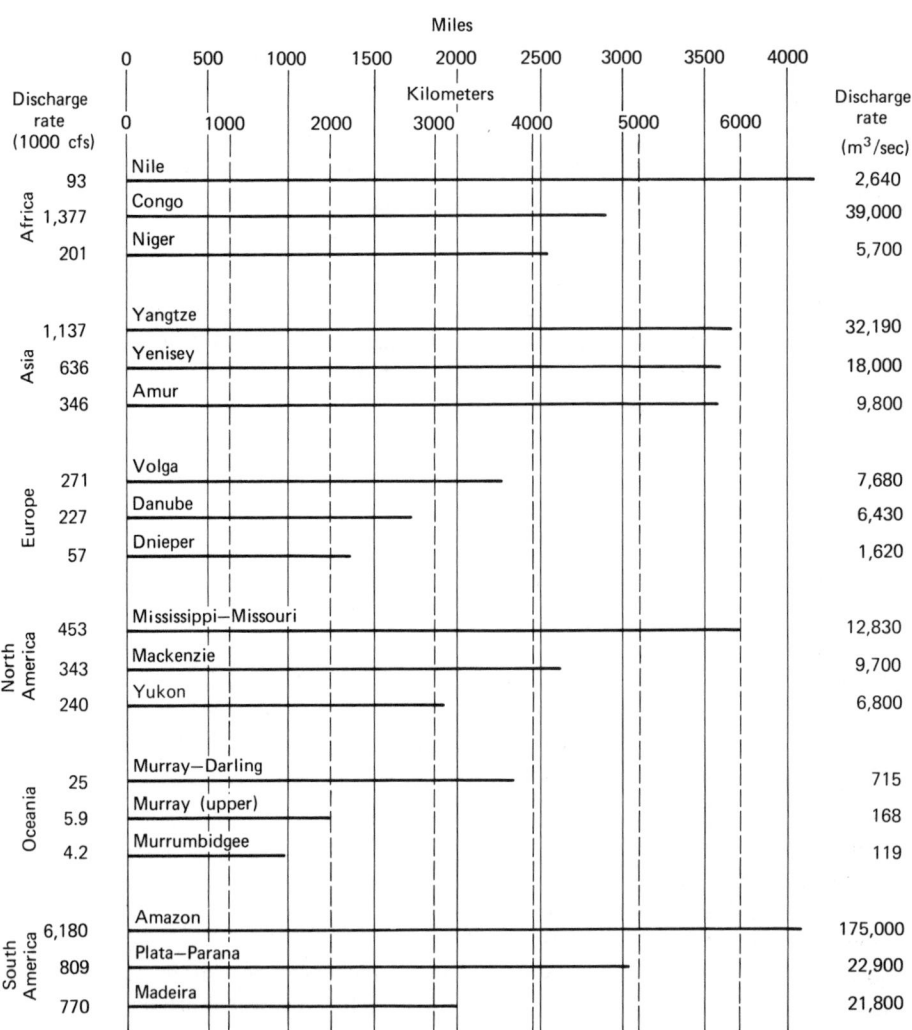

Figure 6.   Three longest rivers of each continent, showing average discharge rate of each river.

TABLE 3. LONGEST RIVERS, BY CONTINENT
AFRICA

Latitude and Longitude	River	Outflow and Location	Length (mi)	Length (km)	Drainage Basin (1000 mi^2)	Drainage Basin (1000 km^2)	Discharge Rate (1000 cfs)	Discharge Rate (m^3/sec)
31.32N, 31.51E[1]	• Nile (off Nil)-Kagera-Ruvuvu-Luvironza	Mediterranean Sea, Egypt	4160	6690	1082	2802	93	2,640
6.04S, 12.24E	Congo (alt Zaire; for Kongo)-Lualaba	Atlantic O, Angola–Zaire	2880	4630	1476	3822	1377	39,000
4.20N, 6.00E	• Niger	G of Guinea, Nigeria	2550	4100	808	2092	201	5,700
18.50S, 36.17E	• Zambezi (alt Zambesi, Zambeze)	Mozambique Channel, Mozambique	1650	2650	514	1331	565	16,000
0.26N, 42.48E	• Shebeli (alt Shabale, Shibeli; for Scebeli)	Balli Swamp, Somalia	1550	2490	77	200	11	320
0.30S, 17.42E	• Ubangi (alt Oubangui)-Uele-Kibali	Congo R, Congo–Zaire	1530	2460	298	773	265	7,500
28.38S, 16.27E	Orange (alt Oranje)	Atlantic O, Namibia–South Africa	1400	2250	261	677	12	345
3.02S, 16.57E	• Kasai (alt Cassai)	Congo R, Zaire	1200	1930	349	904	351	9,950
15.48N, 16.32W	Senegal-Bafing	Atlantic O, Mauritania–Senegal	1050	1700	170	440	29	815
15.38N, 32.31E	• Blue Nile (off Abay, Azraq)	Nile R, Sudan	1000	1610	128	331	57	1,620
18.53S, 22.24E	• Okovanggo (alt Cubango, Okavango)	Okovanggo Basin, Botswana	1000	1610	303	785	9.0	255
5.47N, 0.43E	• Volta-Black Volta (alt Volta Noire)	G of Guinea, Ghana	990	1600	154	398	42	1,180
25.12S, 33.32E	• Limpopo (alt Crocodile)	Indian O, Mozambique	990	1590	170	440		
0.16S, 42.35E	• Juba (alt Ganana; for Giuba)–Ganale-Dorya	Indian O, Somalia	970	1560	76	196	7.1	200
6.46S, 26.58E	• Luvua-Luapula-Chambezi	Lualaba R, Zaire	930	1500	97	250		
0.46N, 24.16E	• Lomami	Congo R, Zaire	900	1450	42	110		

[1] For Damietta (off Dumyat) R distributary; for Nile R proper: 30.10N, 31.06E.

Figure 7. Skylab satellite photograph of the Nile, now generally recognized as the longest river on earth, in the vicinity of the famous Aswan High Dam. (Credit: US National Aeronautics and Space Administration.)

AFRICA

Latitude and Longitude	River	Outflow and Location	Length (mi)	Length (km)	Drainage Basin (1000 mi²)	Drainage Basin (1000 km²)	Discharge Rate (1000 cfs)	Discharge Rate (m³/sec)
12.58N, 14.31E	● Shari (alt Chari)-Sara (alt Ouham)	Chad L, Cameroon–Chad	900	1450	270	700	43	1,230
8.00S, 39.20E	● Rufiji-Luwegu	Indian O, Tanganyika, Tanzania	870	1400	69	178	34	973
1.13S, 16.49E	● Sangha-Kadei	Congo R, Congo	870	1400	70	181	64	1,800
7.50N, 6.50E	● Benue (alt Benoue)	Niger R, Nigeria	810	1300	130	337	112	3,170
1.13N, 23.36E	● Aruwimi-Ituri	Congo R, Zaire	800	1290	45	116		
0.05N, 18.17E	● Ruki-Busira-Tshuapa	Congo R, Zaire	800	1290	67	174		
0.49S, 9.00E	● Ogooue (alt Ogowe)	G of Guinea, Gabon	750	1210	83	216	165	4,670
4.17S, 20.25E	● Sankuru-Lubilash	Kasai R, Zaire	750	1210	60	156	88	2,500
29.04S, 23.38E	● Vaal	Orange R, Cape of Good Hope, South Africa	750	1210			1.3	36
11.12N, 41.40E	Awash (alt Hawash)	Abbe L, Ethiopia	750	1200	21	55	5.7	160
13.27N, 16.37E	Gambia (alt Gambie	Atlantic O, Gambia	700	1130	70	182		
3.14S, 17.22E	● Kwango (alt Cuango)	Kasai R, Zaire	700	1130	102	263	95	2,700
17.40N, 33.56E	● Atbarah (alt Atbara)-Takaze (alt Satit)	Nile R, Sudan	700	1120			25	716
14.30N, 4.12W	Bani-Bagoe	Niger R, Mali	690	1110	50	130	28	796
3.01S, 16.58E	● Fimi-Lukenie	Kasai R, Zaire	660	1070	51	132		
3.22S, 17.22E	● Kwilu (alt Cuilo)	Kwango R, Zaire	650	1050	35	90	2.5	71
13.49N, 10.50W	● Bakoye-Baoule	Senegal R, Mali	620	1000	37	95		
17.47S, 25.10E	● Chobe (alt Linyanti)-Kwando (alt Cuando)	Zambezi R, Botswana–Namibia	620	1000				
15.56S, 28.55E	● Kafue	Zambezi R, Zambia	600	970	58	150		
28.31S, 20.13E	● Molopo	Orange R, Cape of Good Hope, South Africa	600	970				
9.19S, 13.08E	Cuanza	Atlantic O, Angola	600	960	60	156		
12.06N, 15.02E	● Logone-Mbere	Shari R, Chad	600	960	29	76	14	403
17.20S, 11.50E	Cunene (alt Kunene)	Atlantic O, Angola–Namibia	590	940				

AFRICA

Latitude and Longitude	River	Outflow and Location	Length		Drainage Basin		Discharge Rate	
			(mi)	(km)	(1000 mi²)	(1000 km²)	(1000 cfs)	(m³/sec)
3.35N, 9.38E	● Sanaga-Lom	Bight of Bonny of G of Guinea, Cameroon	570	920	52	135	77	2,190
5.02S, 21.07E	● Lulua	Kasai R, Zaire	550	890	25	65		
9.10N, 1.15W	● White Volta (alt Volta Blanche)	Volta R, Ghana	550	890	40	105	8.5	240
3.17N, 9.54E	● Nyong	Bight of Bonny of G of Guinea, Cameroon	530	860	10	26	4.4	125
0.43N, 18.23E	● Lulonga-Lopori	Congo R, Zaire	510	820	25	66		
4.28S, 11.41E	● Kouilou-Niari	Atlantic O, Congo	510	810	22	56	33	940
5.10N, 5.00W	● Bandama	G of Guinea, Ivory Coast	500	800	23	60	12	327
25.46S, 32.43E	● Komati (alt Incomati)	Delagoa B of Indian O, Mozambique	500	800				
0.14S, 20.42E	● Lomela	Busira R, Zaire	500	800				
15.36S, 30.25E	● Luangwa (alt Aruangua)	Zambezi R, Zambia	500	800	56	145		

* * *

4.32N, 36.04E	Omo	Rudolf L, Ethiopia	470	760	26	67		
9.22N, 31.33E	● Sobat-Baro	Nile R, Sudan	460	740	95	245		
10.29S, 40.28E	Ruvuma (alt Rovuma)	Indian O, Mozambique–Tanzania	450	730	58	150		
4.08N, 22.27E	● Bomu (alt Mbomou)	Ubangi R, Central African Empire–Zaire	450	720	97	250		
2.32S, 40.31E	● Tana	Indian O, Kenya	440	710	12	32	4.7	135
5.12N, 3.44W	● Comoe (alt Komoe)	G of Guinea, Ivory Coast	430	700	29	74	9.7	274
36.02N, 0.08E	● Sheliff (alt Cheliff; off Shalaf)	Mediterranean Sea, Algeria	430	700	14	35		
11.24N, 4.07E	Sokoto	Niger R, Nigeria	390	630	32	83	2.5	71
21.00S, 35.02E	Sabi (alt Save)	Mozambique Channel, Mozambique	380	610				
33.19N, 8.20W	● Oum er Rbia	Atlantic O, Morocco	370	600	12	30	3.7	105
4.58N, 6.05W	● Sassandra-Tienba	G of Guinea, Ivory Coast	370	600	19	50	11	325

AFRICA

Latitude and Longitude	River	Outflow and Location	Length (mi)	Length (km)	Drainage Basin (1000 mi²)	Drainage Basin (1000 km²)	Discharge Rate (1000 cfs)	Discharge Rate (m³/sec)
17.42S, 35.19E	● Shire (alt Chire)	Zambezi R, Mozambique	370	600				
11.45N, 15.35W	● Geba-Corubal (alt Koliba)	Atlantic O, Guinea-Bissau	350	560				
7.48N, 0.08E	● Oti	Volta R, Ghana	340	550	28	73	4.6	131
35.06N, 2.20W	● Moulouya	Mediterranean Sea, Morocco	320	520	27	71	1.8	50
4.22N, 7.32W	● Cavally (alt Cavalla)	Atlantic O, Ivory Coast–Liberia	320	510	12	30		
6.16N, 1.49E	Mono	Bight of Benin of G of Guinea, Benin–Togo	310	500	10	25	3.4	96
34.16N, 6.41W	● Sebou	Atlantic O, Morocco	280	450	10	27	4.8	137
16.03S, 46.36E	● Betsiboka	Mozambique Channel, Madagascar	270	440	12	32		
8.30N, 13.15W	● Sierra Leone-Rokel	Atlantic O, Sierra Leone	270	440				
37.07N, 10.13E	● Medjerda (off Majardah)	G of Tunis of Mediterranean Sea, Tunisia	230	360	9	23		

ASIA

Latitude and Longitude	River	Outflow and Location	Length (mi)	Length (km)	Drainage Basin (1000 mi²)	Drainage Basin (1000 km²)	Discharge Rate (1000 cfs)	Discharge Rate (m³/sec)
31.48N, 121.10E	Yangtze (off Chang)	East China Sea, Kiangsu, China	3720	5980	705	1827	1137	32,190
71.50N, 82.40E	Yenisey (alt Yenisei)-Angara[1]-Selenga (alt Selenge)-Ider	Yenisey G of Kara Sea, Russia, USSR	3650	5870	1011	2619	636	18,000
52.56N, 141.10E	Amur (alt Heilung)-Argun (alt Oerhkuna)-Kerulen (off Herlen, Kolulun)	Tatar Strait, Russia, USSR	3590	5780	792	2050	346	9,800
66.45N, 69.30E	Ob-Irtysh[2] (alt Irtish)	G of Ob of Kara Sea, Russia, USSR	3360	5410	1154	2990	360	10,200

[1]Data for Angara-Selenga-Ider: 2120 mi, 3410 km; 401,000 mi², 1,039,000 km²; 147,000 cfs, 4150 m³/sec.

[2]Data for Irtysh: 2640 mi, 4250 km; 634,000 mi², 1,643,000 km²; 147,000 cfs, 4150 m³/sec.

ASIA

Latitude and Longitude	River	Outflow and Location	Length (mi)	Length (km)	Drainage Basin (1000 mi²)	Drainage Basin (1000 km²)	Discharge Rate (1000 cfs)	Discharge Rate (m³/sec)
37.32N, 118.19E	Yellow (alt Hwang; off Huang)	G of Chihli of Yellow Sea, Shantung, China	3010	4840	297	771	54	1,530
72.25N, 126.40E	Lena	Laptev Sea, Russia, USSR	2730	4400	961	2490	569	16,100
10.33N, 105.24E	• Mekong (alt Khong, Lantsang, Mekongk, Tien Giang)	South China Sea, Vietnam	2600	4180	313	811	501	14,200
61.04N, 68.52E	Ob (upper)-Katun	Ob R, Russia, USSR	1970	3180	295	765	174	4,920
46.03N, 61.00E	Syr (Darya)-Naryn	Aral (Sea) L, Kazakhstan, USSR	1880	3020	178	462	17	473
65.48N, 88.04E	Lower Tungu-ska (off Nizhnyaya Tunguska)	Yenisey R, Russia, USSR	1860	2990	183	473	118	3,330
24.20N, 67.47E	Indus (alt Yintu)	Arabian Sea, Pakistan	1790	2880	450	1165	235	6,640
22.50N, 90.50E[1]	Brahmaputra (alt Tsangpo, Yalutsangpu)	B of Bengal, Bangladesh	1770	2840	224	580	575	16,290
16.31N, 97.37E	• Salween (alt Khong, Lu, Nu)	G of Martaban of Andaman Sea, Burma	1750	2820	125	324	353	10,000
41.05N, 86.40E	Tarim (off Talimu)-Yarkand (alt Yeherhchiang)	Tarim Basin, Sinkiang, China	1710	2750	173	447	5.2	146
64.24N, 126.26E	Vilyuy (alt Vilyui)	Lena R, Russia, USSR	1650	2650	175	454	54	1,530
43.40N, 59.01E	Amu (Darya) (alt Oxus)-Panja (alt Pyandzh)-Vakhsh	Aral (Sea) L, Uzbekistan, USSR	1630	2620	180	465	54	1,520
47.00N, 51.48E	Ural	Caspian (Sea) L, Kazakhstan, USSR	1570	2530	89	231	8.7	246
21.55N, 88.05E[2]	• Ganges (alt Ganga)-Bhagirathi	B of Bengal, Bangladesh–India	1560	2510	368	952	441	12,500

[1] For Meghna R distributary; for Brahmaputra R proper: 24.02N, 90.59E.
[2] For Hooghly R distributary; for Ganges R proper: 23.20N, 90.30E.

ASIA

Latitude and Longitude	River	Outflow and Location	Length		Drainage Basin		Discharge Rate	
			(mi)	(km)	(1000 mi^2)	(1000 km^2)	(1000 cfs)	(m^3/sec)
57.42N, 71.12E	Ishim	Irtysh R, Russia, USSR	1520	2450	68	177	1.9	53
29.57N, 48.34E	(Shatt al) Arab-Euphrates (off Firat, Furat)-Kara (Su)	Persian G, Iran–Iraq	1510	2430	427	1105	101	2,860
73.00N, 119.55E	Olenek	Laptev Sea, Russia, USSR	1420	2290	85	220	30	840
63.28N, 129.35E	Aldan	Lena R, Russia, USSR	1410	2270	281	729	177	5,000
69.30N, 161.00E	Kolyma (alt Kolima)	East Siberian Sea, Russia, USSR	1320	2130	248	643	87	2,450
15.50N, 95.06E	● Irrawaddy (off Iyawadi)-Nmai	Andaman Sea, Burma	1300	2090	158	409	447	12,660
70.48N, 148.54E	Indigirka-Khastakh	East Siberian Sea, Russia, USSR	1230	1980	139	360	54	1,520
59.26N, 112.34E	Vitim	Lena R, Russia, USSR	1230	1980	87	225	53	1,510
22.45N, 113.37E	Pearl (alt Canton, Yueh; off Chu)-Si (alt West; off Hsi)-Hung (Shui)-Nanpan	South China Sea, Kwangtung, China	1220	1960	173	448	409	11,590
61.36N, 90.18E	Stony Tunguska (off Podkamennaya Tunguska)	Yenisey R, Russia, USSR	1160	1860	93	240	59	1,680
47.42N, 132.30E	Sungari (off Sunghua)	Amur R, Heilungkiang, China	1160	1860	202	524	71	2,000
31.00N, 47.25E	Tigris (off Dicle, Dijlah)	(Shatt al) Arab R, Iraq	1150	1850	145	373	44	1,250
57.43N, 83.51E	Chulym (alt Chulim)	Ob R, Russia, USSR	1080	1730	52	134	27	775
72.55N, 106.00E	Khatanga-Kotuy	Khatanga G of Laptev Sea, Russia, USSR	1020	1640	141	364	116	3,290
58.06N, 93.00E	Yenisey (upper)	Yenisey R, Russia, USSR	1010	1630	115	299	103	2,910
58.55N, 81.32E	Ket	Ob R, Russia, USSR	1010	1620	36	94	15	415

ASIA

Latitude and Longitude	River	Outflow and Location	Length (mi)	Length (km)	Drainage Basin (1000 mi²)	Drainage Basin (1000 km²)	Discharge Rate (1000 cfs)	Discharge Rate (m³/sec)
70.51N, 153.34E	Alazeya-Nelkan	East Siberian Sea, Russia, USSR	990	1590	25	65	11	320
58.10N, 68.12E	Tobol	Irtysh R, Russia, USSR	990	1590	164	426	29	830
53.20N, 121.26E	Shilka-Onon	Amur R, Russia, USSR	970	1570	78	201	18	504
25.30N, 81.53E	• Yamuna (alt Jumna)-Chambal	Ganges R, Uttar Pradesh, India	950	1530	139	359		
28.57N, 70.30E	Panjnad-Sutlej	Indus R, Pakistan	940	1520	206	533	109	3,080
30.34N, 114.17E	Han (Shui)	Yangtze R, Hupeh, China	930	1500	67	174	42	1,200
71.31N, 136.32E	Yana-Sartang	Laptev Sea, Russia, USSR	930	1490	92	238	32	915
62.38N, 134.32E	Amga	Aldan R, Russia, USSR	910	1460	27	69	6.0	170
17.00N, 81.45E	Godavari	B of Bengal, Andhra Pradesh, India	910	1460	121	313	112	3,180
45.24N, 74.08E	Ili-Tekes	Balkhash L, Kazakhstan, USSR	890	1440	54	140	17	470
60.22N, 120.42E	Olekma	Lena R, Russia, USSR	890	1440	81	210	34	955
15.57N, 80.59E	Krishna (alt Kistna)	B of Bengal, Andhra Pradesh, India	870	1400	100	259	70	1,990
67.32N, 78.40E	Taz	G of Ob of Kara Sea, Russia, USSR	870	1400	58	150	43	1,210
40.40N, 122.15E	Liao	G of Liaotung of Yellow Sea, Liaoning, China	840	1340	83	215	22	624
58.06N, 94.01E	Taseyeva-Chuna	Angara R, Russia, USSR	820	1320	49	128	26	735
26.37N, 101.48E	Yalung	Yangtze R, Szechwan, China	820	1320			88	2,500
21.38N, 72.36E	Narmada (alt Narbada)	G of Cambay of Arabian Sea, Gujarat, India	820	1310	36	93	46	1,290
50.15N, 127.35E	Zeya	Amur R, Russia, USSR	770	1240	90	233	63	1,770
58.48N, 130.35E	Uchur	Aldan R, Russia, USSR	750	1210	44	113	50	1,410
45.00N, 67.44E	Chu	Sauma(kol) L, Kazakhstan, USSR	740	1190	57	148	2.1	61

ASIA

Latitude and Longitude	River	Outflow and Location	Length		Drainage Basin		Discharge Rate	
			(mi)	(km)	(1000 mi^2)	(1000 km^2)	(1000 cfs)	(m^3/sec)
41.45N, 35.59E	Kizil (Irmak)	Black Sea, Turkey	730	1180	30	77	4.7	133
63.28N, 118.50E	Markha	Vilyuy R, Russia, USSR	730	1180	38	99	11	299
45.26N, 124.39E	Nonni (alt Nun; off Nen)	Sungari R, Hei-lungkiang, China	730	1170				
29.23N, 71.02E	Chenab-Chandra	Panjnad R, Pakistan	720	1160	53	138	72	2,050
64.54N, 176.13E	Anadyr (alt Anadir)	G of Anadyr of Bering Sea, Russia, USSR	710	1150	74	191	35	978
37.24N, 60.38E	Hari (Rud) (alt Tedzhen)	Kara (Kum) Des-ert, Turkmenia, USSR	710	1150	27	71	0.9	24
20.17N, 106.34E	Red (alt Coi, Koi; off Hong, Yuan)	G of Tonkin of South China Sea, Vietnam	710	1150	46	120	22[1]	630[1]
29.26N, 113.08E	Siang (off Hsiang)	Tungting L, Hunan, China	710	1150	39	100	88	2,500
50.21N, 106.05E	Orhon (alt Orkhon	Selenga R, Mongolia	700	1120	51	133	4.2	120
31.12N, 61.34E	• Helmand (alt Helmund, Hirmand)	Seistan Basin, Afghanistan–Iran	690	1110	100	259		
68.42N, 158.36E	Omolon	Kolyma R, Russia, USSR	690	1110	44	113		
60.40N, 69.46E	Konda	Irtysh R, Russia, USSR	680	1100	28	73	9.5	270
33.12N, 118.33E	Hwai (off Huai)	Hungtze L, An-hwei, China	680	1090	81	210	38	1,090
54.59N, 73.22E	Om	Irtysh R, Russia, USSR	680	1090	20	53	2.2	61
63.46N, 121.35E	Tyung	Vilyuy R, Russia, USSR	680	1090	19	50		
25.47N, 84.37E	Ghaghara (alt Gogra, Kauriala)	Ganges R, Bihar–Uttar Pradesh, India	670	1080	49	127		
59.07N, 80.46E	Vasyugan	Ob R, Russia, USSR	670	1080	24	62	12	343
60.55N, 73.40E	Bolshoy Yugan	Ob R, Russia, USSR	660	1060	13	35	4.9	138
30.35N, 71.49E	• Ravi	Chenab R, Pakistan	660	1060			8.8	250
54.30N, 134.38E	Maya	Aldan R, Russia, USSR	650	1050	66	171	41	1,170

[1] In China.

ASIA

Latitude and Longitude	River	Outflow and Location	Length (mi)	Length (km)	Drainage Basin (1000 mi²)	Drainage Basin (1000 km²)	Discharge Rate (1000 cfs)	Discharge Rate (m³/sec)
40.30N, 80.48E	• Khotan (off Hotien)–Kara-Kash (off Kalakashih)	Tarim R, Sinkiang, China	640	1030				
57.12N, 66.56E	Tura	Tobol R, Russia, USSR	640	1030	31	80	6.1	174
67.31N, 77.55E	Pur-Pyakupur	G of Ob of Kara Sea, Russia, USSR	640	1020	43	112	31	875
57.43N, 95.24E	Biryusa (alt Ona)	Taseyeva R, Russia, USSR	630	1010	22	56	12	344
0.25S, 109.40E	Kapuas (alt Kapuas-Besar; for Kapoeas)	South China Sea, Borneo I, Indonesia	630	1010	39	102		
41.48N, 86.47E	• Konche (Darya) [alt Kuruk (Darya); off Kungchiao]-Khaydyk(gol) (off Kaitu)	Tarim Basin, Sinkiang, China	630	1010				
37.23N, 50.11E	• Safid-Qezel Owzan	Caspian (Sea) L, Iran	620	1000	22	58	4.6	130
29.34N, 106.35E	Kialing (off Chialing)	Yangtze R, Szechwan, China	620	1000	62	160	88	2,500
13.32N, 100.36E	Chao Phraya (alt Menam)-Nan	G of Siam, Thailand	620	990	58	150	31	883
38.18N, 61.12E	Murgab (alt Morghab)	Kara (Kum) Desert, Turkmenia, USSR	610	980	18	47	1.8	52
50.30N, 69.59E	Nura	Tengiz L, Kazakhstan, USSR	610	980	23	61	0.6	18
38.57N, 117.43E	Hai-Pai (alt Pei)	G of Chihli of Yellow Sea, Hopeh, China	600	970	80	208	8.3	234
60.45N, 76.45E	Vakh	Ob R, Russia, USSR	600	960	30	77	16	448
73.08N, 113.36E	Anabar	Laptev Sea, Russia, USSR	580	940	39	100	18	498
60.22N, 120.50E	Chara	Olekma R, Russia, USSR	570	920	33	86	22	626
71.54N, 102.06E	Kheta	Khatanga R, Russia, USSR	570	920	46	120		
29.43N, 107.24E	Wu	Yangtze R, Szechwan, China	570	920	34	88	53	1,500

ASIA

Latitude and Longitude	River	Outflow and Location	Length (mi)	(km)	Drainage Basin (1000 mi²)	(1000 km²)	Discharge Rate (1000 cfs)	(m³/sec)
48.28N, 135.02E	Ussuri (alt Wusuli)-Ulakhe	Amur R, China–USSR	560	910	72	187	34	953
56.36N, 66.24E	Iset-Miass	Tobol R, Russia, USSR	560	900	32	82	2.5	69
28.46N, 104.38E	● Min-Tatu	Yangtze R, Szechwan, China	560	900	52	134		
66.30N, 87.12E	Kureyka (alt Kureika)	Yenisey R, Russia, USSR	550	890	17	45	25	710
39.20N, 119.10E	Luan	G of Chihli of Yellow Sea, Hopeh, China	540	880	18	46	4.9	140
39.32N, 63.45E	Zeravshan	Kyzyl(kum) Desert, Uzbekistan, USSR	540	880	7	18	5.7	162
51.19N, 106.59E	Khilok	Selenga R, Russia, USSR	540	870	15	38	3.7	105
29.12N, 116.00E	Kan-Kung	Poyang L, Kiangsi, China	540	860			88	2,500
34.41N, 110.10E	Wei	Yellow R, Shensi, China	540	860	24	63		
28.58N, 111.49E	Yuan	Tungting L, Hunan, China	540	860			88	2,500
52.52N, 83.36E	Aley (alt Alei)	Ob R, Russia, USSR	530	860	8	21	1.2	34
20.19N, 86.45E	Mahanadi	B of Bengal, Orissa, India	530	860	51	132	75	2,120
30.25N, 48.12E	● Karun	(Shatt al) Arab R, Iran	530	850	23	61	18	522
21.26N, 95.15E	● Chindwin	Irrawaddy R, Burma	520	840	44	114		
75.41N, 99.20E	Taymyra (alt Taimyra)	Kara Sea, Russia, USSR	520	840	48	124	35	988
56.50N, 84.27E	Tom	Ob R, Russia, USSR	510	830	24	62	40	1,120
73.50N, 87.10E	Pyasina	Kara Sea, Russia, USSR	510	820	70	182	94	2,650
41.07N, 30.39E	Sakarya	Black Sea, Turkey	510	820	21	55	9.1	257
48.01N, 62.45E	Turgay (alt Turgai)	Chelkar-Tengiz Marsh, Kazakhstan, USSR	510	820	61	157		
48.54N, 93.23E	Dzavhan (alt Dzabkhan)	Ayrag L, Mongolia	500	810	27	71	2.1	60
39.55N, 124.20E	Yalu (alt Amnok)	Korea B of Yellow Sea, China–North Korea	500	810	24	63	37	1,040

ASIA

Latitude and Longitude	River	Outflow and Location	Length		Drainage Basin		Discharge Rate	
			(mi)	(km)	(1000 mi^2)	(1000 km^2)	(1000 cfs)	(m^3/sec)
21.15N, 105.20E	Black (off Da, Lihsien)	Red R, Vietnam	500	800				
62.54N, 111.06E	Chona	Vilyuy R, Russia, USSR	500	800	16	41	3.8	109
25.32N, 83.10E	● Gomati (alt Gumti)	Ganges R, Uttar Pradesh, India	500	800	7	19		
1.16S, 104.05E	● Hari (for Djambi)	Berhala Strait, Sumatra I, Indonesia	500	800	20	53		
11.09N, 78.52E	● Kaveri (alt Cauvery)	B of Bengal, Tamil Nadu, India	500	800	31	80	33	934
48.36N, 52.30E	Uil	Aralsor L, Kazakhstan, USSR	500	800	12	31		

* * *

Latitude and Longitude	River	Outflow and Location	Length (mi)	Length (km)	Drainage (1000 mi^2)	Drainage (1000 km^2)	Discharge (1000 cfs)	Discharge (m^3/sec)
25.42N, 84.52E	Son	Ganges R, Bihar, India	480	780	28	72		
56.15N, 162.30E	Kamchatka	Bering Sea, Russia, USSR	470	760	22	56	34	966
64.10N, 65.28E	Sosva (off Severnaya Sosva)	Ob R, Russia, USSR	470	750	38	98	29	834
28.39N, 112.30E	Tze (off Tzu)	Tungting L, Hunan, China	470	750				
23.24N, 110.16E	Yu	Si R, Kwangsi, China	470	750			88	2,500
57.47N, 108.07E	Kirenga	Lena R, Russia, USSR	460	750	18	47	23	648
49.27N, 129.30E	Bureya	Amur R, Russia, USSR	460	740	27	71	32	905
25.34N, 86.42E	Kosi	Ganges R, Bihar, India	450	730	34	87		
31.12N, 72.08E	Jhelum	Chenab R, Pakistan	450	720			31	883
0.35S, 117.17E	Mahakam (alt Kutai)	Makassar Strait, Borneo I, Indonesia	450	720				
38.52N, 38.48E	Murat	Euphrates R, Turkey	450	720				
21.06N, 72.41E	Tapti (alt Tapi)	G of Cambay of Arabian Sea, Gujarat, India	450	720	26	67	9.7	274
46.38N, 53.14E	Emba	Caspian (Sea) L, Kazakhstan, USSR	440	710	16	40	0.5	15

ASIA

Latitude and Longitude	River	Outflow and Location	Length (mi)	Length (km)	Drainage Basin (1000 mi^2)	Drainage Basin (1000 km^2)	Discharge Rate (1000 cfs)	Discharge Rate (m^3/sec)
48.57N, 104.48E	Tuul (alt Tola)	Orhon R, Mongolia	440	700	19	50		
35.36N, 110.42E	Fen	Yellow R, Shansi, China	430	690	15	39		
27.05N, 79.58E	Ramganga	Ganges R, Uttar Pradesh, India	430	690	13	33		
37.28N, 54.03E	Atrek (alt Atrak)	Caspian (Sea) L, Iran-USSR	420	670	11	27	0.3	9
15.19N, 105.30E	Mun	Mekong R, Thailand	420	670				
3.32S, 114.29E	Barito (alt Dusun)	Java Sea, Borneo I, Indonesia	400	650	39	100		
26.05N, 119.32E	Min-Kien (off Chien)	East China Sea, Fukien, China	360	580	22	56	69	1,960
5.42N, 118.23E	Kinabatangan	Sulu Sea, Sabah, Malaysia	350	560				
2.07N, 111.12E	Pajang	South China Sea, Sarawak, Malaysia	350	560				
22.17N, 88.05E	Damodar	Hooghly R, West Bengal, India	340	540	8	22		
6.47S, 112.33E	Solo	Java Sea, Java I, Indonesia	340	540	6	15	31	880
35.07N, 128.57E	Naktong (for Rakuti)	Korea Strait, South Korea	330	520	9	24		
23.48N, 109.31E	Liu	Hung (Shui) R, Kwangsi, China	320	520			88	2,500
2.20S, 104.56E	• Musi (for Moesi)	Bangka Strait, Sumatra I, Indonesia	320	520	22	57		
42.18N, 130.41E	Tumen (alt Tuman, Tumyntszyan)	Sea of Japan, North Korea-USSR	320	520	9	23	2.2	62
30.15N, 120.15E	Tsientang (off Fuchun)	East China Sea, Chekiang, China	310	490	16	43	53	1,490
36.02N, 35.58E	Orontes (off Asi)	Mediterranean Sea, Turkey	300[1]	470[1]	9	23	2.2	62
31.10N, 74.59E	Beas	Sutlej R, Punjab, India	290	460	10	26	18	497
33.55N, 72.14E	• Kabul (for Cabul)	Indus R, Pakistan	270	430	32	83		
22.18N, 72.22E	Sabarmati	G of Cambay of Arabian Sea, Gujarat, India	260	420	21	55	1.4	39

[1]Average of lengths officially given by Syria (350 mi, 570 km) and Turkey (240 mi, 380 km).

ASIA

Latitude and Longitude	River	Outflow and Location	Length (mi)	Length (km)	Drainage Basin (1000 mi²)	Drainage Basin (1000 km²)	Discharge Rate (1000 cfs)	Discharge Rate (m³/sec)
37.57N, 139.04E	Shinano	Sea of Japan, Honshu I, Japan	230	370	5	12		
18.22N, 121.37E	• Cagayan (alt Grande de Cagayan)	Babuyan Channel, Luzon I, Philippines	220	350	14	37	9.3	263
28.01N, 120.44E	Ou (alt Wu)	East China Sea, Chekiang, China	210	340	7	18	21	608
23.41N, 116.38E	Han	South China Sea, Kwangtung, China	200	320	11	29	33	922
7.07N, 124.24E	• Mindanao	Moro G of Celebes Sea, Mindanao I, Philippines	200	320	2	5		
35.44N, 140.51E	Tone	Pacific O, Honshu I, Japan	200	320	7	17	6.5	184
43.15N, 141.23E	Ishikari	Sea of Japan, Hokkaido I, Japan	160	260	6	15		
44.53N, 141.45E	Teshio	Sea of Japan, Hokkaido I, Japan	160	260				
31.46N, 35.33E	• Jordan (off Urdunn, Yarden	Dead (Sea) L, Jordan	160	250	6	16	1.3	37

EUROPE

Latitude and Longitude	River	Outflow and Location	Length (mi)	Length (km)	Drainage Basin (1000 mi²)	Drainage Basin (1000 km²)	Discharge Rate (1000 cfs)	Discharge Rate (m³/sec)
45.45N, 47.52E	Volga	Caspian (Sea) L, Russia, USSR	2290	3690	525	1360	271	7,680
45.20N, 29.40E	Danube (off Donau, Duna, Dunaj, Dunarea, Dunav, Dunay)	Black Sea, Romania–USSR	1770	2850	298	773	227	6,430
46.30N, 32.18E	Dnieper (off Dnepr)	Black Sea, Ukraine, USSR	1420	2280	195	504	57	1,620
47.04N, 39.18E	Don	Sea of Azov of Black Sea, Russia, USSR	1160	1870	163	422	30	851
64.32N, 40.30E	Northern Dvina (off Severnaya Dvina)-Vychegda (alt Vichegda)	White Sea, Russia, USSR	1160	1860	139	360	120	3,400

EUROPE

Latitude and Longitude	River	Outflow and Location	Length (mi)	Length (km)	Drainage Basin (1000 mi²)	Drainage Basin (1000 km²)	Discharge Rate (1000 cfs)	Discharge Rate (m³/sec)
68.13N, 54.15E	Pechora (for Petchora)	Barents Sea, Russia, USSR	1120	1810	124	322	141	4,000
55.25N, 50.40E	Kama	Volga R, Russia, USSR	1120	1800	196	507	128	3,620
56.20N, 43.59E	Oka	Volga R, Russia, USSR	930	1500	95	245	39	1,110
55.54N, 53.33E	Belaya	Kama R, Russia, USSR	890	1430	55	142	30	850
39.24N, 49.19E	Kura	Caspian (Sea) L, Azerbaijan, USSR	850	1360	73	188	18	514
46.18N, 30.17E	Dniester (off Dnestr)	Black Sea, Moldavia, USSR	840	1350	28	72	11	310
51.47N, 4.10E[1]	Rhine (off Rhein, Rhin, Rijn)	North Sea, Netherlands	820	1320	97	252	88	2,490
55.36N, 51.30E	Vyatka (alt Viatka)	Kama R, Russia, USSR	820	1310	50	129	29	830
54.21N, 18.56E	Vistula (for Visla, Weichsel; off Wisla)-Bug (alt Zapadnyy Bug)	G of Danzig of Baltic Sea, Poland	750	1200	75	194	34	967
53.50N, 9.00E	Elbe (alt Labe)	North Sea, Niedersachsen-Schleswig-Holstein, West Germany	720	1160	56	144	25	703
50.33N, 30.32E	Desna	Dnieper R, Ukraine, USSR	700	1130	34	89	12	330
39.56N, 48.20E	Araks (alt Aras)	Kura R, Azerbaijan, USSR	670	1070	39	102	7.5	212
47.35N, 40.54E	Donets (alt Northern Donets; off Severnyy Donets)	Don R, Russia, USSR	650	1050	38	99	5.4	153
47.16N, 2.11W	Loire	B of Biscay, Pays de la Loire, France	630	1020	46	120	31	871
57.00N, 24.00E	Western Dvina (alt Daugava; for Duna; off Zapadnaya Dvina)	G of Riga of Baltic Sea, Latvia, USSR	630	1020	34	88	25	697

[1] For Haringvliet Estuary distributary; for Rhine R proper: 51.52N, 6.02E.

EUROPE

Latitude and Longitude	River	Outflow and Location	Length (mi)	Length (km)	Drainage Basin (1000 mi²)	Drainage Basin (1000 km²)	Discharge Rate (1000 cfs)	Discharge Rate (m³/sec)
49.36N, 42.19E	Khoper	Don R, Russia, USSR	630	1010	24	61	5.3	151
38.40N, 9.24W	Tagus (off Tajo, Tejo)	Atlantic O, Portugal	630	1010	31	81	4.5	128
59.57N, 30.20E	Neva-Volkhov-Lovat	G of Finland of Baltic Sea, Russia, USSR	620	1000	108	281	89	2,530
66.11N, 43.59E	Mezen	White Sea, Russia, USSR	600	970	30	78	23	648
45.15N, 20.17E	Tisza (alt Tisa, Tissa; for Theiss)	Danube R, Serbia, Yugoslavia	600	970	61	157	30	844
51.47N, 4.10E[1]	Meuse (alt Maas)	North Sea, Netherlands	590	950	19	49	9.5	269
53.32N, 14.38E	Oder (alt Odra)-Warta (for Warthe)	Baltic Sea, East Germany–Poland	590	950	46	119	19	526
45.30N, 28.12E	Prut (alt Prutul; for Pruth)	Danube R, Romania–USSR	590	950	11	27	2.5	70
55.18N, 21.23E	Neman (alt Nemunas; for Memel, Niemen)	Baltic Sea, Lithuania–Russia, USSR	580	940	38	98	22	629
44.50N, 20.28E	Sava (alt Save; for Sau, Szava)	Danube R, Serbia, Yugoslavia	580	940	37	96	60	1,700
54.40N, 56.00E	Ufa	Belaya R, Russia, USSR	580	930	20	53	12	351
40.43N, 0.54E	Ebro	Mediterranean Sea, Cataluna, Spain	570	910	33	85	6.1	173
45.20N, 37.22E	Kuban	Sea of Azov of Black Sea, Russia, USSR	560	910	22	58	8.4	238
41.08N, 8.40W	Douro (alt Duero)	Atlantic O, Portugal	560	890	38	98	11	312
56.18N, 46.24E	Vetluga	Volga R, Russia, USSR	550	890	15	39	8.2	231
46.59N, 31.58E	Southern Bug (off Yuzhnyy Bug)	Dnieper R, Ukraine, USSR	530	860	25	64	2.9	83
56.06N, 46.00E	Sura	Volga R, Russia, USSR	520	840	26	67	7.6	215

[1] For Haringvliet Estuary distributary; for Meuse R proper: 51.49N, 5.01E.

EUROPE

Latitude and Longitude	River	Outflow and Location	Length		Drainage Basin		Discharge Rate	
			(mi)	(km)	(1000 mi^2)	(1000 km^2)	(1000 cfs)	(m^3/sec)
43.20N, 4.50E	Rhone	Mediterranean Sea, Langue-doc–Provence–Cote d'Azur, France	500	810	38	99	53	1,500
58.13N, 56.22E	Chusovaya	Kama R, Russia, USSR	500	800	18	48	8.0	226
46.15N, 20.12E	Maros (alt Mu-res, Muresul)	Tisza R, Hungary	500	800	12	30	5.4	154
51.46N, 55.01E	Sakmara	Ural R, Russia, USSR	500	800	12	30	4.7	133
47.31N, 40.45E	Sal	Don R, Russia, USSR	500	800	8	21	0.4	12
37.14N, 7.22W	Guadiana	G of Cadiz of Atlantic O, Portugal–Spain	480	780	26	68	3.2	91
64.08N, 41.54E	Pinega	Northern Dvina R, Russia, USSR	480	780	16	43	12	353
49.26N, 0.26E	Seine	English Channel, Haute-Norman-die, France	480	780	30	79	9.6	272
51.10N, 30.30E	Pripet (off Pripyat)	Dnieper R, Ukraine, USSR	480	770	44	114	13	372
51.27N, 32.34E	Seym (alt Seim)	Desna R, Ukraine, USSR	460	750	11	27	3.6	103
49.35N, 42.41E	Medveditsa	Don R, Russia, USSR	460	740	13	35	2.5	71
43.43N, 24.51E	Olt (for Aluta)	Danube R, Romania	460	740	9	24	5.7	160
53.32N, 8.34E	Weser-Werra	North Sea, Nie-dersachsen, West Germany	460	730	18	46	12	334
57.42N, 11.52E	Gota-Klar	Kattegat Strait, Sweden	450	720	19	50	23	640
52.35N, 14.39E	Oder (upper)	Oder R, Poland	450	720	21	54		
49.01N, 33.32E	Psel	Dnieper R, Ukraine, USSR	450	720	9	23	1.9	54
45.33N, 18.55E	Drava (alt Drau, Drave)	Danube R, Cro-atia, Yugoslavia	440	710	15	40	22	611
45.24N, 28.01E	Siret (for Sereth; off Seret, Siretul)	Danube R, Romania	440	710	18	48	14	400
59.27N, 28.02E	Narva-Velikaya	G of Finland of Baltic Sea, Estonia, USSR	430	700	22	56	15	416

EUROPE

Latitude and Longitude	River	Outflow and Location	Length (mi)	Length (km)	Drainage Basin (1000 mi²)	Drainage Basin (1000 km²)	Discharge Rate (1000 cfs)	Discharge Rate (m³/sec)
56.10N, 42.58E	Klyazma	Oka R, Russia, USSR	430	690	16	42	7.1	202
52.31N, 21.05E	Vistula (upper)	Vistula R, Poland	420	680	33	85		
52.01N, 47.24E	Bolshoy Irgiz	Volga R, Russia, USSR	420	670	9	24	0.8	24
52.08N, 27.17E	Goryn	Pripet R, White Russia, USSR	410	660	11	28	3.1	88
36.47N, 6.22W	Guadalquivir	G of Cadiz of Atlantic O, Andalucia, Spain	410	660	22	57	6.4	182
54.44N, 41.53E	Moksha	Oka R, Russia, USSR	410	660	20	51	5.4	154
65.57N, 56.55E	Usa	Pechora R, Russia, USSR	410	650	36	94	36	1,030
45.35N, 1.03W	Gironde-Garonne (alt Garona)	B of Biscay, Aquitaine-Poitou-Charentes, France	400	650	33	85	21	590
51.57N, 30.48E	Sozh	Dnieper R, White Russia, USSR	400	650	16	42	7.3	207
51.30N, 53.22E	Ilek	Ural R, Russia, USSR	390	620	16	41	1.3	36
43.44N, 46.33E	Terek	Caspian (Sea) L, Russia, USSR	390	620	17	43	11	305
44.57N, 12.04E	Po	Adriatic Sea, Veneto, Italy	380	620	29	75	54	1,540
52.33N, 30.14E	Berezina	Dnieper R, White Russia, USSR	380	610	9	24	4.5	127
59.12N, 10.57E	Glomma (off Glama)	Skagerrak Strait, Norway	380	610	16	41	25	697
60.30N, 32.48E	Svir-Suna	Ladoga L, Russia, USSR	370	600	32	83	22	617
53.10N, 50.04E	Samara	Volga R, Russia, USSR	370	590	18	46	1.7	47
62.48N, 42.56E	Vaga	Northern Dvina R, Russia, USSR	360	570	17	45	15	428
60.45N, 46.20E	Yug	Northern Dvina R, Russia, USSR	360	570	14	37	11	300
65.48N, 24.08E	Torne (alt Tornio)-Muonio-Konkama (alt Kongama)	G of Bothnia of Baltic Sea, Finland–Sweden	350	570	15	39	13	366
60.46N, 46.24E	Sukhona	Northern Dvina R, Russia, USSR	350	560	19	50	15	437

EUROPE

Latitude and Longitude	River	Outflow and Location	Length (mi)	Length (km)	Drainage Basin (1000 mi²)	Drainage Basin (1000 km²)	Discharge Rate (1000 cfs)	Discharge Rate (m³/sec)
52.53N, 11.58E	Havel-Spree	Elbe R, East Germany	340	550	9	24	3.2	90
46.41N, 32.50E	Ingulets	Dnieper R, Ukraine, USSR	340	550	5	14	0.3	9
65.47N, 24.30E	Kemi	G of Bothnia of Baltic Sea, Finland	340	550	20	51	20	578
50.22N, 7.36E	Moselle (alt Mosel)	Rhine R, Rheinland-Pfalz, West Germany	340	550	11	28	10	292
57.20N, 43.08E	Unzha	Volga R, Russia, USSR	340	550	11	27	6.2	176
44.43N, 21.03E	Morava (off Velika Morava)-Southern Morava (off Juzna Morava)	Danube R, Serbia, Yugoslavia	330	540	14	37	8.9	253
65.19N, 52.54E	Izhma	Pechora R, Russia, USSR	330	530	12	31	7.2	203
50.00N, 8.18E	Main	Rhine R, Hessen, West Germany	330	520	10	27	3.5	100
48.49N, 2.24E	Marne	Seine R, Region Parisienne, France	330	520	5	14	3.5	98
60.38N, 17.27E	Dal	G of Bothnia of Baltic Sea, Sweden	320	520	11	29	12	333
48.35N, 13.28E	Inn	Danube R, Bayern, West Germany	320	510	10	26	26	735
54.54N, 23.53E	Viliya (alt Neris; for Wilja)	Neman R, Lithuania, USSR	320	510	10	25	6.6	188
39.09N, 0.14W	Jucar	Mediterranean Sea, Valencia, Spain	310	500	8	21	2.1	60
55.05N, 38.51E	Moscow (off Moskva)	Oka R, Russia, USSR	310	500	7	18	2.3	64
40.52N, 26.12E	Maritsa (alt Evros, Meric)	Aegean Sea, Greece–Turkey	300	490	14	35		
44.18N, 0.20E	Lot	Garonne R, Aquitaine, France	300	480	4	10	4.5	128
52.26N, 20.42E	Narew (alt Narev)	Bug R, Poland	300	480	29	75	10	289
45.44N, 4.50E	Saone	Rhone R, Rhone-Alpes, France	300	480	12	30	15	424

* * *

EUROPE

Latitude and Longitude	River	Outflow and Location	Length (mi)	Length (km)	Drainage Basin (1000 mi²)	Drainage Basin (1000 km²)	Discharge Rate (1000 cfs)	Discharge Rate (m³/sec)
45.02N, 0.35W	Dordogne	Garonne R, Aquitaine, France	290	470	9	23	10	286
62.48N, 17.56E	Angerman	G of Bothnia of Baltic Sea, Sweden	290	460	12	32	16	448
63.47N, 20.16E	Ume	G of Bothnia of Baltic Sea, Sweden	290	460	10	27	16	450
48.50N, 34.05E	Vorskla	Dnieper R, Ukraine, USSR	290	460	6	15	1.1	30
58.50N, 37.11E	Mologa	Volga R, Russia, USSR	280	460	11	30	11	314
65.35N, 22.03E	Lule	G of Bothnia of Baltic Sea, Sweden	280	450	10	25	16	444
58.25N, 31.20E	Msta	Ilmen L, Russia, USSR	280	440	9	23	5.7	161
50.45N, 21.51E	San	Vistula R, Poland	280	440	7	17	4.1	116
46.18N, 16.55E	Mur (alt Mura)	Drava R, Croatia, Yugoslavia	270	440	5	14	5.9	166
51.57N, 11.55E	Saale (alt Sachsische Saale)	Elbe R, East Germany	270	430	9	24	3.7	105
51.22N, 4.15E	Scheldt (alt Escaut, Schelde)	North Sea, Netherlands	270	430	8	20	5.5	155
48.07N, 22.20E	Somes (off Somesul, Szamos)	Tisza R, Romania	270	430	7	19	2.8	80
50.22N, 14.28E	Vltava (alt Moldau)	Elbe R, Bohemia, Czechoslovakia	270	430	11	28	5.1	145
63.58N, 38.02E	Onega	Onega B of White Sea, Russia, USSR	260	420	22	57	17	476
40.35N, 22.50E	Vardar (alt Axios, Vardaris)	G of Salonika of Aegean Sea, Greece	260	420	11	28		
45.10N, 12.20E	Adige (for Etsch)	Adriatic Sea, Veneto, Italy	250	410	6	15	9.3	262
44.42N, 27.51E	Ialomita	Danube R, Romania	250	410	5	12	2.5	70
41.44N, 12.14E	Tiber (off Tevere)	Tyrrhenian Sea, Lazio, Italy	250	410	7	17	8.4	239
47.55N, 18.00E	Vah (for Vag, Waag)	Danube R, Slovakia, Czechoslovakia	240	390	4	11	5.6	158

EUROPE

Latitude and Longitude	River	Outflow and Location	Length (mi)	Length (km)	Drainage Basin (1000 mi^2)	Drainage Basin (1000 km^2)	Discharge Rate (1000 cfs)	Discharge Rate (m^3/sec)
52.30N, 9.55W	Shannon	Atlantic O, Ireland	230	370	6	16	7.0	198
51.25N, 3.00W	Severn	Bristol Channel, England, UK	210	340	8	21	2.2	62
53.32N, 0.08E	● Humber-Trent	North Sea, England, UK	210	330	9	23	7.0	198
51.30N, 0.45E	Thames	North Sea, England, UK	200	320	6	16	2.4	67
47.36N, 8.13E	Aare (alt Aar)	Rhine R, Switzerland	180	290	7	18	19	552
41.45N, 19.34E	Drin	Adriatic Sea, Albania	170	280	2	6	10	290
43.41N, 10.17E	● Arno	Ligurian Sea, Toscana, Italy	150	240	3	8	4.9	140
63.47N, 20.48W	● Thjorsa	Atlantic O, Iceland	140	230	3	7	14	395
56.22N, 3.21W	Tay	North Sea, Scotland, UK	120	190	2	6	5.5	156
56.29N, 10.13E	Gudena	Kattegat Strait, Denmark	100	160	1	3	0.6	16

NORTH AMERICA

Latitude and Longitude	River	Outflow and Location	Length (mi)	Length (km)	Drainage Basin (1000 mi^2)	Drainage Basin (1000 km^2)	Discharge Rate (1000 cfs)	Discharge Rate (m^3/sec)
29.02N, 89.15W	Mississippi-Missouri[1]-Jefferson-Beaverhead-Red Rock	G of Mexico, Louisiana, USA	3740	6020	1247	3230	453	12,830
69.15N, 134.08W	Mackenzie-Slave-Peace-Finlay	Beaufort Sea, NWT, Canada	2630	4240	697	1805	343	9,700
62.32N, 163.54W	Yukon-Lewes-Teslin-Nisutlin	Bering Sea, Alaska, USA	1980	3180	328	850	240	6,800
48.09N, 67.10W	Saint Lawrence (alt Saint-Laurent)-(Great Lakes)-Saint Louis	G of Saint Lawrence, Quebec, Canada	1900[2]	3060[2]	503	1303	460	13,030
25.58N, 97.09W	Rio Grande (alt Bravo)	G of Mexico, Mexico–USA	1880	3030	172	445	2.7	77

[1] Data for Missouri-Jefferson-Beaverhead-Red Rock: 2560 mi, 4130 km; 529,000 mi^2, 1,370,000 km^2; 76,000 cfs, 2160 m^3/sec.

[2] For Saint Lawrence R proper: 590 mi, 960 km.

NORTH AMERICA

Latitude and Longitude	River	Outflow and Location	Length (mi)	Length (km)	Drainage Basin (1000 mi²)	Drainage Basin (1000 km²)	Discharge Rate (1000 cfs)	Discharge Rate (m³/sec)
57.04N, 92.30W	Nelson-Sas-katchewan-South Sas-katchewan-Bow	Hudson B, Manitoba, Canada	1600	2570	414	1072	92	2,600
33.47N, 91.04W	Arkansas	Mississippi R, Arkansas, USA	1450	2330	161	417	42	1,190
31.54N, 114.57W	Colorado	G of California, Baja Califor-nia–Sonora, Mexico	1450	2330	246	637	3.7	104
29.53N, 91.28W	Atchafalaya-Red	G of Mexico, Louisiana, USA	1400	2260	95	246	181[1]	5,120[1]
46.15N, 124.03W	Columbia-Snake[2]	Pacific O, Ore-gon–Washing-ton, USA	1320	2130	253	655	189	5,350
28.52N, 95.22W	Brazos	G of Mexico, Texas, USA	1310	2110	44	115	5.3	149
36.59N, 89.08W	Ohio-Allegheny	Mississippi R, Illinois–Ken-tucky, USA	1310	2100	258	668	257	7,280
58.47N, 94.12W	Churchill-Beaver	Hudson B, Man-itoba, Canada	1300	2100	108	281	36	1,030
38.49N, 90.07W	Mississippi (upper)	Mississippi R, Missouri, USA	1170	1880	172	446	98	2,790
41.03N, 95.53W	Platte-North Platte	Missouri R, Nebraska, USA	990	1590	90	233	5.0	142
29.42N, 101.22W	Pecos	Rio Grande R, Texas, USA	930	1490	38	99	0.1	3
35.27N, 95.03W	Canadian	Arkansas R, Oklahoma, USA	910	1460	30	77	6.3	179
37.04N, 88.34W	Tennessee-Holston	Ohio R, Ken-tucky, USA	900	1450	41	106	64	1,810
28.36N, 95.59W	Colorado	Matagorda B of G of Mexico, Texas, USA	890	1440	41	107	2.4	67
46.12N, 119.02W	Columbia (upper)	Columbia R, Washington, USA	890	1430	96	249	120	3,410
49.04N, 123.07W	Fraser	Strait of Georgia, British Colum-bia, Canada	850	1370	85	220	137	3,880

[1] Inc approximately 25% of the Mississippi R flow, diverted to control flooding; the average discharge rate of the Mississippi at Vicksburg, above the point of diversion, is 565,000 cfs, 16,000 m³/sec. The average discharge rate of the Red R is 31,000 cfs, 880 m³/sec.

[2] Data for Snake: 1000 mi, 1610 km; 109,000 mi², 282,000 km²; 49,000 cfs, 1,390 m³/sec.

NORTH AMERICA

Latitude and Longitude	River	Outflow and Location	Length (mi)	Length (km)	Drainage Basin (1000 mi²)	Drainage Basin (1000 km²)	Discharge Rate (1000 cfs)	Discharge Rate (m³/sec)
53.15N, 105.05W	North Saskatchewan	Saskatchewan R, Saskatchewan, Canada	800	1290	46	119	7.7	219
45.25N, 74.00W	Ottawa (alt Outaouais)	Saint Lawrence R, Quebec, Canada	790	1270	56	146	69	1,950
30.41N, 88.00W	Mobile-Alabama-Coosa-Etowah	Mobile B of G of Mexico, Alabama, USA	780	1260	44	113	61	1,730
35.16N, 95.31W	North Canadian	Canadian R, Oklahoma, USA	780	1260	14	36	0.8	22
58.40N, 110.50W	Athabasca	Athabasca L, Alberta, Canada	760	1230	63	163	23	640
38.11N, 109.53W	Green	Colorado R, Utah, USA	730	1170	44	115	6.5	185
48.03N, 106.19W	Milk	Missouri R, Montana, USA	730	1170	23	58	0.7	20
37.09N, 88.24W	Cumberland	Ohio R, Kentucky, USA	720	1160	18	47	28	793
29.45N, 94.43W	Trinity	Galveston B of G of Mexico, Texas, USA	710	1150	18	47	7.4	209
42.52N, 97.18W	James (alt Dakota)	Missouri R, South Dakota, USA	710	1140	22	57	0.3	9
39.07N, 94.37W	Kansas (alt Kaw)-Smoky Hill	Missouri R, Kansas, USA	710	1140	61	159	6.6	186
60.05N, 162.25W	Kuskokwim	Kuskokwim B of Bering Sea, Alaska, USA	710	1140	49	127	44	1,250
50.24N, 96.48W	Red-Assiniboine	Winnipeg L, Manitoba, Canada	710	1140	111	287	8.6	243
36.07N, 96.30W	Cimarron	Arkansas R, Oklahoma, USA	700	1120	19	49	1.8	50
61.51N, 121.18W	Liard	Mackenzie R, NWT, Canada	690	1120	86	223	78	2,210
18.24N, 92.38W	Usumacinta-Chixoy	B of Campeche of G of Mexico, Tabasco, Mexico	690	1110	40	103	61	1,730
33.57N, 91.05W	White	Mississippi R, Arkansas, USA	680	1100	28	73	29	832
47.58N, 103.59W	Yellowstone	Missouri R, North Dakota, USA	670	1080	70	181	13	365

NORTH AMERICA

Latitude and Longitude	River	Outflow and Location	Length (mi)	Length (km)	Drainage Basin (1000 mi²)	Drainage Basin (1000 km²)	Discharge Rate (1000 cfs)	Discharge Rate (m³/sec)
50.37N, 96.20W	Winnipeg-English	Winnipeg L, Manitoba, Canada	650	1050	44	114	31	881
49.53N, 97.07W	Red (upper)-Otter Tail	Red R, Manitoba, Canada	640	1030	48	124	5.9	167
32.43N, 114.33W	Gila	Colorado R, Arizona, USA	630	1010	58	150	0.03	0.8
21.36N, 105.26W	Santiago (alt Grande de Santiago)-Lerma	Pacific O, Nayarit, Mexico	630	1010	48	125	13	363
34.08N, 96.36W	Washita	Red R, Oklahoma, USA	630	1010	8	21	1.5	44
52.17N, 81.31W	Albany-Cat	James B of Hudson B, Ontario, Canada	610	980	26	69	44	1,230
56.02N, 87.36W	Severn-Black Birch	Hudson B, Ontario, Canada	610	980	39	102	15	439
67.15N, 95.15W	Back	Arctic O, NWT, Canada	600	970	35	91	18	521
31.16N, 91.50W	Black-Ouachita	Red R, Louisiana, USA	600	970	18	47	18	515
66.34N, 145.19W	Porcupine	Yukon R, Alaska, USA	580[1]	940[1]	46	120	20	566
64.33N, 100.06W	Dubawnt	Beverly L, NWT, Canada	580	930	27	69	13	365
47.36N, 102.25W	Little Missouri	Missouri R, North Dakota, USA	560	900	9	23	0.6	16
64.16N, 96.05W	Thelon	Baker L, NWT, Canada	560	900	25	65	4	113
53.50N, 79.00W	Grande (alt Fort George)	James B of Hudson B, Quebec, Canada	550	890	38	98	61	1,720
39.03N, 96.48W	Republican-Arikaree	Kansas R, Kansas, USA	550	890	25	65	1.1	31
65.09N, 151.57W	Tanana-Chisana	Yukon R, Alaska, USA	550	890	44	115	41	1,160
58.30N, 68.10W	Koksoak-Caniapiscau	Ungava B of Hudson Strait, Quebec, Canada	540	870	53	137	85	2,410
33.07N, 79.17W	Santee-Wateree-Catawba	Atlantic O, South Carolina, USA	540	870	15	39	2.3	64

[1]Average of lengths officially given by Canada (710 mi, 1150 km) and USA (460 mi, 740 km).

NORTH AMERICA

Latitude and Longitude	River	Outflow and Location	Length (mi)	Length (km)	Drainage Basin (1000 mi²)	Drainage Basin (1000 km²)	Discharge Rate (1000 cfs)	Discharge Rate (m³/sec)
53.20N, 60.20W	Churchill (for Hamilton)-Ashuanipi	Atlantic O, Newfoundland, Canada	530	860	31	80	56	1,580
40.23N, 91.25W	Des Moines	Mississippi R, Iowa–Missouri, USA	530	860	16	41	5.1	144
44.41N, 101.18W	Cheyenne	Missouri R, South Dakota, USA	530	850	25	66	0.9	26
37.48N, 88.02W	Wabash	Ohio R, Illinois–Indiana, USA	530	850	33	86	27	753
29.43N, 84.58W	Apalachicola-Chattahoochee	G of Mexico, Florida, USA	520	840	20	51	22	614
67.49N, 115.04W	Coppermine	Arctic O, NWT, Canada	520	840	16	40	4.1	116
64.55N, 157.32W	Koyukuk	Yukon R, Alaska, USA	520	840	33	84	22	623
31.08N, 87.57W	Tombigbee	Mobile R, Alabama, USA	520	840	20	52	25	711
43.42N, 99.27W	White	Missouri R, South Dakota, USA	510	820	10	26	0.5	15
49.00N, 117.36W	Pend Oreille-Clark Fork	Columbia R, British Columbia, Canada	500	810	26	67	28	804
62.12N, 159.43W	Innoko	Yukon R, Alaska, USA	500	800				
38.35N, 91.58W	Osage-Marais des Cygnes	Missouri R, Missouri, USA	500	800	15	39	10	283
30.11N, 89.31W	Pearl	G of Mexico, Louisiana–Mississippi, USA	490	790	8	20	8.8	249
32.22N, 90.54W	Yazoo-Tallahatchie	Mississippi R, Mississippi, USA	490	790	9	23	9.6	272
46.44N, 105.26W	Powder	Yellowstone R, Montana, USA	490	780	13	34	0.6	17
49.15N, 117.39W	Kootenay (alt Kootenai)	Columbia R, British Columbia, Canada	480	780	15	39	28	801
51.25N, 78.55W	Nottaway-Bell-Megiscane	James B of Hudson B, Quebec, Canada	480	780	25	66	43	1,210
17.55N, 102.10W	Balsas	Pacific O, Guerrero–Michoacan, Mexico	480	770	43	112	16	439

NORTH AMERICA

Latitude and Longitude	River	Outflow and Location	Length (mi)	Length (km)	Drainage Basin (1000 mi²)	Drainage Basin (1000 km²)	Discharge Rate (1000 cfs)	Discharge Rate (m³/sec)
52.15N, 78.32W	Eastmain	James B of Hudson B, Quebec, Canada	470	760	18	46	32	895
51.30N, 78.48W	Rupert-Temiscamie	James B of Hudson B, Quebec, Canada	470	760	17	43	32	903
15.00N, 83.10W	• Coco (alt Segovia)	Caribbean Sea, Honduras–Nicaragua	470	750			18	500
52.57N, 82.18W	Attawapiskat	James B of Hudson B, Ontario, Canada	460	750	15	39	11	309
28.27N, 96.47W	Guadalupe	San Antonio B of G of Mexico, Texas, USA	460	740	9	23	2.1	59
35.48N, 95.18W	Neosho (alt Grand)	Arkansas R, Oklahoma, USA	460	740	13	33	6.5	184
64.03N, 95.35W	Kazan	Baker L, NWT, Canada	450	730	23	59	13	368
31.20N, 81.20W	Altamaha-Ocmulgee-South	Atlantic O, Georgia, USA	450	720	14	36	13	362
46.09N, 107.28W	Bighorn-Wind	Yellowstone R, Montana, USA	450	720	21	54	3.6	102
55.16N, 77.48W	Grande (Riviere) de la Baleine (alt Great Whale)	Hudson B, Quebec, Canada	450	720	16	43	23	665
50.56N, 109.54W	Red Deer	South Saskatchewan, R, Saskatchewan, Canada	450	720	17	44	3.1	87
49.39N, 99.34W	Souris	Assiniboine R, Manitoba, Canada	450	720	24	62	0.5	13
41.07N, 100.41W	South Platte	Platte R, Nebraska, USA	440	710	24	63	0.2	6
39.32N, 76.04W	Susquehanna	Chesapeake B of Atlantic O, Maryland, USA	440	710	28	71	35	994
60.51N, 115.44W	Hay	Great Slave L, NWT, Canada	440	700	19	49	2.9	83
18.36N, 92.39W	Grijalva (alt Mezcalapa)	B of Campeche of G of Mexico, Tabasco, Mexico	430	700	20	52	7.1	200

NORTH AMERICA

Latitude and Longitude	River	Outflow and Location	Length (mi)	Length (km)	Drainage Basin (1000 mi²)	Drainage Basin (1000 km²)	Discharge Rate (1000 cfs)	Discharge Rate (m³/sec)
33.22N, 79.16W	Pee Dee-Yadkin	Winyah B of Atlantic O, South Carolina, USA	430	700	9	23	9.0	256
48.10N, 69.45W	Saguenay-Peribonca	Saint Lawrence R, Quebec, Canada	430	700	34	88	63	1,780
21.45N, 105.30W	San Pedro (alt Mezquital)	Pacific O, Nayarit, Mexico	430	700	7	18	3.9	110
69.42N, 129.01W	Anderson	Wood B of Beaufort Sea, NWT, Canada	430	690	24	62	7.0	198
38.41N, 85.11W	Kentucky	Ohio R, Kentucky, USA	430	690	7	18	8.1	229
42.46N, 98.03W	Niobrara	Missouri R, Nebraska, USA	430	690	12	31	1.4	40
42.59N, 91.09W	Wisconsin	Mississippi R, Wisconsin, USA	430	690	12	31	8.5	241
42.29N, 96.27W	Big Sioux	Missouri R, Iowa–South Dakota, USA	420	680	9	23	0.8	23
51.58N, 98.04W	Dauphin-Fairford-Red Deer	Winnipeg L, Manitoba, Canada	420	680	31	80	2.4	68
38.58N, 90.28W	Illinois-Kankakee	Mississippi R, Illinois, USA	420	680	28	72	21	586
67.41N, 134.32W	Peel-Ogilvie	Mackenzie R, NWT, Canada	420	680	27	71	16	450
34.37N, 90.35W	Saint Francis	Mississippi R, Arkansas, USA	420	680	8	22	5.9	167
45.51N, 116.47W	Salmon	Snake R, Idaho, USA	420	680	14	36	11	309
27.37N, 110.39W	Yaqui-Bavispe	G of California, Sonora, Mexico	420	680	25	66	3.8	108
45.15N, 66.04W	Saint John	B of Fundy of Atlantic O, New Brunswick, Canada	420	670	22	56	26	736
36.56N, 76.27W	James-Jackson	Chesapeake B of Atlantic O, Virginia, USA	410	670	7	18	7.5	213
41.16N, 72.20W	Connecticut	Long Island Sound of Atlantic O, Connecticut, USA	410	660	11	28	17	471
63.18N, 139.24W	Stewart	Yukon R, Yukon Territory, Canada	400	640	20	51	16	453

NORTH AMERICA

Latitude and Longitude	River	Outflow and Location	Length (mi)	Length (km)	Drainage Basin (1000 mi²)	Drainage Basin (1000 km²)	Discharge Rate (1000 cfs)	Discharge Rate (m³/sec)
39.15N, 75.20W	Delaware	Delaware B of Atlantic O, Delaware–New Jersey, USA	390	630	11	30	14	411
38.50N, 82.08W	Kanawha-New	Ohio R, West Virginia, USA	390	630	12	31	14	408
70.01N, 126.42W	Horton	Amundsen G of Arctic O, NWT, Canada	380	620				
38.00N, 76.23W	Potomac	Chesapeake B of Atlantic O, Maryland–Virginia, USA	380	620	14	38	11	314
51.20N, 80.24W	Moose-Abitibi	James B of Hudson B, Ontario, Canada	380	610	35	91	44	1,250
62.47N, 137.20W	Pelly	Yukon R, Yukon Territory, Canada	380	610	20	52	14	399
35.56N, 76.42W	Roanoke	Albemarle Sound of Atlantic O, North Carolina, USA	380	610	10	26	8.2	232
29.59N, 93.47W	Sabine	Sabine L, Louisiana–Texas, USA	380	610	10	27	8.8	249
38.03N, 121.56W	Sacramento	Suisun B of Pacific O, California, USA	380	610	27	70	27	770
16.30N, 97.31W	• Atoyac (alt Verde)	Pacific O, Oaxaca, Mexico	370	600	7	19	3.2	90
70.27N, 150.07W	Colville	Beaufort Sea, Alaska, USA	370	600	24	62		
29.35N, 104.25W	• Conchos	Rio Grande R, Chihuahua, Mexico	370	590	25	64	0.8	24
37.54N, 87.30W	Green	Ohio R, Kentucky, USA	360	580	9	24	11	306
41.10N, 91.01W	Iowa-Cedar	Missouri R, Iowa, USA	360	580	13	34	6.0	170
37.16N, 110.26W	San Juan	Colorado R, Utah, USA	360	580	23	60	2.8	79
32.02N, 80.53W	Savannah-Seneca	Atlantic O, Georgia–South Carolina, USA	360	580	10	26	11	320
54.09N, 130.05W	Skeena	Chatham Sound of Pacific O, British Columbia, Canada	360	580	17	43	32	917

NORTH AMERICA

Latitude and Longitude	River	Outflow and Location	Length (mi)	Length (km)	Drainage Basin (1000 mi²)	Drainage Basin (1000 km²)	Discharge Rate (1000 cfs)	Discharge Rate (m³/sec)
52.43N, 108.15W	Battle	North Saskatch-ewan R, Sas-katchewan, Canada	350	570	12	31	0.5	14
39.04N, 113.07W	Sevier	Sevier L, Utah, USA	350	570	6	16	0.2	6
41.27N, 112.15W	Bear	Great Salt L, Utah, USA	350	560	7	18	1.7	48
25.54N, 109.22W	• Fuerte-Verde	G of California, Sinaloa, Mexico	350	560	14	36	6.0	171
58.50N, 66.10W	George	Ungava B of Hudson Strait, Quebec, Canada	350	560	16	42	32	909
45.40N, 100.45W	Grand	Missouri R, South Dakota, USA	350	560	6	16	0.3	8
39.06N, 84.30W	Licking	Ohio R, Ken-tucky, USA	350	560	4	10	4.1	116
49.10N, 68.15W	Manicouagan	Saint Lawrence R, Quebec, Canada	350	560	18	46	36	1,030
67.00N, 162.30W	Noatak	Kotzebue Sound of Chukchi Sea, Alaska, USA	350	560	13	33		
44.30N, 68.48W	Penobscot	Atlantic O, Maine, USA	350	560	10	26	12	340
46.21N, 72.31W	Saint-Maurice	Saint Lawrence R, Quebec, Canada	350	560	17	43	26	731
38.04N, 121.51W	San Joaquin	Suisun B of Pa-cific O, Cali-fornia, USA	350	560	14	36	4.7	133
61.03N, 123.22W	South Nahanni	Liard R, NWT, Canada	350	560	13	34	11	317
35.48N, 95.19W	Verdigris	Arkansas R, Oklahoma, USA	350	560	5	13	2.1	59
27.50N, 97.29W	Nueces	Nueces B of G of Mexico, Texas, USA	340	550	19	49	0.9	24
54.45N, 114.17W	Pembina	Athabasca R, Alberta, Canada	340	550	5	13	1.2	34
50.39N, 59.25W	Petit Mecatina (alt Little Mecatina)	G of Saint Law-rence, Quebec, Canada	340	550	8	20	18	518
41.07N, 96.18W	Elkhorn	Platte R, Ne-braska, USA	330	540	7	18	1.1	31
56.31N, 132.24W	Stikine	Stikine Strait, Alaska, USA	330	540	20	51	26	736

NORTH AMERICA

Latitude and Longitude	River	Outflow and Location	Length (mi)	Length (km)	Drainage Basin (1000 mi²)	Drainage Basin (1000 km²)	Discharge Rate (1000 cfs)	Discharge Rate (m³/sec)
38.25N, 87.44W	White	Wabash R, Indiana, USA	330	540	12	31	11	323
32.03N, 91.04W	Big Black	Mississippi R, Mississippi, USA	330	530	3	8	3.3	93
51.10N, 79.45W	Harricana	James B of Hudson B, Ontario, Canada	330	530	11	29	9.0	255
44.54N, 93.09W	Minnesota	Mississippi R, Minnesota, USA	330	530	17	44	3.1	88
33.23N, 112.18W	Salt-Black	Gila R, Arizona, USA	330	530	7	18	0.8	24
33.53N, 78.01W	Cape Fear-Deep	Atlantic O, North Carolina, USA	320	520	6	16	4.7	133
59.33N, 124.01W	Fort Nelson-Sikanni Chief	Liard R, British Columbia, Canada	320	520	18	47	10	286
47.01N, 96.49W	Sheyenne	Red R, North Dakota, USA	320	520	10	26	0.1	4
37.58N, 89.57W	Kaskaskia	Mississippi R, Illinois, USA	320	510	6	16	3.8	108
22.16N, 97.47W	Panuco-Santa Maria	G of Mexico, Tamaulipas–Veracruz, Mexico	320	510	25	66	19	548
67.27N, 133.45W	Arctic Red	Mackenzie R, NWT, Canada	310	500	6	16	4.8	136
25.45N, 102.50W	• Nazas	(Laguna de) Mayran L, Coahuila, Mexico	310	500	14	36	3.4	95
49.05N, 68.23W	Outardes	Saint Lawrence R, Quebec, Canada	310	500	7	19	14	396
43.49N, 117.02W	Owyhee	Snake R, Oregon, USA	310	500	12	31	0.4	11
50.18N, 63.48W	Romaine	G of Saint Lawrence, Quebec, Canada	310	500	6	14	12	340
40.42N, 74.01W	Hudson	New York B of Atlantic O, New Jersey–New York, USA	310	490	13	35	12	354
56.11N, 117.19W	Smoky	Peace R, Alberta, Canada	310	490	19	49	13	365
50.14N, 121.34W	Thompson-North Thompson	Fraser R, British Columbia, Canada	300	490	22	57	28	784
39.35N, 96.34W	Big Blue	Kansas R, Kansas, USA	300	480	10	26	1.6	45

NORTH AMERICA

Latitude and Longitude	River	Outflow and Location	Length (mi)	Length (km)	Drainage Basin (1000 mi^2)	Drainage Basin (1000 km^2)	Discharge Rate (1000 cfs)	Discharge Rate (m^3/sec)
35.53N, 84.29W	Clinch	Tennessee R, Tennessee, USA	300	480	3	9	4.5	127
58.46N, 70.05W	Feuilles (alt Leaf)	Ungava B of Hudson Strait, Quebec, Canada	300	480	16	42	19	547
39.23N, 93.06W	Grand	Missouri R, Missouri, USA	300	480	7	18	3.5	98
57.00N, 92.15W	Hayes	Hudson B, Manitoba, Canada	300	480	37	95		
36.12N, 111.48W	Little Colorado	Colorado R, Arizona, USA	300	480	27	70	0.2	6
51.03N, 80.55W	Moose (upper)-Mattagami	Moose R, Ontario, Canada	300	480	25	65	29	821
30.23N, 88.37W	Pascagoula-Chickasawhay	Mississippi Sound of G of Mexico, Mississippi, USA	300	480	7	18	9.3	263
41.29N, 90.37W	Rock	Mississippi R, Illinois, USA	300	480	11	28	5.3	150
28.48N, 111.49W	● Sonora	G of California, Sonora, Mexico	300	480	11	29		

* * *

44.26N, 102.18W	Belle Fourche	Cheyenne R, South Dakota, USA	290	470	7	18	0.4	11
39.59N, 118.36W	Humboldt	Humboldt L, Nevada, USA	290	470	17	44	0.07	2
41.24N, 97.19W	Loup-Middle Loup	Platte R, Nebraska, USA	290	470	15	39	0.7	20
45.18N, 100.43W	Moreau	Missouri R, South Dakota, USA	290	470	6	16	0.3	8
47.21N, 107.57W	Musselshell	Missouri, R, Montana, USA	290	470	9	23	0.2	6
32.32N, 87.51W	Black Warrior	Tombigbee R, Alabama, USA	290	460	6	16	7.7	218
35.38N, 91.19W	Black	White R, Arkansas, USA	280	450	8	21	8.3	235
39.19N, 92.57W	Chariton	Missouri R, Missouri, USA	280	450	2	5	1.0	28
60.18N, 145.03W	Copper	G of Alaska, Alaska, USA	280	450	24	63	37	1,040
45.44N, 120.39W	John Day	Columbia R, Oregon, USA	280	450	8	21	2.0	57
66.54N, 160.38W	Kobuk	Kotzebue Sound of Chukchi Sea, Alaska, USA	280	450	12	31	8.8	249

NORTH AMERICA

Latitude and Longitude	River	Outflow and Location	Length (mi)	Length (km)	Drainage Basin (1000 mi²)	Drainage Basin (1000 km²)	Discharge Rate (1000 cfs)	Discharge Rate (m³/sec)
29.58N, 93.51W	Neches	Sabine L, Texas, USA	280	450	8	21	6.4	182
35.06N, 76.29W	Neuse	Pamlico Sound of Atlantic O, North Carolina, USA	280	450	3	8	2.9	82
59.03N, 158.23W	Nushagak-Mulchatna	Bristol B of Bering Sea, Alaska, USA	280	450	14	37		
31.58N, 82.32W	Oconee	Altamaha R, Georgia, USA	280	450	5	13	4.8	136
18.42N, 95.38W	Papaloapan-Tuxtepec	Alvarado Lagoon, Veracruz, Mexico	280	440	9	23	44	1,240
30.24N, 81.24W	Saint Johns	Atlantic O, Florida, USA	280	440	3	8	3.3	94
33.44N, 80.38W	Congaree-Broad	Santee R, South Carolina, USA	270	440	8	21	8.3	235
32.30N, 86.16W	Tallapoosa	Alabama R, Alabama, USA	270	430	4	12	4.6	130
45.39N, 122.46W	Willamette	Columbia R, Oregon, USA	270	430	11	29	31	878
30.57N, 84.34W	Flint	Apalachicola R, Georgia, USA	260	430	8	21	8.2	232
38.40N, 91.33W	Gasconade	Missouri R, Missouri, USA	260	430	3	8	3.0	85
46.25N, 105.52W	Tongue	Yellowstone R, Montana, USA	260	430	6	16	0.4	10
43.03N, 86.15W	Grand	Michigan L, Michigan, USA	260	420	5	13	3.5	100
41.17N, 91.21W	Iowa (upper)	Iowa R, Iowa, USA	260	420	4	10	1.5	42
61.15N, 150.36W	Susitna	G of Alaska, Alaska, USA	260	420	19	50	9.8	279
45.38N, 120.55W	Deschutes	Columbia R, Oregon, USA	250	400	11	28	5.8	164
41.33N, 124.05W	Klamath	Pacific O, California, USA	250	400	12	31	17	490
31.50N, 81.03W	Ogeechee	Atlantic O, Georgia, USA	250	400	3	8	2.2	62
29.17N, 83.10W	Suwannee	G of Mexico, Florida, USA	250	400	10	26	6.9	195
44.06N, 77.34W	Trent-Otonabee-Irondale	Ontario L, Ontario, Canada	250	400	5	12	4.4	125
40.04N, 109.40W	White	Green R, Utah, USA	250	400	4	10	0.7	20
34.06N, 98.10W	Wichita	Red R, Texas, USA	250	400	4	10	0.3	8

NORTH AMERICA

Latitude and Longitude	River	Outflow and Location	Length (mi)	Length (km)	Drainage Basin (1000 mi²)	Drainage Basin (1000 km²)	Discharge Rate (1000 cfs)	Discharge Rate (m³/sec)
40.32N, 108.59W	Yampa	Green R, Colorado, USA	250	400	4	10	1.6	45
20.17N, 75.56W	Cauto	G of Guacanayabo of Caribbean Sea, Cuba	230	370	4	11		
43.58N, 69.52W	Androscoggin-Magalloway	Atlantic O, Maine, USA	220	360	4	10	6.0	170
13.14N, 88.49W	Lempa	Pacific O, El Salvador	190	300	7	18	12	350
17.32N, 88.14W	● Belize	Caribbean Sea, Belize	180	290				
19.15N, 72.47W	Artibonite	G of Gonave of Caribbean Sea, Haiti	170	280	4	10		
19.51N, 71.41W	Yaque del Norte	Atlantic O, Dominican Republic	170	280				
46.03N, 73.08W	Richelieu	Saint Lawrence R, Quebec, Canada	110	170	9	24	11	309

OCEANIA

Latitude and Longitude	River	Outflow and Location	Length (mi)	Length (km)	Drainage Basin (1000 mi²)	Drainage Basin (1000 km²)	Discharge Rate (1000 cfs)	Discharge Rate (m³/sec)
35.22S, 139.22E	Murray-Darling[1]-Culgoa-Balonne-Condamine	Indian O, South Australia, Australia	2330	3750	408	1057	25	715
34.07S, 141.55E	Murray (upper)	Murray R, New South Wales, Australia	1090	1750	103	267	5.9	168
29.56S, 146.20E	Barwon-Macintyre-Dumaresq-Severn	Darling R, New South Wales, Australia	980	1580	87	225	5.2	146
34.43S, 143.12E	Murrumbidgee	Murray R, New South Wales, Australia	980	1580	37	97	4.2	119
34.21S, 143.57E	Lachlan	Murrumbidgee R, New South Wales, Australia	920	1480	33	85	1.5	42

[1] Data for Darling-Culgoa-Balonne-Condamine: 1810 mi, 2910 km; 247,000 mi², 640,000 km²; 7,900 cfs, 225 m³/sec.

OCEANIA

Latitude and Longitude	River	Outflow and Location	Length		Drainage Basin		Discharge Rate	
			(mi)	(km)	(1000 mi^2)	(1000 km^2)	(1000 cfs)	(m^3/sec)
8.25S, 143.10E	• Fly-Strickland	G of Papua of Coral Sea, Papua New Guinea	800	1290			318	9,000
3.51S, 144.34E	• Sepik	Pacific O, Papua New Guinea	700	1130				
23.32S, 150.52E	• Fitzroy-Dawson	Pacific O, Queensland, Australia	690	1110	55	143	5.9	166
30.07S, 147.24E	Macquarie	Barwon R, New South Wales, Australia	590	950	18	47	1.5	43
30.00S, 148.07E	Namoi	Barwon R, New South Wales, Australia	530	850	17	43	0.9	25
17.36S, 140.36E	Flinders	G of Carpentaria, Queensland, Australia	520	840	42	108	0.6	16
29.57S, 146.21E	• Bogan	Barwon R, New South Wales, Australia	450	720	10	26	0.07	2
19.39S, 147.30E	Burdekin	Pacific O, Queensland, Australia	440	710	51	131	9.9	281
1.26S, 137.53E	Mamberamo-Taritatu (for Idenburg)	Pacific O, Irian Jaya, Indonesia	420	670				
29.27S, 149.48E	Gwydir	Barwon R, New South Wales, Australia	410	670	10	26	0.9	25
15.12S, 129.43E	• Victoria	Timor Sea, Northern Territory, Australia	400	650	30	78	4.7	133
4.02S, 144.40E	• Ramu (for Ottilien)	Pacific O, Papua New Guinea	400	640				
30.12S, 147.32E	Castlereagh	Barwon R, New South Wales, Australia	340	550	7	18	0.3	8
7.07S, 138.42E	Digul (for Digoel)	Arafura Sea, Irian Jaya, Indonesia	340	540				
		* * *						
33.30S, 151.10E	Hawkesbury	Pacific O, New South Wales, Australia	290	470	8	22	2.3	66
32.50S, 151.42E	Hunter	Pacific O, New South Wales, Australia	290	470	8	20	1.6	46

OCEANIA

Latitude and Longitude	River	Outflow and Location	Length		Drainage Basin		Discharge Rate	
			(mi)	(km)	(1000 mi^2)	(1000 km^2)	(1000 cfs)	(m^3/sec)
37.23S, 174.42E	Waikato	Tasman Sea, North Island, New Zealand	260	420	6	14	12	334
29.25S, 153.22E	Clarence	Pacific O, New South Wales, Australia	240	390	9	23	4.8	136
46.21S, 169.48E	Clutha-Makarora	Pacific O, South Island, New Zealand	200	320	8	22	23	651

SOUTH AMERICA

Latitude and Longitude	River	Outflow and Location	Length		Drainage Basin		Discharge Rate	
			(mi)	(km)	(1000 mi^2)	(1000 km^2)	(1000 cfs)	(m^3/sec)
0.10S, 49.00W	• Amazon (off Amazonas)- Ucayali- Tambo-Ene- Apurimac	Atlantic O, Amapa–Para, Brazil	4080	6570	2375	6150	6180	175,000
35.00S, 57.00W	Plata-Parana- Grande	Atlantic O, Argentina– Uruguay	3030	4880	1197	3100	809	22,900
3.22S, 58.45W	• Madeira- Mamore- Grande (alt Guapay)	Amazon R, Amazonas, Brazil	1990	3200	463	1200	770	21,800
2.37S, 65.44W	• Jurua (alt Yurua)	Amazon R, Amazonas, Brazil	1860	3000	93	240	141	4,000
3.42S, 61.28W	• Purus	Amazon R, Amazonas, Brazil	1860	3000	154	400		
10.30S, 36.24W	Sao Francisco	Atlantic O, Alagoas–Ser- gipe, Brazil	1730	2780	236	611	95	2,690
1.00S, 48.30W	Para-Tocantins	Atlantic O, Para, Brazil	1710	2750	323	836	305	8,630
27.18S, 58.38W	Paraguay	Parana R, Argentina– Paraguay	1610	2600	425	1100	155	4,400
3.08S, 64.46W	• Caqueta (alt Japura, Yapura)	Amazon R, Amazonas, Brazil	1420	2280	120	310	247	7,000
2.24S, 54.41W	• Tapajos- Juruena	Amazon R, Para, Brazil	1380	2220	179	463	212	6,000
5.21S, 48.41W	• Araguaia	Tocantins R, Para, Brazil	1370	2200				

SOUTH AMERICA

Latitude and Longitude	River	Outflow and Location	Length		Drainage Basin		Discharge Rate	
			(mi)	(km)	(1000 mi²)	(1000 km²)	(1000 cfs)	(m³/sec)
34.12S, 58.18W	Uruguay-Canoas	Plata R, Argentina-Uruguay	1370	2200	119	307	194	5,500
1.30S, 51.53W	• Xingu	Amazon R, Para, Brazil	1300	2100	174	450	73	2,060
8.37N, 62.15W	Orinoco	Atlantic O, Venezuela	1280	2060	340	880	890	25,200
3.08S, 59.55W	• Negro (alt Guainia)	Amazon R, Amazonas, Brazil	1240	2000	386	1000	1236	35,000
3.07S, 67.58W	• Putumayo (alt Ica)	Amazon R, Amazonas, Brazil	1240	2000	43	112	177	5,000
11.54S, 65.01W	• Guapore (alt Itenez)	Mamore R, Bolivia-Brazil	1120	1800	232	600	71	2,000
3.00S, 41.50W	• Parnaiba	Atlantic O, Maranhao-Piaui, Brazil	1060	1700	135	350		
11.06N, 74.51W	• Magdalena	Caribbean Sea, Colombia	960	1540	100	260	283	8,000
31.42S, 60.44W	Salado (alt Salado del Norte)	Parana R, Santa Fe, Argentina	930	1500	309	800	1.3	38
1.24S, 61.51W	• Branco-Uraricoera	Negro R, Roraima, Brazil	920	1470				
39.50S, 62.08W	Colorado-Salado-Desaguadero-Bermejo	Atlantic O, Buenos Aires, Argentina	890	1430	42	110	4.7	133
4.30S, 73.27W	Maranon	Amazon R, Peru	880	1410				
7.21S, 58.03W	• Teles Pires (alt Sao Manuel, Tres Barras)	Tapajos R, Mato Grosso-Para, Brazil	870	1400			62	1,750
8.54N, 74.28W	• Cauca	Magdalena R, Colombia	840	1350	24	63	78	2,200
4.03N, 67.44W	• Guaviare	Orinoco R, Colombia	840	1350	56	145		
25.36S, 54.36W	Iguazu (alt Iguacu; for Iguassu)	Parana R, Argentina-Brazil	820	1320	24	62	62	1,750
7.24N, 66.35W	Arauca	Orinoco R, Venezuela	810	1300	7	18		
5.07S, 60.24W	• Aripuana-Roosevelt	Madeira R, Amazonas, Brazil	800	1290				
10.23S, 65.24W	• Beni-Madre de Dios	Madeira R, Bolivia	800	1290				

SOUTH AMERICA

Latitude and Longitude	River	Outflow and Location	Length (mi)	Length (km)	Drainage Basin (1000 mi²)	Drainage Basin (1000 km²)	Discharge Rate (1000 cfs)	Discharge Rate (m³/sec)
20.07S, 51.05W	Paranaiba	Parana R, Mato Grosso–Minas Gerais, Brazil	790	1270				
41.02S, 62.47W	Negro-Neuquen	Atlantic O, Buenos Aires–Rio Negro, Argentina	750	1210	48	125	36	1,010
2.52S, 44.12W	• Itapecuru	Sao Jose B of Atlantic O, Maranhao, Brazil	750	1200	17	45		
2.43S, 66.57W	• Jutai	Amazon R, Amazonas, Brazil	750	1200	12	31	18	500
4.21S, 70.02W	Javari (alt Yacarana, Yavari)	Amazon R, Brazil-Peru	740	1180	35	91	4.2	120
5.10S, 75.32W	Huallaga	Maranon R, Peru	710	1140	37	95		
21.37S, 41.03W	Paraiba (alt Paraiba do Sul)	Atlantic O, Rio de Janeiro, Brazil	710	1140	22	57	12	331
20.40S, 51.35W	• Tiete	Parana R, Sao Paulo, Brazil	700	1130	28	72	13	378
3.04S, 44.35W	• Mearim	Sao Marcos B of Atlantic O, Maranhao, Brazil	680	1100	39	100		
25.21S, 57.42W	Pilcomayo	Paraguay R, Paraguay	680	1100	74	192	3.3	93
15.51S, 38.53W	• Jequitinhonha	Atlantic O, Bahia, Brazil	680	1090	24	62	20	557
26.52S, 58.23W	Bermejo	Paraguay R, Chaco–Formosa, Argentina	660	1060	36	94	11	325
7.37N, 66.25W	Apure-Uribante	Orinoco R, Venezuela	620	1000	50	130	7.4	210
19.37S, 39.49W	• Doce	Atlantic O, Espirito Santo, Brazil	620	1000	32	83	34	969
3.52S, 52.37W	• Iriri	Xingu R, Para, Brazil	620	1000				
6.12N, 67.28W	Meta	Orinoco R, Colombia-Venezuela	620	1000	40	104		
0.02N, 67.16W	• Vaupes (alt Uaupes)	Negro R, Amazonas, Brazil	620	1000				
6.58N, 58.23W	• Essequibo	Atlantic O, Guyana	600	970	27	69	77	2,190

SOUTH AMERICA

Latitude and Longitude	River	Outflow and Location	Length (mi)	Length (km)	Drainage Basin (1000 mi²)	Drainage Basin (1000 km²)	Discharge Rate (1000 cfs)	Discharge Rate (m³/sec)
8.21N, 62.43W	Caroni	Orinoco R, Venezuela	570	920	37	95	177	5,000
38.49S, 64.57W	Colorado (upper)-Grande	Colorado R, La Pampa, Argentina	570	920	10	25		
1.29S, 48.30W	• Guama-Capim	Para R, Para, Brazil	560	900				
22.40S, 53.09W	• Paranapanema	Parana R, Parana–Sao Paulo, Brazil	560	900	22	56	12	348
3.35S, 64.47W	• Tefe	Amazon R, Amazonas, Brazil	560	900				
1.23S, 69.25W	• Apaporis	Caqueta R, Colombia	550	880				
6.25N, 58.37W	• Mazaruni-Cuyuni	Essequibo R, Guyana	550	880				
3.20S, 72.40W	• Napo	Amazon R, Peru	550	880				
10.44S, 73.45W	Urubamba	Ucayali R, Peru	540	860				
6.15S, 42.52W	• Caninde	Parnaiba R, Piaui, Brazil	530	860				
23.18S, 53.42W	• Ivai	Parana R, Parana, Brazil	530	860	14	36	12	340
4.26S, 74.05W	Tigre	Maranon R, Peru	520	840				
43.20S, 65.03W	Chubut	Atlantic O, Chubut, Argentina	500	810	12	31	1.7	49
10.25S, 58.20W	• Arinos	Juruena R, Mato Grosso, Brazil	500	800			45	1,280
10.58S, 66.09W	• Beni (upper)	Beni R, Bolivia	500	800				
3.41S, 44.48W	• Grajau	Mearim R, Maranhao, Brazil	500	800				
1.13S, 46.06W	• Gurupi	Atlantic O, Maranhao-Para, Brazil	500	800	24	61		
11.47S, 37.32W	• Itapicuru	Atlantic O, Bahia, Brazil	500	800	15	39	0.6	17
11.45S, 50.44W	• Mortes	Araguaia R, Mato Grosso, Brazil	500	800				
33.24S, 58.22W	• Negro	Uruguay R, Uruguay	500	800	27	70	22	637
15.39S, 38.57W	• Pardo	Atlantic O, Bahia, Brazil	500	800	17	45	2.2	62
1.33S, 52.38W	• Paru	Amazon R, Para, Brazil	500	800				

* * *

SOUTH AMERICA

Latitude and Longitude	River	Outflow and Location	Length (mi)	Length (km)	Drainage Basin (1000 mi²)	Drainage Basin (1000 km²)	Discharge Rate (1000 cfs)	Discharge Rate (m³/sec)
12.28S, 64.24W	● Itonamas-San Miguel	Guapore R, Bolivia	470	760				
17.13S, 44.49W	● Velhas	Sao Francisco R, Minas Gerais, Brazil	470	760				
5.43N, 53.58W	● Maroni (alt Marowyne)	Atlantic O, French Guiana– Surinam	420	680	24	62	66	1,850
5.55N, 55.10W	● Suriname	Atlantic O, Surinam	370	600				
1.55S, 55.35W	● Trombetas	Amazon R, Para, Brazil	340	550	48	124	53	1,500
47.49S, 73.37W	Baker	G of Penas of Pacific O, Chile	270	440	10	25	21	600
6.00N, 57.04W	● Courantyne (alt Corantijn)	Atlantic O, Guyana– Surinam	270	440				
21.26S, 70.04W	Loa	Pacific O, Chile	270	440	13	34	0.06	1.6
36.49S, 73.10W	Bio–Bio	Arauco G of Pacific O, Chile	240	380	9	24	35	1,000

3a
LONGEST RIVERS,
IRRESPECTIVE OF CONTINENT

Contents

Rank.

Conventional and other (alternate, former, and official) names of river and of its tributaries (if any) constituting the longest watercourse.

Outflow (sea, lake, river, etc.), and subdivision and country in which located.

Total length of watercourse, in miles and kilometers.

Arrangement

By length of watercourse.

Coverage

50 longest rivers.

Rounding

Lengths are rounded to the nearest 10 mi (km).

Entries

50.

TABLE 3a. LONGEST RIVERS, IRRESPECTIVE OF CONTINENT

Rank	River	Outflow and Location	Length (mi)	Length (km)
1.	• Nile (off Nil)-Kagera-Ruvuvu-Luvironza	Mediterranean Sea, Egypt	4160	6690
2.	• Amazon (off Amazonas)-Ucayali-Tambo-Ene-Apurimac	Atlantic O, Amapa–Para, Brazil	4080	6570
3.	Mississippi-Missouri-Jefferson-Beaverhead-Red Rock	G of Mexico, Louisiana, USA	3740	6020
4.	Yangtze (off Chang)	East China Sea, Kiangsu, China	3720	5980
5.	Yenisey (alt Yenisei)-Angara-Selenga (alt Selenge)-Ider	Yenisey G of Kara Sea, Russia, USSR	3650	5870
6.	Amur (alt Heilung)-Argun (alt Oerhkuna)-Kerulen (off Herlen, Kolulun)	Tatar Strait, Russia, USSR	3590	5780
7.	Ob-Irtysh (alt Irtish)	G of Ob of Kara Sea, Russia, USSR	3360	5410
8.	Plata-Parana-Grande	Atlantic O, Argentina–Uruguay	3030	4880
9.	Yellow (alt Hwang; off Huang)	G of Chihli of Yellow Sea, Shantung, China	3010	4840
10.	Congo (alt Zaire; for Kongo)-Lualaba	Atlantic O, Angola–Zaire	2880	4630
11.	Lena	Laptev Sea, Russia, USSR	2730	4400
12.	Mackenzie-Slave-Peace-Finlay	Beaufort Sea, NWT, Canada	2630	4240
13.	• Mekong (alt Khong, Lantsang, Mekongk, Tien Giang)	South China Sea, Vietnam	2600	4180
14.	• Niger	G of Guinea, Nigeria	2550	4100
15.	Murray-Darling-Culgoa-Balonne-Condamine	Indian O, South Australia, Australia	2330	3750
16.	Volga	Caspian (Sea) L, Russia, USSR	2290	3690
17.	• Madeira-Mamore-Grande (alt Guapay)	Amazon R, Amazonas, Brazil	1990	3200
18.	Yukon-Lewes-Teslin-Nisutlin	Bering Sea, Alaska, USA	1980	3180
19.	Ob (upper)-Katun	Ob R, Russia, USSR	1970	3180
20.	Saint Lawrence (alt Saint-Laurent)-(Great Lakes)-Saint Louis	G of Saint Lawrence, Quebec, Canada	1900	3060
21.	Rio Grande (alt Bravo)	G of Mexico, Mexico–USA	1880	3030
22.	Syr (Darya)-Naryn	Aral (Sea) L, Kazakhstan, USSR	1880	3020
23.	• Jurua (alt Yurua)	Amazon R, Amazonas, Brazil	1860	3000
23.	• Purus	Amazon R, Amazonas, Brazil	1860	3000
25.	Lower Tunguska (off Nizhnyaya Tunguska)	Yenisey R, Russia, USSR	1860	2990
26.	Indus (alt Yintu)	Arabian Sea, Pakistan	1790	2880
27.	Danube (off Donau, Duna, Dunaj, Dunarea, Dunav, Dunay)	Black Sea, Romania–USSR	1770	2850
28.	Brahmaputra (alt Tsangpo, Yalutsangpu)	B of Bengal, Bangladesh	1770	2840
29.	• Salween (alt Khong, Lu, Nu)	G of Martaban of Andaman Sea, Burma	1750	2820

Rank	River	Outflow and Location	Length (mi)	Length (km)
30.	Sao Francisco	Atlantic O, Alagoas–Sergipe, Brazil	1730	2780
31.	Para-Tocantins	Atlantic O, Para, Brazil	1710	2750
31.	Tarim (off Talimu)-Yarkand (alt Yeherhchiang)	Tarim Basin, Sinkiang, China	1710	2750
33.	Vilyuy (alt Vilyui)	Lena R, Russia, USSR	1650	2650
33.	• Zambezi (alt Zambesi, Zambeze)	Mozambique Channel, Mozambique	1650	2650
35.	Amu (Darya) (alt Oxus)-Panja (alt Pyandzh)-Vakhsh	Aral (Sea) L, Uzbekistan, USSR	1630	2620
36.	Paraguay	Parana R, Argentina–Paraguay	1610	2600
37.	Nelson-Saskatchewan-South Saskatchewan-Bow	Hudson B, Manitoba, Canada	1600	2570
38.	Ural	Caspian (Sea) L, Kazakhstan, USSR	1570	2530
39.	• Ganges (alt Ganga)-Bhagirathi	B of Bengal, Bangladesh–India	1560	2510
40.	• Shebeli (alt Shabale, Shibeli; for Scebeli)	Balli Swamp, Somalia	1550	2490
41.	• Ubangi (alt Oubangui)-Uele-Kibali	Congo R, Congo–Zaire	1530	2460
42.	Ishim	Irtysh R, Russia, USSR	1520	2450
43.	(Shatt al) Arab-Euphrates (off Firat, Furat)-Kara (Su)	Persian G, Iran–Iraq	1510	2430
44.	Arkansas	Mississippi R, Arkansas, USA	1450	2330
44.	Colorado	G of California, Baja California–Sonora, Mexico	1450	2330
46.	Olenek	Laptev Sea, Russia, USSR	1420	2290
47.	• Caqueta (alt Japura, Yapura)	Amazon R, Amazonas, Brazil	1420	2280
47.	Dnieper (off Dnepr)	Black Sea, Ukraine, USSR	1420	2280
49.	Aldan	Lena R, Russia, USSR	1410	2270
50.	Atchafalaya-Red	G of Mexico, Louisiana, USA	1400	2260

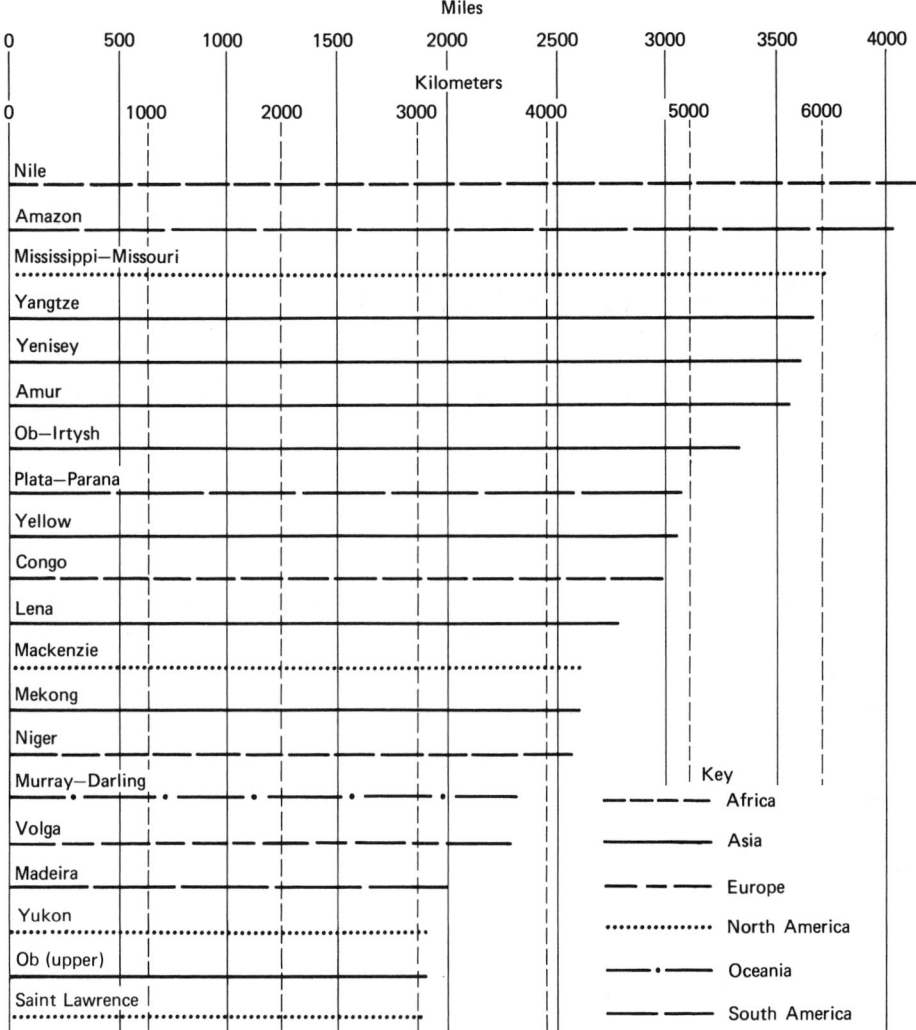

Figure 8. Twenty longest rivers.

3b
RIVERS WITH LARGEST DRAINAGE BASINS

Contents

Rank.

Conventional and other (alternate, former, and official) names of river and of its tributaries (if any) constituting the longest watercourse.

Outflow (sea, lake, river, etc.), and subdivision and country in which located.

Area of drainage basin of river and all its tributaries, in thousands of square miles and square kilometers.

Arrangement

By area of drainage basin.

Coverage

50 rivers with the largest drainage basins.

Rounding

Areas are rounded to the nearest 1000 mi^2 (km^2).

Entries

50.

TABLE 3b. RIVERS WITH LARGEST DRAINAGE BASINS

Rank	River	Outflow and Location	Drainage Basin (1000 mi²)	Drainage Basin (1000 km²)
1.	• Amazon (off Amazonas)-Ucayali-Tambo-Ene-Apurimac	Atlantic O, Amapa–Para, Brazil	2375	6150
2.	Congo (alt Zaire; for Kongo)-Lualaba	Atlantic O, Angola–Zaire	1476	3822
3.	Mississippi-Missouri-Jefferson-Beaverhead-Red Rock	G of Mexico, Louisiana, USA	1247	3230
4.	Plata-Parana-Grande	Atlantic O, Argentina–Uruguay	1197	3100
5.	Ob-Irtysh (alt Irtish)	G of Ob of Kara Sea, Russia, USSR	1154	2990
6.	• Nile (off Nil)-Kagera-Ruvuvu-Luvironza	Mediterranean Sea, Egypt	1082	2802
7.	Yenisey (alt Yenisei)-Angara-Selenga (alt Selenge)-Ider	Yenisey G of Kara Sea, Russia, USSR	1011	2619
8.	Lena	Laptev Sea, Russia, USSR	961	2490
9.	• Niger	G of Guinea, Nigeria	808	2092
10.	Amur (alt Heilung)-Argun (alt Oerhkuna)-Kerulen (off Herlen, Kolulun)	Tatar Strait, Russia, USSR	792	2050
11.	Yangtze (off Chang)	East China Sea, Kiangsu, China	705	1827
12.	Mackenzie-Slave-Peace-Finlay	Beaufort Sea, NWT, Canada	697	1805
13.	Volga	Caspian (Sea) L, Russia, USSR	525	1360
14.	• Zambezi (alt Zambesi, Zambeze)	Mozambique Channel, Mozambique	514	1331
15.	Saint Lawrence (alt Saint-Laurent)-(Great Lakes)-Saint Louis	G of Saint Lawrence, Quebec, Canada	503	1303
16.	• Madeira-Mamore-Grande (alt Guapay)	Amazon R, Amazonas, Brazil	463	1200
17.	Indus (alt Yintu)	Arabian Sea, Pakistan	450	1165
18.	(Shatt al) Arab-Euphrates (off Firat, Furat)-Kara (Su)	Persian G, Iran–Iraq	427	1105
19.	Paraguay	Parana R, Argentina–Paraguay	425	1100
20.	Nelson-Saskatchewan-South Saskatchewan-Bow	Hudson B, Manitoba, Canada	414	1072
21.	Murray-Darling-Culgoa-Balonne-Condamine	Indian O, South Australia, Australia	408	1057
22.	• Negro (alt Guainia)	Amazon R, Amazonas, Brazil	386	1000
23.	• Ganges (alt Ganga)-Bhagirathi	B of Bengal, Bangladesh–India	368	952
24.	• Kasai (alt Cassai)	Congo R, Zaire	349	904
25.	Orinoco	Atlantic O, Venezuela	340	880
26.	Yukon-Lewes-Teslin-Nisutlin	Bering Sea, Alaska, USA	328	850
27.	Para-Tocantins	Atlantic O, Para, Brazil	323	836
28.	• Mekong (alt Khong, Lantsang, Mekongk, Tien Giang)	South China Sea, Vietnam	313	811

Rank	River	Outflow and Location	Drainage Basin	
			(1000 mi^2)	(1000 km^2)
29.	Salado (alt Salado del Norte)	Parana R, Santa Fe, Argentina	309	800
30.	● Okovanggo (alt Cubango, Okavango)	Okovanggo Basin, Botswana	303	785
31.	Danube (off Donau, Duna, Dunaj, Dunarea, Dunav, Dunay)	Black Sea, Romania–USSR	298	773
31.	● Ubangi (alt Oubangui)-Uele-Kibali	Congo R, Congo–Zaire	298	773
33.	Yellow (alt Hwang; off Huang)	G of Chihli of Yellow Sea, Shantung, China	297	771
34.	Ob (upper)-Katun	Ob R, Russia, USSR	295	765
35.	Aldan	Lena R, Russia, USSR	281	729
36.	● Shari (alt Chari)-Sara (alt Ouham)	Chad L, Cameroon–Chad	270	700
37.	Orange (alt Oranje)	Atlantic O, Namibia–South Africa	261	677
38.	Ohio-Allegheny	Mississippi R, Illinois–Kentucky, USA	258	668
39.	Columbia-Snake	Pacific O, Oregon–Washington, USA	253	655
40.	Kolyma (alt Kolima)	East Siberian Sea, Russia, USSR	248	643
41.	Colorado	G of California, Baja California–Sonora, Mexico	246	637
42.	Sao Francisco	Atlantic O, Alagoas–Sergipe, Brazil	236	611
43.	● Guapore (alt Itenez)	Mamore R, Bolivia–Brazil	232	600
44.	Brahmaputra (alt Tsangpo, Yalutsangpu)	B of Bengal, Bangladesh	224	580
45.	Panjnad-Sutlej	Indus R, Pakistan	206	533
46.	Sungari (alt Sunghua)	Amur R, Heilungkiang, China	202	524
47.	Kama	Volga R, Russia, USSR	196	507
48.	Dnieper (off Dnepr)	Black Sea, Ukraine, USSR	195	504
49.	Lower Tunguska (off Nizhnyaya Tunguska)	Yenisey R, Russia, USSR	183	473
50.	Amu (Darya) (alt Oxus)-Panja (alt Pyandzh)-Vakhsh	Aral (Sea) L, Uzbekistan, USSR	180	465

Figure 9. Rubber harvesters on the Amazon, the river with the largest drainage basin and the highest discharge rate. (Credit: Organization of American States.)

3c
RIVERS WITH HIGHEST
DISCHARGE RATES

Contents

Rank.

Conventional and other (alternate, former, and official) names of river and of its tributaries (if any) constituting the longest watercourse.

Outflow (sea, lake, river, etc.), and subdivision and country in which located.

Average discharge rate, in thousands of cubic feet per second and in cubic meters per second. The discharge rate given is that of the main stream, preferably averaged over several recent years and measured at the gauging station registering the greatest discharge (or nearest the mouth of the river).

Arrangement

By discharge rate.

Coverage

50 rivers with the highest average discharge rates.

Rounding

Discharge rates are rounded to the nearest 1000 cfs and to the nearest 10 m^3/sec.

Entries

50.

TABLE 3c. RIVERS WITH HIGHEST DISCHARGE RATES

Rank	River	Outflow and Location	Discharge Rate (1000 cfs)	(m³/sec)
1.	● Amazon (off Amazonas)-Ucayali-Tambo-Ene-Apurimac	Atlantic O, Amapa–Para, Brazil	6180	175,000
2.	Congo (alt Zaire; for Kongo)-Lualaba	Atlantic O, Angola–Zaire	1377	39,000
3.	● Negro (alt Guainia)	Amazon R, Amazonas, Brazil	1236	35,000
4.	Yangtze (off Chang)	East China Sea, Kiangsu, China	1137	32,190
5.	Orinoco	Atlantic O, Venezuela	890	25,200
6.	Plata-Parana-Grande	Atlantic O, Argentina–Uruguay	809	22,900
7.	● Madeira-Mamore-Grande (alt Guapay)	Amazon R, Amazonas, Brazil	770	21,800
8.	Yenisey (alt Yenisei)-Angara-Selenga (alt Selenge)-Ider	Yenisey G of Kara Sea, Russia, USSR	636	18,000
9.	Brahmaputra (alt Tsangpo, Yalutsangpu)	B of Bengal, Bangladesh	575	16,290
10.	Lena	Laptev Sea, Russia, USSR	569	16,100
11.	● Zambezi (alt Zambesi, Zambeze)	Mozambique Channel, Mozambique	565	16,000
12.	● Mekong (alt Khong, Lantsang, Mekongk, Tien Giang)	South China Sea, Vietnam	501	14,200
13.	Saint Lawrence (alt Saint-Laurent)-(Great Lakes)-Saint Louis	G of Saint Lawrence, Quebec, Canada	460	13,030
14.	Mississippi-Missouri-Jefferson-Beaverhead-Red Rock	G of Mexico, Louisiana, USA	453	12,830
15.	● Irrawaddy (off Iyawadi)-Nmai	Andaman Sea, Burma	447	12,660
16.	● Ganges (alt Ganga)-Bhagirathi	B of Bengal, Bangladesh–India	441	12,500
17.	Pearl (alt Canton, Yueh; off Chu)-Si (alt West; off Hsi)-Hung (Shui)-Nanpan	South China Sea, Kwangtung, China	409	11,590
18.	Ob-Irtysh (alt Irtish)	G of Ob of Kara Sea, Russia, USSR	360	10,200
19.	● Salween (alt Khong, Lu, Nu)	G of Martaban of Andaman Sea, Burma	353	10,000
20.	● Kasai (alt Cassai)	Congo R, Zaire	351	9,950
21.	Amur (alt Heilung)-Argun (alt Oerhkuna)-Kerulen (off Herlen, Kolulun)	Tatar Strait, Russia, USSR	346	9,800
22.	Mackenzie-Slave-Peace-Finlay	Beaufort Sea, NWT, Canada	343	9,700
23.	● Fly-Strickland	G of Papua of Coral Sea, Papua New Guinea	318	9,000
24.	Para-Tocantins	Atlantic O, Para, Brazil	305	8,630
25.	● Magdalena	Caribbean Sea, Colombia	283	8,000
26.	Volga	Caspian (Sea) L, Russia, USSR	271	7,680
27.	● Ubangi (alt Oubangui)-Uele-Kibali	Congo R, Congo–Zaire	265	7,500
28.	Ohio-Allegheny	Mississippi R, Illinois–Kentucky, USA	257	7,280

Rank	River	Outflow and Location	Discharge Rate (1000 cfs)	Discharge Rate (m³/sec)
29.	● Caqueta (alt Japura, Yapura)	Amazon R, Amazonas, Brazil	247	7,000
30.	Yukon-Lewes-Teslin-Nisutlin	Bering Sea, Alaska, USA	240	6,800
31.	Indus (alt Yintu)	Arabian Sea, Pakistan	235	6,640
32.	Danube (off Donau, Duna, Dunaj, Dunarea, Dunav, Dunay)	Black Sea, Romania–USSR	227	6,430
33.	● Tapajos-Juruena	Amazon R, Para, Brazil	212	6,000
34.	● Niger	G of Guinea, Nigeria	201	5,700
35.	Uruguay-Canoas	Plata R, Argentina–Uruguay	194	5,500
36.	Columbia-Snake	Pacific O, Oregon–Washington, USA	189	5,350
37.	Atchafalaya-Red	G of Mexico, Louisiana, USA	181	5,120
38.	Aldan	Lena R, Russia, USSR	177	5,000
38.	Caroni	Orinoco R, Venezuela	177	5,000
38.	● Putumayo (alt Ica)	Amazon R, Amazonas, Brazil	177	5,000
41.	Ob (upper)-Katun	Ob R, Russia, USSR	174	4,920
42.	● Ogooue (alt Ogowe)	G of Guinea, Gabon	165	4,670
43.	Paraguay	Parana R, Argentina–Paraguay	155	4,400
44.	● Jurua (alt Yurua)	Amazon R, Amazonas, Brazil	141	4,000
44.	Pechora (for Petchora)	Barents Sea, Russia, USSR	141	4,000
46.	Fraser	Strait of Georgia, British Columbia, Canada	137	3,880
47.	Kama	Volga R, Russia, USSR	128	3,620
48.	Columbia (upper)	Columbia R, Washington, USA	120	3,410
49.	Northern Dvina (off Severnaya Dvina)-Vychegda (alt Vichegda)	White Sea, Russia, USSR	120	3,400
50.	Lower Tunguska (off Nizhnyaya Tunguska)	Yenisey R, Russia, USSR	118	3,330

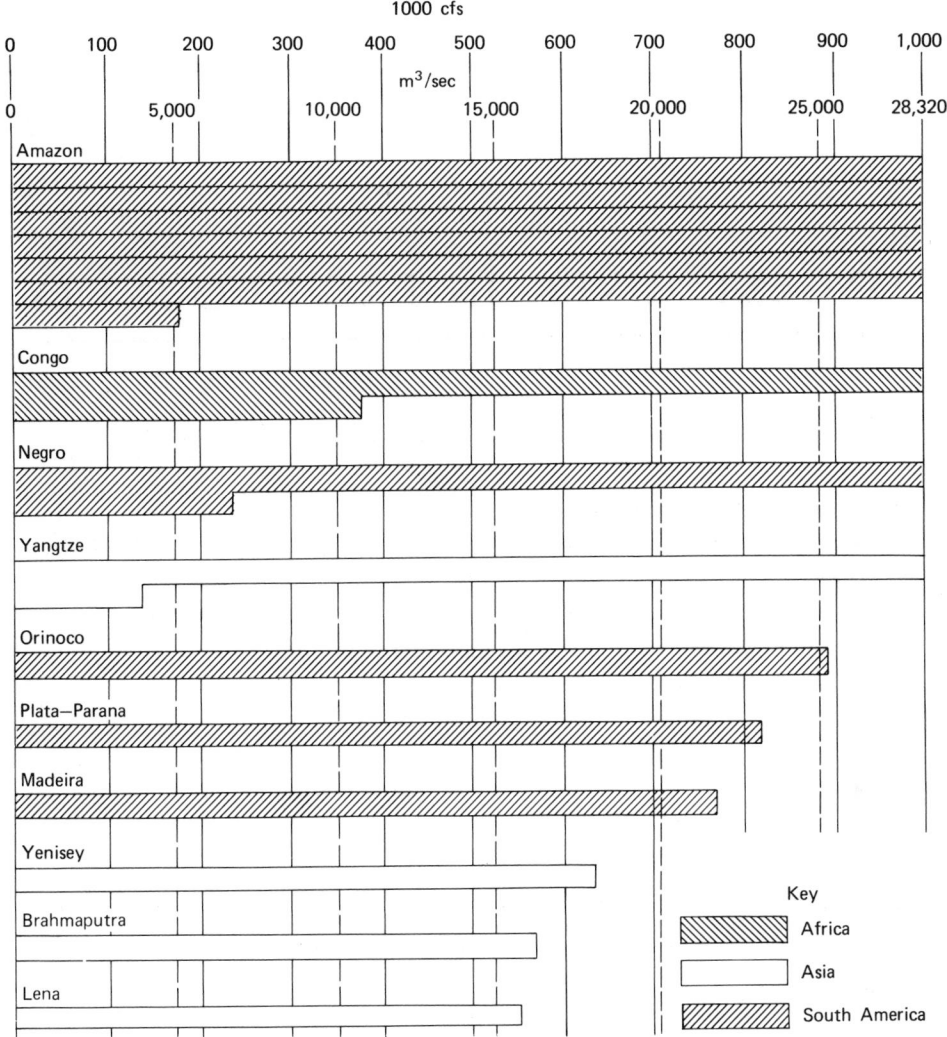

Figure 10. Ten rivers with highest average discharge rates.

3d
LONGEST RIVER IN EACH COUNTRY

Contents

Conventional name of country.

Conventional and other (alternate, former, and official) names of river and of its tributaries (if any) constituting the longest watercourse.

Outflow (sea, lake, river, etc.), and subdivision and country in which located.

Total length of watercourse, in miles and kilometers.

Arrangement

By country, alphabetically.

Coverage

All countries with an area of at least 1000 mi^2 (2590 km^2) having a river of significant length.

Rounding

Lengths are rounded to the nearest 10 mi (km).

Entries

144.

TABLE 3d. LONGEST RIVER IN EACH COUNTRY

Country	River	Outflow and Location	Length (mi)	(km)
Afghanistan	Amu (Darya) (alt Oxus)-Panja (alt Pyandzh)-Vakhsh	Aral (Sea) L, Uzbekistan, USSR	1630	2620
Albania	Drin	Adriatic Sea, Albania	170	280
Algeria	• Sheliff (alt Cheliff; off Shalaf)	Mediterranean Sea, Algeria	430	700
Angola	Congo (alt Zaire; for Kongo)-Lualaba	Atlantic O, Angola–Zaire	2880	4630
Argentina	Plata-Parana-Grande	Atlantic O, Argentina–Uruguay	3030	4880
Australia	Murray-Darling-Culgoa-Balonne-Condamine	Indian O, South Australia, Australia	2330	3750
Austria	Danube (off in Austria Donau)	Black Sea, Romania–USSR	1770	2850
Bangladesh	Brahmaputra (alt Tsangpo, Yalutsangpu)	B of Bengal, Bangladesh	1770	2840
Belgium	Meuse (alt Maas)	North Sea, Netherlands	590	950
Belize	• Belize	Caribbean Sea, Belize	180	290
Benin	• Niger	G of Guinea, Nigeria	2550	4100
Bhutan	• Manas	Brahmaputra R, Assam, India	220	350
Bolivia	• Guapore (alt Itenez)	Mamore R, Bolivia–Brazil	1120	1800
Botswana	• Zambezi (alt Zambesi, Zambeze)	Mozambique Channel, Mozambique	1650	2650
Brazil	• Amazon (off Amazonas)-Ucayali-Tambo-Ene-Apurimac	Atlantic O, Amapa–Para, Brazil	4080	6570
Brunei	• Limbang	South China Sea, Brunei–Malaysia	120	200
Bulgaria	Danube (off in Bulgaria Dunav)	Black Sea, Romania–USSR	1770	2850
Burma	• Mekong (alt Khong, Lantsang, Mekongk, Tien Giang)	South China Sea, Vietnam	2600	4180
Burundi	• Nile (off Nil)-Kagera-Ruvuvu-Luvironza	Mediterranean Sea, Egypt	4160	6690
Cambodia	• Mekong (off in Cambodia Mekongk)	South China Sea, Vietnam	2600	4180
Cameroon	• Shari (alt Chari)-Sara (alt Ouham)	Chad L, Cameroon–Chad	900	1450
Canada	Mackenzie-Slave-Peace-Finlay	Beaufort Sea, NWT, Canada	2630	4240
Central African Empire	• Ubangi (alt Oubangui)-Uele-Kibali	Congo R, Congo–Zaire	1530	2460
Chad	• Shari (alt Chari)-Sara (alt Ouham)	Chad L, Cameroon–Chad	900	1450
Chile	Baker	G of Penas of Pacific O, Chile	270	440
	Loa	Pacific O, Chile	270	440
China	Yangtze (off Chang)	East China Sea, Kiangsu, China	3720	5980
Colombia	• Amazon (off Amazonas)-Ucayali-Tambo-Ene-Apurimac	Atlantic O, Amapa–Para, Brazil	4080	6570
Congo	Congo (alt Zaire; for Kongo)-Lualaba	Atlantic O, Angola–Zaire	2880	4630
Costa Rica	• San Juan	Caribbean Sea, Costa Rica–Nicaragua	140	220

Country	River	Outflow and Location	Length (mi)	Length (km)
Cuba	Cauto	G of Guacanayabo of Caribbean Sea, Cuba	230	370
Cyprus	• Pedias	Mediterranean Sea, Cyprus	62	100
Czechoslovakia	Danube (off in Czechoslovakia Dunaj)	Black Sea, Romania–USSR	1770	2850
Denmark	Gudena	Kattegat Strait, Denmark	100	160
Dominican Republic	Yaque del Norte	Atlantic O, Dominican Republic	170	280
Ecuador	• Putumayo (alt Ica)	Amazon R, Amazonas, Brazil	1240	2000
Egypt	• Nile (off Nil)-Kagera-Ruvuvu-Luvironza	Mediterranean Sea, Egypt	4160	6690
El Salvador	Lempa	Pacific O, El Salvador	190	300
Equatorial Guinea	• Benito	G of Guinea, Equatorial Guinea	200	320
Ethiopia	• Shebeli (alt Shabale, Shibeli; for Scebeli)	Balli Swamp, Somalia	1550	2490
Fiji	Rewa	Pacific O, Viti Levu I, Fiji	95	150
Finland	Torne (alt Tornio)-Muonio-Konkama (alt Kongama)	G of Bothnia of Baltic Sea, Finland–Sweden	350	570
France	Rhine (off in France Rhin)	North Sea, Netherlands	820	1320
French Guiana	• Maroni (alt Marowyne)	Atlantic O, French Guiana–Surinam	420	680
Gabon	• Ogooue (alt Ogowe)	G of Guinea, Gabon	750	1210
Gambia	Gambia (alt Gambie)	Atlantic O, Gambia	700	1130
Germany				
East	Elbe (alt Labe)	North Sea, Niedersachsen–Schleswig-Holstein, West Germany	720	1160
West	Danube (off in Germany Donau)	Black Sea, Romania–USSR	1770	2850
Ghana	• Volta-Black Volta (alt Volta Noire)	G of Guinea, Ghana	990	1600
Greece	Maritsa (off in Greece Evros)	Aegean Sea, Greece–Turkey	300	490
Guatemala	Usumacinta-Chixoy	B of Campeche of G of Mexico, Tabasco, Mexico	690	1110
Guinea	• Niger	G of Guinea, Nigeria	2550	4100
Guinea-Bissau	• Geba-Corubal (alt Koliba)	Atlantic O, Guinea-Bissau	350	560
Guyana	• Essequibo	Atlantic O, Guyana	600	970
Haiti	Artibonite	G of Gonave of Caribbean Sea, Haiti	170	280
Honduras	• Coco (alt Segovia)	Caribbean Sea, Honduras–Nicaragua	470	750
Hungary	Danube (off in Hungary Duna)	Black Sea, Romania–USSR	1770	2850
Iceland	• Thjorsa	Atlantic O, Iceland	140	230
India	Brahmaputra (alt Tsangpo, Yalutsangpu)	B of Bengal, Bangladesh	1770	2840
Indonesia				
in Asia	Kapuas (alt Kapuas-Besar; for Kapoeas)	South China Sea, Borneo I, Indonesia	630	1010
in Oceania (Irian Jaya)	Mamberamo-Taritatu (for Idenburg)	Pacific O, Irian Jaya, Indonesia	420	670

Country	River	Outflow and Location	Length (mi)	Length (km)
Iran	(Shatt al) Arab-Euphrates (off in Iran Furat)-Kara (Su)	Persian G, Iran–Iraq	1510	2430
Iraq	(Shatt al) Arab-Euphrates (off in Iraq Furat)-Kara (Su)	Persian G, Iran–Iraq	1510	2430
Ireland	Shannon	Atlantic O, Ireland	230	370
Israel	● Jordan (off in Israel Yarden)	Dead (Sea) L, Jordan	160	250
Italy	Po	Adraitic Sea, Veneto, Italy	380	620
Ivory Coast	● Volta-Black Volta (alt Volta Noire)	G of Guinea, Ghana	990	1600
Jamaica	Black	Caribbean Sea, Jamaica	44	71
Japan	Shinano	Sea of Japan, Honshu I, Japan	230	370
Jordan	● Jordan (off in Jordan Urdunn)	Dead (Sea) L, Jordan	160	250
Kashmir-Jammu	Indus (alt Yintu)	Arabian Sea, Pakistan	1790	2880
Kenya	● Tana	Indian O, Kenya	440	710
Korea				
North	Yalu (alt Amnok)	Korea B of Yellow Sea, China-North Korea	500	810
South	Naktong (for Rakuti)	Korea Strait, South Korea	330	520
Laos	● Mekong (off in Laos Khong)	South China Sea, Vietnam	2600	4180
Lebanon	Orontes (off Asi)	Mediterranean Sea, Turkey	300	470
Lesotho	Orange (alt Oranje)	Atlantic O, Namibia-South Africa	1400	2250
Liberia	● Cavally (alt Cavalla)	Atlantic O, Ivory Coast–Liberia	320	510
Luxembourg	Moselle (alt Mosel)	Rhine R, Rheinland-Pfalz, West Germany	340	550
Madagascar	● Betsiboka	Mozambique Channel, Madagascar	270	440
Malawi	● Shire (alt Chire)	Zambezi R, Mozambique	370	600
Malaysia				
Peninsular	● Pahang	South China Sea, Peninsular Malaysia, Malaysia	200	320
Sabah	Kinabatangan	Sulu Sea, Sabah, Malaysia	350	560
Sarawak	Rajang	South China Sea, Sarawak, Malaysia	350	560
Mali	● Niger	G of Guinea, Nigeria	2550	4100
Mauritania	Senegal-Bafing	Atlantic O, Mauritania–Senegal	1050	1700
Mexico	Rio Grande (off in Mexico Bravo)	G of Mexico, Mexico-USA	1880	3030
Mongolia	Yenisey (alt Yenisei)-Angara-Selenga (alt Selenge)-Ider	Yenisey G of Kara Sea, Russia, USSR	3650	5870
Morocco	● Oum er Rbia	Atlantic O, Morocco	370	600
Mozambique	● Zambezi (off in Mozambique Zambeze)	Mozambique Channel, Mozambique	1650	2650
Namibia	● Zambezi (alt Zambesi, Zambeze)	Mozambique Channel, Mozambique	1650	2650
Nepal	Ghaghara (alt Gogra, Kauriala)	Ganges R, Bihar–Uttar Pradesh, India	670	1080
Netherlands	Rhine (off in Netherlands Rijn)	North Sea, Netherlands	820	1320
New Caledonia	Diahot	Coral Sea, New Caledonia	56	90

Country	River	Outflow and Location	Length (mi)	Length (km)
New Zealand	Waikato	Tasman Sea, North I, New Zealand	260	420
Nicaragua	• Coco (alt Segovia)	Caribbean Sea, Honduras–Nicaragua	470	750
Niger	• Niger	G of Guinea, Nigeria	2550	4100
Nigeria	• Niger	G of Guinea, Nigeria	2550	4100
Norway	Glomma (off Glama)	Skagerrak Strait, Norway	380	610
Pakistan	Indus (alt Yintu)	Arabian Sea, Pakistan	1790	2880
Panama	• Chepo (alt Bayano)	G of Panama of Pacific O, Panama	100	160
Papua New Guinea	• Fly-Strickland	G of Papua of Coral Sea, Papua New Guinea	800	1290
Paraguay	Plata-Parana-Grande	Atlantic O, Argentina–Uruguay	3030	4880
Peru	• Amazon (off Amazonas)-Ucayali-Tambo-Ene-Apurimac	Atlantic O, Amapa–Para, Brazil	4080	6570
Philippines	• Cagayan (alt Grande de Cagayan)	Babuyan Channel, Luzon I, Philippines	220	350
Poland	Vistula (for Visla, Weichsel; off Wisla)-Bug (alt Zapadnyy Bug)	G of Danzig of Baltic Sea, Poland	750	1200
Portugal (in Europe)	Tagus (off in Portugal Tejo)	Atlantic O, Portugal	630	1010
Puerto Rico	• Plata	Atlantic O, Puerto Rico	45	72
Rhodesia	• Zambezi (alt Zambesi, Zambeze)	Mozambique Channel, Mozambique	1650	2650
Romania	Danube (off in Romania Dunarea)	Black Sea, Romania–USSR	1770	2850
Rwanda	• Nile (off Nil)-Kagera-Ruvuvu-Luvironza	Mediterranean Sea, Egypt	4160	6690
Senegal	Senegal-Bafing	Atlantic O, Mauritania–Senegal	1050	1700
Sierra Leone	• Sierra Leone-Rokel	Atlantic O, Sierra Leone	270	440
Somalia	• Shebeli (alt Shabale, Shibeli; for Scebeli)	Balli Swamp, Somalia	1550	2490
South Africa	Orange (alt Oranje)	Atlantic O, Namibia–South Africa	1400	2250
Spain (in Europe)	Tagus (off in Spain Tajo)	Atlantic O, Portugal	630	1010
Sri Lanka	Mahaweli Ganga	Indian O, Sri Lanka	210	330
Sudan	• Nile (off Nil)-Kagera-Ruvuvu-Luvironza	Mediterranean Sea, Egypt	4160	6690
Surinam	• Maroni (off in Surinam Marowyne)	Atlantic O, French Guiana–Surinam	420	680
Swaziland	• Komati (alt Incomati)	Delagoa B of Indian O, Mozambique	500	800
Sweden	Gota-Klar	Kattegat Strait, Sweden	450	720
Switzerland	Rhine (off in Switzerland Rhein, Rhin)	North Sea, Netherlands	820	1320
Syria	(Shatt al) Arab-Euphrates (off in Syria Furat)-Kara (Su)	Persian G, Iran–Iraq	1510	2430

Country	River	Outflow and Location	Length (mi)	(km)
Taiwan	Choshui	Formosa Strait, Taiwan	120	190
Tanzania	• Nile (off Nil)-Kagera-Ruvuvu-Luvironza	Mediterranean Sea, Egypt	4160	6690
Thailand	• Mekong (off in Thailand Khong)	South China Sea, Vietnam	2600	4180
Togo	• Oti	Volta R, Ghana	340	550
Trinidad and Tobago	• Ortoire	Atlantic O, Trinidad I, Trinidad and Tobago	31	50
Tunisia	• Medjerda (off Majardah)	G of Tunis of Mediterranean Sea, Tunisia	230	360
Turkey				
in Asia	(Shatt al) Arab-Euphrates (off in Turkey Firat)-Kara (Su)	Persian G, Iran–Iraq	1510	2430
in Europe	Maritsa (off in Turkey Meric)	Aegean Sea, Greece–Turkey	300	490
Uganda	• Nile (off Nil)-Kagera-Ruvuvu-Luvironza	Mediterranean Sea, Egypt	4160	6690
UK	Severn	Bristol Channel, England, UK	210	340
Upper Volta	• Volta-Black Volta (alt Volta Noire)	G of Guinea, Ghana	990	1600
Uruguay	Plata-Parana-Grande	Atlantic O, Argentina–Uruguay	3030	4880
USA				
in North America	Mississippi-Missouri-Jefferson-Beaverhead-Red Rock	G of Mexico, Louisiana, USA	3740	6020
in Oceania (Hawaii)	• Kaukonahua (Stream)	Pacific O, Oahu I, Hawaii, USA	63	100
USSR				
in Asia	Yenisey (alt Yenisei)-Angara-Selenga (alt Selenge)-Ider	Yenisey G of Kara Sea, Russia, USSR	3650	5870
in Europe	Volga	Caspian (Sea) L, Russia, USSR	2290	3690
Venezuela	Orinoco	Atlantic O, Venezuela	1280	2060
Vietnam	• Mekong (off in Vietnam Tien Giang)	South China Sea, Vietnam	2600	4180
Yugoslavia	Danube (off in Yugoslavia Dunav)	Black Sea, Romania–USSR	1770	2850
Zaire	Congo (for Kongo; off in Zaire Zaire)-Lualaba	Atlantic O, Angola–Zaire	2880	4630
Zambia	• Zambezi (alt Zambesi, Zambeze)	Mozambique Channel, Mozambique	1650	2650

4
HIGHEST MOUNTAINS (PEAKS), BY CONTINENT

Contents

Latitude and longitude of mountain peak, in degrees and minutes.

Conventional and other (alternate, former, and official) names of peak and of massif, if any (in brackets).

Mountain range or system (unconventional names are given only for initial entry), and subdivision and country in which located.

Elevation (i.e., altitude of summit above sea level), in feet and meters.

Date of first successful ascent.

Arrangement

By elevation, under continent.

Coverage

Above * * *, all named peaks rising to at least the following elevations:

Africa—15,000 ft (4570 m)
Antarctica—14,000 ft (4270 m)
Asia—25,500 ft (7770 m)
Europe—14,800 ft (4510 m)
North America—14,250 ft (4340 m)
Oceania—14,000 ft (4270 m)
South America—20,700 ft (6310 m)

Below * * *, other well-known peaks, including the highest in certain countries and in important ranges.

Rounding

Elevations are rounded to the nearest 10 ft (m).

Special Features

When peak is the highest in a particular range, the name of the range is italicized.
Active volcanoes are indicated by the abbreviation "volc."

Entries

449.

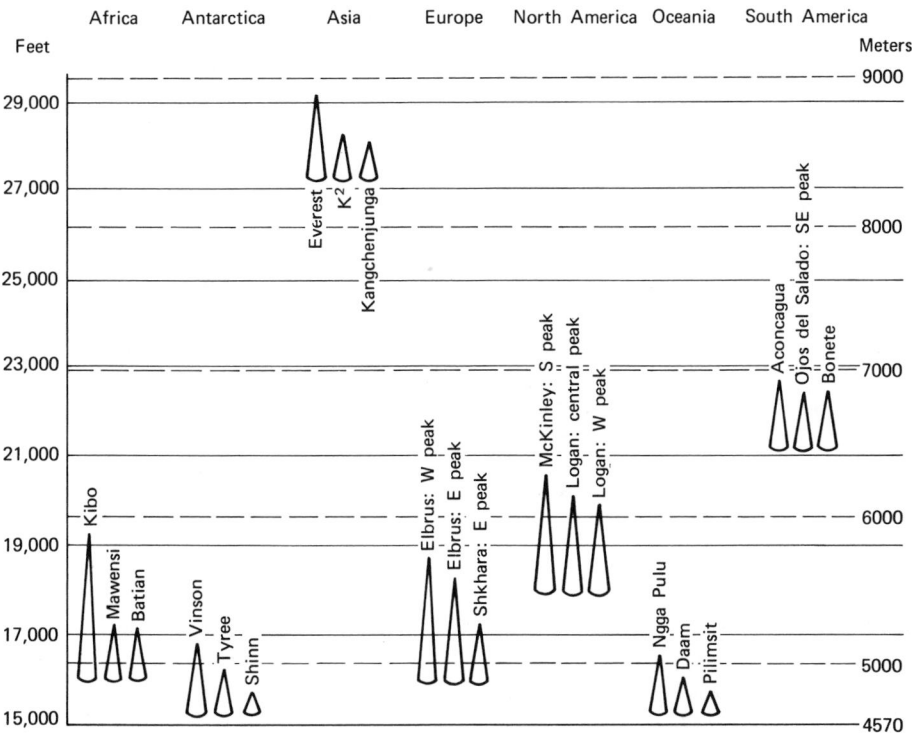

Figure 11.   Three highest mountains of each continent.

TABLE 4. HIGHEST MOUNTAINS (PEAKS), BY CONTINENT
AFRICA

Latitude and Longitude		Peak and [Massif]	Mountain Range or System	Location	Elevation		First Ascent
					(ft)	(m)	
3.04S,	37.21E	Kibo (volc[1]) [Kilimanjaro]	–	Tanganyika, Tanzania	19,340	5890	1889
3.06S,	37.27E	Mawensi (volc[1]) [Kilimanjaro]	–	Tanganyika, Tanzania	17,100	5210	1912
0.09S,	37.18E	Batian [Kenya]	–	Kenya	17,050	5200	1899
0.09S,	37.18E	• Nelion [Kenya]	–	Kenya	17,020	5190	1929
0.23N,	29.52E	Margherita [Stanley]	*Ruwenzori*	Uganda–Zaire	16,760	5110	1906
0.23N,	29.52E	Alexandra [Stanley]	Ruwenzori	Uganda–Zaire	16,700	5090	1906
0.23N,	29.52E	Albert [Stanley]	Ruwenzori	Zaire	16,690	5090	1932
0.23N,	29.52E	Savoia [Stanley]	Ruwenzori	Uganda	16,330	4980	1906
0.23N,	29.52E	Elena [Stanley]	Ruwenzori	Uganda	16,300	4970	1906
0.23N,	29.52E	Elizabeth [Stanley]	Ruwenzori	Uganda	16,170	4930	1953
0.23N,	29.52E	Philip [Stanley]	Ruwenzori	Uganda	16,140	4920	1954
0.23N,	29.52E	Moebius [Stanley]	Ruwenzori	Uganda	16,130	4920	1906
0.24N,	29.53E	Vittorio Emanuele [Speke]	Ruwenzori	Uganda	16,040	4890	1906
0.24N,	29.53E	Ensonga [Speke]	Ruwenzori	Uganda	15,960	4860	1926
0.22N,	29.53E	Edward [Baker]	Ruwenzori	Uganda	15,890	4840	1906
0.24N,	29.53E	Johnston [Speke]	Ruwenzori	Uganda	15,860	4830	1926
0.26N,	29.54E	Umberto [Emin]	Ruwenzori	Zaire	15,740	4800	1906
0.22N,	29.53E	Semper [Baker]	Ruwenzori	Uganda	15,730	4790	1906
0.26N,	29.54E	Kraepelin [Emin]	Ruwenzori	Zaire	15,720	4790	1932
0.26N,	29.55E	Iolanda [Gessi]	Ruwenzori	Uganda	15,470	4720	1906
0.26N,	29.55E	Bottego [Gessi]	Ruwenzori	Uganda	15,420	4700	1906
0.20N,	29.53E	Sella [Luigi di Savoia]	Ruwenzori	Uganda	15,180	4630	1906
0.22N,	29.53E	Wollaston [Baker]	Ruwanzori	Uganda	15,180	4630	1906
0.22N,	29.53E	Moore [Baker]	Ruwenzori	Uganda	15,170	4620	1906
0.20N,	29.53E	Weismann [Luigi di Savoia]	Ruwenzori	Uganda	15,160	4620	1935
0.23N,	29.52E	Great Tooth [Stanley]	Ruwenzori	Uganda	15,100	4600	
0.22N,	29.51E	Wasuwameso [Mugule]	Ruwenzori	Zaire	15,030	4580	
0.19N,	29.52E	Okusoma [Luigi di Savoia]	Ruwenzori	Uganda	15,020	4580	

[1] No known eruption.

AFRICA

Latitude and Longitude	Peak and [Massif]	Mountain Range or System	Location	Elevation (ft)	(m)	First Ascent
0.24N, 29.53E	Trident [Speke]	Ruwenzori	Uganda	15,000	4570	

* * *

Latitude and Longitude	Peak and [Massif]	Mountain Range or System	Location	Elevation (ft)	(m)	First Ascent
3.14S, 36.45E	• Meru (volc)	–	Tanganyika, Tanzania	14,980	4570	
13.15N, 38.24E	Ancua [Ras Dashan (off Rasdajan)]	*Semien*	Ethiopia	14,930	4550	
13.18N, 38.21E	Lagada	Semien	Ethiopia	14,870	4530	
1.30S, 29.27E	Karisimbi (volc)	*Virunga*	Rwanda-Zaire	14,790	4510	
1.27S, 29.26E	Mikeno (volc)	Virunga	Zaire	14,560	4440	
	Karra	*Arusi*	Ethiopia	14,240	4340	
1.08N, 34.33E	Elgon	–	Kenya-Uganda	14,140	4310	1911
6.55N, 39.49E	Batu	*Mendebo*	Ethiopia	14,130	4310	
11.10N, 39.10E	Kollo (alt Collo)	*Lasta*	Ethiopia	14,110	4300	
31.03N, 7.57W	Toubkal: W peak [Toubkal]	*Atlas*	Morocco	13,670	4160	
10.45N, 37.55E	Birhan	*Choke*	Ethiopia	13,630	4150	
1.23S, 29.40E	Muhavura (volc)	Virunga	Rwanda-Uganda	13,540	4130	
4.12N, 9.11E	Fako (volc) [Cameroon (alt Cameroun)]	*Cameroon (alt Cameroun)*	Cameroon	13,350	4070	1861
0.19S, 36.37E	Lesatima	*Aberdare*	Kenya	13,120	4000	
28.16N, 16.38W	Teide (volc)	–	(Canary Islands), Spain[1]	12,200	3720	
19.50N, 18.30E	(Emi) Koussi (volc)	*Tibesti*	Chad	11,470	3490	
29.28S, 29.16E	Thabana Ntlenyana (alt Thabantshonyana)	*Drakens(berg)*	Lesotho	11,420	3480	
29.12S, 29.22E	• Injasuti	Drakens(berg)	Lesotho–South Africa	11,310	3450	
29.06S, 29.20E	Champagne Castle	Drakens(berg)	Lesotho–South Africa	11,080	3380	
3.57N, 32.54E	• Kinyeti	*Imatong*	Sudan	10,460	3190	
21.25N, 18.42E	• Keguer Terbi (alt Chegor Tedi, Hessi)	Tibesti	Chad–Libya	10,330	3150	
21.05S, 55.29E	Neiges (volc)	–	Reunion	10,070	3070	
15.59S, 35.36E	Mlanje	*Mlanje*	Malawi	10,000	3050	

[1] Tenerife I.

AFRICA

Latitude and Longitude		Peak and [Massif]	Mountain Range or System	Location	Elevation		First Ascent
					(ft)	(m)	
3.35N,	8.46E	Malabo (for Santa Isabel) (volc)	—	Equatorial Guinea[1]	9,840	3000	
23.18N,	5.32E	Tahat [Atakor]	*Ahaggar* (alt Hoggar)	Algeria	9,570	2920	1912?
14.01S,	48.58E	Maromokotro	*Tsaratanana*	Madagascar	9,470	2890	
14.56N,	24.21W	Cano (volc)	—	Cape Verde[2]	9,280	2830	
28.31N,	33.57E	Katrinah (alt Catherine) [Musa]	*Sinai*	Egypt	8,650	2640	
12.30S,	15.19E	Moco	*Upanda*	Angola	8,600	2620	
21.10S,	14.33E	Konigstein [Brand(berg)]	*Kaokoveld*	Namibia	8,550	2610	
18.18S,	32.54E	Inyangani	*Inyanga*	Rhodesia	8,510	2590	
11.12N,	49.30E	Faddisome [Hor Bogor]	*Carcar*	Somalia	8,500	2590	
19.47S,	33.09E	Binga	*Chimanimani*	Mozambique–Rhodesia	7,990	2440	
35.20N,	6.40E	Chelia	*Aures* (off Awras)	Algeria	7,640	2330	
20.00N,	8.35E	• Greboun	*Air* (alt Azbine)	Niger	7,550	2300	
8.20N,	11.45E	Vogel	*Banglang*	Nigeria	6,700	2040	
9.13N,	11.07W	• Bintimani	*Loma*	Sierra Leone	6,390	1940	
32.45N,	16.56W	Ruivo de Santana	—	(Madeira Islands), Portugal[3]	6,110	1860	
25.57S,	31.11E	• Emlembe	Drakens(berg)	Swaziland	6,100	1860	
7.37N,	8.25W	Nimba	*Nimba*	Guinea–Ivory Coast–Liberia	5,780	1760	
35.13N,	8.41E	Shanabi (alt Chambi)	*Dorsale*	Tunisia	5,070	1540	

ANTARCTICA

Latitude and Longitude		Peak and [Massif]	Mountain Range or System	Location	Elevation		First Ascent
					(ft)	(m)	
78.35S,	85.25W	• –[Vinson]	*Sentinel*	Antarctica	16,860	5140	1966
78.24S,	85.55W	• Tyree	Sentinel	Antarctica	16,290	4970	1967
78.27S,	85.46W	• Shinn	Sentinel	Antarctica	15,750	4800	1966
78.23S,	86.02W	• Gardner	Sentinel	Antarctica	15,370	4690	1966
78.26S,	85.53W	• Epperly	Sentinel	Antarctica	15,100	4600	
84.20S,	166.19E	• Kirkpatrick	*Queen Alexandra*	Antarctica	14,850	4530	
83.54S,	168.23E	• Elizabeth	Queen Alexandra	Antarctica	14,700	4480	
82.51S,	161.21E	• Markham	*Queen Elizabeth*	Antarctica	14,290	4360	

[1] Macias Nguema I.
[2] Fogo I.
[3] Madeira I.

ANTARCTICA

Latitude and Longitude	Peak and [Massif]	Mountain Range or System	Location	Elevation (ft)	(m)	First Ascent
84.04S, 167.30E	• Bell	Queen Alexandra	Antarctica	14,120	4300	
83.59S, 166.39E	• Mackellar	Queen Alexandra	Antarctica	14,100	4300	

* * *

Latitude and Longitude	Peak and [Massif]	Mountain Range or System	Location	Elevation (ft)	(m)	First Ascent
84.33S, 175.18E	• Kaplan	*Hughes*	Antarctica	13,880	4230	
71.23S, 63.22W	• Jackson	*Gutenko*	Antarctica	13,750	4190	
77.02S, 126.00W	• Sidley	*Executive Committee*	Antarctica	13,720	4180	
71.47S, 168.45E	• Minto	*Admiralty*	Antarctica	13,670	4170	
77.32S, 167.09E	• Erebus (volc)	—	Antarctica	12,450	3790	1908

ASIA

Latitude and Longitude	Peak and [Massif]	Mountain Range or System	Location	Elevation (ft)	(m)	First Ascent
27.59N, 86.56E	Everest (alt Chum-ulangma) [Everest]	*Nepal Himalaya*	China–Nepal	29,030	8850	1953
35.53N, 76.31E	K^2 (alt Chogori, Dapsang, Godwin Austen)	*Karakoram*	Kashmir-Jammu	28,250	8610	1954
27.42N, 88.09E	Kangchenjunga (alt Kanchen-junga): highest peak [Kangchen-junga]	Nepal Himalaya	India–Nepal	28,170	8590	1955
27.58N, 86.56E	Lhotse (alt E^1, Lotzu) [Everest]	Nepal Himalaya	China–Nepal	27,890	8500	1956
27.41N, 88.09E	Kangchenjunga: S peak [Kang-chenjunga]	Nepal Himalaya	India–Nepal	27,800	8470	
27.53N, 87.05E	Makalu I [Makalu]	Nepal Himalaya	China–Nepal	27,790	8470	1955
27.42N, 88.09E	• Kangchenjunga: W peak [Kangchen-junga]	Nepal Himalaya	India–Nepal	27,620	8420	
27.57N, 86.57E	• Lhotse Shar (alt Lhotse: E peak) [Everest]	Nepal Himalaya	China–Nepal	27,500	8380	1970
28.42N, 83.30E	Dhaulagiri I (alt Daulagiri I)	Nepal Himalaya	Nepal	26,810	8170	1960

Figure 12. Located on the border of China and Nepal, Everest, or Chumulangma, is the world's highest mountain peak. (Credit: Department of Tourism, Ministry of Industry and Commerce of Nepal.)

ASIA

Latitude and Longitude	Peak and [Massif]	Mountain Range or System	Location	Elevation (ft)	(m)	First Ascent
28.06N, 86.39E	Cho Oyu (alt Choaoyu): highest peak	Nepal Himalaya	China–Nepal	26,750	8150	1954
28.33N, 84.34E	Manaslu (alt Kutang I): highest peak [Manaslu]	Nepal Himalaya	Nepal	26,660	8130	1956
35.14N, 74.35E	Nanga Parbat: highest peak	*Punjab Himalaya*	Kashmir-Jammu	26,660	8130	1953
28.36N, 83.49E	Annapurna I	Nepal Himalaya	Nepal	26,500	8080	1950
35.43N, 76.42E	Gasherbrum I (alt Hidden) [Gasherbrum]	Karakoram	Kashmir-Jammu	26,470	8070	1958
35.49N, 76.34E	Broad: highest peak [Gasherbrum]	Karakoram	Kashmir-Jammu	26,400	8050	1957
35.46N, 76.39E	Gasherbrum II: highest peak [Gasherbrum]	Karakoram	Kashmir-Jammu	26,360	8030	1956
28.21N, 85.47E	Gosainthan (alt Shishma Pangma; off Kaosengtsan)	Nepal Himalaya	Tibet, China	26,290	8010	1964
35.49N, 76.34E	• Broad: mittel (gipfel) peak [Gasherbrum]	Karakoram	Kashmir-Jammu	26,250	8000	
35.46N, 76.39E	Gasherbrum III [Gasherbrum]	Karakoram	Kashmir-Jammu	26,090	7950	
28.32N, 84.07E	Annapurna II	Nepal Himalaya	Nepal	26,040	7940	1960
35.46N, 76.37E	Gasherbrum IV [Gasherbrum]	Karakoram	Kashmir-Jammu	26,000	7920	1958
28.06N, 86.45E	Gyachung Kang	Nepal Himalaya	China–Nepal	25,990	7920	1964
35.14N, 74.35E	• Nanga Parbat: vor(gipfel) peak	Punjab Himalaya	Kashmir-Jammu	25,950	7910	
27.43N, 88.07E	Kangbachen [Kangchenjunga]	Nepal Himalaya	India–Nepal	25,930	7900	
28.33N, 84.34E	• Manaslu: E pinnacle [Manaslu]	Nepal Himalaya	Nepal	25,900	7900	
36.20N, 75.11E	Distaghil Sar	Karakoram	Kashmir-Jammu	25,870	7890	1960
27.58N, 86.53E	Nuptse (alt E^2) [Everest]	Nepal Himalaya	Nepal	25,850	7880	1961
28.26N, 84.39E	Himalchuli: highest peak	Nepal Himalaya	Nepal	25,800	7860	1956

ASIA

Latitude and Longitude	Peak and [Massif]	Mountain Range or System	Location	Elevation (ft)	(m)	First Ascent
36.11N, 75.13E	• Khiangyang Kish (alt Khinyang Chhish)	Karakoram	Kashmir-Jammu	25,760	7850	
28.06N, 86.41E	Ngojumba Ri (alt Cho Oyu: E peak)	Nepal Himalaya	China–Nepal	25,720	7840	1965
28.30N, 84.34E	Dakura (alt Kutang II) [Manaslu]	Nepal Himalaya	Nepal	25,710	7840	
35.39N, 76.19E	Masherbrum: E peak	Karakoram	Kashmir-Jammu	25,660	7820	1960
30.23N, 79.58E	Nanda Devi: W peak	*Kumaun Himalaya*	Uttar Pradesh, India	25,650	7820	1936
35.15N, 74.35E	Nanga Parbat: N peak	Punjab Himalaya	Kashmir-Jammu	25,650	7820	
27.55N, 87.08E	Chomo Lonzo (alt Makalu) [Makalu]	Nepal Himalaya	China–Nepal	25,640	7820	1954
35.38N, 76.18E	Masherbrum: W peak	Karakoram	Kashmir-Jammu	25,610	7810	
36.09N, 74.29E	Rakaposhi	*Haramosh (Ridge)*	Kashmir-Jammu	25,550	7790	1958
36.31N, 74.31E	Batura Mustagh I (alt Hunza-Kunji I)	Karakoram	Kashmir-Jammu	25,540	7790	1959?
35.46N, 76.39E	• Gasherbrum II: E peak [Gasherbrum]	Karakoram	Kashmir-Jammu	25,500	7770	

* * *

Latitude and Longitude	Peak and [Massif]	Mountain Range or System	Location	Elevation (ft)	(m)	First Ascent
30.56N, 79.35E	Kamet (alt Kameite)	*Zaskar*	China–India	25,440	7760	1931
29.40N, 95.10E	Namcha Barwa (off Namu-chopaerhwa)	*Assam Himalaya*	Tibet, China	25,440	7760	
30.26N, 81.18E	Gurla Mandhata (off Kuala-mantata)	*Nepal-Tibet (Watershed)*	Tibet, China	25,350	7730	
36.25N, 87.25E	• Ulugh Mustagh (off Wulu-komushih)	*Kunlun*	Sinkiang-Tibet, China	25,340	7720	
38.40N, 75.21E	Kungur II (off Kungkoerh II)	*Mustagh Ata (off Mussu-takoate)*	Sinkiang, China	25,330	7720	1955
36.15N, 71.50E	• Tirich Mir: W peak	*Hindu Kush*	Pakistan	25,260	7700	1950
36.15N, 71.50E	• Tirich Mir: E peak	Hindu Kush	Pakistan	25,230	7690	1964

ASIA

Latitude and Longitude	Peak and [Massif]	Mountain Range or System	Location	Elevation (ft)	Elevation (m)	First Ascent
34.52N, 77.45E	• Saser Kangri I	*Saser (Ridge)*	Kashmir-Jammu	25,170	7670	
29.34N, 101.53E	• Minya Konka (alt Minyag Gangkar; off Kungka)	*Tahsueh*	Szechwan, China	24,890	7590	1932
28.14N, 90.36E	Khula Kangri I (alt Kula Gangri I)	Assam Himalaya	Bhutan–China	24,780	7550	
38.16N, 75.09E	• Mustagh Ata (off Mussu-takoate)	Mustagh Ata	Sinkiang, China	24,760	7550	1956
28.01N, 86.54E	Changtse (alt Changtzu, E³) [Everest]	Nepal Himalaya	China–Nepal	24,730	7540	
38.56N, 72.02E	Kommunizma (for Garmo, Stalina)	*Pamir-Alai* (off Pamir-Alay)	Tadzhikistan, USSR	24,590	7490	1933
36.26N, 71.50E	• Noshaq: highest peak	Hindu Kush	Afghanistan-Pakistan	24,580	7490	1960
42.03N, 80.11E	Pobedy (alt Shengli)	*Tien* (alt Tyan)	China–USSR	24,410	7440	1956
32.46N, 81.02E	• Alung Gangri (off Aling)	*Alung* (off Aling)	Tibet, China	24,000	7320	
27.50N, 89.16E	Chomo Lhari (alt Chonolali)	Assam Himalaya	Bhutan–China	24,000	7310	1937
39.20N, 72.55E	Lenina (for Kaufmann)	Pamir-Alai	Kirgizia-Tadzhikistan, USSR	23,380	7130	1928
34.24N, 100.10E	Amne Machin (alt Ani-maching; off Chishih)	*Amne Machin* (off Chishih)	Tsinghai, China	23,300	7100	1960
30.27N, 90.33E	• Nyenchhen Thanglha (off Nienching-tangkula)	*Nyenchhen Thanglha* (off Nienchingtang-kula)	Tibet, China	23,250	7090	
29.56N, 84.33E	• Lombo Kangra (off Lungpu)	*Kailas* (off Kangtissu)	Tibet, China	23,160	7060	
42.15N, 80.10E	Khan-Tengri	Tien	Kirgizia, USSR	22,950	6990	1931
31.04N, 81.19E	Kailas (off Kangtissu)	Kailas	Tibet, China	22,030	6710	
37.09N, 72.26E	Karla Marksa	*Vakhan*	Tadzhikistan, USSR	21,980	6700	1946
38.35N, 97.45E	• Sulo	*Nan*	Tsinghai, China	20,820	6350	
28.17N, 97.46E	• Hkakabo Razi	*Kumon*	Burma	19,300	5880	
35.56N, 52.08E	• Damavand (alt Demavend)	*Alborz* (alt Elburz)	Iran	18,610	5670	1837

ASIA

Latitude and Longitude	Peak and [Massif]	Mountain Range or System	Location	Elevation (ft)	Elevation (m)	First Ascent
39.42N, 44.18E	Great Ararat (off Buyuk-agri) (volc[1])	—	Turkey	16,950	5160	1829
30.50N, 51.35E	• Dinar	*Zagros*	Iran	16,400	5000	
56.04N, 160.38E	Klyuchevskaya (volc)	*Kamchatka*	Russia, USSR	15,670	4770	1931
49.10N, 87.55E	Khuitun (off Huyten) [Tavan Bogd]	*Altai* (alt Altay)	Mongolia	15,270	4650	
49.48N, 86.35E	Belukha: W peak	Altai	Kazakhstan- Russia, USSR	15,160	4620	1903
38.54N, 42.48E	Suphan	—	Turkey	14,550	4430	
37.30N, 44.00E	Geliasin [Resko (alt Cilo)]	*Hakkari*	Turkey	13,680	4170	
6.05N, 116.33E	Kinabalu: highest peak	*Crocker*	Sabah, Malaysia	13,450	4100	1924
23.28N, 120.57E	Yu (alt Hsinkao, Morrison)	*Central* (off Chungyang)	Taiwan	13,110	4000	1896?
38.32N, 35.28E	Erciyas (volc)	—	Turkey	12,850	3920	
1.42S, 101.16E	Kerinci (for Indrapura, Kerintji) (volc)	*Barisan*	Sumatra I, Indonesia	12,470	3800	
35.22N, 138.44E	Fuji (alt Huzi) (volc)	—	Honshu I, Japan	12,390	3780	7th c?
15.20N, 43.55E	• Hadur Shuayb	*Yemen (High-lands)*	North Yemen	12,340	3760	
36.43N, 44.50E	• Algurd (alt Halgurd)	Zagros	Iraq	12,250	3730	
8.24S, 116.28E	Sangkariyan (for Sangkari-jan) (volc[1]) [Rinjani (for Rindjani)]	—	Lombok I, Indonesia	12,220	3730	
8.06S, 112.55E	Mahameru (volc)	*Semeru*	Java I, Indonesia	12,060	3680	
16.52N, 43.22E	• Razikh	Yemen (High-lands)	Saudi Arabia- North Yemen	11,990	3650	
51.45N, 100.20E	Munku-Sardyk	*Sayan*	Mongolia- USSR	11,450	3490	1868
3.21S, 120.01E	Rante Kombola	*Quarles*	Celebes I, Indonesia	11,340	3450	
15.04N, 107.59E	• (Ngoc) Linh	*Annamese* (off Trungphan)	Vietnam	10,500	3200	
35.40N, 138.15E	Kitadake [Shirane]	*Akaishi*	Honshu I, Japan	10,470	3190	
36.48N, 139.08E	Oku-Hotaka [Hotaka]	*Mikuni*	Honshu I, Japan	10,470	3190	

[1] No known eruption.

ASIA

Latitude and Longitude	Peak and [Massif]	Mountain Range or System	Location	Elevation (ft)	Elevation (m)	First Ascent
8.21S, 115.30E	Agung (volc)	–	Bali I, Indonesia	10,310	3140	
22.18N, 103.46E	Fan Si Pan	*Fan Si Pan*	Vietnam	10,310	3140	
23.13N, 57.16E	• Sham	*Akhdar*	Oman	10,190	3110	
34.18N, 36.07E	Makmel [Qurnat al Sawda]	*Lebanon* (off Lubnan)	Lebanon	10,120	3080	
8.55S, 125.30E	Tata Mailau (alt Ramelau)	–	Timor I, Indonesia	9,720	2960	
6.59N, 125.16E	Apo (volc[1])	–	Mindanao I, Philippines	9,690	2950	
16.36N, 120.54E	Pulog	*Central*	Luzon I, Philippines	9,610	2930	
18.59N, 103.10E	• Bia	*Tranninh*	Laos	9,240	2820	
33.24N, 35.50E	Hermon (off Shaykh)	*Anti-Lebanon* (off Sharqi)	Lebanon-Syria	9,230	2810	
41.59N, 128.04E	• Paitou (alt Paektu)	*Changpai* (alt Changbaek)	China–North Korea	8,900	2710	1886
18.35N, 98.29E	Inthanon (alt Angka)	*Phi Pan Nam*	Thailand	8,450	2580	
7.00N, 80.46E	Pidurutalagala	*Piduru (Ridges)*	Sri Lanka	8,280	2520	
13.53N, 45.12E	• Thamar (off Thamir)	*Yemen* (Highlands)	South Yemen	8,240	2510	
4.38N, 102.14E	Tahan	*Cameron (Highlands)*	Peninsular Malaysia, Malaysia	7,190	2190	
34.55N, 32.52E	Khionistra (alt Olympus)	*Troodos*	Cyprus	6,410	1950	
33.22N, 126.32E	• Halla	–	South Korea[2]	6,400	1950	
12.02N, 104.10E	• Aural	*Cardamomes*	Cambodia	5,950	1810	
29.36N, 35.24E	• Ramm	*Sharah*	Jordan	5,750	1750	
33.00N, 35.25E	• Meron (for Jarmaq, Sharqi)	*Galilee*	Israel	3,960	1210	

EUROPE

Latitude and Longitude	Peak and [Massif]	Mountain Range or System	Location	Elevation (ft)	Elevation (m)	First Ascent
43.21N, 42.26E	Elbrus (for Elborus): W peak (volc[1]) [Elbrus (for Elborus)]	*Caucasus* (off Kavkaz)	Russia, USSR	18,480	5630	1874
43.21N, 42.26E	Elbrus: E peak (volc[1]) [Elbrus]	Caucasus	Russia, USSR	18,360	5590	1829

[1] No known eruption.
[2] Cheju I.

EUROPE

Latitude and Longitude	Peak and [Massif]	Mountain Range or System	Location	Elevation (ft)	(m)	First Ascent
43.00N, 43.06E	Shkhara: E peak	Caucasus	Georgia-Russia, USSR	17,060	5200	1888
43.03N, 43.08E	Dykh(-Tau): W peak	Caucasus	Russia, USSR	17,050	5200	1888
43.03N, 43.08E	Dykh(-Tau): E peak	Caucasus	Russia, USSR	16,900	5150	1938
43.03N, 43.13E	Koshtan(-Tau)	Caucasus	Russia, USSR	16,880	5140	1888
43.00N, 43.06E	Shkhara: W peak	Caucasus	Georgia-Russia, USSR	16,880	5140	
43.03N, 43.10E	Pushkina	Caucasus	Russia, USSR	16,730	5100	1938
43.02N, 43.03E	Dzhangi(-Tau): NW peak	Caucasus	Georgia, USSR	16,570	5050	1903
42.42N, 44.31E	Kazbek: E peak	Caucasus	Georgia, USSR	16,560	5050	1868
43.02N, 43.03E	Dzhangi(-Tau): SE peak	Caucasus	Georgia, USSR	16,520	5030	1888
43.02N, 43.02E	Katyn(-Tau)	Caucasus	Georgia-Russia, USSR	16,310	4970	1888
43.02N, 43.05E	Shota Rustaveli	Caucasus	Georgia-Russia, USSR	16,270	4960	1937
43.02N, 43.09E	Mizhirgi: W peak	Caucasus	Russia, USSR	16,170	4930	1934
43.02N, 43.09E	Mizhirgi: E peak	Caucasus	Russia, USSR	16,140	4920	1889
43.04N, 43.13E	Kundyum-Mizhirgi	Caucasus	Russia, USSR	16,010	4880	1946
43.03N, 43.01E	Gestola	Caucasus	Georgia-Russia, USSR	15,930	4860	1886
43.02N, 42.58E	Tetnuld	Caucasus	Georgia, USSR	15,920	4850	1887
45.50N, 6.52E	(Mont-)Blanc (alt Bianco) [Blanc (alt Bianco)]	*Alps* (off Alpe, Alpen, Alpes, Alpi)	France-Italy	15,770	4810	1786
42.43N, 44.25E	Dzhimariy (-Khokh)	Caucasus	Georgia, USSR	15,680	4780	1890
43.02N, 43.04E	Adish	Caucasus	Georgia-Russia, USSR	15,570	4750	1931
43.08N, 42.40E	Ushba: SW peak	Caucasus	Georgia, USSR	15,450	4710	1903
43.08N, 42.40E	Ushba: NE peak	Caucasus	Georgia, USSR	15,400	4690	1888
43.05N, 43.13E	Ullu-Auz(-Bashi)	Caucasus	Russia, USSR	15,360	4680	1888
43.05N, 43.12E	Panoramnyy	Caucasus	Russia, USSR	15,350	4680	1946
43.02N, 43.10E	Krumkol	Caucasus	Russia, USSR	15,320	4670	1937
42.42N, 44.31E	Kazbek: W peak	Caucasus	Georgia, USSR	15,250	4650	1890
42.46N, 43.48E	Uilpata [for Aday(-Khokh)]	Caucasus	Russia, USSR	15,240	4650	1890
42.45N, 44.25E	Shau(-Khokh)	Caucasus	Georgia, USSR	15,240	4640	1936
45.56N, 7.52E	Dufour(spitze) [Rosa]	Alps	Italy-Switzerland	15,200	4630	1855
45.56N, 7.52E	Grenz(gipfel) [Rosa]	Alps	Italy-Switzerland	15,190	4630	1851

EUROPE

Latitude and Longitude	Peak and [Massif]	Mountain Range or System	Location	Elevation (ft)	Elevation (m)	First Ascent
43.20N, 42.25E	Kyukyurtlyukol (-Bashi) [Elbrus]	Caucasus	Russia, USSR	15,170	4620	1936
45.56N, 7.51E	Nordend [Rosa]	Alps	Italy–Switzerland	15,130	4610	1861
43.08N, 42.59E	Tikhtengen: S peak	Caucasus	Georgia–Russia, USSR	15,130	4610	1935
43.08N, 42.59E	Tikhtengen: N peak	Caucasus	Georgia–Russia, USSR	15,130	4610	1936
42.42N, 44.28E	Mayli(-Khokh)	Caucasus	Georgia, USSR	15,100	4600	1903
42.45N, 43.47E	Dubl: N peak	Caucasus	Russia, USSR	15,030	4580	1933
45.56N, 7.52E	Zumstein-(spitze) [Rosa]	Alps	Italy–Switzerland	15,000	4570	1820
45.55N, 7.52E	Signal(kuppe) [Rosa]	Alps	Italy–Switzerland	14,960	4560	1842
43.06N, 43.13E	Dumala(-Tau)	Caucasus	Russia, USSR	14,950	4560	1930
43.03N, 43.15E	Tyutyun (-Bashi)	Caucasus	Russia, USSR	14,930	4550	1933
46.06N, 7.51E	Dom [Mischabel]	Alps	Switzerland	14,910	4540	1858
42.57N, 43.11E	Aylama	Caucasus	Georgia–Russia, USSR	14,890	4540	1889
43.12N, 42.57E	Dzhaylyk (-Bashi)	Caucasus	Russia, USSR	14,890	4540	1936
45.55N, 7.50E	Lyskamm [Rosa]	Alps	Italy–Switzerland	14,890	4540	1861
43.01N, 43.15E	Tyutyun(-Tau)	Caucasus	Russia, USSR	14,890	4540	1933
42.45N, 43.47E	Dubl: S peak	Caucasus	Russia, USSR	14,880	4530	1933
43.01N, 43.03E	Lakutsa	Caucasus	Georgia, USSR	14,830	4520	1931
42.47N, 43.46E	Karaugom: E peak	Caucasus	Russia, USSR	14,810	4510	1890
42.20N, 46.15E	Addala Shukhgelmeer	Caucasus	Russia, USSR	14,800	4510	1935
42.47N, 43.46E	Karaugom: W peak	Caucasus	Russia, USSR	14,800	4510	1937
42.42N, 44.29E	Spartak	Caucasus	Georgia, USSR	14,800	4510	1940

* * *

Latitude and Longitude	Peak and [Massif]	Mountain Range or System	Location	Elevation (ft)	Elevation (m)	First Ascent
46.07N, 7.43E	Weiss(horn)	Alps	Switzerland	14,780	4500	1861
46.05N, 7.51E	Tasch(horn) [Mischabel]	Alps	Switzerland	14,730	4490	1862
45.58N, 7.39E	Matter(horn) [Cervino]	Alps	Italy–Switzerland	14,690	4480	1865
45.51N, 6.53E	Maudit [Blanc]	Alps	France–Italy	14,650	4470	1878
41.13N, 47.51E	Bazar-Dyuzi	Caucasus	Azerbaijan–Russia, USSR	14,550	4430	1873
46.02N, 7.37E	(Dent) Blanche	Alps	Switzerland	14,300	4360	1862

EUROPE

Latitude and Longitude		Peak and [Massif]	Mountain Range or System	Location	Elevation (ft)	(m)	First Ascent
46.32N,	8.08E	Finsteraar(horn)	Alps	Switzerland	14,020	4270	1812
46.28N,	8.00E	Aletsch(horn)	Alps	Switzerland	13,760	4190	1859
46.33N,	7.58E	Jungfrau	Alps	Switzerland	13,640	4160	1811
46.33N,	8.01E	Monch	Alps	Switzerland	13,450	4100	1857
46.34N,	8.01E	Eiger	Alps	Switzerland	13,020	3970	1858
47.04N,	12.42E	Grossglockner	Alps	Austria	12,460	3800	1800
37.03N,	3.19W	Mulhacen	*Nevada*	Andalucia, Spain	11,410	3480	
42.38N,	0.40E	Aneto	*Pyrenees* (alt Pirineos)	Aragon, Spain	11,170	3400	1842
37.50N,	14.55E	Etna (volc)	—	Sicilia, Italy[1]	10,760	3280	
47.25N,	10.59E	Zug(spitze)	Alps	Austria–West Germany	9,720	2960	1820
42.11N,	23.34E	Musala	*Rhodope* (off Rodhopis, Rodopi)	Bulgaria	9,600	2920	
40.05N,	22.21E	Mytikas [Olympus (off Olimbos)]	*Olympus* (off Olimbos)	Greece	9,570	2920	1915?
42.28N,	13.34E	Corno Grande	*Apennines* (off Appennino)	Abruzzi e Molise, Italy	9,560	2910	
41.46N,	23.24E	Vikhren	Rhodope	Bulgaria	9,560	2910	
46.23N,	13.50E	Triglav	Alps	Slovenia, Yugoslavia	9,400	2860	1778
41.44N,	20.32E	Korab	*Korab*	Albania–Yugoslavia	9,030	2750	
42.23N,	8.56E	Cinto	—	Corse, France[2]	8,890	2710	
49.09N,	20.05E	Gerlachovsky (alt Gerlach-ovka; for Stalin)	*Tatra* (off Tatry)	Slovakia, Czechoslovakia	8,710	2650	
43.12N,	4.48W	Cerredo [Europa]	*Cantabrian* (off Cantabrica)	Asturias–Castilla la Vieja–Leon, Spain	8,690	2650	1892
45.36N,	24.44E	Moldoveanu	*Transylvanian Alps* (off Carpatii Meridionali)	Romania	8,340	2540	
45.35N,	24.34E	Negoiu	Transylvanian Alps	Romania	8,320	2530	
49.12N,	20.04E	Rysy	Tatra	Czechoslovakia–Poland	8,200	2500	
61.39N,	8.33E	Glittertinden	*Jotunheimen*	Norway	8,110	2470	

[1] Sicily I.
[2] Corsica I.

EUROPE

Latitude and Longitude	Peak and [Massif]	Mountain Range or System	Location	Elevation (ft)	Elevation (m)	First Ascent
61.37N, 8.17E	Galdhopiggen	Jotunheimen	Norway	8,100	2470	
38.32N, 22.35E	Parnassus (off Parnassos) [Parnassus]	—	Greece	8,060	2460	
42.43N, 24.55E	Botev	Balkan (off Stara)	Bulgaria	7,800	2380	
38.28N, 28.25W	Ponta do Pico (alt Pico) (volc)	—	(Azores), Portugal[1]	7,710	2350	
64.01N, 16.41W	Hvannadal-shnukur	—	Iceland	6,950	2120	
67.53N, 18.31E	Kebnekaise	Kolen	Sweden	6,930	2110	1883
40.19N, 7.37W	Estrela	Estrela	Portugal	6,530	1990	
65.04N, 60.09E	Narodnaya	Ural	Russia, USSR	6,210	1890	
45.32N, 2.50E	Sancy	Auvergne	Auvergne, France	6,190	1890	
79.02N, 17.30E	Newton	—	Svalbard[2]	5,630	1720	
56.48N, 5.00W	(Ben) Nevis	Grampian	Scotland, UK	4,410	1340	
69.18N, 21.16E	Haltiatunturi (alt Reisduod-darhaldde)	Haltia (alt Halddia)	Finland-Norway	4,360	1330	
40.49N, 14.26E	Vesuvius (off Vesuvio) (volc)	—	Campania, Italy	4,190	1280	
50.26N, 12.57E	Fichtel(-berg)	Erz(gebirge)	East Germany	3,980	1210	
52.00N, 9.45W	Carrantuohill (alt Carran-tual)	Macgillicuddy's Reeks	Ireland	3,410	1040	
47.52N, 20.01E	Kekes	Matra	Hungary	3,330	1010	

NORTH AMERICA

Latitude and Longitude	Peak and [Massif]	Mountain Range or System	Location	Elevation (ft)	Elevation (m)	First Ascent
63.04N, 151.00W	McKinley: S peak [McKinley]	Alaska	Alaska, USA	20,320	6190	1913
60.34N, 140.24W	Logan: central peak [Logan]	Saint Elias	Yukon Territory, Canada	19,520	5950	1925
60.34N, 140.25W	• Logan: W peak [Logan]	Saint Elias	Yukon Territory, Canada	19,470	5930	1925
63.04N, 151.00W	McKinley: N peak [McKinley]	Alaska	Alaska, USA	19,470	5930	1910
60.34N, 140.22W	• Logan: E peak [Logan]	Saint Elias	Yukon Territory, Canada	19,420	5920	1957

[1] Pico I.

[2] West Spitsbergen I.

NORTH AMERICA

Latitude and Longitude	Peak and [Massif]	Mountain Range or System	Location	Elevation (ft)	Elevation (m)	First Ascent
19.02N, 97.16W	Citlatepetl (alt Orizaba) (volc[1])	*Neovolcanica*	Puebla-Vera-cruz, Mexico	18,700	5700	1848
60.35N, 140.24W	• Logan: N peak [Logan]	Saint Elias	Yukon Territory, Canada	18,270	5570	1959
60.17N, 140.55W	Saint Elias	Saint Elias	Canada–USA	18,010	5490	1897
19.01N, 98.32W	Popocatepetl (volc)	Neovolcanica	Puebla, Mexico	17,890	5450	1520
62.58N, 151.24W	Foraker	Alaska	Alaska, USA	17,400	5300	1934
19.11N, 98.38W	Ixtacihuatl (alt Iztaccihuatl)	Neovolcanica	Puebla, Mexico	17,340	5290	16th c?
60.35N, 140.39W?	• Queen [Logan]	Saint Elias	Yukon Territory, Canada	17,300	5270	1966
61.01N, 140.28W	Lucania	Saint Elias	Yukon Territory, Canada	17,150	5230	1937
60.35N, 140.39W	King [Logan]	Saint Elias	Yukon Territory, Canada	16,970	5170	1952
61.06N, 140.23W	Steele	Saint Elias	Yukon Territory, Canada	16,640	5070	1935
61.23N, 141.45W	Bona	Saint Elias	Alaska, USA	16,500	5030	1930
61.44N, 143.26W	Blackburn: highest peak	*Wrangell*	Alaska, USA	16,390	5000	1958
61.44N, 143.26W	Blackburn: SE peak	Wrangell	Alaska, USA	16,290	4960	1912
62.13N, 144.08W	Sanford	Wrangell	Alaska, USA	16,240	4950	1938
61.14N, 140.30W	Wood	Saint Elias	Yukon Territory, Canada	15,880	4840	1941
60.20N, 139.42W	Vancouver	Saint Elias	Canada–USA	15,700	4790	1949
61.25N, 141.43W	Churchill	Saint Elias	Alaska, USA	15,640	4770	1951
61.11N, 140.33W	• Slaggard	Saint Elias	Yukon Territory, Canada	15,570	4750	1959
61.12N, 140.30W	• McCauley (alt Macauly)	Saint Elias	Yukon Territory, Canada	15,470	4720	1959
58.54N, 137.31W	Fairweather	Saint Elias	Canada–USA	15,300	4660	1931
61.20N, 141.48W	University	Saint Elias	Alaska, USA	15,030	4580	1955
60.19N, 139.04W	Hubbard	Saint Elias	Canada–USA	15,010	4580	1951
61.17N, 141.09W	Bear	Saint Elias	Alaska, USA	14,850	4530	1951
61.00N, 140.01W	Walsh	Saint Elias	Yukon Territory, Canada	14,780	4500	1941
19.14N, 98.02W	Malinche (alt Matlalcueyetl)	Neovolcanica	Puebla-Tlaxcala, Mexico	14,640	4460	
62.57N, 151.05W	Hunter	Alaska	Alaska, USA	14,570	4440	1954
60.21N, 139.04W	Alverstone	Saint Elias	Canada–USA	14,530[2]	4430[2]	1951
63.06N, 150.56W	Browne Tower [McKinley]	Alaska	Alaska, USA	14,530	4430	1913

[1] Dormant since 1687.

[2] Average of elevations officially given by Canada (14,500 ft, 4420 m) and USA (14,560 ft, 4440 m).

NORTH AMERICA

Latitude and Longitude	Peak and [Massif]	Mountain Range or System	Location	Elevation (ft)	Elevation (m)	First Ascent
36.35N, 118.17W	Whitney	*Sierra Nevada*	California, USA	14,490	4420	1873
61.22N, 141.54W	● Aello	Saint Elias	Alaska, USA	14,440	4400	1967
39.07N, 106.26W	Elbert	*Rocky*	Colorado, USA	14,430	4400	1874
38.55N, 106.19W	Harvard	Rocky	Colorado, USA	14,420	4390	1869
39.11N, 106.28W	Massive	Rocky	Colorado, USA	14,420	4390	1874
46.51N, 121.46W	Rainier (volc)	*Cascade*	Washington, USA	14,410	4390	1870
19.06N, 99.46W	Toluca (alt Zinantecatl)	Neovolcanica	Mexico, Mexico	14,410	4390	
36.39N, 118.21W	Williamson	Sierra Nevada	California, USA	14,370	4380	1884?
39.02N, 106.28W	La Plata	Rocky	Colorado, USA	14,370	4380	
37.35N, 105.29W	Blanca	Rocky	Colorado, USA	14,320	4360	1874
38.04N, 107.28W	Uncompahgre	Rocky	Colorado, USA	14,310	4360	1874
37.59N, 105.35W	Crestone	Rocky	Colorado, USA	14,290	4360	1916
39.21N, 106.06W	Lincoln	Rocky	Colorado, USA	14,290	4350	1861
38.40N, 106.15W	Antero	Rocky	Colorado, USA	14,270	4350	
39.38N, 105.49W	Grays	Rocky	Colorado, USA	14,270	4350	1869
39.39N, 105.49W	Torreys	Rocky	Colorado, USA	14,270	4350	
39.01N, 106.52W	Castle	Rocky	Colorado, USA	14,260	4350	bef 1875
39.35N, 105.38W	Evans	Rocky	Colorado, USA	14,260	4350	
40.15N, 105.37W	Longs	Rocky	Colorado, USA	14,260	4350	1868
39.24N, 106.12W	Quandary	Rocky	Colorado, USA	14,260	4350	
60.37N, 140.11W	McArthur [Logan]	Saint Elias	Yukon Territory, Canada	14,250	4340	1961

* * *

Latitude and Longitude	Peak and [Massif]	Mountain Range or System	Location	Elevation (ft)	Elevation (m)	First Ascent
41.25N, 122.12W	Shasta (volc)	Cascade	California, USA	14,160	4320	1854
62.00N, 144.00W	Wrangell (volc)	Wrangell	Alaska, USA	14,160	4320	1908
38.50N, 105.03W	Pikes (Peak)	Rocky	Colorado, USA	14,110	4300	1820
60.19N, 139.00W	Kennedy	Saint Elias	Yukon Territory, Canada	13,900	4240	1965
15.02N, 91.55W	Tajumulco (volc)	Madre	Guatemala	13,850	4220	
43.44N, 110.48W	Grand Teton	*Teton*	Wyoming, USA	13,770	4200	1872
51.22N, 125.14W	Waddington	*Coast*	British Columbia, Canada	13,100	3990	1936
53.07N, 119.08W	Robson	*Canadian Rocky*	British Columbia, Canada	12,970	3950	1913
9.29N, 83.30W	Chirripo Grande	*Talamanca*	Costa Rica	12,530	3820	
68.50N, 29.45W	Gunnbjorn	—	Greenland	12,140	3700	1935
8.48N, 82.38W	● Chiriqui (alt Baru) (volc)	*Central*	Panama	11,410	3480	
45.22N, 121.42W	Hood	Cascade	Oregon, USA	11,240	3430	1854
40.29N, 121.30W	Lassen (volc)	Cascade	California, USA	10,460	3190	
19.02N, 70.59W	Duarte (for Trujillo)	*Central*	Dominican Republic	10,420	3170	
18.22N, 71.59W	Selle	*Selle*	Haiti	8,790	2680	

NORTH AMERICA

Latitude and Longitude	Peak and [Massif]	Mountain Range or System	Location	Elevation (ft)	Elevation (m)	First Ascent
13.50 N, 89.38W	Santa Ana (volc)	*Apaneca Lamatepeque*	El Salvador	7,730	2360	
18.03N, 76.35W	Blue (Mountain)	*Eastern*	Jamaica	7,400	2260	
13.45N, 86.23W	• Mogoton	*Dipilto*	Honduras–Nicaragua	6,910	2110	
35.46N, 82.16W	Mitchell	*Blue (Ridge) (Appalachian)*	North Carolina, USA	6,680	2040	
19.59N, 76.50W	Turquino	*Maestra*	Cuba	6,480	1970	
44.16N, 71.18W	Washington	*White (Appalachian)*	New Hampshire, USA	6,290	1920	1642
16.03N, 61.40W	Soufriere (volc)	–	Guadeloupe[1]	4,870	1480	
14.48N, 61.10W	Pelee (volc)	–	Martinique	4,580	1400	
18.10N, 66.36W	Punta	*Central*	Puerto Rico	4,390	1340	

OCEANIA

Latitude and Longitude	Peak and [Massif]	Mountain Range or System	Location	Elevation (ft)	Elevation (m)	First Ascent
4.05S, 137.11E	Ngga Pulu [Jaya (for Carstensz, Djaja, Sukarno)]	*Sudirman* (for Nassau)	Irian Jaya, Indonesia	16,500	5030	1936
4.21S, 138.26E	Daam	*Jayawijaya* (for Djajawidjaja, Orange)	Irian Jaya, Indonesia	16,150	4920	
4.03S, 137.02E	Pilimsit (for Idenburg)	Sudirman	Irian Jaya, Indonesia	15,750	4800	
4.15S, 138.45E	Trikora (for Wilhelmina)	Jayawijaya	Irian Jaya, Indonesia	15,580	4750	1913
4.44S, 140.20E	Mandala (for Juliana)	Jayawijaya	Irian Jaya, Indonesia	15,420	4700	1959
5.43S, 145.03E	• Wilhelm	*Bismarck*	Papua New Guinea	15,400	4690	
4.25S, 139.56E	Wisnumurti (for Jan Pieterszoon Coen)	Jayawijaya	Irian Jaya, Indonesia	15,080	4590	
4.42S, 140.06E	Yamin (for Prins Hendrik)	Jayawijaya	Irian Jaya, Indonesia	14,860	4530	
6.07S, 144.42E	• Kubor	*Kubor*	Papua New Guinea	14,300	4360	
5.38S, 145.01E	• Herbert	Bismarck	Papua New Guinea	14,000	4270	

* * *

Latitude and Longitude	Peak and [Massif]	Mountain Range or System	Location	Elevation (ft)	Elevation (m)	First Ascent
19.50N, 155.28W	(Mauna) Kea: highest peak	–	Hawaii, USA[2]	13,800	4210	

[1] Basse-Terre I.
[2] Hawaii I.

OCEANIA

Latitude and Longitude	Peak and [Massif]	Mountain Range or System	Location	Elevation (ft)	Elevation (m)	First Ascent
19.28N, 155.36W	(Mauna) Loa (volc)	–	Hawaii, USA[1]	13,680	4170	
6.04S, 143.53E	• Giluwe	*Hagen*	Papua New Guinea	13,660	4160	
6.16S, 147.04E	• Bangeta	*Saruwaged*	Papua New Guinea	13,470	4110	
8.53S, 147.33E	• Victoria	*Owen Stanley*	Papua New Guinea	13,360	4070	
43.37S, 170.08E	Cook (alt Aorangi): highest peak	*Southern Alps*	South I, New Zealand	12,350	3760	1894
5.55S, 154.59E	• Balbi (volc[2])	*Emperor*	Papua New Guinea[3]	10,210	3110	
20.43N, 156.13W	Haleakala	–	Hawaii, USA[4]	10,020	3060	
9.43S, 160.02E	Makarakomburu	*Kavo*	Solomon Islands[5]	8,030	2450	
17.37S, 149.28W	Orohena	–	French Polynesia[6]	7,340	2240	
36.27S, 148.16E	Kosciusko	*Great Dividing*	New South Wales, Australia	7,330	2230	1840?
13.35S, 172.27W	• Silisili (alt Hertha)	–	Samoa[7]	6,090	1860	
20.36S, 164.46E	Panie	–	New Caledonia	5,340	1630	
17.37S, 178.01E	Tomaniivi (alt Victoria)	–	Fiji[8]	4,340	1320	

SOUTH AMERICA

Latitude and Longitude	Peak and [Massif]	Mountain Range or System	Location	Elevation (ft)	Elevation (m)	First Ascent
32.39S, 70.01W	Aconcagua	*Andes*	Mendoza, Argentina	22,840	6960	1897
27.06S, 68.32W	Ojos del Salado: SE peak (volc[2])	Andes	Argentina–Chile	22,560[9]	6870[9]	1937
27.51S, 68.47W	Bonete	Andes	La Rioja, Argentina	22,550	6870	1913

[1] Hawaii I.
[2] No known eruption.
[3] Bougainville I.
[4] Maui I.
[5] Guadalcanal I.
[6] Tahiti I.
[7] Savaii I.
[8] Viti Levu I.
[9] Average of elevations officially given by Argentina (22,540 ft, 6870 m) and Chile (22,570 ft, 6880 m).

Figure 13. Snow-covered Aconcagua, towering above a desolate and forbidding section of the Argentinian Andes, is the highest mountain in the Western Hemisphere. (Credit: Organization of American States.)

SOUTH AMERICA

Latitude and Longitude	Peak and [Massif]	Mountain Range or System	Location	Elevation (ft)	Elevation (m)	First Ascent
27.47S, 68.51W	Pissis	Andes	Catamarca–La Rioja, Argentina	22,240	6780	1937
9.07S, 77.37W	Huascaran: S peak	*Blanca* (Andes)	Peru	22,210	6770	1932
31.59S, 70.07W	Mercedario	Andes	San Juan, Argentina	22,210	6770	1934
24.43S, 68.33W	Llullaillaco (volc)	Andes	Argentina–Chile	22,100[1]	6730[1]	bef 1550
24.58S, 66.22W	Libertador (for Cachi: N peak) [Cachi]	Andes	Salta, Argentina	22,050	6720	1950
27.06S, 68.32W	• Ojos del Salado: NW peak (volc[2])	Andes	Argentina–Chile	22,050	6720	1937
33.22S, 69.47W	Tupungato	Andes	Argentina–Chile	21,900[3]	6670[3]	1897
27.03S, 68.27W	• Gonzalez: highest peak	Andes	Argentina–Chile	21,850	6660	
9.07S, 77.37W	Huascaran: N peak	Blanca (Andes)	Peru	21,840	6650	1908 or 1939
27.04S, 68.29W	Muerto	Andes	Argentina–Chile	21,820[4]	6650[4]	1950?
10.16S, 76.54W	Yerupaja: N peak	*Huayhuash* (Andes)	Peru	21,760	6630	1950
27.02S, 68.18W	Incahuasi (alt Incaguasi)	Andes	Argentina–Chile	21,700[5]	6610[5]	1859 or 1913
25.55S, 66.52W	Galan	Andes	Catamarca, Argentina	21,650	6600	
27.06S, 68.47W	Tres Cruces: central peak	Andes	Argentina–Chile	21,540[6]	6560[6]	1937
27.03S, 68.27W	• Gonzalez: N peak	Andes	Argentina–Chile	21,490	6550	1955
18.06S, 68.54W	Sajama	*Occidental* (Andes)	Bolivia	21,460	6540	1939
10.16S, 76.54W	• Yerupaja: S peak	Huayhuash (Andes)	Peru	21,380	6510	1958

[1] Average of elevations officially given by Argentina (22,060 ft, 6720 m) and Chile (22,150 ft, 6750 m).

[2] No known eruption.

[3] Average of elevations officially given by Argentina (22,310 ft, 6800 m) and Chile (21,490 ft, 6550 m).

[4] Average of elevations officially given by Argentina (21,460 ft, 6540 m) and Chile (22,200 ft, 6760 m).

[5] Average of elevations officially given by Argentina (21,720 ft, 6620 m) and Chile (21,690 ft, 6610 m).

[6] Average of elevations officially given by Argentina (20,850 ft, 6360 m) and Chile (22,210 ft, 6770 m).

SOUTH AMERICA

Latitude and Longitude		Peak and [Massif]	Mountain Range or System	Location	Elevation		First Ascent
					(ft)	(m)	
27.16S,	68.32W	Nacimiento	Andes	Catamarca, Argentina	21,300	6490	1937
15.31S,	72.42W	Coropuna: highest peak	*Occidental* (Andes)	Peru	21,080	6420	1911
15.50S,	68.34W	Illampu [Sorata]	*Real* (Andes)	Bolivia	21,070	6420	1928
27.08S,	68.49W	Puntiagudo	Andes	Argentina–Chile	21,060[1]	6420[1]	
32.05S,	69.59W	Ramada	Andes	San Juan, Argentina	21,030	6410	1934
9.07S,	77.36W	Chopicalqui (alt Huascaran: E peak)	Blanca (Andes)	Peru	21,000	6400	1932
26.30S,	68.32W	Laudo	Andes	Catamarca, Argentina	21,000	6400	
9.32S,	77.18W	Huantsan: S peak	Blanca (Andes)	Peru	20,980	6390	1952
13.48S,	71.14W	Ausangate: highest peak	*Vilcanota* (Andes)	Peru	20,950	6380	1953
15.51S,	68.36W	Ancohuma [Sorata]	Real (Andes)	Bolivia	20,930	6380	1919
29.08S,	69.48W	Toro	Andes	Argentina–Chile	20,930	6380	
9.02S,	77.41W	Huandoy: central peak	Blanca (Andes)	Peru	20,850	6360	1932
9.02S,	77.41W	• Huandoy: W peak	Blanca (Andes)	Peru	20,850	6360	1954
10.17S,	76.54W	Siula: N peak	Huayhuash (Andes)	Peru	20,850	6360	1936
27.06S,	68.47W	• Tres Cruces: S peak	Andes	Argentina–Chile	20,850	6360	1937
15.31S,	72.42W	• Coropuna: NW peak	Occidental (Andes)	Peru	20,790	6340	1952
13.48S,	71.14W	• Ausangate: E peak	Vilcanota (Andes)	Peru	20,770	6330	1952
29.56S,	69.54W	Tortolas	Andes	Argentina–Chile	20,760[2]	6330[2]	1924 or 1952
16.39S,	67.48W	Illimani: S peak	Real (Andes)	Bolivia	20,740	6320	1898
15.50S,	71.52W	Ampato: highest peak	Occidental (Andes)	Peru	20,700	6310	1950

* * *

16.39S,	67.48W	• Illimani: N peak	Real (Andes)	Bolivia	20,670	6300	1915

[1] Average of elevations officially given by Argentina (19,360 ft, 5900 m) and Chile (22,770 ft, 6940 m).

[2] Average of elevations officially given by Argentina (20,740 ft, 6320 m) and Chile (20,770 ft, 6330 m).

SOUTH AMERICA

Latitude and Longitude		Peak and [Massif]	Mountain Range or System	Location	Elevation (ft)	Elevation (m)	First Ascent
1.28S,	78.48W	● Chimborazo	*Occidental* (Andes)	Ecuador	20,580	6270	1880
18.10S,	69.09W	Parinacota [Payachata]	Andes	Bolivia–Chile	20,440[1]	6230[1]	1928
18.25S,	69.08W	Guallatiri (volc)	Andes	Chile	19,880	6060	1926?
0.40S,	78.26W	● Cotopaxi (volc)	*Oriental* (Andes)	Ecuador	19,340	5900	1872
22.26S,	67.55W	Tocorpuri: highest peak	Andes	Bolivia–Chile	19,140	5830	1939
16.18S,	71.24W	Misti (volc)	Occidental (Andes)	Peru	19,100	5820	1677?
3.00N,	76.00W	Huila (volc[2])	*Central* (Andes)	Colombia	18,870	5750	1944
23.23S,	67.45W	Lascar (volc)	Andes	Chile	18,420	5610	
6.26N,	72.18W	Guican [Cocuy]	*Oriental* (Andes)	Colombia	18,020	5490	
10.48N,	73.41W	Bolivar [Horqueta]	*Santa Marta*	Colombia	17,390	5300	1939
10.49N,	73.41W	Colon (alt Cristobal Colon) [Horqueta]	*Santa Marta*	Colombia	17,390	5300	1939
8.33N,	71.03W	Bolivar (alt Columna)	*Merida*	Venezuela	16,430	5010	1935 or 1936
0.50N,	65.25W	Neblina	*Imeri*	Amazonas, Brazil	9,890	3010	
20.26S,	41.47W	Bandeira	*Caparao*	Espirito Santo– Minas Gerais, Brazil	9,480	2890	
31.59S,	64.59W	Champaqui	*Cordoba*	Cordoba, Argentina	9,460	2880	
5.12N,	60.44W	Roraima	*Pacaraima*	Brazil–Guyana– Venezuela	9,430	2870	1884
22.23S,	44.38W	Agulhas Negras [Itatiaia]	*Mantiqueira*	Minas Gerais– Rio de Janeiro, Brazil	9,140	2790	

[1] Average of elevations officially given by Bolivia (20,120 ft, 6130 m) and Chile (20,770 ft, 6330 m).
[2] No known eruption.

4a
HIGHEST MOUNTAIN,
OR ELEVATION, IN EACH COUNTRY

Contents

Conventional name of country.

Conventional and other (alternate, former, and official) names of mountain peak and of massif, if any (in brackets).

Conventional and other names of mountain range or system.

Country (if any) with which mountain is shared.

Elevation (i.e., altitude of summit above sea level), in feet and meters.

Arrangement

By country, alphabetically.

Coverage

All countries with an area of at least 1000 mi^2 (2590 km^2), except a few countries consisting of a number of small islands.

Rounding

Elevations are rounded to the nearest 10 ft (m).

Entries

159.

TABLE 4a. HIGHEST MOUNTAIN, OR ELEVATION, IN EACH COUNTRY

Country	Peak and [Massif]	Mountain Range or System	Country with which Shared	Elevation (ft)	(m)
Afghanistan	● Noshaq: highest peak	Hindu Kush	Pakistan	24,580	7490
Albania	Korab	Korab	Yugoslavia	9,050	2760
Algeria	Tahat [Atakor]	Ahaggar (alt Hoggar)		9,570	2920
Angola	Moco	Upanda		8,600	2620
Argentina	Aconcagua	Andes		22,840	6960
Australia	Kosciusko	Great Dividing		7,330	2230
Austria	Grossglockner	Alps (off in Austria Alpen)		12,460	3800
Bangladesh	● Keokradong	Chittagong (Hills)		4,030	1230
Belgium	● Botrange	Ardennes		2,270	690
Belize	Victoria	Maya		3,680	1120
Benin	– [Tanekas]	Togo		2,150	650
Bhutan	Khula Kangri I (alt Kula Gangri I)	Assam Himalaya	China	24,780	7550
Bolivia	Sajama	Occidental (Andes)		21,390	6520
Botswana	?	?		5,920	1810
Brazil	Neblina	Imeri		9,890	3010
Brunei	● Pagon	Crocker	Sarawak, Malaysia	6,070	1850
Bulgaria	Musala	Rhodope (off in Bulgaria Rodopi)		9,600	2920
Burma	● Hkakabo Razi	Kumon		19,300	5880
Burundi	● Heha	?		8,750	2670
Cambodia	● Aural	Cardamomes		5,950	1810
Cameroon	Fako [Cameroon (alt Cameroun)]	Cameroon (alt Cameroun)		13,350	4070
Canada	Logan: central peak [Logan]	Saint Elias		19,520	5950
Central African Empire	Gaou	Yade		4,660	1420
Chad	(Emi) Koussi	Tibesti		11,470	3490
Chile	Ojos del Salado: SE peak	Andes	Argentina	22,560	6870
China	Everest (alt Chumulangma) [Everest]	Nepal Himalaya	Nepal	29,030	8850
Colombia	Huila	Central (Andes)		18,870	5750
Congo	?	Mayombe		2,620	800
Costa Rica	Chirripo Grande	Talamanca		12,530	3820
Cuba	Turquino	Maestra		6,480	1970
Cyprus	Khionistra (alt Olympus)	Troodos		6,410	1950
Czechoslovakia	Gerlachovsky (alt Gerlachovka; for Stalin)	Tatra (off Tatry)		8,710	2650
Denmark	● Ejer Bavnehoj	–		560	170
Djibouti	Gouda (alt Goudah)	?		5,500	1670

Country	Peak and [Massif]	Mountain Range or System	Country with which Shared	Elevation (ft)	(m)
Dominican Republic	Duarte (for Trujillo)	Central		10,420	3170
Ecuador	• Chimborazo	Occidental (Andes)		20,580	6270
Egypt	Katrinah (alt Catherine) [Musa]	Sinai		8,650	2640
El Salvador	Santa Ana	Apaneca Lamatepeque		7,730	2360
Equatorial Guinea	Malabo (for Santa Isabel)	−		9,840	3000
Ethiopia	Ancua [Ras Dashan (off Rasdajan)]	Semien		14,930	4550
Fiji	Tomaniivi (alt Victoria)	−		4,340	1320
Finland	Haltiatunturi (alt Reisduoddarhaldde)	Haltia (alt Halddia)	Norway	4,360	1330
France	(Mont-)Blanc (alt Bianco) [Blanc (alt Bianco)]	Alps (off in France Alpes)	Italy	15,770	4810
French Guiana	−[Timotakem]	Tumuc-Humac		2,620	800
Gabon	Iboundji	Chaillu		5,160	1570
Gambia	• ?	−		230	70
Germany					
East	Fichtel(-berg)	Erz(gebirge)		3,980	1220
West	Zug(spitze)	Alps (off in Germany Alpen)	Austria	9,720	2960
Ghana	• Afadjoto	Akwapim-Togo		2,900	890
Greece	Mytikas [Olympus (off Olimbos)]	Olympus (off Olimbos)		9,550	2910
Greenland	Gunnbjorn	−		12,140	3700
Guatemala	Tajumulco	Madre		13,850	4220
Guinea	Nimba	Nimba	Ivory Coast, Liberia	5,780	1760
Guinea-Bissau	?	Fouta Djallon		980	300
Guyana	Roraima	Pacaraima	Brazil, Venezuela	9,430	2870
Haiti	Selle	Selle		8,790	2680
Honduras	• ?	Celaque		9,400	2870
Hungary	Kekes	Matra		3,330	1010
Iceland	Hvannadalshnukur	−		6,950	2120
India	Kangchenjunga (alt Kanchenjunga): highest peak [Kangchenjunga]	Nepal Himalaya	Nepal	28,170	8590
Indonesia					
in Asia	Kerinci (for Indrapura, Kerintji)	Barisan		12,470	3800
in Oceania (Irian Jaya)	Ngga Pulu [Jaya (for Carstensz, Djaja, Sukarno)]	Sudirman (for Nassau)		16,500	5030

Country	Peak and [Massif]	Mountain Range or System	Country with which Shared	Elevation (ft)	(m)
Iran	● Damavand (alt Demavend)	Alborz (alt Elburz)		18,610	5670
Iraq	● Algurd (alt Halgurd)	Zagros		12,250	3730
Ireland	Carrantuohill (alt Carrantual)	Macgillicuddy's Reeks		3,410	1040
Israel	● Meron (for Jarmaq, Sharqi)	Galilee		3,960	1210
Italy	(Mont-)Blanc (alt Bianco) [Blanc (alt Bianco)]	Alps (off in Italy Alpi)	France	15,770	4810
Ivory Coast	Nimba	Nimba	Guinea, Liberia	5,780	1760
Jamaica	Blue (Mountain)	Eastern		7,400	2260
Japan	Fuji (alt Huzi)	–		12,390	3780
Jordan	● Ramm	Sharah		5,750	1750
Kashmir-Jammu	K^2 (alt Chogori, Dapsang, Godwin Austen)	Karakoram		28,250	8610
Kenya	Batian [Kenya]	–		17,050	5200
Korea					
North	● Paitou (alt Paektu)	Changpai (alt Changbaek)	China	8,900	2710
South	● Halla	–		6,400	1950
Kuwait	● ?	–		950	290
Laos	● Bia	Tranninh		9,240	2820
Lebanon	Makmel [Qurnat al Sawda]	Lebanon (off Lubnan)		10,120	3080
Lesotho	Thabana Ntlenyana (alt Thabantshonyana)	Drakens(berg)		11,420	3480
Liberia	Nimba	Nimba	Guinea, Ivory Coast	5,780	1760
Libya	● Kegueur Terbi (alt Chegor Tedi, Hessi)	Tibesti	Chad	10,330	3150
Luxembourg	● Burgplatz	Ardennes		1,840	560
Madagascar	Maromokotro	Tsaratanana		9,470	2890
Malawi	Mlanje	Mlanje		10,000	3050
Malaysia					
Peninsular	Tahan	Cameron (Highlands)		7,190	2190
Sabah	Kinabalu: highest peak	Crocker		13,450	4100
Sarawak	Murud	Tama Abu		7,950	2420
Mali	● ?	Iforas		3,150	960
Mauritania	● Ijill (alt Idjil)	?		2,900	880
Mauritius	Petite Riviere Noire	–		2,710	830
Mexico	Citlaltepetl (alt Orizaba)	Neovolcanica		18,700	5700
Mongolia	Khuitun (off Huyten) [Tavan Bogd]	Altai (alt Altay)		15,270	4650

Country	Peak and [Massif]	Mountain Range or System	Country with which Shared	Elevation (ft)	(m)
Morocco	Toubkal: W peak [Toubkal]	Atlas		13,670	4160
Mozambique	Binga	Chimanimani	Rhodesia	7,990	2440
Namibia	Konigstein [Brand(berg)]	Kaokoveld		8,550	2610
Nepal	Everest (alt Chumulangma) [Everest]	Nepal Himalaya	China	29,030	8850
Netherlands	● Vaalser(berg)	–		1,060	320
New Caledonia	Panie	–		5,340	1630
New Zealand	Cook (alt Aorangi): highest peak	Southern Alps		12,350	3760
Nicaragua	● Mogoton	Dipilto	Honduras	6,910	2110
Niger	● Greboun	Air (alt Azbine)		7,550	2300
Nigeria	Vogel	Banglang		6,700	2040
Norway	Glittertinden	Jotunheimen		8,110	2470
Oman	● Sham	Akhdar		10,190	3110
Pakistan	● Tirich Mir: W peak	Hindu Kush		25,260	7700
Panama	● Chiriqui (alt Baru)	Central		11,410	3480
Papua New Guinea	● Wilhelm	Bismarck		15,400	4690
Paraguay	–[Villa Rica]	?		2,790	850
Peru	Huascaran: S peak	Blanca (Andes)		22,210	6770
Philippines	Apo	–		9,690	2950
Poland	Rysy	Tatra (off Tatry)	Czechoslovakia	8,200	2500
Portugal					
in Africa (Madeira Islands)	Ruivo de Santana	–		6,110	1860
in Europe	Ponta do Pico (alt Pico)	–		7,710	2350
Puerto Rico	Punta	Central		4,390	1340
Qatar	● ?	–		250	80
Reunion	Neiges	–		10,070	3070
Rhodesia	Inyangani	Inyanga		8,510	2590
Romania	Moldoveanu	Transylvanian Alps (off Carpatii Meridionali)		8,340	2540
Rwanda	Karisimbi	Virunga	Zaire	14,790	4510
Saudi Arabia	● Razikh	Yemen (Highlands)	Yemen	11,990	3650
Senegal	?	Fouta Djallon		1,640	500
Sierra Leone	● Bintimani	Loma		6,390	1940
Solomon Islands	Makarakomburu	Kavo		8,030	2450
Somalia	Faddisome [Hor Bogor]	Carcar		8,500	2590
South Africa	● Injasuti	Drakens(berg)	Lesotho	11,310	3450
Spain					
in Africa (Canary Islands)	Teide	–		12,200	3720
in Europe	Mulhacen	Nevada		11,410	3480
Sri Lanka	Pidurutalagala	Piduru (Ridges)		8,280	2520
Sudan	● Kinyeti	Imatong		10,460	3190
Surinam	● –[Wilhelmina]	Wilhelmina		4,200	1280

Country	Peak and [Massif]	Mountain Range or System	Country with which Shared	Elevation (ft)	(m)
Svalbard	Newton	–		5,630	1720
Swaziland	● Emlembe	Drakens(berg)		6,100	1860
Sweden	Kebnekaise	Kolen		6,930	2110
Switzerland	Dufour(spitze) [Rosa]	Alps (off in Switzerland Alpen, Alpes, Alpi)	Italy	15,200	4630
Syria	Hermon (off Shaykh)	Anti-Lebanon (off Sharqi)	Lebanon	9,230	2810
Taiwan	Yu (alt Hsinkao, Morrison)	Central (off Chungyang)		13,110	4000
Tanzania	Kibo [Kilimanjaro]	–		19,340	5890
Thailand	Inthanon (alt Angka)	Phi Pan Nam		8,450	2580
Togo	Baumann (alt Agou)	Togo		3,240	990
Trinidad and Tobago	Aripo	–		3,080	940
Tunisa	Shanabi (alt Chambi)	Dorsale		5,070	1540
Turkey					
in Asia	Great Ararat (off Buyukagri)	–		16,950	5160
in Europe	Mahya	Istranca		3,380	1030
Uganda	Margherita [Stanley]	Ruwenzori	Zaire	16,760	5110
UK	(Ben) Nevis	Grampian		4,410	1340
United Arab Emirates	● Adhan	?		3,700	1130
Upper Volta	?	–		1,000	300
Uruguay	● Animas	Animas		1,640	500
USA					
in North America	McKinley: S peak [McKinley]	Alaska		20,320	6190
in Oceania (Hawaii)	(Mauna) Kea: highest peak	–		13,800	4210
USSR					
in Asia	Kommunizma (for Garmo, Stalina)	Pamir-Alai (off Pamir-Alay)		24,590	7490
in Europe	Elbrus (for Elborus): W peak [Elbrus (for Elborus)]	Caucasus (off Kavkaz)		18,480	5630
Venezuela	Bolivar (alt Columna)	Merida		16,430	5010
Vietnam	● (Ngoc) Linh	Annamese (off Trungphan)		10,500	3200
Western Sahara	?	–		1,640	500
Yemen					
North	● Hadur Shuayb	Yemen (Highlands)		12,340	3760
South	● Thamar (off Thamir)	Yemen (Highlands)		8,240	2510
Yugoslavia	Triglav	Alps (off in Yugoslavia Alpe)		9,390	2860
Zaire	Margherita [Stanley]	Ruwenzori	Uganda	16,760	5110
Zambia	?	Mapinga		7,800	2380

5
LARGEST NATURAL LAKES, BY CONTINENT

Contents

Latitude and longitude, in degrees and minutes, at center of lake.
Conventional and other (alternate, former, and official) names of lake.
Subdivision and country (or countries) in which located.
Area in square miles and square kilometers.
Elevation (i.e., altitude of surface above sea level), in feet and meters.
Greatest recorded depth of water, in feet and meters.

Arrangement

By area, under continent.

Coverage

Above * * *, all natural lakes with an area of at least 300 mi² (780 km²).
Below * * *, other well-known natural lakes, including the largest in certain countries.

Rounding

Areas are rounded to the nearest 10 mi² (km²).

Special Features

Saltwater lakes are indicated by the word "salt," and lagoons, by the abbreviation "lag."
When area and/or depth vary from season to season, the normal minima and maxima are shown as follows: 4,000/10,000.

Entries

290.

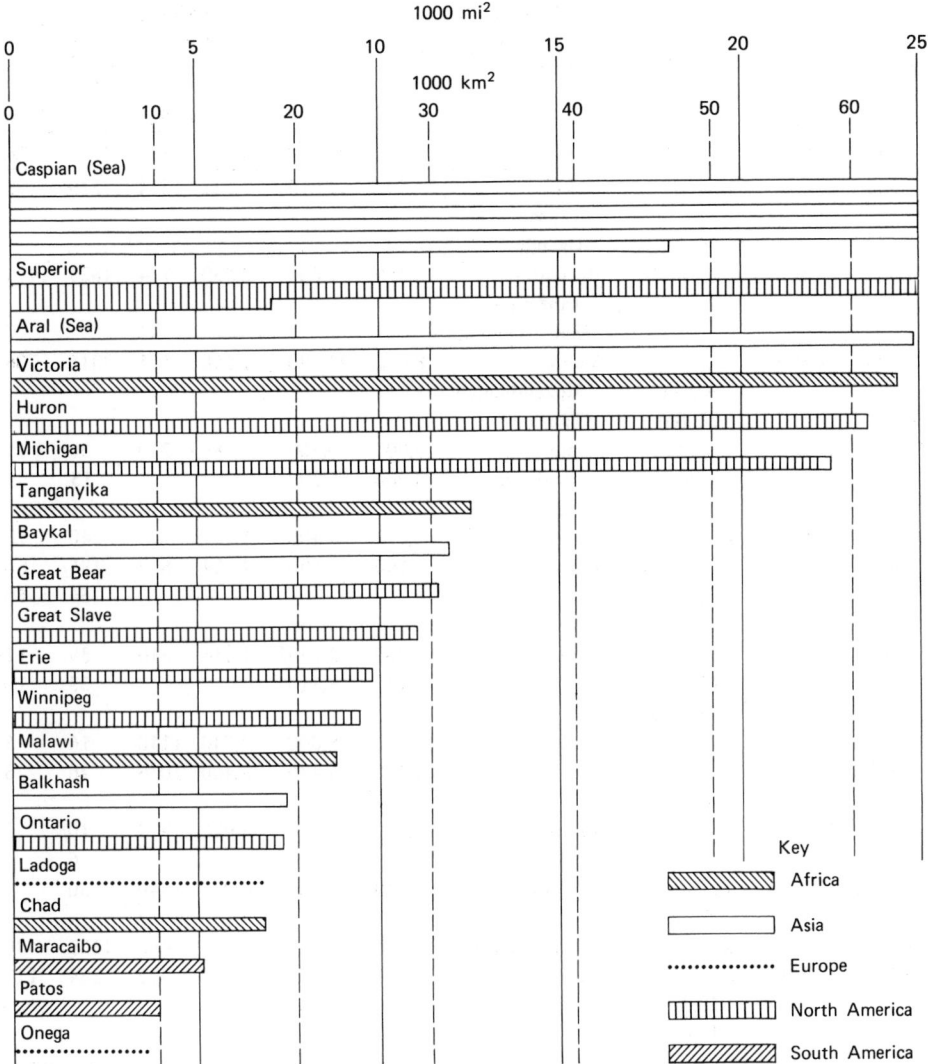

Figure 14. Twenty largest natural lakes.

TABLE 5. LARGEST NATURAL LAKES, BY CONTINENT
AFRICA

Latitude and Longitude	Lake	Location	Area (mi²)	Area (km²)	Elevation (ft)	Elevation (m)	Greatest Depth (ft)	Greatest Depth (m)
1.00S, 33.00E	Victoria	Kenya–Tanzania–Uganda	24,300	62,940	3,721	1134	279	85
6.00S, 29.30E	Tanganyika	Burundi–Tanzania–Zaire–Zambia	12,350	32,000	2,539	774	4825	1471
12.00S, 34.30E	Malawi (alt Niassa, Nyasa)	Malawi–Mozambique–Tanzania	8,680	22,490	1,558	475	2316	706
13.20N, 14.00E	Chad (alt Tchad)	Cameroon–Chad–Niger–Nigeria	4,000/ 10,000	10,360/ 25,900	787	240	36/ 13	11/ 4
3.30N, 36.00E	Rudolf (salt)	Ethiopia–Kenya	2,470	6,400	1,401	427	240	73
1.40N, 31.00E	Mobutu Sese Seko (for Albert)	Uganda–Zaire	2,160	5,590	2,024	617	197/ 164	60/ 50
2.00S, 18.20E	Mai-Ndombe (for Leopold II)	Zaire	900/ 3,170	2,070/ 8,210	1,116	340	39/ 33	12/ 10
11.05S, 29.45E	Bangweulu	Zambia	1,930	5,000	3,740	1140	16	5
1.30N, 33.00E	● Kyoga (alt Kioga)	Uganda	1,710	4,430	3,400	1036	26	8
9.00S, 28.45E	Mweru	Zaire–Zambia	1,680	4,350	3,025	922	10/ 6	3/ 2
12.10N, 37.20E	Tana (alt Tsana)	Ethiopia	1,390	3,600	6,037	1840	30	9
2.00S, 29.10E	Kivu	Rwanda–Zaire	860	2,220	4,790	1460	1575	480
0.21S, 29.35E	Idi Amin Dada (for Edward)	Uganda–Zaire	830	2,150	2,992	912	384	117
8.00S, 32.25E	● Rukwa (salt)	Tanganyika, Tanzania	290/ 1,160	750/ 3,000	2,602	793	shallow	
31.15N, 32.00E	● Manzala (off Manzilah) (salt, lag)	Egypt	530	1,360	0	0	shallow	
6.20N, 37.55E	Abaya (for Margherita)	Ethiopia	450	1,160	4,160	1268	43	13
15.12S, 35.50E	Chilwa (alt Chirua, Shirwa) (salt)	Malawi–Mozambique	400	1,040	1,805	550	shallow	
11.10N, 41.45E	● Abbe (alt Abe) (salt)	Djibouti–Ethiopia	300	780				
5.15N, 3.14W	● Aby (salt, lag)	Ivory Coast	300	780	0	0		

* * *

AFRICA

Latitude and Longitude	Lake	Location	Area (mi^2)	(km^2)	Elevation (ft)	(m)	Greatest Depth (ft)	(m)
16.45N, 3.54W	Faguibine	Mali	230	590			34	10
5.50N, 37.40E	Chama (alt Chamo; for Ruspoli)	Ethiopia	210	550	4,052	1235	33	10
0.48S, 18.03E	Tumba	Zaire	190	500	1,115	340	39/ 33	12/ 10
8.00N, 38.48E	Zeway (alt Zwai)	Ethiopia	170	430	6,057	1846	13	4

ASIA

Latitude and Longitude	Lake	Location	Area (mi^2)	(km^2)	Elevation (ft)	(m)	Greatest Depth (ft)	(m)
42.00N, 50.00E	Caspian (Sea) (off Kaspiyskoye, Khazar) (salt)	Iran–USSR	143,240	371,000	−91	−28	3363	1025
45.00N, 60.00E	Aral (Sea) (off Aralskoye) (salt)	Kazakhstan–Uzbekistan, USSR	24,900	64,500	174	53	220	67
54.00N, 109.00E	Baykal (alt Baikal)	Russia, USSR	12,160	31,500	1,496	456	5315	1620
46.00N, 74.00E	Balkhash (alt Balkash) (salt)	Kazakhstan, USSR	6,560/ 8,490	17,000/ 22,000	1,125	343	87	26
13.00N, 104.00E	● (Tonle) Sap	Cambodia	1,040/ 3,860	2,700/ 10,000			39	12
42.25N, 77.15E	Issyk(-Kul) (salt)	Kirgizia, USSR	2,410	6,240	5,276	1608	2303	702
37.40N, 45.30E	● Rezaiyeh (alt Rizaiyeh, Urmia) (salt)	Iran	1,500/ 2,300	3,880/ 5,960	4,183	1275	52	16
74.30N, 102.30E	Taymyr (alt Taimyr)	Russia, USSR	1,540/ 1,930	4,000/ 5,000	9	3	85	26
37.00N, 100.20E	Koko (alt Tsing; off Ching) (salt)	Tsinghai, China	1,720	4,460	10,489	3197	125	38
45.00N, 132.24E	Khanka (alt Hsingkai)	China–USSR	1,540/ 1,700	4,000/ 4,400	226	69	33	10
29.18N, 112.45E	Tungting	Hunan, China	1,200/ 2,010	3,100/ 5,200	36	11		
54.50N, 77.30E	Chany (salt)	Russia, USSR	970/ 1,930	2,500/ 5,000	345	105	39/ 23	12/ 7

Figure 15. Beach at Bandar-e Pahlavi, Iran, on the southern shore of the Caspian Sea, the largest lake in the world. (Credit: Iran National Tourist Organization.)

ASIA

Latitude and Longitude	Lake	Location	Area (mi²)	(km²)	Elevation (ft)	(m)	Greatest Depth (ft)	(m)
38.33N, 42.46E	Van (salt)	Turkey	1,440	3,740	5,401	1646	82	25
29.00N, 116.25E	Poyang	Kiangsi, China	1,290	3,350	5,906	1800	66	20
50.20N, 92.45E	Uvs (alt Ubsa, Ubsu) (salt)	Mongolia	1,290	3,350	2,490	759	shallow	
40.30N, 90.30E	Lop (alt Lob; off Lopu) (salt)	Sinkiang, China	1,160	3,010	2,520	768	7	2
33.18N, 118.41E	Hungtze (off Hungtse)	Anhwei–Kiangsu, China	1,040	2,700	49	15		
46.10N, 81.50E	Ala(kol) (salt)	Kazakhstan, USSR	1,020	2,650	1,139	347	177	54
51.00N, 100.30E	Hovsgol (alt Hobsogol, Khubsugul, Kosogol)	Mongolia	1,010	2,620	5,328	1624	807	246
30.45N, 90.30E	Nam (alt Tengri; off Namu) (salt)	Tibet, China	970	2,500	15,181	4627		
31.15N, 120.10E	Tai	Chekiang–Kiangsu, China	850	2,210			16	5
30.50N, 47.10E	● Hammar	Iraq	750	1,940			7	2
31.50N, 89.00E	● Zilling (alt Goring, Seling; off Chilin) (salt)	Tibet, China	720	1,860	14,748	4495	26	8
48.00N, 84.00E	Zaysan (alt Zaisan)	Kazakhstan, USSR	690	1,800	1,266	386	33	10
48.00N, 92.10E	Har Us (alt Hara Usa, Khara-Us)	Mongolia	680	1,760	3,783	1153		
38.45N, 33.25E	Tuz (salt)	Turkey	630	1,640	3,035	925	shallow	
49.00N, 117.27E	● Hulun (alt Dalai) (salt)	Inner Mongolia, China	620	1,590	4,183	1275	5	2
50.24N, 68.57E	Tengiz (salt)	Kazakhstan, USSR	610	1,590	997	304	26	8
31.00N, 86.22E	● Tangra (alt Dangrayum; off Tang-kulayumu) (salt)	Tibet, China	540	1,400	15,499	4724		

ASIA

Latitude and Longitude	Lake	Location	Area (mi²)	Area (km²)	Elevation (ft)	Elevation (m)	Greatest Depth (ft)	Greatest Depth (m)
42.00N, 87.00E	Possuteng (alt Baghrash) (salt)	Sinkiang, China	530	1,380	3,406	1038		
49.12N, 93.24E	Hyargas (alt Hirgis, Khirgis) (salt)	Mongolia	530	1,360	3,373	1028		
7.30N, 100.15E	• (Thale) Luang (alt Sap) (salt, lag)	Thailand	500	1,290	0	0	shallow	
2.35N, 98.40E	Toba	Sumatra I, Indonesia	440	1,150	2,973	906	1736	529
44.55N, 82.55E	• Ebi (off Aipi) (salt)	Sinkiang, China	410	1,070	699	213		
19.45N, 85.25E	• Chilka (salt, lag)	Orissa, India	350/ 450	910/ 1,170	0	0	shallow	
31.30N, 35.30E	Dead (Sea) (off Lut, Mayyit, Melah) (salt)	Israel–Jordan	390	1,020	−1,289	−393	1421	433
34.35N, 117.13E	• Weishan	Kiangsu– Shantung, China	390	1,000				
31.31N, 117.33E	• Chao	Anhwei, China	350	900				
14.23N, 121.15E	Bay (salt)	Luzon I, Philippines	340	890	7	2	21	6
28.35N, 90.20E	• Pomo (off Pumuchang)	Tibet, China	340	880	16,194	4936		
47.20N, 87.10E	• Ulyungur (alt Pulunto, Urungu; off Wulunku) (salt)	Sinkiang, China	320	830	1,536	468		
31.06N, 85.35E	• Terinam (off Tiehlinan- mu) (salt)	Tibet, China	310	810	15,368	4684		
29.00N, 90.40E	• Yamdrok (off Yangchoy- ung)	Tibet, China	310	800	14,350	4374		
53.15N, 73.15E	Seletyteniz (salt)	Kazakhstan, USSR	300	780	210	64	10	3

* * *

ASIA

Latitude and Longitude	Lake	Location	Area (mi^2)	Area (km^2)	Elevation (ft)	Elevation (m)	Greatest Depth (ft)	Greatest Depth (m)
34.45N, 51.36E	● Namak (salt)	Iran	290	750			shallow	
46.35N, 81.00E	Sasyk(kol) (salt)	Kazakhstan, USSR	280	740	1,148	350	15	5
32.50N, 119.15E	● Kaoyu (off Kaopao)	Anhwei-Kiangsu, China	270	700				
35.15N, 136.05E	Biwa	Honshu I, Japan	270	690	285	87	314	96
31.10N, 88.15E	● Dzharing (for Kyaring; off Chalin)	Tibet, China	260	670	15,447	4708		
37.40N, 31.30E	Beysehir	Turkey	250	650	3,662	1116	30	9
34.55N, 98.00E	● Ngoring (off Chaling)	Tsinghai, China	250	650	14,010	4270		
47.48N, 117.42E	Buyr (alt Bor, Pei-erh)	China–Mongolia	240	610	1,913	583	36	11
33.45N, 78.43E	● Pangong (alt Pankung) (salt)	China–Kashmir-Jammu	230	600	13,936	4248	142	43
34.52N, 97.30E	● Tsaring (off Oling)	Tsinghai, China	220	570	14,010	4270		
48.06N, 93.12E	Har (alt Hara, Khara)	Mongolia	200	530	3,622	1104		
38.02N, 30.53E	Egridir (alt Egirdir)	Turkey	200	520	3,032	924	43	13
24.50N, 102.43E	Tien	Yunnan, China	150	400	6,400	1950		
39.00N, 73.30E	Kara(kul)	Tadzhikistan, USSR	150	380	12,840	3914	774	236
51.35N, 87.40E	Teletskoye [for Altyn(-Kol)]	Russia, USSR	89	230	1,431	436	1066	325
2.28S, 121.20E	● Matana	Celebes I, Indonesia	75	190	1,253	382	1936	590
38.13N, 72.50E	Sarez (off Sarezskoye)	Tadzhikistan, USSR	33	86	10,627	3239	1657	505
51.27N, 157.05E	Kurile (off Kurilskoye)	Russia, USSR	30	77			1004	306
42.45N, 141.20E	Shikotsu	Hokkaido I, Japan	30	77	814	248	1191	363
40.28N, 140.55E	Towada	Honshu I, Japan	23	59			1096	334
39.43N, 140.40E	Tazawa	Honshu I, Japan	10	26			1394	425

EUROPE

Latitude and Longitude	Lake	Location	Area (mi²)	Area (km²)	Elevation (ft)	Elevation (m)	Greatest Depth (ft)	Greatest Depth (m)
61.00N, 31.30E	Ladoga (off Ladozh-skoye)	Russia, USSR	7,000	18,130	13	4	755	230
61.30N, 35.45E	Onega (off Onezhskoye)	Russia, USSR	3,750	9,700	108	33	394	120
58.55N, 13.30E	Vanern (alt Vaner, Vener)	Sweden	2,160	5,580	144	44	322	98
57.19N, 30.52E	Peipus (off Chudskoye)	Estonia–Russia, USSR	1,660	4,300	98	30	49	15
58.24N, 14.36E	Vattern (alt Vatter, Vetter)	Sweden	740	1,910	289	88	420	128
61.15N, 28.15E	Saimaa (alt Saima)	Finland	680	1,760	249	76	269	82
55.00N, 21.00E	● Kurisches (Haff) (off Kurskiy) (salt, lag)	Lithuania–Russia, USSR	630	1,620	0	0	33	10
40.20N, 45.20E	Sevan (for Gokcha)	Armenia, USSR	530	1,360	6,234	1900	272	83
58.17N, 31.20E	Ilmen	Russia, USSR	230/ 810	600/ 2,100	59	18	36/ 11	11/ 3
63.40N, 34.40E	Vyg(ozero)	Russia, USSR	480	1,250	292	89	79	24
52.35N, 5.30E	● IJssel(meer) [for Zuider(Zee)]	Netherlands	470	1,210	26	8	shallow	
59.30N, 17.12E	Malaren (alt Malar)	Sweden	440	1,140	1	0.3	210	64
60.15N, 37.40E	White (off Beloye)	Russia, USSR	430	1,120	371	113	20	6
61.35N, 25.30E	Paijanne	Finland	420	1,090	256	78	305	93
69.00N, 28.00E	Inari (alt Enare)	Finland	390	1,000	374	114	197	60
65.40N, 32.00E	Top(ozero)	Russia, USSR	380	990	358	109	184	56
63.18N, 33.45E	Seg(ozero)	Russia, USSR	350	910	374	114	318	97
53.46N, 14.14E	Oder(-Haff) [alt Szczecinski; for Stettiner (Haff)] (salt, lag)	East Germany–Poland	350	900	0	0	30	9
64.20N, 27.15E	Oulu (alt Ule)	Finland	350	900	400	122	125	38
63.15N, 29.40E	Pielinen	Finland	330	850	308	94	161	49
67.30N, 33.00E	Imandra	Russia, USSR	310	810	417	127	220	67

* * *

EUROPE

Latitude and Longitude	Lake	Location	Area (mi²)	Area (km²)	Elevation (ft)	Elevation (m)	Greatest Depth (ft)	Greatest Depth (m)
66.05N, 30.58E	Pya(ozero)	Russia, USSR	250	660	331	101	161	49
46.50N, 17.45E	Balaton (alt Platten)	Hungary	230	590	341	104	36	11
46.25N, 6.30E	Geneva (off Geneve, Ginevra, Leman)	France–Switzerland	220	580	1,221	372	1,017	310
47.35N, 9.25E	Constance [alt Boden-(see), Costanza]	Austria–Switzerland–West Germany	210	540	1,299	396	827	252
59.15N, 15.45E	Hjalmaren	Sweden	190	480	72	22	59	18
63.12N, 14.18E	Storsjon i Jamtland	Sweden	180	460	958	292	243	74
54.40N, 6.25W	• Neagh	Northern Ireland, UK	150	400	49	15	102	31
44.54N, 28.57E	Razelm (alt Razim) (salt)	Romania	150	390	10	3	10	3
42.10N, 19.20E	Scutari (off Shkodres, Skadarsko) (salt)	Albania–Yugoslavia	150[1]	380[1]	20	6	144	44
45.40N, 10.41E	• Garda	Lombardia–Trentino-Alto Adige–Veneto, Italy	140	370	213	65	1135	346
60.40N, 11.00E	Mjosa	Norway	140	370	400	122	1473	449
41.00N, 20.45E	Ohrid (off Ohridsko, Ohrit)	Albania–Yugoslavia	140[2]	360[2]	2,280	695	938	286
47.50N, 16.45E	Neusiedler (alt Ferto)	Austria–Hungary	120	320	377	115	5	1
40.55N, 21.00E	Prespa (off Megal Prespa, Prespansko, Prespes)	Albania–Greece–Yugoslavia	110[3]	280[3]	2,799	853	177	54
45.25N, 12.19E	• Venice (off Veneta) (salt, lag)	Veneto, Italy	110	280	0	0	shallow	
66.14N, 17.30E	Hornavan	Sweden	89/110	230/280	1,394	425	725	221

[1] Average of areas officially given by Albania (140 mi², 370 km²) and Yugoslavia (150 mi², 390 km²).
[2] Average of areas officially given by Albania (140 mi², 370 km²) and Yugoslavia (130 mi², 350 km²).
[3] Average of areas officially given by Albania (110 mi², 280 km²) and Yugoslavia (110 mi², 270 km²).

EUROPE

Latitude and Longitude		Lake	Location	Area		Elevation		Greatest Depth	
				(mi^2)	(km^2)	(ft)	(m)	(ft)	(m)
46.54N,	6.53E	Neuchatel [alt Neuen-burger(see)]	Switzerland	84	220	1,408	429	502	153
45.57N,	8.39E	Maggiore [alt Langen-(see), Majeur]	Italy–Switzerland	82	210	633	193	1221	372
53.26N,	9.14W	Corrib	Ireland	66	170	29	9	151	46
46.00N,	9.17E	• Como	Lombardia, Italy	56	150	653	199	1352	412
60.02N,	10.08E	Tyrifjorden	Norway	52	130	207	63	968	295
53.25N,	12.42E	Muritz	East Germany	45	120	203	62	108	33
47.00N,	8.28E	Lucerne (alt Lucerna, Vierwald-statter)	Switzerland	44	110	1,424	434	702	214
53.46N,	21.44E	Sniardwy (for Spirding)	Poland	44	110	381	116	75	23
47.14N,	8.42E	Zurich(see) (alt Zurigo)	Switzerland	35	90	1,332	406	469	143
56.08N,	4.38W	• Lomond	Scotland, UK	27	70	27	8	623	190
57.18N,	4.27W	• Ness	Scotland, UK	22	57	53	16	754	230
45.58N,	9.00E	Lugano [alt Ceresio, Luganer-(see)]	Italy–Switzerland	19	49	886	270	945	288
46.42N,	7.44E	Thun [alt Thuner-(see)]	Switzerland	19	48	1,831	558	712	217
45.44N,	5.52E	Bourget	Rhone-Alpes, France	17	45	758	231	476	145
56.57N,	5.43W	• Morar	Scotland, UK	10	26	31	9	1017	310

NORTH AMERICA

Latitude and Longitude		Lake	Location	Area		Elevation		Greatest Depth	
				(mi^2)	(km^2)	(ft)	(m)	(ft)	(m)
48.00N,	88.00W	Superior	Canada–USA	31,760[1]	82,260[1]	602	183	1333	406
44.30N,	82.15W	Huron	Canada–USA	23,000[2]	59,580[2]	581	177	750	229

[1] Average of areas officially given by Canada (31,700 mi^2, 82,100 km^2) and USA (31,820 mi^2, 82,410 km^2).

[2] Average of areas officially given by Canada (23,000 mi^2, 59,570 km^2) and USA (23,010 mi^2, 59,600 km^2).

NORTH AMERICA

Latitude and Longitude	Lake	Location	Area (mi^2)	Area (km^2)	Elevation (ft)	Elevation (m)	Greatest Depth (ft)	Greatest Depth (m)
44.00N, 87.00W	Michigan	Illinois–Indiana–Michigan–Wisconsin, USA	22,400	58,020	581	177	923	281
66.00N, 121.00W	Great Bear	NWT, Canada	12,100	31,330	512	156	1356	413
61.30N, 114.00W	Great Slave	NWT, Canada	11,030	28,570	513	156	2015	614
42.15N, 81.00W	Erie	Canada–USA	9,920[1]	25,710[1]	572	174	210	64
52.00N, 97.30W	Winnipeg	Manitoba, Canada	9,420	24,390	713	217	92	28
43.45N, 78.00W	Ontario	Canada–USA	7,440[2]	19,270[2]	246	75	802	244
11.30N, 85.30W	Nicaragua (alt Cocibolca)	Nicaragua	3,150	8,150	105	32	230	70
59.05N, 109.30W	Athabasca	Alberta–Saskatchewan, Canada	3,060	7,940	700	213	407	124
57.15N, 102.40W	Reindeer	Manitoba–Saskatchewan, Canada	2,570	6,650	1,106	337	720	219
66.30N, 70.40W	Netilling	NWT, Canada[3]	2,140	5,540	95	29		
52.30N, 100.00W	Winnipegosis	Manitoba, Canada	2,070	5,370	830	253	39	12
49.50N, 88.30W	Nipigon	Ontario, Canada	1,870	4,850	1,050	320	541	165
51.00N, 98.45W	Manitoba	Manitoba, Canada	1,800	4,660	813	248	92	28
41.10N, 112.30W	Great Salt (salt)	Utah, USA	1,680	4,360	4,200	1280	48	15
49.15N, 94.45W	Woods	Canada–USA	1,580[4]	4,100[4]	1,060	323	69	21
63.08N, 101.30W	Dubawnt	NWT, Canada	1,480	3,830	774	236		
65.00N, 71.00W	Amadjuak	NWT, Canada[3]	1,200	3,120	370	113		
53.45N, 59.30W	Melville (salt)	Newfoundland, Canada	1,180	3,070	0	0	840	256
58.15N, 103.20W	Wollaston	Saskatchewan, Canada	1,030	2,680	1,306	398	233	71
59.30N, 155.00W	Iliamna	Alaska, USA	1,000	2,590	50	15	980	299
51.00N, 73.30W	Mistassini	Quebec, Canada	900	2,340	1,220	372	600	183
60.30N, 99.30W	Nueltin	Manitoba–NWT, Canada	880	2,280	911	278		

[1] Average of areas officially given by Canada (9910 mi^2, 25,670 km^2) and USA (9940 mi^2, 25,740 km^2).

[2] Average of areas officially given by Canada (7340 mi^2, 19,010 km^2) and USA (7540 mi^2, 19,530 km^2).

[3] Baffin I.

[4] Average of areas officially given by Canada (1680 mi^2, 4350 km^2) and USA (1480 mi^2, 3850 km^2).

NORTH AMERICA

Latitude and Longitude	Lake	Location	Area (mi²)	Area (km²)	Elevation (ft)	Elevation (m)	Greatest Depth (ft)	Greatest Depth (m)
57.10N, 98.40W	Southern Indian	Manitoba, Canada	870	2,250	835	255	59	18
54.00N, 64.00W	Michikamau	Newfoundland, Canada	780	2,030	1,510	460	262	80
64.10N, 95.20W	Baker	NWT, Canada	730	1,890	8	2	756	230
26.57N, 80.52W	Okeechobee	Florida, USA	700	1,810	19	6	20	6
63.15N, 116.55W	La Martre	NWT, Canada	690	1,780	870	265		
56.00N, 124.00W	Williston	British Columbia, Canada	680	1,760	2,180	664	550	168
50.20N, 92.30W	Seul	Ontario, Canada	640	1,660	1,170	357	112	34
30.13N, 90.07W	Pontchar-train (salt, lag)	Louisiana, USA	620	1,620	0	0	15	5
18.37N, 91.33W	• Terminos (salt, lag)	Campeche, Mexico	600	1,550	0	0	shallow	
62.41N, 98.00W	Yathkyed	NWT, Canada	560	1,450	461	141		
58.35N, 112.05W	Claire	Alberta, Canada	550	1,440	700	213	8	2
57.30N, 106.30W	Cree	Saskatchewan, Canada	550	1,430	1,597	487	148	45
55.10N, 105.00W	La Ronge	Saskatchewan, Canada	550	1,410	1,193	364	135	41
56.00N, 74.30W	Eau Claire (for Clearwater)	Quebec, Canada	530	1,380	790	241		
54.00N, 100.10W	Moose	Manitoba, Canada	530	1,370	838	255		
53.20N, 100.00W	Cedar	Manitoba, Canada	520	1,350	830	253		
60.20N, 102.10W	Kasba	NWT, Canada	520	1,340	1,102	336		
55.10N, 73.15W	Bienville	Quebec, Canada	480	1,250	1,400	427		
53.47N, 94.25W	Island	Manitoba, Canada	470	1,220	744	227		
57.56N, 156.23W	Becharof	Alaska, USA	460	1,190	14	4		
55.25N, 115.25W	Lesser Slave	Alberta, Canada	450	1,170	1,892	577	70	21
48.02N, 94.55W	Red	Minnesota, USA	450	1,170	1,175	358	31	9
42.28N, 82.40W	Saint Clair	Canada–USA	450[1]	1,160[1]	575	175	21	6
54.45N, 94.00W	Gods	Manitoba, Canada	440	1,150	585	178		
20.15N, 103.00W	Chapala	Jalisco-Michoacan, Mexico	440	1,140	5,004	1525	42	13
15.23N, 83.55W	Caratasca (salt, lag)	Honduras	430	1,110	0	0	16	5

[1] Average of areas officially given by Canada (430 mi², 1110 km²) and USA (460 mi², 1190 km²).

NORTH AMERICA

Latitude and Longitude	Lake	Location	Area (mi²)	Area (km²)	Elevation (ft)	Elevation (m)	Greatest Depth (ft)	Greatest Depth (m)
64.27N, 99.00W	Aberdeen	NWT, Canada	420	1,100	261	80		
45.50N, 60.50W	Bras d'Or (salt, lag)	Nova Scotia, Canada[1]	420	1,100	0	0	230	70
44.35N, 73.20W	Champlain	Canada–USA	420[2]	1,100[2]	100	30	400	122
66.30N, 113.27W	Takiyuak	NWT, Canada	420	1,080	1,250	381		
63.55N, 111.00W	MacKay	NWT, Canada	410	1,060	1,415	431		
12.21N, 86.21W	Managua (alt Xolotlan)	Nicaragua	400	1,040	120	37	262	80
48.35N, 72.05W	Saint-Jean	Quebec, Canada	390	1,000	321	98	204	62
66.00N, 100.00W	Garry	NWT, Canada	380	980	487	148		
49.34N, 70.22W	Pipmuacan	Quebec, Canada	380	980	1,300	396		
65.40N, 110.40W	Contwoyto	NWT, Canada	370	960	1,850	564		
33.13N, 115.51W	Salton (Sea) (salt)	California, USA	370	950	−231	−70	48	15
48.42N, 79.45W	Abitibi	Ontario–Quebec, Canada	360	930	868	265		
65.04N, 118.29W	Hottah	NWT, Canada	350	920	592	180		
48.42N, 93.10W	Rainy	Canada–USA	350[3]	910[3]	1,108	338	112	34
9.05N, 82.05W	● Chiriqui (salt, lag[4])	Panama	350	900	0	0	deep	
64.05N, 108.30W	Aylmer	NWT, Canada	330	850	1,230	375		
69.30N, 132.00W	Eskimo North	NWT, Canada	320	840	1	0.3		
46.17N, 79.45W	Nipissing	Ontario, Canada	320	830	644	196	72	22
70.35N, 153.26W	Teshekpuk	Alaska, USA	310	820	5	2		
62.40N, 109.30W	Nonacho	NWT, Canada	300	780	1,160	354		
55.55N, 108.44W	Peter Pond	Saskatchewan, Canada	300	780	1,382	421	79	24
59.30N, 133.45W	Atlin	British Columbia–Yukon Territory, Canada	300	770	2,190	668	930	283

* * *

54.47N, 97.22W	Cross	Manitoba, Canada	290	760	679	207		
57.30N, 75.00W	Minto	Quebec, Canada	290	760	550	168		
44.25N, 79.20W	Simcoe	Ontario, Canada	290	740	718	219	136	41

[1] Cape Breton I.
[2] Average of areas officially given by Canada (360 mi², 930 km²) and USA (490 mi², 1270 km²).
[3] Average of areas officially given by Canada (360 mi², 940 km²) and USA (340 mi², 890 km²).
[4] Actually a bay.

NORTH AMERICA

Latitude and Longitude	Lake	Location	Area (mi²)	Area (km²)	Elevation (ft)	Elevation (m)	Greatest Depth (ft)	Greatest Depth (m)
56.15N, 76.20W	Guillaume-Delisle (for Richmond G) (salt, lag)	Quebec, Canada	270	700	0	0		
53.45N, 90.00W	Big Trout	Ontario, Canada	250	660	698	213		
53.52N, 98.05W	Playgreen	Manitoba, Canada	250	660	711	217		
54.46N, 107.17W	Dore	Saskatchewan, Canada	250	640	1,506	459	67	20
58.38N, 155.52W	Naknek	Alaska, USA	240	630	34	10		
52.45N, 66.15W	Ashuanipi	Newfoundland, Canada	230	600	1,735	529		
15.30N, 89.10W	Izabal	Guatemala	230	590	26	8	59	18
53.15N, 76.45W	Sakami	Quebec, Canada	230	590	640	195	361	110
44.00N, 88.25W	Winnebago	Wisconsin, USA	210	560	747	228	22	7
49.00N, 57.20W	Grand	Newfoundland, Canada[1]	210	540	284	87	360	110
46.14N, 93.39W	Mille Lacs	Minnesota, USA	210	540	1,249	381	36	11
47.51N, 114.07W	Flathead	Montana, USA	200	510	2,892	881	220	67
18.27N, 71.39W	Enriquillo (salt)	Dominican Republic	190	500	-144	-44		
39.06N, 120.02W	Tahoe	California–Nevada, USA	190	500	6,229	1899	1645	501
19.55N, 101.05W	Cuitzeo	Guanajuato–Michoacan, Mexico	180	460	5,975	1821	11	3
47.09N, 94.24W	Leech	Minnesota, USA	180	460	1,290	393		
40.01N, 119.35W	Pyramid (salt)	Nevada, USA	170	450	3,802	1159	330	101
48.10N, 116.21W	Pend Oreille	Idaho, USA	150	380	2,063	629	1200	366
42.24N, 121.54W	Upper Klamath	Oregon, USA	140	370	4,139	1262	45	14
40.12N, 111.48W	Utah	Utah, USA	140	360	4,487	1368	16	5
44.27N, 110.22W	Yellowstone	Wyoming, USA	140	350	7,733	2357	300	91
45.37N, 69.40W	Moosehead	Maine, USA	120	300	1,058	322	246	75
60.10N, 150.50W	Tustumena	Alaska, USA	120	300	90	27		
41.59N, 111.20W	Bear	Idaho–Utah, USA	110	280	5,943	1811	175	53
60.13N, 154.22W	Clark	Alaska, USA	110	280			606	185
38.42N, 118.43W	Walker (salt)	Nevada, USA	110	280	4,000	1219	1000	305
47.26N, 94.12W	Winnibigo-shish	Minnesota, USA	110	280	1,300	396	25	8

[1] Newfoundland I.

NORTH AMERICA

Latitude and Longitude	Lake	Location	Area (mi^2)	Area (km^2)	Elevation (ft)	Elevation (m)	Greatest Depth (ft)	Greatest Depth (m)
53.18N, 126.42W	Eutsuk	British Columbia, Canada	110	270	2,817	859	1060	323
60.18N, 163.43W	Dall	Alaska, USA	100	260				
41.55N, 120.25W	Goose (salt)	California–Oregon, USA	100	260	4,716	1437	24	7
52.33N, 120.59W	Quesnel	British Columbia, Canada	100	260	2,380	725	1560	475
29.52N, 93.50W	Sabine (salt, lag)	Louisiana–Texas, USA	95	250	0	0	shallow	
49.31N, 121.52W	Harrison	British Columbia, Canada	92	240	34	10	916	279
55.22N, 125.54W	Takla	British Columbia, Canada	92	240	2,260	689	941	287
30.15N, 90.30W	Maurepas (salt, lag)	Louisiana, USA	91	240	0	0	shallow	
29.45N, 92.30W	White	Louisiana, USA	81	210	1	0.3	shallow	
43.12N, 75.54W	Oneida	New York, USA	80	210	369	112	55	17
50.03N, 124.27W	Powell	British Columbia, Canada	72	190	175	53	1174	358
43.37N, 71.21W	Winnipesaukee	New Hampshire, USA	72	190	504	154	169	52
42.40N, 76.41W	Cayuga	New York, USA	67	170	382	116	435	133
45.57N, 66.02W	Grand	New Brunswick, Canada	67	170	4	1		
42.39N, 76.53W	Seneca	New York, USA	67	170	445	136	618	188
59.50N, 158.50W	Nuyakuk	Alaska, USA	64	170			930	283
51.17N, 124.00W	Chilko	British Columbia, Canada	61	160	3,860	1177	1200	366
47.50N, 120.01W	Chelan	Washington, USA	55	140	950	290	1605	489
51.12N, 119.35W	Adams	British Columbia, Canada	51	130	1,356	413	1500	457
43.37N, 73.33W	George	New York, USA	44	110	319	97	200	61
42.56N, 122.00W	Crater	Oregon, USA	21	54	6,176	1882	1932	589

OCEANIA

Latitude and Longitude	Lake	Location	Area (mi²)	Area (km²)	Elevation (ft)	Elevation (m)	Greatest Depth (ft)	Greatest Depth (m)
28.30S, 137.20E	Eyre (salt)	South Australia, Australia	0/2,970	0/7,690	−39	−12	4/0	1/0
31.00S, 137.50E	Torrens (salt)	South Australia, Australia	0/2,230	0/5,780	98	30	shallow	
31.35S, 136.00E	Gairdner (salt)	South Australia, Australia	0/1,840	0/4,770	112	34	shallow	
30.44S, 139.48E	Frome (salt)	South Australia, Australia	0/930	0/2,410	160	49	4/0	1/0

* * *

Latitude and Longitude	Lake	Location	Area (mi²)	Area (km²)	Elevation (ft)	Elevation (m)	Greatest Depth (ft)	Greatest Depth (m)
38.50S, 175.56E	Taupo	North I, New Zealand	230	610	1,172	357	522	159
35.26S, 139.10E	Alexandrina (lag)[1]	South Australia, Australia	220	570	0	0	15	5
45.12S, 167.48E	Te Anau	South I, New Zealand	130	340	686	209	906	276
45.05S, 168.34E	Wakatipu	South I, New Zealand	110	290	1,017	310	1240	378
45.30S, 167.30E	Manapouri	South I, New Zealand	55	140	607	185	1453	443
44.30S, 169.17E	Hawea	South I, New Zealand	46	120	1,132	345	1286	392

SOUTH AMERICA

Latitude and Longitude	Lake	Location	Area (mi²)	Area (km²)	Elevation (ft)	Elevation (m)	Greatest Depth (ft)	Greatest Depth (m)
9.40N, 71.30W	Maracaibo (salt, lag)	Venezuela	5,020	13,010	0	0	197	60
31.06S, 51.15W	● Patos (salt, lag)	Rio Grande do Sul, Brazil	3,920	10,140	0	0	15	5
15.48S, 69.24W	Titicaca	Bolivia–Peru	3,100	8,030	12,497	3809	997	304
32.45S, 52.50W	● Mirim (alt Merin) (salt, lag)	Brazil–Uruguay	1,150	2,970	0	0	33	10
46.30S, 72.00W	Buenos Aires (alt General Carrera)	Argentina–Chile	860	2,240	712	217		

[1]Freshwater.

SOUTH AMERICA

Latitude and Longitude	Lake	Location	Area (mi²)	(km²)	Elevation (ft)	(m)	Greatest Depth (ft)	(m)
30.42S, 62.36W	(Mar) Chiquita (salt)	Cordoba, Argentina	720	1,850	230	70	13/10	4/3
50.13S, 72.25W	Argentino	Santa Cruz, Argentina	550	1,410	656	200	984	300
18.45S, 67.07W	Poopo	Bolivia	520	1,340	12,094	3686	10	3
49.35S, 72.35W	Viedma	Santa Cruz, Argentina	420	1,090	820	250		
48.52S, 72.40W	San Martin (alt O'Higgins)	Argentina–Chile	390	1,010	656	200	558	170
45.30S, 68.48W	Colhue Huapi	Chubut, Argentina	310	800	869	265	13	4
41.08S, 72.48W	Llanquihue	Chile	310	800	171	52	1148	350

* * *

Latitude and Longitude	Lake	Location	Area (mi²)	(km²)	Elevation (ft)	(m)	Greatest Depth (ft)	(m)
54.38S, 68.00W	Fagnano (alt Cami)	Argentina–Chile	230	590	827	252	656	200
40.58S, 71.30W	Nahuel Huapi	Neuquen–Rio Negro, Argentina	210	550	2,517	767	1437	438
45.27S, 69.13W	Musters	Chubut, Argentina	170	430	889	271	328	100
40.14S, 72.24W	Ranco	Chile	150	400	230	70	262	80
39.15S, 72.06W	• Villarrica	Chile	66	170	755	230		

5a
DEEPEST NATURAL LAKES

Contents

Rank.
Conventional and other (alternate, former, and official) names of lake.
Subdivision and country (or countries) in which located.
Greatest recorded depth of water, in feet and meters.

Arrangement

By depth.

Coverage

50 deepest natural lakes.

Special Feature

Saltwater lakes are indicated by the word "salt."

Entries

50.

TABLE 5a. DEEPEST NATURAL LAKES

Rank	Lake	Location	Greatest Depth (ft)	(m)
1.	Baykal (alt Baikal)	Russia, USSR	5315	1620
2.	Tanganyika	Burundi–Tanzania–Zaire–Zambia	4825	1471
3.	Caspian (Sea) (off Kaspiyskoye, Khazar) (salt)	Iran–USSR	3363	1025
4.	Malawi (alt Niassa, Nyasa)	Malawi–Mozambique–Tanzania	2316	706
5.	Issyk(-Kul) (salt)	Kirgizia, USSR	2303	702
6.	Great Slave	NWT, Canada	2015	614
7.	● Matana	Celebes I, Indonesia	1936	590
8.	Crater	Oregon, USA	1932	589
9.	Toba	Sumatra I, Indonesia	1736	529
10.	Sarez (off Sarezskoye)	Tadzhikistan, USSR	1657	505
11.	Tahoe	California–Nevada, USA	1645	501
12.	Chelan	Washington, USA	1605	489
13.	Kivu	Rwanda–Zaire	1575	480
14.	Quesnel	British Columbia, Canada	1560	475
15.	Adams	British Columbia, Canada	1500	457
16.	Mjosa	Norway	1473	449
17.	Manapouri	South I, New Zealand	1453	443
18.	Nahuel Huapi	Neuquen–Rio Negro, Argentina	1437	438
19.	Dead (Sea) (off Lut, Mayyit, Melah) (salt)	Israel–Jordan	1421	433
20.	Tazawa	Honshu I, Japan	1394	425
21.	Great Bear	NWT, Canada	1356	413
22.	● Como	Lombardia, Italy	1352	413
23.	Superior	Canada–USA	1333	406
24.	Hawea	South I, New Zealand	1286	392
25.	Maggiore [alt Langen(see), Majeur]	Italy–Switzerland	1221	372
26.	Chilko	British Columbia, Canada	1200	366
26.	Pend Oreille	Idaho, USA	1200	366
28.	Shikotsu	Hokkaido I, Japan	1191	363
29.	Powell	British Columbia, Canada	1174	358
30.	Llanquihue	Chile	1148	350
31.	● Garda	Lombardia–Trentino-Alto Adige–Veneto, Italy	1135	346
32.	Towada	Honshu I, Japan	1096	334
33.	Teletskoye [for Altyn(-Kol)]	Russia, USSR	1066	325
34.	Eutsuk	British Columbia, Canada	1060	323
35.	Geneva (off Geneva, Ginevra, Leman)	France–Switzerland	1017	310
35.	● Morar	Scotland, UK	1017	310
37.	Kurile (off Kurilskoye)	Russia, USSR	1004	306
38.	Walker (salt)	Nevada, USA	1000	305
39.	Titicaca	Bolivia–Peru	997	304
40.	Argentino	Santa Cruz, Argentina	984	300
41.	Iliamna	Alaska, USA	980	299
42.	Tyrifjorden	Norway	968	295

Rank	Lake	Location	Greatest Depth (ft)	(m)
43.	Lugano [alt Ceresio, Luganer(see)]	Italy-Switzerland	945	288
44.	Takla	British Columbia, Canada	941	287
45.	Ohrid (off Ohridsko, Ohrit)	Albania-Yugoslavia	938	286
46.	Atlin	British Columbia-Yukon Territory, Canada	930	283
46.	Nuyakuk	Alaska, USA	930	283
48.	Michigan	Illinois-Indiana-Michigan-Wisconsin, USA	923	281
49.	Harrison	British Columbia, Canada	916	279
50.	Te Anau	South I, New Zealand	906	276

Figure 16. Cape Shamanka, a picturesque spot on Lake Baykal in the Siberian region of the USSR. Baykal is the deepest lake on earth and the one with the greatest volume of fresh water. (Credit: Embassy of the USSR, Washington.)

5b
LAKES WITH GREATEST VOLUME OF WATER

Contents

Rank.

Conventional and other (alternate, former, and official) names of lake.

Subdivision and country (or countries) in which located.

Estimated volume of water (area multiplied by average depth), in cubic miles and cubic kilometers.

Arrangement

By volume of water.

Coverage

20 natural lakes with greatest volume of water.

Special Feature

Saltwater lakes are indicated by the word "salt," and lagoons by the abbreviation "lag."

Entries

20.

TABLE 5b.   LAKES WITH GREATEST VOLUME OF WATER

Rank	Lake	Location	Volume of Water (mi^3)	(km^3)
1.	Caspian (Sea) (off Kaspiyskoye, Khazar) (salt)	Iran–USSR	16,021	66,780
2.	Baykal (alt Baikal)	Russia, USSR	5,517	22,995
3.	• Tanganyika	Burundi–Tanzania–Zaire–Zambia	4,277	17,827
4.	Superior	Canada–USA	2,941	12,258
5.	• Malawi (alt Niassa, Nyasa)	Malawi–Mozambique–Tanzania	1,473	6,140
6.	Michigan	Illinois–Indiana–Michigan–Wisconsin, USA	1,185	4,940
7.	Huron	Canada–USA	849	3,539
8.	• Victoria	Kenya–Tanzania–Uganda	604	2,518
9.	• Great Bear	NWT, Canada	550	2,292
10.	• Great Slave	NWT, Canada	501	2,088
11.	Issyk(-Kul) (salt)	Kirgizia, USSR	417	1,738
12.	Ontario	Canada–USA	398	1,657
13.	Aral (Sea) (off Aralskoye) (salt)	Kazakhstan–Uzbekistan, USSR	348	1,451
14.	Ladoga (off Ladozhskoye)	Russia, USSR	218	908
15.	• Titicaca	Bolivia–Peru	198	827
16.	Erie	Canada–USA	116	483
17.	Winnipeg	Manitoba, Canada	89	371
18.	• Kivu	Rwanda–Zaire	80	333
19.	Onega (off Onezhskoye)	Russia, USSR	70	292
20.	• Maracaibo (salt, lag)	Venezuela	67	280

5c
LARGEST NATURAL LAKE
IN EACH COUNTRY

Contents

Conventional name of country.
Conventional and other (alternate, former, and official) names of lake.
Country (if any) with which lake is shared.
Area in square miles and square kilometers.

Arrangement

By country, alphabetically.

Coverage

All countries with an area of at least 5000 mi^2 (12,950 km^2) having or sharing a natural lake of significant size.

Rounding

Areas are rounded to the nearest 10 mi^2 (km^2) [to the nearest square mile and square kilometer if less than 100 mi^2 (km^2)].

Special Features

Saltwater lakes are indicated by the word "salt," and lagoons, by the abbreviation "lag."
When area varies from season to season, the normal minimum and maximum are shown as follows: 4,000/10,000.

Entries

99.

TABLE 5c.   LARGEST NATURAL LAKE IN EACH COUNTRY

Country	Lake	Country with which Shared	Area (mi²)	Area (km²)
Afghanistan	• (Ab-i-)Istada (salt)		200?	520?
Albania	Scutari (off in Albania Shkodres)	Yugoslavia	150	380
Algeria	• Oran (off Ouahran) (salt, lag)		120	320
Angola	• Dilolo		8	20
Argentina	Buenos Aires (alt General Carrera)	Chile	860	2,240
Australia	Eyre (salt)		0/2,970	0/7,690
Austria	Constance [off in Austria Boden(see)]	Switzerland, West Germany	210	540
Bangladesh	• Chalan (Bil)		20/150	52/390
Benin	Nokoue (salt, lag)		53	140
Bolivia	Titicaca	Peru	3,100	8,030
Botswana	Dow		70?	180?
Brazil	• Patos (salt, lag)		3,920	10,140
Burma	• Indawgyi		80	210
Burundi	Tanganyika	Tanzania, Zaire, Zambia	12,350	32,000
Cambodia	• (Tonle) Sap		1,040/ 3,860	2,700/ 10,000
Cameroon	Chad (alt Tchad)	Chad, Niger, Nigeria	4,000/ 10,000	10,360/ 25,900
Canada	Superior	USA	31,760	82,260
Chad	Chad (alt Tchad)	Cameroon, Niger, Nigeria	4,000/ 10,000	10,360/ 25,900
Chile	Buenos Aires (off in Chile General Carrera)	Argentina	860	2,240
China	Koko (alt Tsing; off Ching) (salt)		1,720	4,460
Colombia	• Tota		23	59
Cuba	• Leche (salt, lag)		30	78
Denmark	• Arre		16	41
Djibouti	• Abbe (alt Abe) (salt)	Ethiopia	300	780
Dominican Republic	Enriquillo (salt)		190	500
Egypt	• Manzala (off Manzilah) (salt, lag)		530	1,360
El Salvador	• Guija		120	310
Ethiopia	Rudolf (salt)	Kenya	2,470	6,400
Finland	Saimaa (alt Saima)		680	1,760
France	Geneva (off in France Geneve)	Switzerland	220	580
Gabon	• Onangue		90?	250?
Germany				
East	(Oder(-Haff) [alt Szczecinski; for Stettiner (Haff)] (salt, lag)	Poland	350	900
West	Constance [off in Germany Boden(see)]	Austria, Switzerland	210	540

Country	Lake	Country with which Shared	Area (mi²)	Area (km²)
Ghana	• Bosumtwi		18	48
Greece	Prespa (off in Greece Megal Prespa)	Albania, Yugoslavia	110	280
Guatemala	Izabal		230	590
Haiti	Saumatre (alt Azuel)		65	170
Honduras	Caratasca (salt, lag)		430	1,110
Hungary	Balaton (alt Platten)		230	590
Iceland	• Thingvallavatn		32	83
India	• Chilka (salt, lag)		350/ 450	910/ 1,170
Indonesia (in Asia)	Toba		440	1,150
Iran	Caspian (Sea) (off in Iran Khazar) (salt)	USSR	143,240	371,000
Iraq	• Hammar		750	1,940
Ireland	Corrib		66	170
Israel	Dead (Sea) (off in Israel Melah) (salt)	Jordan	390	1,020
Italy	• Garda		140	370
Ivory Coast	• Aby (salt, lag)		300	780
Japan	Biwa		270	690
Jordan	Dead (Sea) (off in Jordan Lut, Mayyit) (salt)	Israel	390	1,020
Kashmir-Jammu	• Pangong (alt Pankung) (salt)	China	230	600
Kenya	Victoria	Tanzania, Uganda	24,300	62,940
Korea (North)	Kwangpo (salt, lag)		5	13
Liberia	• Fisherman's		30	78
Madagascar	• Alaotra		70	180
Malawi	Malawi (alt Niassa, Nyasa)	Mozambique, Tanzania	8,680	22,490
Malaysia	• (Tasek) Dampar		40?	100?
Mali	Faguibine		230	590
Mauritania	• Rkiz (alt Cayar)		60?	160?
Mexico	• Terminos (salt, lag)		600	1,550
Mongolia	Uvs (alt Ubsa, Ubsu) (salt)		1,290	3,350
Mozambique	Malawi (off in Mozam- bique Niassa)	Malawi, Tanzania	8,680	22,490
Netherlands	• IJssel(meer) [for Zuider (Zee)]		470	1,210
New Zealand	Taupo		230	610
Nicaragua	Nicaragua (alt Cocibolca)		3,150	8,150
Niger	Chad (alt Tchad)	Cameroon, Chad, Nigeria	4,000/ 10,000	10,360/ 25,900
Nigeria	Chad (alt Tchad)	Cameroon, Chad, Niger	4,000/ 10,000	10,360/ 25,900
Norway	Mjosa		140	370
Pakistan	• Manchhar		30/100	78/260

Country	Lake	Country with which Shared	Area (mi²)	Area (km²)
Panama	● Chiriqui (salt, lag[1])		350	900
Papua New Guinea	● Murray		120?	300?
Paraguay	● Ipoa		100	260
Peru	Titicaca	Bolivia	3,100	8,030
Philippines	Bay (salt)		340	890
Poland	Oder(-Haff) [off in Poland Szczecinski; for Stettiner (Haff)] (salt, lag)	East Germany	350	900
Romania	Razelm (alt Razim) (salt)		150	390
Rwanda	Kivu	Zaire	860	2,220
Senegal	● Guiers		60?	150?
South Africa	● Saint Lucia (salt, lag)		150?	390?
Sri Lanka	● Batticaloa (salt, lag)		46	120
Sweden	Vanern (alt Vaner, Vener)		2,160	5,580
Switzerland	Geneva (off in Switzerland Geneve, Ginevra, Leman)	France	220	580
Syria	Jabbul		58	150
Tanzania	Victoria	Kenya, Uganda	24,300	62,940
Thailand	● (Thale) Luang (alt Sap) (salt, lag)		500	1,290
Tunisia	● Bizerta (alt Bizerte; off Banzart) (salt, lag)		42	110
Turkey				
in Asia	Van (salt)		1,440	3,740
in Europe	Terkos		9	24
Uganda	Victoria	Kenya, Tanzania	24,300	62,940
UK	● Neagh		150	400
Uruguay	● Mirim (off in Uruguay Merin) (salt, lag)	Brazil	1,140	2,970
USA (in North America)	Superior	Canada	31,760	82,260
USSR				
in Asia	Caspian (Sea) (off in USSR Kaspiyskoye)	Iran	143,240	371,000
in Europe	Ladoga (off Ladozhskoye)		7,000	18,130
Venezuela	Maracaibo (salt, lag)		5,020	13,010
Vietnam	● Cau Hai (salt, lag)		40	100
Yugoslavia	Scutari (off in Yugoslavia Skadarsko)	Albania	150	380
Zaire	Tanganyika	Burundi, Tanzania, Zambia	12,350	32,000
Zambia	Tanganyika	Burundi, Tanzania, Zaire	12,350	32,000

[1] Actually a bay.

6
HIGHEST WATERFALLS
(INDIVIDUAL LEAPS),
BY CONTINENT

Contents

Latitude and longitude, in degrees and minutes.
Conventional and alternate names of waterfall.
River, and subdivision and country in which located.
Height of greatest individual leap, in feet and meters.

Arrangement

By height, under continent.

Coverage

Above * * *, all known and named waterfalls with a height of at least 300 ft (91 m).
Below * * *, other well-known waterfalls.

Entries

162.

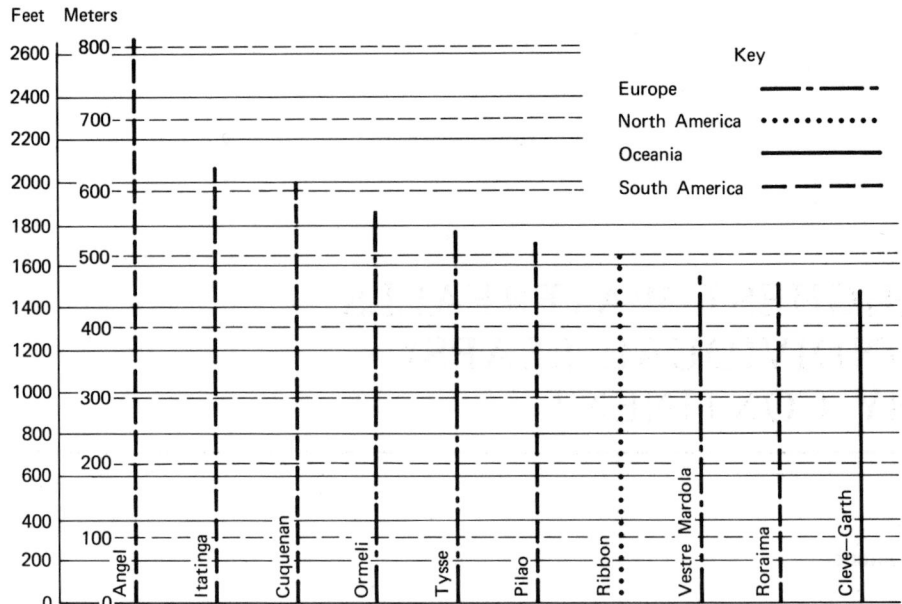

Figure 17. Ten highest waterfalls.

TABLE 6. HIGHEST WATERFALLS (INDIVIDUAL LEAPS), BY CONTINENT

AFRICA

Latitude and Longitude		Waterfall	River and Location	Height (ft)	(m)
28.45S,	28.56E	Tugela: highest fall	Tugela R, Natal, South Africa	1350	411
		Shire (alt Chire)	Shire (alt Chire) R, Malawi	1200?	366?
10.12S,	27.27E	Kaloba (alt Lofoi)	Lofoi R, Zaire	1115	340
18.36S,	32.42E	Mtarazi	Mtarazi R, Mozambique–Rhodesia	1000	305
8.36S,	31.14E	Kalambo	Kalambo R, Tanzania–Zambia	704	215
29.52S,	28.04E	Maletsunyane	Maletsunyane R, Lesotho	630	192
		Fincha	Fincha R, Ethiopia	508	155
28.35S,	20.23E	Aughrabies (alt King George's)	Orange (alt Oranje) R, Cape of Good Hope, South Africa	482	147
5.22N,	40.03E	Baratieri	Ganale–Dorya R, Ethiopia	459	140
31.26S,	29.38E	Magwa	Magwa R, Cape of Good Hope, South Africa	450	137
8.50S,	31.22E	Chirombo	Ieisa (alt Iza) R, Zambia	440	134
14.35S,	29.07E	Lunsemfwa	Lunsemfwa R, Zambia	400	122
31.15S,	28.57E	Tsitsa	Tsitsa R, Cape of Good Hope, South Africa	375	114
17.23S,	14.15E	Ruacana	Cunene (alt Kunene) R, Angola–Namibia	352	107
29.29S,	30.14E	Howick	Umgeni R, Natal, South Africa	311	95
17.55S,	25.51E	Victoria (alt Mosi-oa-Toenja)	Zambezi (alt Zambesi, Zambeze) R, Rhodesia–Zambia	304	92
20.26S,	57.23E	Chamarel	Cap R, Mauritius	300	91

* * *

Latitude and Longitude		Waterfall	River and Location	Height (ft)	(m)
9.06S,	15.57E	Dianzundu (alt Duque de Braganca)	Lucala R, Angola	200?	61?
2.17N,	31.41E	Kabalega (for Murchison)	Nile (off Nil) R, Uganda	130	40

ASIA

Latitude and Longitude		Waterfall	River and Location	Height (ft)	(m)
25.10N,	91.45E	Mawsmai	? R, Assam, India	1148?	350?
6.04N,	116.29E	Kalupis	? R, Sabah, Malaysia	1100?	335?
25.30N,	91.40E	Thylliejlongwa	? R, Assam, India	997?	304?
14.14N,	74.50E	Gersoppa (alt Jog): highest fall	Sharavati R, Karnataka–Maharashtra, India	829	253
7.05N,	80.50E	Kurundu Oya	Kurundu (Oya) R, Sri Lanka	620	189
6.44N,	81.02E	Diyaluma	? R, Sri Lanka	560	171
1.19N,	124.54E	Tondano	Menado (alt Manado) R, Celebes I, Indonesia	492	150
45.31N,	148.53E	Ilya Muromets	? R, Iturup I, Russia, USSR	463	141
6.46N,	80.50E	Bambarakanda: highest fall	? R, Sri Lanka	461	141
33.40N,	135.53E	Nachi	Nachi R, Honshu I, Japan	430	131
6.54N,	80.30E	Laksapana	? R, Sri Lanka	377	115

ASIA

Latitude and Longitude		Waterfall	River and Location	Height (ft)	(m)
24.32N,	81.18E	Bihar	Bihar R, Madhya Pradesh, India	370	113
7.22N,	80.55E	Ratna Ella	Ratna (Ella) R, Sri Lanka	365	111
36.44N,	139.27E	Kegon	Daiya R, Honshu I, Japan	348	106
6.38N,	80.34E	Kirindi Ela	Kirindi (Ela) R, Sri Lanka	347	106
36.48N,	139.26E	Yudaki	? R, Honshu I, Japan	335	102
7.04N,	80.42E	Ramboda	? R, Sri Lanka	329	100
12.15N,	77.10E	Kaveri (alt Cauvery)	Kaveri (alt Cauvery) R, Karnataka–Tamil Nadu, India	320	98
11.27N,	76.35E	Pykara	? R, Tamil Nadu, India	300	91

* * *

Latitude and Longitude		Waterfall	River and Location	Height (ft)	(m)
13.56N,	105.56E	Khone	Mekong (alt Khong, Lantsang, Mekongk, Tien Giang) R, Cambodia–Laos	70	21

EUROPE

Latitude and Longitude		Waterfall	River and Location	Height (ft)	(m)
		Ormeli	? R, Norway	1847	563
		Tysse	Tyssa R, Norway	1749	533
62.34N,	8.11E	Vestre Mardola (alt Western Mardola)	? R, Norway	1535	468
42.42N,	0.00	Gavarnie: highest fall	Pau R, Midi-Pyrenees, France	1385	422
62.21N,	8.04E	Verma	? R, Norway	1250	381
		Austerbo	? R, Norway	1247	380
		Serio	Serio R, Italy	1034	315
60.31N,	7.15E	Rembesdals: highest fall	? R, Norway	984	300
60.07N,	6.44E	Tyssestrengene	Tysso R, Norway	984	300
46.36N,	7.54E	Staubbach	Staub(bach) Creek, Switzerland	980	299
62.34N,	8.11E	Ostre Mardola (alt Eastern Mardola): upper fall	? R, Norway	974	297
61.22N,	7.55E	Vettis: highest fall	Morkedola R, Norway	902	275
60.22N,	7.08E	Valur	? R, Norway	892	272
		Mollius	? R, Norway	883	269
		Austerkrok	? R, Norway	843	257
		Seculejo	? R, Spain	820	250
60.29N,	7.15E	Skykkje	Skykkjua R, Norway	820	250
		Kjos	? R, Norway	738	225
62.34N,	8.11E	Ostre Mardola: lower fall	? R, Norway	722	220
61.23N,	7.26E	Feigum	? R, Norway	715	218
		Rogaland: highest fall	? R, Norway	689	210
		Teverone	? R, Italy	680?	207?
61.27N,	7.59E	Maradals	? R, Norway	656	200
		Aurstaupet	? R, Norway	633	193
60.26N,	7.15E	Voring	Bjoreia R, Norway	597	182
		Stauber	Stauber R, Switzerland	590?	180?
		Sote: highest fall	? R, Norway	577	176

EUROPE

Latitude and Longitude		Waterfall	River and Location	Height (ft)	Height (m)
61.07N,	10.30E	Mesna	? R, Norway	525	160
60.07N,	6.44E	Skjeggedals	Tysso R, Norway	525	160
58.13N,	4.52W	Eas Coul Aulin	? R, Scotland, UK	511	156
59.02N,	7.51E	Kjel (alt Kile): highest fall	Kjel (alt Kile) R, Norway	502	153
46.24N,	8.24E	Frua (alt Toce)	Toce R, Piemonte, Italy	470?	143?
		Hundkastet: highest fall	? R, Norway	459	140
60.44N,	6.53E	Rjoande: highest fall	? R, Norway	459	140
47.12N,	12.10E	Krimmler: lower fall	Krimmler R, Austria	459	140
47.12N,	12.10E	Krimmler: upper fall	Krimmler R, Austria	459	140
60.50N,	6.40E	Stalheims	? R, Norway	413	126
46.24N,	7.26E	Iffigen	Iffigen(bach) Creek, Switzerland	394	120
57.16N,	5.18W	Glomach	Glomach R, Scotland, UK	370	113
59.52N,	8.34E	Rjukan	Mane R, Norway	345	105
		Fagerbakk	Fager(bakk) Creek, Norway	328	100
		Heis	? R, Norway	328	100
47.12N,	12.10E	Krimmler: middle fall	Krimmler R, Austria	328	100

* * *

47.07N,	13.08E	Gastein: lower fall	Gasteiner Ache R, Austria	280	85

NORTH AMERICA

Latitude and Longitude		Waterfall	River and Location	Height (ft)	Height (m)
37.44N,	119.39W	Ribbon	Ribbon Creek, California, USA	1612	491
49.15N,	125.45W	Della	? R, British Columbia, Canada	1443	440
37.45N,	119.36W	Yosemite: upper fall	Yosemite Creek, California, USA	1430	436
51.30N,	116.29W	Takakkaw: highest fall	Yoho R tributary, British Columbia, Canada	1200	366
37.42N,	119.40W	Silver Strand (alt Widow's Tears)	Meadow Brook, California, USA	1170	357
28.13N,	108.14W	Basaseachic	(Arroyo) Basaseachic Creek, Chihuahua, Mexico	1020	311
51.33N,	116.33W	Twin	Twin Falls Creek, British Columbia, Canada	900	274
52.17N,	125.46W	Hunlen	Atnarko Creek, British Columbia, Canada	830	253
46.47N,	121.42W	Fairy	Stevens Creek, Washington, USA	700	213
39.34N,	121.17W	Feather	Fall R, California, USA	640	195
37.43N,	119.39W	Bridalveil	Bridalveil Creek, California, USA	620	189
52.11N,	117.03W	Panther	Nigel Creek, Alberta, Canada	600	183
37.43N,	119.32W	Nevada	Merced R, California, USA	594	181
45.34N,	122.06W	Multnomah: highest fall	Multnomah R, Oregon, USA	542	165
37.43N,	119.36W	Sentinel: lower fall	Sentinel Creek, California, USA	500?	152?
51.57N,	120.11W	Helmcken	Murtle R, British Columbia, Canada	450	137
49.11N,	121.44W	Bridal Veil	Bridal Creek, British Columbia, Canada	400	122

NORTH AMERICA

Latitude and Longitude		Waterfall	River and Location	Height	
				(ft)	(m)
37.43N,	119.34W	Illilouette	Illilouette Creek, California, USA	370	113
46.47N,	121.47W	Comet	Van Trump Creek, Washington, USA	320	98
37.45N,	119.36W	Yosemite: lower fall	Yosemite Creek, California, USA	320	98
37.44N,	119.33W	Vernal	Merced R, California, USA	317	97
44.43N,	110.28W	Yellowstone: lower fall	Yellowstone R, Wyoming, USA	308	94
10.44N,	61.24W	Maracas	? R, Trinidad I, Trinidad and Tobago	300	91
46.47N,	121.43W	Sluiskin	Paradise R, Washington, USA	300	91

* * *

Latitude and Longitude		Waterfall	River and Location	Height	
61.38N,	125.42W	Virginia	South Nahanni R, NWT, Canada	294	90
46.55N,	71.10W	Montmorency	Montmorency R, Quebec, Canada	273	83
47.33N,	121.49W	Snoqualmie	Snoqualmie R, Washington, USA	270	82
35.39N,	85.22W	Fall Creek	Fall Creek, Tennessee, USA	256	78
53.30N,	64.10W	Churchill (for Grand)	Churchill (for Hamilton) R, Newfoundland, Canada	245	75
42.33N,	76.34W	Taughannock	Taughannock Creek, New York, USA	215	66
42.36N,	114.26W	Shoshone	Snake R, Idaho, USA	195	59
43.04N,	79.04W	Niagara	Niagara R, Canada–USA	186	57

OCEANIA

Latitude and Longitude		Waterfall	River and Location	Height	
				(ft)	(m)
21.10N,	156.49W	Kahiwa	? R, Molokai I, Hawaii, USA	1750[1]	533[1]
		Cleve-Garth	? R, New Zealand	1476?	450?
8.54S,	140.06W	Ahui	? R, Nuku-Hiva I, French Polynesia	1148	350
30.32S,	152.03E	Wollomombi: highest fall	Wollomombi R, New South Wales, Australia	1100	334
18.17S,	146.03E	Wallaman (alt Stony Creek)	Stony Creek, Queensland, Australia	970	296
17.43S,	145.35E	Elizabeth Grant	Tully R, Queensland, Australia	900	274
45.28S,	167.10E	Helena	Helena R, South I, New Zealand	830	253
44.48S,	167.44E	Sutherland: upper fall	Arthur R, South I, New Zealand	815	248
16.50S,	145.39E	Barron	Barron R, Queensland, Australia	770	235
44.48S,	167.44E	Sutherland: middle fall	Arthur R, South I, New Zealand	751	229
13.55S,	171.45W	Tiavi	? R, Upolu I, Samoa	600	183
17.47S,	145.35E	Tully: highest fall	Tully R, Queensland, Australia	550?	168?
44.40S,	167.55E	Bowen	Bowen R, South I, New Zealand	520	158
44.36S,	167.52E	Stirling	Stirling R, South I, New Zealand	480	146
19.51N,	155.09W	Akaka	Kolekole Stream, Hawaii I, Hawaii, USA	418	127
34.39S,	150.29E	Fitzroy: highest fall	? R, New South Wales, Australia	400	122

[1] Total fall.

OCEANIA

Latitude and Longitude		Waterfall	River and Location	Height (ft)	(m)
33.43S,	150.23E	Wentworth: upper fall	Wentworth R, New South Wales, Australia	360	110
44.48S,	167.44E	Sutherland: lower fall	Arthur R, South I, New Zealand	338	103
		Waiilikahi	? R, Hawaii I, Hawaii, USA	320	98

SOUTH AMERICA

Latitude and Longitude		Waterfall	River and Location	Height (ft)	(m)
5.57N,	62.30W	Angel: upper fall	Churun R, Venezuela	2648	807
23.07S,	48.36W	Itatinga	Itatinga R, Sao Paulo, Brazil	2060	628
5.13N,	60.51W	Cuquenan (alt Kukenaam)	Cuquenan (alt Kukenaam) R, Guyana–Venezuela	2000	610
		Pilao	Itajai R, Santa Catarina, Brazil	1719	524
		Roraima	Potaro R, Guyana	1500?	457?
5.46N,	61.08W	King George VI	Utshi R, Guyana	1200?	366?
5.22N,	72.45W	Candelas	Cusiana R, Colombia	984	300
12.15S,	73.45W	Sewerd	Cutibireni R, Peru	877	267
		King Edward VIII	Semang R, Guyana	840	256
5.09N,	59.29W	Kaieteur	Potaro R, Guyana	741	226
		Casca d'Anta	Sao Francisco R, Minas Gerais, Brazil	666	203
		Sakaika	? R, Guyana	629	192
5.57N,	62.30W	Angel: lower fall	Churun R, Venezuela	564	172
4.35N,	74.18W	Tequendama	Bogota (alt Funza) R, Colombia	515	157
		Wakowaieng	? R, Guyana	440	134
		Fagundes	Piabanha R, Rio de Janeiro, Brazil	413	126
		Kumarow	Kamarang R, Guyana	400	122
		Itiquira	Itiquira R, Brazil	394	120
5.25N,	59.30W	Marina	Ipobe R, Guyana	360	110
		Papagaio	Papagaio R, Mato Grosso, Brazil	350?	107?
2.22N,	52.40W	Manoa (alt Manaua)	Oyapock (alt Oiapoque) R, Brazil–French Guiana	345	105
5.42N,	61.07W	Great	? R, Guyana	300	91

* * *

Latitude and Longitude		Waterfall	River and Location	Height (ft)	(m)
9.24S,	38.13W	Paulo Afonso	Sao Francisco R, Alagoas-Bahia, Brazil	262	80
25.41S,	54.26W	Iguazu (alt Iguacu; for Iguassu)	Iguazu (alt Iguacu; for Iguassu) R, Argentina–Brazil	230	70
24.02S,	54.16W	Sete Quedas (alt Guaira)	Parana R, Brazil–Paraguay	213	65
20.18S,	49.10W	Marimbondo	Grande R, Minas Gerais-Sao Paulo, Brazil	115	35
31.14S,	57.55W	Grande	Uruguay R, Argentina–Uruguay	75	23
20.36S,	51.33W	Urubupunga	Parana R, Mato Grosso-Sao Paulo, Brazil	27	9

Figure 18. Angel Falls on the Churun River, Venezuela, the highest waterfall on earth. Originating on the Auyan-Tepui, or Devil Mountain, plateau, it was named after Jimmy Angel, a US pilot, who brought news of it to the outside world in 1937. (Credit: Embassy of Venezuela Information Service, Washington.)

6a
GREATEST WATERFALLS
(VOLUME OF WATER)

Contents

Rank.

Conventional and alternate names of waterfall.

River, and subdivision and country in which located.

Average rate of flow, in thousands of cubic feet per second and in cubic meters per second.

Height of greatest individual leap, in feet and meters.

Arrangement

By average rate of flow.

Coverage

All waterfalls with an average rate of flow of at least 20,000 cfs (566 m³/sec).

Rounding

Rates of flow are rounded to the nearest 1000 cfs.

Entries

12.

TABLE 6a. GREATEST WATERFALLS (VOLUME OF WATER)

Rank	Waterfall	River and Location	Average Flow (1000 cfs)	Average Flow (m³/sec)	Height (ft)	Height (m)
1.	Khone	Mekong (alt Khong, Lantsang, Mekongk, Tien Giang) R, Cambodia–Laos	410	11,610	70	21
2.	Sete Quedas (alt Guaira)	Parana R, Brazil–Paraguay	292	8,260	213	65
3.	Niagara	Niagara R, Canada–USA	212	6,000	186	57
4.	Grande	Uruguay R, Argentina–Uruguay	106	3,000	75	23
5.	Urubupunga	Parana R, Mato Grosso-Sao Paulo, Brazil	97	2,750	27	9
6.	Iguazu (alt Iguacu; for Iguassu)	Iguazu (alt Iguacu; for Iguassu) R, Argentina–Brazil	60	1,700	230	70
7.	Marimbondo	Grande R, Minas Gerais–Sao Paulo, Brazil	53	1,500	115	35
8.	Churchill (for Grand)	Churchill (for Hamilton) R, Newfoundland, Canada	40	1,130	245	75
9.	Victoria (alt Mosi-oa-Toenja)	Zambezi (alt Zambesi, Zambeze) R, Rhodesia–Zambia	38	1,090	304	92
10.	Kaveri (alt Cauvery)	Kaveri (alt Cauvery) R, Karnataka–Tamil Nadu, India	33	934	320	98
11.	Paulo Afonso	Sao Francisco R, Alagoas–Bahia, Brazil	25	700	262	80
12.	Kaieteur	Potaro R, Guyana	23	650	741	226

Figure 19. Sete Quedas, or Guaira, on the Parana River between Brazil and Paraguay. Among the higher waterfalls, it carries the greatest volume of water. (Credit: Organization of American States.)

7a
LARGEST COUNTRIES IN AREA

Contents

Rank.
Conventional name of country.
Area, including inland waters, in thousands of square miles and square kilometers.
Percentage of world land area represented by the given area.

Arrangement

By area.

Coverage

50 largest countries in area.

Rounding

Areas are rounded to the nearest 1000 mi^2 (km^2), percentages to the nearest two significant digits.

Entries

50.

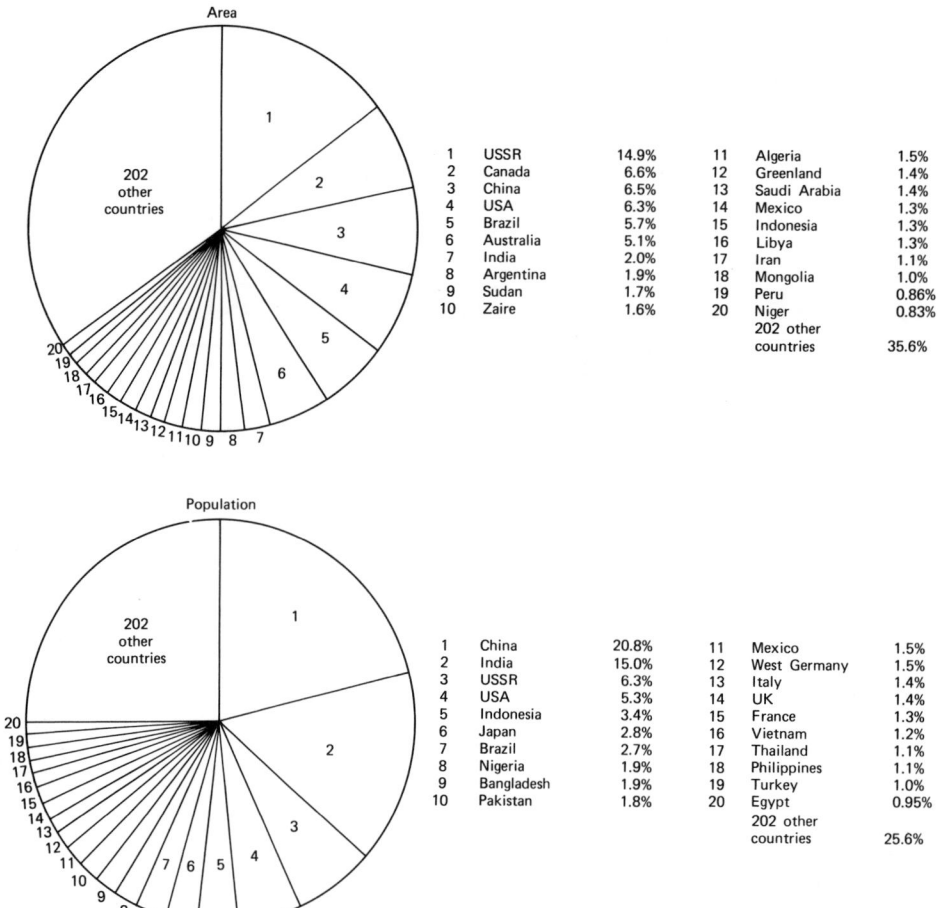

Area

1	USSR	14.9%	11	Algeria	1.5%
2	Canada	6.6%	12	Greenland	1.4%
3	China	6.5%	13	Saudi Arabia	1.4%
4	USA	6.3%	14	Mexico	1.3%
5	Brazil	5.7%	15	Indonesia	1.3%
6	Australia	5.1%	16	Libya	1.3%
7	India	2.0%	17	Iran	1.1%
8	Argentina	1.9%	18	Mongolia	1.0%
9	Sudan	1.7%	19	Peru	0.86%
10	Zaire	1.6%	20	Niger	0.83%
				202 other countries	35.6%

Population

1	China	20.8%	11	Mexico	1.5%
2	India	15.0%	12	West Germany	1.5%
3	USSR	6.3%	13	Italy	1.4%
4	USA	5.3%	14	UK	1.4%
5	Indonesia	3.4%	15	France	1.3%
6	Japan	2.8%	16	Vietnam	1.2%
7	Brazil	2.7%	17	Thailand	1.1%
8	Nigeria	1.9%	18	Philippines	1.1%
9	Bangladesh	1.9%	19	Turkey	1.0%
10	Pakistan	1.8%	20	Egypt	0.95%
				202 other countries	25.6%

Figure 20. Area and population of countries as percentage of world land area and population.

TABLE 7a. LARGEST COUNTRIES IN AREA

Rank	Country	Area (1000 mi^2)	Area (1000 km^2)	% of World Land Area	Rank	Country	Area (1000 mi^2)	Area (1000 km^2)	% of World Land Area
1.	USSR	8,649	22,402	14.9	28.	Mauritania	398	1,031	0.69
2.	Canada	3,852	9,976	6.6	29.	Egypt	387	1,001	0.67
3.	China	3,769	9,761	6.5	30.	Tanzania	365	945	0.63
4.	USA	3,683	9,539	6.3	31.	Nigeria	357	924	0.62
5.	Brazil	3,286	8,512	5.7	32.	Venezuela	352	912	0.61
6.	Australia	2,966	7,682	5.1	33.	Namibia	318	824	0.55
7.	India	1,184	3,066	2.0	34.	Turkey	315	815	0.54
8.	Argentina	1,073	2,780	1.9	35.	Pakistan	310	804	0.54
9.	Sudan	967	2,506	1.7	36.	Mozambique	309	799	0.53
10.	Zaire	906	2,345	1.6	37.	Chile	292	757	0.50
11.	Algeria	885	2,293	1.5	38.	Zambia	291	753	0.50
12.	Greenland	840	2,176	1.4	39.	Burma	262	678	0.45
13.	Saudi Arabia	830	2,150	1.4	40.	Afghanistan	250	647	0.43
14.	Mexico	781	2,022	1.3	41.	Somalia	246	638	0.42
15.	Indonesia	741	1,919	1.3	42.	Central African Empire	241	623	0.41
16.	Libya	731	1,893	1.3					
17.	Iran	636	1,648	1.1	43.	Botswana	232	600	0.40
18.	Mongolia	604	1,565	1.0	44.	Madagascar	227	587	0.39
19.	Peru	496	1,285	0.86	45.	Kenya	225	583	0.39
20.	Niger	482	1,248	0.83	46.	France	213	551	0.37
21.	Angola	481	1,247	0.83	47.	Thailand	198	514	0.34
22.	Mali	479	1,240	0.83	48.	Spain	195	505	0.34
23.	Ethiopia	472	1,222	0.81	49.	Cameroon	184	475	0.32
24.	South Africa	471	1,221	0.81	50.	Papua New Guinea	178	462	0.31
25.	Chad	459	1,189	0.79					
26.	Colombia	440	1,139	0.76	Total		46,932	121,553	80.9
27.	Bolivia	424	1,099	0.73					

7b
LARGEST COUNTRIES IN POPULATION

Contents

Rank.
Conventional name of country.
Latest population in thousands, and year of estimate (e) or census.
Percentage of world population represented by the given population.

Arrangement

By population.

Coverage

50 largest countries in population.

Rounding

Populations are rounded to the nearest 1000, percentages to the nearest two significant digits.

Entries

50.

TABLE 7b. LARGEST COUNTRIES IN POPULATION

Rank	Country	Population (thousands)	Year	% of World Population	Rank	Country	Population (thousands)	Year	% of World Population
1.	China	835,843	1976e	20.8	27.	South Africa	26,129	1976e	0.65
2.	India	604,977	1976e	15.0	28.	Argentina	25,719	1976e	0.64
3.	USSR	255,524	1976e	6.3	29.	Zaire	25,629	1976e	0.64
4.	USA	215,118	1976e	5.3	30.	Canada	22,993	1976	0.57
5.	Indonesia	135,880	1976e	3.4	31.	Colombia	22,540	1976e	0.56
6.	Japan	113,086	1976e	2.8	32.	Yugoslavia	21,560	1976e	0.54
7.	Brazil	109,181	1976e	2.7	33.	Romania	21,559	1977	0.54
8.	Nigeria	77,056	1976e	1.9	34.	Afghanistan	19,803	1976e	0.49
9.	Bangladesh	75,980	1976e	1.9	35.	Morocco	17,828	1976e	0.44
10.	Pakistan	72,368	1976e	1.8	36.	Algeria	17,304	1976e	0.43
11.	Mexico	62,329	1976e	1.5	37.	East Germany	16,786	1976e	0.42
12.	West Germany	61,513	1976e	1.5	38.	Taiwan	16,290	1976e	0.40
13.	Italy	56,189	1976e	1.4	39.	North Korea	16,246	1976e	0.40
14.	UK	55,928	1976e	1.4	40.	Peru	16,090	1976e	0.40
15.	France	52,947	1976e	1.3	41.	Tanzania	15,607	1976e	0.39
16.	Vietnam	47,150	1976e	1.2	42.	Sudan	15,470	1976e	0.38
17.	Thailand	42,960	1976e	1.1	43.	Czechoslovakia	14,918	1976e	0.37
18.	Philippines	42,920	1976e	1.1	44.	Sri Lanka	14,270	1976e	0.35
19.	Turkey	41,120	1976e	1.0	45.	Australia	13,915	1976	0.35
20.	Egypt	38,228	1976	0.95	46.	Kenya	13,847	1976e	0.34
21.	South Korea	35,860	1976e	0.89	47.	Netherlands	13,770	1976e	0.34
22.	Spain	35,849	1976e	0.89	48.	Nepal	12,857	1976e	0.32
23.	Poland	34,362	1976e	0.85	49.	Venezuela	12,360	1976e	0.31
24.	Iran	33,592	1976	0.83	50.	Malaysia	12,300	1976e	0.31
25.	Burma	31,002	1976e	0.77	Total		3,621,430		90.0
26.	Ethiopia	28,678	1976e	0.71					

7c
MOST DENSELY
POPULATED COUNTRIES

Contents

Rank.

Conventional name of country.

Continent in which located.

Year of data, and density of population per square mile and per square kilometer.

Arrangement

By density of population.

Coverage

50 most densely populated countries.

Entries

50.

Figure 21. Panorama of the capital and only city in Macao, the most densely populated country in the world. (Credit: Portuguese National Tourist Office.)

TABLE 7c.   MOST DENSELY POPULATED COUNTRIES

Rank	Country	Continent	Year	Density of Population (per mi²)	(per km²)
1.	Macao	Asia	1976	44,516	17,187
2.	Monaco	Europe	1976	34,533	13,333
3.	Gibraltar	Europe	1976	12,950	5,000
4.	Melilla	Africa	1973	11,100	4,286
5.	Hong Kong	Asia	1976	10,908	4,212
6.	Singapore	Asia	1976	9,883	3,816
7.	Ceuta	Africa	1973	9,353	3,611
8.	Vatican City	Europe	1976	4,120	1,591
9.	Gaza Strip	Asia	1974	2,871	1,108
10.	Bermuda	North America	1976	2,785	1,075
11.	Malta	Europe	1975	2,623	1,013
12.	Johnston and Sand Islands	Oceania	1976	2,590	1,000
13.	Channel Islands	Europe	1976	1,713	662
14.	Barbados	North America	1976	1,508	582
15.	Bangladesh	Asia	1976	1,367	528
16.	Maldives	Asia	1976	1,173	453
17.	Taiwan	Asia	1976	1,167	450
18.	Mauritius	Africa	1976	1,134	438
19.	Midway Islands	Oceania	1976	1,036	400
20.	Bahrain	Asia	1976	1,013	391
21.	South Korea	Asia	1976	943	364
22.	Puerto Rico	North America	1976	933	360
23.	Grenada	North America	1976	896	346
24.	Netherlands	Europe	1976	866	335
25.	Nauru	Oceania	1975	863	333
26.	San Marino	Europe	1976	849	328
27.	Belgium	Europe	1976	839	324
28.	Japan	Asia	1976	776	300
29.	Lebanon	Asia	1976	775	299
30.	Martinique	North America	1976	769	297
31.	Virgin Islands (USA)	North America	1976	715	276
32.	Wake Island	Oceania	1976	647	250
33.	West Germany	Europe	1976	641	247
34.	Tuvalu	Oceania	1976	632	244
35.	Netherlands Antilles	North America	1976	629	243
36.	Saint Vincent	North America	1975	621	240
37.	UK	Europe	1976	594	229
38.	Seychelles	Africa	1977	578	223
39.	Sri Lanka	Asia	1976	563	217
40.	Trinidad and Tobago	North America	1976	552	213
41.	Tokelau Islands	Oceania	1974	518	200
42.	India	Asia	1976	511	197
42.	Reunion	Africa	1976	511	197
44.	El Salvador	North America	1976	508	196
45.	Jamaica	North America	1976	485	187
46.	Italy	Europe	1976	483	187
47.	Guadeloupe	North America	1976	478	184

Rank	Country	Continent	Year	Density of Population (per mi^2)	(per km^2)
48.	Saint Lucia	North America	1976	471	182
49.	Saint Kitts-Nevis	North America	1976	467	180
50.	Israel	Asia	1976	440	170

Figure 22. The princely palace in Monaco, the country that ranks second to Macao in density of population. (Credit: Monaco Government Tourist Office.)

7d
LEAST DENSELY
POPULATED COUNTRIES

Contents

Rank.

Conventional name of country.

Continent in which located.

Year of data, and density of population per square mile and per square kilometer.

Arrangement

By density of population, inversely.

Coverage

50 least densely populated countries.

Rounding

Densities of population are rounded to the nearest two significant digits if between 1 and 100 per square mile or per square kilometer, and to the nearest significant digit if less than 1 per square mile or per square kilometer.

Entries

50.

TABLE 7d. LEAST DENSELY POPULATED COUNTRIES

Rank	Country	Continent	Year	Density of Population	
				(per mi^2)	(per km^2)
1.	Greenland	North America	1976	0.06	0.02
2.	French Southern and Antarctic Lands	Africa	1974	0.07	0.03
3.	Svalbard	Europe	1974	0.1	0.05
4.	Falkland Islands	South America	1975	0.3	0.1
5.	Western Sahara	Africa	1976	1.2	0.5
6.	French Guiana	South America	1976	1.7	0.6
7.	Mongolia	Asia	1976	2.5	1.0
8.	Namibia	Africa	1976	2.9	1.1
9.	Botswana	Africa	1976	3.0	1.2
10.	Libya	Africa	1976	3.5	1.3
11.	Mauritania	Africa	1976	3.7	1.4
12.	Australia	Oceania	1976	4.7	1.8
13.	Iceland	Europe	1976	5.5	2.1
14.	Canada	North America	1976	6.0	2.3
15.	Surinam	South America	1976	6.9	2.7
16.	Chad	Africa	1976	9.0	3.5
17.	Saudi Arabia	Asia	1976	9.2	3.5
18.	Guyana	South America	1976	9.4	3.6
19.	Oman	Asia	1976	9.6	3.7
20.	Niger	Africa	1976	9.8	3.8
21.	Congo	Africa	1976	11	4.1
22.	Bolivia	South America	1976	11	4.2
23.	Central African Empire	Africa	1976	11	4.4
24.	Gabon	Africa	1976	12	4.5
25.	Mali	Africa	1976	13	4.9
26.	Somalia	Africa	1976	13	5.1
27.	South Yemen	Asia	1976	14	5.3
28.	Angola	Africa	1975	14	5.4
29.	Papua New Guinea	Oceania	1976	16	6.1
30.	Sudan	Africa	1976	16	6.2
31.	Belize	North America	1976	16	6.3
32.	New Hebrides	Oceania	1976	17	6.6
33.	Paraguay	South America	1976	17	6.7
34.	Solomon Islands	Oceania	1976	18	6.8
34.	Zambia	Africa	1976	18	6.8
36.	New Caledonia	Oceania	1976	18	7.0
37.	Algeria	Africa	1976	20	7.5
38.	United Arab Emirates	Asia	1976	21	8.1
39.	Djibouti	Africa	1974	21	8.2
40.	Argentina	South America	1976	24	9.3
41.	Zaire	Africa	1976	28	11
42.	Equatorial Guinea	Africa	1976	29	11
43.	USSR	Europe, Asia	1976	30	11
44.	New Zealand	Oceania	1976	30	12
45.	Mozambique	Africa	1976	31	12
46.	Norway	Europe	1976	32	12
47.	Peru	South America	1976	32	13

Rank	Country	Continent	Year	Density of Population	
				(per mi^2)	(per km^2)
48.	Brazil	South America	1976	33	13
49.	Madagascar	Africa	1976	34	13
50.	Venezuela	South America	1976	35	14

7e
COUNTRIES WITH HIGHEST BIRTH RATES

Contents

Rank.
Conventional name of country.
Continent in which located.
Period of data, and number of live births per thousand of population.

Arrangement

By birth rate.

Coverage

50 countries with the highest birth rates.

Entries

50.

TABLE 7e.   COUNTRIES WITH HIGHEST BIRTH RATES

Rank	Country	Continent	Period	Birth Rate	Rank	Country	Continent	Period	Birth Rate
1.	Niger	Africa	1970–75	52.2e	27.	Pakistan	Asia	1970–75	47.4e
2.	Zambia	Africa	1970–75	51.5e	28.	Angola	Africa	1970–75	47.2e
3.	Togo	Africa	1970–75	50.6e	28.	Somalia	Africa	1970–75	47.2e
4.	Malawi	Africa	1970–72	50.5	30.	Kuwait	Asia	1970–75	47.1e
5.	Liberia	Africa	1970–71	50.4	31.	Tanzania	Africa	1970–75	47.0e
6.	Mali	Africa	1970–75	50.1e	32.	Cambodia	Asia	1970–75	46.7e
7.	Rwanda	Africa	1970–75	50.0e	33.	Comoros	Africa	1970–75	46.6e
8.	Benin	Africa	1970–75	49.9e	33.	Guinea	Africa	1970–75	46.6e
9.	North Yemen	Asia	1970–75	49.6e	35.	Sao Tome and Principe	Africa	1968–72	46.4
9.	South Yemen	Asia	1970–75	49.6e					
11.	Bangladesh	Asia	1970–75	49.5e	36.	Morocco	Africa	1970–75	46.2e
11.	Saudi Arabia	Asia	1970–75	49.5e	37.	Madagascar	Africa	1970–75	46.0e
13.	Ethiopia	Africa	1970–75	49.4e	38.	Botswana	Africa	1970–75	45.6e
14.	Honduras	North America	1970–75	49.3e	38.	Ivory Coast	Africa	1970–75	45.6e
14.	Nigeria	Africa	1970–75	49.3e	40.	Syria	Asia	1970–75	45.4e
16.	Afghanistan	Asia	1970–75	49.2e	41.	Iran	Asia	1970–75	45.3e
17.	Swaziland	Africa	1970–75	49.0e	42.	Uganda	Africa	1970–75	45.2e
18.	Ghana	Africa	1970–75	48.8e	42.	Zaire	Africa	1970–75	45.2e
19.	Algeria	Africa	1970–75	48.7e	44.	Congo	Africa	1970–75	45.1e
19.	Kenya	Africa	1970–75	48.7e	45.	Libya	Africa	1970–75	45.0e
21.	Upper Volta	Africa	1970–75	48.5e	45.	Namibia	Africa	1970–75	45.0e
22.	Nicaragua	North America	1970–75	48.3e	45.	New Hebrides	Oceania	1966	45.0e
23.	Iraq	Asia	1970–75	48.1e	48.	Jordan	Asia	1971–75	44.8
24.	Rhodesia	Africa	1970–75	47.9e	48.	Mauritania	Africa	1970–75	44.8e
25.	Sudan	Africa	1970–75	47.8e	50.	Sierra Leone	Africa	1970–75	44.7e
26.	Senegal	Africa	1970–75	47.6e					

7f
COUNTRIES WITH LOWEST
BIRTH RATES

Contents

Rank.
Conventional name of country.
Continent in which located.
Period of data, and number of live births per thousand of population.

Arrangement

By birth rate, inversely.

Coverage

50 countries with the lowest birth rates.

Entries

50.

TABLE 7f. COUNTRIES WITH LOWEST BIRTH RATES

Rank	Country	Continent	Period	Birth Rate	Rank	Country	Continent	Period	Birth Rate
1.	Monaco	Europe	1970–74	8.6	24.	Greece	Europe	1971–75	15.8
2.	West Germany	Europe	1971–75	10.8	25.	Canada	North America	1971–75	15.9
3.	Christmas Island (Australia)	Oceania	1969–73	11.0	25.	Italy	Europe	1971–75	15.9
					27.	France	Europe	1971–75	16.0
4.	East Germany	Europe	1971–75	11.5	28.	Hungary	Europe	1971–75	16.1
5.	Luxembourg	Europe	1971–75	11.7	29.	Bulgaria	Europe	1971–75	16.2
6.	Macao	Asia	1967–70, 72	11.9	30.	Ceuta	Africa	1966–70	16.4
					31.	San Marino	Europe	1969–73	16.8
7.	Pitcairn Island	Oceania	1970–72, 74	12.6	32.	Bermuda	North America	1971–75	17.5
					33.	Malta	Europe	1971–75	17.6
8.	Channel Islands	Europe	1971–75	12.8	34.	USSR	Europe and Asia	1971–75	17.9
9.	Finland	Europe	1971–75	13.1	35.	Poland	Europe	1971–75	18.0
10.	Belgium	Europe	1971–75	13.2	36.	Yugoslavia	Europe	1971–75	18.2
11.	Austria	Europe	1971–75	13.4	37.	Czechoslovakia	Europe	1971–75	18.4
12.	Sweden	Europe	1971–75	13.5	38.	Japan	Asia	1971–75	18.7
13.	Isle of Man	Europe	1971–75	13.7	39.	Spain	Europe	1971–75	19.2
13.	Switzerland	Europe	1971–75	13.7	40.	Australia	Oceania	1971–75	19.3
15.	Canal Zone	North America	1971–75	14.1	40.	Romania	Europe	1971–75	19.3
15.	Norfolk Island	Oceania	1969–72, 74	14.1	42.	Hong Kong	Asia	1971–75	19.4
					43.	Gibraltar	Europe	1971–75	19.7
17.	UK	Europe	1971–75	14.2	44.	Falkland Islands	South America	1970–74	19.8
18.	Denmark	Europe	1971–75	14.6	45.	Portugal	Europe	1971–75	20.2
19.	Netherlands	Europe	1971–75	14.9	46.	Faeroe Islands	Europe	1970–74	20.3
20.	Liechtenstein	Europe	1971–75	15.2	47.	Cocos Islands	Oceania	1969–73	20.4
21.	Melilla	Africa	1966–70	15.3	48.	New Zealand	Oceania	1971–75	20.5
22.	Norway	Europe	1970–74	15.5	48.	Uruguay	South America	1970–74	20.5
22.	USA	North America	1971–75	15.5	50.	Barbados	North America	1971–75	20.6

7g
COUNTRIES WITH HIGHEST
DEATH RATES

Contents

Rank.
Conventional name of country.
Continent in which located.
Period of data, and number of deaths per thousand of population.

Arrangement

By death rate.

Coverage

50 countries with the highest death rates.

Entries

50.

TABLE 7g. COUNTRIES WITH HIGHEST DEATH RATES

Rank	Country	Continent	Period	Death Rate	Rank	Country	Continent	Period	Death Rate
1.	Bangladesh	Asia	1970–75	28.1e	25.	Cameroon	Africa	1970–75	22.0e
2.	Malawi	Africa	1970–72	26.5	26.	Ghana	Africa	1970–75	21.9e
3.	Mali	Africa	1970–75	25.9e	27.	Swaziland	Africa	1970–75	21.8e
4.	Ethiopia	Africa	1970–75	25.8e	28.	Comoros	Africa	1970–75	21.7e
4.	Upper Volta	Africa	1970–75	25.8e	28.	Somalia	Africa	1970–75	21.7e
6.	Niger	Africa	1970–75	25.5e	30.	Madagascar	Africa	1970–75	21.1e
7.	Guinea-Bissau	Africa	1970–75	25.1e	31.	Congo	Africa	1970–75	20.8e
8.	Mauritania	Africa	1970–75	24.9e	32.	Sierra Leone	Africa	1970–75	20.7e
9.	Angola	Africa	1970–75	24.5e	33.	Ivory Coast	Africa	1970–75	20.6e
10.	Gambia	Africa	1970–75	24.1e	33.	North Yemen	Asia	1970–75	20.6e
11.	Chad	Africa	1970–75	24.0e	33.	South Yemen	Asia	1970–75	20.6e
12.	Senegal	Africa	1970–75	23.9e	36.	Bhutan	Asia	1970–75	20.5e
13.	Afghanistan	Asia	1970–75	23.8e	36.	Vietnam	Asia	1970–75	20.5e
14.	Pitcairn Island	Oceania	1965–66, 69, 71, 73	23.6	36.	Zaire	Africa	1970–75	20.5e
					39.	Burundi	Africa	1970–71	20.4e
					40.	Nepal	Asia	1970–75	20.3e
14.	Rwanda	Africa	1970–75	23.6e	40.	Zambia	Africa	1970–75	20.3e
16.	Togo	Africa	1970–75	23.3e	42.	Saudi Arabia	Asia	1970–75	20.2e
17.	Benin	Africa	1970–75	23.0e	43.	Mozambique	Africa	1970–75	20.1e
17.	Botswana	Africa	1970–75	23.0e	43.	Tanzania	Africa	1970–75	20.1e
19.	Guinea	Africa	1970–75	22.9e	45.	New Hebrides	Oceania	1966	20.0e
19.	Maldives	Asia	1963–65	22.9	46.	Equatorial Guinea	Africa	1970–75	19.7e
21.	Laos	Asia	1970–75	22.8e	46.	Lesotho	Africa	1970–75	19.7e
22.	Nigeria	Africa	1970–75	22.7e	48.	Bolivia	South America	1970–75	19.1e
23.	Central African Empire	Africa	1970–75	22.5e	49.	Cambodia	Asia	1970–75	19.0e
24.	Gabon	Africa	1970–75	22.2e	50.	Liberia	Africa	1970–71	18.7

7h
COUNTRIES WITH LOWEST DEATH RATES

Contents

Rank.

Conventional name of country.

Continent in which located.

Period of data, and number of deaths per thousand of population.

Arrangement

By death rate, inversely.

Coverage

50 countries with the lowest death rates.

Entries

50.

TABLE 7h. COUNTRIES WITH LOWEST DEATH RATES

Rank	Country	Continent	Period	Death Rate
1.	Christmas Island (Australia)	Oceania	1969–73	2.1
2.	Canal Zone	North America	1971–75	2.2
3.	Tonga	Oceania	1967–71	2.9
4.	Fiji	Oceania	1970–75	4.3e
5.	Guam	Oceania	1971–75	4.4
6.	Andorra	Europe	1971–75	4.6
6.	Samoa	Oceania	1971–75	4.6
8.	Taiwan	Asia	1971–75	4.7
9.	Western Sahara	Africa	1968–72	4.9
10.	American Samoa	Oceania	1971–75	5.0
11.	Brunei	Asia	1970–74	5.1
11.	Hong Kong	Asia	1971–75	5.1
11.	Netherlands Antilles	North America	1969–73	5.1
14.	Pacific Islands (USA)	Oceania	1971–75	5.2
15.	Kuwait	Asia	1970–75	5.3e
15.	Singapore	Asia	1971–75	5.3
17.	Costa Rica	North America	1971–75	5.4
18.	Cuba	North America	1971–75	5.6
19.	Belize	North America	1970–74	5.8
20.	Bahamas	North America	1971–75	6.0
20.	Macao	Asia	1967–70, 72	6.0
22.	Cayman Islands	North America	1970–74	6.1
23.	Ceuta	Africa	1966–70	6.2
23.	Melilla	Africa	1966–70	6.2
25.	Antigua	North America	1970–74	6.3
25.	Greenland	North America	1969–73	6.3
27.	Cocos Islands	Oceania	1969–73	6.4
27.	Puerto Rico	North America	1971–75	6.4
29.	Japan	Asia	1971–75	6.5
29.	Tokelau Islands	Oceania	1970–72	6.5
29.	Virgin Islands (UK)	North America	1971–75	6.5
29.	Virgin Islands (USA)	North America	1971–75	6.5
33.	Malaysia	Asia	1970–74	6.6
33.	Norfolk Island	Oceania	1969–72, 74	6.6
35.	Cook Islands	Oceania	1970–73	6.7
35.	Grenada	North America	1974–75	6.7
35.	Trinidad and Tobago	North America	1971–75	6.7
38.	Cyprus	Asia	1970–75	6.8e
39.	Iceland	Europe	1971–75	6.9
39.	Martinique	North America	1970–74	6.9
41.	Venezuela	South America	1970–75	7.0e
42.	Bermuda	North America	1971–75	7.1
42.	Guyana	South America	1976	7.1
42.	Israel	Asia	1971–75	7.1
42.	Niue Island	Oceania	1972–75	7.1
42.	Panama	North America	1970–75	7.1e
42.	Wallis and Futuna Islands	Oceania	1966–70	7.1
48.	Jamaica	North America	1971–75	7.2
49.	Guadeloupe	North America	1970–74	7.3
49.	Liechtenstein	Europe	1971–75	7.3

7i
COUNTRIES WITH HIGHEST
LIFE EXPECTANCIES

Contents

Rank.

Conventional name of country.

Continent in which located.

Period of data, and expectation of life at birth, in years, for men, for women, and for both (average).

Arrangement

By life expectancy.

Coverage

50 countries with the highest life expectancies.

Entries

50.

TABLE 7i.   COUNTRIES WITH HIGHEST LIFE EXPECTANCIES

Rank	Country	Continent	Period	Life Expectancy (years)		
				Men	Women	Avg
1.	Norway	Europe	1974–75	71.7	78.0	74.9
1.	Sweden	Europe	1971–75	72.1	77.6	74.9
3.	Iceland	Europe	1971–75	71.6	77.5	74.5
3.	Netherlands	Europe	1975	71.4	77.6	74.5
5.	Japan	Asia	1975	71.8	76.9	74.4
6.	Denmark	Europe	1974–75	71.1	76.8	73.9
7.	Switzerland	Europe	1968–73	70.3	76.2	73.3
8.	France	Europe	1974	69.0	76.9	72.9
9.	Canada	North America	1970–72	69.3	76.4	72.8
10.	USA	North America	1975	68.7	76.5	72.6
11.	Puerto Rico	North America	1971–73	68.9	76.0	72.5
12.	Spain	Europe	1970	69.7	75.0	72.3
13.	Israel	Asia	1975	70.3	73.9	72.1
14.	Italy	Europe	1970–72	69.0	74.9	71.9
14.	UK	Europe	1971–73	68.8	75.1	71.9
16.	Finland	Europe	1975	67.4	75.9	71.6
16.	New Zealand	Oceania	1970–72	68.5	74.6	71.6
18.	Cyprus	Asia	1973	70.0	72.9	71.4
18.	Poland	Europe	1976	68.0	74.9	71.4
20.	Austria	Europe	1975	67.7	74.9	71.3
20.	East Germany	Europe	1975	68.5	74.0	71.3
20.	West Germany	Europe	1973–75	68.0	74.5	71.3
23.	Bulgaria	Europe	1969–71	68.6	73.9	71.2
23.	Hong Kong	Asia	1971	67.4	75.0	71.2
25.	Belgium	Europe	1968–72	67.8	74.2	71.0
25.	Taiwan	Asia	1975	68.4	73.6	71.0
27.	Australia	Oceania	1965–67	67.6	74.1	70.9
28.	Ireland	Europe	1965–67	68.6	72.8	70.7
29.	Malta	Europe	1975	68.5	72.7	70.6
30.	Luxembourg	Europe	1971–73	67.0	73.9	70.4
31.	Cuba	North America	1970	68.5	71.8	70.1
31.	Fiji	Oceania	1970–75	68.5e	71.7e	70.1e
33.	Czechoslovakia	Europe	1973	66.5	73.5	70.0
34.	Romania	Europe	1974–76	67.4	72.0	69.7
35.	Hungary	Europe	1974	66.5	72.4	69.5
36.	Greece	Europe	1960–62	67.5	70.7	69.1
37.	Bermuda	North America	1965–66	65.6	72.3	69.0
37.	USSR	Europe and Asia	1971–72	64.0	74.0	69.0
39.	Kuwait	Asia	1970	66.1	71.8	68.9
40.	Portugal	Europe	1974	65.3	72.0	68.7
41.	Uruguay	South America	1963–64	65.5	71.6	68.5
42.	Jamaica	North America	1969–70	66.7	70.2	68.4
43.	Argentina	South America	1973	65.2	71.4	68.3
44.	Yugoslavia	Europe	1970–72	65.4	70.2	67.8
45.	Albania	Europe	1974	66.5	69.0	67.7
46.	Singapore	Asia	1970	65.1	70.0	67.5
47.	American Samoa	Oceania	1969–71	65.0	69.1	67.0
48.	Trinidad and Tobago	North America	1970	64.1	68.1	66.1
49.	Panama	North America	1970	64.3	67.5	65.9
50.	Sri Lanka	Asia	1967	64.8	66.9	65.8

7j
COUNTRIES WITH LOWEST
LIFE EXPECTANCIES

Contents

Rank.

Conventional name of country.

Continent in which located.

Period of data, and expectation of life at birth, in years, for men, for women, and for both (average).

Arrangement

By life expectancy, inversely.

Coverage

50 countries with the lowest life expectancies.

Entries

50.

TABLE 7j.   COUNTRIES WITH LOWEST LIFE EXPECTANCIES

Rank	Country	Continent	Period	Life Expectancy (years)		
				Men	Women	Avg
1.	Upper Volta	Africa	1960–61	32.1	31.1	31.6
2.	Chad	Africa	1963–64	29.0	35.0	32.0
3.	Central African Empire	Africa	1959–60	33.0	36.0	34.5
4.	Gabon	Africa	1960–61	25.0	45.0	35.0
4.	Togo	Africa	1961	31.6	38.5	35.0
6.	Nigeria	Africa	1965–66	37.2	36.7	36.9
7.	Madagascar	Africa	1966	37.5	38.3	37.9
8.	Ethiopia	Africa	1970–75	36.5e	39.6e	38.0e
8.	Mali	Africa	1970–75	36.5e	39.6e	38.0e
10.	Angola	Africa	1970–75	37.0e	40.1e	38.5e
10.	Guinea-Bissau	Africa	1970–75	37.0e	40.1e	38.5e
10.	Mauritania	Africa	1970–75	37.0e	40.1e	38.5e
10.	Niger	Africa	1970–75	37.0e	40.1e	38.5e
14.	Gambia	Africa	1970–75	38.5e	41.6e	40.0e
14.	Senegal	Africa	1970–75	38.5e	41.6e	40.0e
16.	Afghanistan	Asia	1970–75	39.9e	40.7e	40.3e
17.	Laos	Asia	1970–75	39.1e	41.8e	40.4e
17.	Tanzania	Africa	1967			40.4e
19.	Guinea	Africa	1970–75	39.4e	42.0e	40.7e
20.	Benin	Africa	1970–75	39.4e	42.6e	41.0e
20.	Cameroon	Africa	1970–75	39.4e	42.6e	41.0e
20.	Rwanda	Africa	1970–75	39.4e	42.6e	41.0e
20.	Somalia	Africa	1970–75	39.4e	42.6e	41.0e
24.	Burundi	Africa	1970–71	40.0e	43.0e	41.5e
25.	Indonesia	Asia	1971	40.2	43.0	41.6
26.	Comoros	Africa	1970–75	40.9e	44.1e	42.5e
26.	Malawi	Africa	1970–72	40.9	44.2	42.5
28.	Botswana	Africa	1970–75	41.9e	45.1e	43.5e
28.	Congo	Africa	1970–75	41.9e	45.1e	43.5e
28.	Equatorial Guinea	Africa	1970–75	41.9e	45.1e	43.5e
28.	Ghana	Africa	1970–75	41.9e	45.1e	43.5e
28.	Ivory Coast	Africa	1970–75	41.9e	45.1e	43.5e
28.	Mozambique	Africa	1970–75	41.9e	45.1e	43.5e
28.	Sierra Leone	Africa	1970–75	41.9e	45.1e	43.5e
28.	Swaziland	Africa	1970–75	41.9e	45.1e	43.5e
28.	Zaire	Africa	1970–75	41.9e	45.1e	43.5e
37.	Bhutan	Asia	1970–75	42.2e	45.0e	43.6e
37.	Nepal	Asia	1970–75	42.2e	45.0e	43.6e
39.	Zambia	Africa	1970–75	42.9e	46.1e	44.5e
40.	Vietnam	Asia	1970–75	43.2e	46.0e	44.6e
41.	Bangladesh	Asia	1962–65	44.5	45.0	44.7
42.	North Yemen	Asia	1970–75	43.7e	45.9e	44.8e
42.	South Yemen	Asia	1970–75	43.7e	45.9e	44.8e
44.	Liberia	Africa	1971	45.8	44.0	44.9
45.	Saudi Arabia	Asia	1970–75	44.2e	46.5e	45.3e
46.	Cambodia	Asia	1970–75	44.0e	46.9e	45.4e
47.	Lesotho	Africa	1970–75	44.4e	47.6e	46.0e
48.	India	Asia	1961–70	47.1	45.6	46.3
49.	Bolivia	South America	1970–75	45.7e	47.9e	46.8e
50.	Belize	North America	1944–48	45.0	49.0	47.0

7k
COUNTRIES WITH GREATEST ENERGY PRODUCTION

Contents

Rank.

Conventional name of country.

Continent in which located.

Production of primary commercial energy, expressed in thousands of metric tons of coal equivalent. Primary energy includes coal and lignite, crude petroleum and natural gas liquids, natural gas, and hydro and nuclear electricity. The production estimates are those of the United Nations (see reference 217) but are based generally on official statistics. The United Nations is also responsible for the conversion of each type of energy into coal equivalence.

Percentage of world energy production represented by the given production figure.

Arrangement

By energy production.

Coverage

50 countries with the greatest energy production in 1975, the latest year for which comparable data are available.

Entries

50.

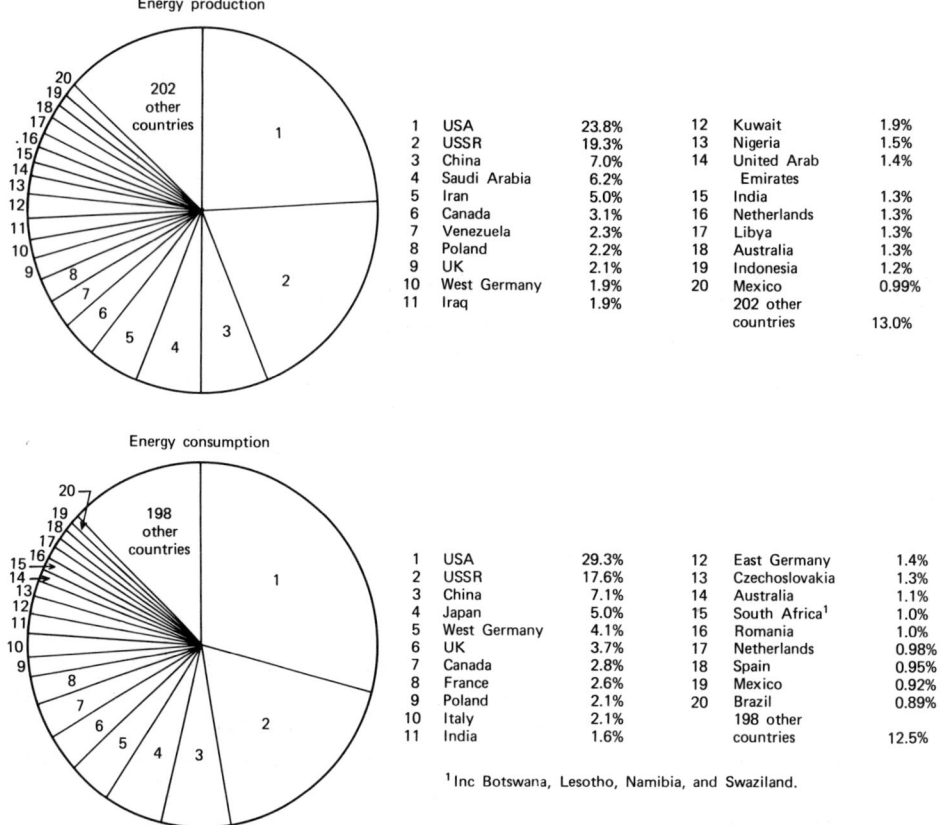

Energy production

1	USA	23.8%	12	Kuwait	1.9%
2	USSR	19.3%	13	Nigeria	1.5%
3	China	7.0%	14	United Arab	1.4%
4	Saudi Arabia	6.2%		Emirates	
5	Iran	5.0%	15	India	1.3%
6	Canada	3.1%	16	Netherlands	1.3%
7	Venezuela	2.3%	17	Libya	1.3%
8	Poland	2.2%	18	Australia	1.3%
9	UK	2.1%	19	Indonesia	1.2%
10	West Germany	1.9%	20	Mexico	0.99%
11	Iraq	1.9%		202 other countries	13.0%

Energy consumption

1	USA	29.3%	12	East Germany	1.4%
2	USSR	17.6%	13	Czechoslovakia	1.3%
3	China	7.1%	14	Australia	1.1%
4	Japan	5.0%	15	South Africa[1]	1.0%
5	West Germany	4.1%	16	Romania	1.0%
6	UK	3.7%	17	Netherlands	0.98%
7	Canada	2.8%	18	Spain	0.95%
8	France	2.6%	19	Mexico	0.92%
9	Poland	2.1%	20	Brazil	0.89%
10	Italy	2.1%		198 other countries	12.5%
11	India	1.6%			

[1] Inc Botswana, Lesotho, Namibia, and Swaziland.

Figure 23. Energy production and consumption of countries as percentage of world energy production and consumption.

TABLE 7k. COUNTRIES WITH GREATEST ENERGY PRODUCTION IN 1975

Rank	Country	Continent	Energy Production[1]	% of World Total	Rank	Country	Continent	Energy Production[1]	% of World Total
1.	USA	North America	2,036,671	23.8	27.	North Korea	Asia	42,430	0.50
2.	USSR	Europe, Asia	1,650,472	19.3	28.	Argentina	South America	41,593	0.49
3.	China	Asia	596,812	7.0	29.	Japan	Asia	37,437	0.44
4.	Saudi Arabia	Asia	530,102	6.2	30.	Qatar	Asia	34,353	0.40
5.	Iran	Asia	426,636	5.0	31.	Yugoslavia	Europe	28,212	0.33
6.	Canada	North America	268,424	3.1	32.	Italy	Europe	27,033	0.32
7.	Venezuela	South America	201,113	2.3	33.	Brazil	South America	25,405	0.30
8.	Poland	Europe	192,417	2.2	34.	Oman	Asia	25,014	0.29
9.	UK	Europe	183,988	2.1	35.	Hungary	Europe	23,710	0.28
10.	West Germany	Europe	165,867	1.9	36.	Norway	Europe	23,557	0.28
11.	Iraq	Asia	165,620	1.9	37.	Brunei	Asia	21,298	0.25
12.	Kuwait	Asia	164,280	1.9	38.	Spain	Europe	20,369	0.24
13.	Nigeria	Africa	131,067	1.5	39.	Colombia	South America	19,111	0.22
14.	United Arab Emirates	Asia	120,522	1.4	40.	Trinidad and Tobago	North America	18,251	0.21
15.	India	Asia	114,624	1.3	41.	South Korea	Asia	17,800	0.21
16.	Netherlands	Europe	111,790	1.3	42.	Gabon	Africa	16,785	0.20
17.	Libya	Africa	111,630	1.3	43.	Bulgaria	Europe	15,015	0.18
18.	Australia	Oceania	111,048	1.3	44.	Syria	Asia	14,359	0.17
19.	Indonesia	Asia	103,076	1.2	45.	Egypt	Africa	13,299	0.16
20.	Mexico	North America	84,436	0.99	46.	Turkey	Asia, Europe	12,434	0.15
21.	Algeria	Africa	83,349	0.97	47.	Ecuador	South America	12,107	0.14
22.	Czechoslovakia	Europe	81,929	0.96	48.	Angola	Africa	11,917	0.14
23.	East Germany	Europe	81,085	0.95	49.	Austria	Europe	10,908	0.13
24.	Romania	Europe	80,659	0.95	50.	Sweden	Europe	8,576	0.10
25.	South Africa[2]	Africa	69,786	0.82	Total			8,435,789	98.5
26.	France	Europe	47,413	0.55	World			8,561,395	100

[1] In thousands of metric tons of coal equivalent.
[2] Inc Botswana, Lesotho, Namibia, and Swaziland.

71
COUNTRIES WITH GREATEST ENERGY CONSUMPTION

Contents

Rank.

Conventional name of country.

Continent in which located.

Consumption of energy, expressed in thousands of metric tons of coal equivalent. The consumption estimates are those of the United Nations (see reference 217) and include coal and lignite, petroleum products, natural gas, and hydro and nuclear electricity. The United Nations is also responsible for the conversion of each type of energy into coal equivalence.

Percentage of world energy consumption represented by the given consumption figure.

Arrangement

By energy consumption.

Coverage

50 countries with the greatest energy consumption in 1975, the latest year for which comparable data are available.

Entries

50.

TABLE 7 1.   COUNTRIES WITH GREATEST ENERGY CONSUMPTION IN 1975

Rank	Country	Continent	Energy Consumption[1]	% of World Total
1.	USA	North America	2,349,549	29.3
2.	USSR	Europe, Asia	1,410,781	17.6
3.	China	Asia	570,467	7.1
4.	Japan	Asia	401,884	5.0
5.	West Germany	Europe	330,490	4.1
6.	UK	Europe	295,329	3.7
7.	Canada	North America	225,568	2.8
8.	France	Europe	208,877	2.6
9.	Poland	Europe	170,338	2.1
10.	Italy	Europe	168,088	2.1
11.	India	Asia	132,054	1.6
12.	East Germany	Europe	115,168	1.4
13.	Czechoslovakia	Europe	105,856	1.3
14.	Australia	Oceania	87,559	1.1
15.	South Africa[2]	Africa	84,231	1.0
16.	Romania	Europe	80,797	1.0
17.	Netherlands	Europe	78,962	0.98
18.	Spain	Europe	76,171	0.95
19.	Mexico	North America	73,431	0.92
20.	Brazil	South America	71,793	0.89
21.	Belgium	Europe	54,698	0.68
22.	Sweden	Europe	50,625	0.63
23.	Iran	Asia	44,662	0.56
24.	North Korea	Asia	44,518	0.55
25.	Argentina	South America	44,517	0.55
26.	Bulgaria	Europe	41,700	0.52
27.	Yugoslavia	Europe	41,208	0.51
28.	Hungary	Europe	38,200	0.48
29.	South Korea	Asia	35,980	0.45
30.	Venezuela	South America	31,653	0.39
31.	Austria	Europe	27,836	0.35
32.	Denmark	Europe	26,652	0.33
33.	Turkey	Asia, Europe	24,701	0.31
34.	Indonesia	Asia	24,224	0.30
35.	Switzerland	Europe	23,407	0.29
36.	Finland	Europe	22,433	0.28
37.	Taiwan	Asia	20,770[3]	0.26
38.	Greece	Europe	18,909	0.24
39.	Norway	Europe	18,458	0.23
40.	Colombia	South America	15,808	0.20
41.	Egypt	Africa	15,092	0.19
42.	Philippines	Asia	13,849	0.17
43.	Pakistan	Asia	12,825	0.16
44.	Algeria	Africa	12,657	0.16
45.	Saudi Arabia	Asia	12,532	0.16
46.	Thailand	Asia	11,897	0.15
47.	Cuba	North America	10,821	0.13
48.	Peru	South America	10,654	0.13
49.	Puerto Rico	North America	9,889	0.12
50.	Ireland	Europe	9,684	0.12
Total			7,808,252	97.3
World			8,022,971	100

[1] In thousands of metric tons of coal equivalent.
[2] Inc Botswana, Lesotho, Namibia, and Swaziland.
[3] Estimated on basis of 1966–70 growth.

7m
COUNTRIES WITH GREATEST
ENERGY CONSUMPTION PER CAPITA

Contents

Rank.

Conventional name of country.

Continent in which located.

Energy consumption per capita, expressed in kilograms of coal equivalent; i.e., total consumption of energy divided by total population. The total consumption estimates are those of the United Nations (see reference 217) and include coal and lignite, petroleum products, natural gas, and hydro and nuclear electricity, but the population figures utilized are either official estimates of the country cited or, if such are unavailable, those determined by the author to be most accurate for the middle of the year 1975.

Arrangement

By consumption of energy per capita.

Coverage

50 countries with the greatest energy consumption per capita in 1975, the latest year for which comparable data are available.

Entries

50.

TABLE 7m. COUNTRIES WITH GREATEST ENERGY CONSUMPTION
PER CAPITA IN 1975

Rank	Country	Continent	Energy Consumption Per Capita[1]	Rank	Country	Continent	Energy Consumption Per Capita[1]
1.	Virgin Islands (USA)	North America	36,374	25.	West Germany	Europe	5,345
				26.	UK	Europe	5,277
2.	Wake Island	Oceania	24,000	27.	Denmark	Europe	5,273
3.	Christmas Island (Australia)	Oceania	20,667	28.	United Arab Emirates	Asia	5,068
				29.	Poland	Europe	5,016
4.	Qatar	Asia	19,118	30.	Bulgaria	Europe	4,777
5.	Luxem- bourg	Europe	15,504	31.	Finland	Europe	4,766
				32.	Iceland	Europe	4,720
6.	Canal Zone	North America	15,195	33.	Norway	Europe	4,606
				34.	Faeroe Islands	Europe	4,317
7.	Netherlands Antilles	North America	12,231	35.	France	Europe	3,975
8.	Bahrain	Asia	11,668	36.	Romania	Europe	3,815
9.	USA	North America	11,003	37.	Austria	Europe	3,700
				38.	Switzerland	Europe	3,656
10.	Canada	North America	9,940	39.	Hungary	Europe	3,624
				40.	Japan	Asia	3,590
11.	New Caledonia	Oceania	9,409	41.	Saint-Pierre and Miquelon	North America	3,500
12.	Kuwait	Asia	8,727				
13.	Brunei	Asia	8,423	42.	Puerto Rico	North America	3,185
14.	Nauru	Oceania	7,571				
15.	Czechoslo- vakia	Europe	7,151	43.	New Zea- land	Oceania	3,123
16.	East Germany	Europe	6,835	43.	Trinidad and Tobago	North America	3,123
17.	Guam	Oceania	6,745				
18.	Australia	Oceania	6,415	45.	Ireland	Europe	3,097
19.	Bahamas	North America	6,279	46.	Bermuda	North America	3,089
20.	Sweden	Europe	6,167	47.	Italy	Europe	3,012
21.	Greenland	North America	5,900	48.	South Africa[2]	Africa	2,934
22.	Netherlands	Europe	5,783	49.	Cayman Islands	North America	2,818
23.	Belgium	Europe	5,584				
24.	USSR	Europe, Asia	5,570	50.	North Korea	Asia	2,808

[1] In kilograms of coal equivalent.
[2] Inc Botswana, Lesotho, Namibia, and Swaziland.

7n
COUNTRIES WITH GREATEST GROSS NATIONAL PRODUCT

Contents

Rank.

Conventional name of country.

Continent in which located.

Gross national product (GNP), expressed in millions of US dollars. The GNP estimates are those of the World Bank (see reference 218) but are based generally on official national account statistics. The Bank has converted these statistics into US dollars, using a three-year base period (1974–76) "to reflect more fully the most current market rates of exchange and prices."

Percentage of world total of GNP's represented by the given GNP figure.

Arrangement

By GNP.

Coverage

50 countries with the greatest GNP in 1976, the latest year for which comparable data are available.

Entries

50.

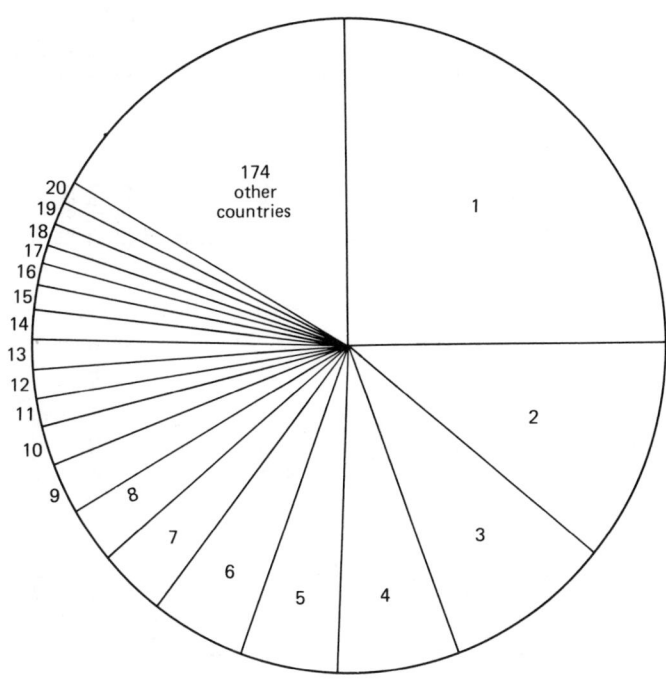

1	USA	25.2%	12	Poland	1.5%
2	USSR	10.5%	13	India	1.4%
3	Japan	8.2%	14	Netherlands	1.3%
4	West Germany	6.8%	15	Australia	1.2%
5	France	5.2%	16	Sweden	1.1%
6	China	5.1%	17	East Germany	1.1%
7	UK	3.3%	18	Mexico	1.0%
8	Canada	2.6%	19	Belgium	0.99%
9	Italy	2.5%	20	Iran	0.98%
10	Brazil	1.9%		174 other countries	16.6 %
11	Spain	1.5%			

Figure 24. Gross national product of countries as percentage of world total of GNPs.

TABLE 7n. COUNTRIES WITH GREATEST GROSS NATIONAL PRODUCT IN 1976

Rank	Country	Continent	Gross National Product[1]	% of World Total	Rank	Country	Continent	Gross National Product[1]	% of World Total
1.	USA	North America	1,698,060	25.2	27.	Denmark	Europe	37,770	0.56
2.	USSR	Europe, Asia	708,170	10.5	28.	Yugoslavia	Europe	36,170	0.54
3.	Japan	Asia	553,140	8.2	29.	South Africa	Africa	34,850	0.52
4.	West Germany	Europe	457,540	6.8	30.	Indonesia	Asia	32,440	0.48
5.	France	Europe	346,730	5.2	31.	Venezuela	South America	31,750	0.47
6.	China	Asia	343,090	5.1	32.	Romania	Europe	31,070	0.46
7.	UK	Europe	225,150	3.3	33.	Norway	Europe	29,920	0.44
8.	Canada	North America	174,120	2.6	34.	Nigeria	Africa	29,320	0.44
9.	Italy	Europe	171,250	2.5	35.	Finland	Europe	26,570	0.39
10.	Brazil	South America	125,570	1.9	36.	Hungary	Europe	24,140	0.36
11.	Spain	Europe	104,090	1.5	37.	South Korea	Asia	24,050	0.36
12.	Poland	Europe	98,130	1.5	38.	Greece	Europe	23,600	0.35
13.	India	Asia	95,880	1.4	39.	Bulgaria	Europe	20,270	0.30
14.	Netherlands	Europe	85,320	1.3	40.	Philippines	Asia	17,810	0.26
15.	Australia	Oceania	83,380	1.2	41.	Taiwan	Asia	17,500	0.26
16.	Sweden	Europe	71,290	1.1	42.	Kuwait	Asia	16,480	0.24
17.	East Germany	Europe	70,880	1.1	42.	Portugal	Europe	16,480	0.24
18.	Mexico	North America	67,640	1.0	44.	Thailand	Asia	16,230	0.24
19.	Belgium	Europe	66,660	0.99	45.	Algeria	Africa	16,060	0.24
20.	Iran	Asia	66,250	0.98	46.	Libya	Africa	16,000	0.24
21.	Czechoslovakia	Europe	57,250	0.85	47.	Iraq	Asia	15,940	0.24
22.	Switzerland	Europe	56,900	0.85	48.	Colombia	South America	15,400	0.23
23.	Turkey	Asia, Europe	40,960	0.61	49.	Israel	Asia	13,980	0.21
24.	Austria	Europe	40,080	0.60	50.	New Zealand	Oceania	13,120	0.19
25.	Argentina	South America	39,920	0.59	Total			6,442,880	95.7
26.	Saudi Arabia	Asia	38,510	0.57	World[2]			6,730,390	100

[1] In millions of US dollars at market prices.
[2] Exc 28 small countries with a total population of 1,113,000 (0.003% of the world's total), for which GNP estimates are unavailable.

7o
COUNTRIES WITH GREATEST GROSS NATIONAL PRODUCT PER CAPITA

Contents

Rank.

Conventional name of country.

Continent in which located.

Year of data, and gross national product (GNP) per capita, expressed in US dollars; i.e., GNP divided by total population. The GNP estimates are those of the World Bank (see reference 218), but the population figures utilized are either official estimates of the country cited or, if such are unavailable, those determined by the author to be most accurate for the middle of the year in question.

Arrangement

By GNP per capita.

Coverage

50 countries with the greatest GNP per capita.

Entries

50.

TABLE 7o. COUNTRIES WITH GREATEST GROSS NATIONAL PRODUCT PER CAPITA

Rank	Country	Continent	Year	Gross National Product Per Capita[1]	Rank	Country	Continent	Year	Gross National Product Per Capita[1]
1.	Kuwait	Asia	1976	15,984	27.	Japan	Asia	1976	4,891
2.	United Arab Emirates	Asia	1976	14,428	28.	Faeroe Islands	Europe	1975	4,878
3.	Qatar	Asia	1976	14,059	29.	Nauru	Oceania	1970	4,667
4.	Switzerland	Europe	1976	8,966	30.	New Caledonia	Oceania	1975	4,545
5.	Sweden	Europe	1976	8,671					
6.	USA	North America	1976	7,894	31.	Greenland	North America	1975	4,400
7.	Canada	North America	1976	7,573	32.	East Germany	Europe	1976	4,223
8.	Denmark	Europe	1976	7,445	33.	New Zealand	Oceania	1976	4,193
9.	West Germany	Europe	1976	7,438	34.	UK	Europe	1976	4,026
10.	Norway	Europe	1976	7,432	35.	Israel	Asia	1976	3,960
11.	Belgium	Europe	1976	6,741	36.	Czechoslovakia	Europe	1976	3,838
12.	France	Europe	1976	6,549					
13.	Luxembourg	Europe	1976	6,508	37.	Gibraltar	Europe	1975	3,333
14.	Guam	Oceania	1975	6,489	38.	Bahamas	North America	1976	3,318
15.	Bermuda	North America	1975	6,429	39.	Italy	Europe	1976	3,048
16.	Libya	Africa	1976	6,324	40.	French Polynesia	Oceania	1975	2,969
17.	Iceland	Europe	1976	6,273					
18.	Netherlands	Europe	1976	6,196	41.	Channel Islands	Europe	1975	2,946
19.	Australia	Oceania	1976	5,992					
20.	Brunei	Asia	1975	5,655	42.	Spain	Europe	1976	2,904
21.	Finland	Europe	1976	5,621	43.	Poland	Europe	1976	2,856
22.	Canal Zone	North America	1975	5,610	44.	USSR	Europe, Asia	1976	2,771
23.	American Samoa	Oceania	1975	5,517	45.	Singapore	Asia	1976	2,700
					46.	Oman	Asia	1976	2,693
24.	Austria	Europe	1976	5,334	47.	Greece	Europe	1976	2,575
25.	Virgin Islands (USA)	North America	1975	5,275	48.	Venezuela	South America	1976	2,569
					49.	Ireland	Europe	1976	2,559
26.	Saudi Arabia	Asia	1976	5,067	50.	Bahrain	Asia	1976	2,548

[1] In US dollars at market prices.

8a
LARGEST CITIES
(WITHIN CITY LIMITS)

Contents

Rank.

Conventional name of city, and subdivision and country in which located.

Year of data. If the year is followed by the letter "e," the population has been estimated; otherwise, the figure given is from census returns.

Latest population, in thousands, of city proper.[1]

Arrangement

By population.

Coverage

All cities with a population of at least 2 million.

Rounding

Populations are rounded to the nearest 1000.

Entries

54.

[1]Occasionally, population data are conventionally given only for a city with suburbs included (indicated by the letter "s").

TABLE 8a.   LARGEST CITIES (WITHIN CITY LIMITS)

Rank	City and Location	Year	Population (thousands)	Rank	City and Location	Year	Population (thousands)
1.	Shanghai, China	1974e	10,820	29.	Berlin, East and West Germany	1976e	3,073
2.	Tokyo, Honshu I, Japan	1975	8,647		East Berlin, East Germany	1976e	1,106
3.	Mexico, Distrito Federal, Mexico	1976e	8,628		West Berlin, West-Berlin, West Germany	1976e	1,967
4.	Peking, China	1974e	7,570	30.	Sydney, New South Wales, Australia	1976	3,021s
5.	Moscow, Russia in Europe, USSR	1976e	7,563	31.	Canton, Kwangtung, China	1965e	3,000
6.	New York, New York, USA	1975e	7,482	31.	Mukden, Liaoning, China	1970e	3,000
7.	Sao Paulo, Sao Paulo, Brazil	1975e	7,199	33.	Buenos Aires, Distrito Federal, Argentina	1975e	2,975
8.	Seoul, South Korea	1975	6,889	34.	Rome, Lazio, Italy	1976e	2,884
9.	Tientsin, China	1974e	6,280	35.	Osaka, Honshu I, Japan	1975	2,779
10.	Bombay, Maharashtra, India	1971	5,971	36.	Los Angeles, California, USA	1975e	2,727
11.	Jakarta, Java I, Indonesia	1975e	5,490	37.	Yokohama, Honshu I, Japan	1975	2,622
12.	Cairo, Egypt	1976	5,084	38.	London, England, UK	1975e	2,553
13.	Rio de Janeiro, Guanabara, Brazil	1975e	4,858	39.	Istanbul, Turkey	1975	2,547
14.	Tehran, Iran	1976	4,496	40.	Melbourne, Victoria, Australia	1976	2,479
15.	Hong Kong, Hong Kong	1976	4,407	41.	Madras, Tamil Nadu, India	1971	2,469
16.	Bangkok, Thailand	1975e	4,130	42.	Pusan, South Korea	1975	2,454
17.	Leningrad, Russia in Europe, USSR	1976e	3,911	43.	Alexandria, Egypt	1976	2,312
18.	Madrid, Castilla la Nueva, Spain	1975e	3,634	44.	Paris, Region Parisienne, France	1975	2,300
19.	Lushun-Talien, Liaoning, China	1965e	3,600	45.	Singapore, Singapore	1976e	2,278
20.	Karachi, Pakistan	1972	3,499	46.	Wuhan, Hupeh, China	1958e	2,226
21.	Ho Chi Minh City, Vietnam	1976e	3,460	47.	Chungking, Szechwan, China	1958e	2,165
22.	Lima, Peru	1972	3,303s	47.	Lahore, Pakistan	1972	2,165
23.	Delhi, Delhi, India	1971	3,288	49.	Taipei, Taiwan	1976e	2,089
24.	Rangoon, Burma	1973	3,187s	50.	Nagoya, Honshu I, Japan	1975	2,080
25.	Santiago, Chile	1975e	3,186s	51.	Budapest, Hungary	1976e	2,071
26.	Calcutta, West Bengal, India	1971	3,149	52.	Kiev, Ukraine, USSR	1976e	2,013
27.	Bogota, Colombia	1976e	3,102	53.	Kinshasa, Zaire	1974e	2,008
28.	Chicago, Illinois, USA	1975e	3,099	54.	Harbin, Heilungkiang, China	1970e	2,000

8b
LARGEST CITIES, INCLUDING ADJACENT SUBURBAN AREAS

Contents

Rank.

Conventional name of city, and subdivision and country in which located.

Year of data. If the year is followed by the letter "e," the population has been estimated; otherwise, the figure given is from census returns.

Latest population, in thousands, of city and adjacent suburban areas.[1] Preferably, this is the population of the urbanized area or conurbation only; some officially reported suburban areas, however, include rural districts.

Arrangement

By population.

Coverage

All cities with a population of at least 1 million, including adjacent suburban areas.

Rounding

Populations are rounded to the nearest 1000.

Entries

163.

---

[1] When population data are not officially supplied for adjacent suburban areas, the letter "s" is omitted, and the population of the city proper is given.

Figure 25. High-altitude aerial photograph of what is still the largest urbanized area in the world—New York City, showing principally Manhattan, Brooklyn, and the nearby New Jersey suburbs. (Credit: US National Aeronautics and Space Administration.)

TABLE 8b. LARGEST CITIES, INCLUDING ADJACENT SUBURBAN AREAS

Rank	City and Location	Year	Population (thousands)
1.	New York, New York, USA	1970	16,208s
2.	Mexico, Distrito Federal, Mexico	1976e	11,943s
3.	Tokyo, Honshu I, Japan	1975	11,540s
4.	Shanghai, China	1974e	10,820
5.	Buenos Aires, Distrito Federal, Argentina	1975e	9,245s
6.	Paris, Region Parisienne, France	1975	8,450s
7.	Los Angeles, California, USA	1970	8,345s
8.	Moscow, Russia in Europe, USSR	1976e	7,734s
9.	Peking, China	1974e	7,570
10.	Sao Paulo, Sao Paulo, Brazil	1975e	7,199s
11.	London, England, UK	1975e	7,111s
12.	Calcutta, West Bengal, India	1971	7,031s
13.	Seoul, South Korea	1975	6,889
14.	Cairo, Egypt	1975e	6,757s
15.	Chicago, Illinois, USA	1970	6,717s
16.	Tientsin, China	1974e	6,280
17.	Bombay, Maharashtra, India	1971	5,971
18.	Jakarta, Java I, Indonesia	1975e	5,490
19.	Manila, Luzon I, Philippines	1975	4,863s
20.	Rio de Janeiro, Guanabara, Brazil	1975e	4,858s
21.	Tehran, Iran	1976	4,496
22.	Hong Kong, Hong Kong	1976	4,407
23.	Leningrad, Russia in Europe, USSR	1976e	4,372s
24.	Bangkok, Thailand	1975e	4,130
25.	Philadelphia, Pennsylvania, USA	1970	4,022s
26.	Detroit, Michigan, USA	1970	3,974s
27.	Delhi, Delhi, India	1971	3,647s
28.	Madrid, Castilla la Nueva, Spain	1975e	3,634
29.	Lushun-Talien, Liaoning, China	1965e	3,600
30.	Karachi, Pakistan	1972	3,499
31.	Ho Chi Minh City, Vietnam	1976e	3,460
32.	Lima, Peru	1972	3,303s
33.	Baghdad, Iraq	1977	3,206s
34.	Rangoon, Burma	1973	3,187s
35.	Santiago, Chile	1975e	3,186s
36.	Madras, Tamil Nadu, India	1971	3,170s
37.	Bogota, Colombia	1976e	3,102
38.	Berlin, East and West Germany	1976e	3,073
	East Berlin, East Germany	1976e	1,106
	West Berlin, West-Berlin, West Germany	1976e	1,967
39.	Sydney, New South Wales, Australia	1976	3,021s
40.	Canton, Kwangtung, China	1965e	3,000
40.	Mukden, Liaoning, China	1970e	3,000
42.	San Francisco, California, USA	1970	2,988s
43.	Rome, Lazio, Italy	1976e	2,884
44.	Toronto, Ontario, Canada	1976	2,803s
45.	Montreal, Quebec, Canada	1976	2,802s
46.	Osaka, Honshu I, Japan	1975	2,779
47.	Boston, Massachusetts, USA	1970	2,653s
48.	Yokohama, Honshu I, Japan	1975	2,622
49.	Istanbul, Turkey	1975	2,547
50.	Athens, Greece	1971	2,540s
51.	Caracas, Venezuela	1975e	2,487s
52.	Washington, District of Columbia, USA	1970	2,481s
53.	Melbourne, Victoria, Australia	1976	2,479s
54.	Pusan, South Korea	1975	2,454

Rank	City and Location	Year	Population (thousands)
55.	Alexandria, Egypt	1976	2,312
56.	Singapore, Singapore	1976e	2,278
57.	Wuhan, Hupeh, China	1958e	2,226
58.	Chungking, Szechwan, China	1958e	2,165
58.	Lahore, Pakistan	1972	2,165
60.	Taipei, Taiwan	1976e	2,089
61.	Nagoya, Honshu I, Japan	1975	2,080
62.	Guadalajara, Jalisco, Mexico	1976e	2,076s
63.	Budapest, Hungary	1976e	2,071
64.	Kiev, Ukraine, USSR	1976e	2,013
65.	Kinshasa, Zaire	1974e	2,008
66.	Harbin, Heilungkiang, China	1970e	2,000
67.	Cleveland, Ohio, USA	1970	1,960s
68.	Bucharest, Romania	1977	1,934s
69.	Saint Louis, Missouri, USA	1970	1,883s
70.	Havana, Cuba	1976e	1,861
71.	Pittsburgh, Pennsylvania, USA	1970	1,846s
72.	Barcelona, Cataluna, Spain	1975e	1,828
73.	Casablanca, Morocco	1974e	1,808
74.	Changchun, Kirin, China	1965e	1,800
75.	Hyderabad, Andhra Pradesh, India	1971	1,796s
76.	Ahmadabad, Gujarat, India	1971	1,742s
77.	Monterrey, Nuevo Leon, Mexico	1976e	1,725s
78.	Hamburg, Hamburg, West Germany	1976e	1,707
79.	Milan, Lombardia, Italy	1976e	1,705
80.	Minneapolis, Minnesota, USA	1970	1,704s
81.	Ankara, Turkey	1975	1,701
82.	Lagos, Nigeria	1972e	1,700s

Rank	City and Location	Year	Population (thousands)
83.	Houston, Texas, USA	1970	1,678s
84.	Bangalore, Karnataka, India	1971	1,654s
85.	Tashkent, Uzbekistan, USSR	1976e	1,643
86.	Vienna, Austria	1975e	1,604
87.	Baltimore, Maryland, USA	1970	1,580s
88.	Belo Horizonte, Minas Gerais, Brazil	1975e	1,557s
89.	Surabaya, Java I, Indonesia	1971	1,552
90.	Algiers, Algeria	1976e	1,504s
91.	Pyongyang, North Korea	1974e	1,500
91.	Sian, Shensi, China	1970e	1,500
93.	Warsaw, Poland	1976e	1,463
94.	Kyoto, Honshu I, Japan	1975	1,461
95.	Nanking, Kiangsu, China	1958e	1,455
96.	Hanoi, Vietnam	1976e	1,443
97.	Johannesburg, Transvaal, South Africa	1970	1,433s
98.	Baku, Azerbaijan, USSR	1976e	1,406s
99.	Kharkov, Ukraine, USSR	1976e	1,385
100.	Stockholm, Sweden	1976e	1,364s
101.	Kobe, Honshu I, Japan	1975	1,361
102.	Dallas, Texas, USA	1970	1,339s
103.	Dacca, Bangladesh	1974	1,320
104.	Munich, Bayern, West Germany	1976e	1,311
104.	Taegu, South Korea	1975	1,311
106.	Gorkiy, Russia in Europe, USSR	1976e	1,305
107.	Copenhagen, Denmark	1975e	1,287s
108.	Novosibirsk, Russia in Asia, USSR	1976e	1,286
109.	Kanpur, Uttar Pradesh, India	1971	1,275s
110.	Milwaukee, Wisconsin, USA	1970	1,252s

Rank	City and Location	Year	Population (thousands)
111.	Recife, Pernambuco, Brazil	1975e	1,250s
112.	Giza, Egypt	1976	1,246
113.	Sapporo, Hokkaido I, Japan	1975	1,241
114.	Seattle, Washington, USA	1970	1,238s
115.	Salvador, Bahia, Brazil	1975e	1,237s
116.	Montevideo, Uruguay	1975	1,230
117.	Naples, Campania, Italy	1976e	1,224
118.	Miami, Florida, USA	1970	1,220s
119.	Bandung, Java I, Indonesia	1971	1,200
119.	Tsinan, Shantung, China	1968?e	1,200
121.	San Diego, California, USA	1970	1,198s
122.	Medellin, Colombia	1976e	1,195
123.	Haiphong, Vietnam	1976e	1,191
123.	Turin, Piemonte, Italy	1976e	1,191
125.	Minsk, White Russia, USSR	1976e	1,189s
126.	Kuybyshev, Russia in Europe, USSR	1976e	1,186
127.	Prague, Bohemia, Czechoslovakia	1976e	1,176
128.	Atlanta, Georgia, USA	1970	1,173s
129.	Beirut, Lebanon	1975e	1,172s
130.	Sverdlovsk, Russia in Asia, USSR	1976e	1,171
131.	Vancouver, British Columbia, Canada	1976	1,166s
132.	Lyons, Rhone-Alpes, France	1975	1,159s
133.	Tel Aviv-Jaffa, Israel	1974e	1,157s
134.	Tsingtao, Shantung, China	1958e	1,144
135.	Chengtu, Szechwan, China	1958e	1,135
135.	Poona, Maharashtra, India	1971	1,135s
137.	Cincinnati, Ohio, USA	1970	1,111s
138.	Fortaleza, Ceara, Brazil	1975e	1,110s
139.	Kansas City, Missouri, USA	1970	1,102s
140.	Cape Town, Cape of Good Hope, South Africa	1970	1,097s
141.	Buffalo, New York, USA	1970	1,087s
142.	Birmingham, England, UK	1975e	1,085
143.	Addis Ababa, Ethiopia	1974	1,083
144.	Kitakyushu, Kyushu I, Japan	1975	1,058
145.	Taiyuan, Shansi, China	1958e	1,053
146.	Brussels, Belgium	1975e	1,051s
147.	Denver, Colorado, USA	1970	1,047s
148.	Porto Alegre, Rio Grande do Sul, Brazil	1975e	1,044s
149.	Damascus, Syria	1975e	1,042
150.	Lisbon, Portugal	1970	1,034s
151.	Rotterdam, Netherlands	1976e	1,031s
152.	Tbilisi, Georgia, USSR	1976e	1,030
153.	San Jose, California, USA	1970	1,025s
154.	Odessa, Ukraine, USSR	1976e	1,023
155.	Kaohsiung, Taiwan	1976e	1,020
156.	Kawasaki, Honshu I, Japan	1975	1,015
157.	Marseilles, Provence-Cote d'Azur, France	1975	1,011s
158.	Cologne, Nordrhein-Westfalen, West Germany	1976e	1,010
159.	Cali, Colombia	1976e	1,003
160.	Fukuoka, Kyushu I, Japan	1975	1,002
160.	Omsk, Russia in Asia, USSR	1976e	1,002
162.	Fushun, Liaoning, China	1965e	1,000
162.	Shiukwan, Kwangtung, China	1972?e	1,000

8c
HIGHEST CITIES

Contents

Rank.

Conventional name of city, and subdivision and country in which located.

Elevation (i.e., altitude above sea level), in feet and meters. When available, the given elevation is that of the city center or principal business district, or is an average of elevations at several points in the city; otherwise, it is that of the meteorological station.

Arrangement

By elevation.

Coverage

All cities listed in Table 11b of this work with an elevation of at least 3000 ft (914 m).

Entries

132.

Figure 26. Panorama of Potosi, Bolivia, the world's highest city, with Cerro Rico, famous for its silver lode, in the background. (Credit: Bolivian Institute of Tourism.)

Figure 27. Panorama of Lhasa, Tibet, China, the second highest city in the world, showing the celebrated Potala Palace. (Credit: Foreign Language Press, Peking.)

TABLE 8c. HIGHEST CITIES

Rank	City and Location	Elevation (ft)	(m)	Rank	City and Location	Elevation (ft)	(m)
1.	Potosi, Bolivia	13,045	3976	34.	Taxco, Guerrero, Mexico	5,840	1780
2.	Lhasa, Tibet, China	12,002	3658	35.	Hamadan, Iran	5,824	1775
3.	La Paz, Bolivia	11,736	3577	36.	Butte, Montana, USA	5,755	1754
4.	Cuzco, Peru	11,152	3399	37.	Johannesburg, Transvaal, South Africa	5,750	1753
5.	Quito, Ecuador	9,249	2819				
6.	Sucre, Bolivia	9,154	2790	38.	Kokiu, Yunnan, China	5,709	1740
7.	Toluca, Mexico, Mexico	8,793	2680	39.	Irapuato, Guanajuato, Mexico	5,656	1724
8.	Bogota, Colombia	8,675	2644				
9.	Cochabamba, Bolivia	8,393	2558	40.	Mixco, Guatemala	5,551	1692
10.	Pachuca de Soto, Hidalgo, Mexico	7,960	2426	41.	Queretaro, Queretaro, Mexico	5,528	1685
11.	Addis Ababa, Ethiopia	7,900	2408	42.	Nairobi, Kenya	5,453	1662
12.	Asmara, Ethiopia	7,789	2374	43.	Germiston, Transvaal, South Africa	5,450	1661
13.	Arequipa, Peru	7,559	2304				
14.	Mexico, Distrito Federal, Mexico	7,546	2300	44.	Boulder, Colorado, USA	5,430	1655
15.	Netzahualcoyotl, Mexico, Mexico	7,474	2278	45.	Windhoek, Namibia	5,428	1654
				46.	Uruapan del Progreso, Michoacan, Mexico	5,361	1634
16.	Sining, Tsinghai, China	7,363	2244				
17.	Sana, North Yemen	7,260	2242	47.	Lakewood, Colorado, USA	5,355	1632
18.	Simla, Himachal Pradesh, India	7,225	2202	48.	Taif, Hejaz, Saudi Arabia	5,348	1630
19.	Puebla, Puebla, Mexico	7,094	2162				
20.	Manizales, Colombia	7,021	2140	49.	Aurora, Colorado, USA	5,342	1628
21.	Santa Fe, New Mexico, USA	6,996	2132	50.	Denver, Colorado, USA	5,280	1609
				51.	Saltillo, Coahuila, Mexico	5,246	1599
22.	Guanajuato, Guanajuato, Mexico	6,726	2050	52.	Isfahan, Iran	5,217	1590
23.	Morelia, Michoacan, Mexico	6,368	1941	53.	Srinagar, Kashmir-Jammu	5,205	1586
24.	Kunming, Yunnan, China	6,211	1893	54.	Guadalajara, Jalisco, Mexico	5,141	1567
25.	Durango, Durango, Mexico	6,198	1889	55.	Casper, Wyoming, USA	5,123	1561
26.	Aguascalientes, Aguascalientes, Mexico	6,195	1888	56.	Siakwan, Yunnan, China	5,118	1560
27.	Leon, Guanajuato, Mexico	6,185	1885	57.	Leninakan, Armenia, USSR	5,105	1556
28.	San Luis Potosi, San Luis Potosi, Mexico	6,158	1877	58.	Oaxaca, Oaxaca, Mexico	5,086	1550
29.	Harar, Ethiopia	6,089	1856	59.	Cuernavaca, Morelos, Mexico	5,059	1542
30.	Cheyenne, Wyoming, USA	6,062	1848	60.	Medellin, Colombia	5,056	1541
31.	Colorado Springs, Colorado, USA	6,012	1833	61.	Kigali, Rwanda	5,053	1540
				62.	Shiraz, Iran	5,049	1539
32.	Celaya, Guanajuato, Mexico	5,932	1808	63.	Albuquerque, New Mexico, USA	4,950	1509
33.	Kabul, Afghanistan	5,903	1799	64.	Lanchow, Kansu, China	4,948	1508
				65.	Guatemala, Guatemala	4,928	1502
				66.	Salisbury, Rhodesia	4,831	1472

Rank	City and Location	Elevation (ft)	(m)	Rank	City and Location	Elevation (ft)	(m)
67.	Vereeniging, Transvaal, South Africa	4,725	1440	100.	El Paso, Texas, USA	3,762	1147
68.	Idaho Falls, Idaho, USA	4,709	1435	101.	Ciudad Juarez, Chihuahua, Mexico	3,734	1138
69.	Pueblo, Colorado, USA	4,695	1431	102.	Torreon, Coahuila, Mexico	3,708	1130
70.	Chihuahua, Chihuahua, Mexico	4,692	1430	103.	Amarillo, Texas, USA	3,676	1120
71.	Jalapa, Veracruz, Mexico	4,682	1427	104.	Yinchwan, Ningsia, China	3,645	1111
72.	Bloemfontein, Orange Free State, South Africa	4,678	1426	105.	Kayseri, Turkey	3,514	1071
				105.	Kweiyang, Kweichow, China	3,514	1071
73.	Pereira, Colombia	4,672	1424	107.	Blantyre, Malawi	3,501	1067
74.	Kashgar, Sinkiang, China	4,629	1411	107.	Lilongwe, Malawi	3,501	1067
75.	Provo, Utah, USA	4,549	1387	109.	Huhehot, Inner Mongolia, China	3,484	1062
76.	Antananarivo, Madagascar	4,531	1381	110.	Kandahar, Afghanistan	3,462	1055
				111.	Tatung, Shansi, China	3,442	1049
77.	Reno, Nevada, USA	4,491	1369	112.	Cali, Colombia	3,432	1046
78.	Tabriz, Iran	4,469	1362	113.	Calgary, Alberta, Canada	3,428	1045
79.	Pocatello, Idaho, USA	4,464	1361				
80.	Kitwe, Zambia	4,429	1350	114.	Paotow, Inner Mongolia, China	3,425	1044
81.	Bulawayo, Rhodesia	4,405	1343				
82.	Kathmandu, Nepal	4,388	1337	115.	Yenan, Shensi, China	3,400	1036
83.	Pretoria, Transvaal, South Africa	4,375	1333	116.	Konya, Turkey	3,366	1026
				117.	Great Falls, Montana, USA	3,330	1015
84.	Kermanshah, Iran	4,337	1322				
85.	Ogden, Utah, USA	4,299	1310	118.	Gaborone, Botswana	3,329	1015
86.	Ulan-Bator, Mongolia	4,295	1309	119.	Osh, Kirgizia, USSR	3,324	1013
87.	Salt Lake City, Utah, USA	4,266	1300	120.	Tegucigalpa, Honduras	3,304	1007
				121.	Lubbock, Texas, USA	3,241	988
88.	Orizaba, Veracruz, Mexico	4,213	1284	122.	Meshed, Iran	3,232	985
				123.	Rapid City, South Dakota, USA	3,231	985
89.	Lusaka, Zambia	4,196	1279				
90.	Ndola, Zambia	4,137	1261	124.	Billings, Montana, USA	3,117	950
91.	Helena, Montana, USA	4,124	1257	124.	Curitiba, Parana, Brazil	3,117	950
92.	Ibague, Colombia	4,098	1249	126.	Qom, Iran	3,045	928
93.	Lubumbashi, Zaire	4,035	1230	127.	Bucaramanga, Colombia	3,035	925
94.	Kampala, Uganda	3,910	1192	128.	Cordoba, Veracruz, Mexico	3,032	924
95.	Tehran, Iran	3,908	1191				
96.	Las Cruces, New Mexico, USA	3,895	1187	129.	Caracas, Venezuela	3,025	922
				129.	Herat, Afghanistan	3,025	922
97.	Salta, Salta, Argentina	3,878	1182	131.	Bangalore, Karnataka, India	3,021	921
98.	San Jose, Costa Rica	3,845	1172				
99.	Brasilia, Distrito Federal, Brazil	3,809	1161	132.	Tepic, Nayarit, Mexico	3,002	915

8d
OLDEST CITIES, BY CONTINENT

Contents

Rank.

Conventional name of city, and subdivision and country in which located.

Year or century in which the first permanent settlement within present city limits was made. A question mark following a given date usually indicates that, although its exact origin is unknown, the city in question was first mentioned in historical records at that date. When the time of settlement cannot be ascertained but the city is known to have existed before a certain date, that date is given, preceded by the abbreviation "bef." It should be observed that the settlement date of a city bears no relation either to the date when the city was formally established (or chartered) or to the date when the present name was adopted.

Arrangement

By date, under continent.

Coverage

All cities listed in Table 11a of this work that were settled as early as the following dates:
Africa—1600 AD
Asia—300 BC
Europe—300 BC
North America—1640 AD
Oceania—1850 AD
South America—1640 AD

Entries

324.

Figure 28. Vista of Gaziantep, Turkey, probably the oldest continuously populated city in the world, which was settled more than 5600 years ago. (Credit: Turkish Tourism and Information Office.)

TABLE 8d.   OLDEST CITIES, BY CONTINENT

AFRICA

Rank	City and Location	Date Settled	Rank	City and Location	Date Settled
1.	Giza, Egypt	bef 2568 BC	23.	Rabat-Sale, Morocco	11th c
2.	Asyut, Egypt	bef 2160 BC	23.	Timbuktu, Mali	11th c
2.	Luxor, Egypt	bef 2160 BC	26.	Marrakesh, Morocco	1062
4.	Tangier, Morocco	15th c BC?	27.	Zaria, Nigeria	1095?
5.	Tripoli, Tripolitania, Libya	7th c BC?	28.	Kano, Nigeria	12th c?
			28.	Tanta, Egypt	12th c?
6.	Aswan, Egypt	6th c BC?	30.	Mansurah, Egypt	1221
6.	Bengasi, Cyrenaica, Libya	6th c BC	31.	Ouagadougou, Upper Volta	14th c?
8.	Tunis, Tunisia	4th c BC?	32.	Funchal, (Madeira Islands), Portugal	1425
9.	Alexandria, Egypt	332 BC			
10.	Constantine, Algeria	3rd c BC?	33.	Suez, Egypt	15th c
11.	Sfax, Tunisia	2nd c AD?	34.	Las Palmas, (Canary Islands), Spain	1478
12.	Cairo, Egypt	641?			
13.	Annaba, Algeria	7th c	35.	Santa Cruz de Tenerife, (Canary Islands), Spain	1494
13.	Harar, Ethiopia	7th c?			
15.	Mombasa, Kenya	8th c?	36.	Casablanca, Morocco	1515
16.	Fez, Morocco	808	37.	Maputo, Mozambique	1544
17.	Mogadishu, Somalia	908?	38.	Accra, Ghana	16th c
18.	Algiers, Algeria	10th c	38.	Iwo, Nigeria	16th c?
18.	Meknes, Morocco	10th c	38.	Niamey, Niger	16th c?
18.	Oran, Algeria	10th c	38.	Porto-Novo, Benin	16th c
21.	Mahalla al Kubra, Egypt	985?	38.	Zanzibar, Zanzibar, Tanzania	16th c
22.	Oujda, Morocco	994	43.	Luanda, Angola	1575
23.	Damanhur, Egypt	11th c?	44.	Oshogbo, Nigeria	1600?

ASIA

Rank	City and Location	Date Settled	Rank	City and Location	Date Settled
1.	Gaziantep, Turkey	3650 BC?	18.	Peking, China	11th c BC?
2.	Jerusalem, Israel	3000 BC?	20.	Aleppo, Syria	bef 1000 BC
2.	Kirkuk, Iraq	3000 BC?	21.	Changchow, Fukien, China	10th c BC?
4.	Konya, Turkey	2600 BC?			
5.	Sian, Shensi, China	2205 BC?	21.	Chengchow, Honan, China	10th c BC?
6.	Shaohing, Chekiang, China	2000 BC?	21.	Foochow, Fukien, China	10th c BC?
7.	Loyang, Honan, China	1900 BC?	21.	Hofei, Anhwei, China	10th c BC?
8.	Ankara, Turkey	17th c BC?	21.	Nanchung, Szechwan, China	10th c BC?
9.	Changchih, Shansi, China	16th c BC?			
10.	Tel Aviv-Jaffa, Israel	bef 1472 BC	21.	Nanning, Kwangsi, China	10th c BC?
11.	Gaza, Gaza Strip	1468 BC?	21.	Tatung, Shansi, China	10th c BC?
12.	Beirut, Lebanon	15th c BC?	28.	Canton, Kwangtung, China	9th c BC?
12.	Liaoyang, Liaoning, China	15th c BC?			
14.	Damascus, Syria	bef 14th c BC	29.	Trebizond, Turkey	756 BC?
15.	Varanasi, Uttar Pradesh, India	12th c BC?	30.	Adana, Turkey	8th c BC?
			31.	Hengyang, Hunan, China	7th c BC?
16.	Pyongyang, North Korea	1122 BC?	31.	Kaifeng, Honan, China	7th c BC?
17.	Hamadan, Iran	1100 BC?	31.	Osaka, Honshu I, Japan	7th c BC?
18.	Izmir, Turkey	11th c BC?	31.	Shaoyang, Hunan, China	7th c BC?

ASIA

Rank	City and Location	Date Settled	Rank	City and Location	Date Settled
31.	Suchow, Kiangsu, China	7th c BC?	45.	Soochow, Kiangsu, China	525 BC?
31.	Tripoli, Lebanon	7th c BC	46.	Isfahan, Iran	5th c BC?
31.	Yangchow, Kiangsu, China	7th c BC?	46.	Madurai, Tamil Nadu, India	5th c BC?
38.	Mathura, Uttar Pradesh, India	600 BC?	46.	Omiya, Honshu I, Japan	5th c BC?
38.	Sialkot, Pakistan	600 BC?	49.	Hantan, Hopeh, China	4th c BC
40.	Rangoon, Burma	585 BC	49.	Herat, Afghanistan	4th c BC?
41.	Patna, Bihar, India	6th c BC?	49.	Kabul, Afghanistan	4th c BC?
41.	Samsun, Turkey	6th c BC	49.	Kweilin, Kwangsi, China	4th c BC?
43.	Gaya, Bihar, India	545 BC?	49.	Samarkand, Uzbekistan, USSR	4th c BC?
44.	Colombo, Sri Lanka	543 BC?	54.	Antioch, Turkey	300 BC?

EUROPE

Rank	City and Location	Date Settled	Rank	City and Location	Date Settled
1.	Zurich, Switzerland	3000 BC?	29.	Naples, Campania, Italy	600 BC?
2.	Lisbon, Portugal	2000 BC?	31.	Besancon, Franche-Comte, France	6th c BC?
2.	Porto, Portugal	2000 BC?	31.	Brescia, Lombardia, Italy	6th c BC
4.	Athens, Greece	bef 13th c BC	31.	Bristol, England, UK	6th c BC
5.	La Coruna, Galicia, Spain	bef 12th c BC	31.	Kerch, Ukraine, USSR	6th c BC
6.	Malaga, Andalucia, Spain	12th c BC	31.	Kutaisi, Georgia, USSR	6th c BC
7.	Cadiz, Andalucia, Spain	1100 BC?	31.	Monaco, Monaco	6th c BC?
8.	Pisa, Toscana, Italy	11th c BC?	31.	Perugia, Umbria, Italy	6th c BC?
9.	Metz, Lorraine, France	1000 BC?	31.	Stara Zagora, Bulgaria	6th c BC?
9.	Rome, Lazio, Italy	1000 BC?	31.	Tarragona, Cataluna, Spain	6th c BC?
11.	Toulon, Provence-Cote d'Azur, France	9th c BC	31.	Varna, Bulgaria	6th c BC
12.	Cordova, Andalucia, Spain	bef 8th c BC	41.	Cagliari, Sardegna, Italy	540 BC
13.	Cannes, Provence-Cote d'Azur, France	8th c BC?	42.	Bologna, Emilia-Romagna, Italy	510 BC?
13.	Catania, Sicilia, Italy	8th c BC	43.	Bergamo, Lombardia, Italy	5th c BC?
13.	Messina, Sicilia, Italy	8th c BC	43.	Genoa, Liguria, Italy	5th c BC?
13.	Palermo, Sicilia, Italy	8th c BC	43.	Granada, Andalucia, Spain	5th c BC?
13.	Ravenna, Emilia-Romagna, Italy	8th c BC	43.	Le Mans, Pays de la Loire, France	5th c BC?
13.	Reggio di Calabria, Calabria, Italy	8th c BC	43.	Lerida, Cataluna, Spain	5th c BC?
19.	Syracuse, Sicilia, Italy	734 BC	43.	Mainz, Rheinland-Pfalz, West Germany	5th c BC?
20.	Lucca, Toscana, Italy	718 BC?	43.	Mantua, Lombardia, Italy	5th c BC?
21.	Taranto, Puglia, Italy	708 BC?	43.	Modena, Emilia-Romagna, Italy	5th c BC?
22.	Terni, Umbria, Italy	672 BC?	43.	Monza, Lombardia, Italy	5th c BC?
23.	Istanbul, Turkey	658 BC	43.	Neuss, Nordrhein-Westfalen, West Germany	5th c BC?
24.	Constanta, Romania	7th c BC?	43.	Nice, Provence-Cote d'Azur, France	5th c BC?
24.	Huelva, Andalucia, Spain	7th c BC?			
24.	Jerez de la Frontera, Andalucia, Spain	7th c BC?			
24.	Seville, Andalucia, Spain	7th c BC?			
24.	Vigo, Galicia, Spain	7th c BC?			
29.	Marseilles, Provence-Cote d'Azur, France	600 BC?			

EUROPE

Rank	City and Location	Date Settled	Rank	City and Location	Date Settled
43.	Patras, Greece	5th c BC?	61.	Trent, Trentino-Alto Adige, Italy	4th c BC?
43.	Piraeus, Greece	5th c BC			
43.	Regensburg, Bayern, West Germany	5th c BC?	61.	Turin, Piemonte, Italy	4th c BC?
			61.	Vienna, Austria	4th c BC?
43.	Salamanca, Leon, Spain	5th c BC?	67.	Plovdiv, Bulgaria	341 BC?
43.	Siena, Toscana, Italy	5th c BC?	68.	Cosenza, Calabria, Italy	bef 331 BC
59.	Rimini, Emilia-Romagna, Italy	400 BC?	69.	Salonika, Greece	315 BC
			70.	Padua, Veneto, Italy	302 BC?
60.	Ancona, Marche, Italy	390 BC?	71.	Rheims, Champagne, France	300 BC?
61.	Arezzo, Toscana, Italy	4th c BC?			
61.	Milan, Lombardia, Italy	4th c BC	71.	Toulouse, Midi-Pyrenees, France	300 BC?
61.	Novara, Piemonte, Italy	4th c BC?			

NORTH AMERICA

Rank	City and Location	Date Settled	Rank	City and Location	Date Settled
1.	Toluca, Mexico, Mexico	1120 AD?	27.	Tallahassee, Florida, USA	bef 1539
2.	Jalapa, Veracruz, Mexico	1313?	28.	Mazatlan, Sinaloa, Mexico	bef 1541
3.	Mexico, Distrito Federal, Mexico	1325	29.	Morelia, Michoacan, Mexico	1541
4.	Guanajuato, Guanajuato, Mexico	bef 1400	30.	Guadalajara, Jalisco, Mexico	1542
5.	Queretaro, Queretaro, Mexico	1440	31.	Irapuato, Guanajuato, Mexico	1547
6.	Orizaba, Veracruz, Mexico	1457	32.	Acapulco, Guerrero, Mexico	1550
7.	Oaxaca, Oaxaca, Mexico	1486			
8.	Santo Domingo, Dominican Republic	1496	32.	Netzahualcoyotl, Mexico, Mexico	16th c?
9.	Santiago de los Caballeros, Dominican Republic	1504	34.	Durango, Durango, Mexico	1563
10.	Santiago de Cuba, Cuba	1514	35.	Saint Augustine, Florida, USA	1565
11.	Havana, Cuba	1519	36.	Celaya, Guanajuato, Mexico	1570
12.	Cuernavaca, Morelos, Mexico	bef 1521	37.	Aguascalientes, Aguascalientes, Mexico	1575
12.	Managua, Nicaragua	bef 1521			
14.	San Juan, Puerto Rico	1521	37.	Saltillo, Coahuila, Mexico	1575
15.	Tepic, Nayarit, Mexico	bef 1524			
16.	San Salvador, El Salvador	1525	39.	Leon, Guanajuato, Mexico	1576
17.	Willemstad, Netherlands Antilles	1527	39.	San Luis Potosi, San Luis Potosi, Mexico	1576
18.	Merida, Yucatan, Mexico	bef 1528			
19.	Camaguey, Cuba	1528	39.	Santa Ana, El Salvador	1576?
20.	Taxco, Guerrero, Mexico	1529	42.	Tegucigalpa, Honduras	1578
21.	San Miguel, El Salvador	1530	43.	Monterrey, Nuevo Leon, Mexico	1579?
22.	Puebla, Puebla, Mexico	1532			
23.	Culiacan, Sinaloa, Mexico	1533	44.	Coatzacoalcos, Veracruz, Mexico	1580
24.	Pachuca de Soto, Hidalgo, Mexico	1534			
25.	San Pedro Sula, Honduras	1536	45.	Saint John's, Newfoundland, Canada	1583?
25.	Uruapan del Progreso, Michoacan, Mexico	1536	46.	Port of Spain, Trinidad and Tobago	bef 1595

NORTH AMERICA

Rank	City and Location	Date Settled	Rank	City and Location	Date Settled
47.	Villahermosa, Tabasco, Mexico	1596	65.	Trois-Rivieres, Quebec, Canada	1634
48.	Veracruz, Veracruz, Mexico	1599	65.	Waltham, Massachusetts, USA	1634
49.	Quebec, Quebec, Canada	1608	67.	Hartford, Connecticut, USA	1635
50.	Santa Fe, New Mexico, USA	1609	68.	Providence, Rhode Island, USA	1636
51.	Hampton, Virginia, USA	1610	68.	Springfield, Massachusetts, USA	1636
52.	Cordoba, Veracruz, Mexico	1617	70.	Belize, Belize	1638?
53.	Newport News, Virginia USA	1621	70.	Cranston, Rhode Island, USA	1638
54.	Albany, New York, USA	1624	70.	New Haven, Connecticut, USA	1638
54.	New York, New York, USA	1624	70.	Wilmington, Delaware, USA	1638
56.	Quincy, Massachusetts, USA	1625	74.	Yonkers, New York, USA	bef 1639
57.	Salem, Massachusetts, USA	1626	75.	Bridgeport, Connecticut, USA	1639
58.	Bridgetown, Barbados	1628	75.	Chihuahua, Chihuahua, Mexico	1639
59.	Jersey City, New Jersey, USA	1629?	75.	Newport, Rhode Island, USA	1639
59.	Lynn, Massachusetts, USA	1629	75.	Newton, Massachusetts, USA	1639
61.	Boston, Massachusetts, USA	1630	79.	New Bedford, Massachusetts, USA	1640
61.	Cambridge, Massachusetts, USA	1630			
61.	Somerville, Massachusetts, USA	1630			
64.	Williamsburg, Virginia, USA	1633			

OCEANIA

Rank	City and Location	Date Settled	Rank	City and Location	Date Settled
1.	Hilo, Hawaii, USA	bef 1778 AD	9.	Perth, Western Australia, Australia	1829
2.	Sydney, New South Wales, Australia	1788	10.	Melbourne, Victoria, Australia	1835
3.	Honolulu, Hawaii, USA	bef 1794	11.	Adelaide, South Australia, Australia	1836
4.	Hobart, Tasmania, Australia	1804	12.	Geelong, Victoria, Australia	1837
4.	Newcastle, New South Wales, Australia	1804	13.	Papeete, French Polynesia	1840?
6.	Wollongong, New South Wales, Australia	1815	13.	Wellington, North I, New Zealand	1840
7.	Brisbane, Queensland, Australia	1824	15.	Auckland, North I, New Zealand	1841
7.	Canberra, Australian Capital Territory, Australia	1824?	16.	Dunedin, South I, New Zealand	1845

OCEANIA

Rank	City and Location	Date Settled	Rank	City and Location	Date Settled
17.	Apia, Samoa	bef 1850	18.	Christchurch, South I, New Zealand	1850

SOUTH AMERICA

Rank	City and Location	Date Settled	Rank	City and Location	Date Settled
1.	Quito, Ecuador	1000 AD?	32.	San Juan, San Juan, Argentina	1562
2.	Cuzco, Peru	11th c			
3.	Arequipa, Peru	bef 1425	33.	Niteroi, Rio de Janeiro, Brazil	1565
4.	Lima, Peru	bef 1532			
5.	Cartagena, Colombia	1533	33.	Rio de Janeiro, Guanabara, Brazil	1565
6.	Trujillo, Peru	1534			
7.	Olinda, Pernambuco, Brazil	1535	35.	Caracas, Venezuela	1567
			35.	El Valle, Venezuela	1567
7.	Recife, Pernambuco, Brazil	1535?	35.	Nova Iguacu, Rio de Janeiro, Brazil	1567?
7.	Vitoria, Espirito Santo, Brazil	1535	38.	Maracaibo, Venezuela	1571
			39.	Cordoba, Cordoba, Argentina	1573
10.	Cali, Colombia	1536			
10.	Santos, Sao Paulo, Brazil	1536	39.	Santa Fe, Santa Fe, Argentina	1573
12.	Asuncion, Paraguay	1537	41.	Cochabamba, Bolivia	1574
12.	Callao, Peru	1537	42.	Buenos Aires, Distrito Federal, Argentina	1580
12.	Guayaquil, Ecuador	1537			
15.	Bogota, Colombia	bef 1538	43.	Salta, Salta, Argentina	1582
15.	Sucre, Bolivia	bef 1538	44.	Joao Pessoa, Paraiba, Brazil	1585
17.	Paramaribo, Surinam	1540?			
18.	Santiago, Chile	1541	45.	Vina del Mar, Chile	1586?
19.	Valparaiso, Chile	1544?	46.	Corrientes, Corrientes, Argentina	1588
20.	Potosi, Bolivia	1545			
21.	La Paz, Bolivia	bef 1548	47.	San Felix de Guayana, Venezuela	1590
22.	Salvador, Bahia, Brazil	1549			
23.	Concepcion, Chile	1550	48.	Santa Cruz, Bolivia	1595
24.	Ibague, Colombia	1551	49.	Moron, Buenos Aires, Argentina	1600
24.	Santo Andre, Sao Paulo, Brazil	1551			
			50.	Fortaleza, Ceara, Brazil	1609
26.	Barquisimeto, Venezuela	1552			
26.	Sao Bernardo do Campo, Sao Paulo, Brazil	1552	51.	Sao Luis, Maranhao, Brazil	1612
28.	Sao Paulo, Sao Paulo, Brazil	1554	52.	Belem, Para, Brazil	1616
			52.	Medellin, Colombia	1616
29.	Valencia, Venezuela	1555	54.	Bucaramanga, Colombia	1622
30.	Guarulhos, Sao Paulo, Brazil	1560	55.	Georgetown, Guyana	1625
			56.	Barranquilla, Colombia	1629
31.	Mendoza, Mendoza, Argentina	1561	57.	Campos, Rio de Janeiro, Brazil	1634

8e
WARMEST CITIES

Contents

Rank.

Conventional name of city, and subdivision and country in which located.

Average temperature throughout the year, in degrees Fahrenheit and in degrees Celsius (centigrade).

Arrangement

By temperature, from high to low.

Coverage

All cities listed in Table 11b of this work with an average temperature of at least 80°F (26.7°C).

Entries

117.

TABLE 8e. WARMEST CITIES

Rank	City and Location	Average Temperature (°F)	(°C)	Rank	City and Location	Average Temperature (°F)	(°C)
1.	Djibouti, Djibouti	86.0	30.0	26.	Myingyan, Burma	81.9[4]	27.7[4]
2.	Timbuktu, Mali	84.7	29.3	26.	Rajahmundry, Andhra	81.9[5]	27.7[5]
2.	Tirunelveli, Tamil Nadu, India	84.7[1]	29.3[1]		Pradesh, India		
				26.	Surat, Gujarat, India	81.9	27.7
2.	Tuticorin, Tamil Nadu, India	84.7[1]	29.3[1]	26.	Warangal, Andhra Pradesh, India	81.9	27.7
5.	Aden, South Yemen	84.0	28.9	26.	Willemstad, Netherlands	81.9	27.7
5.	Madurai, Tamil Nadu, India	84.0	28.9		Antilles		
				33.	Acapulco, Guerrero,	81.7	27.6
5.	Niamey, Niger	84.0	28.9		Mexico		
8.	Tiruchirapalli, Tamil Nadu, India	83.8	28.8	33.	Ho Chi Minh City, Vietnam	81.7	27.6
9.	Khartoum, Sudan	83.7	28.7	33.	Ipoh, Peninsular	81.7	27.6
9.	Omdurman, Sudan	83.7[2]	28.7[2]		Malaysia, Malaysia		
11.	Madras, Tamil Nadu, India	83.5	28.6	36.	Bandar Seri Begawan, Brunei	81.5	27.5
11.	Ouagadougou, Upper Volta	83.5	28.6	36.	Doha, Qatar	81.5[6]	27.5[6]
				36.	Hyderabad, Pakistan	81.5	27.5
13.	Jidda, Hejaz, Saudi Arabia	83.3	28.5	36.	Kuala Lumpur, Peninsular Malaysia, Malaysia	81.5	27.5
14.	Bamako, Mali	82.6	28.1	40.	Cebu, Cebu I,	81.3	27.4
14.	Bangkok, Thailand	82.6	28.1		Philippines		
16.	Maracaibo, Venezuela	82.4	28.0	40.	Pegu, Burma	81.3[7]	27.4[7]
16.	Ndjamena, Chad	82.4	28.0	40.	Pontianak, Borneo I,	81.3	27.4
16.	Salem, Tamil Nadu, India	82.4	28.0		Indonesia		
				40.	Rangoon, Burma	81.3	27.4
19.	Phnom-Penh, Cambodia	82.2	27.9	40.	Teresina, Piaui, Brazil	81.3	27.4
19.	Vellore, Tamil Nadu, India	82.2	27.9	45.	Ahmadabad, Gujarat, India	81.1	27.3
21.	Barranquilla, Colombia	82.0	27.8	45.	Bombay, Maharashtra, India	81.1	27.3
21.	Denpasar, Bali I, Indonesia	82.0	27.8	45.	Calicut, Kerala, India	81.1	27.3
21.	George Town, Peninsular Malaysia, Malaysia	82	27.8	45.	Cirebon, Java I, Indonesia	81.1	27.3
21.	Guntur, Andhra Pradesh, India	82.0[3]	27.8[3]	45.	Cucuta, Colombia	81.1	27.3
				45.	Medina, Hejaz, Saudi Arabia	81.1	27.3
21.	Vijayawada, Andhra Pradesh, India	82.0[3]	27.8[3]	45.	Paramaribo, Surinam	81.1	27.3
26.	Cuttack, Orissa, India	81.9	27.7	45.	Thana, Maharashtra, India	81.1[8]	27.3[8]
26.	Mandalay, Burma	81.9	27.7				

[1] Data for Palayamcottai, nr Tirunelveli and Tuticorin.
[2] Data for Khartoum, nr Omdurman.
[3] Data for Masulipatnam, nr Guntur and Vijayawada.
[4] Data for Mandalay, nr Myingyan.
[5] Data for Kakinada, nr Rajahmundry.
[6] Data for Dhahran, Saudi Arabia, nr Doha.
[7] Data for Rangoon, nr Pegu.
[8] Data for Bombay, nr Thana.

Rank	City and Location	Average Temperature (°F)	(°C)	Rank	City and Location	Average Temperature (°F)	(°C)
45.	Ulhasnagar-Kalyan, Maharashtra, India	81.1[1]	27.3[1]	76.	Calcutta, West Bengal, India	80.4	26.9
54.	Amravati, Maharashtra, India	81.0	27.2	76.	Colombo, Sri Lanka	80.4	26.9
54.	Bhavnagar, Gujarat, India	81.0	27.2	76.	Cotonou, Benin	80.4	26.9
54.	Kupang, Timor I, Indonesia	81.0	27.2	76.	Howrah, West Bengal, India	80.4[3]	26.9[3]
54.	Mogadishu, Somalia	81.0	27.2	76.	Jakarta, Java I, Indonesia	80.4	26.9
54.	Peshawar, Pakistan	81.0	27.2	76.	Lagos, Nigeria	80.4	26.9
54.	Porto-Novo, Benin	81.0	27.2	76.	Managua, Nicaragua	80.4	26.9
54.	Visakhapatnam, Andhra Pradesh, India	81.0	27.2	76.	Manaus, Amazonas, Brazil	80.4	26.9
61.	Cartagena, Colombia	80.8	27.1	76.	Mushin, Nigeria	80.4[4]	26.9[4]
61.	Cochin, Kerala, India	80.8	27.1	76.	Nagpur, Maharashtra, India	80.4	26.9
61.	Colon, Panama	80.8	27.1	76.	Panama, Panama	80.4[5]	26.9[5]
61.	Mangalore, Karnataka, India	80.8	27.1	76.	Port Moresby, Papua New Guinea	80.4	26.9
61.	Sholapur, Maharashtra, India	80.8	27.1	76.	Raipur, Madhya Pradesh, India	80.4	26.9
61.	Trivandrum, Kerala, India	80.8	27.1	76.	Semarang, Java I, Indonesia	80.4	26.9
67.	Bassein, Burma	80.6	27.0	76.	South Suburban, West Bengal, India	80.4[3]	26.9[3]
67.	Davao, Mindanao I, Philippines	80.6	27.0	76.	Tanjungkarang-Telukbetung, Sumatra I, Indonesia	80.4[6]	26.9[6]
67.	Georgetown, Guyana	80.6	27.0	94.	Bacolod, Negros I, Philippines	80.2	26.8
67.	Iloilo, Panay I, Philippines	80.6	27.0	94.	Banjarmasin, Borneo I, Indonesia	80.2	26.8
67.	Jamshedpur, Bihar, India	80.6	27.0	94.	Kediri, Java I, Indonesia	80.2[7]	26.8[7]
67.	Kota, Rajasthan, India	80.6	27.0	94.	Surabaya, Java I, Indonesia	80.2	26.8
67.	Moulmein, Burma	80.6	27.0	98.	Ado-Ekiti, Nigeria	80.1[8]	26.7[8]
67.	Singapore, Singapore	80.6	27.0				
67.	Vadodara, Gujarat, India	80.6	27.0				
76.	Aswan, Egypt	80.4	26.9				
76.	Bhilainagar-Durg, Madhya Pradesh, India	80.4[2]	26.9[2]				

[1] Data for Bombay, nr Ulhasnagar-Kalyan.
[2] Data for Raipur, nr Bhilainagar-Durg.
[3] Data for Calcutta, nr Howrah and South Suburban.
[4] Data for Lagos, nr Mushin.
[5] Data for Balboa Heights, Canal Zone, nr Panama.
[6] Data for Jakarta, nr Tanjungkarang-Telukbetung.
[7] Data for Surabaya, nr Kediri.
[8] Data for Ibadan, nr Ado-Ekiti.

Rank	City and Location	Average Temperature (°F)	(°C)	Rank	City and Location	Average Temperature (°F)	(°C)
98.	Basilan, Basilan I, Philippines	80.1[1]	26.7[1]	98.	Pasay, Luzon I, Philippines	80.1[2]	26.7[2]
98.	Caloocan, Luzon I, Philippines	80.1[2]	26.7[2]	98.	Pasig, Luzon I, Philippines	80.1[2]	26.7[2]
98.	Conakry, Guinea	80.1	26.7	98.	Quezon City, Luzon I, Philippines	80.1	26.7
98.	Enugu, Nigeria	80.1	26.7				
98.	Fortaleza, Ceara, Brazil	80.1	26.7	98.	Rajkot, Gujarat, India	80.1	26.7
98.	Holguin, Cuba	80.1[3]	26.7[3]	98.	San Miguel, El Salvador	80.1	26.7
98.	Ibadan, Nigeria	80.1	26.7	98.	Zamboanga, Mindanao I, Philippines	80.1	26.7
98.	Iwo, Nigeria	80.1[4]	26.7[4]				
98.	Jodhpur, Rajasthan, India	80.1	26.7	116.	Malacca, Peninsular Malaysia, Malaysia	80	26.7
98.	Makati, Luzon I, Philippines	80.1[2]	26.7[2]	116.	Qui Nhon, Vietnam	80	26.7
98.	Manila, Luzon I, Philippines	80.1	26.7				

[1] Data for Zamboanga, nr Basilan.
[2] Data for Manila, nr Caloocan, Makati, Pasay, and Pasig.
[3] Data for Gibara, nr Holguin.
[4] Data for Ibadan, nr Iwo.

8f
COOLEST CITIES

Contents

Rank.

Conventional name of city, and subdivision and country in which located.

Averate temperature throughout the year in degrees Fahrenheit and in degrees Celsius (centigrade).

Arrangement

By temperature, from low to high.

Coverage

All cities listed in Table 11b of this work with an average temperature of 40°F (4.4°C) or lower.

Entries

108.

TABLE 8f. COOLEST CITIES

Rank	City and Location	Average Temperature (°F)	(°C)	Rank	City and Location	Average Temperature (°F)	(°C)
1.	Norilsk, Russia in Asia, USSR	12.4[1]	-10.9[1]	20.	Petropavlovsk, Kazakhstan, USSR	32.9	0.5
2.	Yakutsk, Russia in Asia, USSR	13.8	-10.1	23.	Zlatoust, Russia in Asia, USSR	33.1	0.6
3.	Ulan-Bator, Mongolia	23.9	-4.5	24.	Berezniki, Russia in Europe, USSR	33.3	0.7
4.	Fairbanks, Alaska, USA	25.7	-3.5	24.	Novokuznetsk, Russia in Asia, USSR	33.3	0.7
5.	Chita, Russia in Asia, USSR	27.1	-2.7	26.	Archangel, Russia in Europe, USSR	33.4	0.8
6.	Bratsk, Russia in Asia, USSR	28.0	-2.2	26.	Severodvinsk, Russia in Europe, USSR	33.4	0.8
7.	Ulan-Ude, Russia in Asia, USSR	28.9	-1.7	28.	Nizhniy Tagil, Russia in Asia, USSR	33.6	0.9
8.	Angarsk, Russia in Asia, USSR	29.7	-1.3	29.	Barnaul, Russia in Asia, USSR	34.0	1.1
9.	Irkutsk, Russia in Asia, USSR	30.0	-1.1	30.	Tyumen, Russia in Asia, USSR	34.3	1.3
10.	Komsomolsk-na-Amure, Russia in Asia, USSR	30.7	-0.7	31.	Khabarovsk, Russia in Asia, USSR	34.5	1.4
11.	Tomsk, Russia in Asia, USSR	30.9	-0.6	31.	Tselinograd, Kazakhstan, USSR	34.5	1.4
12.	Kemerovo, Russia in Asia, USSR	31.3	-0.4	33.	Kirov, Russia in Europe, USSR	34.7	1.5
13.	Novosibirsk, Russia in Asia, USSR	31.8	-0.1	34.	Kamensk-Uralskiy, Russia in Asia, USSR	34.9[4]	1.6[4]
14.	Blagoveshchensk, Russia in Asia, USSR	32.0	0.0	34.	Kustanay, Kazakhstan, USSR	34.9	1.6
14.	Omsk, Russia in Asia, USSR	32.0	0.0	34.	Rubtsovsk, Russia in Asia, USSR	34.9	1.6
16.	Kurgan, Russia in Asia, USSR	32.4[2]	0.2[2]	34.	Saskatoon, Saskatchewan, Canada	34.9	1.6
16.	Murmansk, Russia in Europe, USSR	32.4	0.2	34.	Sverdlovsk, Russia in Asia, USSR	34.9	1.6
18.	Prokopyevsk, Russia in Asia, USSR	32.7[3]	0.4[3]	39.	Anchorage, Alaska, USA	35.0	1.7
18.	Syktyvkar, Russia in Europe, USSR	32.7	0.4	40.	Chelyabinsk, Russia in Asia, USSR	35.2	1.8
20.	Biysk, Russia in Asia, USSR	32.9	0.5	40.	Kopeysk, Russia in Asia, USSR	35.2[5]	1.8[5]
20.	Krasnoyarsk, Russia in Asia, USSR	32.9	0.5				

[1] Data for Dudinka, nr Norilsk.
[2] Data for Staro-Siderovo, nr Kurgan.
[3] Data for Kiselevsk, nr Prokopyevsk.
[4] Data for Sverdlovsk, nr Kamensk-Uralskiy.
[5] Data for Chelyabinsk, nr Kopeysk.

Rank	City and Location	Average Temperature (°F)	(°C)	Rank	City and Location	Average Temperature (°F)	(°C)
40.	Perm, Russia in Europe, USSR	35.2	1.8	60.	Tsitsihar, Heilungkiang, China	36.9	2.7
43.	Pavlodar, Kazakhstan, USSR	35.4	1.9	60.	Yaroslavl, Russia in Europe, USSR	36.9[2]	2.7[2]
43.	Petropavlovsk-Kamchatskiy, Russia in Asia, USSR	35.4	1.9	63.	Mutankiang, Heilungkiang, China	37.0	2.8
45.	Oulu, Finland	35.6	2.0	64.	Edmonton, Alberta, Canada	37.1	2.8
46.	Regina, Saskatchewan, Canada	35.7	2.1	65.	Cheboksary, Russia in Europe, USSR	37.2	2.9
47.	Izhevsk, Russia in Europe, USSR	35.8	2.1	66.	Orsk, Russia in Europe, USSR	37.4	3.0
47.	Yuzhno-Sakhalinsk, Russia in Asia, USSR	35.8	2.1	66.	Ust-Kamenogorsk, Kazakhstan, USSR	37.4	3.0
49.	Petrozavodsk, Russia in Europe, USSR	36.0	2.2	68.	Dzerzhinsk, Russia in Europe, USSR	37.6[3]	3.1[3]
50.	Karaganda, Kazakhstan, USSR	36.1	2.3	68.	Gorkiy, Russia in Europe, USSR	37.6	3.1
50.	Temir-Tau, Kazakhstan, USSR	36.1[1]	2.3[1]	70.	Semipalatinsk, Kazakhstan, USSR	37.8	3.2
50.	Yoshkar-Ola, Russia in Europe, USSR	36.1	2.3	70.	Ulyanovsk, Russia in Europe, USSR	37.8	3.2
53.	Winnipeg, Manitoba, Canada	36.2	2.3	72.	Harbin, Heilungkiang, China	37.9	3.3
54.	Naberezhnyye Chelny, Russia in Europe, USSR	36.3	2.4	72.	Ivanovo, Russia in Europe, USSR	37.9	3.3
54.	Thunder Bay, Ontario, Canada	36.3	2.4	72.	Vladimir, Russia in Europe, USSR	37.9	3.3
54.	Vologda, Russia in Europe, USSR	36.3	2.4	75.	Chicoutimi, Quebec, Canada	38.0	3.3
57.	Cherepovets, Russia in Europe, USSR	36.7	2.6	75.	Jonquiere, Quebec, Canada	38.0[4]	3.3[4]
57.	Sterlitamak, Russia in Europe, USSR	36.7	2.6	77.	Kiamusze, Heilungkiang, China	38.1	3.4
57.	Ufa, Russia in Europe, USSR	36.7	2.6	78.	Calgary, Alberta, Canada	38.2	3.4
60.	Kostroma, Russia in Europe, USSR	36.9	2.7	79.	Butte, Montana, USA	38.4	3.6
				80.	Aktyubinsk, Kazakhstan, USSR	38.5	3.6

[1] Data for Karaganda, nr Temir-Tau.
[2] Data for Kostroma, nr Yaroslavl.
[3] Data for Gorkiy, nr Dzerzhinsk.
[4] Data for Chicoutimi, nr Jonquiere.

Rank	City and Location	Average Temperature (°F)	(°C)	Rank	City and Location	Average Temperature (°F)	(°C)
80.	Kaluga, Russia in Europe, USSR	38.5	3.6	96.	Sault Sainte Marie, Ontario, Canada	39.3	4.1
80.	Kazan, Russia in Europe, USSR	38.5	3.6	97.	Grand Forks, North Dakota, USA	39.6	4.2
83.	Duluth, Minnesota, USA	38.6	3.7	98.	Leningrad, Russia in Europe, USSR	39.7	4.3
84.	Saransk, Russia in Europe, USSR	38.7	3.7	98.	Tula, Russia in Europe, USSR	39.7	4.3
85.	Kisi, Heilungkiang, China	38.8	3.8	100.	Vitebsk, White Russia, USSR	39.8[1]	4.3[1]
85.	Kuybyshev, Russia in Europe, USSR	38.8	3.8	101.	Kirin, Kirin, China	39.9	4.4
85.	Tampere, Finland	38.8	3.8	101.	Lyubertsy, Russia in Europe, USSR	39.9[2]	4.4[2]
88.	Kalinin, Russia in Europe, USSR	39.0	3.9	101.	Moscow, Russia in Europe, USSR	39.9	4.4
88.	Novgorod, Russia in Europe, USSR	39.0	3.9	101.	Podolsk, Russia in Europe, USSR	39.9[2]	4.4[2]
88.	Orenburg, Russia, in Europe, USSR	39.0	3.9	101.	Ryazan, Russia in Europe, USSR	39.9	4.4
88.	Penza, Russia in Europe, USSR	39.0	3.9	101.	Syzran, Russia in Europe, USSR	39.9	4.4
88.	Sundsvall, Sweden	39.0	3.9	101.	Tolyatti, Russia in Europe, USSR	39.9	4.4
88.	Yinchwan, Ningsia, China	39.0	3.9	101.	Uralsk, Kazakhstan, USSR	39.9	4.4
94.	Smolensk, Russia in Europe, USSR	39.2	4.0				
94.	Vladivstok, Russia in Asia, USSR	39.2	4.0				

[1] Data for Novoye Korolevo, nr Vitebsk.
[2] Data for Moscow, nr Lyubertsy and Podolsk.

8g
CITIES WITH MOST PRECIPITATION

Contents

Rank.

Conventional name of city, and subdivision and country in which located.

Average annual precipitation (i.e., rainfall plus the rain equivalent of snowfall), in inches and in millimeters.

Arrangement

By precipitation.

Coverage

All cities listed in Table 11b of this work with annual precipitation averaging at least 70 in (1778 mm).

Entries

113.

TABLE 8g. CITIES WITH MOST PRECIPITATION

Rank	City and Location	Annual Precipitation (in)	(mm)	Rank	City and Location	Annual Precipitation (in)	(mm)
1.	Monrovia, Liberia	202.01	5131	35.	Ipoh, Peninsular Malaysia, Malaysia	101.60	2581
2.	Moulmein, Burma	189.76	4820				
3.	Padang, Sumatra I, Indonesia	175.28	4452	36.	Thana, Maharashtra, India	97.52	2477
				37.	Fukui, Honshu I, Japan	97.32	2472
4.	Conakry, Guinea	170.91	4341	38.	Kuala Lumpur, Peninsular Malaysia, Malaysia	96.1	2441
5.	Bogor, Java I, Indonesia	166.34	4225				
				39.	Kagoshima, Kyushu I, Japan	95.79	2433
6.	Douala, Cameroon	161.77	4109				
7.	Cayenne, French Guiana	147.40	3744	40.	Singapore, Singapore	95.08	2415
8.	Freetown, Sierra Leone	143.27	3639	41.	Colombo, Sri Lanka	94.3	2395
9.	Ambon, Ambon I, Indonesia	138.98	3530	41.	Port Harcourt, Nigeria	94.3	2395
				43.	Toyama, Honshu I, Japan	94.02	2388
10.	Mangalore, Karnataka, India	133.78	3398	44.	Rhondda, Wales, UK	93.87	2384
				45.	Palembang, Sumatra I, Indonesia	93.74	2381
11.	Hilo, Hawaii, USA	133.57	3393				
12.	Pontianak, Borneo I, Indonesia	131.69	3345	46.	Shizuoka, Honshu I, Japan	92.72	2355
13.	Bandar Seri Begawan, Brunei	131.0	3327	46.	Ulhasnagar-Kalyan, Maharashtra, India	92.72	2355
14.	Colon, Panama	130.75	3321	48.	Quezon City, Luzon I, Philippines	91.62	2327
15.	Calicut, Kerala, India	125.12	3178				
16.	Suva, Fiji	124.41	3160	49.	Nikko, Honshu I, Japan	89.41	2271
17.	Libreville, Gabon	122.83	3120	50.	Cordoba, Veracruz, Mexico	89.29	2268
18.	Jambi, Sumatra I, Indonesia	122.01	3099				
				51.	Cirebon, Java I, Indonesia	88.90	2258
19.	Cochin, Kerala, India	119.96	3047	51.	Paramaribo, Surinam	88.90	2258
20.	Keelung, Taiwan	119.80	3043	53.	Iloilo, Panay I, Philippines	88.43	2246
21.	Hue, Vietnam	118.78	3017	54.	Santos, Sao Paulo, Brazil	87.87	2232
22.	Coatzacoalcos, Veracruz, Mexico	113.54	2884	55.	Georgetown, Guyana	87.44	2221
23.	Minatitlan, Veracruz, Mexico	113.23	2876	56.	Malacca, Peninsular Malaysia, Malaysia	86.8	2205
24.	Chittagong, Bangladesh	112.52	2858	57.	Belem, Para, Brazil	86.10	2187
25.	Apia, Samoa	111.77	2839	58.	Bacolod, Negros I, Philippines	85.55	2173
26.	Ujung Pandang, Celebes I, Indonesia	109.21	2774	59.	Surakarta, Java I, Indonesia	85.24	2165
27.	Bassein, Burma	108.98	2768	60.	Bandung, Java I, Indonesia	85.00	2159
28.	Banjarmasin, Borneo I, Indonesia	108.46	2755	61.	Hong Kong, Hong Kong	84.92	2157
29.	George Town, Peninsular Malaysia, Malaysia	107.7	2736	62.	Dehra Dun, Uttar Pradesh, India	84.61	2149
30.	Menado, Celebes I, Indonesia	106.38	2702	63.	Abidjan, Ivory Coast	84.41	2144
31.	Kanazawa, Honshu I, Japan	104.80	2662	64.	Tanjungkarang-Telukbetung, Sumatra I, Indonesia	83.70	2126
32.	Rangoon, Burma	104.29	2649				
33.	Kochi, Shikoku I, Japan	104.09	2644	65.	Petropolis, Rio de Janeiro, Brazil	83.54	2122
34.	Miyazaki, Kyushu I, Japan	102.13	2594	66.	Naha, Okinawa I, Japan	83.39	2118
				67.	Orizaba, Veracruz, Mexico	83.27	2115

Rank	City and Location	Annual Precipitation (in)	(mm)	Rank	City and Location	Annual Precipitation (in)	(mm)
68.	Taipei, Taiwan	82.68	2100	90.	San Jose, Costa Rica	76.26	1937
69.	Numazu, Honshu I, Japan	82.44	2094	92.	Davao, Mindanao I, Philippines	75.91	1928
70.	Da Nang, Vietnam	81.61	2073				
71.	Manila, Luzon I, Philippines	81.46	2069	93.	Kaohsiung, Taiwan	75.71	1923
				94.	Villahermosa, Tabasco, Mexico	75.51	1918
72.	Manizales, Colombia	80.91	2055				
73.	Pereira, Colombia	80.87	2054	95.	Hamamatsu, Honshu I, Japan	75.00	1905
74.	Ibague, Colombia	80.83	2053				
74.	Medan, Sumatra I, Indonesia	80.83	2053	96.	Gifu, Honshu I, Japan	74.96	1904
				97.	Panama, Panama	74.92	1903
76.	Semarang, Java I, Indonesia	80.04	2033	98.	Sao Luis, Maranhao, Brazil	74.05	1881
77.	Kandy, Sri Lanka	79.7	2024	99.	Dacca, Bangladesh	73.35	1863
78.	Santa Ana, El Salvador	79.61	2022	100.	Fort-de-France, Martinique	73.19	1859
79.	Narayanganj, Bangladesh	79.21	2012				
80.	Sasebo, Kyushu I, Japan	78.90	2004	101.	Niigata, Honshu I, Japan	72.83	1850
81.	Yogyakarta, Java I, Indonesia	78.82	2002	102.	Macao, Macao	72.68	1846
				103.	Papeete, French Polynesia	72.60	1844
82.	Malang, Java I, Indonesia	78.46	1993	104.	Tainan, Taiwan	72.40	1839
83.	Bissau, Guinea-Bissau	78.19	1986	105.	Ogwr, Wales, UK	72.17	1833
84.	Nagasaki, Kyushu I, Japan	77.80	1976	106.	Salvador, Bahia, Brazil	72.13	1832
85.	Kweilin, Kwangsi, China	77.40	1966	107.	Trivandrum, Kerala, India	71.34	1812
86.	Bergen, Norway	77.09	1958	108.	Manaus, Amazonas, Brazil	71.3	1811
87.	Kumamoto, Kyushu I, Japan	76.34	1939	109.	Akita, Honshu I, Japan	71.14	1807
				110.	Bombay, Maharashtra, India	71.06	1805
87.	Yokosuka, Honshu I, Japan	76.34	1939				
				111.	Kediri, Java I, Indonesia	70.63	1794
89.	Onitsha, Nigeria	76.3	1938	112.	San Salvador, El Salvador	70.55	1792
90.	Ho Chi Minh City, Vietnam	76.26	1937	113.	Taichung, Taiwan	70.24	1784

8h
CITIES WITH LEAST PRECIPITATION

Contents

Rank.

Conventional name of city, and subdivision and country in which located.

Average annual precipitation (i.e., rainfall plus the rain equivalent of snowfall), in inches and in millimeters.

Arrangement

By precipitation, inversely.

Coverage

All cities listed in Table 11b of this work with annual precipitation averaging 12 in (305 mm) or less.

Entries

109.

Figure 29. A Pacific Ocean port on the edge of the Atacama Desert, Antofagasta, Chile, is probably the city with the least precipitation. (Credit: Organization of American States.)

TABLE 8h. CITIES WITH LEAST PRECIPITATION

Rank	City and Location	Annual Precipitation (in)	(mm)	Rank	City and Location	Annual Precipitation (in)	(mm)
1.	Antofagasta, Chile	0.02	0.4	40.	Khartoum, Sudan	6.34	161
2.	Luxor, Egypt	0.02	0.5	42.	Torreon, Coahuila, Mexico	6.50	165
3.	Aswan, Egypt	0.04	1	43.	Multan, Pakistan	6.57	167
4.	Asyut, Egypt	0.20	5	44.	Potosi, Bolivia	6.69	170
5.	Callao, Peru	0.47	12	45.	Astrakhan, Russia in Europe, USSR	6.89	175
6.	Trujillo, Peru	0.54	14				
7.	Suez, Egypt	0.87	22	46.	Phoenix, Arizona, USA	7.05	179
8.	Giza, Egypt	1.10	28	47.	Reno, Nevada, USA	7.20	183
9.	Cairo, Egypt	1.14	29	48.	Sumgait, Azerbaijan, USSR	7.28	185
10.	Zagazig, Egypt	1.18	30	49.	Ciudad Juarez, Chihuahua, Mexico	7.36	187
11.	Lima, Peru	1.22	31				
12.	Aden, South Yemen	1.61	41	50.	Alexandria, Egypt	7.40	188
13.	Chiclayo, Peru	1.63	41	50.	Namangan, Uzbekistan, USSR	7.40	188
14.	Tanta, Egypt	1.65	42				
15.	Jidda, Hejaz, Saudi Arabia	2.09	53	52.	Ahwaz, Iran	7.48	190
16.	Mansurah, Egypt	2.13	54	53.	Mesa, Arizona, USA	7.52	191
17.	Dubayy, United Arab Emirates	2.4	61	54.	Tempe, Arizona, USA	7.63	194
				55.	Mendoza, Mendoza, Argentina	7.76	197
18.	Medina, Hejaz, Saudi Arabia	2.44	62	55.	Sfax, Tunisia	7.76	197
19.	Doha, Qatar	2.5	63	57.	Albuquerque, New Mexico USA	7.77	197
20.	Port Said, Egypt	2.95	75				
21.	Mexicali, Baja California, Mexico	2.99	76	57.	El Paso, Texas, USA	7.77	197
				59.	Ulan-Ude, Russia in Asia, USSR	7.95	202
22.	Kashgar, Sinkiang, China	3.07	78	60.	Yinchwan, Ningsia, China	7.99	203
23.	Manama, Bahrain	3.23	82	61.	Yakima, Washington, USA	8.00	203
24.	San Juan, San Juan, Argentina	3.66	93	62.	Las Cruces, New Mexico, USA	8.01	203
25.	Las Vegas, Nevada, USA	3.76	96	63.	Karachi, Pakistan	8.03	204
26.	Riyadh, Nejd, Saudi Arabia	3.82	97	64.	Scottsdale, Arizona, USA	8.06	205
				65.	Meshed, Iran	8.23	209
27.	Damanhur, Egypt	4.06	103	66.	Andizhan, Uzbekistan, USSR	8.27	210
28.	Arequipa, Peru	4.31	109				
29.	Isfahan, Iran	4.61	117	66.	Ashkhabad, Turkmenia, USSR	8.27	210
30.	Kuwait, Kuwait	5.04	128				
30.	Zarqa, Jordan	5.04	128	68.	Ciudad Obregon, Sonora, Mexico	8.31	211
32.	Djibouti, Djibouti	5.12	130				
33.	Bukhara, Uzbekistan, USSR	5.31	135	69.	Norilsk, Russia in Asia, USSR	8.39	213
34.	Palm Springs, California, USA	5.33	135	69.	Yakutsk, Russia in Asia, USSR	8.39	213
35.	Abadan, Iran	5.67	144	71.	Kandahar, Afghanistan	8.86	225
36.	Bakersfield, California, USA	5.72	145	71.	Timbuktu, Mali	8.86	225
				73.	Idaho Falls, Idaho, USA	8.89	226
37.	Nouakchott, Mauritania	5.90	150	74.	Damascus, Syria	8.98	228
38.	Baghdad, Iraq	6.14	156	75.	Tehran, Iran	9.02	229
39.	Hyderabad, Pakistan	6.18	157	76.	Almeria, Andalucia, Spain	9.09	231
40.	Basra, Iraq	6.34	161				

Rank	City and Location	Annual Precipitation (in)	(mm)	Rank	City and Location	Annual Precipitation (in)	(mm)
76.	Herat, Afghanistan	9.09	231	93.	Semipalatinsk, Kazakhstan, USSR	10.51	267
78.	Urumtsi, Sinkiang, China	9.17	233	94.	Amman, Jordan	10.75	273
79.	Baku, Azerbaijan, USSR	9.37	238	95.	Uralsk, Kazakhstan, USSR	10.79	274
80.	San Diego, California, USA	9.45	240	96.	Pocatello, Idaho, USA	10.80	274
81.	Kirovabad, Azerbaijan, USSR	9.76	248	97.	Tucson, Arizona, USA	11.05	281
81.	Marrakesh, Morocco	9.76	248	98.	Casper, Wyoming, USA	11.22	285
83.	Santa Cruz de Tenerife, (Canary Islands), Spain	9.88	251	98.	Fairbanks, Alaska, USA	11.22	285
84.	Riverside, California, USA	9.92	252	98.	Ismailia, Egypt	11.22	285
85.	Pavlodar, Kazakhstan, USSR	10.00	254	101.	Dzhambul, Kazakhstan, USSR	11.30	287
85.	Ulan-Bator, Mongolia	10.00	254	102.	Helena, Montana, USA	11.38	289
87.	Aktyubinsk, Kazakhstan, USSR	10.20	259	103.	Boise City, Idaho, USA	11.50	292
88.	Fresno, California, USA	10.24	260	104.	Hama, Syria	11.77	299
89.	Long Beach, California, USA	10.25	260	105.	Bratsk, Russia in Asia, USSR	11.85	301
90.	Praia, Cape Verde	10.28	261	105.	Murcia, Murcia, Spain	11.85	301
91.	Tabriz, Iran	10.35	263	107.	Modesto, California, USA	11.87	301
92.	Bengasi, Cyrenaica, Libya	10.47	266	108.	Pueblo, Colorado, USA	11.91	303
				109.	Paotow, Inner Mongolia, China	11.97	304

9a
HIGHEST BUILDINGS

Contents

Conventional and alternate names of building.
City, and subdivision and country in which located.
Year of completion.
Height above street level, in feet and meters.

Arrangement

By height.

Coverage

Above * * *, all buildings standing or under construction, excluding observation and
television towers, with a height of at least 600 ft (183m).
Below * * *, other well-known high buildings.

Entries

141.

Figure 30. Sears Tower, the highest office building ever constructed, dwarfs most of the other skyscrapers in Chicago's business center. (Credit: Sears, Roebuck and Co.)

Figure 31. Twin towers of the World Trade Center, New York, the second highest office building in the world. (Credit: New York Convention and Visitors Bureau, Inc.)

TABLE 9a. HIGHEST BUILDINGS

Rank	Building	Location	Year Completed[1]	Height (ft)	Height (m)
1.	Sears Tower	Chicago, Illinois, USA	1974	1454	443
2.	World Trade Center	New York, New York, USA	1972	1350[2]	411[2]
3.	Empire State	New York, New York, USA	1931	1250	381
4.	Standard Oil (Indiana)	Chicago, Illinois, USA	1971	1136	346
5.	John Hancock Center	Chicago, Illinois, USA	1967	1127	344
6.	Chrysler	New York, New York, USA	1930	1046	319
7.	Eiffel Tower (off Tour Eiffel)[3]	Paris, Region Parisienne, France	1889	984	300
8.	60 Wall Tower (for Cities Service)	New York, New York, USA	1934	950	290
9.	First Bank Tower (alt Bank of Montreal, Banque de Montreal, First Canadian Place)	Toronto, Ontario, Canada	1976	935	285
10.	40 Wall Street (for Bank of Manhattan)	New York, New York, USA	1930	927	283
11.	Citicorp Center	New York, New York, USA	1977	914	279
12.	Water Tower Place	Chicago, Illinois, USA	1976	859	262
13.	United California Bank	Los Angeles, California, USA	1973	858	262
14.	Transamerica Pyramid	San Francisco, California, USA	1973	853	260
15.	First National Bank	Chicago, Illinois, USA	1969	850	259
15.	RCA	New York, New York, USA	1933	850	259
17.	United States Steel	Pittsburgh, Pennsylvania, USA	1969	841	256
18.	Chase Manhattan	New York, New York, USA	1960	813	248
19.	Pan Am	New York, New York, USA	1961	808	246
20.	Woolworth	New York, New York, USA	1913	792	241
21.	John Hancock Tower	Boston, Massachusetts, USA	1970	790	241
22.	Commerce Court West (alt Canadian Imperial Bank of Commerce)	Toronto, Ontario, Canada	1971	784	239
23.	Bank of America	San Francisco, California, USA	1969	778	237
24.	IDS Center	Minneapolis, Minnesota, USA	1971	772	235

[1]UC, under construction.
[2]Two buildings of equal height.
[3]Primarily an observation tower.

Rank	Building	Location	Year Completed[1]	Height (ft)	(m)
25.	Palace of Culture and Science (off Palac Kultury i Nauki)	Warsaw, Poland	1955	768	234
26.	One Penn Plaza	New York, New York, USA	1972	766	233
27.	MLC Centre	Sydney, New South Wales, Australia	1974	760	232
28.	Exxon (alt 1251 Avenue of the Americas	New York, New York, USA	1971	750	229
28.	Prudential Tower	Boston, Massachusetts, USA	1964	750	229
30.	United States Steel (alt One Liberty Plaza)	New York, New York, USA	1971	743	226
31.	Citibank (alt 20 Exchange Place)	New York, New York, USA	1931	741	226
31.	Ikebokoru	Tokyo, Honshu I, Japan	1977?	741	226
33.	Detroit Plaza Hotel	Detroit, Michigan, USA	1977	740	226
34.	Security Pacific National Bank	Los Angeles, California, USA	1973	738	225
34.	Shinjuku Mitsui	Tokyo, Honshu I, Japan	1974	738	225
36.	Toronto-Dominion Bank Tower	Toronto, Ontario, Canada	1967	736	224
37.	One Astor Place	New York, New York, USA	1971	731	223
38.	Peachtree Center Plaza Hotel	Atlanta, Georgia, USA	1976	724	221
39.	Moscow M. V. Lomonosov State University	Moscow, Russia, USSR	1953	720	219
40.	One Shell Plaza	Houston, Texas, USA	1971	715	218
41.	First International	Dallas, Texas, USA	1973	710	216
42.	Terminal Tower	Cleveland, Ohio, USA	1930	708	216
43.	Union Carbide	New York, New York, USA	1960	707	215
44.	General Motors	New York, New York, USA	1967	705	215
45.	Metropolitan Life	New York, New York, USA	1909	700	213
46.	Atlantic Richfield	Los Angeles, California, USA	1971	699	213
46.	Bank of America	Los Angeles, California, USA	1971	699	213
48.	500 Fifth Avenue	New York, New York, USA	1930	697	212
48.	One Shell Square	New Orleans, Louisiana, USA	1972	697	212
50.	Avon Products (alt Nine West 57th Street)	New York, New York, USA	1972	688	210
51.	Chemical Bank	New York, New York, USA	1964	687	209

[1]UC, under construction.

Rank	Building	Location	Year Completed[1]	Height (ft)	Height (m)
51.	55 Water Street	New York, New York, USA	1972	687	209
53.	Chanin	New York, New York, USA	1928	680	207
53.	Tour Maine-Montparnasse	Paris, Region Parisienne, France	1972	680	207
55.	Gulf and Western	New York, New York, USA	1969	679	207
56.	One Houston Center	Houston, Texas, USA	1977	678	207
57.	Marine Midland	New York, New York, USA	1966	677	206
58.	McGraw-Hill (alt 1221 Avenue of the Americas)	New York, New York, USA	1972	674	205
59.	Lincoln	New York, New York, USA	1929	673	205
60.	IBM	Chicago, Illinois, USA	1971	670	204
60.	1633 Broadway	New York, New York, USA	1971	670	204
62.	Bank of Oklahoma Tower	Tulsa, Oklahoma, USA	1975	667	203
63.	Civic Center (alt City Hall)	Chicago, Illinois, USA	1965	662	202
64.	Carlton Centre Tower	Johannesburg, Transvaal, South Africa	1971	656	200
64.	OCBC Centre (alt Overseas Chinese Banking Corporation)	Singapore, Singapore	1976	656	200
64.	Shinjuku Sumitomo	Tokyo, Honshu I, Japan	1973	656	200
64.	Torre Parque Central	Caracas, Venezuela		656	200
68.	1100 Milam	Houston, Texas, USA	1974	651	198
69.	Ukraina Hotel	Moscow, Russia, USSR	bef 1961	650	198
70.	American Brands	New York, New York, USA	1967	648	198
71.	Lake Point Tower	Chicago, Illinois, USA	1968	645	197
72.	Irving Trust	New York, New York, USA	1930	640	195
73.	345 Park Avenue	New York, New York, USA	1968	634	193
74.	Yasuda Kasai	Tokyo, Honshu I, Japan	1975?	633	193
75.	Gateway Arch	Saint Louis, Missouri, USA	1965	630	192
75.	Grace Plaza (alt 1114 Avenue of the Americas)	New York, New York, USA	1972	630	192
75.	Home Insurance Company	New York, New York, USA	1966	630	192
75.	One New York Plaza	New York, New York, USA	1969	630	192
79.	One Hammarskjold Plaza	New York, New York, USA	1972	628	191
80.	Burlington House	New York, New York, USA	1969	625	190

[1]UC, under construction.

Rank	Building	Location	Year Completed[1]	Height (ft)	Height (m)
80.	First National Bank	Dallas, Texas, USA	1965	625	190
80.	First Wisconsin Center and Office Tower	Milwaukee, Wisconsin, USA	1973	625	190
80.	Waldorf-Astoria Hotel	New York, New York, USA	1931	625	190
84.	Place Victoria (alt Bourse, Stock Exchange)	Montreal, Quebec, Canada	1965	624	190
84.	State Office Tower	Columbus, Ohio, USA	1973	624	190
86.	Crocker-Citizens Plaza	Los Angeles, California, USA	1967	620	189
86.	Olympic Tower	New York, New York, USA	1974	620	189
86.	Ten East 40th Street	New York, New York, USA	1928	620	189
89.	General Electric	New York, New York, USA	1939	616	188
90.	New York Life	New York, New York, USA	1928	615	187
91.	Place Ville-Marie (alt Banque Royale du Canada, Royal Bank of Canada)	Montreal, Quebec, Canada	1962	612	187
92.	Development Bank of Singapore (alt DBS)	Singapore, Singapore	1973	610	186
93.	Penney	New York, New York, USA	1964	609	186
93.	Royal Trust Tower	Toronto, Ontario, Canada	1969	609	186
93.	Seattle-First National Bank	Seattle, Washington, USA	1969	609	186
96.	Qantas	Sydney, New South Wales, Australia	1971	607	185
97.	Exxon (for Humble Oil)	Houston, Texas, USA	1961	606	185
98.	Board of Trade	Chicago, Illinois, USA	1930	605	184
99.	Canadian Imperial Bank of Commerce (alt Banque de Commerce Canadienne Imperiale)	Montreal, Quebec, Canada	1962	604	184
99.	Federal Reserve	Boston, Massachusetts, USA	1977	604	184
101.	Australia Square Tower	Sydney, New South Wales, Australia	1968	602	183
102.	Boston Company	Boston, Massachusetts, USA	1970	601	183
102.	Prudential	Chicago, Illinois, USA	1954	601	183
104.	560 Lexington Avenue	New York, New York, USA	1977	600	183
104.	National Westminster Bank	London, England, UK	UC	600	183
	Two Dallas Centre	Dallas, Texas, USA	UC	(51 stories)	

[1]UC, under construction.

Rank	Building	Location	Year Completed[1]	Height (ft)	Height (m)
	(Edificio) Italia	Sao Paulo, Sao Paulo, Brazil	1966	(45 stories)	
		* * *			
	Republic Bank Tower	Dallas, Texas, USA	1964	598	182
	Embarcadero Center	San Francisco, California, USA	1973	595	181
	Empire State Plaza Tower	Albany, New York, USA	1969	589	180
	Marina City Apartments	Chicago, Illinois, USA	1962	588[2]	179[2]
	Connaught Centre	Hong Kong, Hong Kong	1973	585	178
	Harbour Centre	Vancouver, British Columbia, Canada	1974	581	177
	Post Office Tower	London, England, UK	1966	580	177
	Carew Tower	Cincinnati, Ohio, USA	1930	574	175
	Keio Plaza Hotel	Tokyo, Honshu I, Japan	1970	558	170
	City National Bank (alt Penobscot)	Detroit, Michigan, USA	1928	557	170
	Washington Monument[3]	Washington, District of Columbia, USA	1884	555	169
	Vehicle Assembly	Cape Canaveral, Florida, USA	1966	552	168
	United Nations Secretariat	New York, New York, USA	1950	550	168
	City Hall	Philadelphia, Pennsylvania, USA	1901	548	167
	Mole Antonelliana	Turin, Piemonte, Italy	1863	548	167
	Torre de Madrid	Madrid, Castilla la Nueva, Spain	1959	541	165
	Cathedral of Learning (alt University of Pittsburgh)	Pittsburgh, Pennsylvania, USA	1956	535	163
	Cathedral (off Munster)	Ulm, Baden-Wurttemberg, West Germany	1890	529	161
	Banco do Estado de Sao Paulo	Sao Paulo, Sao Paulo, Brazil	1946	528	161
	Seagram	New York, New York, USA	1958	525	160
	Cathedral (off Dom Sankt Peter)	Cologne, Nordrhein-Westfalen, West Germany	1880	515	157
	Cathedrale Notre-Dame	Rouen, Haute-Normandie, France	1530	512	156
	Karl-Marx-Universitat	Leipzig, East Germany	1971	502	153
	World Trade Center	Tokyo, Honshu I, Japan	1969	499	152
	CBS	New York, New York, USA	1965	491	150

[1] UC, under construction.
[2] Two buildings of equal height.
[3] Primarily an observation tower.

Rank	Building	Location	Year Completed[1]	Height (ft)	Height (m)
	Kasumigaseki	Tokyo, Honshu I, Japan	1968	482	147
	Tribune Tower	Chicago, Illinois, USA	1925	462	141
	Pyramid of Khufu (alt Pyramid of Cheops)	Giza, Egypt	2568 BC?	450	137
	Saint Peter's Basilica (off Basilica di San Pietro)	Vatican City	1615	435	132
	Bayer A. G.	Leverkusen, Nordrhein-Westfalen, West Germany	1962	434	132
	Pirelli	Milan, Lombardia, Italy	1960	414	126
	Shwedagon Pagoda	Rangoon, Burma	1774	326	99
	Statue of Liberty	New York, New York, USA	1886	305	93
	Leaning Tower (alt Torre Pendente; off Campanile)	Pisa, Toscana, Italy	1350	179	54

[1]UC, under construction.

Figure 32. Styles in contemporary skyscraper architecture. Top left: MLC Centre, Sydney (1974); top right: Transamerica Pyramid, San Francisco (1973); bottom left: United States Steel Building, Pittsburgh (1969); bottom right: Detroit Plaza Hotel, the world's highest hostelry, surrounded by four office towers of Renaissance Center, Detroit (1977). (Credit, respectively: Australian Information Service; Transamerica Corporation; United States Steel Corporation; Metropolitan Detroit Convention & Visitors Bureau.)

9b
LONGEST BRIDGES (SPAN)

Contents

Conventional and alternate names of bridge.
Body of water spanned, and subdivision and country in which located.
Type of bridge.
Use to which bridge is put.
Year of completion.
Longest span, in feet and meters.

Arrangement

By length of span.

Coverage

All bridges standing or under construction with spans of at least the following lengths:
Suspension bridges—1200 ft (366 m)
Cantilever bridges—1100 ft (335 m)
Cable-stayed girder bridges—900 ft (274 m)
Continuous truss bridges—900 ft (274 m)
Steel arch bridges—900 ft (274 m)
Concrete arch bridges—750 ft (229 m)
Concrete cable-stayed girder bridges—750 ft (229 m)
Continuous plate and box girder bridges—750 ft (229 m)
Additionally, the longest span bridge is listed for each of the following types: bascule, simple truss, swing span, and vertical lift.

Entries

157.

Figure 33. Artist's impression of the Humber suspension bridge near Hull, England, which, when completed, will have the longest span on record. (Credit: British Information Services.)

Figure 34. Verrazano-Narrows suspension bridge, New York City. This bridge, with the longest span thus far completed, crosses the channel between the upper and lower sections of New York Bay, connecting Brooklyn with Staten Island. (Credit: New York Convention and Visitors Bureau, Inc.)

TABLE 9b. LONGEST BRIDGES (SPAN)

Bridge	Body of Water and Location	Type[1]	Use[2]	Year Completed[3]	Span (ft)	Span (m)
Humber	Humber R, England, UK	S	H	UC (1979?)	4626	1410
Verrazano-Narrows	New York B of Atlantic O, New York, USA	S	H	1964	4260	1298
Golden Gate	San Francisco B of Pacific O, California, USA	S	H	1937	4200	1280
Mackinac	Straits of Mackinac, Michigan, USA	S	H	1957	3800	1158
Bosporus	Bosporus Strait, Turkey	S	H	1973	3524	1074
George Washington	Hudson R, New Jersey–New York, USA	S	H	1931	3500	1067
25th of April (for Salazar; off 25 de Abril)	Tagus R, Portugal	S	H; RR	1966	3323	1013
Forth Road	Forth R, Scotland, UK	S	H	1964	3300	1006
Severn	Severn R, England–Wales, UK	S	H	1966	3240	988
Tacoma Narrows II	Puget Sound of Pacific O, Washington, USA	S	H	1950	2800	853
Angostura	Orinoco R, Venezuela	S	H	1967	2336	712
Kammon	Kammon Strait, Honshu I-Kyushu I, Japan	S	H	1973	2336	712
San Francisco-Oakland Bay (alt Transbay)	San Francisco B of Pacific O, California, USA	S	H; RT	1936	2310[4]	704[4]
Bronx-Whitestone	East R, New York, USA	S	H	1939	2300	701
Pierre Laporte (alt Frontenac, Quebec Road)	Saint Lawrence R, Quebec, Canada	S	H	1970	2190	668
Delaware Memorial I	Delaware R, Delaware–New Jersey, USA	S	H	1951	2150	655
Delaware Memorial II	Delaware R, Delaware–New Jersey, USA	S	H	1968	2150	655
Seaway Skyway	Saint Lawrence R, Canada–USA	S	H	1960	2150	655
Walt Whitman	Delaware R, New Jersey–Pennsylvania, USA	S	H	1957	2000	610
Tancarville	Seine R, Haute-Normandie, France	S	H	1959	1995	608
Little Belt (off Lillebaelt)	Little Belt Strait, Denmark	S	H	1969	1969	600
Ambassador	Detroit R, Canada–USA	S	H	1929	1850	564

[1] B, bascule; C, cantilever; CA, concrete arch; CCSG, concrete cable-stayed girder; CPBG, continuous plate and box girder; CSG, cable-stayed girder; CT, continuous truss; S, suspension; SA, steel arch; SS, swing span; ST, simple truss; VL, vertical lift.

[2] H, highway; RR, railroad; RT, rapid transit.

[3] UC, under construction.

[4] Two spans of this length, and two additional spans each of 1160 ft, 354 m.

Bridge	Body of Water and Location	Type[1]	Use[2]	Year Completed[3]	Span (ft)	Span (m)
Quebec	Saint Lawrence R, Quebec, Canada	C	H; RR[4]	1917	1800	549
Throgs Neck	East R, New York, USA	S	H	1961	1800	549
Benjamin Franklin	Delaware R, New Jersey–Pennsylvania, USA	S	H; RT	1926	1750	533
Kvalsund	Kvalsund R, Norway	S	H	1977	1723	525
Skjomen	Skjomen Fjord of Norwegian Sea, Norway	S	H	1971	1723	525
Forth	Forth R, Scotland, UK	C	RR	1890	1710[5]	521[5]
New River Gorge	New R, West Virginia, USA	SA	H	1977	1700	518
Osaka Port	Yodo R, Honshu I, Japan	C	H	1974	1673	510
Bayonne (alt Kill van Kull)	Kill van Kull, New Jersey–New York, USA	SA	H	1931	1652	504
Sydney Harbour	Sydney Harbor of Pacific O, New South Wales, Australia	SA	H; RR	1932	1650	503
Commodore John Barry (alt Chester)	Delaware R, New Jersey–Pennsylvania, USA	C	H	1973	1644	501
Kleve-Emmerich	Rhine R, Nordrhein-Westfalen, West Germany	S	H	1965	1640	500
Bear Mountain	Hudson R, New York, USA	S	H	1924	1632	497
Chesapeake Bay I (alt William Preston Lane, Jr, Memorial I)	Chesapeake B of Atlantic O, Maryland, USA	S	H	1952	1600	488
Chesapeake Bay II (alt William Preston Lane, Jr, Memorial II)	Chesapeake B of Atlantic O, Maryland, USA	S	H	1973	1600	488
Newport	Narragansett B of Atlantic O, Rhode Island, USA	S	H	1969	1600	488
Williamsburg	East R, New York, USA	S	H; RT	1903	1600	488
Brooklyn	East R, New York, USA	S	H; RT	1883	1595	486
Greater New Orleans	Mississippi R, Louisiana, USA	C	H	1958	1575	480
Lions Gate	Burrard Inlet of Strait of Georgia, British Columbia, Canada	S	H	1939	1550	472
Sotra	Vatle(straumen) Channel, Norway	S	H	1971	1535	468
Hooghly I (alt Howrah)	Hooghly R, West Bengal, India	C	H; RT	1943	1500	457

[1] B, bascule; C, cantilever; CA, concrete arch; CCSG, concrete cable-stayed girder; CPBG, continuous plate and box girder; CSG, cable-stayed girder; CT, continuous truss; S, suspension; SA, steel arch; SS, swing span; ST, simple truss; VL, vertical lift.

[2] H, highway; RR, railroad; RT, rapid transit.

[3] UC, under construction.

[4] Railroad traffic only since opening of Pierre Laporte Bridge in 1970.

[5] Two spans of this length.

Bridge	Body of Water and Location	Type[1]	Use[2]	Year Completed[3]	Span (ft)	Span (m)
Hooghly II	Hooghly R, West Bengal, India	CSG	H	1977	1500	457
Mid-Hudson (alt Poughkeepsie)	Hudson R, New York, USA	S	H	1930	1500	457
Vincent Thomas (alt San Pedro-Terminal Island)	San Pedro B of Pacific O, California, USA	S	H	1964	1500	457
Manhattan	East R, New York, USA	S	H; RT	1909	1470	448
Angus L. Mac-Donald (alt Halifax Harbour)	Halifax Harbor of Atlantic O, Nova Scotia, Canada	S	H	1955	1447	441
A. Murray Mac-Kay	Halifax Harbor of Atlantic O, Nova Scotia, Canada	S	H	1970	1400	427
San Francisco-Oakland Bay (alt Transbay)	San Francisco B of Pacific O, California, USA	C	H; RT	1936	1400	427
Triborough	East R, New York, USA	S	H	1936	1380	421
Alvsborg	Gota R, Sweden	S	H	1966	1368	417
Saint-Nazaire	Loire R, Pays de la Loire, France	CSG	H	1975	1325	404
Namhae Island	Cheju Strait, South Korea	S	H	1973	1312	400
Rande	Rande Strait, Galicia, Spain	CSG	H	1977	1312	400
Dames Point	Saint Johns R, Florida, USA	CSG	H	UC	1300	396
Aquitaine	Garonne R, Aquitaine, France	S	H	1967	1293	394
Cologne-Roden-kirchen (off Koln-Roden-kirchen)	Rhine R, Nordrhein-Westfalen, West Germany	S	H	1955	1240	378
Baton Rouge	Mississippi R, Louisiana, USA	C	H	1969	1235	376
Astoria	Columbia R, Oregon–Washington, USA	CT	H	1966	1232	376
Fremont	Willamette R, Oregon, USA	SA	H	1973	1225	373
Luling	Mississippi R, Louisiana, USA	CSG	H	UC (1980?)	1222	372
Zdakov (alt Orlik)	Vltava R, Bohemia, Czechoslovakia	SA	H	1967	1214	370
Tappan Zee (alt Nyack-Tarrytown)	Hudson R, New York, USA	C	H	1955	1212	369
Saint Johns	Willamette R, Oregon, USA	S	H	1931	1207	368
Dusseldorf-Flehe	Rhine R, Nordrhein-Westfalen, West Germany	CSG	H	UC (1979?)	1205	367

[1] B, bascule; C, cantilever; CA, concrete arch; CCSG, concrete cable-stayed girder; CPBG, continuous plate and box girder; CSG, cable-stayed girder; CT, continuous truss; S, suspension; SA, steel arch; SS, swing span; ST, simple truss; VL, vertical lift.
[2] H, highway; RR, railroad; RT, rapid transit.
[3] UC, under construction.

Bridge	Body of Water and Location	Type[1]	Use[2]	Year Completed[3]	Span (ft)	Span (m)
Wakato	Dokai B of Korea Strait, Kyushu I, Japan	S	H	1962	1204	367
Longview	Columbia R, Oregon–Washington, USA	C	H	1930	1200	366
Mount Hope	Mount Hope B of Atlantic O, Rhode Island, USA	S	H	1929	1200	366
Francis Scott Key	Patapsco R, Maryland, USA	C	H	1976	1200	366
Port Mann	Fraser R, British Columbia, Canada	SA	H	1964	1200	366

* * *

Bridge	Body of Water and Location	Type[1]	Use[2]	Year Completed[3]	Span (ft)	Span (m)
Queensboro	East R, New York, USA	C	H; RT	1909	1182[4]	360[4]
Duisburg-Neuenkamp	Rhine R, Nordrhein-Westfalen, West Germany	CSG	H	1970	1148	350
Thatcher Ferry	Panama Canal, Canal Zone	SA	H	1962	1128	344
Hercilio Luz (alt Florianopolis)	Strait between Norte B and Sul B of Atlantic O, Santa Catarina, Brazil	S	H; RT	1926	1114	340
West Gate	Yarra R, Victoria, Australia	CSG	H	1977[5]	1102	336
Carquinez I	Carquinez Strait, California, USA	C	H	1927	1100[6]	335[6]
Carquinez II	Carquinez Strait, California, USA	C	H	1958	1100[6]	335[6]
Laviolette (alt Trois-Rivieres)	Saint Lawrence R, Quebec, Canada	SA	H	1967	1100	335
Second Narrows	Burrard Inlet of Strait of Georgia, British Columbia, Canada	C	H	1960	1100	335
Jacques Cartier (alt Montreal Harbour)	Saint Lawrence R, Quebec, Canada	C	H; RR	1930	1097	334
Brazo Largo	Parana Guazu R, Entre Rios, Argentina	CSG	H; RR	1976	1083	330
Runcorn-Widnes	Mersey R, England, UK	SA	H	1961	1083	330
Zarate	Parana de las Palmas R, Buenos Aires, Argentina	CSG	H; RR	1975	1083	330
Birchenough	Sabi R, Rhodesia	SA	H	1935	1080	329
Richmond-San Rafael	San Francisco B of Pacific O, California, USA	C	H	1957	1070[6]	326[6]

[1] B, bascule; C, cantilever; CA, concrete arch; CCSG, concrete cable-stayed girder; CPBG, continuous plate and box girder; CSG, cable-stayed girder; CT, continuous truss; S, suspension; SA, steel arch; SS, swing span; ST, simple truss; VL, vertical lift.
[2] H, highway; RR, railroad; RT, rapid transit.
[3] UC, under construction.
[4] A second span of 984 ft, 300 m.
[5] Completion originally scheduled for 1972 but delayed by collapse of side span in October 1970.
[6] Two spans of this length.

Bridge	Body of Water and Location	Type[1]	Use[2]	Year Completed[3]	Span (ft)	Span (m)
Kohlbrand	Elbe R, Hamburg, West Germany	CSG	H	1974	1066	325
Oshima	Oshima (Naruto) Strait, Honshu I, Japan	CT	H	1976	1066	325
Cincinnati-Covington	Ohio R, Kentucky–Ohio, USA	S	H	1867[4]	1057	322
Knie	Rhine R, Nordrhein-Westfalen, West Germany	CSG	H	1969	1050	320
Brotonne	Seine R, Haute-Normandie, France	CCSG	H	1976	1050	320
Glen Canyon	Colorado R, Arizona, USA	SA	H	1959	1028	313
Mannheim-Seckenheim	Rhine R, Baden-Wurttemberg, West Germany	CCSG	H	UC	1011	308
Wheeling	Ohio R, Ohio–West Virginia, USA	S	H	1849[5]	1010	308
Erskine	Clyde R, Scotland, UK	CSG	H	1971	1000	305
Gladesville	Parramatta R, New South Wales, Australia	CA	H	1964	1000	305
Lewiston-Queenston	Niagara R, Canada–USA	SA	H	1962	1000	305
Bratislava	Danube R, Slovakia, Czechoslovakia	CSG	H	1972	994	303
Perrine	Snake R, Idaho, USA	SA	H	1976	993	303
Severin	Rhine R, Nordrhein-Westfalen, West Germany	CSG	H	1960	991	302
Kiev	Dnieper R, Ukraine, USSR	CSG	H	1975	984	300
Kuronoseto	Kurono Strait, Kyushu I, Japan	CT	H	1974	984	300
Costa e Silva (alt Rio de Janeiro-Niteroi)	Guanabara B of Atlantic O, Guanabara–Rio de Janeiro, Brazil	CPBG	H	1974	984	300
Temmon-Kyo	Misumi Strait, Kyushu I, Japan	CT	H	1966	984	300
Pasco-Kennewick	Columbia R, Washington, USA	CCSG	H	1978	981	299
Hell Gate	East R, New York, USA	SA	RR	1917	977	298
Amizade (alt Foz do Iguacu)	Parana R, Brazil–Paraguay	CA	H	1965	951	290
Deggenau	Danube R, Bayern, West Germany	CSG	H	1975	951	290
Rainbow	Niagara R, Canada–USA	SA	H	1941	950	290

[1] B, bascule; C, cantilever; CA, concrete arch; CCSG, concrete cable-stayed girder; CPBG, continuous plate and box girder; CSG, cable-stayed girder; CT, continuous truss; S, suspension; SA, steel arch; SS, swing span; ST, simple truss; VL, vertical lift.
[2] H, highway; RR, railroad; RT, rapid transit.
[3] UC, under construction.
[4] Reconstructed in 1898.
[5] Reconstructed in 1856.

Bridge	Body of Water and Location	Type[1]	Use[2]	Year Completed[3]	Span (ft)	Span (m)
Kurt Schumacher (alt Mannheim-Nord)	Rhine R, Baden-Wurttemberg-Rheinland-Pfalz, West Germany	CSG	H	1969	945	288
Wadi al Kuf Gorge	(Wadi al) Kuf R, Cyrenaica, Libya	CCSG	H	1971	925	282
Friedrich Ebert (alt Bonn-Nord)	Rhine R, Nordrhein-Westfalen, West Germany	CSG	H	1967	919	280
Leverkusen	Rhine R, Nordrhein-Westfalen, West Germany	CSG	H	1964	919	280
Tjorn	Askero Fjord of Skagerrak Strait, Sweden	SA	H	1960	912	278
Ravenswood	Ohio R, Ohio–West Virginia, USA	CT	H	UC	902	275
Memphis	Mississippi R, Arkansas–Tennessee, USA	SA	H	1972	900[4]	274[4]
Arrabida	Douro R, Portugal	CA	H	1963	885	270
Royal Gorge[5]	Arkansas R, Colorado, USA	S	H	1929	880	268
Sando	Angerman R, Sweden	CA	H	1943	866	264
Sava I	Sava R, Serbia, Yugoslavia	CPBG	H	1956	856	261
Zoo	Rhine R, Nordrhein-Westfalen, West Germany	CPBG	H	1966	850	259
Dubuque	Mississippi R, Illinois–Iowa, USA	CT	H	1943	845	258
Sava II	Sava R, Serbia, Yugoslavia	CPBG	H	1970	820	250
Sibenik	Krka R, Croatia, Yugoslavia	CA	H	1967	808	246
Chaco-Corrientes	Parana R, Chaco–Corrientes, Argentina	CCSG	H	1973	804	245
Auckland Harbour	Waitemata Harbor, North I, New Zealand	CPBG	H	1969	800	244
Saikai	Inoura Strait, Kyushu I, Japan	SA	H	1955	800	244
Koror-Babelthuap	(Toagel) Mid Channel, Pacific Islands (USA)	CCSG	H	1978	790	241
Hamana	Hamana Lagoon, Honshu I, Japan	CCSG	H	1975?	787	240
Koblenz-Sud	Rhine R, Rheinland-Pfalz, West Germany	CPBG	H	1973	774	236
Shimonoseki	Kammon Strait, Honshu I, Japan	CCSG	H	1975	774	236
General Rafael Urdaneta	Maracaibo L, Venezuela	CCSG	H	1962	771[6]	235[6]

[1] B, bascule; C, cantilever; CA, concrete arch; CCSG, concrete cable-stayed girder; CPBG, continuous plate and box girder; CSG, cable-stayed girder; CT, continuous truss; S, suspension; SA, steel arch; SS, swing span; ST, simple truss; VL, vertical lift.

[2] H, highway; RR, railroad; RT, rapid transit.

[3] UC, under construction.

[4] Two spans of this length.

[5] Height of deck above water: 1053 ft, 321 m—a record.

[6] Five spans of this length.

Bridge	Body of Water and Location	Type[1]	Use[2]	Year Completed[3]	Span (ft)	Span (m)
Grand Duchess Charlotte	Alzette R, Luxembourg	CPBG	H	1966	768	234
Fiumarella	Fiumarella R, Calabria, Italy	CA	H	1961	758	231
Bonn-Sud	Rhine R, Nordrhein-Westfalen, West Germany	CPBG	H	1971	755	230
Urado	Urado B of Pacific O, Shikoku I, Japan	CCSG	H	1972	755	230
San Mateo-Hayward II	San Francisco B of Pacific O, California, USA	CPBG	H	1967	750	229
Metropolis	Ohio R, Illinois-Kentucky, USA	ST	RR	1917	720	219
Arthur Kill	Arthur Kill, New Jersey–New York, USA	VL	RR	1959	558	170
Ferdan	Suez Canal, Egypt	SS	H; RR	1965	550	168
Sault Sainte Marie	Sault Sainte Marie Canal, Canada–USA	B	RR	1914	336	102

In addition to the bridges listed above, Japan has projected the construction of 18 main bridges to provide both a highway and a railroad connection between the main islands of Honshu and Shikoku by taking advantage of the existence of a number of smaller, intervening islands. Calling as it does for the construction, along three separate routes, of no less than fifteen long-span bridges that qualify for inclusion in this list, including the longest and sixth longest suspension and the eighth and ninth longest cantilever bridges ever designed, this project can justly be called the greatest bridge-building effort in the history of engineering. Completion of the first route, a combined highway and railroad link, is scheduled for 1987, and three of the bridges on the other two routes are already under construction. Following are the fifteen qualifying bridges:

Bridge	Body of Water and Location	Type[1]	Use[2]	Year Completed[3]	Span (ft)	Span (m)
Akashi Strait	Akashi Strait, Honshu I, Japan	S	H; RR	[4]	5840	1780
South Bisaneto	Bisan Strait, Shikoku I, Japan	S	H; RR	UC (1987?)	3609	1100
Third Kurushima	Kurushima Strait, Shikoku I, Japan	S	H	[4]	3281	1000
North Bisaneto	Bisan Strait, Shikoku I, Japan	S	H; RR	UC (1987?)	3248	990
Shimotsuiseto	Shimotsui Strait, Honshu I–Shikoku I, Japan	S	H; RR	UC (1987?)	3084	940
Tatara	Tatara Strait, Honshu I–Shikoku I, Japan	S	H	[4]	2920	890
Onaruto	Naruto Strait, Honshu I, Japan	S	H; RR	UC (1983?)	2874	876
First Kurushima	Kurushima Strait, Shikoku I, Japan	S	H	[4]	2822	860
Innoshima	Innoshima Strait, Honshu I, Japan	S	H	UC (1982?)	2526	770
Oshima	Hakata Strait, Shikoku I, Japan	S	H	[4]	1805	550

[1] B, bascule; C, cantilever; CA, concrete arch; CCSG, concrete cable-stayed girder; CPBG, continuous plate and box girder; CSG, cable-stayed girder; CT, continuous truss; S, suspension; SA, steel arch; SS, swing span; ST, simple truss; VL, vertical lift.
[2] H, highway, RR, railroad; RT, rapid transit.
[3] UC, under construction.
[4] Projected.

Bridge	Body of Water and Location	Type[1]	Use[2]	Year Completed[3]	Span (ft)	Span (m)
Second Kurushima	Kurushima Strait, Shikoku I, Japan	S	H	[4]	1805	550
Hitsuishijima	Shimotsui Strait, Shikoku I, Japan	C	H; RR	UC (1987?)	1312	400
Igurojima	Shimotsui Strait, Shikoku I, Japan	C	H; RR	UC (1987?)	1312	400
Omishima	Hakata Strait, Shikoku I, Japan	SA	H	UC (1979?)	974	297
Iguchi	Innoshima Strait, Honshu I, Japan	CCSG	H	[4]	820	250

[1] B, bascule; C, cantilever; CA, concrete arch; CCSG, concrete cable-stayed girder; CPBG, continuous plate and box girder; CSG, cable-stayed girder; CT, continuous truss; S, suspension; SA, steel arch; SS, swing span; ST, simple truss; VL, vertical lift.
[2] H, highway; RR, railroad; RT, rapid transit.
[3] UC, under construction.
[4] Projected.

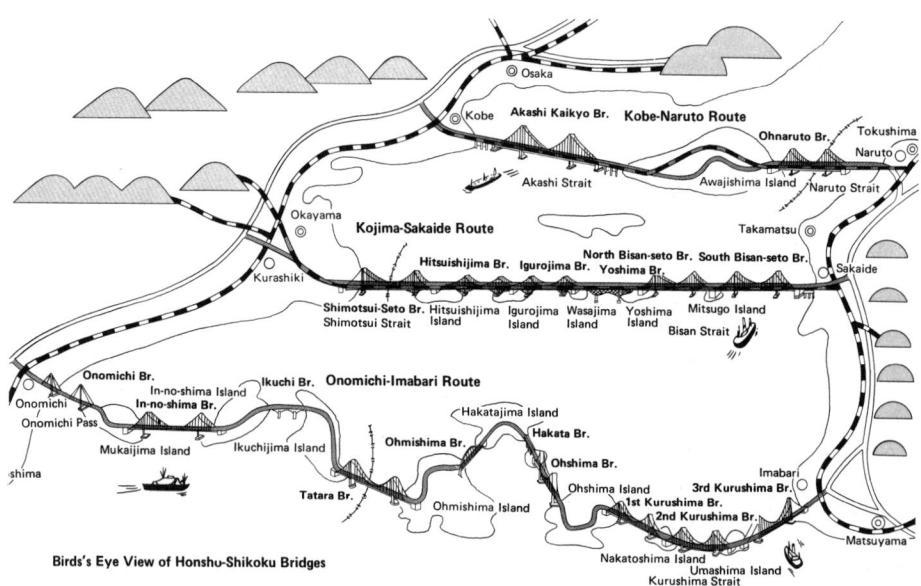

Figure 35. Outline of the Honshu-Shikoku Bridge Project, designed to provide a triple link between those two main Japanese islands. (Credit: Honshu-Shikoku Bridge Authority.)

9c
LONGEST RAILROAD, HIGHWAY, AND CANAL TUNNELS

Contents

Conventional and alternate names of tunnel.
Subdivision and country in which located.
Use to which tunnel is put.
Year of completion.
Length from portal to portal, in miles and kilometers.

Arrangement

By length.

Coverage

Above * * *, all railroad, highway, and canal tunnels, built or under construction, with a length of at least 4.5 mi (7.24 km).
Below * * *, other such tunnels that are long and well known.

Entries

85.

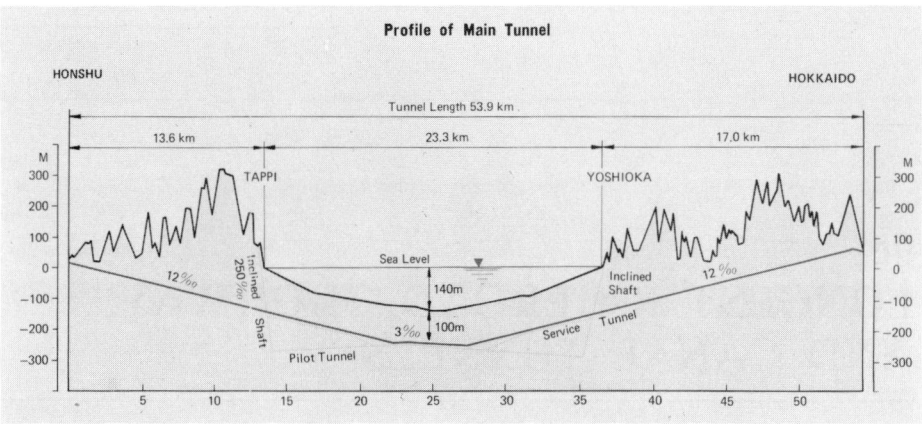

Profile of Main Tunnel

HONSHU   HOKKAIDO

Figure 36. Seikan Tunnel, which is expected to connect Honshu and Hokkaido Islands in Japan by a record-breaking 33.5-mi (54-km) bore under Tsugaru Strait by 1982. Top: profile of the main tunnel; bottom: site of the Honshu Island approach. (Credit: Japanese National Railways.)

276

TABLE 9c. LONGEST RAILROAD, HIGHWAY, AND CANAL TUNNELS

Tunnel	Location	Use[1]	Year Completed[2]	Length (mi)	Length (km)
Seikan (underwater)	Hokkaido I–Honshu I, Japan	RR	UC (1982?)	33.46	53.85
Shimizu III	Honshu I, Japan	RR	UC	13.86	22.3
Simplon II (alt Sempione II)	Italy–Switzerland	RR	1922	12.32	19.82
Simplon I (alt Sempione I)	Italy–Switzerland	RR	1906	12.30	19.80
New Kammon (off Shin Kammon) (underwater)	Honshu I–Kyushu I, Japan	RR	1974	11.63	18.71
Apennine (off Appennino)	Emilia-Romagna–Toscana, Italy	RR	1934	11.51	18.52
Gotthard (alt Saint-Gotthard)	Switzerland	H	UC (1978?)	10.14	16.32
Rokko	Honshu I, Japan	RR	1972	10.10	16.25
Furka	Switzerland	RR	UC (1978?)	9.57	15.4
Gotthard (alt Saint-Gotthard)	Switzerland	RR	1882	9.32	15.00
Nakayama	Honshu I, Japan	RR	UC	9.13	14.7
Lotschberg	Switzerland	RR	1913	9.08	14.61
Haruna	Honshu I, Japan	RR	UC	8.95	14.4
Hokuriku	Honshu I, Japan	RR	1962	8.62	13.87
Mont-Cenis (alt Frejus, Monte Cenisio)	France–Italy	RR	1871	8.49	13.66
Shimizu II	Honshu I, Japan	RR	1967	8.38	13.49
Aki	Shikoku I, Japan	RR	1974?	8.10	13.04
Mont-Cenis (alt Frejus, Monte Cenisio)	France–Italy	H	UC (1978?)	7.89	12.7
Cascade	Washington, USA	RR	1929	7.79	12.54
Mont-Blanc (alt Monte Bianco)	France–Italy	H	1965	7.25	11.67
Kubiki	Honshu I, Japan	RR	1969	7.05	11.35
Flathead	Montana, USA	RR	1970	7.0	11.27
Lierasen	Norway	RR	1973	6.65	10.7
Santa Lucia	Catania, Italy	RR	1974	6.38	10.26
Arlberg	Austria	RR	1884	6.37	10.25
Gran Sasso	Abruzzi e Molise, Italy	H	1976	6.21	10.0
Shimizu I	Honshu I, Japan	RR	1930	6.03	9.70
Moffat	Colorado, USA	RR	1928	5.97	9.61
North Kyushu	Kyushu I, Japan	RR	1974	5.97	9.60
Seelisberg	Switzerland	H	UC	5.75[3]	9.25[3]
Kvineshei	Norway	RR	1943	5.63	9.06
Bingo	Japan	RR	1974	5.54	8.92
Kaimai	North I, New Zealand	RR	UC	5.49	8.84
Rimutaka	North I, New Zealand	RR	1955	5.47	8.80
Otira	South I, New Zealand	RR	1923	5.37	8.65
Ricken	Switzerland	RR	1910	5.35	8.60
Grenchenberg	Switzerland	RR	1915	5.33	8.58
Tauern	Austria	RR	1909	5.31	8.55

[1]C, canal; H, highway; RR, railroad.

[2]UC, under construction.

[3]Two parallel tubes of this length.

Tunnel	Location	Use[1]	Year Completed[2]	Length (mi)	Length (km)
Fukuoka	Kyushu I, Japan	RR	1974	5.28	8.50
Haegebostad	Norway	RR	1943	5.27	8.47
Ena	Honshu I, Japan	H	1975	5.25	8.45
Ronco (alt Giovi)	Liguria, Italy	RR	1889	5.16	8.30
Hauenstein	Switzerland	RR	1916	5.05	8.13
Colle di Tenda (alt Col de Tende)	France–Italy	RR	1900[3]	5.03	8.10
Connaught	British Columbia, Canada	RR	1916	5.02	8.08
Karawanken (alt Karavanke)	Austria–Yugoslavia	RR	1906	4.96	7.98
Kobe	Honshu I, Japan	RR	1970	4.95	7.97
Tanna II	Honshu I, Japan	RR	1964	4.94	7.96
Somport	France–Spain	RR	1928	4.89	7.87
Tanna I	Honshu I, Japan	RR	1934	4.85	7.80
Ulriken	Norway	RR	1964	4.76	7.66
Hosaka	Honshu I, Japan	RR	1970	4.72	7.59
Hoosac	Massachusetts, USA	RR	1875	4.70	7.56
Monte Orso	Lazio, Italy	RR	1927	4.68	7.53
Lupacino	Italy	RR	1958	4.67	7.51
Vivola	Lazio, Italy	RR	1927	4.57	7.35

* * *

Tunnel	Location	Use[1]	Year Completed[2]	Length (mi)	Length (km)
Jungfrau	Switzerland	RR	1912	4.43	7.12
Rove	Provence-Cote d'Azur, France	C	1927	4.42	7.11
Severn (underwater)	England–Wales, UK	RR	1886	4.36	7.01
Rokko II	Honshu I, Japan	H	1974	4.29	6.90
San Bernardino (alt Bernhardin)	Switzerland	H	1967	4.10	6.6
Tauern	Austria	H	1974	3.98	6.4
Haneda (underwater)	Honshu I, Japan	RR	1971	3.72	5.98
Grand Saint-Bernard (alt Gran San Bernardo)	Italy–Switzerland	H	1964	3.64	5.85
Felber-Tauern	Austria	H	1967	3.48	5.6
Katschberg	Austria	H	1974	3.36	5.4
Viella	Cataluna, Spain	H	1941	3.13	5.04
New Sasago (off Shin Sasago)	Honshu I, Japan	H	UC	2.74	4.42
Kammon (underwater)	Honshu I–Kyushu I, Japan	RR	1942	2.24	3.6
Kammon (underwater)	Honshu I–Kyushu I, Japan	H	1958	2.15	3.46
Mersey (alt Queensway) (underwater)	England, UK	H	1934	2.13	3.43
Elbe (underwater)	Hamburg-Niedersachsen, West Germany	H	1973	2.06	3.32
Transandine Summit (off Cumbre)	Argentina–Chile	RR	1910	1.97	3.17

[1]C, canal; H, highway; RR, railroad.
[2]UC, under construction.
[3]Closed since World War II.

Tunnel	Location	Use[1]	Year Completed[2]	Length (mi)	Length (km)
Tsuruga	Honshu I, Japan	H	UC	1.97	3.17
Pyrenees (alt Pirineos)	France–Spain	H	1970	1.87	3.01
Sasago II	Honshu I, Japan	H	1958	1.83	2.95
Rokko I	Honshu I, Japan	H	1967	1.76	2.84
Brooklyn-Battery (underwater)	New York, USA	H	1950	1.73	2.78
Eisenhower Memorial	Colorado, USA	H	1971	1.70	2.74
Kuriko II	Honshu I, Japan	H	1966	1.66	2.67
Holland (underwater)	New Jersey–New York, USA	H	1927	1.62	2.61
Lincoln I (underwater)	New Jersey–New York, USA	H	1937	1.56	2.51
Lincoln III (underwater)	New Jersey–New York, USA	H	1957	1.52	2.45
Baltimore Harbor (underwater)	Maryland, USA	H	1958	1.45	2.33
Lincoln II (underwater)	New Jersey–New York, USA	H	1945	1.42	2.29

[1]C, canal; H, highway; RR, railroad.
[2]UC, under construction.

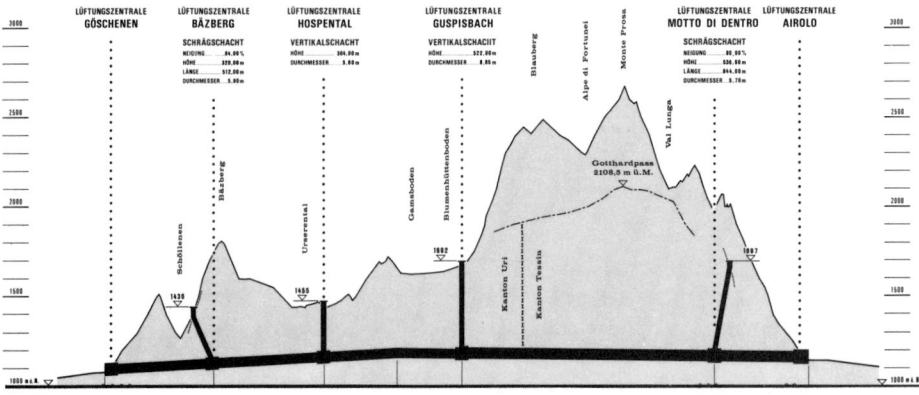

Figure 37. Gotthard Tunnel, recently bored through the Sankt Gotthard Massif in Switzerland, has a record length for highway tunnels of more than 10 mi (16 km). Top: longitudinal profile of the tunnel; bottom: site of the southern approach at Airolo. (Credit: Swiss Federal Office of Highways and Rivers.)

9d
HIGHEST DAMS

Contents

Rank.
Conventional and alternate names of dam.
River dammed, and subdivision and country in which located.
Type of dam.
Year of completion.
Height of dam, excluding projections, in feet and meters.
Volume of structure, in millions of cubic yards and cubic meters.

Arrangement

By height.

Coverage

Above * * *, all dams in place or under construction with a height of at least 500 ft (152 m).
Below * * *, other well-known high dams.

Rounding

Volumes are rounded to the nearest 100,000 yd^3 (m^3).

Entries

87.

Figure 38. Hydropower station at Nurek Dam on the Vakhsh River in Tadzhikistan, USSR. Although the Nurek is the highest dam already in place, an even higher one, the Rogun, is being built on the same river. (Credit: Embassy of the USSR, Washington.)

TABLE 9d. HIGHEST DAMS

Rank	Dam	River and Location	Type[1]	Year Completed[2]	Height (ft)	Height (m)	Volume[3] (yd³)	Volume[3] (m³)
1.	Rogun	Vakhsh R, Tadzhiki-stan, USSR	R	UC	1066	325	91.6	70.0
2.	Nurek	Vakhsh R, Tadzhiki-stan, USSR	R	1974	984	300	75.9	58.0
3.	Grande Dixence	Dixence R, Switzer-land	CG	1962	932	284	7.8	6.0
4.	Inguri	Inguri R, Georgia, USSR	CA	UC	892	272	5.0	3.8
5.	Vaiont	Vaiont R, Veneto, Italy	CA	1961	858	262	0.5	0.4
6.	Mica	Columbia R, British Columbia, Canada	R	1974	794	242	42.0	32.1
6.	Sayano-Shushenskaya (alt Sayansk)	Yenisey R, Russia, USSR	CA	UC	794	242	11.9	9.1
8.	Chicoasen	Grijalva R, Chiapas, Mexico	E	UC	787	240	15.7	12.0
8.	Patia	Patia R, Colombia	R	UC	787	240	30.9	23.6
10.	Chivor (for Esmeralda)	Bata R, Colombia	R	1975	778	237	14.1	10.8
11.	Mauvoisin	Drance de Bagnes R, Switzerland	CA	1958	777	237	2.7	2.0
12.	Oroville	Feather R, California, USA	E	1968	770	235	78.0	59.6
13.	Chirkey	Sulak R, Russia, USSR	CA	1975	764	233	1.6	1.2
14.	Contra	Verzasca R, Switzer-land	CA	1965	754	230	0.9	0.7
15.	Bhakra	Sutlej R, Punjab, India	CG	1962	742	226	5.4	4.1
16.	Hoover (for Boulder)	Colorado R, Arizona–Nevada, USA	CA	1936	726	221	4.4	3.4
17.	Mratinje	Piva R, Montenegro, Yugoslavia	CA	1975	722	220	1.0	0.8
18.	Dworshak	North Fork Clear-water R, Idaho, USA	CG	1974	717	219	6.5	5.0
19.	Glen Canyon	Colorado R, Arizona, USA	CA	1964	710	216	4.9	3.7
20.	Toktogul	Naryn R, Kirgizia, USSR	CA	UC	705	215	3.5	2.7
21.	Daniel Johnson	Manicouagan R, Quebec, Canada	CA	1967	703	214	2.9	2.2
22.	Liuchia	Yellow R, China	E	1974	699	213		
23.	Keban	Euphrates R, Turkey	R	1973	696	212	19.6	15.0

[1] A, arch; C, concrete; E, earth; G, gravity; R, rock.
[2] UC, under construction.
[3] In millions of cubic yards and cubic meters.

Rank	Dam	River and Location	Type[1]	Year Completed[2]	Height (ft)	Height (m)	Volume[3] (yd³)	Volume[3] (m³)
24.	Auburn	North Fork American R, California, USA	CA	UC	695	212	6.0	4.6
25.	Luzzone	Brenno di Luzzone R, Switzerland	CA	1963	682	208	1.8	1.4
26.	Mohammed Reza Shah Pahlavi	Dez R, Iran	CA	1963	668	204	0.6	0.5
27.	Reza Shah Kabir	Karun R, Iran	CA	1973	656	200	1.6	1.2
28.	Almendra (alt Vallarino)	Tormes R, Leon, Spain	CA	1971	646	197	2.9	2.2
29.	Bullard's Bar	North Yuba R, California, USA	CA	1970	635	194	2.7	2.1
30.	Melones	Stanislaus R, California, USA	R	1975	625	190	16.0	12.2
31.	Kurobe Number 4	Kurobe R, Honshu I, Japan	CA	1964	610	186	1.8	1.4
31.	Swift	Lewis R, Washington, USA	E	1958	610	186	15.8	12.1
33.	Kolnbrein	Malta (Bach) Creek, Austria	CA	UC	607	185	1.8	1.4
33.	Oymapinar	Manavgat R, Turkey	CA	UC	607	185	0.7	0.5
35.	Mossyrock	Cowlitz R, Washington, USA	CA	1969	605	184	1.2	0.9
36.	Shasta	Sacramento R, California, USA	CG	1954	602	183	8.7	6.7
37.	W. A. C. Bennett	Peace R, British Columbia, Canada	E	1967	600	183	57.2	43.7
38.	Tignes	Isere R, Rhone-Alpes, France	CA	1952	592	180	0.8	0.6
39.	Amir Kabir	Karaj R, Iran	CA	1962	591	180	0.9	0.7
39.	Dartmouth	Mitta Mitta R, Victoria, Australia	E	1976	591	180	16.9	12.9
39.	Grande Maison	Eau d'Olle (Ruisseau) Creek, Rhone-Alpes, France	E	UC	591	180	24.9	19.0
39.	Itaipu	Parana R, Brazil–Paraguay	E	UC	591	180	35.3	27.0
39.	Karakaya	Euphrates R, Turkey	CA	UC	591	180	1.8	1.4
39.	Tachien (alt Tehchi)	Tachia R, Taiwan	CA	1974	591	180	0.6	0.4
45.	Emosson	Barberine R, Switzerland	CA	1973	590	180	1.4	1.1
46.	Don Pedro	Tuolumne R, California, USA	R	1971	585	178	16.8	12.8
47.	Alpe Gera	Cormor R, Friuli-Venezia Giulia, Italy	CG	1965	584	178	2.3	1.7

[1] A, arch; C, concrete; E, earth; G, gravity; R, rock.
[2] UC, under construction.
[3] In millions of cubic yards and cubic meters.

Rank	Dam	River and Location	Type[1]	Year Completed[2]	Height (ft)	Height (m)	Volume[3] (yd³)	Volume[3] (m³)
48.	Takase	Takase R, Honshu I, Japan	E	UC	577	176	14.0	10.7
49.	Hasan Ugurlu	Yesil (Irmak) R, Turkey	E	UC	574	175	11.8	9.0
49.	Nader Shah	Marun R, Iran	R	UC	574	175	9.4	7.2
51.	Gura Apelor Retezat	(Riul) Mare R, Romania	R	UC	568	173	11.8	9.0
52.	Hungry Horse	South Fork Flathead R, Montana, USA	CA	1953	564	172	3.1	2.4
53.	Idikki	Periyar R, Kerala, India	CA	1974	561	171	0.6	0.5
54.	Amaluza	Paute R, Ecuador	CA	UC	558	170	1.5	1.2
55.	Charvak	Chirchik R, Uzbekistan, USSR	E	1970	551	168	25.0	19.1
56.	Cabora Bassa	Zambezi R, Mozambique	CA	1974	550	168	0.6	0.5
56.	Grand Coulee	Columbia R, Washington, USA	CG	1942	550	168	10.6	8.1
58.	Vidraru	Arges R, Romania	CA	1963	544	166	0.7	0.5
59.	King Paul (alt Kremasta)	Akheloos R, Greece	E	1965	541	165	10.7	8.2
60.	Ross	Skagit R, Washington, USA	CA	1949	540	165	0.9	0.7
61.	Trinity	Trinity R, California, USA	E	1962	537	164	29.3	22.4
62.	Raul Leoni	Caroni R, Venezuela	G	UC	532	162	100.0	76.5
63.	Talbingo (alt Yalwal)	Tumut R, New South Wales, Australia	R	1971	530	162	18.5	14.1
64.	La Grande Number 2	Grande R, Quebec, Canada	E	UC	525	160	30.0	22.9
64.	Thomson	Thomson R, Victoria, Australia	R	UC	525	160	12.6	9.6
64.	Yellowtail	Bighorn R, Montana, USA	CA	1966	525	160	1.5	1.1
67.	Cougar	South Fork McKenzie R, Oregon, USA	R	1964	519	158	13.0	9.9
68.	Fierze	Drin R, Albania	E	1973	518	158	0.9	0.7
68.	Finstertal	Finster(bach) Creek, Austria	R	UC	518	158	5.7	4.4
68.	Gokcekaya	Sakarya R, Turkey	CA	1973	518	158	0.8	0.6
71.	Oku-Tadami	Tadami R, Honshu I, Japan	CG	1961	515	157	2.1	1.6
72.	Speccheri	Leno di Vallarsa R, Trentino-Alto Adige, Italy	CA	1957	514	157	0.2	0.1
73.	Tseuzier (alt Zeuzier)	Liene (alt Lienne) R, Switzerland	CA	1957	512	156	0.4	0.3

[1] A, arch; C, concrete; E, earth; G, gravity; R, rock.
[2] UC, under construction.
[3] In millions of cubic yards and cubic meters.

Rank	Dam	River and Location	Type[1]	Year Completed[2]	Height (ft)	Height (m)	Volume[3] (yd³)	Volume[3] (m³)
74.	Sakuma	Tenryu R, Honshu I, Japan	CG	1956	510	155	1.5	1.1
75.	Monteynard	Drac R, Rhone-Alpes, France	CA	1962	509	155	0.6	0.5
76.	Goscheneralp	Goschenerreuss R, Switzerland	E	1960	508	155	12.2	9.4
76.	Nagawado	Azusa R, Honshu I, Japan	CA	UC	508	155	0.9	0.7
78.	Bhumiphol (alt Yanhee)	Ping R, Thailand	CG	1964	505	154	1.3	1.0
78.	Miyagase	Nakatsu R, Honshu I, Japan	CG	UC	505	154	2.6	2.0
80.	Flaming Gorge	Green R, Utah, USA	CA	1964	502	153	1.0	0.8
80.	Place Moulin	Buthier R, Valle d'Aosta, Italy	CG	1965	502	153	2.0	1.5
80.	Tedorigawa	Tedori R, Honshu I, Japan	R	UC	502	153	12.6	9.6
83.	Gepatsch	Faggen(bach) Creek, Austria	E	1964	500	152	9.2	7.1
83.	Santa Giustina	Noce R, Trentino-Alto Adige, Italy	CA	1950	500	152	0.1	0.1

* * *

Rank	Dam	River and Location	Type[1]	Year Completed[2]	Height (ft)	Height (m)	Volume[3] (yd³)	Volume[3] (m³)
	Akosombo	Volta R, Ghana	R	1965	463	141	10.4	8.0
	Kariba	Zambezi R, Rhodesia–Zambia	CG	1959	420	128	1.3	1.0
	Guri	Caroni R, Venezuela	E	1968	348	106	4.9	3.8

[1] A, arch; C, concrete; E, earth; G, gravity; R, rock.
[2] UC, under construction.
[3] In millions of cubic yards and cubic meters.

9e
LARGEST DAMS
(VOLUME OF STRUCTURE)

Contents

Rank.

Conventional and alternate names of dam.

River dammed, and subdivision and country in which located.

Type of dam.

Year of completion.

Volume of structure, in millions of cubic yards and cubic meters.

Height of dam, excluding projections, in feet and meters.

Arrangement

By volume of structure.

Coverage

All dams in place or under construction with a volume of at least 20,000,000 yd^3 (15,291,100 m^3).

Rounding

Volumes are rounded to the nearest 100,000 yd^3 (m^3).

Entries

65.

TABLE 9e. LARGEST DAMS (VOLUME OF STRUCTURE)

Rank	Dam	River and Location	Type[1]	Year Completed[2]	Volume[3] (yd³)	(m³)	Height (ft)	(m)
1.	Tarbela	Indus R, Pakistan	E	1975	186.0	142.3	486	148
2.	Fort Peck	Missouri R, Montana, USA	E	1940	125.6	96.1	250	76
3.	Raul Leoni	Caroni R, Venezuela	G	UC	100.0	76.5	532	162
4.	Oahe	Missouri R, South Dakota, USA	E	1963	92.0	70.3	245	75
5.	Oosterschelde	Vense Gat Oosterschelde R, Netherlands	E	UC	91.6	70.0	148	45
5.	Rogun	Vakhsh R, Tadzhikistan, USSR	R	UC	91.6	70.0	1066	325
7.	Mangla	Jhelum R, Pakistan	E	1968	85.9	65.6	380	116
8.	Gardiner	South Saskatchewan R, Saskatchewan, Canada	E	1966	85.7	65.5	223	68
9.	Oroville	Feather R, California, USA	E	1968	78.0	59.6	770	235
10.	San Luis	San Luis Creek, California, USA	E	1967	77.7	59.4	382	116
11.	Nurek	Vakhsh R, Tadzhikistan, USSR	R	1974	75.9	58.0	984	300
12.	Nagajunasagar	Krishna R, Andhra Pradesh, India	E	1966	73.6	56.3	409	125
13.	Garrison	Missouri R, North Dakota, USA	E	1956	66.5	50.8	210	64
14.	Cochiti	Rio Grande R, New Mexico, USA	E	1975	64.6	49.4	253	77
15.	Tabka (alt Thawra)	Euphrates R, Syria	E	1974	60.2	46.0	197	60
16.	Gorkiy	Volga R, Russia, USSR	E	1955	58.0	44.3	105	32
17.	Kiev	Dnieper R, Ukraine, USSR	E	1964	57.5	44.0	72	22
18.	W. A. C. Bennett	Peace R, British Columbia, Canada	E	1967	57.2	43.7	600	183
19.	Aswan High [off (Sadd al) Aali]	Nile R, Egypt	R	1970	55.7	42.6	364	111
20.	Saratov	Volga R, Russia, USSR	E	1967	52.8	40.4	131	40
21.	Fort Randall	Missouri R, South Dakota, USA	E	1956	50.2	38.4	165	50
22.	Kanev	Dnieper R, Ukraine, USSR	E	1974	49.5	37.9	82	25
23.	Itumbiara	Paranaiba R, Goias–Minas Gerais, Brazil	E	UC	47.1	36.0	328	100
24.	Kakhovka	Dnieper R, Ukraine, USSR	E	1955	46.6	35.6	121	37
25.	Beas Pong	Beas R, Punjab, India	E	1975	45.8	35.0	381	116
26.	Tsimlyansk	Don R, Russia, USSR	E	1952	44.3	33.9	128	39
26.	Volga–V. I. Lenin	Volga R, Russia, USSR	E	1955	44.3	33.9	148	45
28.	Castaic	Castaic Creek, California, USA	E	1973	44.0	33.6	340	104
29.	Jari	Jari R, Pakistan	E	1967	42.4	32.4	234	71

[1] A, arch; C, concrete; E, earth; G, gravity; R, rock.
[2] UC, under construction.
[3] In millions of cubic yards and cubic meters.

Rank	Dam	River and Location	Type[1]	Year Completed[2]	Volume[3] (yd³)	Volume[3] (m³)	Height (ft)	Height (m)
30.	Mica	Columbia R, British Columbia, Canada	R	1974	42.0	32.1	794	242
31.	Kremenchug	Dnieper R, Ukraine, USSR	E	1961	41.2	31.5	98	30
32.	Yacyreta-Apipe	Parana R, Argentina–Paraguay	E	UC	37.5	28.7	125	38
33.	Dneprodzerzhinsk	Dnieper R, Ukraine, USSR	E	1964	35.9	27.4	115	35
34.	Sao Simao	Paranaiba R, Goias–Minas Gerais, Brazil	R	UC	35.8	27.4	394	120
35.	Itaipu	Parana R, Brazil–Paraguay	E	UC	35.3	27.0	591	180
36.	Ukai	Tapti R, Guharat, India	E	UC	33.4	25.5	225	69
37.	Volga–22nd Congress	Volga R, Russia, USSR	E	1958	33.0	25.2	144	44
38.	Suapiti	Konkoure R, Guinea	E	1960?	32.7	25.0	394	120
39.	Kingsley	North Platte R, Nebraska, USA	E	1942	32.0	24.5	170	52
40.	Patia	Patia R, Colombia	R	UC	30.9	23.6	787	240
41.	Warm Springs	Dry Creek, California, USA	E	UC	30.4	23.2	318	97
42.	Diablo	Rio Grande R, Texas, USA	E	UC	30.0	22.9	280	85
42.	La Grande Number 3	Grande R, Quebec, Canada	E	UC	30.0	22.9	320	98
42.	La Grande Number 2	Grande R, Quebec, Canada	E	UC	30.0	22.9	525	160
45.	Balimela	Sileru R, Orissa, India	E	UC	29.6	22.6	230	70
46.	Ilha Solteira	Parana R, Sao Paulo, Brazil	E	1973	29.5	22.6	295	90
47.	Trinity	Trinity R, California, USA	E	1962	29.3	22.4	537	164
48.	Bilandi Tank	Bilandi R, India	E	UC	26.9	20.6	105	32
49.	Navajo	San Juan R, New Mexico, USA	E	1963	26.8	20.5	402	123
50.	Haringvliet	Haringvliet Estuary, Netherlands	E	1970	26.2	20.0	79	24
51.	Agua Vermelha	Grande R, Sao Paulo, Brazil	E	UC	25.7	19.6	295	90
52.	Hirakud	Mahanadi R, Orissa, India	E	1956	25.1	19.2	200	61
53.	Charvak	Chirchik R, Uzbekistan, USSR	E	1970	25.0	19.1	551	168
53.	La Grande Number 4	Grande R, Quebec, Canada	E	UC	25.0	19.1	385	117
55.	Grande Maison	Eau d'Olle (Ruisseau) Creek, Rhone-Alpes, France	E	UC	24.9	19.0	591	180
56.	Marimbondo	Grande R, Minas Gerais, Brazil	E	1975	24.3	18.6	295	90
57.	Gatun	Chagres R, Canal Zone	E	1912	23.0	17.6	115	35
58.	Poechos	Chira R, Peru	E	UC	22.9	17.5	164	50
59.	Bratsk	Angara R, Russia, USSR	E	1964	22.2	17.0	410	125

[1] A, arch; C, concrete; E, earth; G, gravity; R, rock.
[2] UC, under construction.
[3] In millions of cubic yards and cubic meters.

Rank	Dam	River and Location	Type[1]	Year Completed[2]	Volume[3] (yd³)	(m³)	Height (ft)	(m)
59.	Mornos	Mornos R, Greece	E	UC	22.2	17.0	414	126
59.	Sterkfontein	Nuwe Jaar(spruit) R, South Africa	E	UC	22.2	17.0	305	93
62.	Twin Buttes	Concho R, Texas, USA	E	1963	21.4	16.4	134	41
63.	Tuttle Creek	Big Blue R, Kansas, USA	E	1962	21.0	16.1	157	48
64.	Mingechaur	Kura R, Azerbaijan, USSR	E	1953	20.4	15.6	262	80
65.	Ivankovo	Volga R, Russia, USSR	E	1937	20.2	15.4	98	30

[1] A, arch; C, concrete; E, earth; G, gravity; R, rock.
[2] UC, under construction.
[3] In millions of cubic yards and cubic meters.

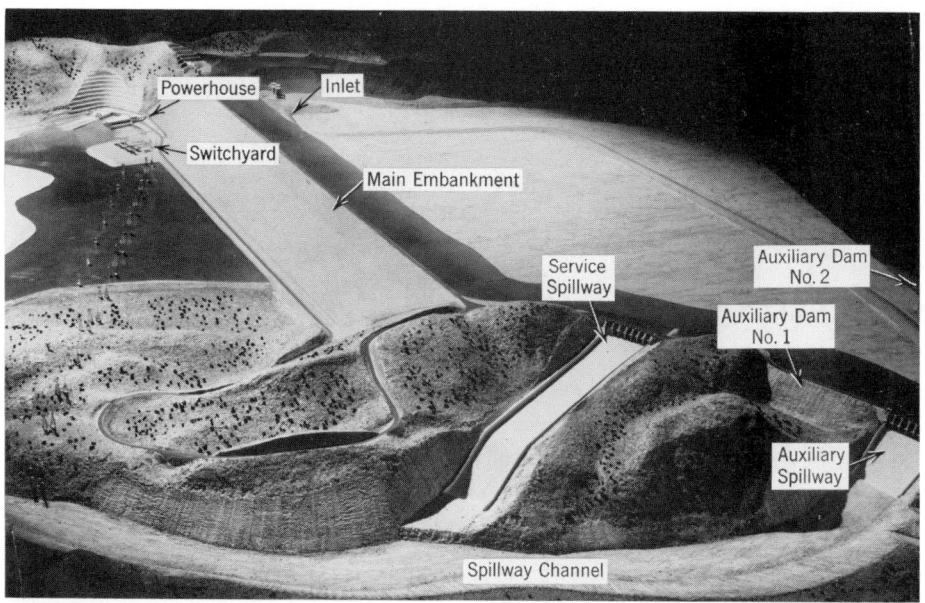

Figure 39. Tarbela Dam on the Indus River in Pakistan, the world's largest in volume of structure. Top: scale model of the dam; bottom: service spillway. (Credit, respectively: Tippetts-Abbett-McCarthy-Stratton, New York; Tomas Sennett for the World Bank.)

9f
LARGEST UNIVERSITIES

Contents

Rank.

Name of university.

City, and subdivision and country in which located.

Student enrollment (i.e., thousands of full-time and part-time students, excluding correspondence students, enrolled in the university and its affiliated institutions[1]).

Academic staff (i.e., number of full-time and part-time teachers).

Date of foundation. The date given is the year in which the institution itself or (preferably) its oldest constituent college or the educational institution from which it directly evolved was established (or chartered). This date may precede by several years the date when the first instruction was given. Also, it bears no necessary relation to the length of time during which the university has been operative, since many universities have been closed for periods of years or even decades.

Arrangement

By student enrollment.

Coverage

100 largest universities (as defined in the Introduction, under General Explanatory Notes) in student enrollment.

Rounding

Enrollments are rounded to the nearest 1000 students.

Entries

105.

[1] In India most universities include as affiliates numerous educational institutions of lower level, such as preparatory schools and other secondary schools; thus the size of these universities may be exaggerated with respect to that of universities in other countries, where only college-level students are included in the enrollment statistics.

Figure 40. Albany campus of the State University of New York, the university with the largest student enrollment. The Academic Podium (center) is 1600 ft (488 m) long and 600 ft (183 m) wide and comprises 13 interconnected buildings. All but one of the four 23-story towers and all of the low-rise units surrounding them are residence space. (Credit: State University of New York at Albany.)

TABLE 9f.   LARGEST UNIVERSITIES

Rank	University	Location	Students (thousands)	Teachers	Date Founded
1.	State University of New York	Albany, New York, USA	350	25,500	1844
	Albany campus	Albany, New York, USA	15	1,514	1844
	Campuses at or nr Buffalo	Buffalo, New York, USA	51	3,607	1846
2.	City University of New York	New York, New York, USA	251	18,121	1817
3.	Universite de Paris	Paris, Region Parisienne, France	233		1200
4.	University of Calcutta	Calcutta, West Bengal, India	227		1817
5.	University of California	Berkeley, California, USA	182	12,500	1855
	Berkeley campus	Berkeley, California, USA	35	2,429	1873
	Los Angeles campus	Los Angeles, California, USA	60	3,134	1881
6.	Universidad de Buenos Aires	Buenos Aires, Distrito Federal, Argentina	174	9,101	1821
7.	University of Madras	Madras, Tamil Nadu, India	152		1794
8.	University of Wisconsin	Madison, Wisconsin, USA	143	9,202	1849
	Madison campus	Madison, Wisconsin, USA	39	3,103	1849
9.	Universita degli Studi di Roma	Rome, Lazio, Italy	122	6,806	1303
10.	Universidad Nacional Autonoma de Mexico	Mexico, Distrito Federal, Mexico	120	15,964	1551
11.	University of Bombay	Bombay, Maharashtra, India	119		1832
12.	University of Kerala	Trivandrum, Kerala, India	116		1937
13.	University of North Carolina	Chapel Hill, North Carolina, USA	111	9,681	1789
	Chapel Hill campus	Chapel Hill, North Carolina, USA	21	2,891	1789
14.	Nihon University	Tokyo, Honshu I, Japan	102	3,505	1889
15.	Gujarat University	Ahmadabad, Gujarat, India	101		1949
16.	Universidad Complutense de Madrid	Madrid, Castilla la Nueva, Spain	97	4,570	1498
17.	University of Mysore	Mysore, Karnataka, India	95		1833
18.	Kurukshetra University	Kurukshetra, Haryana, India	94		1956
19.	University of Poona	Poona, Maharashtra, India	93		1885
20.	Agra University	Agra, Uttar Pradesh, India	92		1927
20.	Nagpur University	Nagpur, Maharashtra, India	92	4,124	1923
20.	University of Texas	Austin, Texas, USA	92	7,709	1881
	Austin campus	Austin, Texas, USA	49	3,468	1881
23.	University of London	London, England, UK	82	6,630	13th c
23.	University of Rajasthan	Jaipur, Rajasthan, India	82		1873
25.	Universita degli Studi di Napoli	Naples, Campania, Italy	79	3,256	1224
25.	University of Minnesota	Minneapolis, Minnesota, USA	79	5,981	1851
	Minneapolis campus	Minneapolis, Minnesota, USA	67	5,420	1851
27.	Indiana University	Bloomington, Indiana, USA	77	4,470	1820
	Bloomington campus	Bloomington, Indiana, USA	33	1,541	1820
27.	Madurai University	Madurai, Tamil Nadu, India	77		1958
29.	Panjab University	Chandigarh, Chandigarh, India	75		1947
29.	University of Alexandria	Alexandria, Egypt	75	2,955	1942
29.	University of Delhi	Delhi, Delhi, India	75		1881

Rank	University	Location	Students (thousands)	Teachers	Date Founded
32.	Shivaji University	Kolhapur, Maharashtra, India	74	2,568	1962
33.	Universidad Nacional de La Plata	La Plata, Buenos Aires, Argentina	73	4,216	1884
34.	Ain Shams University	Cairo, Egypt	69	1,262	1950
34.	Osmania University	Hyderabad, Andhra Pradesh, India	69		1887
36.	Pennsylvania State University	State College, Pennsylvania, USA	68	5,784	1855
	State College campus	State College, Pennsylvania, USA	40	4,538	1855
37.	Universidad de Chile	Santiago, Chile	64	14,000	1738
37.	University of the East	Manila, Luzon I, Philippines	64	1,548	1946
	Manila campus	Manila, Luzon I, Philippines		1,483	1946
39.	Instituto Politecnico Nacional (univ)	Mexico, Distrito Federal, Mexico	62	11,000	1931
39.	University of Cairo	Giza, Egypt	62	3,402	1908
39.	University of Illinois	Urbana and Champaign, Illinois, USA	62	9,119	1867
	Urbana-Champaign campus	Urbana and Champaign, Illinois, USA	37	6,785	1867
42.	University of Gorakhpur	Gorakhpur, Uttar Pradesh, India	61		1933
43.	Marathwada University	Aurangabad, Maharashtra, India	59	2,001	1958
43.	University of Maryland	College Park, Maryland, USA	59	4,718	1807
	College Park campus	College Park, Maryland, USA	48	3,454	1856
45.	University of Dacca	Dacca, Bangladesh	58		1910
46.	Gauhati University	Jhalukbari, Assam, India	55	2,457	1914
46.	Lalit Narayan Mithila University	Darbhanga, Bihar, India	55		1947
46.	Universidad Nacional de Cordoba	Cordoba, Cordoba, Argentina	55	4,212	1613
46.	University of Missouri	Columbia, Missouri, USA	55	6,220	1839
	Columbia campus	Columbia, Missouri, USA	26	3,515	1839
50.	Ohio State University	Columbus, Ohio, USA	54	3,637	1870
	Columbus campus	Columbus, Ohio, USA	50		1870
50.	Universidad de La Habana	Havana, Cuba	54	3,066	1728
52.	Universita degli Studi di Bologna	Bologna, Emilia-Romagna, Italy	53	2,951	1088?
53.	Universidad de Puerto Rico (alt University of Puerto Rico)	San Juan, Puerto Rico	52	3,337	1903
	San Juan campus	San Juan, Puerto Rico	28		1903
53.	Universite d'Aix-Marseille	Aix-en-Provence, Provence-Cote d'Azur, France	52	3,825	1409
	Aix-en-Provence campus	Aix-en-Provence, Provence-Cote d'Azur, France	11	393	1409
	Marseilles campus	Marseilles, Provence-Cote d'Azur, France	41	3,432	1854
55.	University of Bihar	Muzaffarpur, Bihar, India	51		1889

Rank	University	Location	Students (thousands)	Teachers	Date Founded
55.	Univerzitet u Beogradu	Belgrade, Serbia, Yugoslavia	51	3,684	1808
57.	Universidad Central del Ecuador	Quito, Ecuador	50	2,350	1586
57.	Universidad Central de Venezuela	Caracas, Venezuela	50	2,700	1696
57.	Universita degli Studi di Milano	Milan, Lombardia, Italy	50	2,304	1923
57.	University of Calicut	Calicut, Kerala, India	50		1968
61.	Magadh University	Bodh Gaya, Bihar, India	49		1944
61.	University of Burdwan	Burdwan, West Bengal, India	49		1960
61.	University of Tennessee	Knoxville, Tennessee, USA	49	3,257	1794
	Knoxville campus	Knoxville, Tennessee, USA	30	1,668	1794
64.	Michigan State University	East Lansing, Michigan, USA	48	4,586	1855
64.	Ranchi University	Ranchi, Bihar, India	48	1,571	1899
64.	Waseda University	Tokyo, Honshu I, Japan	48	2,159	1882
67.	Louisiana State University and Agricultural and Mechanical College	Baton Rouge, Louisiana, USA	47	4,671	1853
	Baton Rouge campus	Baton Rouge, Louisiana, USA	26	1,960	1869
68.	Rutgers, the State University of New Jersey	New Brunswick, New Jersey, USA	46	5,161	1766
	New Brunswick campus	New Brunswick, New Jersey, USA	31		1766
68.	University of Michigan	Ann Arbor, Michigan, USA	46	5,362	1817
	Ann Arbor campus	Ann Arbor, Michigan, USA	38		1837
68.	University of Toronto	Toronto, Ontario, Canada	46	4,969	1827
71.	Guru Nanak University	Amritsar, Punjab, India	45		1969
71.	Utkal University	Bhubaneswar, Orissa, India	45	1,673	1943
73.	Saurashtra University	Rajkot, Gujarat, India	44		1965
74.	Karnatak University	Hubli-Dharwar, Karnataka, India	43	3,122	1917
74.	Pontifical University of Santo Tomas	Manila, Luzon I, Philippines	43	1,606	1611
74.	Universita degli Studi di Bari	Bari, Puglia, Italy	43	2,236	1924
74.	University of Hawaii	Honolulu, Hawaii, USA	43	2,327	1907
	Campuses at or nr Honolulu	Honolulu, Hawaii, USA	37	2,106	1907
78.	Universita degli Studi di Padova	Padua, Veneto, Italy	42	2,379	1222
78.	Universite de Toulouse	Toulouse, Midi-Pyrenees, France	42	1,463	1229
78.	University of Damascus	Damascus, Syria	42	745	1903
81.	Far Eastern University	Manila, Luzon I, Philippines	40	1,150	1928
81.	Universidad de Barcelona	Barcelona, Cataluna, Spain	40	2,100	bef 1377
81.	Universidad de la Republica	Montevideo, Uruguay	40	2,149	1833
81.	University of Kentucky	Lexington, Kentucky, USA	40	3,137	1865
	Lexington campus	Lexington, Kentucky, USA	22	2,479	1865
81.	Univerzitet Kiril i Metodij vo Skopje	Skopje, Macedonia, Yugoslavia	40	1,500	1946
86.	Kanpur University	Kanpur, Uttar Pradesh, India	39		1955

Rank	University	Location	Students (thousands)	Teachers	Date Founded
86.	Universita degli Studi di Firenze	Florence, Toscana, Italy	39	2,295	1321
86.	Universita degli Studi di Palermo	Palermo, Sicilia, Italy	39	1,686	1777
86.	University of Cincinnati	Cincinnati, Ohio, USA	39	3,161	1819
	Cincinnati campus	Cincinnati, Ohio, USA	35		1819
86.	University of Nebraska	Lincoln, Nebraska, USA	39	3,179	1869
	Lincoln campus	Lincoln, Nebraska, USA	22	1,525	1869
91.	Purdue University	West Lafayette, Indiana, USA	38	4,988	1869
	West Lafayette campus	West Lafayette, Indiana, USA	29	4,477	1869
91.	Universite de Lyon	Lyons, Rhone-Alpes, France	38		1809
91.	University of Houston	Houston, Texas, USA	38	1,769	1927
91.	Wayne State University	Detroit, Michigan, USA	38	2,142	1868
95.	Arizona State University	Tempe, Arizona, USA	37	2,204	1885
96.	Chuo University	Tokyo, Honshu I, Japan	36	1,356	1885
96.	Ludwig-Maximilians- Universitat Munchen	Munich, Bayern, West Germany	36	1,650	1472
96.	Northeastern University	Boston, Massachusetts, USA	36	2,139	1898
96.	University of Washington	Seattle, Washington, USA	36	3,775	1861
100.	San Diego State University	San Diego, California, USA	35	1,996	1897
100.	Southern Illinois University	Carbondale, Illinois, USA	35	3,076	1869
	Carbondale campus	Carbondale, Illinois, USA	21		1869
100.	Temple University	Philadelphia, Pennsylvania, USA	35	3,057	1884
100.	Universita degli Studi di Torino	Turin, Piemonte, Italy	35	2,281	1404
100.	Universite de Lille	Lille, Nord, France	35	1,836	1560
100.	University of Pittsburgh	Pittsburgh, Pennsylvania, USA	35	4,029	1787
	Pittsburgh campus	Pittsburgh, Pennsylvania, USA	30	3,843	1787

9g
LARGEST LIBRARIES

Contents

Rank.

Name of library.

City, and subdivision and country in which located.

Size of collection (i.e., thousands of volumes of books and pamphlets, usually excluding bound and unbound periodicals).

Date of foundation. The date given is preferably the year in which the library was opened on a permanent basis, that is, the one since which it has been continuously operative. By "library," however, is meant the nucleus of the present book collection, not necessarily the institution under its name today.

Arrangement

By size of collection.

Coverage

100 largest libraries in size of collection.

Rounding

Collections are rounded to the nearest 1000 volumes.

Entries

100.

Figure 41. Main building of the Library of Congress, the world's largest in number of volumes. The building has been a landmark in Washington since 1897. (Credit: Library of Congress.)

Figure 42. Urquhart Building of the Lending Division of the British Library at Wetherby, West Yorkshire, England. Second in size to the Library of Congress, the British Library ceased to be part of the British Museum in 1973, although the greater portion of its reference collections is still housed there. (Credit: British Library.)

TABLE 9g. LARGEST LIBRARIES

Rank	Library	Location	Volumes (thousands)	Date Founded
1.	Library of Congress	Washington, District of Columbia, USA	17,889	1800
2.	British Library (inc British Museum Library)	London, England, UK	12,150	1753
3.	V. I. Lenin State Library of the USSR	Moscow, Russia in Europe, USSR	11,750	1828
4.	Harvard University Library	Cambridge, Massachusetts, USA	9,207	1638
5.	New York Public Library	New York, New York, USA	8,898	1848
6.	M. E. Saltykov-Shchedrin State Public Library	Leningrad, Russia in Europe, USSR	8,000	1795
7.	Bibliotheque Nationale	Paris, Region Parisienne, France	7,000	1480
8.	Biblioteca Academiei Republicii Socialiste Romania	Bucharest, Romania	6,624	1867
9.	National Diet Library	Tokyo, Honshu I, Japan	6,560	1948
10.	Yale University Library	New Haven, Connecticut, USA	6,519	1701
11.	Shanghai Library	Shanghai, China	6,500	1952
12.	Deutsche Bucherei	Leipzig, East Germany	6,078	1912
13.	Chicago Public Library	Chicago, Illinois, USA	5,948	1873
14.	Biblioteca Centrala de Stat	Bucharest, Romania	5,642	1955
15.	USSR Academy of Sciences Library	Leningrad, Russia in Europe, USSR	5,526	1714
16.	Deutsche Staatsbibliothek	Berlin, East Germany	5,238	1661
17.	Urbana-Champaign Campus Library of the University of Illinois	Urbana, Illinois, USA	5,227	1868
18.	New York State Library	Albany, New York, USA	5,000	1818
19.	Bibliotheque de l'Universite de Paris	Paris, Region Parisienne, France	4,876	1570?
20.	Statni Knihovna	Prague, Bohemia, Czechoslovakia	4,672	1366?
21.	Columbia University Library	New York, New York, USA	4,662	1761
22.	Berkeley Campus Library of the University of California	Berkeley, California, USA	4,658	1868
23.	National Library of Peking	Peking, China	4,600	1909
24.	Los Angeles Public Library	Los Angeles, California, USA	4,533	1872
25.	National Library of New Zealand	Wellington, North I, New Zealand	4,418	1856
26.	University of Toronto Library	Toronto, Ontario, Canada	4,382	1842
27.	Cornell University Library	Ithaca, New York, USA	4,273	1865
28.	Biblioteca Nazionale Centrale	Florence, Toscana, Italy	4,210	1747
29.	Stanford University Library	Stanford, California, USA	4,092	1892
30.	Cape Provincial Library Service (alt Kaapse Provinsiale Biblioteekdiens)	Cape Town, Cape of Good Hope, South Africa	4,073	1945
31.	Los Angeles County Public Library	Los Angeles, California, USA	4,058	1912
32.	University of Tokyo Library	Tokyo, Honshu I, Japan	4,040	1893
33.	Central Library of the Chinese Academy of Sciences	Peking, China	4,000	1951

Rank	Library	Location	Volumes (thousands)	Date Founded
34.	Ann Arbor Campus Library of the University of Michigan	Ann Arbor, Michigan, USA	3,924	1838
35.	Boston Public Library	Boston, Massachusetts, USA	3,864	1854
36.	Central Library of Natural Sciences of the USSR Academy of Sciences	Moscow, Russia in Europe, USSR	3,784	1934
37.	Mirabeau B. Lamar Austin Campus Library of the University of Texas	Austin, Texas, USA	3,726	1883
38.	Bayerische Staatsbibliothek	Munich, Bayern, West Germany	3,700	1558
39.	Joseph Regenstein Library of the University of Chicago	Chicago, Illinois, USA	3,622	1891
40.	Universitatsbibliothek der Humboldt-Universitat zu Berlin	Berlin, East Germany	3,598	1831
41.	Central Scientific Library of The Ukrainian SSR Academy of Sciences	Kiev, Ukraine, USSR	3,594	1919
42.	Brooklyn Public Library	New York, New York, USA	3,520	1897
43.	Los Angeles Campus Library of the University of California	Los Angeles, California, USA	3,519	1919
44.	Bodleian Library (alt University of Oxford Library)	Oxford, England, UK	3,500	1602
45.	A. M. Gorkiy Scientific Library of the Moscow M. V. Lomonosov State University	Moscow, Russia in Europe, USSR	3,485	1755
46.	Kyoto University Library	Kyoto, Honshu I, Japan	3,438	1899
47.	Columbus Campus Library of Ohio State University	Columbus, Ohio, USA	3,413	1873
48.	Universitats- und Landes-bibliothek Sachsen-Anhalt	Halle, East Germany	3,300	1696
49.	McGill University Library	Montreal, Quebec, Canada	3,297	1821
50.	Kent County Library	Maidstone, England, UK	3,286	1921
51.	Transvaal Provincial Library Service (alt Transvaalse Provinsiale Biblioteek-diens)	Pretoria, Transvaal, South Africa	3,270	1943
52.	Lancashire County Library	Preston, England, UK	3,191	1925
53.	National A. F. Myashnikyan State Library of the Armenian SSR	Yerevan, Armenia, USSR	3,163	1832
54.	Universitetsbiblioteket	Oslo, Norway	3,162	1811
55.	Cleveland Public Library	Cleveland, Ohio, USA	3,092	1869
56.	Public Library of Cincinnati and Hamilton County	Cincinnati, Ohio, USA	3,055	1856
57.	O. Meredith Wilson Minneapolis Campus Library of the University of Minnesota	Minneapolis, Minnesota, USA	3,047	1851
58.	Bibliotheque Nationale et Universitaire	Strasbourg, Alsace, France	3,020	1872

Rank	Library	Location	Volumes (thousands)	Date Founded
59.	Universitatsbibliothek	Leipzig, East Germany	3,019	1543
60.	Queens Borough Public Library	New York, New York, USA	3,013	1896
61.	Center for Research Library	Chicago, Illinois, USA	3,000	1949
61.	Central House of the Soviet Army Library	Moscow, Russia in Europe, USSR	3,000	
61.	Narodniho Muzea Knihovna	Prague, Bohemia, Czechoslovakia	3,000	1818
61.	National Library of Scotland	Edinburgh, Scotland, UK	3,000	1682
61.	Peking University Library	Peking, China	3,000	1902
61.	University of Cambridge	Cambridge, England, UK	3,000	1400
67.	Philadelphia Free Library	Philadelphia, Pennsylvania, USA	2,974	1891
68.	Memorial Madison Campus Library of the University of Wisconsin	Madison, Wisconsin, USA	2,973	1850
69.	Buffalo and Erie County Public Library	Buffalo, New York, USA	2,931	1836
70.	Biblioteca Nacional	Madrid, Castilla la Nueva, Spain	2,923	1712
71.	Alisher Navoi State Library of the Uzbek SSR	Tashkent, Uzbekistan, USSR	2,900	1870
72.	Institute of Scientific Information on Social Sciences of the USSR Academy of Sciences	Moscow, Russia in Europe, USSR	2,872	1918
73.	California State Library	Sacramento, California, USA	2,844	1850
74.	Lvov V. Stefanik State Scientific Library of the Ukrainian SSR Academy of Sciences	Lvov, Ukraine, USSR	2,824	1940
75.	Princeton University Library	Princeton, New Jersey, USA	2,812	1746
76.	V. I. Lenin State Library of the White Russian SSR	Minsk, White Russia, USSR	2,800	1922
77.	Scientific Library of the Tbilisi State University	Tbilisi, Georgia, USSR	2,790	1918
78.	Bloomington Campus Library of Indiana University	Bloomington, Indiana, USA	2,763	1824
79.	M. Gorkiy Scientific Library of the Leningrad A. A. Zhdanov State University	Leningrad, Russia in Europe, USSR	2,759	1819
80.	University of Pennsylvania Library	Philadelphia, Pennsylvania, USA	2,700	1750
81.	Hampshire County Library	Winchester, England, UK	2,647	1925
82.	State Public Historical Library of the Russian SFSR	Moscow, Russia in Europe, USSR	2,634	1938
83.	William R. Perkins Library of Duke University	Durham, North Carolina, USA	2,622	1838
84.	Biblioteka Narodowa	Warsaw, Poland	2,600	1928
85.	Birmingham Public Library	Birmingham, England, UK	2,562	1861
86.	Staatsbibliothek Preussischer Kulturbesitz	Berlin, West-Berlin, West Germany	2,560	1661
87.	Library Service of Western Australia	Perth, Western Australia, Australia	2,541	1887

Rank	Library	Location	Volumes (thousands)	Date Founded
88.	Fovarosi Szabo Ervin Konyvtar	Budapest, Hungary	2,535	1904
89.	John Rylands University Library of Manchester	Manchester, England, UK	2,500	1851
89.	National Karl Marx State Library of the Georgian SSR	Tbilisi, Georgia, USSR	2,500	1846
91.	Liverpool Public Library	Liverpool, England, UK	2,498	1852
92.	Northwestern University Library	Evanston, Illinois, USA	2,475	1856
93.	Statni Vedecka Knihovna	Brno, Moravia, Czechoslovakia	2,438	1770
94.	Essex County Library	Chelmsford, England, UK	2,423	1926
95.	V. G. Korolenko State Scientific Library	Kharkov, Ukraine, USSR	2,407	1886
96.	Biblioteca Centrala Universitara Cluj	Cluj, Romania	2,401	1872
97.	Nanking Library	Nanking, Kiangsu, China	2,400	1908
98.	Bibliotheque Royale Albert Ier	Brussels, Belgium	2,391	1837
99.	Osterreichische National-bibliothek	Vienna, Austria	2,390	1526
100.	Milwaukee Public Library	Milwaukee, Wisconsin, USA	2,389	1847

COUNTRY GAZETTEER,
BY CONTINENT

Contents

This gazetteer consists of four tables.

10a. Latest Area, Population, and Density of Population

Conventional name of country.

Area, including inland waters, in thousands of square miles and square kilometers.

Latest population, in thousands, by census and by estimate, with year of data for each. If the census date is 1976 or 1977, no estimate is given. It should be noted also that most of the estimates are official ones based either on registration data or on birth, death, and migration records. Only occasionally are they simple projections from past growth rates.

Density of population, based on latest available data, per square mile and per square kilometer.[1]

10b. Evolution of Population

Conventional name of country.

Population, in thousands, for the years 1960, 1950, 1930, 1900, 1850, and 1800, or for the times nearest to those years for which reliable data are available, with the exact year stated. All populations are for the country as it existed at the time when the population was determined, not for its present-day area. For populations "circa 1960," a 1956 census figure is normally given in preference to a 1960 estimate. Question marks are used when the exact year is unknown; for example, the notation "(00?e)" below the tabular heading "Circa 1900" indicates that the population preceding it was estimated about but not necessarily during the year 1900.

10c. Additional Statistics

Conventional name of country.

Period of record for birth and death rates.

---

[1] Computed on the basis of actual areas rather than the rounded areas given in the table.

Number of live births per thousand of population.

Number of deaths per thousand of population.

Period of record for life expectancies.

Expectation of life at birth, in years, for men, for women, and for both (average).

Year of data for energy production and consumption.

Production of energy, in thousands of metric tons of coal equivalent. See introduction to Table 7k for further explanation.

Consumption of energy, in thousands of metric tons of coal equivalent. See introduction to Table 7l for further explanation.

Consumption of energy per capita, in kilograms of coal equivalent. See introduction to Table 7m for further explanation.

Year of data for gross national product.

Gross national product, in millions of US dollars at market prices. See introduction to Table 7n for further explanation.

Gross national product per capita, in US dollars at market prices. See introduction to Table 7o for further explanation.

10d. *Supplemental Information*

Conventional name of country.

[a] Alternate, former, and official names of country. See General Explanatory Notes in the Introduction for explanation. In addition, formerly separate countries now included in the country cited and component parts of the country that are sometimes regarded as separate countries (e.g., Guernsey, a part of the Channel Islands) are listed after the abbreviation "inc."

[b] Conventional and other (alternate, former, and official) names of capital (or capitals), and its latest population in thousands, with year of census or estimate (e).

[c] Conventional and other (alternate, former, and official) names of the largest city other than the capital (or capitals), and its latest population, in thousands, with year of census or estimate (e)[1]

Arrangement

By country, alphabetically under continent.

Coverage

All countries, regardless of size or sovereignty, but excluding provinces, states, and regions generally recognized as subdivisions of another country, for example, Tibet, a subdivision of China, and Ukraine, a subdivision of the USSR. For historical reasons, however, some states and so forth that once were separate countries are entered under the names of the countries of which they are now subdivisions. All countries located in more than one continent are listed under each applicable continent.

[1] In some countries the "largest city" is actually a small town or even a village. No attempt has been made to differentiate in the index, where each is identified simply as a city.

Rounding

Areas are rounded to the nearest 1000 mi^2 and km^2 if more than 100,000 mi^2 or km^2, to the nearest 100 if between 10,000 and 100,000, to the nearest 10 if between 1000 and 10,000, and to the nearest square mile and square kilometer if less than 1000 mi^2 or km^2.

Populations are rounded to the nearest 1000, except that when the original figure is less than 1000 the rounding is done to the nearest 100.

Densities of population are rounded to the nearest two significant digits if between 1 and 100 per square mile or per square kilometer, and to the nearest significant digit if less than 1 per square mile or per square kilometer.

Entries

259.

AFRICA

Table 10a. Latest Area, Population, and Density of Population

Country	Area (1000 mi²)	Area (1000 km²)	Latest Population (thousands) — Census	Latest Population (thousands) — Estimate	Density of Population (per mi²)	Density of Population (per km²)
Algeria	885	2293	11,822 (1966)	17,304 (1976)	20	7.5
Angola	481	1247	5,646 (1970)	6,700 (1975)	14	5.4
Benin	43.5	113	2,106 (1961)	3,197 (1976)	74	28
Botswana	232	600	609 (1971)	693 (1976)	3.0	1.2
British Indian Ocean Territory	0.030	0.078	1 (1960–62)	2 (1976)	66	26
Burundi	10.7	27.8	3,350 (1970–71)	3,864 (1976)	360	139
Cameroon	184	475	7,663 (1976)		42	16
Cape Verde	1.56	4.03	272 (1970)	303 (1976)	195	75
Central African Empire	241	623	1,203 (1959–60); 2,256 (1968)	2,740 (1976)	11	4.4
Ceuta	0.007	0.018	67 (1970)	65 (1973)	9,353	3,611
Chad	459	1189	3,254 (1963–64)	4,116 (1976)	9.0	3.5
Comoros	0.720	1.87	244[1] (1966)	274 (1976)	380	147
Congo	132	342	1,089 (1970); 1,300 (1974)	1,390 (1976)	11	4.1
Djibouti	8.49	22.0	81 (1960–61)	180 (1974)	21	8.2
Egypt	387	1001	30,076 (1966); 38,228 (1976)		99	38
Equatorial Guinea	10.8	28.1	246 (1960)	316 (1976)	29	11
Ethiopia	472	1222		24,069 (1970); 28,678 (1976)	61	23
Eritrea	45.4	118	598 (1931)	2,125 (1975)	47	18
French Southern and Antarctic Lands	2.92	7.56	0.2 (1970)	0.2 (1974)	0.07	0.03

[1] Inc Mayotte, listed below.

AFRICA

Country	Area (1000 mi²)	Area (1000 km²)	Latest Population (thousands) Census		Latest Population (thousands) Estimate		Density of Population (per mi²)	Density of Population (per km²)
Gabon	103	268	950	(1969–70)	1,200	(1976)	12	4.5
Gambia	4.36	11.3	493	(1973)	538	(1976)	123	48
Ghana	92.1	239	8,559	(1970)	10,309	(1976)	112	43
Guinea	95.0	246	5,143	(1972)	6,050	(1976)	64	25
Guinea-Bissau	13.9	36.1	487	(1970)	534	(1976)	38	15
Ivory Coast	125	322	6,673	(1975)	7,028	(1976)	56	22
Kenya	225	583	10,943	(1969)	13,847	(1976)	62	24
Lesotho	11.7	30.4	852	(1966)			104	40
			1,214	(1976)				
Liberia	43.0	111	1,523	(1970)	1,600	(1976)	37	14
			1,503	(1974)				
Libya	731	1893	2,291	(1973)	2,530	(1976)	3.5	1.3
Madagascar	227	587	6,200	(1966)	7,660	(1976)	34	13
			7,520	(1975)				
Malawi	45.7	118	4,040	(1966)	5,175	(1976)	113	44
Mali	479	1240	6,035	(1976)			13	4.9
Mauritania[1]	398	1031	1,050	(1964–65)			3.7	1.4
			1,481	(1976)				
Mauritius	0.790	2.04	851	(1972)	895	(1976)	1,134	438
Mayotte	0.144	0.374	32	(1966)	36	(1976)	249	96
Melilla	0.005	0.014	65	(1970)	60	(1973)	11,100	4,286
Morocco[1]	177	459	15,379	(1971)	17,828	(1976)	101	39
Mozambique	309	799	8,234	(1970)	9,444	(1976)	31	12
Namibia	318	824	762	(1970)	910	(1976)	2.9	1.1
Niger	482	1248	2,766	(1959–60)	4,727	(1976)	9.8	3.8
Nigeria	357	924	79,759[2]	(1973)	77,056	(1976)	216	83
Portugal (in Africa: Madeira Islands)	0.307	0.796	253	(1970)	277	(1975)	901	348

[1] Exc Western Sahara (for Spanish Sahara), listed below.
[2] Figure repudiated by Nigerian Government.

AFRICA

Country	Area (1000 mi²)	Area (1000 km²)	Latest Population (thousands) — Census	Latest Population (thousands) — Estimate	Density of Population (per mi²)	Density of Population (per km²)
Reunion	0.969	2.51	417 (1967) 477 (1974)	495 (1976)	511	197
Rhodesia	151	391	5,099 (1969)	6,530 (1976)	43	17
Rwanda	10.2	26.3	3,573 (1970)	4,289 (1976)	422	163
Saint Helena	0.162	0.419	6 (1966) 6 (1975)	6 (1976)	37	14
Sao Tome and Principe	0.372	0.964	74 (1970)	81 (1976)	218	84
Senegal	75.7	196	3,906 (1970–71) 5,085 (1976)		67	26
Seychelles	0.107	0.278	53 (1971) 62 (1977)		578	223
Sierra Leone	27.7	71.7	2,729 (1974)	3,111 (1976)	112	43
Somalia	246	638	1,022 (1931)	3,261 (1976)	13	5.1
South Africa	471	1221	21,794 (1970)	26,129 (1976)	55	21
Spain (in Africa: Canary Islands)	2.81	7.27	1,170 (1970)	1,299 (1976)	463	179
Sudan	967	2506	14,172 (1973)	15,470 (1976)	16	6.2
Swaziland	6.70	17.4	395 (1966) 499 (1976)		74	29
Tanzania	365	945	12,313 (1967)	15,607 (1976)	43	17
Tanganyika	364	943	11,959 (1967)	15,176 (1976)	42	16
Zanzibar	0.950	2.46	355 (1967)	431 (1976)	454	175
Togo	21.9	56.8	1,951 (1970)	2,283 (1976)	104	40
Tunisia	63.4	164	4,533 (1966) 5,588 (1975)	5,737 (1976)	91	35
Uganda	93.1	241	9,549 (1966)	11,943 (1976)	128	50
Upper Volta	106	274	6,144 (1975)	6,280 (1976)	59	23
Western Sahara	103	266	76 (1970) 75 (1974)	128 (1976)	1.2	0.5

AFRICA

Country	Area (1000 mi²)	Area (1000 km²)	Latest Population (thousands) Census	Latest Population (thousands) Estimate	Density of Population (per mi²)	Density of Population (per km²)
Zaire	906	2345	21,638 (1970)	25,629 (1976)	28	11
Zambia	291	753	4,057 (1969)	5,138 (1976)	18	6.8
			4,696 (1974)			

ASIA

Country	Area (1000 mi²)	Area (1000 km²)	Latest Population (thousands) Census	Latest Population (thousands) Estimate	Density of Population (per mi²)	Density of Population (per km²)
Afghanistan	250	647		17,086 (1970)	79	31
				19,803 (1976)		
Bahrain	0.256	0.662	216 (1971)	259 (1976)	1,013	391
Bangladesh	55.6	144	71,479 (1974)	75,980 (1976)	1,367	528
Bhutan	18.1	47.0	1,035 (1969)	1,202 (1976)	66	26
Brunei	2.23	5.76	136 (1971)	177 (1976)	80	31
Burma	262	678	28,886 (1973)	31,002 (1976)	118	46
Cambodia	69.9	181	5,729 (1962)	7,735 (1976)	111	43
China[1]	3769	9761	582,603 (1953)	835,843 (1976)	222	86
Cyprus	3.57	9.25	632 (1973)	639 (1976)	179	69
Gaza Strip	0.146	0.378	356 (1967)	419 (1974)	2,871	1,108
Hong Kong	0.404	1.05	3,948 (1971)		10,908	4,212
			4,407 (1976)			
India[2]	1184	3066	543,333[3] (1971)	604,977 (1976)	511	197
Sikkim	2.82	7.30	210 (1971)		75	29
Indonesia						
Total	741	1919	119,232[4] (1971)	135,880 (1976)	183	71
in Asia	578	1498	118,309[4] (1971)	134,895 (1976)	233	90
East Timor	5.76	14.9	609 (1970)	688 (1976)	119	46

[1] Exc Kashmir-Jammu and Taiwan, listed below.
[2] Exc Kashmir-Jammu, listed below.
[3] Exc Sikkim.
[4] Exc East Timor.

ASIA

Country	Area (1000 mi²)	Area (1000 km²)	Latest Population (thousands) Census	Latest Population (thousands) Estimate	Density of Population (per mi²)	Density of Population (per km²)
Iran	636	1648	25,789 (1966); 33,592 (1976)		53	20
Iraq	168	435	8,047 (1965); 12,171 (1977)		72	28
Israel[1]	8.02	20.8	3,148 (1972)	3,530 (1976)	440	170
Japan	146	378	103,720[2] (1970); 111,940 (1975)	113,086 (1976)	776	300
Ryukyu Islands	0.867	2.25	945 (1970); 1,043 (1975)	1,059 (1976)	1,221	472
Jordan	35.6[3]	92.1[3]	1,706 (1961)	2,079[3] (1976)	58[3]	23[3]
Kashmir-Jammu	85.8	222	4,022 (1941)	6,960 (1976)	81	31
Korea						
North Korea	46.5	121		16,246 (1976)	349	135
South Korea	38.0	98.5	30,882 (1970); 34,709 (1975)	35,860 (1976)	943	364
Kuwait	6.88	17.8	739 (1970); 995 (1975)	1,031 (1976)	150	58
Laos	91.4	237	944 (1931)	3,383 (1976)	37	14
Lebanon	4.02	10.4	2,790 (1970)	3,110 (1976)	775	299
Macao	0.006	0.016	249 (1970)	275 (1976)	44,516	17,187
Malaysia	127	330	10,440 (1970)	12,300 (1976)	97	37
Peninsular Malaysia	50.8	132	8,810 (1970)	10,331 (1976)	203	79
Sabah	28.5	73.7	653 (1970)	830 (1976)	29	11
Sarawak	48.1	124	976 (1970)	1,138 (1976)	24	9.1

[1] Exc Gaza Strip, listed above, and West Bank, listed below.
[2] Exc Ryukyu Islands.
[3] Exc West Bank, listed below.

ASIA

Country	Area		Latest Population (thousands)		Density of Population	
	(1000 mi²)	(1000 km²)	Census	Estimate	(per mi²)	(per km²)
Maldives	0.115	0.298	114 (1970) 123 (1972) 129 (1974)	135 (1976)	1,173	453
Mongolia	604	1565	1,198 (1969)	1,488 (1976)	2.5	1.0
Nepal	56.1	145	11,556 (1971)	12,857 (1976)	229	88
Oman	82.0	212		791 (1976)	9.6	3.7
Pakistan¹	310	804	64,980 (1972)	72,368 (1976)	233	90
Philippines	116	300	36,684 (1970) 41,831 (1975)	42,920 (1976)	371	143
Qatar	4.25	11.0		170 (1975)	40	15
Saudi Arabia	830	2150	7,013 (1974)	7,600 (1976)	9.2	3.5
Singapore	0.231	0.597	2,075 (1970)	2,278 (1976)	9,883	3,816
Sri Lanka	25.3	65.6	12,711 (1971)	14,270 (1976)	563	217
Syria	71.5	185	6,305 (1970)	7,596 (1976)	106	41
Taiwan	14.0	36.2	13,505 (1966) 14,676 (1970)	16,290 (1976)	1,167	450
Thailand	198	514	34,397 (1970)	42,960 (1976)	216	84
Turkey	315	815				
Total			35,605 (1970) 40,348 (1975)	41,020 (1976)	130	50
in Asia	305	790	32,394 (1970) 36,548 (1975)	37,157 (1976)	122	47
United Arab Emirates	32.3	83.6	179 (1968) 673 (1976)		21	8.1
USSR (in Asia)	6608	17,114	66,848 (1970)	73,697 (1976)	11	4.3
Vietnam	129	333	17,702 (1931)	47,150 (1976)	367	142
North Vietnam	61.3	159	23,787 (1974)		388	150
South Vietnam	67.3	174	9,606 (1931)	19,582 (1974)	291	112

¹ Exc Kashmir-Jammu, listed above.

ASIA

Country	Area (1000 mi²)	Area (1000 km²)	Latest Population (thousands) Census	Estimate	Density of Population (per mi²)	(per km²)
West Bank	2.18	5.65	599 (1967)	700 (1976)	321	124
Yemen						
North Yemen	75.3	195	5,238 (1975)		70	27
South Yemen	129	333	1,590 (1973)	1,749 (1976)	14	5.3

EUROPE

Country	Area (1000 mi²)	Area (1000 km²)	Latest Population (thousands) Census	Estimate	Density of Population (per mi²)	(per km²)
Albania	11.1	28.7	1,626 (1960)	2,548 (1976)	230	89
Andorra	0.175	0.453	27 (1975)	29 (1976)	166	64
Austria	32.4	83.9	7,456 (1971)	7,514 (1976)	232	90
Belgium	11.8	30.5	9,651 (1970)	9,889 (1976)	839	324
Bulgaria	42.8	111	8,228 (1965)	8,761 (1976)	205	79
			8,728 (1975)			
Channel Islands	0.075	0.195	123 (1971)	129 (1976)	1,713	662
Czechoslovakia	49.4	128	14,345 (1970)	14,918 (1976)	302	117
Denmark	16.6	43.1	4,938 (1970)	5,073 (1976)	305	118
Faeroe Islands	0.540	1.40	39 (1970)	40 (1976)	74	29
Finland	130	337	4,598 (1970)	4,727 (1976)	36	14
France	213	551	49,797 (1968)	52,947 (1976)	249	96
			52,656 (1975)			
Germany						
East Germany	41.8	108	17,068 (1971)	16,786 (1976)	402	155
West Germany	96.0	249	60,651 (1970)	61,513 (1976)	641	247
Gibraltar	0.002	0.006	27 (1970)	30 (1976)	12,950	5,000
Greece	51.0	132	8,769 (1971)	9,165 (1976)	180	69
Hungary	35.9	93.0	10,322 (1970)	10,596 (1976)	295	114
Iceland	39.8	103	205 (1970)	220 (1976)	5.5	2.1
Ireland	27.1	70.3	2,978 (1971)	3,162 (1976)	117	45

EUROPE

Country	Area		Latest Population (thousands)		Density of Population	
	(1000 mi²)	(1000 km²)	Census	Estimate	(per mi²)	(per km²)
Isle of Man	0.227	0.588	56 (1971)		264	102
			60 (1976)			
Italy	116	301	53,745 (1971)	56,189 (1976)	483	187
Liechtenstein	0.061	0.157	21 (1970)	24 (1976)	396	153
			24 (1974)			
Luxembourg	0.998	2.59	340 (1970)	358 (1976)	359	138
Malta	0.122	0.316	316 (1967)	320 (1975)	2,623	1,013
Monaco	0.0007	0.0018	23 (1968)	24 (1976)	34,533	13,333
Netherlands	15.9	41.2	13,046 (1971)	13,770 (1976)	866	335
Norway	125	324	3,874 (1970)	4,026 (1976)	32	12
Poland	121	313	32,642 (1970)	34,362 (1976)	285	110
Portugal						
Total	35.5	92.1	8,663 (1970)	9,694 (1976)	273	105
in Europe	35.2	91.3	8,410 (1970)	9,417 (1976)	267	103
Azores	0.902	2.33	287 (1970)	302 (1975)	335	129
Romania	91.7	237	19,103 (1966)		235	91
			21,559 (1977)			
San Marino	0.024	0.061	18 (1971)	20 (1976)	849	328
Spain						
Total	195	505	33,824 (1970)	35,849 (1976)	184	71
in Europe	192	497	32,654 (1970)	34,550 (1976)	180	69
Balearic Islands	1.94	5.01	558 (1970)	625 (1976)	323	125
Svalbard	24.1	62.4	4 (1971)	3 (1974)	0.1	0.05
Sweden	174	450	8,077 (1970)	8,222 (1976)	47	18
			8,209 (1975)			
Switzerland	15.9	41.3	6,270 (1970)	6,346 (1976)	398	154
Turkey (in Europe)	9.41	24.4	3,211 (1970)	3,863 (1976)	410	158
			3,800 (1975)			

EUROPE

Country	Area (1000 mi²)	Area (1000 km²)	Latest Population (thousands) Census	Estimate	Density of Population (per mi²)	(per km²)
UK	94.2	244	55,515 (1971)	55,928 (1976)	594	229
England and Wales	58.3	151	48,750 (1971)	49,184 (1976)	843	325
Northern Ireland	5.46	14.1	1,536 (1971)	1,538 (1976)	282	109
Scotland	30.4	78.8	5,229 (1971)	5,205 (1976)	171	66
USSR						
Total	8649	22,402	241,720 (1970)	255,524 (1976)	30	11
in Europe	2042	5288	174,872 (1970)	181,827 (1976)	89	34
Vatican City	0.0002	0.0004	0.9 (1948)	0.7 (1976)	4,120	1,591
Yugoslavia	98.8	256	20,523 (1971)	21,560 (1976)	218	84

NORTH AMERICA

Country	Area (1000 mi²)	Area (1000 km²)	Latest Population (thousands) Census	Estimate	Density of Population (per mi²)	(per km²)
Anguilla	0.035	0.091	6 (1960)	6 (1976)	171	66
Antigua	0.170	0.440	65 (1970)	71 (1976)	416	161
Bahamas	5.38	13.9	169 (1970)	211 (1976)	39	15
Barbados	0.166	0.431	238 (1970)	251 (1976)	1,508	582
Belize	8.87	23.0	121 (1970)	144 (1976)	16	6.3
Bermuda	0.020	0.053	53 (1970)	57 (1976)	2,785	1,075
Canada	3852	9976	21,568 (1971) 22,993 (1976)		6.0	2.3
Newfoundland	156	405	522 (1971) 558 (1976)		3.6	1.4
Canal Zone	0.553	1.43	44 (1970)	40 (1976)	72	28
Cayman Islands	0.100	0.259	10 (1970)	11 (1976)	110	43
Costa Rica	19.6	50.7	1,872 (1973)	2,012 (1976)	103	40
Cuba	42.8	111	8,569 (1970)	9,460 (1976)	221	85
Dominica	0.290	0.751	71 (1970)	78 (1976)	269	104
Dominican Republic	18.7	48.4	4,009 (1970)	4,835 (1976)	259	100

NORTH AMERICA

Country	Area		Latest Population (thousands)		Density of Population	
	(1000 mi²)	(1000 km²)	Census	Estimate	(per mi²)	(per km²)
El Salvador	8.12	21.0	3,555 (1971)	4,123 (1976)	508	196
Greenland	840	2176	47 (1970)	50 (1976)	0.06	0.02
Grenada	0.133	0.344	94 (1970)	119 (1976)	896	346
Guadeloupe	0.687	1.78	313 (1967)	328 (1976)	478	184
			324 (1974)			
Guatemala	42.0	109	5,160 (1973)	6,256 (1976)	149	57
Haiti	10.7	27.7	4,330 (1971)	4,668 (1976)	436	168
Honduras	43.3	112	2,654 (1974)	2,830 (1976)	65	25
Jamaica	4.24	11.0	1,814 (1970)	2,057 (1976)	485	187
Martinique	0.425	1.10	320 (1967)	327 (1976)	769	297
			325 (1974)			
Mexico	781	2022	48,225 (1970)	62,329 (1976)	80	31
Montserrat	0.039	0.102	12 (1970)	13 (1976)	329	127
Netherlands Antilles	0.383	0.993	218 (1971)	241 (1976)	629	243
Nicaragua	50.2	130	1,878 (1971)	2,233 (1976)	44	17
Panama	29.2	75.6	1,428 (1970)	1,719 (1976)	59	23
Puerto Rico	3.44	8.90	2,712 (1970)	3,205 (1976)	933	360
Saint Kitts-Nevis	0.103	0.266	45 (1970)	48 (1976)	467	180
Saint Lucia	0.238	0.616	101 (1970)	112 (1976)	471	182
Saint-Pierre and Miquelon	0.093	0.242	5 (1967)	6 (1976)	64	25
			6 (1974)			
Saint Vincent	0.150	0.388	87 (1970)	93 (1975)	621	240
Trinidad and Tobago	1.98	5.13	941 (1970)	1,093 (1976)	552	213
Turks and Caicos Islands	0.166	0.430	6 (1970)	6 (1976)	36	14
USA						
Total	3683	9539	204,839 (1970)	215,118 (1976)	58	23
in North America	3677	9522	204,069 (1970)	214,287 (1976)	58	23
Alaska	586	1519	302 (1970)	357 (1976)	0.6	0.2
Virgin Islands (UK)	0.059	0.153	10 (1970)	11 (1975)	186	72
Virgin Islands (USA)	0.133	0.344	62 (1970)	95 (1976)	715	276

OCEANIA

Country	Area (1000 mi²)	Area (1000 km²)	Latest Population (thousands) Census	Latest Population (thousands) Estimate	Density of Population (per mi²)	Density of Population (per km²)
American Samoa	0.076	0.197	27 (1970), 29 (1974)	29 (1975)	381	147
Australia	2966	7682	12,756 (1971), 13,915 (1976)		4.7	1.8
Tasmania	26.2	67.8	390 (1971), 407 (1976)		16	6.0
Christmas Island (Australia)	0.054	0.140	3 (1971)	3 (1976)	55	21
Cocos Islands	0.005	0.014	0.6 (1971)	0.5 (1976)	92	36
Cook Islands	0.093	0.241	21 (1971), 20 (1973), 18 (1976)		193	75
Fiji	7.07	18.3	477 (1966), 588 (1976)		83	32
French Polynesia	1.54	4.00	119 (1971), 137 (1977)		89	34
Gilbert Islands	0.308	0.798	54[1] (1968), 58[1] (1973)	58 (1976)	188	73
Guam	0.212	0.549	85 (1970)	93 (1976)	439	169
Indonesia (in Oceania: Irian Jaya)	163	422	923 (1971)	985 (1976)	6.0	2.3
Johnston and Sand Islands	0.0004	0.001	1 (1970)	1 (1976)	2,590	1,000
Midway Islands	0.002	0.005	2 (1970)	2 (1976)	1,036	400
Nauru	0.008	0.021	6 (1966), 7 (1975)		863	333
New Caledonia	7.36	19.1	101 (1969), 132 (1974), 133 (1976)		18	7.0

[1] Inc Tuvalu (for Ellice Islands), listed below.

OCEANIA

Country	Area		Latest Population (thousands)				Density of Population	
	(1000 mi²)	(1000 km²)	Census		Estimate		(per mi²)	(per km²)
New Hebrides	5.70	14.8	78	(1967)	97	(1976)	17	6.6
New Zealand	104	269	2,863	(1971)			30	12
			3,129	(1976)				
Niue Island	0.100	0.259	5	(1971)			40	15
			4	(1976)				
Norfolk Island	0.014	0.036	2	(1971)	2	(1976)	144	56
Northern Mariana Islands	0.184	0.477	10	(1970)	15	(1975)	81	31
			14	(1973)				
Pacific Islands (USA)	0.533	1.38	95¹	(1970)	126¹	(1976)	208	80
			115¹	(1973)				
Papua New Guinea	178	462	2,490	(1971)	2,829	(1976)	16	6.1
New Guinea (Australia)	92.2	239	1,798	(1971)	1,898	(1973)	21	8.0
Papua	86.1	223	692	(1971)	662	(1973)	7.7	3.0
Pitcairn Island	0.002	0.005	0.1	(1971)	0.1	(1975)	52	20
Samoa	1.13	2.93	147	(1971)			134	52
			152	(1976)				
Solomon Islands	11.2	28.9	161	(1970)			18	6.8
			197	(1976)				
Tokelau Islands	0.004	0.010	2	(1972)			518	200
			2	(1974)				
Tonga	0.270	0.699	77	(1966)			333	129
			90	(1976)				
Tuvalu	0.010	0.025	6	(1968)	6	(1976)	632	244
			6	(1973)				
USA (in Oceania: Hawaii)	6.45	16.7	770	(1970)	831	(1976)	129	50
Wake Island	0.003	0.008	2	(1970)	2	(1976)	647	250

¹Inc Northern Mariana Islands.

OCEANIA

Country	Area (1000 mi²)	Area (1000 km²)	Latest Population (thousands) Census	Estimate	Density of Population (per mi²)	(per km²)
Wallis and Futuna Islands	0.077	0.200	9 (1969) 9 (1976)		117	45

SOUTH AMERICA

Country	Area (1000 mi²)	Area (1000 km²)	Latest Population (thousands) Census	Estimate	Density of Population (per mi²)	(per km²)
Argentina	1073	2780	23,362 (1970)	25,719 (1976)	24	9.3
Bolivia	424	1099	4,648 (1976)		11	4.2
Brazil	3286	8512	93,289 (1970)	109,181 (1976)	33	13
Chile	292	757	8,885 (1970)	10,454 (1976)	36	14
Colombia	440	1139	21,070 (1973)	22,540 (1976)	51	20
Ecuador	109	284	6,522 (1974)	6,890 (1976)	63	24
Galapagos Islands	3.03	7.84	4 (1974)		1.3	0.5
Falkland Islands	6.28	16.3	2 (1972) 2 (1975)		0.3	0.1
French Guiana	35.1	91.0	44 (1967) 55 (1974)	58 (1976)	1.7	0.6
Guyana	83.0	215	702 (1970)	783 (1976)	9.4	3.6
Paraguay	157	407	2,358 (1972)	2,724 (1976)	17	6.7
Peru	496	1285	14,161 (1972)	16,090 (1976)	32	13
Surinam	63.0	163	385 (1971)	435 (1976)	6.9	2.7
Uruguay	68.5	178	2,782 (1975)	2,800 (1976)	41	16
Venezuela	352	912	10,722 (1971)	12,360 (1976)	35	14

Table 10b. Evolution of Population
AFRICA

Population (thousands)

Country	Circa 1960		Circa 1950		Circa 1930		Circa 1900		Circa 1850		Circa 1800	
Algeria	10,784	(60)	8,682	(48)	6,553	(31)	4,429	(96)	2,496	(56)	1,500	(00?e)
Angola	4,841	(60)	4,145	(50)	2,615	(29)	4,119	(00?e)	2,500	(50?e)		
Benin	2,106	(61)	1,535	(51e)	980	(26)	749	(06)				
Botswana	543	(64)	296	(46)	153	(21)	121	(04)				
British Indian Ocean Territory	1	(60–62)										
Burundi[1]	2,224	(60e)	1,908	(52)								
Cameroon	5,017	(60–65)	4,085	(50e)	2,996	(31)	3,500	(00?e)				
Cape Verde	200	(60)	148	(50)	146	(30)	147	(00)	83	(73)	42	(00?e)
Central African Empire	1,203	(59–60)	1,072	(50e)	1,066	(26)	2,130	(06e)				
Ceuta	73	(60)	60	(50)	51	(30)	13	(00)	7	(57)		
Chad	3,254	(63–64)	2,241	(50e)	974	(26)	885	(06e)				
Comoros[2]	183	(58)	166	(51)	129	(36)	96	(06)	28	(50)	21	(00?e)
Congo	797	(60–61)	684	(50e)	699	(26)	259	(06e)				
Djibouti	81	(60–61)	56	(51e)	70	(31)	208	(06)				
Egypt	26,085	(60)	19,022	(47)	14,218	(27)	9,794	(97)	4,476	(46)	2,460	(00e)
Equatorial Guinea	246	(60)	199	(50)	167	(32)	161	(00e)				
Ethiopia	20,600	(60e)	15,000	(50e)	5,500	(30?e)	4,500	(00?e)	3,000	(31e)	1,800	(00?e)
Eritrea	[1,422	(62e)]	1,104	(50e)	598	(31)	330	(99)				
French Southern and Antarctic Lands	0.1	(63e)										
Gabon	630	(60–61)	409	(50e)	389	(26)	376	(06e)				
Gambia	315	(63)	280	(51)	200	(31)	103	(01)				

[1] Burundi with Rwanda: 3,406 (1935e).
[2] Inc Mayotte.

AFRICA

Population (thousands)

Country	Circa 1960		Circa 1950		Circa 1930		Circa 1900		Circa 1850		Circa 1800	
Ghana	6,727	(60)	3,736	(48)	2,870	(31)	1,503	(11)	408	(71e)		
Guinea	3,072	(60e)	2,570	(55)	2,096	(26)	1,498	(06)				
Guinea-Bissau	521	(60)	511	(50)	377	(31)	820	(00?e)				
Ivory Coast	3,100	(57–58)	2,169	(51e)	1,725	(26)	889	(06)				
Kenya	8,636	(62)	5,406	(48)	3,025	(31)	3,000	(00?e)				
Lesotho	642	(56)	564	(46)	562	(36)	349	(04)	128	(75)		
Liberia	1,016	(62)	1,648	(49e)	2,500	(30?e)	2,060	(00?e)	250	(50e)		
Libya	1,564	(64)	1,089	(54)	704	(31)	1,000	(00?e)	750	(50?e)	1,000	(00?e)
Madagascar	5,393	(60e)	4,256	(50e)	3,759	(31)	2,505	(01)	3,000	(50?e)	4,000	(00?e)
Malawi	3,460	(60e)	2,050	(45)	1,603	(31)	737	(01)				
Mali[1]	4,100	(60–61)	3,347	(51e)	2,635	(26)						
Mauritania	1,050	(64–65)	657	(51e)	289	(26)	223	(06)				
Mauritius	701	(62)	517	(52)	403	(31)	371	(01)	181	(51)	59	(97)
Mayotte	[23	(58)]	[17	(48e)]	[17	(36)]	[12	(01?)]	2	(57)		
Melilla	79	(60)	81	(50)	63	(30)	9	(00)				
Morocco	11,626	(60)	9,125	(52)	6,235	(31)	7,000	(00?e)	8,500	(50?e)	5,000	(00?e)
Mozambique	6,604	(60)	5,732	(50)	4,006	(30)	3,120	(00?e)	300	(50?e)		
Namibia	526	(60)	418	(51)	259	(26)	207	(04e)				
Niger[2]	2,876	(59–60)	2,127	(51e)	1,219	(26)						
Nigeria	55,670	(63)	29,731	(52–53)	19,158	(31)	17,133	(11e)				
Portugal (in Africa: Madeira Islands)	268	(60)	267	(50)	212	(30)	151	(00)	107	(54)	90	(00?e)
Reunion	349	(61)	242	(46)	198	(31)	173	(02)	106	(52)	65	(04)
Rhodesia	3,857	(62)	1,765	(48)	1,109	(31)	613	(04e)				
Rwanda[3]	2,665	(60e)	2,148	(52)								
Saint Helena	5	(56)	5	(46)	4	(31)	4	(04)	7	(61)	2	(05)

[1] Mali with Niger: 5,059 (1906).
[2] Niger with Mali: 5,059 (1906).
[3] Rwanda with Burundi: 3,406 (1935e).

AFRICA

Population (thousands)

Country	Circa 1960		Circa 1950		Circa 1930		Circa 1900		Circa 1850		Circa 1800	
Sao Tome and Principe	64	(60)	60	(50)	59	(21)	42	(00)	17	(50?e)	5	(00?e)
Senegal	3,110	(60–61)	2,093	(51e)	1,358	(26)	1,247	(11)				
Seychelles	41	(60)	35	(47)	27	(31)	19	(01)	7	(50)	7	(25)
Sierra Leone	2,180	(63)	1,858	(48)	1,769	(31)	1,403	(11)				
Somalia	2,010	(60e)	1,747	(51e)	1,022	(31)	400	(00?e)				
South Africa	16,003	(60)	12,668	(51)	9,588	(36)	5,176	(04)	267[1]	(56)	62[1]	(98)
Spain (in Africa: Canary Islands)	944	(60)	793	(50)	555	(30)	359	(00)	234	(57)	174	(97)
Sudan	10,263	(55–56)	8,350	(50e)	5,508	(31e)	3,000	(10e)				
Swaziland	237	(56)	185	(46)	157	(36)	85	(11)				
Tanzania												
Tanganyika	8,788	(57)	7,408	(52)	5,023	(31)	4,145	(13)				
Zanzibar	299	(58)	264	(48)	235	(31)	197	(10)	130	(46e)	200	(11e)
Togo	1,440	(58–60)	1,395	(50e)	1,044	(31e)	2,250	(00?e)				
Tunisia	4,113	(56)	3,231	(46)	2,411	(31)	1,939	(11)	1,520	(81)	1,000	(00?e)
Uganda	6,537	(59)	4,959	(48)	3,553	(31)	2,843	(11)				
Upper Volta	4,400	(60–61)	3,109	(51e)	3,240	(26)						
Western Sahara	24	(60)	37	(50e)	32	(30?e)	115	(00?e)				
Zaire	12,769	(55–58)	11,258	(50e)	8,764	(30e)	9,000	(10e)				
Zambia	3,490	(63)	1,930	(50–51)	1,345	(31)	497	(00?e)				

ASIA

Population (thousands)

Country	Circa 1960		Circa 1950		Circa 1930		Circa 1900		Circa 1850		Circa 1800	
Afghanistan	14,483	(61e)	12,000	(50e)	7,000	(30?e)	4,550	(00?e)	4,000	(50?e)	3,000	(00?e)
Bahrain	143	(59)	110	(50)	120	(30?e)	70	(00?e)				

[1] For Cape of Good Hope Province only.

ASIA

Population (thousands)

Country	Circa 1960	Circa 1950	Circa 1930	Circa 1900	Circa 1850	Circa 1800
Bangladesh[1]						
Bhutan	670 (60e)	300 (50e)	250 (30?e)	200 (00?e)	20 (64e)	
Brunei	84 (60)	41 (47)	30 (31)	22 (11)		
Burma	22,325 (60e)	16,824 (41)	14,667 (31)	10,491 (01)	7,722 (91)	4,231 (26)
Cambodia	5,729 (62)	3,640 (51e)	2,806 (31)	1,194 (06)	1,000 (50?e)	1,000 (00?e)
China	646,530 (57e)	582,603 (53)	438,933 (31e)	372,563 (10e)	429,931 (50e)	295,273 (00e)
Cyprus	578 (60)	450 (46)	348 (31)	237 (01)	186 (81)	[84 (00?e)]
Gaza Strip	377 (60e)	198 (50e)				
Hong Kong	3,133 (61)	2,015 (51e)	840 (31)	399 (01)	33 (50)	
India	435,512[2] (61)	357,303[2] (51)	338,061[3] (31)	283,870[3] (01)	203,415[3] (67–72)	131,000[3] (20e)
Sikkim	162 (61)	138 (51)	110 (31)	59 (01)		
Indonesia						
Total[4]	97,069 (61)	77,271 (50e)	60,727 (30)	37,694 (00e)	19,319 (63e)	13,476 (00?e)
in Asia[4]	96,319 (61)	76,571 (50e)	60,413 (30)	37,494 (00e)	19,119 (63e)	
East Timor	517 (60)	442 (50)	442 (26)	300 (00?e)		
Iran	18,955 (56)	16,550 (40)	15,055 (33)	9,000 (97e)	5,000 (50?e)	
Iraq	6,340 (57)	4,816 (47)	3,300 (30?e)	[1,398 (10e)]		
Israel	2,183 (61)	1,258 (50e)	1,036 (31)			
Japan	93,419[5] (60)	83,419[5] (50)	64,450 (30)	43,756 (98)	33,111 (72)	25,471 (98e)
Ryukyu Islands	883 (60)	918 (50)	[785 (30)]	[454 (98)]		
Jordan[6]	1,706 (61)	1,329 (52)	300 (29e)			
Kashmir-Jammu	4,835 (61e)	4,022 (41)	[3,646 (31)]	[2,906 (01)]		

[1] East Pakistan until December 1971 (see under Pakistan).
[2] Exc Burma, Kashmir-Jammu, and Sikkim.
[3] Exc Burma and Sikkim.
[4] Exc East Timor.
[5] Exc Ryukyu Islands.
[6] Inc West Bank.

ASIA

Population (thousands)

Country	Circa 1960		Circa 1950		Circa 1930		Circa 1900		Circa 1850		Circa 1800	
Korea					20,438	(30)	12,934	(09)	10,519	(83)	3,000	(00?e)
North Korea	10,789	(60e)	9,102	(49e)								
South Korea	24,989	(60)	20,167	(49)								
Kuwait	322	(61)	170	(50e)	51	(30?e)						
Laos	1,805	(60e)	1,360	(51e)	944	(31)	664	(06)	1,000	(50?e)		
Lebanon	2,152	(61e)	1,257	(50e)	629	(21–22)	[200	(00?e)]				
Macao	169	(60)	188	(50)	158	(27)	64	(99)	52	(62e)	34	(22e)
Malaysia												
Peninsular Malaysia	6,279	(57)	4,908	(47)	3,788	(31)	2,351	(11)				
Sabah	454	(60)	334	(51)	270	(31)	105	(01)				
Sarawak	745	(60)	546	(47)	440	(37e)	500	(01e)				
Maldives	82	(56)	82	(46)	79	(31)	72	(11)	175	(50?e)		
Mongolia	1,017	(63)	732	(50)	648	(18)						
Nepal	9,413	(61)	8,432	(52–54)	5,574	(20)	5,639	(11)	2,000	(50?e)	2,000	(20e)
Oman	565	(60e)	550	(50e)	500	(30?e)	1,000	(00?e)				
Pakistan	93,832[1]	(61)	75,842[1]	(51)	56,887[2]	(31)	[45,504[2]	(01)]				
East Pakistan	50,854	(61)	42,063	(51)	35,604[2]	(31)	[28,928[2]	(01)]				
West Pakistan	42,978[1]	(61)	33,779[1]	(51)	21,283[2]	(31)	[16,577[2]	(01)]				
Philippines	27,088	(60)	19,234	(48)	16,000	(39)	7,635	(03)	6,171	(77)	1,522	(99e)
Qatar	55	(63e)	17	(51e)	26	(30?e)						
Saudi Arabia	6,400	(62e)	6,000	(51e)	4,200	(30?e)	[3,000	(00?e)]	[4,000	(50?e)]		
Singapore	1,446	(57)	938	(47)	560	(31)	229	(01)	54	(50)	0.2	(00?e)
Sri Lanka	10,582	(63)	8,098	(53)	5,307	(31)	3,566	(01)	2,400	(71)	852	(24)
Syria	4,565	(60)	3,503	(50e)	1,506	(21–22)	[2,690	(00?e)]				
Taiwan	9,368	(56)	7,618	(50)	4,593	(30)	2,925	(01)				
Thailand	26,258	(60)	17,443	(47)	11,506	(29)	8,266	(11)	5,000	(54e)	1,900	(00?e)

[1] Exc Kashmir-Jammu.
[2] Included with India.

ASIA

Population (thousands)

Country	Circa 1960		Circa 1950		Circa 1930		Circa 1900		Circa 1850		Circa 1800	
Turkey												
Total	27,755	(60)	20,947	(50)	13,648	(27)	23,813	(10e)	26,636	(44–50?e)	20,912	(00?e)
in Asia	25,470	(60)	19,363	(50)	12,608	(27)	17,683¹	(10e)	16,050¹	(50?e)	11,090¹	(00?e)
United Arab Emirates	111	(62e)	80	(49e)								
USSR (in Asia)	53,559	(59)	38,614	(39)	26,752	(26)	13,506	(97)	4,103	(56)	2,500	(95?e)
Vietnam					17,702	(31)	14,281	(06)	10,000	(50?e)	17,000	(00?e)
North Vietnam	15,917	(60)	16,114	(53e)								
South Vietnam	14,100	(60e)	9,766	(53e)								
West Bank	[805]	(61)]	[742]	(52)]								
Yemen												
North Yemen	5,000	(60e)	4,500	(50e)	2,000	(30?e)	[750]	(10e)]				
South Yemen	1,000	(60e)	750	(50e)	650	(37e)	194	(00?e)				

EUROPE

Population (thousands)

Country	Circa 1960		Circa 1950		Circa 1930		Circa 1900		Circa 1850		Circa 1800	
Albania	1,626	(60)	1,122	(45)	1,003	(30)						
Andorra	14	(65)	6	(54)	5	(21)	8	(01)	7	(50?e)		
Austria	7,074	(61)	6,934	(51)	6,760	(34)	26,151	(00)	17,535	(50)	8,511	(00)
Belgium	9,190	(61)	8,512	(47)	8,092	(30)	6,694	(00)	4,337	(46)	3,008	(00)
Bulgaria	7,614	(56)	7,029	(46)	5,479	(26)	3,744	(00)	[2,008]	(80)]	[1,800]	(00?e)]
Channel Islands	111	(61)	103	(51)	93	(31)	96	(01)	91	(51)	49	(21)
Czechoslovakia	13,746	(61)	12,338	(50)	14,730	(30)						
Denmark	4,585	(60)	4,281	(50)	3,544	(30)	2,450	(01)	1,408	(50)	926	(01)
Faeroe Islands	35	(60)	32	(50)	24	(30)	15	(01)	8	(50)	5	(01)
Finland	4,446	(60)	4,030	(50)	3,463	(30)	2,713	(00)	1,637	(50)	833	(00)

¹ Inc Iraq, Lebanon, Saudi Arabia, Syria, and North Yemen.

EUROPE

Population (thousands)

Country	Circa 1960	Circa 1950	Circa 1930	Circa 1900	Circa 1850	Circa 1800
France	46,528 (62)	42,844 (54)	41,835 (31)	38,962 (01)	35,783 (51)	27,349 (01)
Germany			65,218 (33)	56,367 (00)	29,800 (49)	24,833 (16)
East Germany	17,004 (64)	18,388 (50)				
West Germany	56,175 (61)	49,843 (50)				
Gibraltar	25 (61)	23 (51)	17 (31)	20 (01)	16 (44)	3 (87)
Greece	8,389 (61)	7,633 (51)	6,205 (28)	2,434 (96)	987 (48)	[753 (28)]
Hungary	9,961 (60)	9,205 (49)	8,685 (30)	19,255 (00)	13,192 (50)	8,003 (85)
Iceland	176 (60)	144 (50)	109 (30)	78 (01)	59 (50)	47 (01)
Ireland	2,818[1] (61)	2,961[1] (51)	2,972[1] (26)	4,459 (01)	6,552 (51)	6,802 (21)
Isle of Man	48 (61)	55 (51)	49 (31)	55 (01)	52 (51)	28 (92)
Italy	49,904 (61)	47,159 (51)	40,310 (31)	33,172 (01)	25,017 (61)	14,134 (00?e)
Liechtenstein	17 (60)	14 (50)	8 (30)	10 (01)	8 (52)	6 (12)
Luxembourg	315 (60)	291 (47)	300 (30)	235 (00)	195 (51)	134 (21)
Malta	320 (57)	306 (48)	242 (31)	185 (01)	123 (51)	114 (98)
Monaco	22 (62)	20 (51)	25 (28)	15 (97)	8 (57)	6 (00?e)
Netherlands	11,462 (60)	9,625 (47)	7,936 (30)	5,104 (99)	3,057 (49)	1,880 (95)
Norway	3,591 (60)	3,279 (50)	2,814 (30)	2,240 (00)	1,490 (55)	883 (01)
Poland	29,776 (60)	25,008 (50)	32,107 (31)	9,402 (97)	4,852 (51e)	2,600 (15e)
Portugal						
Total	8,851 (60)	8,441 (50)	6,826 (30)	5,423 (00)	3,844 (54)	3,115 (01e)
in Europe[2]	8,583 (60)	8,174 (50)	6,614 (30)	5,272 (00)	3,737 (54)	3,025 (01e)
Azores	[328 (60)]	[319 (50)]	[255 (30)]	[256 (00)]	[238 (54)]	[142 (00?e)]
Romania	17,489 (56)	15,873 (48)	14,281 (30)	5,957 (99)	[3,865 (59)]	[2,200 (00?e)]
San Marino	15 (60e)	12 (47)	13 (30e)	10 (06)	7 (64)	7 (00?e)

[1] Exc Northern Ireland.
[2] Inc Azores.

EUROPE

Population (thousands)

Country	Circa 1960		Circa 1950		Circa 1930		Circa 1900		Circa 1850		Circa 1800	
Spain												
Total	30,431	(60)	27,977	(50)	23,564	(30)	18,594	(00)	15,455	(57)	10,541	(97)
in Europe	29,486	(60)	27,183	(50)	23,009	(30)	18,235	(00)	15,220	(57)	10,367	(97)
Balearic Islands	443	(60)	422	(50)	366	(30)	312	(00)	263	(57)	187	(97)
Svalbard	3	(60)	2	(46)	0.6	(30)						
Sweden	7,495	(60)	7,042	(50)	6,142	(30)	5,136	(00)	3,483	(50)	2,347	(00)
Switzerland	5,429	(60)	4,715	(50)	4,066	(30)	3,315	(00)	2,393	(50)	1,843	(95)
Turkey (in Europe)	2,285	(60)	1,584	(50)	1,041	(27)	6,130	(10e)	10,586[1]	(44e)	9,822[2]	(00?e)
UK	52,709	(61)	50,225	(51)	46,052	(31)	37,000[3]	(01)	20,817[3]	(51)	10,501[3]	(01)
England and Wales	46,105	(61)	43,758	(51)	39,952	(31)	32,528	(01)	17,928	(51)	8,893	(01)
Northern Ireland	1,425	(61)	1,371	(51)	1,257	(26)						
Scotland	5,179	(61)	5,096	(51)	4,843	(31)	4,472	(01)	2,889	(51)	1,608	(01)
USSR												
Total	208,827	(59)	170,557	(39)	147,028	(26)	116,238	(97)	64,903	(56)	33,000	(95?e)
in Europe	155,268	(59)	131,943	(39)	120,276	(26)	102,732	(97)	60,800	(56)	30,500	(95?e)
Vatican City	0.9	(60e)	0.9	(48)	0.6	(30)						
Yugoslavia	18,549	(61)	15,772	(48)	13,934	(31)						

NORTH AMERICA

Population (thousands)

Country	Circa 1960		Circa 1950		Circa 1930		Circa 1900		Circa 1850		Circa 1800	
Anguilla	6	(60)	5	(46)	4	(21)	4	(01)	3	(71)	3	(25)
Antigua	54	(60)	42	(46)	30	(21)	35	(01)	36	(44)	36	(17)
Bahamas	130	(63)	85	(53)	60	(31)	54	(01)	28	(51)	14	(03)
Barbados	232	(60)	193	(46)	156	(21)	183	(91)	136	(51)	82	(11)

[1] Inc Bulgaria and Romania.
[2] Inc Bulgaria, Greece, and Romania.
[3] Exc Ireland.

NORTH AMERICA

Population (thousands)

Country	Circa 1800		Circa 1850		Circa 1900		Circa 1930		Circa 1950		Circa 1960	
Belize	10	(90)	11	(45)	37	(01)	51	(31)	59	(46)	90	(60)
Bermuda			11	(51)	18	(01)	28	(31)	37	(50)	43	(60)
Canada	430¹	(14e)	2,561¹	(52)	5,371¹	(01)	10,377¹	(31)	14,009	(51)	18,238	(61)
Newfoundland	70	(16e)	96	(45)	221	(01)	290	(35)	[361	(51)]	[458	(61)]
Canal Zone					63	(12)	39	(30)	53	(50)	42	(60)
Cayman Islands	0.9	(02)	2	(34e)	4	(91)	5	(21)	7	(43)	8	(60)
Costa Rica	25	(1778)	80	(44)	243	(92)	472	(27)	801	(50)	1,336	(63)
Cuba	272	(92)	1,008	(41)	1,573	(99)	3,962	(31)	5,829	(53)	6,826	(60e)
Dominica	26	(05)	22	(44)	29	(01)	37	(21)	48	(46)	60	(60)
Dominican Republic	153	(85)	136	(50?e)	610	(00?e)	1,479	(35)	2,136	(50)	3,047	(60)
El Salvador	165	(07)	300	(50?e)	1,007	(01)	1,434	(30)	1,856	(50)	2,511	(61)
Greenland	6	(05)	10	(55)	12	(01)	17	(30)	24	(51)	33	(60)
Grenada	31	(11)	33	(51)	63	(01)	66	(21)	72	(46)	89	(60)
Guadeloupe	115	(12)	125	(52)	182	(01)	267	(31)	278	(46)	283	(61)
Guatemala	396	(1778)	850	(55e)	1,365	(93)	2,005	(21)	2,791	(50)	4,288	(64)
Haiti	511	(89e)	572	(50?e)	1,294	(01)	2,291	(18)	3,097	(50)	4,156	(60e)
Honduras	93	(91)	202	(50?e)	544	(01)	854	(30)	1,369	(50)	1,885	(61)
Jamaica	219	(91e)	377	(44)	639	(91)	858	(21)	1,237	(43)	1,610	(60)
Martinique	84	(89)	130	(53)	204	(01)	234	(31)	262	(46)	291	(61)
Mexico	5,800	(03e)	6,382	(31)	13,607	(00)	16,553	(30)	25,791	(50)	34,923	(60)
Montserrat	11	(00?e)	7	(46)	12	(01)	12	(21)	14	(46)	12	(60)
Netherlands Antilles	37	(00?e)	32	(63)	55	(11)	72	(30)	162	(50e)	189	(60)
Nicaragua	107	(1778)	257	(67)	505	(06)	638	(20)	1,057	(50)	1,536	(63)
Panama					337	(11)	467	(30)	805	(50)	1,076	(60)
Puerto Rico	155	(00)	448	(46)	953	(99)	1,544	(30)	2,211	(50)	2,350	(60)
Saint Kitts-Nevis	44	(00?e)	33	(50)	43	(01)	34	(21)	41	(46)	51	(60)
Saint Lucia	17	(03)	24	(51)	50	(01)	52	(21)	70	(46)	86	(60)

¹ Exc Newfoundland.

NORTH AMERICA

Population (thousands)

Country	Circa 1800	Circa 1850	Circa 1900	Circa 1930	Circa 1950	Circa 1960
Saint-Pierre and Miquelon	2 (00?e)	3 (61)	6 (06)	4 (31)	5 (51)	5 (62)
Saint Vincent	24 (12?)	30 (51)	41 (91)	48 (31)	62 (46)	80 (60)
Trinidad and Tobago	34 (02)	83 (51)	274 (01)	413 (31)	563 (46)	834 (60)
Turks and Caicos Islands	2 (03e)	4 (61)	5 (01)	6 (21)	6 (43)	6 (60)
USA Total	5,308[1] (00)	23,192[1] (50)	75,995[1] (00)	122,775[1] (30)	151,132[1] (50)	180,698 (60)
in North America	5,308[2] (00)	23,192[2] (50)	75,995[2] (00)	122,775[2] (30)	151,132[2] (50)	180,065 (60)
Alaska	0.8 (00?e)	11 (56)	64 (00)	59 (29)	129 (50)	[226 (60)]
Virgin Islands (UK)		7 (41)	5 (01)	5 (21)	7 (46)	7 (60)
Virgin Islands (USA)	37 (96)	40 (50)	31 (01)	22 (30)	27 (50)	32 (60)

OCEANIA

Population (thousands)

Country	Circa 1800	Circa 1850	Circa 1900	Circa 1930	Circa 1950	Circa 1960
American Samoa			6 (00)	10 (30)	19 (50)	20 (60)
Australia	37[3] (28)	1,063[3] (61)	3,954 (01)	6,690 (33)	7,626 (47)	10,596 (61)
Tasmania	6 (21)	70 (51)	[172 (01)]	[228 (33)]	[257 (47)]	[350 (61)]
Christmas Island (Australia)			0.04 (98e)	1 (26)	0.9 (47)	3 (61)
Cocos Islands			0.6 (98)	0.9 (24)	2 (47)	0.6 (61)
Cook Islands		16 (50?e)	8 (02)	10 (26)	15 (51)	18 (61)
Fiji		127 (81)	120 (01)	198 (36)	260 (46)	346 (56)

[1] Exc Alaska and Hawaii.
[2] Exc Alaska.
[3] Exc Tasmania and exc aborigines in Australia.

OCEANIA

Population (thousands)

Country	Circa 1960	Circa 1950	Circa 1930	Circa 1900	Circa 1850	Circa 1800
French Polynesia	85 (62)	63 (51)	40 (31)	31 (06)	16 (76)	120 (00?e)
Gilbert Islands[1]	49 (63)	36 (47)	34 (31)	31 (11)		
Guam	67 (60)	59 (50)	19 (30)	10 (01)		
Indonesia (in Oceania: Irian Jaya)	750 (60e)	700 (50e)	314 (30)	200 (00?e)	200 (65e)	
Johnston and Sand Islands	0.2 (60)	0.05 (50)	0.07 (40)			
Midway Islands	2 (60)	0.4 (50)	0.04 (30)			
Nauru	5 (61)	3 (49)	3 (33)			
New Caledonia	87 (63)	65 (51)	57 (31)	54 (01)	42 (50?e)	
New Hebrides	63 (60e)	49 (50e)	60 (30e)	50 (00?e)		110 (00?e)
New Zealand	2,415 (61)	1,939 (51)	1,408 (26)	816 (01)	102 (61)	100 (00?e)
Niue Island	5 (61)	5 (51)	4 (31)	4 (02)	1 (54e)	
Norfolk Island	0.8 (61)	0.9 (47)	1 (33)	0.9 (01)	0.5 (71)	0.9 (05)
Northern Mariana Islands	[8 (60)]	[6 (50)]	[19 (30)]			
Pacific Islands (USA)[2]	71 (58)	55 (50)	70 (30)			
Papua New Guinea						
New Guinea (Australia)	1,402 (60e)	1,080 (50e)	520 (33e)			
Papua	503 (60e)	373 (50e)	280 (30?e)	350 (91e)		
Pitcairn Island	0.1 (63e)	0.1 (47)	0.2 (21)	0.2 (05)	0.2 (51)	0.03 (00)
Samoa	114 (61)	85 (51)	40 (26)	33 (00)	34 (50?e)	
Solomon Islands	124 (59)	100 (50e)	94 (31)	150 (11e)		

[1] Inc Tuvalu.
[2] Inc Northern Mariana Islands.

OCEANIA

Population (thousands)

Country	Circa 1960	Circa 1950	Circa 1930	Circa 1900	Circa 1850	Circa 1800
Tokelau Islands	2 (61)	2 (51)	1 (26)	0.9 (00)		
Tonga	57 (56)	34 (39)	29 (31)	21 (01)	18 (50?e)	200 (00?e)
Tuvalu	[5 (63)]	[4 (47)]	[‹ (31)]	[3 (11)]		
USA (in Oceania: Hawaii)	[633 (60)]	500 (50)	368 (30)	154 (00)	84 (50)	130 (32)
Wake Island	1 (60)	0.3 (50)				
Wallis and Futuna Islands	11 (62e)	8 (55)	6 (31)			

SOUTH AMERICA

Population (thousands)

Country	Circa 1960	Circa 1950	Circa 1930	Circa 1900	Circa 1850	Circa 1800
Argentina	20,011 (60)	15,894 (47)	7,924 (14)	4,045 (95)	1,830 (69)	311 (97e)
Bolivia	3,824 (60e)	3,019 (50)	2,397 (30e)	1,696 (00)	1,544 (54)	1,019 (31)
Brazil	70,269 (60)	51,989 (50)	30,636 (20)	17,438 (00)	10,112 (72)	3,200 (06e)
Chile	7,375 (60)	5,933 (52)	4,287 (30)	2,712 (95)	1,439 (54)	1,010 (35)
Colombia	17,485 (64)	11,356 (51)	7,851 (28)	4,144 (05)	2,243 (51)	1,224 (25)
Ecuador	4,650 (62)	3,203 (50)	2,500 (30e)	1,272 (92e)	870 (46e)	558 (25)
Galapagos Islands	2 (62)	1 (50)	2 (34e)	0.2 (00?e)	0.2 (35e)	0 (00)
Falkland Islands	3 (62)	2 (53)	3 (31)	2 (01)	0.3 (47)	
French Guiana	34 (61)	29 (46)	28 (31)	33 (01)	21 (41)	16 (15e)
Guyana	560 (60)	376 (46)	311 (31)	278 (91)	136 (51)	98 (31)
Paraguay	1,854 (62)	1,360 (50)	992 (36)	636 (99–00)	1,337 (57)	560 (00?e)
Peru	10,521 (61)	7,969 (50e)	7,373 (40)	4,610 (96e)	2,001 (50)	1,076 (91–95)
Surinam	328 (64)	185 (50)	119 (21)	87 (01)	58 (63)	
Uruguay	2,596 (63)	2,193 (50e)	1,877 (30e)	916 (00)	132 (52)	31 (96e)
Venezuela	7,556 (61)	5,092 (50)	3,027 (26)	2,324 (91)	1,564 (54)	660 (25)

Table 10c. Additional Statistics
AFRICA

Country	Birth and Death Rates			Life Expectancy (years)				Energy Production and Consumption		Consumption		Gross National Product		
	Period	Births	Deaths	Period	Men	Women	Avg	Year	Production[1]	Total[1]	Per Capita[2]	Year	Total[3]	Per Capita[4]
Algeria	1970–75	48.7e	15.4e	1970–75	51.7e	54.8e	53.2e	1975	83,349	12,657	754	1976	16,060	928
Angola	1970–75	47.2e	24.5e	1970–75	37.0e	40.1e	38.5e	1975	11,917	1,113	166	1976	1,830	273
Benin	1970–75	49.9e	23.0e	1970–75	39.4e	42.6e	41.0e	1975		162	52	1976	430	135
Botswana	1970–75	45.6e	23.0e	1970–75	41.9e	45.1e	43.5e					1976	280	404
British Indian Ocean Territory														
Burundi	1970–71	42.0e	20.4e	1970–71	40.0e	43.0e	41.5e	1975		48	13	1976	460	119
Cameroon	1970–75	40.4e	22.0e	1970–75	39.4e	42.6e	41.0e	1975	140	668	90	1976	2,240	292
Cape Verde	1970–74	32.0	11.4	1970–75	48.3e	51.7e	50.0e	1975		18	61	1975	80	272
Central African Empire	1970–75	43.4e	22.5e	1959–60	33.0	36.0	34.5	1975	6	61	23	1976	420	153
Ceuta	1966–70	16.4	6.2									1974	90[5]	720[5]
Chad	1970–75	44.0e	24.0e	1963–64	29.0	35.0	32.0	1975		158	39	1976	510	124
Comoros[6]	1970–75	46.6e	21.7e	1970–75	40.9e	44.1e	42.5e	1975		16	52	1976	60	194
Congo	1970–75	45.1e	20.8e	1970–75	41.9e	45.1e	43.5e	1975	2,660	281	208	1976	700	504
Djibouti	1970	42.0	7.6					1975		48	267	1975	200	1,111
Egypt	1970–74	34.9	13.5	1970	50.2	53.3	51.7	1975	13,299	15,092	405	1976	10,530	275
Equatorial Guinea	1970–75	36.8e	19.7e	1970–75	41.9e	45.1e	43.5e	1975		31	100	1976	110	348
Ethiopia	1970–75	49.4e	25.8e	1970–75	36.5e	39.6e	38.0e	1975	43	800	29	1976	2,960	103
Eritrea														

[1] In thousands of metric tons of coal equivalent.

[2] In kilograms of coal equivalent.

[3] In millions of US dollars at market prices.

[4] In US dollars at market prices.

[5] Inc Melilla.

[6] Inc Mayotte.

334   Countries: Additional Statistics

AFRICA

Country	Birth and Death Rates			Life Expectancy (years)				Energy Production and Consumption				Gross National Product		
	Period	Births	Deaths	Period	Men	Women	Avg	Year	Production[1]	Consumption Total[1]	Consumption Per Capita[2]	Year	Total[3]	Per Capita[4]
French Southern and Antarctic Lands														
Gabon	1970–75	32.2e	22.2e	1960–61	25.0	45.0	35.0	1975	16,785	540	470	1976	1,410	1,175
Gambia	1970–75	43.3e	24.1e	1970–75	38.5e	41.6e	40.0e	1975		34	65	1976	100	186
Ghana	1970–75	48.8e	21.9e	1970–75	41.9e	45.1e	43.5e	1975	492	1,792	182	1976	5,920	574
Guinea	1970–75	46.6e	22.9e	1970–75	39.4e	42.0e	40.7e	1975	4	405	70	1976	880	145
Guinea-Bissau	1970–75	40.1e	25.1e	1970–75	37.0e	40.1e	38.5e	1975		43	82	1976	70	131
Ivory Coast	1970–75	45.6e	20.6e	1970–75	41.9e	45.1e	43.5e	1975	34	1,787	268	1976	4,280	609
Kenya	1970–75	48.7e	16.0e	1969	46.9	51.2	49.0	1975	80	2,325	174	1976	3,280	237
Lesotho	1970–75	39.0e	19.7e	1970–75	44.4e	47.6e	46.0e					1976	210	173
Liberia	1970–71	50.4	18.7	1971	45.8	44.0	44.9	1975	39	690	445	1976	720	450
Libya	1970–75	45.0e	14.7e	1970–75	51.4e	54.5e	52.9e	1975	111,630	3,175	1,301	1976	16,000	6,324
Madagascar	1970–75	46.0e	21.1e	1966	37.5	38.3	37.9	1975	21	539	72	1976	1,870	244
Malawi	1970–72	50.5	26.5	1970–72	40.9	44.2	42.5	1975	31	285	57	1976	700	135
Mali	1970–75	50.1e	25.9e	1970–75	36.5e	39.6e	38.0e	1975	4	141	24	1976	590	98
Mauritania	1970–75	44.8e	24.9e	1970–75	37.0e	40.1e	38.5e	1975		143	108	1976	460	311
Mauritius	1971–75	25.4	7.8	1971–73	60.7	65.3	63.0	1975	5	239	266	1976	600	670
Mayotte														
Melilla	1966–70	15.3	6.2									1974	90[5]	720[5]
Morocco	1970–75	46.2e	15.7e	1970–75	51.4e	54.5e	52.9e	1975	906	4,750	275	1976	9,220	517
Mozambique	1970–75	43.1e	20.1e	1970–75	41.9e	45.1e	43.5e	1975	536	1,714	185	1976	1,600	169

[1] In thousands of metric tons of coal equivalent.
[2] In kilograms of coal equivalent.
[3] In millions of US dollars at market prices.
[4] In US dollars at market prices.
[5] Inc Ceuta.

AFRICA

Country	Birth and Death Rates			Life Expectancy (years)				Energy Production and Consumption				Gross National Product		
										Consumption				
	Period	Births	Deaths	Period	Men	Women	Avg	Year	Production[1]	Total[1]	Per Capita[2]	Year	Total[3]	Per Capita[4]
Namibia	1970–75	45.0e	16.7e	1970–75	47.5e	50.0e	48.7e	1975				1975	860	968
Niger	1970–75	52.2e	25.5e	1970–75	37.0e	40.1e	38.5e	1975		163	35	1976	740	157
Nigeria	1970–75	49.3e	22.7e	1965–66	37.2	36.7	36.9	1975	131,067	5,644	75	1976	29,320	381
Portugal (in Africa: Madeira Islands)	1971–75	23.5	10.5	1971–75										
Reunion	1970–74	29.6	7.4	1963–67	55.8	62.4	59.1	1975	16	217	447	1975	960	1,979
Rhodesia	1970–75	47.9e	14.4e	1970–75	49.8e	53.3e	51.5e	1975	4,154	4,819	751	1976	3,560	545
Rwanda	1970–75	50.0e	23.6e	1970–75	39.4e	42.6e	41.0e	1975	18	57	14	1976	480	112
Saint Helena	1971–75	23.2	10.5					1968		2	333			
Sao Tome and Principe	1968–72	46.4	12.9					1975		8	100	1976	40	494
Senegal	1970–75	47.6e	23.9e	1970–75	38.5e	41.6e	40.0e	1975		807	165	1976	1,980	389
Seychelles	1971–75	31.8	8.6	1970–72	61.9	68.0	64.9	1975		28	475	1975	30	508
Sierra Leone	1970–75	44.7e	20.7e	1970–75	41.9e	45.1e	43.5e	1975		318	104	1976	610	196
Somalia	1970–75	47.2e	21.7e	1970–75	39.4e	42.6e	41.0e	1975		113	36	1976	370	113
South Africa	1970–75	42.9e	15.5e	1970–75	49.8e	53.3e	51.5e	1975	69,786[5]	84,231[5]	2,934[5]	1976	34,850	1,334
Spain (in Africa: Canary Islands)	1971–75	23.6	6.4	1971–75										
Sudan	1970–75	47.8e	17.5e	1970–75	47.3e	49.9e	48.6e	1975	15	2,479	167	1976	4,610	298
Swaziland	1970–75	49.0e	21.8e	1970–75	41.9e	45.1e	43.5e					1976	240	481

[1] In thousands of metric tons of coal equivalent.
[2] In kilograms of coal equivalent.
[3] In millions of US dollars at market prices.
[4] In US dollars at market prices.
[5] Inc Botswana, Lesotho, Namibia, and Swaziland.

AFRICA

Country	Birth and Death Rates			Life Expectancy (years)				Energy Production and Consumption				Gross National Product		
	Period	Births	Deaths	Period	Men	Women	Avg	Year	Production[1]	Consumption Total[1]	Per Capita[2]	Year	Total[3]	Per Capita[4]
Tanzania	1970–75	47.0e	20.1e	1967			40.4e	1975	42	1,051	69			
Tanganyika	1970–75	47.0e	20.1e	1957			37.5e	1975	42	1,026	69	1976	2,700	178
Zanzibar	1970	47.0e	21.0e	1958			42.8	1975		25	59			
Togo	1970–75	50.6e	23.3e	1961	31.6	38.5	35.0	1975		145	65	1976	600	263
Tunisia	1971–75	36.4		1970–75	52.5e	55.7e	54.1e	1975	7,111	2,579	462	1976	4,790	835
	1970–75		13.8e											
Uganda	1970–75	45.2e	15.9e	1970–75	48.3e	51.7e	50.0e	1975	100	630	55	1976	2,820	236
Upper Volta	1970–75	48.5e	25.8e	1960–61	32.1	31.1	31.6	1975		118	19	1976	710	113
Western Sahara	1968–72	22.8	4.9											
Zaire	1970–75	45.2e	20.5e	1970–75	41.9e	45.1e	43.5e	1975	646	1,949	78	1976	3,510	137
Zambia	1970–75	51.5e	20.3e	1970–75	42.9e	46.1e	44.5e	1975	1,545	2,470	504	1976	2,200	428

ASIA

Country	Birth and Death Rates			Life Expectancy (years)				Energy Production and Consumption				Gross National Product		
	Period	Births	Deaths	Period	Men	Women	Avg	Year	Production[1]	Consumption Total[1]	Per Capita[2]	Year	Total[3]	Per Capita[4]
Afghanistan	1970–75	49.2e	23.8e	1970–75	39.9e	40.7e	40.3e	1975	4,236	1,011	52	1976	2,300	116
Bahrain	1970–75	30.0e						1975	7,347	3,092	11,668	1976	660	2,548
Bangladesh	1970–75	49.5e	28.1e	1962–65	44.5	45.0	44.7	1975	693	2,179	30	1976	8,470	111
Bhutan	1970–75	43.6e	20.5e	1970–75	42.2e	45.0e	43.6e					1976	90	75

[1] In thousands of metric tons of coal equivalent.
[2] In kilograms of coal equivalent.
[3] In millions of US dollars at market prices.
[4] In US dollars at market prices.

ASIA

Country	Birth and Death Rates			Life Expectancy (years)				Energy Production and Consumption				Gross National Product		
	Period	Births	Deaths	Period	Men	Women	Avg	Year	Production[1]	Consumption Total[1]	Consumption Per Capita[2]	Year	Total[3]	Per Capita[4]
Brunei	1970–74	35.7	5.1					1975	21,298	1,415	8,423	1975	950	5,655
Burma	1970–75	39.5e	15.8e	1970–75	48.6e	51.5e	50.0e	1975	1,497	1,595	53	1976	3,730	120
Cambodia	1970–75	46.7e	19.0e	1970–75	44.0e	46.9e	45.4e	1975		131	16	1974	570	72
China	1970–75	26.9e	10.3e	1970–75	59.9e	63.3e	61.6e	1975	596,812	570,467	693	1976	343,090	410
Cyprus	1970–75	22.2e	6.8e	1973	70.0	72.9	71.4	1975		817	1,279	1976	930	1,455
Gaza Strip	1966	44.3	8.1											
Hong Kong	1971–75	19.4	5.1	1971	67.4	75.0	71.2	1975		4,885	1,133	1976	9,410	2,135
India	1972–74	35.2	15.6	1961–70	47.1	45.6	46.3	1975	114,624	132,054	221	1976	95,880	157
Sikkim	1954	28.8e	15.9e									1973	20	95
Indonesia														
Total	1970–75	42.9e	16.9e	1971	40.2	43.0	41.6	1975	103,076	24,213	183	1976	32,440	239
in Asia	1970–75	42.9e	16.9e	1971	40.2	43.0	41.6							
East Timor	1970–75	44.3e	23.0e	1970–75	39.2e	40.7e	39.9e	1975		11	16	1974	100	152
Iran	1970–75	45.3e	15.6e	1970–75	50.7e	51.3e	51.0e	1975	426,636	44,662	1,362	1976	66,250	1,972
Iraq	1970–75	48.1e	14.6e	1970–75	51.2e	54.3e	52.7e	1975	165,620	7,929	713	1976	15,940	1,385
Israel	1971–75	27.5	7.1	1975	70.3	73.9	72.1	1975	7,282	9,459	2,807	1976	13,980	3,960
Japan	1971–75	18.7	6.5	1975	71.8	76.9	74.4	1975	37,437	401,884	3,590	1976	553,140	4,891
Ryukyu Islands	1974–75	22.4	5.4	1965	68.9	75.6	72.3							
Jordan[5]	1971–75	44.8		1959–63	52.6	52.0	52.3	1975		1,102	408	1976	1,710	615
Kashmir-Jammu	1970–75		14.7e											

[1] In thousands of metric tons of coal equivalent.
[2] In kilograms of coal equivalent.
[3] In millions of US dollars at market prices.
[4] In US dollars at market prices.
[5] Inc West Bank.

ASIA

Country	Birth and Death Rates			Life Expectancy (years)				Energy Production and Consumption				Gross National Product		
										Consumption				
	Period	Births	Deaths	Period	Men	Women	Avg	Year	Production[1]	Total[1]	Per Capita[2]	Year	Total[3]	Per Capita[4]
Korea														
North Korea	1970–75	35.7e	9.4e	1970–75	58.8e	62.5e	60.6e	1975	42,430	44,518	2,808	1976	7,610	468
South Korea	1970–75	28.8e	8.9e	1970	63.0	67.0	65.0	1975	17,800	35,980	1,037	1976	24,050	671
Kuwait	1970–75	47.1e	5.3e	1970	66.1	71.8	68.9	1975	164,280	8,683	8,727	1976	16,480	15,984
Laos	1970–75	44.6e	22.8e	1970–75	39.1e	41.8e	40.4e	1975	31	208	63	1976	310	92
Lebanon	1970–75	39.8e	9.9e	1970–75	61.4e	65.1e	63.2e	1975	98	2,663	871	1974	3,290	1,107
Macao	1967–70, 72	11.9	6.0					1975		72	266	1975	220	815
Malaysia	1970–74	33.0	6.6	1970	60.4	63.9	62.1	1975	7,008	6,661	551	1976	10,900	886
Peninsular Malaysia	1971–75	32.6	6.8	1974	65.0	70.3	67.6	1975	124	5,523	541			
Sabah	1970–74	36.9	5.1	1970	48.8	45.4	47.1	1975	684	462	611			
Sarawak	1970–74	29.9	4.9	1970	51.1	52.7	51.9	1975	6,200	676	598			
Maldives	1963–65	41.3	22.9					1975				1975	10	76
Mongolia	1970–75	38.8e	9.3e	1970–75	59.1e	62.3e	60.7e	1975	1,012	1,576	1,099	1976	1,280	860
Nepal	1970–75	42.9e	20.3e	1970–75	42.2e	45.0e	43.6e	1975	12	129	10	1976	1,490	116
Oman								1975	25,014	259	338	1976	2,130	2,693
Pakistan	1970–75	47.4e	16.5e	1962	53.7	48.8	51.2	1975	8,198	12,825	183	1976	12,190	168
Philippines	1970–75	43.8e	10.5e	1970–75	56.9e	60.0e	58.4e	1975	562	13,849	331	1976	17,810	415
Qatar								1975	34,353	3,250	19,118	1976	2,390	14,059
Saudi Arabia	1970–75	49.5e	20.2e	1970–75	44.2e	46.5e	45.3e	1975	530,102	12,532	1,717	1976	38,510	5,067
Singapore	1971–75	21.2	5.3	1970	65.1	70.0	67.5	1975		4,839	2,151	1976	6,150	2,700
Sri Lanka	1968–72	30.2	7.8	1967	64.8	66.9	65.8	1975	136	1,770	127	1976	2,750	193

[1] In thousands of metric tons of coal equivalent.
[2] In kilograms of coal equivalent.
[3] In millions of US dollars at market prices.
[4] In US dollars at market prices.

ASIA

Country	Birth and Death Rates			Life Expectancy (years)				Energy Production and Consumption				Gross National Product		
	Period	Births	Deaths	Period	Men	Women	Avg	Year	Production[1]	Consumption Total[1]	Consumption Per Capita[2]	Year	Total[3]	Per Capita[4]
Syria	1970–75	45.4e	15.4e	1975	54.9	58.7	56.8	1975	14,359	3,505	476	1976	5,970	786
Taiwan	1971–75	24.0	4.7	1975	68.4	73.6	71.0	1975	6,630[5]	20,770[5]	1,295	1976	17,500	1,074
Thailand	1970–75	43.4e	10.8e	1960	53.6	58.7	56.1	1975	458	11,897	281	1976	16,230	378
Turkey														
Total	1970–75	39.6e	12.5e	1966			53.7	1975	12,434	24,701	612	1976	40,960	999
				1950–51	46.0	50.4	48.2							
in Asia														
United Arab Emirates								1975	120,522	3,041	5,068	1976	9,710	14,428
USSR (in Asia)														
Vietnam	1970–75	41.5e	20.5e	1970–75	43.2e	46.0e	44.6e	1975	4,301	8,406	186	1975	7,100	157
North Vietnam	1970–75	41.4e	17.9e	1970–75	46.6e	49.5e	48.0e	1975	4,299	4,618	186	1974	3,000	126
South Vietnam	1970–75	42.7e	23.6e	1970–75	39.1e	41.9e	40.5e	1975	2	3,788	185	1974	3,440	176
West Bank														
Yemen														
North Yemen	1970–75	49.6e	20.6e	1970–75	43.7e	45.9e	44.8e	1975		328	63	1976	1,540	294
South Yemen	1970–75	49.6e	20.6e	1970–75	43.7e	45.9e	44.8e	1975		555	328	1976	480	274

EUROPE

Country	Birth and Death Rates			Life Expectancy (years)				Energy Production and Consumption				Gross National Product		
	Period	Births	Deaths	Period	Men	Women	Avg	Year	Production[1]	Consumption Total[1]	Consumption Per Capita[2]	Year	Total[3]	Per Capita[4]
Albania	1967–71	34.4	8.3	1974	66.5	69.0	67.7	1975	4,172	1,838	741	1976	1,330	522

[1] In thousands of metric tons of coal equivalent.
[2] In kilograms of coal equivalent.
[3] In millions of US dollars at market prices.
[4] In US dollars at market prices.
[5] Estimated on basis of 1966–70 growth.

EUROPE

Country	Birth and Death Rates			Life Expectancy (years)				Energy Production and Consumption		Consumption		Gross National Product		
	Period	Births	Deaths	Period	Men	Women	Avg	Year	Production[1]	Total[1]	Per Capita[2]	Year	Total[3]	Per Capita[4]
Andorra		21.0	4.6											
Austria	1971–75	13.4	12.7	1975	67.7	74.9	71.3	1975	10,908	27,836	3,700	1976	40,080	5,334
Belgium	1971–75	13.2	12.1	1968–72	67.8	74.2	71.0	1975	8,427	54,698	5,584	1976	66,660	6,741
Bulgaria	1971–75	16.2	9.8	1969–71	68.6	73.9	71.2	1975	15,015	41,700	4,777	1976	20,270	2,314
Channel Islands	1971–75	12.8	12.4									1975	380	2,946
Czechoslovakia	1971–75	18.4	11.5	1973	66.5	73.5	70.0	1975	81,929	105,856	7,151	1976	57,250	3,838
Denmark	1971–75	14.6	10.0	1974–75	71.1	76.8	73.9	1975	248	26,652	5,273	1976	37,770	7,445
Faeroe Islands	1970–74	20.3	7.4					1975	7	177	4,317	1975	200	4,878
Finland	1971–75	13.1	9.5	1975	67.4	75.9	71.6	1975	1,524	22,433	4,766	1976	26,570	5,621
France	1971–75	16.0	10.6	1974	69.0	76.9	72.9	1975	47,413	208,877	3,975	1976	346,730	6,549
Germany														
East Germany	1971–75	11.5	13.8	1975	68.5	74.0	71.3	1975	81,085	115,168	6,835	1976	70,880	4,223
West Germany	1971–75	10.8	11.9	1973–75	68.0	74.5	71.3	1975	165,867	330,490	5,345	1976	457,540	7,438
Gibraltar	1971–75	19.7	8.0					1975		34	1,259	1975	90	3,333
Greece	1971–75	15.8	8.6	1960–62	67.5	70.7	69.1	1975	6,156	18,909	2,089	1976	23,600	2,575
Hungary	1971–75	16.1	11.9	1974	66.5	72.4	69.5	1975	23,710	38,200	3,624	1976	24,140	2,278
Iceland	1971–75	21.0	6.9	1971–75	71.6	77.5	74.5	1975	274	1,029	4,720	1976	1,380	6,273
Ireland	1971–75	22.4	11.1	1965–67	68.6	72.8	70.7	1975	2,333	9,684	3,097	1976	8,090	2,559
Isle of Man	1971–75	13.7	17.4									1975	120	2,034
Italy	1971–75	15.9	9.8	1970–72	69.0	74.9	71.9	1975	27,033	168,088	3,012	1976	171,250	3,048
Liechtenstein	1971–75	15.2	7.3											

[1] In thousands of metric tons of coal equivalent.
[2] In kilograms of coal equivalent.
[3] In millions of US dollars at market prices.
[4] In US dollars at market prices.

EUROPE

Country	Birth and Death Rates			Life Expectancy (years)				Energy Production and Consumption				Gross National Product		
										Consumption				
	Period	Births	Deaths	Period	Men	Women	Avg	Year	Production[1]	Total[1]	Per Capita[2]	Year	Total[3]	Per Capita[4]
Luxembourg	1971–75	11.7	12.2	1971–73	67.0	73.9	70.4	1975	61	5,535	15,504	1976	2,330	6,508
Malta	1971–75	17.6	9.5	1975	68.5	72.7	70.6	1975		310	969	1975	460	1,437
Monaco	1970–74	8.6	11.8											
Netherlands	1971–75	14.9	8.3	1975	71.4	77.6	74.5	1975	111,790	78,962	5,783	1976	85,320	6,196
Norway	1971–75	15.5	10.0	1974–75	71.7	78.0	74.9	1975	23,557	18,458	4,606	1976	29,920	7,432
Poland	1971–75	18.0	8.4	1976	68.0	74.9	71.4	1975	192,417	170,338	5,016	1976	98,130	2,856
Portugal														
Total	1971–75	20.2	10.8	1974	65.3	72.0	68.7	1975	1,014	8,609	894	1976	16,480	1,700
in Europe	1971–75	20.1	10.8											
Azores	1971–75	21.8	11.2											
Romania	1971–75	19.3	9.4	1974–76	67.4	72.0	69.7	1975	80,659	80,797	3,815	1976	31,070	1,448
San Marino	1969–73	16.8	7.5											
Spain														
Total	1971–75	19.2	8.4	1970	69.7	75.0	72.3	1975	20,369	76,171	2,154	1976	104,090	2,904
in Europe	1971–75	19.0	8.5											
Balearic Islands	1971–75	20.2	10.4											
Svalbard														
Sweden	1971–75	13.5	10.5	1971–75	72.1	77.6	74.9	1975	8,576	50,625	6,167	1976	71,290	8,671
Switzerland	1971–75	13.7	8.9	1968–73	70.3	76.2	73.3	1975	4,974	23,407	3,656	1976	56,900	8,966
Turkey (in Europe)														
UK	1971–75	14.2	11.9	1971–73	68.8	75.1	71.9	1975	183,988	295,329	5,277	1976	225,150	4,026
England and Wales	1971–75	14.0	11.9	1972–74	69.2	75.6	72.4							
Northern Ireland	1971–75	18.7	11.0	1972–74	65.0	73.6	69.3							
Scotland	1971–75	14.5	12.2	1973–75	67.4	73.9	70.7							

[1] In thousands of metric tons of coal equivalent.
[2] In kilograms of coal equivalent.
[3] In millions of US dollars at market prices.
[4] In US dollars at market prices.

EUROPE

Country	Birth and Death Rates			Life Expectancy (years)				Energy Production and Consumption		Consumption		Gross National Product		
	Period	Births	Deaths	Period	Men	Women	Avg	Year	Production[1]	Total[1]	Per Capita[2]	Year	Total[3]	Per Capita[4]
USSR														
Total in Europe														
Vatican City														
	1971–75	17.9	8.7	1971–72	64.0	74.0	69.0	1975	1,650,472	1,410,781	5,570	1976	708,170	2,771
Yugoslavia	1971–75	18.2	8.7	1970–72	65.4	70.2	67.8	1975	28,212	41,208	1,930	1976	36,170	1,678

NORTH AMERICA

Country	Birth and Death Rates			Life Expectancy (years)				Energy Production and Consumption		Consumption		Gross National Product		
	Period	Births	Deaths	Period	Men	Women	Avg	Year	Production[1]	Total[1]	Per Capita[2]	Year	Total[3]	Per Capita[4]
Anguilla														
Antigua	1970–74	21.7	6.3	1959–61	60.5	64.3	62.4	1975		153	2,155	1975	60	845
Bahamas	1971–75	23.1	6.0	1969–71	64.0	67.3	65.6	1975		1,281	6,279	1976	700	3,318
Barbados	1971–75	20.6	8.6	1959–61	62.7	67.4	65.1	1975	3	264	1,056	1976	380	1,514
Belize	1970–74	39.0	5.8	1944–48	45.0	49.0	47.0	1975		73	521	1975	90	643
Bermuda	1971–75	17.5	7.1	1965–66	65.6	72.3	69.0	1975		173	3,089	1975	360	6,429
Canada	1971–75	15.9	7.4	1970–72	69.3	76.4	72.8	1975	268,424	225,568	9,940	1976	174,120	7,573
Newfoundland	1971–75	22.0	6.1	1970–72	69.3	75.7	72.5							

[1] In thousands of metric tons of coal equivalent.
[2] In kilograms of coal equivalent.
[3] In millions of US dollars at market prices.
[4] In US dollars at market prices.

NORTH AMERICA

Country	Birth and Death Rates			Life Expectancy (years)				Energy Production and Consumption				Gross National Product		
	Period	Births	Deaths	Period	Men	Women	Avg	Year	Production[1]	Consumption Total[1]	Consumption Per Capita[2]	Year	Total[3]	Per Capita[4]
Canal Zone	1971–75	14.1	2.2					1975	28	623	15,195	1975	230	5,610
Cayman Islands	1970–74	28.5	6.1					1975		31	2,818			
Costa Rica	1971–75	30.0	5.4	1962–64	61.9	64.8	63.3	1975	161	1,070	544	1976	2,090	1,039
Cuba	1971–75	25.1	5.6	1970	68.5	71.8	70.1	1975	267	10,821	1,168	1976	8,120	858
Dominica	1965–69	40.4	8.9	1958–62	57.0	59.2	58.1	1975	1	16	211	1975	40	526
Dominican Republic	1970–73	41.7		1959–61	57.1	58.6	57.9	1975	24	2,151	458	1976	3,750	776
	1970–75		11.0e											
El Salvador	1971–75	41.3	8.2	1960–61	56.6	60.4	58.5	1975	69	993	248	1976	2,030	492
Greenland	1969–73	22.7	6.3	1966–70	58.9	65.7	62.3	1975		295	5,900	1975	220	4,400
Grenada	1974–75	26.8	6.7	1959–61	60.1	65.6	62.8	1975		31	272	1976	50	420
Guadeloupe	1970–74	28.5	7.3	1963–67	62.5	67.3	64.9	1975		200	613	1975	490	1,503
Guatemala	1970–74	42.6	13.1	1963–65	48.3	49.7	49.0	1975	49	1,444	268	1976	4,070	651
Haiti	1970–75	35.8e	16.3e	1970–75	49.0e	51.0e	50.0e	1975	14	136	30	1976	930	199
Honduras	1970–75	49.3e	14.6e	1970–75	52.1e	55.0e	53.5e	1975	49	706	258	1976	1,160	410
Jamaica	1971–75	32.3	7.2	1969–70	66.7	70.2	68.4	1975	16	2,895	1,427	1976	2,230	1,084
Martinique	1970–74	24.6	6.9	1963–67	63.3	67.4	65.3	1975		386	1,184	1975	770	2,362
Mexico	1970–75	42.0e		1975	62.8	66.6	64.7	1975	84,436	73,431	1,221	1976	67,640	1,085
	1971–75		8.0											
Montserrat	1969–73	23.2	10.0	1966–74	61.6	68.2	64.9	1975		9	692			
Netherlands Antilles	1969–73	21.9	5.1	1966–70	58.9	65.7	62.3	1975		2,960	12,231	1975	410	1,694
Nicaragua	1970–75	48.3e	13.9e	1970–75	51.2e	54.6e	52.9e	1975	48	1,033	479	1976	1,760	788
Panama	1971–75	34.0		1970	64.3	67.5	65.9	1975	14	1,442	865	1976	2,260	1,315
	1970–75		7.1e											

[1] In thousands of metric tons of coal equivalent.
[2] In kilograms of coal equivalent.
[3] In millions of US dollars at market prices.
[4] In US dollars at market prices.

NORTH AMERICA

Country	Birth and Death Rates			Life Expectancy (years)				Energy Production and Consumption		Consumption		Gross National Product		
	Period	Births	Deaths	Period	Men	Women	Avg	Year	Production[1]	Total[1]	Per Capita[2]	Year	Total[3]	Per Capita[4]
Puerto Rico	1971–75	23.9	6.4	1971–73	68.9	76.0	72.5	1975	39	9,889	3,185	1976	7,670	2,393
Saint Kitts-Nevis[5]	1974–75	23.5	9.8	1959–61	58.0	61.9	59.9	1975		19	358	1975	30	566
Saint Lucia	1968–71	41.2	8.3	1959–61	55.1	58.5	56.8	1975		37	343	1975	60	556
Saint-Pierre and Miquelon	1966–70	27.9	9.6					1975		21	3,500			
Saint Vincent	1969–73	36.3	9.2	1959–61	58.5	59.7	59.1	1975	1	15	161	1975	40	430
Trinidad and Tobago	1971–75	25.1	6.7	1970	64.1	68.1	66.1	1975	18,251	3,373	3,123	1976	2,450	2,242
Turks and Caicos Islands	1970–71, 73	32.9	9.6											
USA														
Total	1971–75	15.5	9.2	1975	68.7	76.5	72.6	1975	2,036,671	2,349,549	11,003	1976	1,698,060	7,894
in North America	1971–75	15.5	9.3	1969–71	67.0	74.6	70.8							
Alaska	1971–75	21.3	4.4	1969–71	66.0	74.0	70.0							
Virgin Islands (UK)	1971–75	26.3	6.5	1946	49.5	54.8	52.1	1975		12	1,091			
Virgin Islands (USA)	1971–75	36.1	6.5					1975		3,310	36,374	1975	480	5,275

[1] In thousands of metric tons of coal equivalent.
[2] In kilograms of coal equivalent.
[3] In millions of US dollars at market prices.
[4] In US dollars at market prices.
[5] Inc Anguilla.

OCEANIA

Country	Birth and Death Rates			Life Expectancy (years)				Energy Production and Consumption				Gross National Product		
										Consumption				
	Period	Births	Deaths	Period	Men	Women	Avg	Year	Production[1]	Total[1]	Per Capita[2]	Year	Total[3]	Per Capita[4]
American Samoa	1971–75	36.9	5.0	1969–71	65.0	69.1	67.0	1975		49	1,690	1975	160	5,517
Australia	1971–75	19.3	8.4	1965–67	67.6	74.1	60.9	1975	111,048	87,559	6,415	1976	83,380	5,992
Tasmania	1971–75	19.0	8.4											
Christmas Island (Australia)	1969–73	11.0	2.1					1975		62	20,667			
Cocos Islands	1969–73	20.4	6.4											
Cook Islands	1970–73	34.4	6.7					1975		9	474	1973	8	400
Fiji	1971–75	29.1		1970–75	68.5e	71.7e	70.1e	1975		334	581	1976	670	1,139
	1970–75		4.3e											
French Polynesia	1967–71	42.2	9.3					1975		112	875	1975	380	2,969
Gilbert Islands[5]	1970–71	22.5		1958–62	56.9	59.0	57.9	1975		23	371	1975	60	968
Guam	1971–75	32.8	4.4					1975		634	6,745	1975	610	6,489
Indonesia (in Oceania: Irian Jaya)														
Johnston and Sand Islands														
Midway Islands														
Nauru	1966–68	40.3	7.5					1975	44	53	7,571	1970	28	4,667
New Caledonia	1970–74	35.5	9.2					1975		1,242	9,409	1975	600	4,545
New Hebrides	1966	45.0e	20.0e					1975		53	546	1975	50	521

[1] In thousands of metric tons of coal equivalent.
[2] In kilograms of coal equivalent.
[3] In millions of US dollars at market prices.
[4] In US dollars at market prices.
[5] Inc Tuvalu.

OCEANIA

Country	Birth and Death Rates			Life Expectancy (years)				Energy Production and Consumption				Gross National Product		
										Consumption				
	Period	Births	Deaths	Period	Men	Women	Avg	Year	Production[1]	Total[1]	Per Capita[2]	Year	Total[3]	Per Capita[4]
New Zealand	1971–75	20.5	8.4	1970–72	68.5	74.6	71.6	1975	4,673	9,603	3,123	1976	13,120	4,193
Niue Island	1972–75	28.1	7.1											
Norfolk Island	1969–72, 74	14.1	6.6											
Northern Mariana Islands														
Pacific Islands (USA)[5]	1971–75	33.8	5.2					1975		109	893	1975	120	984
Papua New Guinea	1970–75	40.6e	17.1e	1970–75	47.7e	47.6e	47.6e	1975	22	766	278	1976	1,400	495
New Guinea (Australia)														
Papua														
Pitcairn Island	1970–72, 74	12.6												
	1965–66, 69, 71, 73		23.6											
Samoa	1971–75	29.5	4.6	1961–66	60.8	65.2	63.0	1975	1	24	159	1976	50	329
Solomon Islands	1969	36.1	13.0					1975		46	242	1975	50	263
Tokelau Islands	1970–72	27.7	6.5											
Tonga	1967–71	31.6	2.9					1975		16	182	1975	40	455

[1] In thousands of metric tons of coal equivalent.
[2] In kilograms of coal equivalent.
[3] In millions of US dollars at market prices.
[4] In US dollars at market prices.
[5] Inc Northern Mariana Islands.

OCEANIA

Country	Birth and Death Rates Period	Births	Deaths	Life Expectancy (years) Period	Men	Women	Avg	Energy Production and Consumption Year	Production[1]	Consumption Total[1]	Consumption Per Capita[2]	Gross National Product Year	Total[3]	Per Capita[4]
Tuvalu														
USA (in Oceania: Hawaii)	1971–75	18.8	5.2	1969–71	71.0	76.8	73.9							
Wake Island								1975		48	24,000			
Wallis and Futuna Islands	1966–70	32.7	7.1											

SOUTH AMERICA

Country	Birth and Death Rates Period	Births	Deaths	Life Expectancy (years) Period	Men	Women	Avg	Energy Production and Consumption Year	Production[1]	Consumption Total[1]	Consumption Per Capita[2]	Gross National Product Year	Total[3]	Per Capita[4]
Argentina	1969–73 / 1970–75	23.4	8.8e	1973	65.2	71.4	68.3	1975	41,593	44,517	1,754	1976	39,920	1,552
Bolivia	1970–75	44.0e	19.1e	1970–75	45.7e	47.9e	46.8e	1975	4,949	1,707	374	1976	2,280	491
Brazil	1974–75 / 1970–75	40.5	8.8e	1960–70	57.6	61.1	59.4	1975	25,405	71,793	670	1976	125,570	1,150
Chile	1968–72	26.3	8.9	1969–70	60.5	66.0	63.2	1975	4,716	7,843	765	1976	10,980	1,050
Colombia	1970–75	40.6e	8.8e	1970–75	59.2e	62.7e	60.9e	1975	19,111	15,808	717	1976	15,400	683
Ecuador	1970–75	41.8e	9.5e	1961–63	51.0	53.7	52.4	1975	12,107	2,975	442	1976	4,690	681
Galapagos Islands														

[1] In thousands of metric tons of coal equivalent.
[2] In kilograms of coal equivalent.
[3] In millions of US dollars at market prices.
[4] In US dollars at market prices.

SOUTH AMERICA

Country	Birth and Death Rates			Life Expectancy (years)				Energy Production and Consumption				Gross National Product		
										Consumption				
	Period	Births	Deaths	Period	Men	Women	Avg	Year	Production[1]	Total[1]	Per Capita[2]	Year	Total[3]	Per Capita[4]
Falkland Islands	1970–74	19.8	9.0					1975		2	1,000	1975	100	1,754
French Guiana	1970–74	31.7	8.0					1975		57	1,000	1976	430	549
Guyana	1976	26.7	7.1	1959–61	59.0	63.0	61.0	1975		881	1,147			
Paraguay	1970–75	39.8e	8.9e	1970–75	60.3e	63.6e	61.9e	1975	55	406	153	1976	1,680	617
Peru	1970–75	41.0e	11.9e	1960–65	52.6	55.5	54.0	1975	6,793	10,654	671	1976	12,610	784
Surinam	1965–66 1970–75	38.9	7.5e	1963	62.5	66.7	64.6	1975	125	870	2,062	1975	500	1,185
Uruguay	1970–74	20.5	9.5	1963–64	65.5	71.6	68.5	1975	139	2,886	1,037	1976	3,900	1,393
Venezuela	1970–75	36.1e	7.0e	1961	61.2	64.7	62.9	1975	201,113	31,653	2,639	1976	31,750	2,569

[1] In thousands of metric tons of coal equivalent.
[2] In kilograms of coal equivalent.
[3] In millions of US dollars at market prices.
[4] In US dollars at market prices.

Table 10d.   Supplemental Information
AFRICA

Algeria: [a]off Jazair. [b]Algiers (alt Alger; off Jazair): 904 (1966), 1504s (1974e). [c]Oran (off Ouahran): 327 (1966), 485s (1974e).

Angola: [a]inc Cabinda. [b]Luanda (for Loanda, Sao Paulo de Loanda): 475 (1970). [c]Huambo (for Nova Lisboa): 62 (1970); Lobito: 60 (1970).

Benin: [a]for Dahomey. [b]Porto-Novo: 104 (1975e). [c]Cotonou (for Kotonu): 178 (1975e).

Botswana: [a]for Bechuanaland. [b]Gaborone (for Gaberones): 37 (1976e). [c]Francistown: 25 (1976e); Serowe: 24 (1976e).

British Indian Ocean Territory: [a]none. [b]Victoria (in Seychelles). [c]none.

Burundi: [a]for Urundi. [b]Bujumbura (for Usumbura): 157s (1976e). [c]Muhinga: 22 (1971e).

Cameroon: [a]alt Cameroun; for Kamerun. [b]Yaounde (alt Yaunde): 274s (1975e). [c]Douala (alt Duala): 486s (1975e).

Cape Verde: [a]off Cabo Verde. [b]Praia (on Sao Tiago I): 21 (1970). [c]Mindelo (on Sao Vicente I): 29 (1970).

Central African Empire: [a]for Central African Republic, Ubangi-Shari; off Empire Centrafricain. [b]Bangui: 302s (1971e). [c]Berberati: 40 (1968e); Bossangoa: 36 (1968e).

Ceuta: [a]alt Spanish North Africa (in part). [b]Ceuta: 65 (1973e). [c]none.

Chad: [a]off Tchad. [b]Ndjamena (for Fort-Lamy): 224 (1975). [c]Sarh (for Fort-Archambault): 44 (1972e); Moundou: 40 (1972e).

Comoros: [a]alt Comoro Islands; for Mayotte; off Comores. [b]Moroni (on Grande Comore I): 12 (1974e). [c]Mutsamudu: (on Anjouan I): 8 (1966).

Congo: [a]for Congo (Brazzaville), Middle Congo. [b]Brazzaville: 290 (1974). [c]Pointe-Noire: 142 (1974).

Djibouti: [a]alt Jibuti; for French Somaliland, French Territory of the Afars and the Issas. [b]Djibouti (alt Jibuti): 120s (1973e). [c]Tadjoura: 1 (1948e).

Egypt: [a]alt Arab Republic of Egypt; for United Arab Republic; off Jumhuriyat Misr al Arabiyah, Misr. [b]Cairo (off Qahirah): 5084 (1976), 6757s (1975e). [c]Alexandria (off Iskandariyah): 2312 (1976).

Equatorial Guinea: [a]for Spanish Guinea; inc Macias Nguema (for Fernando Po), Rio Muni; off Guinea Ecuatorial. [b]Malabo (for Santa Isabel): (on Macias Nguema I): 19 (1970e). [c]Bata: 4 (1960).

Ethiopia: [a]for Abyssinia; inc Eritrea; off Ityopya. [b]Addis Ababa: 1083 (1974). [c]Asmara: 296 (1974).

French Southern and Antarctic Lands: [a]inc Kerguelen; off Terres Australes et Antarctiques Francaises. [b]Port-aux-Francais (on Kerguelen I): 0.08 (1973e). [c]none.

Gabon: [a]for Gabun. [b]Libreville: 105 (1969-70), 112s (1973e). [c]Port-Gentil: 48 (1969-70).

Gambia: [a]none. [b]Banjul (for Bathurst): 44 (1976e). [c]Serrekunda: 26 (1973).

Ghana: [a]for Gold Coast. [b]Accra (for Akkra): 717 (1975e), 738s (1970). [c]Kumasi (for Coomassie): 260, 345s (1970).

Guinea: [a]for French Guinea; off Guinee. [b]Conakry (for Konakry): 526s (1972). [c]Kankan: 29 (1964e), 176s (1970e).

Guinea-Bissau: [a]for Portuguese Guinea; off Guine-Bissau. [b]Bissau: 71 (1970). [c]Bafata: 8 (1970).

Ivory Coast: [a]off Cote d'Ivoire. [b]Abidjan: 904s (1974e). [c]Bouake: 156 (1974e).

Kenya: [a]for British East Africa. [b]Nairobi: 736 (1976e). [c]Mombasa (alt Mvita): 351 (1976e).

Lesotho: [a]for Basutoland. [b]Maseru: 15 (1976). [c]Mohale's Hoek: 3 (1966).

Liberia: [a]none. [b]Monrovia: 167, 204s (1974). [c]Buchanan: 24 (1974).

Libya: [a]inc Cyrenaica, Tripolitania; off Libiyah. [b]Tripoli (off Tarabulus): 551 (1973). [c]Bengasi (alt Benghazi; off Banghazi): 140 (1973).

Madagascar: [a]alt Madagaskar; for Malagasy Republic. [b]Antananarivo (for Tananarive): 452 (1975). [c]Antsirabe: 79 (1975); Tamatave: 77 (1975).

Malawi: [a]for Nyasaland. [b]Lilongwe: 103 (1977e). [c]Blantyre (inc Limbe since 1949): 229 (1977e).

Mali: [a]for French Sudan. [b]Bamako: 170s (1974e). [c]Mopti: 32 (1974e).

Mauritania: [a]alt Islamic Republic of Mauritania, Mauretania; off Jumhuriyat Muritaniyah al Islamiyah, Mauritanie, Muritaniyah, Republique Islamique de Mauritanie. [b]Nouakchott: 70 (1974e), 134s (1976). [c]Nouadhibou (for Port-Etienne): 22 (1976); Kaedi: 21 (1976).

Mauritius: [a]for Ile de France; inc Rodrigues (alt Rodriguez). [b]Port Louis: 141 (1976e). [c]Beau Bassin-Rose Hill: 84 (1976e).

Mayotte: [a]none. [b]Labattoir (on Pamanzi I): 3 (1966). [c]Sada (on Mayotte I): 2 (1966).

Melilla: [a]alt Spanish North Africa (in part). [b]Melilla: 60 (1973e). [c]none.

Morocco: [a]for Marrocco; off Maghrib. [b]Rabat-Sale (Sale: for Sallee; off Sla): 616 (1974e). [c]Casablanca (off Dar al Baida): 1808 (1974e).

Mozambique: [a]off Mocambique. [b]Maputo (for Lourenco Marques): 102, 355s (1970). [c]Nampula: 126 (1970).

Namibia: [a]alt South-West Africa, Suidwes-Afrika. [b]Windhoek (for Aigams, Windhuk): 65 (1970). [c]Walvis Bay (alt Walvisbaai): 23 (1970).

Niger: [a]none. [b]Niamey: 130 (1975e). [c]Maradi: 40 (1973e); Zinder: 35 (1973e).

Nigeria: [a]none. [b]Lagos: 1061 (1975e), 1700s (1972e). [c]Ibadan: 847 (1975e).

Reunion: [a]for Bourbon. [b]Saint-Denis: 66 (1967), 104s (1974). [c]Saint-Pierre: 19, 40s (1967).

Rhodesia: [a]alt Zimbabwe; for Southern Rhodesia. [b]Salisbury: 568s (1976e). [c]Bulawayo: 340s (1976e).

Rwanda: [a]alt Ruanda. [b]Kigali: 90 (1977). [c]Butare (for Astrida): 25 (1970).

Saint Helena: [a]inc Ascension, Tristan da Cunha. [b]Jamestown (on Saint Helena I): 1 (1975). [c]Georgetown (on Ascension I): ? [Ascension I: 1 (1975)].

Sao Tome and Principe: [a]off Sao Tome e Principe (Sao Tome: for Sao Thome). [b]Sao Tome (for Sao Thome) (on Sao Tome I): 17 (1970). [c]Santo Antonio (on Principe I): 2 (1970).

Senegal: [a]none. [b]Dakar: 799s (1976). [c]Thies: 117 (1976).

Seychelles: [a]none. [b]Victoria (alt Port Victoria) (on Mahe I): 23s (1977). [c]Anse aux Pins (on Mahe I): 3s (1960).

Sierra Leone: [a]none. [b]Freetown: 274 (1974). [c]Bo (alt Bo Town): 30 (1974).

Somalia: [a]alt Somali Republic; inc British Somaliland, Italian Somaliland. [b]Mogadishu (alt Mogadiscio, Muqdisho; off Hamar): 285 (1976e). [c]Hargeisa: 80 (1969e).

South Africa: [a]alt Suid-Afrika; for Union of South Africa. [b]Cape Town (alt Kaapstad): 691, 1097s (1970); Pretoria: 544, 562s (1970). [c]Durban: 730, 843s (1970); Johannesburg: 655, 1433s (1970).

Sudan: [a]for Anglo-Egyptian Sudan. [b]Khartoum (off Khurtum): 334 (1973). [c]Omdurman (off Umm Durman): 299 (1973).

Swaziland: [a]none. [b]Mbabane: 21 (1973e). [c]Manzini (for Bremersdorp): 6s (1966).

Tanzania: [a]inc Tanganyika (for German East Africa), Zanzibar (alt Unguja). [b]Dar es Salaam: 517 (1975e). [c]Zanzibar (alt Unguja) (on Zanzibar I): 68 (1967). *Tanganyika:* [b]Dar es Salaam: 517 (1975e). [c]Tanga: 61 (1967). *Zanzibar:* [b]Zanzibar (alt Unguja) (on Zanzibar I): 68 (1967). [c]Wete (on Pemba I): 8 (1967).

Togo: [a]alt Togoland; for French Togo. [b]Lome: 214 (1975e). [c]Sokode: 31 (1974e).

Tunisia: [a]off Tunisiyah. [b]Tunis: 550, 944s (1975). [c]Sfax (off Safaqis): 171 (1975), 216s (1966).

Uganda: [a]inc Buganda. [b]Kampala: 331 (1969). [c]Jinja-Njeru: 53 (1969).

Upper Volta: [a]alt Voltaic Republic; off Haute-Volta. [b]Ouagadougou (for Wagadugu): 169 (1975e). [c]Bobo Dioulasso: 113 (1975e).

Western Sahara: [a]for Spanish Sahara. [b]Aaiun (alt Aiun): 24 (1974). [c]Villa Cisneros (alt Cisneros): 5 (1974).

Zaire: [a]for Belgian Congo, Congo (Kinshasa), Kongo. [b]Kinshasa (for Leopoldville): 2008 (1974e). [c]Kananga (for Luluabourg): 601 (1974e).

Zambia: [a]for Northern Rhodesia. [b]Lusaka: 401s (1974). [c]Kitwe: 251s (1974).

ASIA

Afghanistan: [a]none. [b]Kabul (for Cabul): 749 (1975e). [c]Kandahar (alt Qandahar): 209 (1975e).

Bahrain: [a]alt Bahrein; off Bahrayn. [b]Manama (off Manamah) (on Bahrain I): 110 (1975e). [c]Muharraq (on Muharraq I): 51 (1975e).

Bangladesh: [a]for East Pakistan. [b]Dacca (for Jahangirnagar): 1320 (1974). [c]Chittagong (for Islamabad, Porto Grande): 458 (1974).

Bhutan: [a]off Druk-yul. [b]Thimbu [alt Thimphu; off Tashi Chho (Dzong)]: 10 (1971e); Paro (Dzong): 3 (1969e). [c]Punakha: ?

Brunei: [a]none. [b]Bandar Seri Begawan (for Brunei, Brunei Town): 17 (1971), 75s (1976e). [c]Seria: 21 (1971).

Burma: [a]alt Union of Burma; for Birmah; off Myanma. [b]Rangoon (off Yangon): 3187s (1973). [c]Mandalay (off Mandale): 417s (1973).

Cambodia: [a]alt Khmer Republic; for French Indo-China (in part); off Kampuchea. [b]Phnom-Penh: 50 (1975e). [c]Battambang: 45 (1969e).

China: [a]alt Mainland China; off Chunghua. [b]Peking (for Peiping; off Peiching): 7570 (1974e). [c]Shanghai: 10,820 (1974e).

Cyprus: [a]off Kibris, Kypros. [b]Nicosia (off Levkosia): 117s (1974e). [c]Limassol (off Lemesos): 79 (1973e).

Gaza Strip: [a]off Qita Ghazzah. [b]Gaza (alt Azzah; off Ghazzah): 118 (1967). [c]Khan Yunus: 53 (1967).

Hong Kong: [a]alt Hsiangchiang, Hsiangkang. [b]Hong Kong (alt Hsiangchiang, Hsiangkang; inc Kowloon, New Kowloon, Victoria): 4407 (1976). [c]none.

India: [a]inc Sikkim; off Bharat. [b]New Delhi: 302 (1971). [c]Bombay: 5971 (1971); Calcutta: 3149, 7031s (1971).

Indonesia: [a]for Dutch East Indies, Netherlands Indies; inc East Timor (for Portuguese Timor), Irian Jaya (for Netherlands New Guinea, West Irian, West New Guinea). [b]Jakarta (alt Djakarta, for Batavia): 5490 (1975e). [c]Surabaya (alt Surabaja, for Soerabaja): 1552 (1971).

Iran: [a]for Persia. [b]Tehran (alt Teheran): 4496 (1976). [c]Isfahan (alt Ispahan; off Esfahan): 672 (1976); Meshed (off Mashhad): 670 (1976).

Iraq: [a]alt Irak; for Mesopotamia. [b]Baghdad (alt Bagdad): 1491 (1965), 3206s (1977). [c]Basra (off Basrah): 311, 311s (1965).

Israel: [a]for Palestine. [b]Jerusalem (alt Quds ash Sharif; inc Jordanian Jerusalem; off Yerushalayim): 355 (1975e). [c]Tel Aviv-Jaffa (Jaffa: off Yafo): 354 (1975e), 1157s (1974e).

Japan: [a]inc Ryukyu Islands; off Nihon, Nippon. [b]Tokyo (alt Tokio; for Edo, Yedo): 8647, 11,540s (1975). [c]Osaka: 2779 (1975).

Jordan: [a]for Trans-Jordan; off Urdunn. [b]Amman: 598 (1974e). [c]Zarqa (alt Zerka, Zerqa): 226 (1974e).

Kashmir-Jammu: [a]alt Jammu and Kashmir (Kashmir: for Cashmere). [b]Srinagar (for Cashmere): 415, 423s (1971); Jammu: 158, 164s (1971). [c]Anantnag: 28 (1971); Sopore: 28 (1971).

Korea: [a]for Chosen, Choson, Corea; off Tae Han Min Guk. *North Korea:* [b]Pyongyang (for Heijo): 1500 (1974e). [c]Chongjin (for Seishin): 300 (1974e). *South Korea:* [b]Seoul (alt Kyongsong; for Hanyang, Keijo; off Soul): 6889 (1975). [c]Pusan (alt Busan; for Fusan): 2454 (1975).

Kuwait: [a]alt Kuweit; off Kuwayt. [b]Kuwait (alt Kuweit; off Kuwayt): 78, 275s (1975). [c]Hawalli: 130 (1975).

Laos: [a]for French Indo-China (in part); off Lao. [b]Vientiane: 177 (1973). [c]Savannakhet: 51 (1973).

Lebanon: [a]off Lubnan. [b]Beirut (for Beyrouth; off Bayrut): 800 (1972e), 1172s (1975e). [c]Tripoli (off Tarabulus): 150 (1972e).

Macao: [a]alt Aomen; off Macau. [b]Macao (alt Aomen; off Macau): 241 (1970). [c]none.

Malaysia: [a]inc Peninsular Malaysia (for British Malaya, Malaya, West Malaysia; off Semenanjung Malaysia), Sabah (for British North Borneo, North Borneo), Sarawak. [b]Kuala Lumpur: 557 (1975e), 708s (1970). [c]George Town (alt Penang; off Pinang): 269, 332s (1970). *Peninsular Malaysia:* [b]Kuala Lumpur: 557 (1975e), 708s (1970). [c]George Town (alt Penang; off Pinang): 269, 332s (1970). *Sabah:* [b]Kota Kinabalu (for Jesselton): 41 (1970). [c]Sandakan: 42 (1970). *Sarawak:* [b]Kuching: 64 (1970). [c]Sibu: 51 (1970).

Maldives: [a]off Divehi Raaje. [b]Male [on Male (alt King's) I]: 16 (1974). [c]Gan (on Addu I): ?

Mongolia: [a]for Outer Mongolia; off Mongol. [b]Ulan-Bator (for Kulun, Urga; off Ulaanbaatar): 334 (1976e). [c]Darhan (alt Darkhan): 55 (1977e).

Nepal: [a]for Nepaul. [b]Kathmandu (alt Katmandu): 150 (1971). [c]Patan (alt Lalitpur): 59 (1971).

Oman: [a]for Masqat wah Oman, Muscat and Oman. [b]Muscat (off Masqat): 15 (1973e). [c]Matrah: 18 (1973e).

Pakistan: [a]none. [b]Islamabad: 77 (1972). [c]Karachi: 3499 (1972).

Philippines: [a]off Pilipinas. [b]Manila (on Luzon I): 1454, 4863s (1975). [c]Quezon City (on Luzon I): 960 (1975).

Qatar: [a]alt Katar. [b]Doha (off Dawhah): 130 (1975e). [c]Wakrah (alt Waqra): 5 (1950?e).

Saudi Arabia: [a]inc Asir, Hejaz, Nejd (alt Najd); off Arabiyah as Suudiyah. [b]Riyadh (off Riyad): 667 (1974). [c]Jidda (alt Jedda; off Juddah): 561 (1974).

Singapore: [a]off Singapura. [b]Singapore (off Singapura): 2278 (1976e). [c]none.

Sri Lanka: [a]for Ceylon, Serendib. [b]Colombo: 592, 850s (1974e). [c]Dehiwala-Mount Lavinia: 136 (1973e).

Syria: [a]off Sham, Suriyah. [b]Damascus (off Dimashq): 1042 (1975e). [c]Aleppo (off Halab): 779 (1975e).

Taiwan: [a]alt Formosa, Republic of China. [b]Taipei [for Taihoku; inc Daitotei, Moko (alt Banka, Manka)]: 2089 (1976e). [c]Kaohsiung (for Takao; inc Kigo): 1020 (1976e).

Thailand: [a]for Siam; off Prathet Thai. [b]Bangkok (off Krungthep Mahanakhon): 4130 (1975e). [c]Chiengmai (off Chiang Mai): 93 (1972e).

Turkey: [a]for Ottoman Empire; off Turkiye. [b]Ankara (for Angora): 1701 (1975). [c]Istanbul (for Constantinople) (in Europe): 2547 (1975).

United Arab Emirates: [a]for Pirate Coast, Trucial Oman, Trucial States; inc Abu Dhabi (off Abu Zaby), Ajman, Dubai (off Dubayy), Fujairah (off Fujayrah), Ras al Khaimah (off Ras al Khaymah), Sharjah (alt Sharjah and Kalba; inc Kalba; off Shariqah), Umm al Qaiwain (off Umm al Qaywayn); off Amiiraat al Arabiyah al Muttahidah. [b]Abu Zaby (alt Abu Dhabi) (on Abu Zaby I): 95 (1974e). [c]Dubayy (alt Dubai): 100 (1974e).

USSR (see under Europe).

Vietnam: [a]for French Indo-China (in part); inc North Vietnam, South Vietnam; off Viet Nam. [b]Hanoi (for Kecho): 1443 (1976e). [c]Ho Chi Minh City (for Saigon; inc Cholon): 3460 (1976e).

West Bank: [a]alt Judaea and Samaria. [b]Jerusalem (in Israel). [c]Nabulus: 64 (1970e).

Yemen: [a]off Yaman. *North Yemen:* [a]for Yemen. [b]Sana (alt Sanaa): 135 (1975). [c]Hudaydah (alt Hodeida): 92 (1973e). *South Yemen:* [a]for Hadhramaut, South Arabia; inc Aden. [b]Aden: 132 (1973). [c]Mukalla: 50 (1970e).

EUROPE

Albania: ^{a}off Shqiperi. ^{b}Tirane (alt Tirana): 192 (1975e). ^{c}Shkoder (alt Scutari): 62 (1975e); Durres (alt Durazzo): 60 (1975e).

Andorra: ^{a}none. ^{b}Andorra (alt Andorra la Vella): 11s (1975). ^{c}Les Escaldes: 4 (1965).

Austria: ^{a}off Osterreich. ^{b}Vienna (off Wien): 1604 (1975e). ^{c}Graz: 248 (1971).

Belgium: ^{a}off Belgie, Belgique. ^{b}Brussels (off Brussel, Bruxelles): 153, 1051s (1975e). ^{c}Antwerp (off Antwerpen, Anvers: 209, 662s (1975e).

Bulgaria: ^{a}off Bulgariya. ^{b}Sofia (off Sofiya): 966 (1975). ^{c}Plovdiv (for Philippopolis) 300 (1975).

Channel Islands: ^{a}inc Guernsey, Jersey. ^{b}Saint Helier (on Jersey I): 28 (1971); Saint Peter Port (on Guernsey I): 16 (1971). ^{c}Saint Saviour (on Jersey I): 11 (1971).

Czechoslovakia: ^{a}off Ceskoslovensko. ^{b}Prague (for Prag; off Praha): 1176 (1976e). ^{c}Brno (for Brunn): 363 (1976e).

Denmark: ^{a}off Danmark. ^{b}Copenhagen (off Kobenhavn): 562, 1287s (1975e). ^{c}Arhus (alt Aarhus): 199 (1970), 246s (1975e).

Faeroe Islands: ^{a}alt Faroe Islands; off Faeroerne. ^{b}Torshavn (alt Thorshavn) (on Streymoy I): 11 (1975e). ^{c}Klaksvik (alt Klakksvik) (on Bordhoy I): 4 (1966).

Finland: ^{a}alt Suomi. ^{b}Helsinki (alt Helsingfors): 493 (1976e), 861s (1974e). ^{c}Tampere (alt Tammerfors): 166 (1976e), 236s (1974e); Turku (alt Abo): 165 (1976e), 233s (1974e).

France: ^{a}none. ^{b}Paris: 2300, 8450s (1975). ^{c}Marseilles (off Marseille): 909, 1011s (1975); Lyons (off Lyon): 457, 1159s (1975).

Germany: ^{a}off Deutschland. *East Germany:* ^{b}East Berlin (off Ost-Berlin): 1106 (1976e). ^{c}Leipzig: 565 (1976e). *West Germany:* ^{b}Bonn [inc Bad Godesberg (alt Godesberg) since 1968]: 283 (1976e). ^{c}West Berlin (off West-Berlin): 1967 (1976e).

Gibraltar: ^{a}none. ^{b}Gibraltar: 30 (1976e). ^{c}none.

Greece: ^{a}off Ellas, Hellas. ^{b}Athens (off Athinai): 867, 2540s (1971). ^{c}Salonika (alt Salonica; off Thessaloniki): 346, 557s (1971).

Hungary: ^{a}off Magyarorszag. ^{b}Budapest [inc Buda (for Ofen), Pest]: 2071 (1976e). ^{c}Miskolc: 200 (1976e).

Iceland: ^{a}off Island. ^{b}Reykjavik: 85, 99s (1974e). ^{c}Akureyri: 12 (1974e); Kopavogur: 12 (1974e).

Ireland: ^{a}off Eire. ^{b}Dublin (off Baile Atha Cliath): 568, 680s (1971). ^{c}Cork (off Corcaigh): 129, 134s (1971).

Isle of Man: ^{a}none. ^{b}Douglas: 20 (1976). ^{c}Onchan: 6 (1976); Ramsey: 5 (1976).

Italy: ^{a}off Italia. ^{b}Rome (off Roma): 2884 (1976e). ^{c}Milan (for Mailand; off Milano): 1705 (1976e).

Liechtenstein: ^{a}none. ^{b}Vaduz: 5 (1976e). ^{c}Schaan: 4 (1973e).

Luxembourg: ^{a}alt Lutzelburg, Luxemburg. ^{b}Luxembourg (alt Lutzelburg, Luxemburg): 80 (1976e). ^{c}Esch-sur-Alzette: 27 (1975e).

Malta: ^{a}inc Gozo. ^{b}Valletta (alt Valetta) (on Malta I): 14 (1975e). ^{c}Sliema (on Malta I): 20 (1975e).

Monaco: ^{a}none. ^{b}Monaco (inc Monte-Carlo): 24 (1976e). ^{c}none.

Netherlands: ^{a}alt Holland; off Nederland. ^{b}Amsterdam: 751, 987s (1976e); Hague (alt The Hague; off 's Gravenhage): 479, 682s (1976e). ^{c}Rotterdam: 615, 1031s (1976e).

Norway: ^{a}off Norge. ^{b}Oslo (for Christiania, Kristiania): 463 (1976e), 701s (1970). ^{c}Trondheim (for Nidaros, Trondhjem): 119 (1970), 135s (1976e); Bergen: 113 (1970), 213s (1976e).

Poland: ^{a}off Polska. ^{b}Warsaw (off Warszawa): 1463 (1976e). ^{c}Lodz: 810 (1976e).

Portugal: ^{a}inc Azores (off Acores), Madeira Islands (off Madeira). ^{b}Lisbon (off Lisboa): 830 (1975e), 1034s (1970). ^{c}Porto (alt Oporto): 336 (1975e), 693s (1970).

Romania: [a]alt Roumania, Rumania. [b]Bucharest (off Bucuresti): 1807, 1934s (1977). [c]Timisoara (for Temesvar): 269 (1977), 229s (1975e); Constanta (alt Constantsa; for Kustendje): 257 (1977), 268s (1975e).

San Marino: [a]none. [b]San Marino: 5 (1975e). [c]Borgo Maggiore: 0.8 (1937); Serravalle: 0.7 (1937).

Spain: [a]inc Balearic Islands (off Islas Baleares), Canary Islands (off Islas Canarias); off Espana. [b]Madrid: 3634 (1975e). [c]Barcelona: 1828 (1975e).

Svalbard: [a]alt Spitsbergen. [b]Longyearbyen [on West Spitsbergen (off Vestspitsbergen) I]: 0.9 (1970?e). [c]Ny-Alesund (on West Spitsbergen I): 0.2 (1946e).

Sweden: [a]off Sverige. [b]Stockholm: 661, 1364s (1976e). [c]Goteborg (alt Gothenburg): 442, 692s (1976e).

Switzerland: [a]off Schweiz, Suisse, Svizzera. [b]Bern (alt Berna, Berne): 149, 283s (1976e). [c]Zurich (alt Zurigo): 388, 708s (1976e).

Turkey (see under Asia).

UK: [a]alt Great Britain and Northern Ireland, United Kingdom; inc England and Wales (Wales: alt Cymru), Northern Ireland, Scotland; off United Kingdom of Great Britain and Northern Ireland. [b]London: 2553, 7111s (1975e). [c]Birmingham: 1085 (1975e). *England and Wales:* [b]London: 2553, 7111s (1975e). [c]Birmingham: 1085 (1975e). *Northern Ireland:* [b]Belfast: 374 (1975e). [c]Londonderry (alt Derry): 84 (1975e). *Scotland:* [b]Edinburgh: 470 (1975e). [c]Glasgow: 881 (1975e).

USSR: [a]alt Soviet Union, Union of Soviet Socialist Republics; for Russia; off Soyuz Sovetskikh Sotsialisticheskikh Respublik. [b]Moscow (off Moskva): 7563, 7734s (1976e). [c]Leningrad (for Petrograd, Saint Petersburg): 3911, 4372s (1976e).

Vatican City: [a]alt Holy See; off Citta del Vaticano. [b]Vatican City (off Citta del Vaticano): 0.7 (1976e). [c]none.

Yugoslavia: [a]alt Jugoslavia; inc Bosnia and Herzegovina, Montenegro, Serbia; off Jugoslavija. [b]Belgrade (off Beograd): 870 (1975e), 775s (1971). [c]Zagreb (for Agram, Zagrab): 566 (1971).

NORTH AMERICA

Anguilla: [a] for Snake Island. [b]London (in UK). [c]Road Bay: ?

Antigua: [a]inc Barbuda. [b]Saint John's (on Antigua I): 24 (1975e). [c]All Saints' Village (on Antigua I): 2 (1960).

Bahamas: [a]alt Bahama Islands. [b]Nassau (on New Providence I): 102s (1970). [c]Freeport (on Grand Bahama I): 15 (1970).

Barbados: [a]none. [b]Bridgetown: 9, 115s (1970). [c]Speightstown: 2 (1960).

Belize: [a]for British Honduras. [b]Belmopan: 5 (1973e), 0.3 (1970). [c]Belize: 41 (1972e).

Bermuda: [a]for Somers Islands. [b]Hamilton (on Bermuda I): 3, 14s (1970). [c]Saint George's (on Saint George's I): 2 (1970).

Canada: [a]inc Newfoundland. [b]Ottawa (for Bytown): 304, 693s (1976). [c]Montreal (for Ville-Marie): 1081, 2802s (1976); Toronto (for York): 633, 2803s (1976).

Canal Zone: [a]alt Panama Canal Zone. [b]Balboa Heights: 0.2 (1970). [c]Balboa: 3 (1970).

Cayman Islands: [a]none. [b]George Town (on Grand Cayman I): 4 (1970). [c]West Bay (on Grand Cayman I): 3 (1970).

Costa Rica: [a]none. [b]San Jose: 215, 406s (1973). [c]Alajuela: 30 (1973).

Cuba: [a]none. [b]Havana (alt Habana; off La Habana): 1861 (1976e). [c]Santiago de Cuba (alt Santiago): 316 (1976e).

Dominica: [a]none. [b]Roseau: 18 (1976e), 12 (1970). [c]Portsmouth: 3 (1970).

Dominican Republic: [a]off Republica Dominicana. [b]Santo Domingo (for Ciudad Trujillo): 923 (1975e). [c]Santiago de los Caballeros (alt Santiago): 209 (1975e).

El Salvador: [a]alt Salvador. [b]San Salvador: 388 (1973e). [c]Santa Ana: 96, 159s (1971).

Greenland: [a]off Gronland. [b]Godthab: 9 (1976e). [c]Holsteinsborg: 4 (1976e).

Grenada: [a]none. [b]Saint George's (on Grenada I): 7 (1970), 30s (1975e). [c]Grenville (on Grenada I): 15s (1975e).

Guadeloupe: [a]none. [b]Basse-Terre (on Basse-Terre I): 15 (1974). [c]Pointe-a-Pitre (on Grande-Terre I): 24 (1974).

Guatemala: [a]none. [b]Guatemala (alt Guatemala City): 701 (1973). [c]Mixco: 115 (1973).

Haiti: [a]for Hayti. [b]Port-au-Prince: 459 (1975e), 494s (1971). [c]Cap-Haitien (for Cap-Francais): 46 (1971).

Honduras: [a]none. [b]Tegucigalpa: 271 (1974). [c]San Pedro Sula: 148 (1974).

Jamaica: [a]none. [b]Kingston: 170 (1973e), 626s (1975e). [c]Montego Bay: 44 (1970); Spanish Town: 41 (1970).

Martinique: [a]none. [b]Fort-de-France (for Fort-Royal): 94 (1967), 99s (1974). [c]Schoelcher: 11, 13s (1967).

Mexico: [a]none. [b]Mexico (alt Mexico City; for Tenochtitlan): 8628, 11,943s (1976e). [c]Guadalajara: 1641, 2076s (1976e).

Montserrat: [a]none. [b]Plymouth: 1 (1970). [c]none.

Netherlands Antilles: [a]alt Dutch West Indies; for Curacao; off Nederlandse Antillen. [b]Willemstad (on Curacao I): 50 (1970e), 94s (1960). [c]Oranjestad (on Aruba I): 16 (1968e).

Nicaragua: [a]none. [b]Managua: 313 (1974e). [c]Leon: 56 (1971).

Panama: [a]none. [b]Panama (alt Panama City): 416 (1976e). [c]Colon (for Aspinwall): 95 (1975e).

Puerto Rico: [a]for Porto Rico. [b]San Juan (inc Rio Piedras since 1950): 472 (1972e), 820s (1970). [c]Bayamon: 148 (1970).

Saint Kitts-Nevis: [a]Saint Kitts: alt Saint Christopher. [b]Basseterre (on Saint Kitts I): 16 (1970). [c]Sandy Point (on Saint Kitts I): 4 (1960).

Saint Lucia: [a]none. [b]Castries: 4, 39s (1970). [c]Vieux Fort: 3 (1960).

Saint-Pierre and Miquelon: [a]off Saint-Pierre et Miquelon. [b]Saint-Pierre (on Saint-Pierre I): 5 (1974). [c]Miquelon (on Miquelon I): 0.6 (1967).

Saint Vincent: [a]none. [b]Kingstown (on Saint Vincent I): 22 (1973e). [c]Georgetown (on Saint Vincent I): 1 (1960).

Trinidad and Tobago: [a]none. [b]Port of Spain (on Trinidad I): 60 (1973e). [c]San Fernando (on Trinidad I): 37 (1970).

Turks and Caicos Islands: [a]none. [b]Grand Turk (on Grand Turk I): 2 (1970). [c]Cockburn Harbour (on South Caicos I): 1 (1970).

USA: [a]alt United States, US; inc Alaska (for Russian America), Hawaii (for Hawaiian Islands, Sandwich Islands); off United States of America. [b]Washington: 693 (1976e), 2481s (1970). [c]New York (alt New York City; for New Amsterdam; inc Brooklyn since 1898): 7482 (1975e), 16,208s (1970).

Virgin Islands (UK): [a]alt British Virgin Islands. [b]Road Town (on Tortola I): 3 (1973e). [c]East End-Long Look (on Tortola I): 2 (1970).

Virgin Islands (USA): [a]alt Virgin Islands of the United States; for Danish West Indies. [b]Charlotte Amalie (for Saint Thomas) (on Saint Thomas I): 12 (1970). [c]Christiansted (on Saint Croix I): 3 (1970).

OCEANIA

American Samoa: [a]alt Eastern Samoa. [b]Pago Pago (on Tutuila I): 3, 11s (1974). [c]Nu'uuli (on Tutuila I): 2 (1974).

Australia: [a]inc Tasmania (for Van Diemen's Land). [b]Canberra: 208s (1976). [c]Sydney: 3021s (1976).

Christmas Island (Australia): [a]none. [b]Flying Fish Cove: ? [c]none.

Cocos Islands: [a]alt Keeling Islands. [b]Bantam Village (on Home I): ? [c]none.

Cook Islands: [a]for Harvey Islands. [b]Avarua (on Rarotonga I): 6 (1971). [c]Arorangi (on Rarotonga I): 2 (1966).

Fiji: [a]alt Fiji Islands; for Fidji, Viti. [b]Suva (on Viti Levu I): 64 (1976), 80s (1966). [c]Lautoka (on Viti Levu I): 12 (1966).

French Polynesia: [a]for French Oceania; inc Marquesas Islands, Society Islands (inc Tahiti); off Polynesie Francaise. [b]Papeete (on Tahiti I): 25, 65s (1971). [c]Uturoa (on Raiatea I): 3 (1971).

Gilbert Islands: [a]none. [b]Bairiki (on Tarawa I): 2 (1973). [c]Bikenibeu (on Tarawa I): 3 (1973).

Guam: [a]inc Mariana Islands (for Ladrone Islands) (in part). [b]Agana: 2 (1974e). [c]Tamuning: 8 (1970).

Johnston and Sand Islands: [a]none. [b]Washington (in USA). [c]none.

Midway Islands: [a]none. [b]Washington (in USA). [c]none.

Nauru: [a]for Pleasant Island; off Naoero. [b]Yaren: ? [c]none.

New Caledonia: [a]off Nouvelle-Caledonie. [b]Noumea (for Port-de-France) [on New Caledonia (off Nouvelle-Caledonie) I]: 42 (1969), 56s (1976). [c]Mont-Dore (on New Caledonia I): 11s (1976).

New Hebrides: [a]alt Nouvelles-Hebrides. [b]Vila (alt Port-Vila) [on Efate (alt Vate) I]: 3 (1967), 13s (1973e). [c]Santo (alt Luganville) [on Espiritu Santo (alt Santo; for Marina) I]: 4 (1972e), 5s (1967).

New Zealand: [a]none. [b]Wellington (on North I): 140, 327s (1976). [c]Christchurch (on South I): 172, 295s (1976); Auckland (on North I): 151, 743s (1976).

Niue Island: [a]for Savage Island. [b]Alofi: 1 (1971), 4s (1976e). [c]none.

Norfolk Island: [a]none. [b]Kingston: 1 (1970?e). [c]none.

Northern Mariana Islands: [a]for Ladrone Islands (in part). [b]Saipan (on Saipan I): 13 (1973). [c]Rota (on Rota I): 1 (1973); Tinian (on Tinian I): 0.7 (1973).

Pacific Islands (USA): [a]inc Caroline Islands (for New Philippines), Marshall Islands; off Trust Territory of the Pacific Islands. [b]Saipan (in Northern Mariana Islands). [c]Moen (on Moen I): 10 (1973).

Papua New Guinea: [a]inc New Guinea (Australia) [inc Bismarck Archipelago (inc New Britain)], Papua. [b]Port Moresby (on New Guinea I): 113 (1976). [c]Lae (on New Guinea I): 39 (1971). *New Guinea:* [c]Lae (on New Guinea I): 39 (1971); Rabaul (on New Britain I): 27 (1971). *Papua:* [b]Port Moresby (on New Guinea I): 113 (1976). [c]Daru (on Daru I): 6 (1971).

Pitcairn Island: [a]none. [b]Adamstown: ? [c]none.

Samoa: [a]for Navigators' Islands, Western Samoa. [b]Apia (on Upolu I): 32s (1976). [c]Faleasiu (on Upolu I): 2 (1971).

Solomon Islands: [a]for British Solomon Islands. [b]Honiara (on Guadalcanal I): 15 (1976). [c]Auki Station (on Malaita I): 0.9 (1970).

Tokelau Islands: [a]alt Union Islands. [b]Wellington (in New Zealand). [c]Fakaofo (on Fakaofo I): ? [Fakaofo I: 0.6 (1974)].

Tonga: [a]alt Tonga Islands; for Friendly Islands. [b]Nukualofa (on Tongatapu I): 18 (1976). [c]Neiafu (on Vavau I): 3 (1956).

Tuvalu: [a]for Ellice Islands. [b]Fongafale (on Funafuti I): 0.9 (1973). [c]Lolua (on Nanumea I): 0.6 (1973); Savave (on Nukufetau I): 0.6 (1973).

Wake Island: [a]none. [b]Washington (in USA). [c]none.

Wallis and Futuna Islands: [a]off Wallis et Futuna (Futuna Islands: for Hoorn Islands). [b]Mata-Utu [on Uvea (alt Wallis) I]: 0.4 (1976). [c]Vaitupu [on Uvea (alt Wallis) I]: 0.6 (1969).

SOUTH AMERICA

Argentina: [a]none. [b]Buenos Aires (for Buenos Ayres): 2975, 9245s (1975e). [c]Cordoba: 903 (1975e).

Bolivia: [a]none. [b]La Paz (alt Paz; for Choqueyapu; off La Paz de Ayacucho): 655 (1976); Sucre (for Charcas, Chuquisaca): 62 (1976). [c]Santa Cruz (off Santa Cruz de la Sierra): 257 (1976).

Brazil: [a]off Brasil. [b]Brasilia: 282 (1970), 763s (1975e). [c]Sao Paulo: 7199, 7199s (1975e).

Chile: [a]for Chili. [b]Santiago (alt Santiago de Chile): 3186s (1975e). [c]Valparaiso: 249, 592s (1975e).

Colombia: [a]for New Granada. [b]Bogota (for Santa Fe de Bogota, Teusaquillo): 3102 (1976e). [c]Medellin: 1195 (1976e).

Ecuador: [a]inc Galapagos (off Colon) Islands. [b]Quito: 600 (1974). [c]Guayaquil (for Santiago de Guayaquil): 823 (1974).

Falkland Islands: [a]alt Islas Malvinas; inc South Georgia. [b]Stanley (alt Port Stanley) (on East Falkland I): 1 (1974e). [c]Goose Green (on East Falkland I): 0.1 (1972).

French Guiana: [a]inc Inini; off Guyane Francaise. [b]Cayenne: 20 (1967), 30s (1974). [c]Saint-Laurent-du-Maroni: 3, 5s (1967).

Guyana:[a]for British Guiana. [b]Georgetown: 66, 167s (1970). [c]New Amsterdam: 18 (1970).

Paraguay: [a]none. [b]Asuncion (for Nuestra Senora de la Asuncion): 442 (1976e). [c]Fernando de la Mora: 37 (1972).

Peru: [a]none. [b]Lima (for Rimac): 3303s (1972). [c]Arequipa: 302 (1972); Callao: 297 (1972).

Surinam: [a]for Dutch Guiana, Netherlands Guiana; off Suriname. [b]Paramaribo: 102 (1971). [c]Nieuw Nickerie: 35 (1971).

Uruguay: [a]alt Republica Oriental del Uruguay; for Banda Oriental. [b]Montevideo: 1230 (1975). [c]Paysandu: 80 (1975); Salto: 80 (1975).

Venezuela: [a]none. [b]Caracas (for Santiago de Leon de Caracas): 1035 (1971), 2487s (1975e). [c]Maracaibo: 652 (1971).

CITY GAZETTEER,
BY CONTINENT AND COUNTRY

Contents

This gazetteer consists of three tables.

11a. Date of Settlement and Evolution of Population

Conventional name of city.

Year or century in which the first permanent settlement within present city limits was made. A question mark following a given date usually indicates that, although its exact origin is unknown, the city in question was first mentioned in historical records at that date. When the time of settlement cannot be ascertained but the city is known to have existed before a certain date, that date is given, preceded by the abbreviation "bef." It should be observed that the settlement date of a city bears no relation either to the date when the city was formally established (or chartered) or to the date when the present name was adopted.

Latest population, in thousands, and year of data. If the year is followed by the letter "e," the population has been estimated; otherwise, the figure given is from census returns. For many cities, two population figures are listed for a given year; the first is for the city proper, and the second, followed by the letter "s," for the city and adjacent suburban areas. Preferably, this is the population of the urbanized area or conurbation only; some officially reported suburban areas, however, include rural districts.

Similar population data for the years 1970, 1960, 1950, 1930, 1900, 1850, 1800 (and sometimes earlier), or the times nearest to those years for which reliable figures are available, with the exact year stated. All populations are for the city as it existed at the time the population was determined, not for its present-day area. For populations "circa 1960," a 1956 census figure is normally given in preference to a 1960 estimate. Question marks are used when the exact year is unknown; for example, the notation "(00?e)" below the tabular heading "circa 1900" indicates that the population preceding it was estimated about, but not necessarily during, the year 1900.

11b. Location, Elevation, and Climatic Data[1]

Conventional name of city.

Latitude and longitude, in degrees and minutes. The notation "50.46N, 6.06E" means 50 degrees 46 minutes north of the equator and 6 degrees 6 minutes east of Greenwich, a section of London.

Elevation (i.e., altitude above sea level), in feet. When available, the given elevation is that of the city center or principal business district, or it is an average of elevations at several points in the city; otherwise, it is that of the meteorological station.

Period of record for climatic data. The period cited refers to the average temperature statistics; extreme temperature and precipitation data frequently cover a longer period.

Average temperature throughout the year. This and all other temperatures in this table are expressed in degrees Fahrenheit.

Average temperatures during the warmest and coolest months, and the specific months.

Highest and lowest temperatures of record (absolute high and low).

Highest and lowest average annual temperatures. These are, in effect, the extreme temperatures likely to occur during a given year.

Average annual precipitation (i.e., rainfall plus the rain equivalent of snowfall), in inches.

11c. Supplemental Information

Conventional name of city.

[a] Division of country in which city is located. This information is given for cities in the larger countries; see General Explanatory Notes in the Introduction for further information.

[b] Alternate, former, and official names of city. See General Explanatory Notes for explanation. In addition, important formerly separate cities that have been incorporated into the city cited and certain well-known sections of the city sometimes regarded as separate cities (e.g., Cholon, a section of Ho Chi Minh City) are listed after the abbreviation "inc." If a date is given after the "inc," this is related to the populations entered in Table 11a. Thus the notation (under New York, USA) that reads "inc Brooklyn since 1898" means that the populations given for New York before 1898 exclude Brooklyn.

[c] Hydrographic features within present city limits. These are listed in the following order: seas (including oceans, gulfs, bays, straits, etc.), by size; rivers, by length, except that main rivers are always listed before their tributaries; natural lakes (including lagoons), by size. Creeks and other small streams are cited only if there is no river. If there is no hydrographic feature within the city limits, the nearest sea, river, or lake within approximately 10 mi (16 km) is given, preceded by the abbreviation "nr." Unconventional (i.e., alternate, former, and official) names of rivers and lakes are not

---

[1] Because of the widespread use of electronic calculators, which greatly facilitate conversion, metric data have been omitted from the City Gazetteer to save space. The conversions, however, are quite simple. To change feet of elevation into meters, it is only necessary to multiply the number of feet by 0.3048. To change Fahrenheit degrees of temperature into Celsius, or centigrade, degrees, simply subtract 32 from the number of Fahrenheit degrees and multiply the remainder by 0.5555. And to change inches of precipitation into millimeters, just multiply the number of inches by 25.4.

entered here if the river or lake in question is listed in Table 3 or 5, and thus in the index.

[d] Universities, with date of foundation (in parentheses), student enrollment (i.e., thousands of full-time and part-time students, excluding correspondence students, enrolled in the university and its affiliated institutions), and academic staff (i.e., number of full-time and part-time teachers). Universities are listed in order of student enrollment.

Except as noted in the gazetteer itself,[1] all universities, as defined in the General Explanatory Notes, within approximately 10 mi (16 km) of the city are listed.[2] Those outside the city limits, however, are described like this example under Visakhapatnam, India: "Andhra University (at Waltair, nr Visakhapatnam)." If a university was founded other than at its present location, the notation reads as in this example under Macon, USA: "Mercer University (1833 at Penfield, relocated 1871)."

The date given for the foundation of a university is the year in which the institution itself, or (preferably) its oldest constituent college or the educational institution from which it directly evolved, was established (or chartered). This date may precede by several years the date when instruction was first given. Also, it bears no necessary relation to the length of time during which the university has been operative, since many universities have been closed for periods of years or even decades.

If no university exists at or near a given city, the notation "none" appears after [d].

[e] Largest libraries, with date of foundation (in parentheses) and size of collection (i.e., thousands of volumes of books and pamphlets, usually excluding bound and unbound periodicals) for each library. Libraries are listed by size of collection.

Although only the largest libraries are cited, if the city has a public library of any importance, it is listed along with all other libraries that are larger. The name of the city is omitted from the name of each city (or municipal) public library; otherwise, the names of libraries are given in full. To save space, however, the names of universities already entered under [d] are not repeated when a university library is listed unless it is impossible without the name to determine to which university the library belongs; instead, the library is entered simply as ". . . University Library" or "University . . . Library," depending on whether the university in question is called, for example, New York University or University of New York.

As with universities, large libraries located within approximately 10 mi (16 km) of the city are also listed, but are identified as being "nr" the city. Again, as with universities, those founded elsewhere and subsequently moved to the city under which they are cited are entered as in this example under Harrisburg, USA: "State Library of Pennsylvania (1745 at Philadelphia, relocated 1812?)."

The date of foundation given for libraries is preferably the year in which the library was opened on a permanent basis, that is, the one since which it has been continuously operative. By "library," however, is meant the nucleus of the present book collection, not necessarily the institution under its name today.

[f] Supplemental population data. These most often include populations at various dates for the cities listed after "inc" in Section [b] of this table. All data are in thousands, followed by the year of record in parentheses.

---

[1] For a few large cities with many universities, only universities above a certain size are listed.
[2] Provided, of course, that the university in question is not located in another city entered in the gazetteer.

Arrangement

By city, alphabetically under country, the countries being arranged alphabetically under continent.

Coverage

All cities with at least the following populations (including suburbs if the suburban population is officially determined):
Africa—200,000
Asia, except USSR—200,000
Europe, except USSR—100,000[1]
North America—100,000
Oceania—100,000
South America—200,000
USSR—150,000
In addition, other well-known cities almost as large, as well as many smaller cities important historically or attractive to tourists, are included.

Rounding

Populations are rounded to the nearest 1000, except that when the original figure is less than 1000 the rounding is done to the nearest 100.

Highest and lowest temperatures of record and highest and lowest average annual temperatures are rounded to the nearest degree.

University student enrollments are rounded to the nearest 1000 students, except that enrollments of less than 1000 are rounded to the nearest 100 students.

Library collections are rounded to the nearest 1000 volumes.

Entries

2045.

[1] In the UK local government has been radically reorganized. In England and Wales, excluding London, the previously existing 83 county boroughs, 259 non-county boroughs, 522 urban districts, and 468 rural districts were replaced by 36 metropolitan districts and 333 non-metropolitan districts, effective in 1974. A somewhat similar reorganization took effect in Northern Ireland in 1973 and in Scotland in 1975. As a result of this consolidation of urban and rural districts, nearly all the new local districts include both urban and rural areas. In this work, those districts that have been accorded city or borough status, as well as a few others known to be predominantly urban, are treated as cities. Some of these new districts also have new names; the familiar cities and boroughs included within them are listed in Table 11c.

Table 11a. Date of Settlement and Evolution of Population

AFRICA

City	Date Settled	Latest	ca 1970	ca 1960	ca 1950	ca 1930	ca 1900	ca 1850	ca 1800	18th c
					Population (thousands)					
ALGERIA										
*Algiers	10th c	1504s(74e)	904 (66)	722 (60)	266 (48)	221 (31)	97 (01)	97 (49)	73 (08e)	100 (1634e)
			943s(66)	884s(60)	489s(48)	257s(31)	137s(01)			
Annaba	7th c	313s(74e)	152 (66)	135 (60)	78 (48)	66 (31)	32 (01)	10 (47)		
			169s(66)	165s(60)	103s(48)	69s(31)				
Constantine	3rd c BC?	350s(74e)	244 (66)	169 (60)	80 (48)	79 (31)	41 (01)	23 (49)	20 (08e)	
			254s(66)	223s(60)	119s(48)	105s(31)	48s (01)			
Oran	10th c	485s(74e)	327 (66)	357 (60)	245 (48)	154 (31)	88 (01)	25 (49)	16 (00?e)	
			328s(66)	426s(60)	273s(48)	164s(31)				
ANGOLA										
*Luanda	1575	475 (70)	475 (70)	225 (60)	142 (50)	67 (40)	20 (98)	16 (70)		
BENIN										
Cotonou	bef 1868	178 (75e)	139 (70e)	78 (61)	23 (46–49e)	8 (32)	4 (11)			
*Porto-Novo	16th c	104 (75e)	87 (70)	64 (61)	31 (46–49e)	27 (32)	30 (00?e)			
BOTSWANA										
*Gaborone	1895?	37 (76e)	18 (71)	4 (64)	12s(46)	8s (36)				
BURUNDI										
*Bujumbura	1897	157s (76e)	79s (70e)	47s (60e)	17 (49)					
CAMEROON										
Douala	bef 1884	486s (75e)	250s (70e)	128s (62)	125s(56)	28 (31)	23 (02e)			

AFRICA

Population (thousands)

City	Date Settled	18th c	ca 1800	ca 1850	ca 1900	ca 1930	ca 1950	ca 1960	ca 1970	Latest
*Yaounde	1888					6 (36)	30 (50e)	93 (62)	166s(69e)	274s (75e)
(CANARY ISLANDS), SPAIN										
*Las Palmas	1478		9 (00?e)	13 (57)	45 (00)	78 (30)	153 (50)	194 (60)	287 (70)	341 (75e)
Santa Cruz de Tenerife	1494	7 (1768)		11 (57)	38 (00)	62 (30)	103 (50)	133 (60)	151 (70)	160 (75e)
CAPE VERDE *Praia	1652?			2 (50?e) village	4 (00?)	6 (30?)	10 (50)	13 (60)	21 (70)	21 (70)
CENTRAL AFRICAN EMPIRE *Bangui	1890					13 (26)	41 (50e)	80 (59–60)	299s(68)	302s(71e)
CHAD *Ndjamena	1900					6 (26)	23 (50e)	88s(62)	157s(70e)	224s(75e)
CONGO *Brazzaville	1880				5 (00?e)	4 (26)	83 (50e)	136s (61–62)	175 (70)	290 (74)
DJIBOUTI *Djibouti	1888				15 (00e)	9 (28)	22 (48e)	31s(57e)	62s(70e)	120s (73e)
EGYPT Alexandria	332 BC	15 (1700?e)	15 (00e)	164 (62)	320 (97)	573 (27)	919 (47)	1516 (60)	1801 (66)	2312 (76)

Figure 43. Buildings clustered around Liberation Square in Cairo, the largest city and the largest urbanized area in Africa. The Nile River and part of Zamalik Island can be seen in the background. (Credit: Government of Egypt Ministry of Culture.)

AFRICA

Population (thousands)

City	Date Settled	Latest	ca 1970	ca 1960	ca 1950	ca 1930	ca 1900	ca 1850	ca 1800	18th c
Aswan	6th c BC?	230 (76)	128 (66)	48 (60)	26 (47)	16 (27)	13 (97)	6 (82)		
Asyut	bef 2160 BC	470 (76)	154 (66)	127 (60)	90 (47)	57 (27)	42 (97)	26 (62)	12 (00e)	
*Cairo	641?	5084 (76) 6757s (75e)	4220 (66)	3349 (60)	2091 (47)	1065 (27)	570 (97)	257 (62)	260 (00e)	
Damanhur	11th c?	176 (74e)	146 (66)	127 (60)	85 (47)	52 (27)	32 (97)	20 (82)		
Giza	bef 2568 BC	1246 (76)	571 (66)	262 (60)	66 (47)	27 (27)	17 (97)	11 (72)	3 (00e)	
Ismailia	1863	166 (76)	144 (66)	116 (60)	68 (47)	25 (27)	7 (97)	3 (72)		
Luxor	bef 2160 BC	77 (66)	77 (66)	35 (60)	24 (47)	13 (27)	11 (97)	4 (82)		
Mahalla al Kubra	985?	288 (74e)	225 (66)	178 (60)	116 (47)	46 (27)	32 (97)	28 (82)	17 (00e)	
Mansurah	1221	232 (74e)	191 (66)	147 (60)	102 (47)	64 (27)	36 (97)	16 (72)	7 (00e)	
Port Said	1859	263 (76)	283 (66)	245 (60)	178 (47)	101 (27)	42 (97)	9 (72)		
Shubra al Khaymah		394 (76)	173 (66)	101 (60)	15 (47)	7 (27?)	2 (97?)	2 (82)		
Suez	15th c	194 (76)	264 (66)	204 (60)	107 (47)	41 (27)	17 (97)	13 (72)		
Tanta	12th c?	278 (74e)	230 (66)	184 (60)	140 (47)	90 (27)	57 (97)	19 (62)	10 (00e)	
Zagazig	1850?	195 (74e)	151 (66)	124 (60)	82 (47)	53 (27)	36 (97)	20 (82)		
ETHIOPIA										
*Addis Ababa	1887	1083 (74)	796 (70)	449 (61)	400 (51e)	70 (28e)	30 (00e)			
Asmara	bef 1889	296 (74)	218 (70)	130 (60e)	117 (48e)	22 (31)	9 (05)			
Harar	7th c?	48 (72e)	45 (70e)	40 (58e)	40 (48e)	36 (30?e)	40 (00?e)	30 (70?e)		
GABON										
*Libreville	1849	105 (69–70) 112s (73e)	105 (69–70)	31s (61)	10 (50e)	4 (31)	3 (03e)			

AFRICA

City	Date Settled	Population (thousands)								
		Latest	ca 1970	ca 1960	ca 1950	ca 1930	ca 1900	ca 1850	ca 1800	18th c
GAMBIA										
*Banjul	1816	44 (76e)	39 (73)	28 (63) 40s(63)	20 (51)	14 (31)	7 (01)	4 (50?e)		
GHANA										
*Accra	16th c	717 (75e) 738s (70)	564 (70) 738s (70)	338 (60) 388s(60)	134 (48)	51 (31)	15 (01)	3 (50?e)		
Kumasi	17th c	260 (70) 345s (70)	260 (70) 345s (70)	181 (60) 218s(60)	71 (48)	36 (31)	6 (06)	18 (50?e)	13 (17e)	
GUINEA										
*Conakry	1890	526s (72)	526s (72)	43 (60) 112s(60)	38 (46–49e)	9 (31)	7 (11)			
GUINEA-BISSAU										
*Bissau	1758	71 (70)	71 (70)	47 (60)	18 (50)	1 (30)				
IVORY COAST										
*Abidjan	1904	904s(74e)	550s (70e)	180s(60e)	56 (46–49e)	10 (31)	1 (11)			
Bouake	bef 1898	156 (74e)	120 (70e)	45 (61e)	22 (50?e)	5 (37e)				
KENYA										
Mombasa	8th c?	351 (76e)	247 (69)	180 (62)	85 (48)	56 (31)	27 (00?e)	6 (50?e)		
*Nairobi	1899	736 (76e)	509 (69)	267 (62) 315s(62)	75 (48) 119s(48)	48 (31)	5 (07)			
LIBERIA										
*Monrovia	1822	167 (74) 204s(74)	96 (69–70)	81 (62)	41 (56)	10 (30?e)	5 (97e)	2 (52e)		

AFRICA

Population (thousands)

City	Date Settled	Latest	ca 1970	ca 1960	ca 1950	ca 1930	ca 1900	ca 1850	ca 1800	18th c
LIBYA										
Bengasi	6th c BC	140 (73)	140 (73)	137 (64)	70 (54)	31 (28)	15 (00?e)	3 (50?e)	5 (00?e)	
*Tripoli	7th c BC?	551 (73)	551 (73)	214 (64)	130 (54)	72 (31)	42 (00e)	15 (42e)	15 (05e)	
MADAGASCAR										
*Antananarivo	17th c	452 (75)	333 (68e)	248 (60)	174 (49e)	92 (31)	63 (11)	25 (50?e)		
(MADEIRA ISLANDS), PORTUGAL										
*Funchal	1425	38 (70)	38 (70)	43 (60)	37 (50)	31 (30)	21 (00)	17 (45)	15 (00e)	
MALAWI										
Blantyre	1876	229 (77e)	104 (66)	62 (61e)	6 (49e)	8 (30e)	6 (00?e)			
*Lilongwe	1947	103 (77e)	19 (66)	5s (56)	2 (49e)					
MALI										
*Bamako	bef 1883	170s (74e)	182s (68e)	130s (60e)	60 (46–49e)	20 (31)	7 (11)	0.8 (83)		
Timbuktu	11th c	12 (71e)	12 (71e)	7 (60e)	7 (46–49e)	6 (31)	7 (11)	13 (54e)	60 (00?e)	
MAURITANIA										
*Nouakchott	1903	70 (74e) 134s (76)	30 (70e)	6 (61)	0.4 (50?e)					
MAURITIUS										
*Port Louis	1735	141 (76e)	134 (72)	89 (62)	70 (52)	54 (31)	53 (01)	50 (51)	20 (00?e)	

AFRICA

City	Date Settled	Population (thousands)								
		Latest	ca 1970	ca 1960	ca 1950	ca 1930	ca 1900	ca 1850	ca 1800	18th c
MOROCCO										
Casablanca	1515	1808 (74e)	1506 (71)	965 (60)	682 (52)	160 (31)	20 (00?e)	0.7 (50?e)	0.1 (00?e)	
Fez	808	325 (71)	325 (71)	216 (60)	179 (52)	107 (31)	145 (00?e)	88 (50?e)	100 (00?e)	
Marrakesh	1062	333 (71)	333 (71)	243 (60)	215 (52)	192 (31)	50 (00e)	80 (50e)	30 (00?e)	
Meknes	10th c	248 (71)	248 (71)	176 (60)	140 (52)	54 (31)	35 (00?e)	50 (50?e)	15 (00?e)	
Oujda	994	176 (71)	176 (71)	129 (60)	81 (52)	29 (31)	8 (00?e)			
*Rabat-Sale	11th c	616 (74e)	523 (71)	303 (60)	203 (52)	79 (31)	61 (00e)	38 (50e)		
Tangier	15th c BC?	214 (74e)	188 (71)	142 (60)	85 (47e)	80 (26e)	35 (00e)	8 (50e)	12 (00?e)	
MOZAMBIQUE										
*Maputo	1544	102 (70)	102 (70)	79 (60)	94 (50)	43 (30)	6 (00)	1 (67e)		
		355s (70)	355s (70)	184s (60)	100s(50)					
NAMIBIA										
*Windhoek	bef 1890	65 (70)	65 (70)	36 (60)	20 (51)	8 (26)				
NIGER										
*Niamey	16th c?	130 (75e)	79 (68e)	30 (59)	9 (46–49e)	3 (31)				
NIGERIA										
Abeokuta	1830	253 (75e)	226 (71e)	187 (63)	84 (52)	46 (31)	60 (00?e)	60 (50e)		
Ado-Ekiti	19th c?	213 (75e)	190 (71e)	158 (63)	30 (52)					
Enugu	1909	187 (75e)	167 (71e)	138 (63)	63 (53)	13 (31)				
Ibadan	bef 1821	847 (75e)	758 (71e)	627 (63)	459 (52)	387 (31)	345 (11)	70 (51e)		
Ilesha	18th c?	224 (75e)	200 (71e)	166 (63)	72 (52)	22 (31)	40 (00?e)			
Ilorin	18th c	282 (75e)	252 (71e)	209 (63)	41 (52)	47 (31)	36 (11)	70 (51e)		
Iwo	16th c?	214 (75e)	192 (71e)	159 (63)	100 (52)	57 (31)	60 (00?e)			
Kaduna	1913	202 (75e)	181 (71e)	150 (63)	39 (52)	11 (31)				
Kano	12th c?	399 (75e)	357 (71e)	295 (63)	130 (52)	89 (31)	39 (11)	30 (51e)	35 (24e)	

AFRICA

City	Date Settled	Population (thousands)								
		Latest	ca 1970	ca 1960	ca 1950	ca 1930	ca 1900	ca 1850	ca 1800	18th c
*Lagos	1700?	1061 (75e)	900 (71e)	665 (63)	267 (52)	126 (31)	42 (01)	29 (71)	5 (89e)	
		1700s (72e)	1600s (71e)							
Mushin		197 (75e)	176 (71e)	146 (63)	32 (52)					
Ogbomosho	17th c	432 (75e)	387 (71e)	320 (63)	140 (52)	87 (31)	60 (00?e)	25 (51e)		
Onitsha	17th c	220 (75e)	197 (71e)	163 (63)	77 (53)	18 (31)	16 (00?e)			
Oshogbo	1600?	282 (75e)	253 (71e)	209 (63)	123 (52)	50 (31)	35 (00?e)			
Port Harcourt	1912	242 (75e)	217 (71e)	180 (63)	72 (53)	15 (31)				
Zaria	1095?	224 (75e)	201 (71e)	166 (63)	54 (52)	28 (31)	45 (00?e)	50 (50?e)		
REUNION										
*Saint-Denis	1665	104s (74)	66 (67)	37 (61)	26 (54)	27 (31)	27 (02)	18 (51)	7 (04)	
			85s (67)	65s (61)	42s (54)					
RHODESIA										
Bulawayo	1893	340s (76e)	210 (68e)	125 (56e)	43 (46–48)	31 (31e)	5 (00?e)			
					53s (46–48)					
*Salisbury	1890	568s (76e)	245s (69)	190s (59e)	90 (51)	29 (31e)	5 (00?e)			
					119s (51)					
RWANDA										
*Kigali	bef 1935	90 (77)	50 (70)	5 (59e)	2 (49)					
SENEGAL										
*Dakar	1857	799s (76)	583s (70–71)	375s (61)	171 (46e)	54 (31)	18 (04)	3 (65)		
SIERRA LEONE										
*Freetown	1788	274 (74)	179 (70e)	128 (63)	65 (47e)	55 (31)	34 (01)	18 (60)	5s (20)	

AFRICA

City	Date Settled	Population (thousands)								
		Latest	ca 1970	ca 1960	ca 1950	ca 1930	ca 1900	ca 1850	ca 1800	18th c
SOMALIA										
*Mogadishu	908?	285 (76e)	173 (67)	91 (59e)	74 (48e)	36 (31)	7 (00?e)	4 (50?e)	4 (25e)	
SOUTH AFRICA										
Bloemfontein	1846	148 (70)	148 (70)	113 (60)	81 (51)	53 (36)	34 (04)	1 (65?e)		
			180s (70)	145s (60)	109s(51)	64s (36)				
*Cape Town	1652	691 (70)	691 (70)	508 (60)	441 (51)	296 (36)	78 (04)	26 (56)	17 (00?)	
		1097s(70)	1097s(70)	807s(60)	578s(51)	344s (36)	170s (04)			
Durban	1824	730 (70)	730 (70)	560 (60)	435 (51)	240 (36)	68 (04)	5 (66)		
		843s(70)	843s(70)	681s(60)	480s(51)	260s (36)				
Germiston	1887	210 (70)	210 (70)	148 (60)	116 (51)	68 (36)	29 (04)	0.5 (50?e)		
		282s(70)	282s(70)	214s(60)	168s(51)	79s (36)				
Johannesburg	1886	655 (70)	655 (70)	595 (60)	632 (51)	475 (36)	99 (04)			
		1433s(70)	1433s(70)	1153s(60)	884s(51)	519s (36)	156s (04)			
Pietermaritzburg	1839	114 (70)	114 (70)	92 (60)	73 (51)	47 (36)	31 (04)	5 (71?)		
		159s(70)	159s(70)	129s(60)	92s(51)	56s (36)				
Port Elizabeth	1820	387 (70)	387 (70)	249 (60)	169 (51)	99 (36)	33 (04)	12 (65)		
		469s(70)	469s(70)	291s(60)	189s(51)	110s (36)				
*Pretoria	1855	544 (70)	544 (70)	304 (60)	232 (51)	108 (36)	37 (04)	0.3 (67e)		
		562s(70)	562s(70)	423s(60)	285s(51)	129s (36)				
Vereeniging	1892	170 (70)	170 (70)	79 (60)	60 (51)	20 (36)	0.9(04?)			
		304s(70)	304s(70)	135s(60)	123s(51)					
SUDAN										
*Khartoum	1823	334 (73)	194 (68e)	93 (56)	68 (47e)	50 (29e)	8 (00e)	25 (50e)		
				246s(56)	223s(47e)					
Omdurman	18th c?	299 (73)	206 (68e)	114 (55)	123 (47e)	104 (29e)	43 (09)			
TANZANIA										
*Dar es Salaam	1862	517 (75e)	273 (67)	129 (57)	69 (48)	23 (31)	17 (00e)			

AFRICA

City	Date Settled	Population (thousands)								
		Latest	ca 1970	ca 1960	ca 1950	ca 1930	ca 1900	ca 1850	ca 1800	18th c
Zanzibar	16th c	68 (67)	68 (67)	58 (58)	45 (48)	45 (31)	58 (00e)	25 (50?e)		
TOGO										
*Lome	19th c	214 (75e)	148 (70)	77 (58–60)	33 (50)	18 (31)	4 (03)	6 (50?e)		
TUNISIA										
Sfax	2nd c?	171 (75)	88 (66) / 216s (66)	66 (56)	55 (46)	40 (31)	15 (00?e)			
*Tunis	4th c BC?	550 (75) / 944s (75)	469 (66) / 648s (66)	410 (56)	365 (46)	202 (31)	228 (06)	130 (50?e)	125 (00e)	
UGANDA										
*Kampala	1890	331 (69)	331 (69)	47 (59) / 123s(59)	22 (48)	5 (30e)	3 (11)			
UPPER VOLTA										
*Ouagadougou	14th c?	169 (75e)	110 (70e)	59 (61)	21 (46–49e)	11 (31)	5 (00?e)			
ZAIRE										
Kananga	1884	601 (74e)	429 (70)	115 (59e)	11 (48e)	40 (32e)	5 (01e)			
*Kinshasa	1887	2008 (74e)	1323 (70)	402 (59e)	209 (50e)	12 (30e)	0.4(00?)			
Kisangani	1882	311 (74e)	230 (70)	127 (60e)	25 (48e)	17 (32e)				
Lubumbashi	1910	404 (74e)	318 (70)	184 (59e)	103 (50e)					
Mbuji-Mayi		337 (74e)	256 (70)	41 (58e)						
ZAMBIA										
Kitwe	1931?	251s (74)	200s (69)	102s (63)	70s(56)	17s (46)				

AFRICA

City	Date Settled	Latest	ca 1970	ca 1960	ca 1950	ca 1930	ca 1900	ca 1850	ca 1800	18th c
*Lusaka	1910?	401s (74)	262s (69)	105s (63)	71s (56)	2 (31)				
Ndola	19th c?	229 (74)	160s (69)	77s (63)	69s (56)	8 (31)				

ASIA

City	Date Settled	Latest	ca 1970	ca 1960	ca 1950	ca 1930	ca 1900	ca 1850	ca 1800	18th c
AFGHANISTAN										
Herat	4th c BC?	157 (75e)	102 (70e)	35 (60e)	76 (48e)	30	45 (00?e)	45 (50?e)	100 (10e)	
*Kabul	4th c BC?	749 (75e)	307 (70e)	292 (65), 439s (65)	206 (48e)	80 (30?e)	100 (00e)	40 (53e)	200 (00?e)	
Kandahar	1747	209 (75e)	130 (70e)	50 (60e)	77 (48e)	45 (30?e)	40 (00e)	58 (50e)	100 (09e)	
BAHRAIN										
*Manama	1507?	110 (75e)	89 (71)	62 (59)	40 (50)	25 (30?e)	8 (00?e)	5 (50?e)		
BANGLADESH										
Chittagong	8th c?	458 (74)	458 (74)	364 (61)	290 (51)	53 (31)	22 (01)	21 (72)		
*Dacca	1608?	1320 (74)	1320 (74)	557 (61)	336 (51)	139 (31)	90 (01)	69 (72)	68 (38)	
Khulna	18th c?	436 (74)	436 (74)	128 (61)	41 (51)	19 (31)	10 (01)			
Narayanganj	17th c?	177 (74)	177 (74)	162 (61)	68 (51)	34 (31)	24 (01)	11 (72)		
BRUNEI										
*Bandar Seri Begawan	15th c?	17 (71), 75s (76e)	17 (71), 37s (71)	10 (60), 23s (60)	11 (47)	10 (31)	15 (00?e)	22 (48e)	6 (38)	

ASIA

Population (thousands)

City	Date Settled	Latest	ca 1970	ca 1960	ca 1950	ca 1930	ca 1900	ca 1850	ca 1800	18th c
BURMA										
Bassein	12th c?	336s(73)	336s(73)		78 (53)	46 (31)	32 (01)	21 (72)	8 (26e)	
Henzada	16th c?	284s(73)	284s(73)		62 (53)	29 (31)	25 (01)	15 (66)		
Mandalay	1857	417s(73)	417s(73)	195 (58)	186 (53)	148 (31)	184 (01)	189 (91)		
Moulmein	bef 1826	203s(73)	203s(73)	108 (58)	103 (53)	66 (31)	58 (01)	24 (57)		
Myingyan		220s(73)	220s(73)		37 (53)	25 (31)	16 (01)	20 (91)		
Pegu	573	255s(73)	255s(73)		47 (53)	22 (31)	14 (01)	4 (72)	30 (00?e)	150 (1600e)
*Rangoon	585 BC	3187s(73)	3187s(73)	822 (57e)	737 (53)	400 (31)	235 (01)	99 (72)		
CAMBODIA										
*Phnom-Penh	1371	50 (75e)	650 (69e)	394 (62)	111 (48e)	96 (31)	50 (00e)	17 (66e)		
CHINA										
Amoy	16th c?	308 (58e)		308 (58e)	224 (53)	196 (35e)	96 (97)	275 (47e)	145 (32)	
Anking	13th c	129 (58e)		129 (58e)	105 (53)	121 (32e)	40 (00?e)			
Anshan	1908	833 (58e)		833 (58e)	549 (53)	214 (40)				
Antung	1896?	370 (58e)		370 (58e)	360 (53)	315 (40)	20 (08e)			
Canton	9th c BC?	3000 (65e)	3000 (65e)	2200 (58)	1599 (53)	1157 (35e)	900 (05e)	1236 (47e)	1500 (00?e)	
Changchih	16th c BC?	180 (58e)		180 (58e)	98 (53)	56 (22e)				
Changchow (Fukien)	10th c BC?	90 (56e)		90 (56e)	81 (53)			1000 (50e)		
Changchow (Kiangsu)	1st c?	300 (58e)		300 (58e)	296 (53)	79 (35e)	200 (00?e)			
Changchun	11th c?	1800 (65e)	1800 (65e)	988 (58e)	855 (53)	555 (40)	80 (08e)			
Changsha	3rd c BC?	709 (58e)		709 (58e)	651 (53)	480 (35e)	230 (05e)			
Chefoo	bef 1862	140 (58e)		140 (58e)	116 (53)	140 (34e)	82 (05e)	10 (69e)		
Chengchow	10th c BC?	785 (58e)		785 (58e)	595 (53)	80 (31e)				
Chengtu	3rd c BC?	1135 (58e)		1135 (58e)	857 (53)	481 (36?e)	475 (00?e)	800 (72e)		
Chinchow	12th c	400 (58e)		400 (58e)	352 (53)	142 (40)	14 (00?e)			
Chinkiang	12th c?	190 (58e)		190 (58e)	201 (53)	205 (36e)	168 (05e)	137 (50e)		

ASIA

City	Date Settled	Latest	ca 1970	ca 1960	ca 1950	ca 1930	ca 1900	ca 1850	ca 1800	18th c
					Population (thousands)					
Chinwangtao	bef 1879	210 (58e)		210 (58e)	187 (53)	20 (31e)	5 (05e)			
Chuchow	bef 1912	190 (58e)		190 (58e)	127 (53)					
Chungking	250?	2165 (58e)		2165 (58e)	1772 (53)	281 (34e)	620 (05e)	200 (61e)		
Foochow	10th c BC?	623 (58e)		623 (58e)	553 (53)	359 (35e)	624 (05e)	600 (47e)		
Foshan	1669	300 (68?e)	300 (68?e)	120 (58e)	122 (53)	96 (48e)	400 (00?e)	400 (69e)		
Fushun	1900?	1000 (65e)	1000 (65e)	1019 (58e)	679 (53)	270 (40)				
Fusin		290 (58e)		290 (58e)	189 (53)	143 (40)				
Hangchow	606	794 (58e)		794 (58e)	697 (53)	576 (35e)	350 (05e)	700 (50?e)	1000	(00?e)
Hantan	4th c BC	380 (58e)		380 (58e)	90 (53)					
Harbin	1896?	2000 (70e)	2000 (70e)	1595 (58e)	1163 (53)	661 (40)				
Hengyang	7th c BC?	240 (58e)		240 (58e)	235 (53)	102 (35e)	20 (02e)			
Hofei	10th c BC?	500 (63?e)	500 (63?e)	360 (58e)	184 (53)	70 (34e)	20 (00?e)			
Hoihow	bef 1876	402 (58e)		402 (58e)	135 (53)	15	12 (00?e)			
						(30?e)				
Hokang		200 (58e)		200 (58e)	90 (53)	20 (40)				
Huhehot	9th c	320 (58e)		320 (58e)	148 (53)	84 (35e)	200 (05e)			
Hwainan	bef 1940	500 (63?e)	500 (63?e)	280 (58e)	287 (53)					
Hwangshih		200 (63?e)	200 (63?e)	135 (58e)	110 (53)	6				
						(20?e)				
Ichang	1st c?	90 (56e)		90 (56e)	73 (53)	108 (31e)	45 (05e)			
Ichun	1948?	200 (58e)		200 (58e)	35 (53)					
Ipin	1st c?	190 (58e)		190 (58e)	178 (53)	78 (35e)	50 (07e)			
Kaifeng	7th c BC?	318 (58e)		318 (58e)	299 (53)	303 (36e)	200 (08e)			
Kalgan	1429	480 (58e)		480 (58e)	229 (53)	145 (35e)	47 (00e)			
Kashgar	2nd c BC?	100 (58e)		100 (58e)	91 (53)	80	65 (00?e)	30 (50?e)		
						(30?e)				
Kiamusze		232 (58e)		232 (58e)	146 (53)	129 (40)				
Kingtehchen	6th c	266 (58e)		266 (58e)	82 (53)	140 (34e)	200 (05e)	1000 (50?e)	1000	(00?e)
Kirin	1673	583 (58e)		583 (58e)	435 (53)	173 (40)	120 (09e)	135 (64e)	300	(12e)
Kisi	1935?	253 (58e)		253 (58e)	35 (53)					

ASIA

City	Date Settled	Population (thousands)								
		Latest	ca 1970	ca 1960	ca 1950	ca 1930	ca 1900	ca 1850	ca 1800	18th c
Kokiu		180 (58e)		180 (58)	170 (53)	50 (22e)	10 (00?e)			
Kunming	1st c?	900 (58e)		900 (58)	699 (53)	144 (34e)	45 (07)			
Kweilin	4th c BC?	170 (58e)		170 (58)	145 (53)	75 (36e)	150 (07?e)			
Kweiyang	14th c	530 (58e)		530 (58)	271 (53)	117 (35e)	100 (00?e)			
Lanchow	1st c?	732 (58e)		732 (58)	397 (53)	106 (36e)	337 (00e)			
Lhasa	400?	70 (58?e)		70 (58?e)	50 (53)	20 (30?e)	30 (00?e)	50 (50?e)	50 (00?e)	
Liaoyang	15th c BC?	169 (58e)		169 (58)	147 (53)	100 (40)	60 (00?e)	50 (50?e)		
Liaoyuan		177 (58e)		177 (58)	120 (53)	32 (38e)		80 (73e)		
Liuchow	3rd c BC?	300	300 (63?e)	190 (58)	158 (53)	60 (22e)	35 (00?e)			
Loyang	1900 BC?	500 (58e)		500 (58)	171 (53)	77 (35e)	20 (00?e)			
Luchow	2nd c?	130 (58e)		130 (58)	289 (53)	74 (35e)	40 (00?e)			
Lushun-Talien	600?	3600 (65e)	3600 (65e)	1590 (58)	892 (53)	327 (30)	54 (00?e)			
Malipo		196 (53)			196 (53)					
Mukden	4th c?	3000 (70e)	3000 (70e)	2423 (58)	2300 (53)	1134 (40)	158 (08e)	170 (64e)		
Mutankiang	bef 1903	251 (58e)		251 (58)	151 (53)	178 (40)				
Nanchang	12th c?	520 (58e)		520 (58)	398 (53)	277 (35e)	233 (00e)	300 (50?e)		
Nanchung	10th c BC?	206 (58e)		206 (58)	165 (53)	53 (35e)				
Nanking	2nd c BC?	1455 (58e)	1455 (58e)	1455 (58)	1092 (53)	1013 (35e)	270 (05e)	400 (50?e)	1000 (00?e)	
Nanning	10th c BC?	260 (58e)		260 (58)	195 (53)	89 (36e)	25 (05e)			
Nantung	1st c?	240 (58e)		240 (58)	260 (53)	133 (35e)				
Neikiang	1st c?	180 (58e)		180 (58)	190 (53)	32 (48e)				
Ningpo	8th c	280 (58e)		280 (58)	238 (53)	244 (33e)	260 (05e)	280 (50e)		
Paoki	3rd c BC?	180 (58e)		180 (58)	130 (53)	56 (48e)				
Paoting	13th c	380 (58e)		380 (58)	125 (53)	312 (28e)	60 (07e)	135 (66e)		
Paotow	11th c?	800	800 (63?e)	490 (58)	149 (53)	67 (35e)	20 (00?e)			
*Peking	11th c BC?	7570 (74e) (63?e)	7500 (70e)	4148 (58)	2768 (53)	1565 (35e)	700 (05e)	1649 (45)	700 (00?e)	
Pengpu	bef 1912	330 (58e)		330 (58)	253 (53)	105 (34)				

Figure 44. The famous bund and waterfront along the Whangpoo River at Shanghai, the largest city in Asia. (Credit: Foreign Language Press, Peking.)

ASIA

Population (thousands)

City	Date Settled	Latest	ca 1970	ca 1960	ca 1950	ca 1930	ca 1900	ca 1850	ca 1800	18th c
Penki	bef 1726	449 (58e)		449 (58e)	449 (53)	100 (40)	3 (00?e)			48 (1730?)
Shanghai	bef 11th c	10,820 (74e)	10,000 (70e)	6977 (58e)	6204 (53)	3259 (31e)	423 (95)	149 (65)		
Shangkiu	3rd c BC?	165 (58e)		165 (58e)	134 (53)	74 (35e)				
Shaohing	2000 BC?	160 (58e)		160 (58e)	131 (53)	178 (33e)	500 (00?e)	500 (71e)		
Shaoyang	7th c BC?	170 (58e)		170 (58e)	118 (53)	76 (35e)				
Shihkiachwang	bef 1904	623 (58e)		623 (58e)	373 (53)	217 (35e)				
Shiukwan	3rd c BC?	1000 (72?e)	1000 (72?e)	200 (63?e)	82 (53)	208 (35e)	100 (00?e)			
Siakwan		400 (68?e)	400 (68?e)		26 (53)	10 (20e)				
Sian	2205 BC?	1500 (70e)	1500 (70e)	1368 (58e)	787 (53)	155 (36e)	1000 (00?e)	1000 (72e)	232 (12)	
Siangtan	3rd c BC?	300 (68?e)	300 (68?e)	247 (58e)	184 (53)	123 (33e)	300 (00?e)	1000 (70e)		
Sinhailien	3rd c BC?	210 (58e)		210 (58e)	208 (53)					
Sining	2nd c BC?	300 (63?e)	300 (63?e)	150 (58e)	94 (53)	56 (46e)	60 (00?e)			
Sinsiang	1st c?	420 (72e)	420 (72e)	203 (58e)	170 (53)					
Soochow	525 BC?	651 (58e)		651 (58e)	474 (53)	390 (36e)	500 (05e)	2000 (52e)		
Suchow	7th c BC?	710 (58e)		710 (58e)	373 (53)	160 (35e)	40 (00?e)			
Swatow	bef 1860	300 (68?e)	300 (68?e)	250 (58e)	280 (53)	190 (35e)	60 (05e)	45 (72e)		
Szeping	bef 1900	200 (63?e)	200 (63?e)	130 (58e)	126 (53)	68 (40)				
Taichow	1st c?	200 (58e)		200 (58e)	160 (53)	66 (35e)				
Taiyuan	3rd c BC	1053 (58e)		1053 (58e)	721 (53)	139 (34e)	230 (00?e)	250 (66e)		
Tangshan[1]	1878?	812 (58e)		812 (58e)	693 (53)	149 (35e)				
Tatung	10th c BC?	243 (58e)		243 (58e)	228 (53)	50 (22e)				
Tientsin	bef 13th c	6280 (74e)	4000 (68e)	3278 (58e)	2694 (53)	1068 (35e)	750 (05e)	300 (50e)		

[1] Devastated by earthquake in July 1976.

ASIA

Population (thousands)

City	Date Settled	Latest	ca 1970	ca 1960	ca 1950	ca 1930	ca 1900	ca 1850	ca 1800	18th c
Tsamkong	3rd c BC?	170 (58e)		170 (58e)	166 (53)	230s (26)	177s (06)			
Tsiaotso	bef 1953	250 (58e)		250 (58e)	35 (53)					
Tsinan	6th c	1200 (68?e)	1200 (68?e)	882 (58e)	680 (53)	437 (35e)	100 (00?e)			
Tsingtao	bef 1891	1144 (58e)		1144 (58e)	917 (53)	318 (28e)	35 (07)			
Tsitsihar	1691	704 (58e)		704 (58e)	345 (53)	133 (40)	25 (00?e)			
Tzekung	3rd c BC	400 (63?e)	400 (63?e)	280 (58e)	291 (53)	176 (38e)		60 (68e)		
Tzepo	1st c?	875 (58e)		875 (58e)	184 (53)					
Urumtsi		700 (66?e)	700 (66?e)	320 (58e)	141 (53)	60 (22e)	40 (00e)	150 (65e)		
Weifang	1st c?	190 (58e)		190 (58e)	149 (53)	83 (34e)	100 (00?e)	100 (69e)		
Wenchow	4th c?	210 (58e)		210 (58e)	202 (53)	203 (27e)	80 (05e)	500 (70e)		
Wuhan	3rd c	2226 (58e)		2226 (58e)	1427 (53)	1353 (35e)	1500 (00?e)	996 (50e)		
Wuhu	3rd c?	230 (56e)		230 (56e)	242 (53)	150 (33e)	137 (05e)			
Wusih	1st c?	616 (58e)		616 (58e)	582 (53)	272 (36e)	200 (00?e)			
Wutungkiao	bef 1912	140 (58e)		140 (58e)	199 (53)					
Yangchow	7th c BC?	160 (58e)		160 (58e)	180 (53)	138 (36e)	100 (00?e)	360 (68e)		
Yangchuan	bef 283 BC	200 (58e)		200 (58e)	177 (53)					
Yenan	1st c?	15 (62e)		15 (62e)	30 (50e)	30 (30e)				
Yinchwan	bef 2nd c BC	160 (63?e)	160 (63?e)	60 (56e)	84 (53)	85 (22e)	12 (00?e)			
Yingkow	1836	161 (58e)		161 (58e)	131 (53)	181 (40)	60 (00?e)	80 (64e)		
CYPRUS										
*Nicosia	12th c?	117s (74e)	116s (73)	46 (60)	34 (46)	24 (31)	15 (01)	12 (81)	12 (00?e)	
GAZA STRIP										
*Gaza	1468 BC?	118 (67)	118 (67)		38 (46e)	17 (31)	21 (00e)	3 (50?e)	5 (00e)	

378 Cities: Date of Settlement and Evolution of Population

ASIA

Population (thousands)

City	Date Settled	Latest	ca 1970	ca 1960	ca 1950	ca 1930	ca 1900	ca 1850	ca 1800	18th c
HONG KONG										
*Hong Kong	5th c?	4407 (76)	3948 (71)	3133 (61)	2015 (51e)	840 (31)	284 (01)	33 (50)		
INDIA										
Agra	12th c?	592 (71)	592 (71)	462 (61)	334 (51)	230 (31)	188 (01)	149 (72)	60 (13e)	
		635s (71)	635s (71)	509s (61)	376s (51)			125 (53)		
Ahmadabad	bef 1298	1586 (71)	1586 (71)	1150 (61)	788 (51)	314 (31)	186 (01)	120 (72)	100 (20e)	
		1742s (71)	1742s (71)	1206s (61)	794s (51)			97 (51)		
Ajmer	145?	263 (71)	263 (71)	231 (61)	197 (51)	120 (31)	74 (01)	35 (72)	23 (37e)	
		264s (71)	264s (71)							
Aligarh	12th c?	252 (71)	252 (71)	185 (61)	142 (51)	84 (31)	72 (01)	59 (72)		
Allahabad	240 BC?	491 (71)	491 (71)	413 (61)	312 (51)	184 (31)	172 (01)	144 (72)	20 (03e)	
		513s (71)	513s (71)	431s (61)	332s (51)			72 (53)		
Ambala	14th c?	186 (71)	186 (71)	182 (61)	152 (51)	85 (31)	79 (01)	51 (68)		
Amravati	bef 1756	194 (71)	194 (71)	138 (61)	103 (51)	47 (31)	34 (01)	23 (67)		
Amritsar	1574	408 (71)	408 (71)	376 (61)	326 (51)	265 (31)	162 (01)	134 (68)	80 (00e)	
		458s (71)	458s (71)	398s (61)						
Asansol	bef 1881	156 (71)	156 (71)	103 (61)	76 (51)	31 (31)	15 (01)			
		242s (71)	242s (71)	169s (61)	95s (51)					
Bangalore	1537	1541 (71)	1541 (71)	905 (61)	779 (51)	306 (31)	69 (01)	61 (71)	60 (05e)	
		1654s (71)	1654s (71)	1207s (61)			159s (01)	143s (71)		
Bareilly	1537?	296 (71)	296 (71)	260 (61)	195 (51)	144 (31)	131 (01)	103 (72)	66 (22)	
		326s (71)	326s (71)	273s (61)	208s (51)			111 (53)		
Belgaum	bef 1160	192 (71)	192 (71)	128 (61)	104 (51)	41 (31)	26 (01)	32 (72)	8 (20)	
		214s (71)	214s (71)	147s (61)	120s (51)					
Bhagalpur	bef 7th c	172 (71)	172 (71)	144 (61)	115 (51)	84 (31)	76 (01)	70 (72)	30 (10e)	
Bhatpara	bef 1815	205 (71)	205 (71)	148 (61)	135 (51)	85 (31)	22 (01)	10 (81)		
Bhavnagar	1723	225 (71)	225 (71)	176 (61)	138 (51)	76 (31)	56 (01)	36 (72)		
		226s (71)	226s (71)							

ASIA

City	Date Settled	Latest	ca 1970	ca 1960	ca 1950	ca 1930	ca 1900	ca 1850	ca 1800	18th c
Bhilainagar-Durg	10th c?	225 (71)	225 (71)	135 (61)	20 (51)	13 (31)	4 (01)	4 (81)		
		245s (71)	245s (71)							
Bhopal	11th c?	298 (71)	298 (71)	185 (61)	102 (51)	61 (31)	77 (01)	55 (81)		
		385s (71)	385s (71)	223s(61)						
Bikaner	1488	189 (71)	189 (71)	151 (61)	117 (51)	86 (31)	53 (01)	33 (81)		
		209s (71)	209s (71)							
Bombay	150?	5971 (71)	5971 (71)	4152 (61)	2839 (51)	1161 (31)	776 (01)	644 (72)	162 (16)	70 (1744e)
								566 (49)		10 (1698e)
Calcutta	1495?	3149 (71)	3149 (71)	2927 (61)	2549 (51)	1197 (31)	848 (01)	448 (72)	413 (50)	117 (1752e)
		7031s(71)	7031s(71)	4405s(61)	4578s(51)	1486s (31)	949s (01)	795s (72)	230 (37)	12 (1700e)
Calicut	7th c	334 (71)	334 (71)	193 (61)	159 (51)	99 (31)	77 (01)	48 (71)		
				249s (61)						
Chandigarh	1950	219 (71)	219 (71)	89 (61)	5 (51)					
		233s (71)	233s (71)	120s (61)						
Cochin	bef 4th c	439 (71)	439 (71)	35 (61)	26 (51)	23 (31)	19 (01)	14 (71)		
				313s(61)	193s(51)					
Coimbatore	bef 1768	356 (71)	356 (71)	286 (61)	198 (51)	95 (31)	53 (01)	35 (71)		
		736s(71)	736s(71)							
Cuttack	10th c?	194 (71)	194 (71)	146 (61)	103 (51)	65 (31)	51 (01)	43 (72)	40 (22e)	
		206s(71)	206s(71)							
Dehra Dun	1699	166 (71)	166 (71)	127 (61)	116 (51)	40 (31)	28 (01)	7 (72)		
		203s(71)	203s(71)	156s(61)	144s(51)					
Delhi	993	3288 (71)	3288 (71)	2062 (61)	915 (51)	348 (31)	209 (01)	154 (68)	150 (20?e)	
		3647s(71)	3647s(71)	2359s(61)	1384s(51)	447s (31)		152 (53)		
Dhanbad	bef 1860	80 (71)	80 (71)	47 (61)	34 (51)	16 (31)	12 (21)			
		434s(71)	434s(71)	201s(61)	74s(51)					
Durgapur	1960?	207 (71)	207 (71)	42 (61)						
Gauhati	16th c?	124 (71)	124 (71)	101 (61)	44 (51)	22 (31)	12 (01)	11 (72)		
		200s(71)	200s(71)							

Population (thousands)

ASIA

Population (thousands)

City	Date Settled	Latest	ca 1970	ca 1960	ca 1950	ca 1930	ca 1900	ca 1850	ca 1800	18th c
Gaya	545 BC?	180 (71)	180 (71)	151 (61)	134 (51)	88 (31)	71 (01)	67 (72)		
Gorakhpur	1400?	231 (71)	231 (71)	180 (61)	124 (51)	59 (31)	64 (01)	51 (72)		
								55 (53)		
Guntur	18th c	270 (71)	270 (71)	187 (61)	125 (51)	65 (31)	31 (01)	18 (71)		
Gwalior	525?	385 (71)	385 (71)	301 (61)	34 (51)	22 (31)	17 (01)			
		406s (71)	406s (71)		242s(51)	127s (31)	139s(01)	88s (81)		
Howrah	17th c?	738 (71)	738 (71)	513 (61)	434 (51)	225 (31)	158 (01)	84 (72)		
Hubli-Dharwar	11th c?	379 (71)	379 (71)	248 (61)	196 (51)	132 (31)	91 (01)	65 (72)		
Hyderabad	1589	1607 (71)	1607 (71)	1119 (61)	860 (51)	346 (31)	352 (01)	124 (81)	100 (00?e)	
		1796s(71)	1796s(71)	1251s(61)	1129s(51)	467s(31)	448s(01)	367s (81)		
Indore	1715	543 (71)	543 (71)	395 (61)	311 (51)	127 (31)	87 (01)	75 (81)	10 (18e)	
		561s(71)	561s(71)							
Jabalpur	1100?	426 (71)	426 (71)	295 (61)	204 (51)	124 (31)	90 (01)	55 (72)		
		535s(71)	535s(71)	367s(61)	357s(51)					
Jaipur	1728	615 (71)	615 (71)	403 (61)	291 (51)	144 (31)	160 (01)	138 (70)	72 (00e)	
		637s(71)	637s(71)							
Jamnagar	1540	200 (71)	200 (71)	140 (61)	103 (51)	55 (31)	54 (01)	35 (72)		
		228s(71)	228s(71)	149s(61)	104s(51)					
Jamshedpur	1908?	342 (71)	342 (71)	304 (61)	194 (51)	84 (31)	6 (11)			
		456s(71)	456s(71)	328s(61)	218s(51)					
Jhansi	1553?	173 (71)	173 (71)	140 (61)	127 (51)	77 (31)	56 (01)	30 (72e)		
		198s(71)	198s(71)	170s(61)						
Jodhpur	1459	318 (71)	318 (71)	225 (61)	181 (51)	95 (31)	79 (01)	63 (81)		
								60 (50e)		
Jullundur	100?	296 (71)	296 (71)	223 (61)	169 (51)	89 (31)	68 (01)	50 (68)		
				265s(61)						
Kanpur	1778	1154 (71)	1154 (71)	895 (61)	636 (51)	244 (31)	197 (01)	123 (72)		
		1275s(71)	1275s(71)	971s(61)	705s(51)			118 (53)		
Kolhapur	2nd c?	259 (71)	259 (71)	187 (61)	137 (51)	70 (31)	54 (01)	40 (72)		
		268s(71)	268s(71)	193s(61)						

ASIA

Population (thousands)

City	Date Settled	Latest	ca 1970	ca 1960	ca 1950	ca 1930	ca 1900	ca 1850	ca 1800	18th c
Kota	14th c?	213 (71)	213 (71)	120 (61)	65 (51)	38 (31)	34 (01)	40 (81)		
Lucknow	bef 1478	749 (71)	749 (71)	595 (61)	445 (51)	275 (31)	264 (01)	285 (69)	300 (00e)	
		814s(71)	814s(71)	656s(61)	497s(51)					
Ludhiana	1480	398 (71)	398 (71)	244 (61)	154 (51)	69 (31)	49 (01)	40 (68)		
		401s(71)	401s(71)							
Madras	1504	2469 (71)	2469 (71)	1729 (61)	1416 (51)	647 (31)	509 (01)	368 (71)	300s (94e)	
		3170s(71)	3170s(71)							
Madurai	5th c BC?	549 (71)	549 (71)	425 (61)	362 (51)	182 (31)	106 (01)	52 (71)	20 (12)	
		712s(71)	712s(71)					42 (51)		
Malegaon	1740	192 (71)	192 (71)	121 (61)	55 (51)	29 (31)	19 (01)	10 (72)		
Mangalore	6th c?	165 (71)	165 (71)	143 (61)	117 (51)	67 (31)	44 (01)	30 (71)	20 (00?e)	
		215s(71)	215s(71)	170s(61)						
Mathura	600 BC?	132 (71)	132 (71)	117 (61)	99 (51)	61 (31)	60 (01)	59 (72)		
		140s(71)	140s(71)	125s(61)	106s(51)			66 (53)		
Meerut	3rd c BC?	271 (71)	271 (71)	209 (61)	158 (51)	137 (31)	118 (01)	81 (72)		
		368s(71)	368s(71)	284s(61)	233s(51)			82 (53)		
Moradabad	1625	259 (71)	259 (71)	180 (61)	154 (51)	111 (31)	75 (01)	62 (72)		
		273s(71)	273s(71)	192s(61)	162s(51)			57 (53)		
Mysore	10th c?	356 (71)	356 (71)	254 (61)	244 (51)	107 (31)	68 (01)	58 (71)	80 (00?e)	
Nagpur	1700?	866 (71)	866 (71)	644 (61)	449 (51)	215 (31)	128 (01)	84 (72)		
		930s(71)	930s(71)	690s(61)				111 (54)		
Nasik	100 BC?	176 (71)	176 (71)	131 (61)	97 (51)	49 (31)	21 (01)	22 (72)	30 (20e)	
		272s(71)	272s(71)	216s(61)				22 (50)		
*New Delhi	993	302 (71)	302 (71)	262 (61)	276 (51)	65 (31)	31 (21)			
Patna	6th c BC?	473 (71)	473 (71)	365 (61)	250 (51)	160 (31)	135 (01)	159 (72)	266 (22)	
		491s(71)	491s(71)					284 (37)		
Poona	1604?	856 (71)	856 (71)	598 (61)	481 (51)	234 (31)	153 (01)	119 (72)	81 (22)	40 (1760?e)
		1135s(71)	1135s(71)	737s(61)	589s(51)			73 (51)		
Raipur	750?	175 (71)	175 (71)	140 (61)	90 (51)	45 (31)	32 (01)	19 (72)	5 (18e)	
		206s(71)	206s(71)							

ASIA

	Date		Population (thousands)								
City	Settled	Latest	ca 1970	ca 1960	ca 1950	ca 1930	ca 1900	ca 1850	ca 1800	18th c
Rajahmundry	11th c?	166 (71) 189s (71)	166 (71) 189s (71)	130 (61)	105 (51)	64 (31)	36 (01)	20 (71)		
Rajkot	17th c?	301 (71)	301 (71)	194 (61)	132 (51)	47 (31)	36 (01)	12 (72)		
Ranchi	bef 1834	176 (71)	176 (71)	122 (61)	94 (51)	51 (31)	26 (01)	12 (72)		
Saharanpur	1340?	256s (71) 225 (71)	256s (71) 225 (71)	140s (61) 185 (61)	107s (51) 143 (51)	79 (31)	66 (01)	44 (72) 32 (53)		
Salem	16th c?	309 (71) 416s (71)	309 (71) 416s (71)	249 (61)	202 (51)	102 (31)	71 (01)	50 (71)	19 (43)	
Sangli	10th c?	115 (71) 202s (71)	115 (71) 202s (71)	74 (61) 127s (61)	50 (51) 91s (51)	30 (31)	17 (01)	13 (72)		
Sholapur	1345?	398 (71)	398 (71)	338 (61)	266 (51)	145 (31)	75 (01)	53 (72)		
Simla	bef 1819	55 (71)	55 (71)	43 (61)	46 (51)	23 (31)	39 (01)	8 (68)		
South Dum Dum	bef 1756	174 (71)	174 (71)	111 (61)	61 (51)	18 (31)	11 (01)	10 (72)		
South Suburban	1495?	273 (71)	273 (71)	186 (61)	104 (51)	39 (31)	26 (01)			
Surat	bef 150	472 (71) 493s (71)	472 (71) 493s (71)	288 (61)	223 (51)	99 (31)	119 (01)	108 (72) 90 (51)	124 (16)	200 (1705e)
Thana	11th c?	171 (71) 207s (71)	171 (71) 207s (71)	101 (61) 109s (61)	62 (51) 68s (51)	22 (31)	16 (01)	14 (72)		
Tiruchirapalli	bef 200	307 (71) 465s (71)	307 (71) 465s (71)	250 (61)	219 (51)	143 (31)	105 (01)	77 (71)	80 (20e)	
Tirunelveli	7th c?	108 (71) 267s (71)	108 (71) 267s (71)	88 (61) 190s (61)	73 (51) 161s (51)	57 (31)	40 (01)	21 (71)		
Trivandrum	bef 1196	410 (71)	410 (71)	240 (61) 302s (61)	187 (51)	96 (31)	58 (01)	38 (81)		
Tuticorin	1540?	155 (71) 182s (71)	155 (71) 182s (71)	124 (61) 127s (61)	99 (51)	60 (31)	28 (01)	11 (71)		
Ujjain	bef 263 BC	203 (71) 209s (71)	203 (71) 209s (71)	144 (61)	130 (51)	54 (31)	40 (01)	33 (81)	100 (00?e)	
Ulhasnagar-Kalyan	2nd c?	268 (71) 396s (71)	268 (71) 396s (71)	181 (61) 194s (61)	140 (51)	26 (31)	11 (01)	13 (72)		

ASIA

Population (thousands)

City	Date Settled	Latest	ca 1970	ca 1960	ca 1950	ca 1930	ca 1900	ca 1850	ca 1800	18th c
Vadodara	812?	467 (71) / 467s (71)	467 (71) / 467s (71)	298 (61)	211 (51)	113 (31)	104 (01)	112 (72)	100 (18e)	
Varanasi	12th c BC?	584 (71) / 607s(71)	584 (71) / 607s(71)	485 (61) / 490s(61)	342 (51) / 356s(51)	205 (31)	209 (01)	175 (72) / 186 (53)	168 (10?e)	
Vellore	1274?	139 (71) / 179s (71)	139 (71) / 179s (71)	114 (61) / 123s(61)	106 (51)	57 (31)	44 (01)	38 (71)		
Vijayawada	17th c?	317 (71) / 345s (71)	317 (71) / 345s (71)	230 (61)	161 (51)	60 (31)	24 (01)	8 (71)		
Visakhapatnam	14th c?	353 (71) / 363s(71)	353 (71) / 363s (71)	182 (61)	108 (51)	57 (31)	41 (01)	32 (71)	20 (00?e)	
Warangal	12th c	208 (71)	208 (71)	156 (61)	133 (51)	62 (31)	5 (01)	3 (91)		
INDONESIA										
Ambon	1521	80 (71)	80 (71)	55 (61)	26 (55e)	17 (30)	8 (95)	9 (41)	6 (00?e)	
Banjarmasin	14th c	283 (71)	283 (71)	214 (61)	150 (51e)	66 (30)	45 (95)	30 (50?e)		
Bandung	1810	1200 (71)	1200 (71)	973 (61)	686 (52e)	167 (30)	27 (95)			
Bogor	1745	195 (71)	195 (71)	154 (61)	103 (51e)	65 (30)	25 (95)	5 (50?e)		
Cirebon	15th c?	179 (71)	179 (71)	158 (61)	106 (55e)	54 (30)	21 (96)	12 (50e)		
Denpasar	16th c?	98 (71)	98 (71)	57 (61)		17 (30)				
*Jakarta	5th c?	5490 (75e)	4579 (71)	2907 (61)	1861 (52)	435 (30)	115 (95)	54 (42)	14 (1750) / 95s (1750)	20 (1700) / 53s (1700)
Jambi	11th c?	158 (71)	158 (71)	113 (61)	63 (56e)	23 (30)	9 (05)			
Kediri	10th c?	179 (71)	179 (71)	159 (61)	137 (56e)	49 (30)	17 (95)	6 (50?e)		
Kupang	1613	49 (71)	49 (71)	30 (61)		7 (30)	0.3(95?)			
Malang	13th c?	422 (71)	422 (71)	341 (61)	265 (53e)	87 (30)	13 (95)			
Medan	bef 1869	636 (71)	636 (71)	479 (61)	250 (51e)	77 (30)	13 (95)			
Menado	17th c?	170 (71)	170 (71)	130 (61)	64 (54e)	28 (30)	9 (96)	2 (50?e)		
Padang	1637	196 (71)	196 (71)	144 (61)	109 (51e)	52 (30)	32 (95)	12 (50?e)		
Palembang	2nd c?	583 (71)	583 (71)	475 (61)	208 (51e)	108 (30)	54 (95)	42 (55)	27 (00e)	
Pontianak	1772?	218 (71)	218 (71)	150 (61)	121 (55e)	45 (30)	17 (95)	12 (50?e)		

ASIA

Population (thousands)

City	Date Settled	Latest	ca 1970	ca 1960	ca 1950	ca 1930	ca 1900	ca 1850	ca 1800	18th c
Semarang	17th c?	642 (71)	642 (71)	503 (61)	309 (51e)	218 (30)	83 (95)	28 (50?e)	25 (00e)	
Surabaya	14th c?	1552 (71)	1552 (71)	1008 (61)	926 (53e)	342 (30)	125 (95)	85 (57)	25 (15)	
Surakarta	1744	413 (71)	413 (71)	368 (61)	319 (51e)	165 (30)	105 (95)	100 (50?e)	105 (15e)	
Tanjungkarang-Telukbetung	bef 1866	198 (71)	198 (71)	134 (61)	89 (55e)	25 (30)	3 (96?)			
Ujung Pandang	15th c?	434 (71)	434 (71)	384 (61)	214 (51e)	85 (30)	17 (95)	20 (45e)	100 (00?e)	
Yogyakarta	1749	341 (71)	341 (71)	313 (61)	228 (51e)	137 (30)	58 (95)	43 (50?e)	100 (15e)	
IRAN										
Abadan	1047?	296 (76)	273 (66)	226 (56)	40 (40)	40 (33)				
Ahwaz	12th c?	329 (76)	206 (66)	120 (56)	46 (40)	32 (33)	2 (00?e)	2 (50?e)		
Hamadan	1100 BC?	156 (76)	124 (66)	100 (56)	104 (40)	100 (33)	30 (00e)	30 (50e)	40 (00?e)	
Isfahan	5th c BC?	672 (76)	424 (66)	255 (56)	205 (40)	100 (33)	86 (00e)	74 (82)	300 (00e)	600 (1673e)
Kermanshah	4th c	291 (76)	188 (66)	125 (56)	89 (40)	70 (33)	35 (00e)	30 (50e)	9 (00?e)	
Meshed	817?	670 (76)	410 (66)	242 (56)	167 (40)	139 (33)	62 (00e)	45 (50?e)	50 (00?e)	
Qom	7th c?	247 (76)	134 (66)	96 (56)	53 (40)	39 (33)	23 (00e)	10 (60e)		
Rasht	14th c?	187 (76)	144 (66)	109 (56)	122 (40)	90 (33)	35 (00?e)	25 (50?e)	10 (00?e)	
Shiraz	693?	416 (76)	270 (66)	171 (56)	129 (40)	120 (33)	31 (00e)	30 (50?e)	40 (00?e)	
Tabriz	3rd c?	599 (76)	403 (66)	290 (56)	214 (40)	220 (33)	162 (00e)	37 (50e)	55 (11e)	
*Tehran	12th c	4496 (76)	2720 (66)	1512 (56)	540 (40)	360 (33)	204 (00e)	70 (60e)	15 (97e)	
IRAQ										
*Baghdad	762?	1491 (65) 3206s (77)	1491 (65) 1657s (65)	361 (57) 1057s (57)	467 (47)	145 (20e)	156 (00e)	50 (53e)	96 (00?e)	14 (1652e)
Basra	636	311 (65) 311s (65)	311 (65) 311s (65)	165 (57)	102 (47)	50 (20e)	19 (00e)	60 (50?e)	40 (00?e)	
Karbala	680?	84 (65) 123s (65)	84 (65) 123s (65)	61 (57)	44 (47)	65 (20e)	65 (00?e)	20 (50?e)		
Kirkuk	3000 BC?	175 (65) 184s (65)	175 (65) 184s (65)	121 (57)	68 (47)	20 (20?e)	30s (00?e)	13 (50?e)	18 (00?e)	

ASIA

City	Date Settled	Population (thousands)								
		Latest	ca 1970	ca 1960	ca 1950	ca 1930	ca 1900	ca 1850	ca 1800	18th c
Mosul	636?	264 (65) 264s (65)	264 (65) 264s (65)	178 (57)	134 (47)	65 (20e)	60 (00e)	40 (50?e)	35 (00?e)	
Najaf	8th c	134 (65) 136s (65)	134 (65) 136s (65)	74 (56)	56 (47)	25 (20?e)				
ISRAEL										
Haifa	4th c?	227 (75e) 354s (74e)	220 (72) 335s (72)	183 (61) 260s (61)	147 (51e)	51 (31)	11 (00e)	3 (50?e)		
*Jerusalem	3000 BC?	355 (75e)	314 (72)	228 (61)	207 (51e)	91 (31)	55 (00e)	14 (50e)	19 (97e)	
Tel Aviv-Jaffa	bef 1472 BC	354 (75e) 1157s (74e)	364 (72) 1030s (72)	386 (61) 776s (61)	345 (51e)	97 (31)	21 (00e)	5 (50e)		
JAPAN										
Akashi	16th c?	235 (75)	207 (70)	130 (60)	66 (50)	39 (30)	21 (98)	14 (74)		
Akita	733	261 (75)	236 (70)	204 (60)	126 (50)	51 (30)	29 (98)	38 (74)		
Amagasaki	16th c	546 (75)	554 (70)	406 (60)	279 (50)	50 (30)	43 (98)	12 (74)	40 (00e)	39 (1747)
Aomori	16th c	264 (75)	240 (70)	202 (60)	106 (50)	77 (30)	28 (98)	11 (74)		
Asahikawa	1893	321 (75)	288 (70)	188 (60)	123 (50)	83 (30)	24 (03)			
Beppu		134 (75)	124 (70)	108 (60)	93 (50)	43 (30)	6 (98)			
Chiba	1126	659 (75)	482 (70)	242 (60)	134 (50)	49 (30)	26 (98)	3 (74)		
Fuji	17th c?	199 (75)	181 (70)	47 (60)	22 (50)					
Fujisawa	14th c?	266 (75)	229 (70)	125 (60)	85 (50)	25 (30)	6 (98)	6 (74)	4 (43)	
Fukui	1575?	231 (75)	201 (70)	150 (60)	101 (50)	64 (30)	44 (98)	40 (74)	40 (00e)	
Fukuoka	bef 1281	1002 (75)	853 (70)	647 (60)	393 (50)	228 (30)	66 (98)	21 (74)	25 (00e)	
Fukushima	1180?	247 (75)	227 (70)	139 (60)	93 (50)	46 (30)	21 (98)	6 (74)		
Fukuyama	1619	330 (75)	255 (70)	141 (60)	67 (50)	38 (30)	17 (98)	18 (74)		
Funabashi		423 (75)	325 (70)	135 (60)	83 (50)	23 (30)	10 (98)	9 (74)		
Gifu	1565?	409 (75)	386 (70)	304 (60)	212 (50)	90 (30)	32 (98)	11 (74)		
Hachinohe	bef 1664	224 (75)	209 (70)	174 (60)	104 (50)	53 (30)	11 (98)	10 (74)		
Hachioji	bef 1590	323 (75)	254 (70)	158 (60)	83 (50)	52 (30)	23 (98)	8 (74)		

ASIA

Population (thousands)

City	Date Settled	Latest	ca 1970	ca 1960	ca 1950	ca 1930	ca 1900	ca 1850	ca 1800	18th c
Hakodate	15th c?	307 (75)	242 (70)	243 (60)	229 (50)	197 (30)	78 (98)	113 (74)		
Hamamatsu	1505?	469 (75)	432 (70)	333 (60)	152 (50)	109 (30)	12 (98)	11 (74)	6 (43)	2 (1759)
Higashiosaka		525 (75)	500 (70)	213 (60)	150 (50)	24 (30)				
Himeji	10th c?	436 (75)	408 (70)	329 (60)	212 (50)	62 (30)	35 (98)	25 (74)	25 (00e)	
Hirakata		298 (75)	217 (70)	80 (60)	44 (50)	6 (30)	2 (98)			
Hiratsuka	17th c?	196 (75)	164 (70)	108 (60)	52 (50)	3 (30)	1 (98)	2 (43)		
Hiroshima	1559	853 (75)	542 (70)	431 (60)	286 (50)	270 (30)	122 (98)	74 (74)	50 (04)	75 (1633e)
Hitachi		202 (75)	193 (70)	161 (60)	56 (50)	28 (30)				
Ibaraki	16th c?	210 (75)	164 (70)	72 (60)	35 (50)		3 (98)			
Ichihara		194 (75)	156 (70)	14 (60)	5 (50)					
Ichikawa		319 (75)	261 (70)	157 (60)	103 (50)	21 (30)	2 (98)			
Ichinomiya	7th c	238 (75)	219 (70)	183 (60)	71 (50)	42 (30)	10 (98)	7 (74)		
Ise	4 BC	105 (75)	104 (70)	99 (60)	69 (50)	51 (30)	28 (98)	22 (74)	10 (00e)	
Iwaki	14th c?	330 (75)	327 (70)	58 (60)	27 (50)	10 (30)	6 (98)	3 (74)		
Kagoshima	764?	457 (75)	403 (70)	296 (60)	229 (50)	137 (30)	53 (98)	27 (74)	72 (26)	
Kamakura	7th c?	166 (75)	139 (70)	99 (60)	85 (50)	27 (30)	6 (98)	6 (74)		
Kanazawa	bef 1471	395 (75)	361 (70)	299 (60)	252 (50)	157 (30)	84 (98)	110 (74)	115 (95e)	65 (1710)
Kashiwa	17th c?	203 (75)	151 (70)	64 (60)	21 (50)					
Kasugai		214 (75)	162 (70)	77 (60)	48 (50)					
Kawagoe	830	225 (75)	171 (70)	108 (60)	53 (50)	34 (30)	19 (98)			
Kawaguchi	17th c?	346 (75)	306 (70)	170 (60)	125 (50)	22 (30)	3 (98)			
Kawasaki	1150?	1015 (75)	973 (70)	633 (60)	319 (50)	104 (30)	4 (98)	3 (74)	2 (43)	
Kitakyushu	1318	1058 (75)	1042 (70)	986 (60)	711 (50)	473 (30)	88 (98)			
Kobe	1160?	1361 (75)	1289 (70)	1114 (60)	765 (50)	788 (30)	216 (98)			
Kochi	1600?	281 (75)	240 (70)	196 (60)	162 (50)	97 (30)	37 (98)	11 (74)	25 (00e)	
Kofu	13th c?	194 (75)	183 (70)	161 (60)	122 (50)	79 (30)	38 (98)	40 (74)	25 (00e)	
Koriyama	17th c?	265 (75)	242 (70)	103 (60)	71 (50)	51 (30)	6 (98)	16 (74)		
Koshigaya		196 (75)	139 (70)	50 (60)	7 (50)		3 (98)	5 (74)		
Kumamoto	15th c	488 (75)	440 (70)	374 (60)	268 (50)	164 (30)	61 (98)	45 (74)		
Kurashiki	17th c?	393 (75)	340 (70)	125 (60)	53 (50)	30 (30)	7 (98)	6 (74)		

ASIA

Population (thousands)

City	Date Settled	Latest	ca 1970	ca 1960	ca 1950	ca 1930	ca 1900	ca 1850	ca 1800	18th c
Kure	bef 1887	243 (75)	235 (70)	210 (60)	188 (50)	190 (30)	11 (98)			
Kurume	bef 1600	204 (75)	194 (70)	155 (60)	101 (50)	83 (30)	29 (98)	20 (74)		
Kushiro	17th c?	207 (75)	192 (70)	151 (60)	93 (50)	52 (30)	3 (98)			
Kyoto	8th c	1461 (75)	1419 (70)	1285 (60)	1102 (50)	765 (30)	353 (98)	239 (74)	379 (98)	351 (1696)
Machida		255 (75)	203 (70)	71 (60)	21 (50)	7 (30)				
Maebashi	bef 1582	250 (75)	234 (70)	182 (60)	97 (50)	85 (30)	34 (98)	15 (74)		
Matsudo	17th c?	345 (75)	254 (70)	86 (60)	53 (50)	11 (30)	3 (98)			
Matsumoto	1504	186 (75)	163 (70)	149 (60)	86 (50)	72 (30)	31 (98)			
Matsuyama	1603	367 (75)	323 (70)	239 (60)	164 (50)	82 (30)	37 (98)	12 (74)		14 (1725)
Mito	12th c?	198 (75)	174 (70)	139 (60)	67 (50)	51 (30)	34 (98)			
Miyazaki		234 (75)	203 (70)	158 (60)	103 (50)	55 (30)	5 (98)	12 (74)		
Morioka	1596	216 (75)	196 (70)	157 (60)	118 (50)	62 (30)	33 (98)	21 (74)		
Nagano	7th c	307 (75)	285 (70)	161 (60)	101 (50)	74 (30)	31 (98)	7 (74)		
Nagasaki	12th c?	450 (75)	421 (70)	344 (60)	242 (50)	204 (30)	107 (98)	30 (74)	32 (89)	65 (1696)
Nagoya	1612?	2080 (75)	2036 (70)	1592 (60)	1031 (50)	907 (30)	244 (98)	125 (74)	100 (00e)	100 (1692e)
Naha	7th c?	295 (75)	276 (70)	223 (60)	76 (50)	61 (30)	35 (98)	15 (74)		
Nara	7th c?	258 (75)	208 (70)	135 (60)	78 (50)	53 (30)	31 (98)	21 (74)		
Neyagawa		254 (75)	207 (70)	46 (60)	30 (50)	3 (30)				
Niigata	bef 1869	423 (75)	384 (70)	315 (60)	221 (50)	125 (30)	53 (98)	33 (74)		
Nikko	766	26 (75)	29 (70)	33 (60)	29 (50)	20 (30)	3 (98)	3 (74)		
Nishinomiya	15th c?	401 (75)	377 (70)	263 (60)	127 (50)	39 (30)	12 (98)	9 (74)		
Numazu	1579	199 (75)	189 (70)	143 (60)	102 (50)	44 (30)	12 (98)	16 (74)	5 (43)	
Oita	13th c?	320 (75)	261 (70)	125 (60)	94 (50)	57 (30)	9 (98)	7 (74)		
Okayama	16th c?	513 (75)	375 (70)	261 (60)	163 (50)	139 (30)	58 (98)	32 (74)	56 (1764e)	55 (1667e)
Okazaki	1455	235 (75)	211 (70)	166 (60)	96 (50)	66 (30)	15 (98)	13 (74)	6 (43)	
Omiya	5th c BC?	328 (75)	269 (70)	170 (60)	100 (50)	29 (30)	2 (98)	3 (74)		
Osaka	7th c BC?	2779 (75)	2980 (70)	3012 (60)	1956 (50)	2454 (30)	821 (98)	272 (74)	376 (01)	404 (1749)
								317 (54)		280 (1625)
Otaru	bef 1874	184 (75)	192 (70)	199 (60)	178 (50)	145 (30)	57 (98)	4 (74)		
Otsu	2nd c?	191 (75)	172 (70)	114 (60)	85 (50)	34 (30)	34 (98)			

Figure 45. Tokyo is the largest urbanized area in Asia and third largest in the world. Shown here is the Shinjuku area, one of the most rapidly developing business centers. (Credit: Embassy of Japan, Washington.)

ASIA

City	Date Settled	Population (thousands)								
		Latest	ca 1970	ca 1960	ca 1950	ca 1930	ca 1900	ca 1850	ca 1800	18th c
Sagamihara		377 (75)	278 (70)	102 (60)	69 (50)					
Sakai	bef 1336	751 (75)	594 (70)	340 (60)	214 (50)	120 (30)	50 (98)	39 (74)	45 (13)	
Sapporo	1871	1241 (75)	1010 (70)	524 (60)	314 (50)	169 (30)	37 (98)	2 (74)		
Sasebo	bef 1890	251 (75)	248 (70)	262 (60)	194 (50)	133 (30)	37 (98)	10 (90)		
Sendai	1600	615 (75)	545 (70)	425 (60)	342 (50)	190 (30)	83 (98)	52 (74)	67 (1764)	
Shimizu	16th c?	243 (75)	235 (70)	143 (60)	88 (50)	56 (30)	5 (98)	4 (74)		
Shimonoseki	bef 1185	267 (75)	258 (70)	247 (60)	194 (50)	99 (30)	43 (98)	18 (74)		
Shizuoka	bef 1569	447 (75)	416 (70)	329 (60)	239 (50)	136 (30)	42 (98)	32 (74)		
Suita	1335	301 (75)	260 (70)	117 (60)	78 (50)	24 (30)	4 (98)	3 (74)		
Takamatsu	15th c	299 (75)	274 (70)	228 (60)	125 (50)	80 (30)	34 (98)	33 (74)		
Takasaki	16th c?	211 (75)	193 (70)	142 (60)	93 (50)	60 (30)	31 (98)	11 (74)	7 (01)	4 (1745)
Takatsuki	17th c?	331 (75)	231 (70)	79 (60)	43 (50)	6 (30)	3 (98)			
Tokorozawa		197 (75)	137 (70)	66 (60)	43 (50)					
Tokushima	bef 1585	239 (75)	223 (70)	183 (60)	121 (50)	91 (30)	62 (98)	49 (74)	40 (00e)	
*Tokyo	12th c	8647 (75)	8841 (70)	8310 (60)	5385 (50)	2071 (30)	1440 (98)	596 (74)	1000 (00e)	510 (1750)
		11,540s (75)	11,161s (70)	9124s (60)	6277s (50)	4971s (30)				
Toyama	1572?	290 (75)	269 (70)	207 (60)	154 (50)	75 (30)	60 (98)	45 (74)	27 (41)	
Toyohashi	15th c	285 (75)	259 (70)	216 (60)	146 (50)	99 (30)	22 (98)	8 (74)		
Toyonaka	bef 1681	398 (75)	368 (70)	199 (60)	86 (50)	16 (30)	4 (98)			
Toyota	bef 1818	249 (75)	197 (70)	47 (60)	32 (50)	14 (30)	6 (98)			
Urawa		331 (75)	269 (70)	169 (60)	115 (50)	25 (30)		2 (74)		
Utsunomiya	12th c?	344 (75)	301 (70)	239 (60)	107 (50)	81 (30)	32 (98)	15 (74)	6 (43)	
Wakayama	16th c	390 (75)	365 (70)	285 (60)	191 (50)	117 (30)	64 (98)	61 (74)		
Yamagata	764?	220 (75)	204 (70)	189 (60)	105 (50)	63 (30)	35 (98)	18 (74)	25 (00e)	
Yao	bef 1337	262 (75)	228 (70)	123 (60)	67 (50)	11 (30)	4 (98)	3 (74)		
Yokkaichi	bef 17th c	247 (75)	229 (70)	196 (60)	124 (50)	52 (30)	25 (98)	10 (74)	7 (43)	
Yokohama	bef 1854	2622 (75)	2238 (70)	1376 (60)	951 (50)	620 (30)	194 (98)	65 (74)		
Yokosuka	bef 1600	390 (75)	348 (70)	287 (60)	251 (50)	110 (30)	25 (98)	3 (74)		

ASIA

Population (thousands)

City	Date Settled	Latest	ca 1970	ca 1960	ca 1950	ca 1930	ca 1900	ca 1850	ca 1800	18th c
JORDAN										
*Amman	1878	598 (74e)	330 (67)	246 (61)	103 (52)	12 (30e)				
Zarqa	1878?	226 (74e)	121 (67)	96 (61)	3 (47e)					
KASHMIR-JAMMU										
*Jammu	12th c?	158 (71) 164s (71)	158 (71) 164s (71)	103 (61)	50 (41)	39 (31)	36 (01)	8 (44e)	150 (80?e)	
*Srinagar	6th c	415 (71) 423s (71)	415 (71) 423s (71)	285 (61) 295s (61)	210 (41)	174 (31)	123 (01)	133 (73)	175 (09e)	
KOREA: NORTH										
Chongjin		300 (74e)	265 (67e)	200 (61e)	184 (44)	33 (30)				
Hamhung	15th c?	150 (70e)	150 (70e)	200 (55e)	112 (44)	40 (30)	14 (08)			
Hungnam		200 (74e)	200 (74e)	174 (55e)	144 (44)	23 (30)				
Kaesong	bef 919	240 (74e)	238 (65e)	140 (60e)	89 (49)	49 (30)	56 (99)			
Kimchaek		265 (67e)	265 (67e)		68 (44)	11 (30)				
*Pyongyang	1122 BC?	1500 (74e)	840 (67e)	653 (60e)	343 (44)	137 (30)	74 (99)			
Sinuiju	1910?	165 (67e)	165 (67e)		118 (44)	44 (30)				
Wonsan	bef 1880	215 (67e)	215 (67e)		113 (44)	43 (30)	16 (08)			
KOREA: SOUTH										
Chongju	10th c?	193 (75)	141 (70)	92 (60)	65 (49)	20 (30?)	37 (99)			
Chonju	57 BC?	311 (75)	258 (70)	188 (60)	101 (49)	38 (30)	27 (99)			
Inchon	4th c?	800 (75)	634 (70)	401 (60)	266 (49)	64 (30)				
Kwangju	57 BC?	607 (75)	494 (70)	314 (60)	139 (49)	33 (30)	23 (99)			
Masan	13th c?	372 (75)	187 (70)	158 (60)	91 (49)	26 (30)	17 (99)			
Mokpo	bef 1897	193 (75)	174 (70)	130 (60)	111 (49)	32 (30)	9 (99)			

ASIA

City	Date Settled	Population (thousands)								
		Latest	ca 1970	ca 1960	ca 1950	ca 1930	ca 1900	ca 1850	ca 1800	18th c
Pusan	bef 1443	2454 (75)	1842 (70)	1164 (60)	474 (49)	130 (30)	17 (99)			
*Seoul	6 BC?	6889 (75)	5433 (70)	2445 (60)	1446 (49)	355 (30)	201 (99)	90 (50e)	190 (93)	
Songnam	bef 1970	272 (75)	165 (73)	village (70)						
Suwon	12th c?	224 (75)	167 (70)	91 (60)	53 (49)	14 (30?)				
Taegu	bef 8th c	1311 (75)	1064 (70)	677 (60)	314 (49)	101 (30)	45 (99)			
Taejon		507 (75)	407 (70)	229 (60)	127 (49)	21 (30)				
Ulsan		253 (75)	157 (70)	30 (60)	24 (49)	15 (30?)				
KUWAIT										
*Kuwait	17th c	78 (75) 275s(75)	80 (70) 218s(70)	97 (61) 152s(61)	80 (48e)	40 (30?e)	20 (92e)	30 (70e)		
LAOS										
*Vientiane	bef 1350	177 (73)	177 (73)	162s(62e)	10 (48e)	10 (31)				
LEBANON										
*Beirut	15th c BC?	800 (72e) 1172s(75e)	475 (70) 939s(70)	400 (58e)	201 (49e)	162 (31)	126 (00e)	12 (50?e)	7 (00?e)	
Tripoli	7th c BC	150 (72e)	157 (70)	115 (58e)	65 (49e)	37 (31)	29 (00e)	14 (50e)	16 (19e)	
MACAO										
*Macao	1st c?	241 (70)	241 (70)	161 (60)	167 (50)	157 (27)	64 (99)	30 (50)	34s (00?e)	
MALAYSIA										
George Town	1786	269 (70) 332s(70)	269 (70) 332s(70)	235 (57)	189 (47)	149 (31)	94 (01)	10 (50?e)		
Ipoh	19th c	248 (70) 257s(70)	248 (70) 257s(70)	126 (57)	81 (47)	54 (31)	13 (01)			
*Kuala Lumpur	1857	557 (75e) 708s(70)	452 (70) 708s(70)	316 (57)	176 (47)	111 (31)	47 (11)			

ASIA

City	Date Settled	Population (thousands)								
		Latest	ca 1970	ca 1960	ca 1950	ca 1930	ca 1900	ca 1850	ca 1800	18th c
Malacca	1402?	87 (70)	87 (70)	70 (57)	55 (47)	38 (31)	16 (01)	12 (32e)	12 (00?e)	
MONGOLIA										
*Ulan-Bator	1649	334 (76e)	267 (69)	218 (63)	70 (51e)	30 (30?e)	34 (00e)	30 (50?e)		
NEPAL										
*Kathmandu	723	150 (71)	150 (71)	123 (61)	107 (52)	109s (20)	60 (00e)	35 (50?e)	100 (00e)	
PAKISTAN										
Faisalabad	1892	822 (72)	822 (72)	425 (61)	179 (51)	43 (31)	9 (01)			
Gujranwala	630?	360 (72)	360 (72)	196 (61)	121 (51)	59 (31)	29 (01)	19 (68)		
Hyderabad	bef 8th c	628 (72)	628 (72)	435 (61)	242 (51)	102 (31)	69 (01)	43 (72)	15 (00?e)	
*Islamabad	1961	77 (72)	77 (72)							
Karachi	1729	3499 (72)	3499 (72)	1913 (61)	1065 (51)	248 (31)	109 (01)	57 (72) 22s (53)	15 (30e)	
Lahore	1st c?	2165 (72)	2165 (72)	1296 (61)	849 (51)	430 (31)	203 (01)	99 (68) 94 (55)		
Multan	7th c?	542 (72)	542 (72)	358 (61)	190 (51)	119 (31)	87 (01)	46 (68)	100 (09e)	
Peshawar	400?	268 (72)	268 (72)	219 (61)	151 (51)	122 (31)	95 (01)	81 (68) 53 (53)		
Rawalpindi	bef 1765	615 (72)	615 (72)	340 (61)	237 (51)	119 (31)	88 (01)	19 (68)		
Sargodha	1903	201 (72)	201 (72)	129 (61)	78 (51)	27 (31)	9 (11)			
Sialkot	600 BC?	204 (72)	204 (72)	164 (61)	168 (51)	101 (31)	58 (01)	46 (81)		
PHILIPPINES										
Bacolod	bef 1849	223 (75)	187 (70)	119 (60)	101 (48)	57 (39)	12 (03)	5 (77)		
Caloocan	bef 1815	393 (75)	274 (70)	146 (60)	58 (48)	39 (39)	6 (03)	8 (77)		
Cebu	bef 1521	408 (75)	347 (70)	251 (60)	168 (48)	147 (39)	31 (03)	15 (77)		
Davao	1849?	482 (75)	392¹ (70)	226 (60)	111 (48)	96 (39)	9 (03)	2 (77)		

¹ Urban: 148,000.

ASIA

City	Date Settled	Population (thousands)								
		Latest	ca 1970	ca 1960	ca 1950	ca 1930	ca 1900	ca 1850	ca 1800	18th c
Iloilo	bef 1569	227 (75)	210 (70)	151 (60)	110 (48)	90 (39)	19 (03)	7 (77) 4 (49)	2 (18)	
Makati	16th c?	332 (75)	265 (70)	115 (60)	41 (48)	34 (39)	3 (03)	4 (87)		
*Manila	14th c?	1454 (75) 4863s(75)	1331 (70) 4363s(70)	1139 (60)	984 (48)	623 (39)	204 (03)	169 (87) 130 (70)	85 (14e)	54 (1762e) 30 (1697e)
Pasay	bef 1629	187 (75)	206 (70)	133 (60)	89 (48)	55 (39)	7 (03)	8 (77)		
Pasig	bef 1582	211 (75)	156 (70)	62 (60)	35 (48)	28 (39)	11 (03)	16 (77)		
Quezon City	bef 1903	960 (75)	754 (70)	398 (60)	108 (48)	39 (39)	3 (03)			
Zamboanga	1719	262 (75)	200¹(70)	131 (60)	103 (48)	74 (39)	21 (03)	5 (77)	1 (00?e)	
QATAR										
*Doha	1868	130 (75e)	95 (71e)	45 (63e)	10 (50?e)					
SAUDI ARABIA										
*Jidda	17th c	561 (74)	400 (71e)	148 (62–63)	60 (50?e)	40 (30?e)	25 (00e)	22 (50?e)	5 (00?e)	
Mecca	2nd c?	367 (74)	270 (71e)	159 (62–63)	135 (50?e)	130 (30?e)	60 (00e)	35 (50e)	18 (00?e)	
Medina	135?	198 (74)	100 (71e)	72 (62–63)	40 (50?e)	30 (30?e)	44 (00e)	18 (50e)	6 (00?e)	
*Riyadh	1824?	667 (74)	400 (71e)	197 (62–63)	80 (50?e)	30 (30?e)	30 (00?e)	28 (62e)		
Taif	bef 7th c	205 (74)	60 (68e)	54 (62–63)	25 (50?e)	8 (30?e)	8 (00?e)			
SINGAPORE										
*Singapore	11th c?	2278 (76e)	2075 (70)	1446 (57)	680 (47) 938s(47)	446 (31) 560s (31)	193 (01) 229s (01)	26 (50) 54s (50)	0.2 (00?e)	

¹ Urban: 42,000.

ASIA

Population (thousands)

City	Date Settled	Latest	ca 1970	ca 1960	ca 1950	ca 1930	ca 1900	ca 1850	ca 1800	18th c
SRI LANKA										
*Colombo	543 BC?	592 (74e) 850s(74e)	562 (71)	511 (63)	426 (53)	284 (31)	155 (01)	100 (71)	50	(04e)
Kandy	13th c?	98 (74e)	94 (71)	68 (63)	57 (53)	37 (31)	26 (01)			
SYRIA										
Aleppo	bef 1000 BC	779 (75e)	639 (70)	425 (60)	363 (50e)	249 (31)	117 (00e)	100 (50?e)	200	(00?e)
*Damascus	bef 14th c BC	1042 (75e)	837 (70) 923s(70)	530 (60)	335 (50e)	229 (31)	165 (00e)	150 (50?e)	130	(00?e)
Hama	3rd c BC?	162 (75e)	137 (70)	97 (60)	96 (53e)	40 (31)	45 (00e)	40 (50?e)	85	(00?e)
Homs	1st c?	267 (75e)	215 (70)	137 (60)	127 (53e)	53 (31)	48 (00e)	27 (50e)	25	(00?e)
TAIWAN										
Chiayi	17th c	253 (76e)	239 (70)	152 (56)	86 (50)	58 (30)	20 (01)			
Hsinchu	bef 1731	230 (76e)	208 (70)	126 (56)	125 (50)	44 (30)	19 (01)			
Kaohsiung	bef 1864	1020 (76e)	828 (70)	365 (56)	268 (50)	63 (30)	14 (11)			
Keelung	1626?	343 (76e)	324 (70)	194 (56)	145 (50)	75 (30)	13 (01)			
Panchiao		282 (76e)	115 (70)	38 (55e)		5 (35)				
Pingtung		179 (76e)	165 (70)	103 (56)		46 (35)				
Sanchung		285 (76e)	236 (70)	66 (55e)						
Taichung	18th c	561 (76e)	448 (70)	247 (56)	200 (50)	54 (30)	8 (05)			
Tainan	1590	537 (76e)	475 (70)	284 (56)	221 (50)	95 (30)	48 (01)	90 (69e)		
*Taipei	1708	2089 (76e)	1710 (70)	737 (56)	503 (50)	230 (30)	79 (01)			
THAILAND										
*Bangkok	1782	4130 (75e)	1867 (70)	1300 (60)	605 (47)	685 (29)	587 (11)	160 (49)	35	(00?e)
TURKEY										
Adana	8th c BC?	475 (75)	347 (70)	232 (60)	118 (50)	73 (27)	47 (00e)	20 (50?e)	30	(00?e)
*Ankara	17th c BC?	1701 (75)	1236 (70)	650 (60)	289 (50)	75 (27)	32 (00e)	20 (50?e)	20	(00?e)

ASIA

City	Date Settled	Latest	ca 1970	ca 1960	ca 1950	ca 1930	ca 1900	ca 1850	ca 1800	18th c
Antioch	300 BC?	78 (75)	67 (70)	46 (60)	30 (50)	19 (27)	28 (00e)	20 (50e)	18 (00?e)	
Bursa	3rd c BC	346 (75)	276 (70)	154 (60)	104 (50)	62 (27)	76 (00e)	65 (50?e)	60 (02e)	
Diyarbakir	bef 230	170 (75)	150 (70)	80 (60)	45 (50)	32 (27)	36 (00e)	27 (56)	38 (10e)	40 (1750e)
Eskisehir	3rd c?	260 (75)	216 (70)	153 (60)	90 (50)	32 (27)	19 (00?e)	village		
Gaziantep	3650 BC?	301 (75)	228 (70)	124 (60)	72 (50)	40 (27)	43 (00?e)	20 (50?e)	20 (00?e)	
Izmir	11th c BC?	637 (75)	521 (70)	361 (60)	228 (50)	154 (27)	196 (00e)	140 (51e)	120 (14e)	27 (1701e)
Kayseri	4th c	207 (75)	161 (70)	103 (60)	65 (50)	39 (27)	72 (00e)	18 (49)	25 (00?e)	
Konya	2600 BC?	247 (75)	200 (70)	120 (60)	64 (50)	48 (27)	44 (00e)	25 (50?e)	30 (00?e)	
Samsun	6th c BC	168 (75)	134 (70)	88 (60)	44 (50)	30 (27)	11 (00e)	2 (50?e)	2 (00e)	
Trebizond	756 BC?	97 (75)	81 (70)	53 (60)	34 (50)	25 (27)	35 (05e)	27 (50?e)	15 (00?e)	20 (1694e)
UNITED ARAB EMIRATES										
*Abu Zaby	1761	95 (74e)	22 (68)	8 (60?e)	6 (50?e)	5 (30?e)	6 (00?e)			
Dubayy	1799?	100 (74e)	57 (68)	45 (61)	20 (50?e)		8 (00?e)			
USSR										
Aktyubinsk	1869	179 (76e)	150 (70)	97 (59)	49 (39)	21 (26)	3 (97)			
Alma-Ata	1854	851 (76e)	730 (70)	456 (59)	222 (39)	45 (26)	23 (97)	10 (67)		
Andizhan	9th c	220 (76e)	188 (70)	131 (59)	85 (39)	73 (26)	48 (97)			
Angarsk	1948	231 (76e)	203 (70)	135 (59)	10 (51e)					
Ashkhabad	1881	297 (76e)	253 (70)	170 (59)	127 (39)	52 (26)	19 (97)			
Barnaul	1738	514 (76e)	439 (70)	303 (59)	148 (39)	74 (26)	21 (97)	11 (56)	7 (82)	
Biysk	1709	209 (76e)	186 (70)	146 (59)	80 (39)	46 (26)	17 (97)	3 (56)	2 (11)	
Blagoveshchensk	1856	171 (76e)	128 (70)	94 (59)	58 (39)	61 (26)	33 (97)	3 (67)		
Bratsk	1631	195 (76e)	155 (70)	43 (59)	2 (56e)					
Bukhara	630?	144 (76e)	112 (70)	69 (59)	50 (39)	47 (26)	65 (00e)	70 (66e)	150 (32e)	
Chelyabinsk	1736	989 (76e)	875 (70)	689 (59)	273 (39)	59 (26)	20 (97)	4 (56)	2 (11)	
Chimkent	8th c?	296 (76e)	247 (70)	153 (59)	74 (39)	21 (26)	11 (97)	4 (67)		
Chita	1653	290 (76e)	241 (70)	172 (59)	121 (39)	62 (26)	12 (97)	0.9 (56)		

ASIA

Population (thousands)

City	Date Settled	Latest	ca 1970	ca 1960	ca 1950	ca 1930	ca 1900	ca 1850	ca 1800	18th c
Dushanbe	bef 1922	448 (76e)	374 (70)	227 (59)	83 (39)	6 (26)		1 (67)		
Dzhambul	18th c	246 (76e)	187 (70)	113 (59)	64 (39)	25 (26)	12 (97)			
Frunze	1825	498 (76e)	431 (70)	220 (59)	93 (39)	37 (26)	7 (97)			
Irkutsk	1669	519 (76e)	451 (70)	366 (59)	250 (39)	99 (26)	51 (97)	24 (56)	13 (11)	
Kamensk-Uralskiy	1682	185 (76e)	169 (70)	141 (59)	51 (39)	5 (26)	6 (97)	5 (67?e)		
Karaganda	1857	570 (76e)	523 (70)	383 (59)	154 (39)	0.1(26)		0.3 (60?)		
Kemerovo	1720	446 (76e)	385 (70)	289 (59)	137 (39)	22 (26)	6 (20)			
Khabarovsk	1858	513 (76e)	436 (70)	323 (59)	207 (39)	50 (26)	15 (97)	1 (67?e)		
Kokand	10th c?	152 (76e)	133 (70)	105 (59)	85 (39)	69 (26)	81 (97)	45 (50e)	60 (00?e)	
Komsomolsk-na-Amure	1858	246 (76e)	218 (70)	177 (59)	71 (39)	village				
Kopeysk	bef 1917	157 (76e)	156 (70)	162 (59)	60 (39)	9 (26)				
Krasnoyarsk	1628	758 (76e)	648 (70)	412 (59)	190 (39)	72 (26)	27 (97)	6 (56)	3 (11)	
Kurgan	1553	297 (76e)	244 (70)	146 (59)	53 (39)	28 (26)	10 (97)	3 (56)	0.7 (11)	
Kustanay	1883	151 (76e)	124 (70)	86 (59)	34 (39)	25 (26)	14 (97)			
Magnitogorsk	1929	393 (76e)	364 (70)	311 (59)	146 (39)					
Namangan	1610	217 (76e)	175 (70)	123 (59)	80 (39)	74 (26)	62 (97)			
Nizhniy Tagil	1725	396 (76e)	378 (70)	338 (59)	160 (39)	39 (26)	31 (97)	2 (63)		
Norilsk	1935	168 (76e)	135 (70)	118 (59)	14 (39)					
Novokuznetsk	1617	530 (76e)	499 (70)	382 (59)	166 (39)	4 (26)	3 (97)	2 (56)		
Novosibirsk	1893	1286 (76e)	1161 (70)	885 (59)	404 (39)	121 (26)	8 (97)			
Omsk	1716	1002 (76e)	821 (70)	581 (59)	289 (39)	162 (26)	37 (97)	16 (56)	5 (11)	
Osh	9th c?	155 (76e)	120 (70)	65 (59)	33 (39)	31 (26)	34 (97)			
Pavlodar	1720	247 (76e)	187 (70)	90 (59)	29 (39)	18 (26)	8 (97)	0.2 (63)		
Petropavlovsk	1752	196 (76e)	173 (70)	131 (59)	92 (39)	47 (26)	20 (97)	7 (56)		
Petropavlovsk-Kamchatskiy	1740	202 (76e)	154 (70)	86 (59)	35 (39)	2 (26)	0.4(97)	2 (56)		
Prokopyevsk	18th c	267 (76e)	274 (70)	282 (59)	107 (39)	11 (26)				
Rubtsovsk	1888	171 (76e)	145 (70)	111 (59)	38 (39)	16 (26)				

ASIA

City	Date Settled	Latest	ca 1970	ca 1960	ca 1950	ca 1930	ca 1900	ca 1850	ca 1800	18th c
Samarkand	4th c BC?	304 (76e)	267 (70)	196 (59)	136 (39)	105 (26)	55 (97)	18 (50e)	3 (11)	
Semipalatinsk	1718	277 (76e)	236 (70)	156 (59)	110 (39)	57 (26)	26 (97)	7 (56)	4 (11)	
Sverdlovsk	1721	1171 (76e)	1025 (70)	779 (59)	423 (39)	136 (26)	43 (97)	17 (56)	30 (00?e)	
Tashkent	7th c	1643 (76e)	1385 (70)	927 (59)	556 (39)	324 (26)	156 (97)	64 (67)		
Temirtau	1930	200 (76e)	166 (70)	77 (59)	5 (39)					
Tomsk	1604	413 (76e)	338 (70)	249 (59)	145 (39)	92 (26)	52 (97)	20 (56)	9 (11)	
Tselinograd	1830	217 (76e)	180 (70)	99 (59)	31 (39)	13 (26)	10 (97)	5 (62)		
Tyumen	1586	335 (76e)	269 (70)	150 (59)	79 (39)	50 (26)	30 (97)	11 (56)	9 (11)	
Ulan-Ude	1666	302 (76e)	254 (70)	174 (59)	126 (39)	29 (26)	8 (97)	3 (56)	2 (11)	
Uralsk	1617?	157 (76e)	134 (70)	99 (59)	67 (39)	36 (26)	36 (97)	11 (56)		
Ust-Kamenogorsk	1720	262 (76e)	230 (70)	150 (59)	20 (39)	14 (26)	9 (97)	3 (56)	2 (82)	
Vladivostok	1860	526 (76e)	441 (70)	291 (59)	206 (39)	108 (26)	29 (97)	0.5 (73)	3 (00?e)	
Yakutsk	1642	143 (76e)	108 (70)	74 (59)	53 (39)	11 (26)	7 (91)	6 (63)		
Yuzhno-Sakhalinsk	1881	131 (76e)	106 (70)	86 (59)	39 (40)	25 (30)				
Zlatoust	1754	195 (76e)	180 (70)	161 (59)	99 (39)	48 (26)	21 (97)	10 (61)		
VIETNAM										
Da Nang	17th c?	492 (73e)	428 (70e)	105 (60e)	51 (43)	6 (26)	7 (11)			
Haiphong	bef 1882	1191 (76e)	200 (66e)	182 (60)		70 (36)	16 (00)			
			650s (71e)	369s (60)	143s(48e)					
*Hanoi	599	1443 (76e)	1378 (74)	415 (60)		149 (36)	103 (00)	70 (50?e)	40 (00?e)	
				644s (60)	237s(48e)					
Ho Chi Minh City	bef 17th c	3460 (76e)	1761 (70e)	1400 (60e)	1179 (58e)	256 (36)	160 (00)	110 (50e)	35 (00?e)	
Hue	200 BC?	209 (73e)	209 (69e)	104 (60e)	96 (53e)	61 (21)	50 (00)	60 (50?e)	30 (22e)	
Nha Trang	3rd c?	216 (73e)	103 (69e)	49 (60e)	25 (53e)	15 (36)				
Qui Nhon	bef 1874	214 (73e)	117 (69e)	31 (60e)		10 (36)	8 (00?e)			
YEMEN: NORTH										
*Sana	bef 530	135 (75)	89 (72)	89 (60e)	28 (48e)	22 (30?e)	59 (00e)	40 (50?e)		

ASIA

City	Date Settled	Latest	ca 1970	ca 1960	ca 1950	ca 1930	ca 1900	ca 1850	ca 1800	18th c
					Population (thousands)					
YEMEN: SOUTH										
*Aden	3rd c BC?	132 (73)	132 (73)	99 (55)	57 (46)	32 (31)	23 (01?)	16 (42)		

EUROPE

City	Date Settled	Latest	ca 1970	ca 1960	ca 1950	ca 1930	ca 1900	ca 1850	ca 1800	18th c
					Population (thousands)					
ALBANIA										
*Tirane	1604	192 (75e)	171 (70e)	136 (60)	60 (45)	31 (30)	12 (05e)	10 (50?e)	2 (00?e)	
AUSTRIA										
Graz	881?	248 (71)	248 (71)	237 (61)	226 (51)	153 (34)	138 (00)	55 (51)	30 (10)	32 (1770e) 22 (1702e)
Innsbruck	1170	115 (71)	115 (71)	101 (61)	95 (51)	61 (34)	27 (00)	13 (51)	9 (08e)	10 (1761) 6 (1655)
Linz	1st c	203 (71)	203 (71)	196 (61)	185 (51)	109 (34)	59 (00)	27 (51)	16 (84)	10 (1754e)
Salzburg	7th c?	129 (71)	129 (71)	108 (61)	103 (51)	40 (34)	33 (00)	17 (51)	13 (11)	14 (1713)
*Vienna	4th c BC?	1604 (75e)	1615 (71)	1628 (61)	1766 (51)	1875 (34)	1675 (00)	431 (51)	231 (00)	175 (1754) 114 (1710)
BELGIUM										
Anderlecht	11th c?	101 (75e)	104 (70)	95 (61)	86 (47)	80 (30)	48 (00)	6 (46)	2 (02)	2 (1750)
Antwerp	660?	209 (75e)	225 (70)	253 (61)	263 (47)	284 (30)	273 (00)	88 (46)	54 (00)	43 (1755)
		662s(75e)	671s(70)	652s(61)	590s(47)	484s(30)	400s(00)			67 (1699)
Bruges	865	120 (74e)	117 (70)	52 (61) 103s(61)	53 (47) 91s(47)	51 (30)	52 (00)	49 (46)	31 (96)	28 (1738) 35 (1699)
*Brussels	6th c?	153 (75e) 1051s(75e)	161 (70) 1075s(70)	170 (61) 1023s(61)	185 (47) 956s(47)	200 (30) 869s(30)	184 (00) 626s(00)	124 (46) 212s (46)	66 (00) 70s (00)	54 (1755)
Charleroi	bef 1665	22 (75e) 209s(75e)	24 (70) 214s(70)	26 (61) 217s(61)	26 (47) 209s(47)	29 (30)	24 (00)	7 (46)	4 (00?)	1 (1695)

EUROPE

Population (thousands)

City	Date Settled	Latest	ca 1970	ca 1960	ca 1950	ca 1930	ca 1900	ca 1850	ca 1800	18th c
Ghent	7th c	143 (75e)	149 (70)	158 (61)	166 (47)	170 (30)	160 (00)	103 (46)	55 (00)	44 (1741)
La Louviere	1189	219s(75e)	225s(70)	229s(61)	228s(47)	219s (30)				
		23 (75e)	23 (70)	23 (61)	22 (47)	24 (30)	18 (00)	12 (80)		
Liege	558?	113s(75e)	113s(70)	111s(61)	108s(47)					
		139 (75e)	146 (70)	153 (61)	156 (47)	166 (30)	158 (00)	76 (46)	43 (00)	57 (1750e)
		433s(75e)	441s(70)	445s(61)	425s(47)	252s(30)				50 (1650e)
Louvain	9th c	31 (74e)	31 (70)	33 (61)	37 (47)	39 (30)	42 (00)	30 (46)	19 (00)	
Schaerbeek	282?	113 (75e)	119 (70)	117 (61)	124 (47)	119 (30)	64 (00)	6 (46)	1 (00)	
BULGARIA										
Burgas	18th c	144 (75)	106 (65)	73 (56)	44 (46)	31 (26)	12 (00)	6 (50?e)		
Pleven	6th c	108 (75)	79 (65)	58 (56)	39 (46)	29 (26)	19 (00)	15 (69e)		
Plovdiv	341 BC?	300 (75)	223 (65)	162 (56)	125 (46)	85 (26)	43 (00)	30 (50?e)	30 (17e)	
Ruse	17th c	160 (75)	129 (65)	83 (56)	54 (46)	46 (26)	33 (00)	30 (50?e)	24 (00?e)	
Sliven	1153?	90 (75)	68 (65)	46 (56)	36 (46)	29 (26)	25 (01)	21 (88)		
*Sofia	100?	966 (75)	801 (65)	608 (56)	435 (46)	213 (26)	68 (00)	43 (62e)	46 (00?e)	
			894s(65)	645s(56)						
Stara Zagora	6th c BC?	122 (75)	89 (65)	55 (56)	37 (46)	29 (26)	20 (00)	20 (60?e)	20 (00?e)	
Varna	6th c BC	252 (75)	180 (65)	120 (56)	78 (46)	61 (26)	33 (00)	18 (50?e)	16 (00?e)	
CZECHOSLOVAKIA										
Bratislava	892?	350 (76e)	284 (70)	242 (61)	193 (50)	124 (30)	62 (00)	42 (51)	21 (05)	26 (1773)
Brno	9th c	363 (76e)	336 (70)	314 (61)	285 (50)	264 (30)	109 (00)	47 (51)	23 (04)	15 (1754)
Havirov	1955?	93 (76e)	82 (70)	51 (61)						
Karlovy Vary	1349	61 (76e)	44 (70)	43 (61)	36 (50)	24 (30)	15 (00)	3 (51)	2 (02)	
Kosice	1235	181 (76e)	145 (70)	79 (61)	63 (50)	70 (30)	36 (00)	13 (51)	12 (00?e)	
Olomouc	11th c	98 (76e)	80 (70)	70 (61)	64 (50)	66 (30)	22 (00)	13 (45)	9 (08)	
Ostrava	1267	317 (76e)	279 (70)	234 (61)	189 (50)	125 (30)	30 (00)	4 (57)	0.8 (00?e)	
Plzen	976?	163 (76e)	148 (70)	138 (61)	124 (50)	114 (30)	68 (00)	11 (51)	5 (01)	
*Prague	9th c?	1176 (76e)	1078 (70)	1005 (61)	932 (50)	848 (30)	202 (00)	118 (51)	77 (00)	59 (1754)
										48 (1703)

EUROPE

Population (thousands)

City	Date Settled	Latest	ca 1970	ca 1960	ca 1950	ca 1930	ca 1900	ca 1850	ca 1800	18th c
DENMARK										
Alborg	11th c?		100 (70)	86 (60)	80 (50)	44 (30)	31 (01)	10 (60)	6 (01)	4 (1769)
		155s (75e)	143s (70)	96s (60)	88s (50)					4 (1672)
Arhus	10th c?		199 (70)	120 (60)	116 (50)	81 (30)	52 (01)	8 (50)	4 (01)	4 (1769)
		246s (75e)	224s (70)	177s (60)	151s (50)					3 (1672)
*Copenhagen	1043?	562 (75e)	623 (70)	721 (60)	768 (50)	617 (30)	378 (01)	130 (50)	101 (01)	80 (1769)
		1287s (75e)	1380s (70)	1262s (60)	1168s (50)	771s (30)	477s (01)			62 (1699e)
Frederiksberg	bef 1621	95 (75e)	102 (70)	114 (60)	119 (50)	106 (30)	14 (01)	2 (50)	1 (01)	1 (1769)
Odense	10th c		137 (70)	111 (60)	101 (50)	57 (30)	40 (01)	11 (51)	6 (01)	5 (1769)
		169s (75e)	155s (70)							4 (1672)
FINLAND										
Espoo	1458?	124 (76e)	97 (70)	57s (60)	25s (50)	8s (30)	6s (00)			
*Helsinki	1550	493 (76e)	510 (70)	453 (60)	369 (50)	244 (30)	91 (00)	21 (50)	9 (00)	
		861s (74e)		566s (60)	414s (50)					
Lahti	1878	95 (76e)	88 (70)	67 (60)	45 (50)	10 (30)	0.8 (00)			
Oulu	1375?	93 (76e)	85 (70)	58 (60)	38 (50)	24 (30)	15 (00)	6 (50)	3 (00)	
Tampere	1779	166 (76e)	155 (70)	127 (60)	101 (50)	56 (30)	36 (00)	3 (50)	0.5 (00)	
		236s (74e)								
Turku	12th c?	165 (76e)	152 (70)	124 (60)	102 (50)	67 (30)	38 (00)	17 (50)	10 (00)	
		233s (74e)								
Vantaa	bef 1960	122 (76e)	81 (70)	2 (60)						
FRANCE										
Aix-en-Provence	123 BC	111 (75)	90 (68)	68 (62)	54 (54)	38 (31)	29 (01)	27 (51)	24 (01)	46 (1745)
										24 (1720e)
Ajaccio	7th c	51 (75)	42 (68)	41 (62)	33 (54)	24 (31)	22 (01)	12 (51)	6 (01)	
Amiens	54 BC?	131 (75)	118 (68)	105 (62)	93 (54)	90 (31)	91 (01)	52 (51)	40 (01)	33 (1745)
		153s (75)	137s (68)	113s (62)	94s (54)					

EUROPE

Population (thousands)

City	Date Settled	Latest	ca 1970	ca 1960	ca 1950	ca 1930	ca 1900	ca 1850	ca 1800	18th c
Angers	3rd c?	138 (75)	129 (68)	115 (62)	102 (54)	86 (31)	82 (01)	47 (51)	33 (01)	23 (1745)
Angouleme	bef 1st c BC	181s (75)	163s (68)	134s (62)	43 (54)	37 (31)	38 (01)	21 (51)	15 (01)	
		98s (75)	92s (68)	75s (62)						
Annecy	9th c?	53 (75)	54 (68)	43 (62)	33 (54)	20 (31)	13 (01)	5 (57)	5 (01)	
		101s (75)	82s (68)	57s (62)						
Argenteuil	660?	103 (75)	90 (68)	82 (62)	63 (54)	71 (31)	17 (01)	5 (51)	5 (01)	
Avignon	2nd c BC?	91 (75)	86 (68)	73 (62)	63 (54)	57 (31)	47 (01)	36 (51)	21 (01)	27 (1759)
		154s (75)	139s (68)							
Bayonne	3rd c?	43 (75)	43 (68)	37 (62)	33 (54)	32 (31)	28 (01)	18 (51)	13 (01)	
		120s (75)	110s (68)							
Besancon	6th c BC?	120 (75)	113 (68)	96 (62)	73 (54)	60 (31)	55 (01)	41 (51)	30 (01)	12 (1745)
		124s (75)	116s (68)							
Bethune	11th c	27 (75)	27 (68)	23 (62)	22 (54)	20 (31)	12 (01)	8 (51)	5 (01)	
		145s (75)	145s (68)							
Biarritz	12th c	28 (75)	27 (68)	26 (62)	23 (54)	23 (31)	11 (01)	2 (51)		
Bordeaux	3rd c BC	223 (75)	267 (68)	250 (62)	258 (54)	263 (31)	257 (01)	131 (51)	91 (01)	63 (1745)
		594s (75)	555s (68)	462s (62)	416s (54)					
Boulogne-Billancourt	11th c	104 (75)	109 (68)	107 (62)	94 (54)	86 (31)	44 (01)	8 (51)	2 (01)	
Boulogne-sur-Mer	bef 1st c BC	48 (75)	49 (68)	49 (62)	42 (54)	52 (31)	50 (01)	31 (51)	11 (01)	
		95s (75)	93s (68)	90s (62)						
Brest	3rd c?	167 (75)	154 (68)	136 (62)	111 (54)	70 (31)	84 (01)	61 (51)	27 (01)	21 (1726)
		190s (75)	169s (68)							
Bruay-en-Artois	bef 1851	26 (75)	29 (68)	31 (62)	32 (54)	32 (31)	15 (01)	0.7 (51)		
		116s (75)	127s (68)	110s (62)						
Caen	9th c?	119 (75)	110 (68)	91 (62)	68 (54)	58 (31)	45 (01)	45 (51)	31 (01)	35 (1745)
		183s (75)	152s (68)	117s (62)	75s (54)					
Calais	7th c	79 (75)	75 (68)	70 (62)	60 (54)	70 (31)	60 (01)	11 (51)	7 (01?)	
		100s (75)	94s (68)	74s (62)						

EUROPE

Population (thousands)

City	Date Settled	Latest	ca 1970	ca 1960	ca 1950	ca 1930	ca 1900	ca 1850	ca 1800	18th c
Cannes	8th c BC?	71 (75)	67 (68)	58 (62)	50 (54)	47 (31)	30 (01)	6 (51)	3 (01)	
		255s(75)	213s(68)	74s(62)	62s(54)					
Clermont-Ferrand	20 BC?	157 (75)	149 (68)	128 (62)	113 (54)	103 (31)	53 (01)	34 (51)	24 (01)	9 (1745)
		229s(75)	205s(68)	160s(62)	138s(54)					
Denain	8th c	26 (75)	28 (68)	29 (62)	27 (54)	28 (31)	23 (01)	9 (51)	0.9 (01?)	
		126s(75)	127s(68)							
Dijon	bef 273	152 (75)	145 (68)	136 (62)	113 (54)	91 (31)	71 (01)	32 (51)	21 (01)	30 (1753)
		208s(75)	184s(68)	154s(62)	117s(54)					
Douai	7th c?	45 (75)	49 (68)	48 (62)	43 (54)	42 (31)	34 (01)	21 (51)	18 (01)	
		203s(75)	205s(68)	134s(62)	47s(54)					
Dunkerque	7th c?	83 (75)	28 (68)	28 (62)	21 (54)	32 (31)	39 (01)	29 (51)	21 (01)	8 (1726)
		165s(75)	143s(68)	122s(62)	88s(54)					
Grenoble	1st c BC?	166 (75)	162 (68)	157 (62)	116 (54)	91 (31)	69 (01)	31 (51)	23 (01)	23 (1726)
		392s(75)	332s(68)	233s(62)	147s(54)					
La Rochelle	1023?	75 (75)	73 (68)	67 (62)	59 (54)	45 (31)	26 (01)	17 (51)	18 (01)	
		95s(75)	88s(68)	75s(62)						
Le Havre	1517	218 (75)	200 (68)	184 (62)	140 (54)	165 (31)	130 (01)	29 (51)	16 (01)	7 (1726)
		265s(75)	247s(68)	223s(62)	173s(54)					
Le Mans	5th c BC?	152 (75)	143 (68)	132 (62)	112 (54)	77 (31)	63 (01)	27 (51)	17 (01)	16 (1745)
		188s(75)	166s(68)	142s(62)	113s(54)					
Lens	1096?	40 (75)	42 (68)	43 (62)	41 (54)	34 (31)	24 (01)	10 (51)	2 (01)	
		313s(75)	326s(68)	261s(62)						
Lille	1030?	172 (75)	191 (68)	193 (62)	195 (54)	202 (31)	211 (01)	76 (51)	55 (01)	53 (1745)
		931s(75)	881s(68)	431s(62)	359s(54)					45 (1677)
Limoges	bef 52 BC	144 (75)	133 (68)	118 (62)	106 (54)	93 (31)	84 (01)	42 (51)	21 (01)	18 (1745)
		165s(75)	148s(68)	120s(62)						
Lorient	1664	70 (75)	66 (68)	61 (62)	47 (54)	43 (31)	44 (01)	26 (51)	20 (01)	
		106s(75)	99s(68)	77s(62)						
Lyons	43 BC	457 (75)	528 (68)	529 (62)	471 (54)	580 (31)	459 (01)	177 (51)	109 (01)	109 (1745)
		1159s(75)	1075s(68)	886s(62)	650s(54)					

EUROPE

City	Date Settled	Population (thousands) Latest	ca 1970	ca 1960	ca 1950	ca 1930	ca 1900	ca 1850	ca 1800	18th c
Marseilles	600 BC?	909 (75)	889 (68)	778 (62)	661 (54)	801 (31)	491 (01)	195 (51)	111 (01)	88 (1745)
		1011s (75)	954s (68)	807s (62)						
Maubeuge	7th c	35 (75)	32 (68)	27 (62)	24 (54)	24 (31)	21 (01)	8 (51)	5 (01)	
		97s (75)	91s (68)	64s (62)						
Metz	1000 BC?	112 (75)	108 (68)	103 (62)	86 (54)	79 (31)	58 (00)	43 (51)	32 (01)	75 (1745)
		181s (75)	166s (68)	147s (62)	112s (54)					20 (1684)
Montbeliard	843?	30 (75)	24 (68)	22 (62)	17 (54)	13 (31)	10 (01)	6 (51)	4 (01)	
		132s (75)	115s (68)							
Montpellier	8th c?	191 (75)	162 (68)	119 (62)	98 (54)	87 (31)	76 (01)	46 (51)	34 (01)	35 (1745)
		210s (75)	171s (68)	124s (62)						
Montreuil	12th c?	97 (75)	96 (68)	92 (62)	76 (54)	70 (31)	32 (01)	4 (51)	4 (01)	
Mulhouse	717?	117 (75)	116 (68)	109 (62)	99 (54)	100 (31)	89 (00)	30 (51)	9 (06)	
		219s (75)	199s (68)	165s (62)	111s (54)					
Nancy	947?	108 (75)	123 (68)	129 (62)	125 (54)	121 (31)	103 (01)	45 (51)	30 (01)	29 (1745)
		279s (75)	258s (68)	209s (62)	176s (54)					14 (1709e)
Nanterre	5th c?	95 (75)	90 (68)	83 (62)	53 (54)	43 (31)	14 (01)	3 (51)	2 (01)	
Nantes	1st c BC?	257 (75)	260 (68)	240 (62)	223 (54)	187 (31)	133 (01)	96 (51)	74 (01)	39 (1745)
		440s (75)	394s (68)	328s (62)	242s (54)					
Nice	5th c BC?	344 (75)	322 (68)	293 (62)	244 (54)	220 (31)	105 (01)	34 (48)	18 (01)	16 (1758)
		441s (75)	393s (68)	310s (62)						
Nimes	2nd c BC?	128 (75)	123 (68)	100 (62)	89 (54)	89 (31)	81 (01)	54 (51)	39 (01)	18 (1722)
		130s (75)	125s (68)							
Orleans	273	106 (75)	96 (68)	84 (62)	76 (54)	72 (31)	67 (01)	47 (51)	36 (01)	37 (1745)
		205s (75)	168s (68)	126s (62)	96s (54)					
*Paris	3rd c BC?	2300 (75)	2591 (68)	2790 (62)	2850 (54)	2891 (31)	2714 (01)	1053 (51)	548 (01)	577 (1760e)
		8450s (75)	8197s (68)	7369s (62)	6436s (54)			1227s (51)		510 (1718e)
										412 (1637e)
Pau	12th c?	83 (75)	74 (68)	60 (62)	48 (54)	39 (31)	34 (01)	16 (51)	9 (01)	
		126s (75)	110s (68)							

Figure 46. Aerial view of Paris, the largest urbanized area in Europe, showing the Eiffel Tower and the Palais de Chaillot. (Credit: French Government Tourist Office.)

EUROPE

City	Date Settled	Population (thousands)								
		Latest	ca 1970	ca 1960	ca 1950	ca 1930	ca 1900	ca 1850	ca 1800	18th c
Perpignan	10th c	106 (75)	102 (68)	83 (62)	70 (54)	74 (31)	36 (01)	22 (51)	11 (01)	
		114s (75)	107s (68)							
Poitiers	3rd c?	81 (75)	71 (68)	62 (62)	53 (54)	42 (31)	40 (01)	29 (51)	18 (01)	
		93s (75)	80s (68)	69s (62)						
Rennes	1st c BC?	198 (75)	181 (68)	152 (62)	124 (54)	89 (31)	75 (01)	40 (51)	26 (01)	19 (1726)
		222s (75)	193s (68)	157s (62)						
Rheims	300 BC?	178 (75)	153 (68)	134 (62)	121 (54)	113 (31)	108 (01)	46 (51)	20 (01)	23 (1745)
		202s (75)	168s (68)	144s (62)	126s(54)					
Roubaix	9th c?	110[1] (75)	115[1] (68)	113 (62)	110 (54)	117 (31)	124 (01)	35 (51)	8 (01)	
				340s (62)	267s(54)					
Rouen	bef 4th c	115 (75)	120 (68)	121 (62)	117 (54)	123 (31)	116 (01)	100 (51)	87 (01)	92 (1745)
		389s (75)	370s (68)	325s (62)	246s(54)					60 (1694e)
Saint-Denis	3rd c?	96 (75)	99 (68)	95 (62)	81 (54)	82 (31)	61 (01)	16 (51)	4 (01)	
Saint-Etienne	1195?	220 (75)	213 (68)	201 (62)	182 (54)	191 (31)	147 (01)	56 (51)	16 (01)	16 (1726)
		337s (75)	331s (68)	290s (62)	185s(54)					
Saint-Nazaire	1st c BC?	69 (75)	63 (68)	58 (62)	39 (54)	40 (31)	36 (01)	5 (51)	3 (01)	
		119s (75)	111s (68)							
Strasbourg	12 BC	253 (75)	249 (68)	229 (62)	201 (54)	181 (31)	151 (00)	76 (51)	49 (01)	50 (1750)
		359s (75)	335s (68)	302s (62)	239s(54)					26 (1697)
Thionville	870?	43 (75)	37 (68)	32 (62)	23 (54)	17 (31)	10 (00)	5 (51)	5 (01)	
		142s (75)	136s (68)	125s (62)						
Toulon	9th c BC	182 (75)	175 (68)	162 (62)	141 (54)	133 (31)	102 (01)	69 (51)	20 (01)	26 (1720)
		382s (75)	340s (68)	221s (62)						
Toulouse	300 BC?	374 (75)	371 (68)	324 (62)	269 (54)	195 (31)	150 (01)	93 (51)	50 (01)	63 (1745)
		504s (75)	440s (68)	329s (62)						38 (1698)
Tourcoing	1080?	102 (75)	99 (68)	89 (62)	83 (54)	82 (31)	79 (01)	28 (51)	12 (01)	
Tours	2nd c?	141 (75)	128 (68)	93 (62)	84 (54)	79 (31)	65 (01)	11[2] (51)	22 (01)	46 (1745)
		235s (75)	202s (68)	151s (62)	117s(54)					

[1] Urbanized area population for 1968 and 1975 included in Lille urbanized area.
[2] But 25,000 in 1841 and 33,000 in 1856.

EUROPE

Population (thousands)

City	Date Settled	Latest	ca 1970	ca 1960	ca 1950	ca 1930	ca 1900	ca 1850	ca 1800	18th c
Troyes	1st c BC?	72 (75)	75 (68)	67 (62)	59 (54)	59 (31)	53 (01)	27 (51)	24 (01)	18 (1745)
		128s(75)	114s(68)	98s(62)	84s(54)					
Valence	123 BC	69 (75)	62 (68)	53 (62)	41 (54)	34 (31)	27 (01)	16 (51)	8 (01)	
		104s(75)	92s(68)	68s(62)						
Valenciennes	693?	42 (75)	47 (68)	45 (62)	43 (54)	42 (31)	31 (01)	23 (51)	17 (01)	16 (1745)
		224s(75)	224s(68)	172s(62)	65s(54)					17 (1699)
Versailles	1682	94 (75)	91 (68)	87 (62)	84 (54)	67 (31)	55 (01)	37 (51)	25 (01)	5 (1745)
Villeurbanne	18th c?	117 (75)	120 (68)	105 (62)	82 (54)	82 (31)	29 (01)	5 (51)	2 (92)	

GERMANY: EAST
*Berlin (see under Germany: West)

City	Date Settled	Latest	ca 1970	ca 1960	ca 1950	ca 1930	ca 1900	ca 1850	ca 1800	18th c
Brandenburg	928	93 (76e)	94 (71)	90 (64)	82 (50)	64 (33)	49 (00)	18 (49)	13 (00)	9 (1770)
										8 (1722)
Cottbus	1156?	101 (76e)	83 (71)	73 (64)	61 (50)	52 (33)	39 (00)	9 (49)	6 (02)	3 (1750)
Dessau	1213	101 (76e)	98 (71)	95 (64)	92 (50)	79 (33)	51 (00)	14 (52)	9 (02)	
Dresden	1206?	510 (76e)	502 (71)	504 (64)	494 (50)	642 (33)	396 (00)	94 (49)	62 (00)	63 (1755)
										21 (1699)
Erfurt	741?	205 (76e)	197 (71)	190 (64)	189 (50)	145 (33)	85 (00)	32 (49)	17 (02)	14 (1758)
										12 (1664)
Gera	995?	117 (76e)	112 (71)	107 (64)	99 (50)	84 (33)	46 (00)	13 (52)	7 (00)	2 (1647)
Gorlitz	1071?	83 (76e)	87 (71)	89 (64)	100 (50)	94 (33)	81 (00)	25 (58)	9 (15)	5 (1717e)
Gotha	775?	58 (76e)	57 (71)	57 (64)	57 (50)	48 (33)	35 (00)	15 (52)	11 (02)	12 (1760)
										7 (1715e)
Halle	806	234 (76e)	257 (71)	274 (64)	289 (50)	209 (33)	157 (00)	34 (49)	15 (00)	13 (1751)
Jena	9th c?	101 (76e)	88 (71)	84 (64)	80 (50)	58 (33)	21 (00)	7 (52)	4 (18)	4 (1784)
Karl-Marx-Stadt	1136	308 (76e)	299 (71)	293 (64)	293 (50)	351 (33)	207 (00)	31 (49)	11 (06)	5 (1700)
Leipzig	10th c?	565 (76e)	584 (71)	595 (64)	618 (50)	713 (33)	456 (00)	62 (49)	31 (00)	30 (1748)
										22 (1700)
Magdeburg	805?	279 (76e)	272 (71)	265 (64)	260 (50)	307 (33)	230 (00)	56 (49)	23 (00)	11 (1750)
										11 (1722)

EUROPE

Population (thousands)

City	Date Settled	Latest	ca 1970	ca 1960	ca 1950	ca 1930	ca 1900	ca 1850	ca 1800	18th c
Potsdam	10th c?	122 (76e)	111 (71)	110 (64)	118 (50)	74 (33)	60 (00)	40 (49)	18 (01)	15 (1750) / 1 (1713)
Rostock	1160?	217 (76e)	199 (71)	179 (64)	133 (50)	90 (33)	44 (00)	23 (50)	14 (03)	9 (1773)
Schwerin	1018?	110 (76e)	97 (71)	91 (64)	94 (50)	54 (33)	39 (00)	20 (50)	10 (03)	3 (1764)
Zwickau	11th c	122 (76e)	127 (71)	128 (64)	139 (50)	85 (33)	56 (00)	13 (49)	4 (00)	4 (1755) / 4 (1699)
GERMANY: WEST										
Aachen	1st c?	242 (76e)	173 (70)	170 (61)	130 (50)	163 (33)	135 (00)	51 (49)	24 (99)	
Augsburg	11 BC	248 (76e)	212 (70)	209 (61)	185 (50)	177 (33)	89 (00)	38 (49)	29 (07)	26 (1703)
Baden-Baden	214?	49 (76e)	37 (70)	40 (61)	37 (50)	30 (33)	16 (00)	8 (61)	2 (04)	
Bayreuth	1194	67 (76e)	64 (70)	62 (61)	59 (50)	37 (33)	29 (00)	14 (52)	11 (09)	5 (1715)
Bergisch Gladbach	1271?	99 (76e)	50 (70)	42 (61)	33 (50)	20 (33)	11 (00)	5 (52)	2 (97)	2 (1773)
Berlin	1230?	3073 (76e)	3208 (^{1})	3268 (^{2})	3336 (50)	4243 (33)	1889 (00)	424 (49)	172 (00)	113 (1750) / 29 (1700)
East Berlin		1106 (76e)	1086 (71)	1071 (64)	1189 (50)					
West Berlin		1967 (76e)	2122 (70)	2197 (61)	2147 (50)					
Bielefeld	1015?	315 (76e)	169 (70)	175 (61)	154 (50)	121 (33)	63 (00)	10 (49)	3 (89)	3 (1718)
Bochum	900?	413 (76e)	344 (70)	361 (61)	290 (50)	315 (33)	66 (00)	5 (49)	2 (00)	1 (1750)
*Bonn	10 BC?	283 (76e)	274 (70)	144 (61)	115 (50)	99 (33)	51 (00)	18 (49)	9 (01)	2 (1722)
Bottrop	900?	101 (76e)	107 (70)	111 (61)	93 (50)	86 (33)	25 (00)	5 (71)	2 (15)	
Bremen	787?	571 (76e)	582 (70)	565 (61)	445 (50)	323 (33)	163 (00)	53 (49)	36 (12)	
Bremerhaven	1827	143 (76e)	140 (70)	142 (61)	114 (50)	26 (33)	20 (00)	5 (55)	1 (34?)	
Brunswick	861	267 (76e)	224 (70)	246 (61)	224 (50)	157 (33)	128 (00)	39 (49)	28 (11)	
Cologne	38 BC?	1010 (76e)	848 (70)	809 (61)	595 (50)	757 (33)	373 (00)	95 (49)	44 (00)	45 (1754)
Darmstadt	11th c?	136 (76e)	141 (70)	136 (61)	95 (50)	93 (33)	72 (00)	30 (52)	9 (03)	9 (1772)

^{1} East Berlin, 1971; West Berlin, 1970.
^{2} East Berlin, 1964; West Berlin, 1961.

EUROPE

Population (thousands)

City	Date Settled	Latest	ca 1970	ca 1960	ca 1950	ca 1930	ca 1900	ca 1850	ca 1800	18th c
Dortmund	885?	628 (76e)	640 (70)	641 (61)	507 (50)	541 (33)	143 (00)	11 (49)	4 (09)	3 (1700e)
Duisburg	8th c	587 (76e)	455 (70)	503 (61)	411 (50)	440 (33)	93 (00)	9 (49)	4 (00)	3 (1756)
Dusseldorf	1159	658 (76e)	664 (70)	703 (61)	501 (50)	499 (33)	214 (00)	26 (49)	20 (00)	3 (1714); 9 (1746)
Erlangen	1002	101 (76e)	84 (70)	70 (61)	50 (50)	32 (33)	23 (00)	11 (52)	9 (08)	5 (1673)
Essen	852	674 (76e)	698 (70)	727 (61)	605 (50)	654 (33)	119 (00)	9 (49)	4 (00)	8 (1752)
Esslingen	777?	95 (76e)	87 (70)	83 (61)	71 (50)	43 (33)	27 (00)	13 (49)	7 (03)	2 (1698e)
Flensburg	12th c	93 (76e)	95 (70)	99 (61)	103 (50)	67 (33)	49 (00)	16 (47)	11 (03)	7 (1769)
Frankfurt am Main	1st c?	631 (76e)	670 (70)	683 (61)	532 (50)	556 (33)	289 (00)	59 (49)	40 (11)	23 (1700)
Freiburg im Breisgau	1120	174 (76e)	162 (70)	145 (61)	110 (50)	99 (33)	62 (00)	15 (46)	8 (00?e)	
Furth	793	101 (76e)	95 (70)	98 (61)	100 (50)	77 (33)	54 (00)	17 (52)	13 (03)	3 (1689)
Gelsenkirchen	1150?	320 (76e)	348 (70)	383 (61)	315 (50)	333 (33)	37 (00)	0.8 (52)	0.3 (98)	
Gottingen	953?	124 (76e)	110 (70)	81 (61)	79 (50)	47 (33)	30 (00)	11 (52)	9 (07)	1 (1765)
Hagen	14th c	227 (76e)	201 (70)	196 (61)	146 (50)	148 (33)	51 (00)	5 (49)	2 (97)	0.7 (1722)
Hamburg	811?	1707 (76e)	1794 (70)	1832 (61)	1606 (50)	1129 (33)	706 (00)	133 (50)	95 (11)	54 (1756); 50 (1650e)
Hamm	1226	172 (76e)	85 (70)	71 (61)	60 (50)	54 (33)	31 (00)	7 (43)	3 (98)	2 (1756)
Hannover	1163?	549 (76e)	524 (70)	573 (61)	444 (50)	444 (33)	236 (00)	50 (52)	23 (10)	3 (1719)
Heidelberg	1196?	130 (76e)	121 (70)	127 (61)	116 (50)	85 (33)	40 (00)	15 (52)	16 (18)	15 (1766); 11 (1689)
Heilbronn	741?	113 (76e)	102 (70)	89 (61)	65 (50)	60 (33)	38 (00)	12 (49)	6 (03)	8 (1725e); 6 (1769)
Herne	8th c?	189 (76e)	104 (70)	113 (61)	112 (50)	99 (33)	28 (00)	1 (47)	0.6 (09)	
Hildesheim	8th c?	104 (76e)	94 (70)	96 (61)	72 (50)	63 (33)	43 (00)	16 (52)	11 (02)	8 (1762e)
Ingolstadt	806?	88 (76e)	70 (70)	53 (61)	41 (50)	29 (33)	22 (00)	7 (52)	5 (04)	6 (1700e)

EUROPE

City	Date Settled	Population (thousands)								
		Latest	ca 1970	ca 1960	ca 1950	ca 1930	ca 1900	ca 1850	ca 1800	18th c
Iserlohn	1033?	96 (76e)	58 (70)	55 (61)	46 (50)	34 (33)	27 (00)	11 (43)	4 (98)	4 (1765)
Kaiserslautern	882?	100 (76e)	100 (70)	86 (61)	63 (50)	63 (33)	48 (00)	10 (49)	3 (02)	2 (1744), 0.9 (1698)
Karlsruhe	1715	278 (76e)	259 (70)	242 (61)	199 (50)	155 (33)	97 (00)	23 (49)	8 (00)	2 (1738)
Kassel	913?	203 (76e)	214 (70)	208 (61)	162 (50)	175 (33)	106 (00)	36 (49)	18 (95)	19 (1750), 12 (1723)
Kiel	10th c?	261 (76e)	272 (70)	273 (61)	254 (50)	218 (33)	108 (00)	12 (50)	7 (03)	4 (1750)
Koblenz	9 BC	118 (76e)	119 (70)	99 (61)	66 (50)	65 (33)	45 (00)	25 (49)	8 (00)	1 (1663)
Krefeld	1105?	227 (76e)	222 (70)	213 (61)	172 (50)	165 (33)	107 (00)	36 (49)	8 (04)	6 (1756), 1 (1650)
Leverkusen	1107?	165 (76e)	107 (70)	95 (61)	66 (50)	44 (33)	14 (00)	7 (61)	3 (97)	
Lubeck	1143	231 (76e)	239 (70)	235 (61)	238 (50)	129 (33)	82 (00)	26 (51)	25 (07)	19 (1750), 24 (1700)
Ludwigshafen	17th c	168 (76e)	176 (70)	169 (61)	124 (50)	107 (33)	62 (00)	2 (53)	0.1 (43)	
Mainz	5th c BC?	184 (76e)	172 (70)	134 (61)	88 (50)	143 (33)	84 (00)	41 (52)	21 (00)	27 (1771)
Mannheim	766?	311 (76e)	332 (70)	314 (61)	246 (50)	275 (33)	141 (00)	24 (52)	28 (18)	9 (1725e)
Marl	900?	92 (76e)	77 (70)	72 (61)	51 (50)	32 (33)	2 (00)	2 (43)	2 (11)	
Moers	9th c?	101 (76e)	53 (70)	47 (61)	35 (50)	29 (33)	6 (00)	3 (45)	2 (87)	
Monchengladbach	972	261 (76e)	151 (70)	152 (61)	125 (50)	127 (33)	58 (00)	4 (49)	1 (98)	1 (1756)
Mulheim an der Ruhr	11th c?	188 (76e)	191 (70)	186 (61)	150 (50)	133 (33)	38 (00)	11 (49)	5 (07)	1 (1722)
Munich	1158	1311 (76e)	1294 (70)	1085 (61)	832 (50)	735 (33)	500 (00)	94 (52), 127s (52)	40 (01), 49s (01)	29s (1722)
Munster	800?	264 (76e)	198 (70)	183 (61)	118 (50)	122 (33)	64 (00)	25 (49)	14 (95)	7 (1685)
Neuss	5th c BC?	148 (76e)	115 (70)	93 (61)	63 (50)	56 (33)	28 (00)	9 (49)	4 (98)	
Nuremberg	1040?	495 (76e)	474 (70)	455 (61)	362 (50)	410 (33)	261 (00)	51 (49)	25 (06)	30 (1750e)
Oberhausen	1315?	236 (76e)	247 (70)	257 (61)	203 (50)	192 (33)	42 (00)	6 (62)	0.2 (19?)	
Offenbach am Main	977?	114 (76e)	117 (70)	116 (61)	89 (50)	81 (33)	51 (00)	13 (52)	5 (00)	0.8 (1700)

EUROPE

Population (thousands)

City	Date Settled	Latest	ca 1970	ca 1960	ca 1950	ca 1930	ca 1900	ca 1850	ca 1800	18th c
Oldenburg in Oldenburg	12th c?	135 (76e)	131 (70)	125 (61)	123 (50)	67 (33)	27 (00)	7 (52)	4 (00)	5 (1648)
Osnabruck	772	161 (76e)	144 (70)	139 (61)	110 (50)	94 (33)	52 (00)	14 (52)	8 (05e)	3 (1763e)
Paderborn	777	105 (76e)	67 (70)	54 (61)	40 (50)	37 (33)	24 (00)	9 (46)	5 (10)	
Pforzheim	1067?	108 (76e)	90 (70)	82 (61)	54 (50)	80 (33)	43 (00)	9 (52)	5 (00)	
Recklinghausen	1179	122 (76e)	125 (70)	131 (61)	105 (50)	87 (33)	34 (00)	4 (49)	2 (02)	
Regensburg	5th c BC?	131 (76e)	130 (70)	125 (61)	117 (50)	81 (33)	45 (00)	26 (52)	23 (02e)	13 (1698)
Remscheid	1173?	132 (76e)	136 (70)	127 (61)	103 (50)	101 (33)	58 (00)	12 (49)	5 (04)	7 (1733)
Reutlingen	1090?	95 (76e)	80 (70)	67 (61)	46 (50)	33 (33)	21 (00)	13 (49)	8 (03)	13 (1650)
Saarbrucken	999?	204 (76e)	128 (70)	131 (61)	90 (46)	129 (30)	23 (00)	9 (49)	3 (02)	3 (1769)
Salzgitter	1504	117 (76e)	118 (70)	110 (61)	101 (50)	3 (33)	2 (05)	2 (49?)		
Siegen	9th c?	116 (76e)	57 (70)	49 (61)	39 (50)	33 (33)	22 (00)	6 (43)	4 (07)	2 (1651)
Solingen	965?	170 (76e)	176 (70)	170 (61)	148 (50)	140 (33)	45 (00)	7 (49)	3 (04)	
Stuttgart	1150?	594 (76e)	633 (70)	638 (61)	498 (50)	415 (33)	177 (00)	48 (49)	18 (98)	17 (1758)
Trier	15 BC	100 (76e)	104 (70)	87 (61)	76 (50)	77 (33)	43 (00)	19 (49)	9 (02)	8 (1730)
Ulm	854?	98 (76e)	93 (70)	93 (61)	71 (50)	62 (33)	43 (00)	21 (49)	13 (95)	15 (1750)
Velbert	875?	96 (76e)	55 (70)	52 (61)	41 (50)	30 (33)	17 (00)	6 (42)	4 (08)	
Wiesbaden	3rd c BC	249 (76e)	250 (70)	253 (61)	221 (50)	160 (33)	86 (00)	14 (50)	2 (00)	0.7(1699)
Wilhelmshaven	bef 1852	103 (76e)	103 (70)	100 (61)	101 (50)	28 (33)	23 (00)	0.2 (55)		
Witten	1214?	108 (76e)	97 (70)	96 (61)	76 (50)	73 (33)	34 (00)	5 (52)	0.9 (96)	0.6(1739)
Wolfsburg	700?	126 (76e)	89 (70)	65 (61)	25 (50)	7 (39)	0.3(00?)	0.3 (71?)		
Wuppertal	1070?	403 (76e)	418 (70)	421 (61)	363 (50)	409 (33)	299 (00)	75 (49)	35 (10)	2 (1610)
Wurzburg	7th c	112 (76e)	117 (70)	117 (61)	78 (50)	101 (33)	75 (00)	23 (49)	16 (05)	10 (1621)
GIBRALTAR *Gibraltar	711	30 (76e)	27 (70)	25 (61)	23 (51)	17 (31)	20 (01)	16 (44)	3 (87)	

EUROPE

City	Date Settled	Latest	ca 1970	ca 1960	ca 1950	ca 1930	ca 1900	ca 1850	ca 1800	18th c
					Population (thousands)					
GREECE										
*Athens	bef 13th c BC	867 (71)	867 (71)	628 (61)	555 (51)	396 (28)	129 (96)	26 (48)	12 (00e)	
		2540s (71)	2540s (71)	1853s (61)	1379s (51)	802s (28)				
Iraklion	832?	78 (71)	78 (71)	63 (61)	51 (51)	33 (28)	23 (00)	15 (50?e)	13 (00?e)	
		85s (71)	85s (71)	70s (61)	58s (51)					
Patras	5th c BC?	112 (71)	112 (71)	95 (61)	79 (51)	61 (28)	50 (96)	15 (48)	6 (00?e)	
		121s (71)	121s (71)	102s (61)	88s (51)					
Peristeri	1923	118 (71)	118 (71)	79 (61)	36 (51)	7 (28)				
Piraeus	5th c BC	187 (71)	187 (71)	184 (61)	186 (51)	193 (28)	51 (96)	5 (48)		
Salonika	315 BC	346 (71)	346 (71)	251 (61)	217 (51)	237 (28)	115 (00e)	70 (50?e)	70 (00?e)	
		557s (71)	557s (71)	378s (61)	297s (51)	251s (28)				
HUNGARY										
*Budapest	bef 19 AD	2071 (76e)	1945 (70)	1805 (60)	1058 (49)	1006 (30)	716 (00)	156 (51)	47 (87)	12 (1720)
					1571s (49)					
Debrecen	1211	187 (76e)	157 (70)	130 (60)	120 (49)	117 (30)	72 (00)	31 (51)	29 (87)	24 (1770)
Gyor	bef 50 AD	119 (76e)	87 (70)	71 (60)	57 (49)	51 (30)	28 (00)	16 (51)	13 (87)	4 (1711)
Miskolc	13th c	200 (76e)	172 (70)	144 (60)	104 (49)	64 (30)	41 (00)	16 (51)	14 (87)	
Pecs	1st c BC?	163 (76e)	145 (70)	115 (60)	78 (49)	62 (30)	42 (00)	16 (51)	9 (87)	
Szeged	1138?	170 (76e)	119 (70)	99 (60)	133 (49)	135 (30)	100 (00)	51 (51)	22 (87)	5 (1720)
Szekesfehervar	11th c	96 (76e)	72 (70)	56 (60)	42 (49)	41 (30)	32 (00)	15 (51)	12 (08)	11 (1777)
ICELAND										
*Reykjavik	874	85 (74e)	82 (70)	72 (60)	56 (50)	28 (30)	7 (01)	1 (50)	0.3 (01)	0.2(1703)
		99s (74e)	95s (70)	80s (60)	59s (50)	29s (30)	7s (01)	2s (50)	0.9s (01)	0.6s(1703)
IRELAND										
Cork	7th c	129 (71)	129 (71)	78 (61)	75 (51)	78 (26)	76 (01)	88 (51)	101 (21)	5 (1659)
		134s (71)	134s (71)	116s (61)	107s (51)					

EUROPE

Population (thousands)

City	Date Settled	Latest	ca 1970	ca 1960	ca 1950	ca 1930	ca 1900	ca 1850	ca 1800	18th c
*Dublin	836?	568 (71) 680s(71)	568 (71) 680s(71)	537 (61) 593s(61)	522 (51) 570s(51)	317 (26) 419s (26)	291 (01)	262 (51)	182 (98)	129 (1753) 64 (1682) 9 (1659)
Limerick	812?	57 (71)	57 (71)	51 (61)	51 (51)	39 (26)	38 (01)	53 (51)	39 (00?e)	3 (1659)
ITALY										
Alessandria	1168	103 (76e)	102 (71)	93 (61)	82 (51)	80 (31)	72 (01)	52 (61) 42 (48)	19 (1774)	22 (1741)
Ancona	390 BC?	108 (76e)	110 (71)	100 (61)	86 (51)	83 (31)	55 (01)	47 (61) 23 (46)	17 (00?)	10 (1736) 9 (1701)
Arezzo	4th c BC?	91 (76e)	87 (71)	75 (61)	67 (51)	57 (31)	44 (01)	37 (61)	8 (00?e)	7 (1745)
Bari	180 BC?	384 (76e)	357 (71)	312 (61)	268 (51)	170 (31)	78 (01)	34 (61)	18 (94)	
Bergamo	5th c BC?	128 (76e)	127 (71)	115 (61)	103 (51)	79 (31)	47 (01)	41 (61) 34 (51)	36 (85)	30 (1702e)
Bologna	510 BC?	486 (76e)	491 (71)	445 (61)	341 (51)	239 (31)	148 (01)	113 (61) 97 (53)	67 (00)	64 (1741) 63 (1701)
Bolzano	14 BC?	107 (76e)	106 (71)	89 (61)	71 (51)	37 (31)	14 (00)	8 (57)	8 (00?e)	
Brescia	6th c BC	215 (76e)	210 (71)	173 (61)	142 (51)	115 (31)	69 (01)	39 (61) 34 (51)	28 (85)	29 (1764)
Cagliari	540 BC	240 (76e)	223 (71)	184 (61)	139 (51)	101 (31)	53 (01)	32 (61) 27 (48)	19 (83)	20 (1751) 15 (1698)
Catania	8th c BC	399 (76e)	400 (71)	364 (61)	300 (51)	225 (31)	147 (01)	70 (61) 57 (50)	45 (98)	26 (1748) 16 (1714)
Catanzaro	9th c	92 (76e)	86 (71)	74 (61)	60 (51)	41 (31)	32 (01)	22 (61)	13 (93)	3 (1669)
Como	bef 196 BC	97 (76e)	98 (71)	82 (61)	70 (51)	51 (31)	38 (01)	23 (61) 18 (51)	15 (00)	11 (1750) 9 (1723)
Cosenza	bef 331 BC	102 (76e)	102 (71)	79 (61)	57 (51)	35 (31)	21 (01)	17 (61)	9 (94)	24 (1750)
Cremona	218 BC	82 (76e)	82 (71)	74 (61)	69 (51)	62 (31)	37 (01)	32 (61) 28 (51)	21 (00)	23 (1686)

EUROPE

Population (thousands)

City	Date Settled	Latest	ca 1970	ca 1960	ca 1950	ca 1930	ca 1900	ca 1850	ca 1800	18th c
Ferrara	753?	155 (76e)	154 (71)	153 (61)	134 (51)	116 (31)	87 (01)	69 (61) 68 (53)	26 (97)	28 (1740) 25 (1701)
Florence	59 BC?	465 (76e)	458 (71)	437 (61)	375 (51)	305 (31)	198 (01)	96 (61) 106 (49)	78 (00)	74 (1751) 64 (1688)
Foggia	1069?	153 (76e)	142 (71)	119 (61)	98 (51)	56 (31)	53 (01)	32 (61)	13 (94)	11 (1736)
Forli	188 BC?	110 (76e)	105 (71)	92 (61)	78 (51)	60 (31)	43 (01)	37 (61)	16 (06)	12 (1701)
Genoa	5th c BC?	801 (76e)	817 (71)	784 (61)	688 (51)	591 (31)	220 (01)	131 (61) 125 (48)	91 (99)	65 (1681)
La Spezia	12th c?	121 (76e)	125 (71)	122 (61)	112 (51)	115 (31)	66 (01)	12 (61)	4 (06)	
Leghorn	807?	178 (76e)	175 (71)	161 (61)	142 (51)	120 (31)	96 (01)	95 (61) 80 (55)	53 (00)	30 (1745)
Lucca	718 BC?	92 (76e)	91 (71)	88 (61)	88 (51)	80 (31)	73 (01)	66 (61)	17 (06)	21 (1744)
Mantua	5th c BC?	66 (76e)	66 (71)	62 (61)	54 (51)	42 (31)	30 (01)	30 (51)	22 (02)	24 (1751) 21 (1676)
Messina	8th c BC	265 (76e)	251 (71)	255 (61)	221 (51)	180 (31)	147 (01)	103 (61) 97 (50)	44 (98)	27 (1748) 40 (1713)
Milan	4th c BC	1705 (76e)	1732 (71)	1583 (61)	1274 (51)	962 (31)	490 (01)	241 (61) 193 (51)	135 (00)	124 (1750) 115 (1714)
Modena	5th c BC?	179 (76e)	171 (71)	139 (61)	111 (51)	90 (31)	63 (01)	53 (61)	27 (05)	18 (1751)
Monza	5th c BC?	121 (76e)	114 (71)	84 (61)	73 (51)	60 (31)	42 (01)	26 (61) 19 (51)	11 (05)	
Naples	600 BC?	1224 (76e)	1227 (71)	1183 (61)	1011 (51)	832 (31)	548 (01)	440 (61) 416 (50)	427 (96)	305 (1742) 216 (1707)
Novara	4th c BC?	102 (76e)	101 (71)	88 (61)	69 (51)	61 (31)	44 (01)	25 (61)	12 (1774)	10 (1752)
Padua	302 BC?	242 (76e)	232 (71)	198 (61)	168 (51)	127 (31)	81 (01)	54 (51)	32 (02)	41 (1766)
Palermo	8th c BC	673 (76e)	643 (71)	588 (61)	491 (51)	380 (31)	306 (01)	197 (61) 179 (50)	141 (98)	118 (1747) 128 (1625)
Parma	183 BC	178 (76e)	175 (71)	141 (61)	123 (51)	69 (31)	49 (01)	44 (61) 41 (54)	34 (99)	35 (1700e)

EUROPE

City	Date Settled	Population (thousands)								
		Latest	ca 1970	ca 1960	ca 1950	ca 1930	ca 1900	ca 1850	ca 1800	18th c
Pavia	3rd c BC?	88 (76e)	87 (71)	75 (61)	64 (51)	49 (31)	34 (01)	27 (61) 26 (51)	24 (00)	24 (1750)
Perugia	6th c BC?	137 (76e)	130 (71)	113 (61)	95 (51)	77 (31)	61 (01)	43 (61)	16 (00?e)	16 (1701)
Pescara	48 AD?	135 (76e)	122 (71)	87 (61)	65 (51)	44 (31)	7 (01)	4 (61)		
Piacenza	218 BC	109 (76e)	107 (71)	89 (61)	73 (51)	64 (31)	36 (01)	32 (61)	28 (20)	31 (1758)
Pisa	11th c BC?	103 (76e)	103 (71)	91 (61)	78 (51)	71 (31)	60 (01)	49 (61)	17 (00?e)	13 (1745)
Pistoia	bef 62 BC	95 (76e)	93 (71)	85 (61)	78 (51)	73 (31)	65 (01)	12 (61)	10 (00?e)	9 (1745) 8 (1672)
Prato	10th c	154 (76e)	143 (71)	111 (61)	78 (51)	67 (31)	51 (01)	35 (61)	10 (00?e)	7 (1745) 7 (1672)
Ravenna	8th c BC	138 (76e)	132 (71)	116 (61)	92 (51)	77 (31)	63 (01)	57 (61) 54 (53)	26 (05)	14s (1736) 15s (1701)
Reggio di Calabria	8th c BC	178 (76e)	166 (71)	153 (61)	141 (51)	124 (31)	45 (01)	27 (61)	18 (94)	
Reggio nell'Emilia	2nd c BC	130 (76e)	129 (71)	116 (61)	107 (51)	90 (31)	59 (01)	47 (61)	15 (05)	15 (1775)
Rimini	400 BC?	126 (76e)	118 (71)	93 (61)	77 (51)	63 (31)	44 (01)	33 (61)	17 (05)	10 (1736) 8 (1701)
*Rome	1000 BC?	2884 (76e)	2782 (71)	2188 (61)	1652 (51)	937 (31)	425 (01)	176 (53)	153 (00)	158 (1750) 149 (1700)
Salerno	197 BC	162 (76e)	155 (71)	117 (61)	91 (51)	61 (31)	42 (01)	27 (61) 19 (50)	9 (89)	
Sassari	bef 12th c	116 (76e)	107 (71)	90 (61)	70 (51)	52 (31)	38 (01)	26 (61)	16 (83)	14 (1751) 12 (1698)
Sesto San Giovanni	15th c?	99 (76e)	92 (71)	71 (61)	45 (51)	31 (31)	7 (01)	4 (61) 4 (51)		
Siena	5th c BC?	65 (76e)	66 (71)	61 (61)	53 (51)	46 (31)	27 (01)	23 (61) 20 (43)	16 (84)	16 (1717)
Syracuse	734 BC	121 (76e)	109 (71)	89 (61)	71 (51)	49 (31)	32 (01)	20 (61)	16 (98)	18 (1747) 17 (1713)

EUROPE

Population (thousands)

City	Date Settled	Latest	ca 1970	ca 1960	ca 1950	ca 1930	ca 1900	ca 1850	ca 1800	18th c
Taranto	708 BC?	244 (76e)	227 (71)	195 (61)	169 (51)	112 (31)	60 (01)	28 (61)	18 (00?e)	6 (1736)
Terni	672 BC?	113 (76e)	107 (71)	95 (61)	84 (51)	62 (31)	30 (01)	14 (61)	7 (00?e)	5 (1701)
Torre del Greco	12th c?	99 (76e)	92 (71)	78 (61)	64 (51)	50 (31)	35 (01)	9 (61)	16 (89)	
Trent	4th c BC?	98 (76e)	92 (71)	75 (61)	62 (51)	54 (31)	25 (01)	9 (51)	7 (00?e)	
Treviso	3rd c BC?	91 (76e)	90 (71)	75 (61)	63 (51)	52 (31)	33 (01)	18 (51)	11 (80)	10 (1766) 7 (1632)
Trieste	181 BC?	268 (76e)	272 (71)	273 (61)	273 (51)	250 (31)	133 (00)	64 (51)	24 (01)	
Turin	4th c BC?	1191 (76e)	1168 (71)	1026 (61)	719 (51)	591 (31)	330 (01)	173 (61) 143 (48)	78 (00)	69 (1750) 44 (1702)
Udine	983?	104 (76e)	101 (71)	86 (61)	73 (51)	64 (31)	37 (01)	23 (51)	15 (15)	
Venice	6th c	362 (76e)	363 (71)	347 (61)	317 (51)	250 (31)	148 (01)	123 (51)	134 (02)	149 (1761) 133 (1696)
Verona	bef 89 BC	271 (76e)	266 (71)	221 (61)	179 (51)	152 (31)	74 (01)	51 (51)	41 (95)	41 (1738)
Vicenza	1st c BC?	119 (76e)	117 (71)	98 (61)	80 (51)	64 (31)	44 (01)	30 (51)	29 (02)	28 (1766) 26 (1710)
LUXEMBOURG										
*Luxembourg	963	80 (76e)	76 (70)	72 (60)	62 (47)	54 (30)	21 (00)	12 (40)	9 (00e)	
MALTA										
*Valletta	1566	14 (75e)	16 (67)	18 (57)	19 (48)	23 (31)	23 (01)	25 (51)	24 (98)	11 (1632)
MONACO										
*Monaco	6th c BC?	24 (76e)	24 (70e)	22 (62)	20 (51)	25 (28)	15 (97)	8 (57)	6 (00?e)	
NETHERLANDS										
*Amsterdam	1275?	751 (76e) 987s (76e)	820 (71e) 1036s (71e)	865 (60) 911s (60)	804 (47) 838s (47)	757 (30)	511 (99)	224 (49)	217 (95)	241 (1748e) 105 (1622)
Apeldoorn	793?	134 (76e)	126 (71e)	104 (60)	63 (47)	60 (30)	26 (99)	10 (49)	2 (95)	

EUROPE

Population (thousands)

City	Date Settled	Latest	ca 1970	ca 1960	ca 1950	ca 1930	ca 1900	ca 1850	ca 1800	18th c
Arnhem	893?	126 (76e) 281s(76e)	132 (71e) 274s(71e)	125 (60) 152s(60)	97 (47) 114s(47)	78 (30)	57 (99)	19 (49)	10 (95)	
Breda	1198?	118 (76e) 151s(76e)	122 (71e) 150s(71e)	108 (60)	82 (47) 85s(47)	45 (30)	26 (99)	15 (49)	8 (95)	
Delft	1075?	86 (76e)	86 (71e)	74 (60)	62 (47)	51 (30)	32 (99)	18 (49)	14 (95)	
Dordrecht	1018	102 (76e) 187s(76e)	101 (71e) 172s(71e)	82 (60)	68 (47)	56 (30)	38 (99)	21 (49)	18 (95)	20 (1632)
Eindhoven	1232?	193 (76e) 358s(76e)	190 (71e) 341s(71e)	168 (60)	135 (47) 137s(47)	95 (30)	5 (99)	3 (49)	3 (30)	
Enschede	1118?	142 (76e) 239s(76e)	141 (71e) 233s(71e)	124 (60)	80 (47)	52 (30)	24 (99)	4 (49)	3 (30)	
Groningen	1006?	163 (76e) 202s(76e)	171 (71e) 205s(71e)	145 (60)	132 (47) 143s(47)	105 (30)	67 (99)	34 (49)	24 (95)	
Haarlem	960?	165 (76e) 232s(76e)	173 (71e) 239s(71e)	169 (60) 224s(60)	157 (47) 201s(47)	120 (30)	64 (99)	26 (49)	21 (95)	50 (1748e) 39 (1622)
*Hague	1242?	479 (76e) 682s(76e)	538 (71e) 711s(71e)	605 (60) 692s(60)	533 (47) 592s(47)	438 (30)	206 (99)	72 (49)	38 (95)	41 (1748e) 17 (1622)
Heerlen	2nd c?	71 (76e) 265s(76e)	75 (71e) 265s(71e)	72 (60)	57 (47)	47 (30)	6 (99)	5 (49)	3 (01)	
Hilversum	13th c	94 (76e) 110s(76e)	99 (71e) 115s(71e)	101 (60)	85 (47)	57 (30)	19 (99)	5 (49)	3 (95)	
Leiden	9th c?	100 (76e) 168s(76e)	100 (71e) 165s(71e)	97 (60) 117s(60)	87 (47) 102s(47)	71 (30)	54 (99)	36 (49)	31 (95)	63 (1748e) 45 (1622)
Maastricht	50 AD?	111 (76e) 146s(76e)	112 (71e) 144s(71e)	91 (60)	74 (47)	61 (30)	34 (99)	25 (49)	18 (95)	
Nijmegen	70 AD?	148 (76e) 214s(76e)	150 (71e) 206s(71e)	130 (60) 133s(60)	107 (47)	82 (30)	43 (99)	21 (49)	13 (95)	
Rotterdam	9th c?	615 (76e) 1031s(76e)	679 (71e) 1066s(71e)	729 (60) 827s(60)	646 (47) 716s(47)	587 (30)	319 (99)	90 (49)	53 (95)	56 (1748e) 20 (1622)

EUROPE

City	Date Settled	Population (thousands)								
		Latest	ca 1970	ca 1960	ca 1950	ca 1930	ca 1900	ca 1850	ca 1800	18th c
's Hertogenbosch	1134?	86 (76e)	81 (71e)	72 (60)	53 (47)	42 (30)	31 (99)	22 (49)	13 (95)	
		179s(76e)	169s(71e)							
Tilburg	709?	152 (76e)	154 (71e)	138 (60)	114 (47)	79 (30)	34 (99)	15 (49)	9 (95)	
		213s(76e)	206s(71e)							
Utrecht	630?	251 (76e)	278 (71e)	255 (60)	185 (47)	155 (30)	102 (99)	48 (49)	32 (95)	25 (1748e)
		464s(76e)	459s(71e)		223s(47)					
Velsen	10th c?	64 (76e)	68 (71e)	64 (60)	41 (47)	41 (30)	11 (99)	2 (49)		
		133s(76e)	138s(71e)							
Zaanstad	1398?	125 (76e)	66 (71e)	49 (60)	42 (47)	33 (30)	21 (99)	11 (49)	10 (95)	
		137s(76e)	131s(71e)							
NORWAY										
Bergen	1070	113 (70)	113 (70)	116 (60)	113 (50)	98 (30)	72 (00)	22 (45)	17 (01)	14 (1769)
		213s(76e)	206s(70)	156s(60)	144s(50)	118s (30)				
*Oslo	1048	463 (76e)	475 (70)	476 (60)	434 (50)	253 (30)	229 (00)	33 (45)	10 (01)	
		701s(70)	701s(70)	599s(60)	506s(50)	395s(30)				
Stavanger	9th c?	78 (70)	78 (70)	53 (60)	51 (50)	47 (30)	30 (00)	9 (45)	2 (01)	
		87s(76e)	84s(70)	67s(60)	75s(50)	61s(30)				
Trondheim	997	119 (70)	119 (70)	59 (60)	57 (50)	54 (30)	39 (00)	15 (45)	9 (01)	
		135s(76e)	125s(70)	97s(60)	78s(50)	66s(30)				
POLAND										
Bialystok	1310	201 (76e)	167 (70)	121 (60)	69 (50)	91 (31)	64 (97)	14 (58)	6 (11)	
Bielsko-Biala	13th c	124 (76e)	106 (70)	75 (60)	57 (50)	45 (31)	25 (00)	12 (51)	6 (00?)	
Bydgoszcz	13th c	330 (76e)	282 (70)	232 (60)	163 (50)	117 (31)	47 (00)	13 (49)	4 (00?)	
Bytom	9th c?	236 (76e)	187 (70)	183 (60)	174 (50)	101 (33)	51 (00)	6 (49)	2 (04)	1 (1755)
Chorzow	1136?	156 (76e)	152 (70)	147 (60)	129 (50)	81 (31)	58 (00)	4 (52)	2 (00)	
Czestochowa	1220?	203 (76e)	188 (70)	165 (60)	112 (50)	118 (31)	45 (97)	3 (50?e)	2 (00)	
Elblag	1237	99 (76e)	90 (70)	77 (60)	48 (50)	72 (31)	53 (00)	25 (58)	16 (97)	11 (1772)

EUROPE

Population (thousands)

City	Date Settled	Latest	ca 1970	ca 1960	ca 1950	ca 1930	ca 1900	ca 1850	ca 1800	18th c
Gdansk	980?	434 (76e)	366 (70)	287 (60)	195 (50)	256 (29)	141 (00)	64 (49)	41 (00)	48 (1745) / 50 (1705)
Gdynia	1253?	225 (76e)	191 (70)	148 (60)	103 (50)	30 (31)	0.9(10)			
Gliwice	1276?	200 (76e)	172 (70)	150 (60)	133 (50)	111 (33)	52 (00)	9 (49)	3 (00)	1 (1750)
Gorzow Wielkopolski	10th c	91 (76e)	75 (70)	59 (60)	33 (50)	46 (33)	34 (00)	12 (49)	6 (01)	4 (1750)
Jastrzebie Zdroj		96 (76e)	24 (70)	3 (60)	2 (50)					
Kalisz	160?	93 (76e)	78 (70)	71 (60)	55 (50)	55 (31)	22 (97)	14 (67)	6 (02)	
Katowice	1598?	349 (76e)	305 (70)	270 (60)	176 (50)	126 (31)	32 (00)	8 (71)	3 (06)	
Kielce	1084?	157 (76e)	127 (70)	90 (60)	61 (50)	58 (31)	23 (97)	5 (50?e)	2 (00?e)	
Krakow	700?	701 (76e)	589 (70)	481 (60)	344 (50)	219 (31)	91 (00)	50 (51)	25 (98)	30 (1700e)
Lodz	1423	810 (76e)	763 (70)	710 (60)	620 (50)	605 (31)	314 (00)	16 (50)	0.8 (06)	
Lublin	10th c?	282 (76e)	238 (70)	181 (60)	117 (50)	112 (31)	50 (97)	18 (58)	9 (87)	
Olsztyn	1348	122 (76e)	95 (70)	68 (60)	44 (50)	43 (33)	24 (00)	4 (49)	2 (02)	2 (1772)
Opole	10th c?	108 (76e)	87 (70)	63 (60)	38 (50)	45 (33)	30 (00)	8 (52)	3 (99)	1 (1751)
Plock	bef 10th c	91 (76e)	72 (70)	44 (60)	33 (50)	33 (31)	24 (97)	22 (67)	3 (97)	
Poznan	9th c	527 (76e)	472 (70)	408 (60)	321 (50)	245 (31)	117 (00)	44 (49)	12 (97)	
Radom	1154?	180 (76e)	159 (70)	130 (60)	80 (50)	78 (31)	29 (97)	6 (50?e)	0.4 (50?)	
Ruda Slaska	1303?	152 (76e)	143 (70)	132 (60)	110 (50)	24 (29)	12 (00)	0.4 (50?)		
Rybnik	10th c	103 (76e)	44 (70)	34 (60)	27 (50)	23 (31)	7 (00)	3 (43)	1 (00?)	
Rzeszow	12th c?	100 (76e)	83 (70)	62 (60)	28 (50)	27 (31)	13 (00)	6 (51)	3 (00?e)	
Sosnowiec	1279?	198 (76e)	145 (70)	132 (60)	96 (50)	109 (31)	57 (00)	12 (90)		
Szczecin	9th c	376 (76e)	338 (70)	269 (60)	179 (50)	271 (33)	211 (00)	47 (49)	18 (00)	13 (1750) / 11 (1709)
Tarnow	1105?	100 (76e)	86 (70)	71 (60)	37 (50)	45 (31)	29 (00)	7 (51)	4 (00?e)	
Torun	1231	158 (76e)	130 (70)	105 (60)	81 (50)	54 (31)	30 (00)	13 (49)	8 (02)	
Tychy	bef 1629	140 (76e)	71 (70)	50 (60)	13 (50)	6 (31)	5 (00)	3 (71)		
Walbrzych	12th c?	129 (76e)	125 (70)	117 (60)	94 (50)	47 (33)	15 (00)	4 (49)	2 (99)	
*Warsaw	10th c	1463 (76e)	1316 (70)	1139 (60)	804 (50)	1172 (31)	684 (00)	164 (51)	63 (99)	18 (1676e)
Wloclawek	10th c?	93 (76e)	78 (70)	63 (60)	52 (50)	56 (31)	23 (97)	9 (67)		

EUROPE

City	Date Settled	Population (thousands)								
		Latest	ca 1970	ca 1960	ca 1950	ca 1930	ca 1900	ca 1850	ca 1800	18th c
Wodzislaw Slaski	1257?	104 (76e)	26 (70)	9 (60)	6 (50)	5 (31)	3 (00)	2 (71)	1 (00?)	50 (1747)
Wroclaw	980?	584 (76e)	526 (70)	431 (60)	309 (50)	625 (33)	423 (00)	111 (49)	65 (00)	41 (1710)
Zabrze	13th c	204 (76e)	197 (70)	190 (60)	172 (50)	130 (33)	20 (00)	6 (71)	0.9 (19?)	
PORTUGAL										
Coimbra	2nd c BC?	56 (70)	56 (70)	46 (60)	43 (50)	27 (30)	18 (00)	18 (65)	12 (00?e)	12 (1732)
*Lisbon	2000 BC?	830 (75e) 1034s (70)	769 (70) 1034s (70)	817 (60) 1335s (60)	790 (50)	594 (30)	356 (00)	251 (58)	230 (02e)	126 (1626)
Ponta Delgada	1439?	20 (70)	20 (70)	24 (60)	23 (50)	15 (30)	18 (00)	17 (78)	8 (00?e)	41 (1766)
Porto	2000 BC?	336 (75e) 693s (70)	307 (70) 693s (70)	305 (60) 746s (60)	285 (50)	232 (30)	168 (00)	81 (58)	74 (02e)	16 (1706)
ROMANIA										
Arad	11th c	171 (77) 169s (75e)	126 (66) 137s (66)	106 (56)	87 (48)	77 (30)	54 (00)	20 (51)	8 (87)	
Bacau	1408?	127 (77) 135s (75e)	73 (66) 87s (66)	54 (56)	34 (48)	31 (30)	16 (99)	15 (73)		
Baia Mare	12th c	101 (77) 104s (73e)	63 (66) 109s (66)	36 (56)	21 (48)	14 (30)	11 (00)	6 (51)	4 (87)	
Braila	1368?	195 (77) 177s (75e)	139 (66) 144s (66)	102 (56)	96 (48)	68 (30)	56 (99)	9 (50?e)	30 (00?e)	
Brasov	1225	257 (77) 206s (75e)	163 (66) 263s (66)	124 (56)	83 (48)	59 (30)	35 (00)	24 (51)	18 (89)	
*Bucharest	630?	1807 (77) 1934s (77)	1366 (66) 1511s (66)	1178 (56) 1291s (56)	886 (48) 1042s (48)	639 (30)	276 (99)	122 (59)	32 (98)	
Buzau	1431?	98 (77) 78s (73e)	56 (66) 82s (66)	48 (56)	43 (48)	36 (30)	22 (99)	4 (50?e)		
Cluj	12th c	262 (77) 223s (75e)	186 (66) 223s (66)	155 (56)	118 (48)	101 (30)	47 (00)	19 (51)	15 (97)	

EUROPE

City	Date Settled	Latest	ca 1970	ca 1960	ca 1950	ca 1930	ca 1900	ca 1850	ca 1800	18th c
				Population (thousands)						
Constanta	7th c BC?	257 (77) / 268s(66)	150 (66) / 199s(66)	100 (56) / 126s(56)	79 (48)	59 (30)	15 (99)	5 (50?e)		
Craiova	15th c	222 (77) / 226s(75e)	149 (66) / 173s(66)	97 (56)	85 (48)	63 (30)	46 (99)	8 (50?e)	2 (00?e)	
Galati	1418?	239 (77) / 209s(75e)	151 (66) / 151s(66)	96 (56)	80 (48)	101 (30)	63 (99)	36 (39?e)	5 (00?e)	
Hunedoara	1267?	80 (77) / 89s(73e)	68 (66) / 101s(66)	36 (56)	7 (48)	5 (30)	4 (00)	2 (57)		
Iasi	1408?	265 (77) / 238s(75e)	161 (66) / 195s(66)	113 (56)	94 (48)	103 (30)	78 (99)	66 (59)	15 (90)	
Oradea	1080	171 (77) / 169s(75e)	123 (66) / 135s(66)	99 (56)	82 (48)	83 (30)	47 (00)	21 (51)	5 (87)	
Pitesti	16th c?	124 (77) / 126s(73e)	60 (66) / 79s(66)	38 (56)	29 (48)	20 (30)	16 (99)	7 (59)		
Ploiesti	16th c	199 (77) / 236s(75e)	147 (66) / 191s(66)	115 (56)	96 (48)	79 (30)	45 (99)	27 (59)	3 (25?e)	
Resita	1768	85 (77) / 72s(73e)	57 (66) / 121s(66)	41 (56)	25 (48)	20 (30)	12 (00)	1 (51?)		
Satu Mare	1006?	104 (77) / 86s(73e)	68 (66) / 68s(66)	52 (56)	47 (48)	50 (30)	27 (00)	13 (51)	10 (00?e)	
Sibiu	12th c	151 (77) / 150s(75e)	110 (66) / 110s(66)	90 (56)	61 (48)	49 (30)	26 (00)	16 (51)	14 (86)	
Timisoara	1212	269 (77) / 229s(75e)	174 (66) / 193s(66)	142 (56)	112 (48)	92 (30)	50 (00)	21 (51)	9 (87)	
Tirgu Mures	1332?	130 (77) / 127s(75e)	86 (66) / 105s(66)	65 (56)	47 (48)	39 (30)	18 (00)	9 (51)	5 (87)	
SPAIN										
Albacete	713	103 (75e)	93 (70)	74 (60)	72 (50)	42 (30)	22 (00)	17 (57)	6 (00?e)	
Alicante	713?	222 (75e)	185 (70)	122 (60)	104 (50)	73 (30)	50 (00)	17 (57)	13 (04)	

EUROPE

City	Date Settled	Latest	ca 1970	ca 1960	ca 1950	Population (thousands) ca 1930	ca 1900	ca 1850	ca 1800	18th c
Almeria	238 BC?	129 (75e)	115 (70)	87 (60)	76 (50)	54 (30)	47 (00)	23 (57)	7 (00?e)	
Badajoz	3rd c BC?	104 (75e)	102 (70)	96 (60)	79 (50)	44 (30)	31 (00)	22 (57)	10 (00?e)	
Badalona	3rd c BC?	163 (70)	163 (70)	92 (60)	62 (50)	44 (30)	19 (00)	10 (57)	2 (16)	
Baracaldo	12th c?	109 (70)	109 (70)	78 (60)	42 (50)	34 (30)	15 (00)	0.3 (57)		
Barcelona	200 BC?	1828 (75e)	1745 (70)	1558 (60)	1280 (50)	1006 (30)	533 (00)	160 (57)	115 (97)	70 (1759) 37 (1715e)
Bilbao	1300?	472 (75e)	410 (70)	298 (60)	229 (50)	162 (30)	83 (00)	18 (57)	11 (02)	
Burgos	884	141 (75e)	120 (70)	82 (60)	74 (50)	40 (30)	30 (00)	26 (57)	12 (22)	
Cadiz	1100 BC?	144 (75e)	136 (70)	118 (60)	100 (50)	76 (30)	69 (00)	62 (57)	66 (87)	
Cartagena	225 BC?	147 (70)	147 (70)	124 (60)	113 (50)	103 (30)	100 (00)	22 (57)	25 (00?e)	
Castellon de la Plana	1251	112 (75e)	94 (70)	62 (60)	53 (50)	37 (30)	30 (00)	20 (57)	13 (94e)	
Cordova	bef 8th c BC	254 (75e)	236 (70)	198 (60)	165 (50)	103 (30)	58 (00)	36 (57)	25 (00e)	
Elche	3rd c BC?	123 (70)	123 (70)	73 (60)	56 (50)	38 (30)	27 (00)	20 (57)	17 (00?e)	
Gijon	1st c BC	188 (70)	188 (70)	125 (60)	111 (50)	78 (30)	48 (00)	10 (57)	3 (00?e)	
Granada	5th c BC?	207 (75e)	190 (70)	157 (60)	154 (50)	118 (30)	76 (00)	62 (57)	70 (97)	
Hospitalet	987?	242 (70)	242 (70)	123 (60)	72 (50)	38 (30)	5 (00)	3 (57)		
Huelva	7th c BC?	108 (75e)	97 (70)	74 (60)	64 (50)	45 (30)	21 (00)	8 (57)	5 (00?e)	
Jerez de la Frontera	7th c BC?	150 (70)	150 (70)	131 (60)	108 (50)	72 (30)	63 (00)	39 (57)	8 (00?e)	
La Coruna	bef 12th c BC	194 (75e)	190 (70)	178 (60)	134 (50)	74 (30)	44 (00)	27 (57)	4 (00?e)	
Leon	70 AD	123 (75e)	105 (70)	73 (60)	60 (50)	29 (30)	16 (00)	10 (57)	6 (86?)	
Lerida	5th c BC?	106 (75e)	91 (70)	64 (60)	53 (50)	39 (30)	21 (00)	20 (57)	17 (00?e)	
Logrono	1st c?	97 (75e)	84 (70)	61 (60)	52 (50)	34 (30)	19 (00)	10 (57)	7 (00?e)	
*Madrid	931?	3634 (75e)	3146 (70)	2260 (60)	1618 (50)	953 (30)	540 (00)	271 (57)	168 (97)	160 (1750e) 75 (1650?e)
Malaga	12th c BC	411 (75e)	374 (70)	301 (60)	276 (50)	188 (30)	130 (00)	93 (57)	52 (05)	
Murcia	825	240 (75e)	244 (70)	250 (60)	218 (50)	159 (30)	112 (00)	27 (57)	40 (07)	
Oviedo	761	167 (75e)	154 (70)	127 (60)	106 (50)	75 (30)	48 (00)	14 (57)	7 (00?)	

EUROPE

City	Date Settled	Population (thousands)								
		Latest	ca 1970	ca 1960	ca 1950	ca 1930	ca 1900	ca 1850	ca 1800	18th c
Palma	123 BC	277 (75e)	234 (70)	159 (60)	137 (50)	88 (30)	64 (00)	40 (57)	30 (87)	39 (1700e)
Pamplona	1st c BC?	176 (75e)	147 (70)	98 (60)	72 (50)	42 (30)	29 (00)	23 (57)	14 (00?)	
Sabadell	13th c	159 (70)	159 (70)	105 (60)	59 (50)	46 (30)	23 (00)	14 (57)	5 (25?e)	
Salamanca	5th c BC?	144 (75e)	125 (70)	90 (60)	80 (50)	47 (30)	26 (00)	15 (57)	15 (00?e)	
San Sebastian	1014?	181 (75e)	166 (70)	135 (60)	114 (50)	78 (30)	38 (00)	9 (57)	5 (00?)	
Santa Coloma de Gramanet	11th c	107 (70)	107 (70)	33 (60)	15 (50)	13 (30)	2 (00)	1 (57)		
Santander	1068?	166 (75e)	150 (70)	118 (60)	102 (50)	85 (30)	55 (00)	25 (57)	4 (00?e)	
Saragossa	bef 25 BC	568 (75e)	480 (70)	326 (60)	264 (50)	174 (30)	99 (00)	56 (57)	43 (87)	
Seville	7th c BC?	601 (75e)	548 (70)	442 (60)	377 (50)	229 (30)	148 (00)	82 (57)	96 (87)	66 (1746)
Tarragona	6th c BC?	101 (75e)	78 (70)	49 (60)	39 (50)	31 (30)	23 (00)	18 (57)	7 (00?e)	
Tarrasa	3rd c BC?	139 (70)	139 (70)	92 (60)	59 (50)	40 (30)	16 (00)	9 (57)	3 (00e)	
Toledo	193 BC?	46 (75e)	44 (70)	41 (60)	40 (50)	27 (30)	23 (00)	15 (57)	22 (00?e)	5 (1752)
Valencia	137 BC	731 (75e)	654 (70)	505 (60)	509 (50)	320 (30)	214 (00)	87 (57)	61 (00)	
Valladolid	1074?	287 (75e)	236 (70)	152 (60)	124 (50)	91 (30)	69 (00)	40 (57)	20 (00?e)	
Vigo	7th c BC?	197 (70)	197 (70)	145 (60)	138 (50)	65 (30)	23 (00)	8 (57)	5 (00?e)	
Vitoria	581	180 (75e)	137 (70)	74 (60)	52 (50)	41 (30)	31 (00)	19 (57)	6 (00?e)	
SWEDEN										
Boras	1622	104 (76e)	75 (70)	67 (60)	58 (50)	38 (30)	16 (00)	3 (50)	2 (00)	
Eskilstuna	12th c?	92 (76e)	94 (70)	59 (60)	53 (50)	33 (30)	14 (00)	4 (50)	1 (00)	
Goteborg	1619	442 (76e)	452 (70)	405 (60)	354 (50)	244 (30)	131 (00)	26 (50)	13 (00)	
		692s (76e)	618s (70)	487s (60)	380s(50)					
Helsingborg	1085?	101 (76e)	101 (70)	77 (60)	72 (50)	56 (30)	25 (00)	4 (50)	2 (00)	
Jonkoping	13th c	108 (76e)	108 (70)	51 (60)	44 (50)	31 (30)	23 (00)	6 (50)	3 (00)	
Linkoping	12th c	110 (76e)	105 (70)	65 (60)	55 (50)	30 (30)	15 (00)	5 (50)	3 (00)	
Malmo	1150?	240 (76e)	266 (70)	229 (60)	192 (50)	128 (30)	61 (00)	13 (50)	4 (00)	
		453s (76e)	445s (70)	246s(60)	196s(50)					
Norrkoping	bef 1384	120 (76e)	116 (70)	91 (60)	85 (50)	61 (30)	41 (00)	17 (50)	9 (00)	
Orebro	13th c	117 (76e)	116 (70)	75 (60)	67 (50)	38 (30)	22 (00)	5 (50)	3 (00)	

EUROPE

City	Date Settled	Latest	ca 1970	ca 1960	ca 1950	ca 1930	ca 1900	ca 1850	ca 1800	18th c
*Stockholm	1255	661 (76e)	740 (70)	808 (60)	744 (50)	502 (30)	301 (00)	93 (50)	76 (00)	54 (1750)
		1364s(76e)	1345s(70)	1149s(60)	928s(50)					15 (1663)
Sundsvall	1621	94 (76e)	65 (70)	29 (60)	26 (50)	18 (30)	15 (00)	3 (50)	1 (00)	
Uppsala	9th c?	140 (76e)	127 (70)	78 (60)	63 (50)	30 (30)	23 (00)	7 (50)	5 (00)	
Vasteras	1120?	118 (76e)	117 (70)	78 (60)	60 (50)	30 (30)	12 (00)	4 (50)	3 (00)	
SWITZERLAND										
Basel	374?	191 (76e)	213 (70)	207 (60)	184 (50)	148 (30)	109 (00)	27 (50)	16 (95)	15 (1779)
		369s(76e)	373s(70)	300s(60)	258s(50)					
*Bern	1191	149 (76e)	162 (70)	163 (60)	146 (50)	112 (30)	64 (00)	28 (50)	11 (98)	14 (1764)
		283s(76e)	259s(70)	221s(60)	195s(50)					
Biel	11th c	60 (76e)	64 (70)	59 (60)	48 (50)	38 (30)	30 (00)	6 (50)	3 (00)	
		89s(76e)								
Geneva	58 BC?	155 (76e)	174 (70)	176 (60)	145 (50)	124 (30)	105 (00)	29 (50)	23 (02)	22 (1755)
		323s(76e)	321s(70)	238s(60)	195s(50)					17 (1698)
Lausanne	3rd c?	134 (76e)	137 (70)	126 (60)	107 (50)	76 (30)	47 (00)	17 (50)	10 (00?)	
		227s(76e)	219s(70)	163s(60)	137s(50)					
Luzern	8th c	65 (76e)	70 (70)	67 (60)	61 (50)	47 (30)	29 (00)	10 (50)	4 (99)	
		156s(76e)	149s(70)	120s(60)	98s(50)					
Saint-Gall	614	77 (76e)	81 (70)	76 (60)	68 (50)	64 (30)	54 (00)	18 (50)	8 (08)	
		87s(76e)								
Winterthur	1180	89 (76e)	93 (70)	80 (60)	67 (50)	54 (30)	22 (00)	5 (50)	3 (00?e)	
		107s(76e)	106s(70)	91s(60)	75s(50)					
Zurich	3000 BC?	388 (76e)	423 (70)	440 (60)	390 (50)	250 (30)	151 (00)	17 (50)	10 (90)	11 (1756)
		708s(76e)	675s(70)	537s(60)	495s(50)					9 (1637)
TURKEY										
Edirne	125?	63 (75)	54 (70)	39 (60)	30 (50)	35 (27)	80 (05e)	130 (50e)	100 (00?e)	96 (1750e)
										100 (1693e)

EUROPE

Population (thousands)

City	Date Settled	Latest	ca 1970	ca 1960	ca 1950	ca 1930	ca 1900	ca 1850	ca 1800	18th c
Istanbul	658 BC	2547 (75)	2132 (70)	1467 (60)	983 (50)	691 (27)	874 (00?e) / 1125s (00?e)	900 (50?e) / 891s (44)	400 (00?e) / 598s (00?e)	700 (1690e)
UK (see note 1, p. 361)										
Aberdeen	700?	210 (75e)	182 (71)	185 (61)	183 (51)	167 (31)	154 (01)	72 (51)	27 (01)	16 (1755) / 6 (1707)
Barnsley	1086?	224 (75e)	75 (71)	75 (61)	76 (51)	72 (31)	41 (01)	13 (51)	4s (01)	
Basildon	bef 1510	138 (75e)	129 (71)	89 (61)	43 (51)	40 (31)	18 (01)	2 (51)		
Bath	bef 1st c	84 (75e)	85 (71)	81 (61)	79 (51)	69 (31)	50 (01)	54 (51)	32 (01)	
Belfast	1177	374 (75e)	360 (71)	416 (61) / 529s (61)	444 (51)	415 (26)	349 (01)	100 (51)	18 (98)	9 (1758) / 0.6(1659)
Beverley	700?	107 (75e)	17 (71)	16 (61)	16 (51)	14 (31)	13 (01)	9 (51)	6 (01)	
Birmingham	11th c?	1085 (75e)	1015 (71)	1107 (61)	1113 (51)	1003 (31)	522 (01)	233 (51)	61 (01)	25 (1741) / 15 (1700e)
Blackburn	6th c?	142 (75e)	102 (71)	106 (61)	111 (51)	123 (31)	128 (01)	47 (51)	12 (01)	
Blackpool	16th c?	147 (75e)	152 (71)	153 (61)	147 (51)	102 (31)	47 (01)	2 (51)	0.5s(01)	
Bolton	11th c?	263 (75e)	154 (71)	161 (61)	167 (51)	177 (31)	168 (01)	61 (51)	13 (01)	5 (1773)
Bournemouth	18th c	146 (75e)	154 (71)	154 (61)	145 (51)	117 (31)	47 (01)	6 (71)		
Bradford	1066?	461 (75e)	294 (71)	296 (61)	292 (51)	299 (31)	280 (01)	104 (51)	6 (01)	
Brighton	1086?	159 (75e)	161 (71)	163 (61)	156 (51)	147 (31)	123 (01)	70 (51)	7s (01)	
Bristol	6th c BC	420 (75e)	427 (71)	436 (61)	443 (51)	404 (31)	329 (01)	137 (51)	41 (01)	47 (1753) / 11 (1607)
Burnley	8th c	93 (75e)	76 (71)	81 (61)	85 (51)	98 (31)	97 (01)	21 (51)	3s (01)	
Bury	10th c?	180 (75e)	68 (71)	60 (61)	59 (51)	56 (31)	58 (01)	31 (51)	7 (01)	
Calderdale	bef 1066	192 (75e)	91 (71)	96 (61)	98 (51)	98 (31)	105 (01)	34 (51)	9 (01)	
Cambridge	730?	104 (75e)	99 (71)	95 (61)	81 (51)	70 (31)	38 (01)	28 (51)	10 (01)	6 (1749)
Canterbury	43 AD?	115 (75e)	33 (71)	30 (61)	28 (51)	25 (31)	25 (01)	18 (51)	9 (01)	
Cardiff	.1st c	284 (75e)	279 (71)	257 (61)	244 (51)	227 (31)	164 (01)	18 (51)	2 (01)	
Carlisle	bef 1st c	100 (75e)	72 (71)	71 (61)	68 (51)	57 (31)	45 (01)	26 (51)	10 (01)	
Charnwood	11th c?	132 (75e)	46 (71)	39 (61)	35 (51)	27 (31)	22 (01)	11 (51)	5s (01)	

EUROPE

City	Date Settled	Population (thousands)								
		Latest	ca 1970	ca 1960	ca 1950	ca 1930	ca 1900	ca 1850	ca 1800	18th c
Cheltenham	773?	86 (75e)	74 (71)	72 (61)	63 (51)	49 (31)	49 (01)	35 (51)	3s (01)	
Chester	48 AD?	117 (75e)	63 (71)	59 (61)	48 (51)	41 (31)	46 (01)	28 (51)	15 (01)	
Chesterfield	2nd c?	94 (75e)	70 (71)	68 (61)	69 (51)	64 (31)	27 (01)	7 (51)	4 (01)	
Chichester	bef 1st c	95 (75e)	21 (71)	20 (61)	19 (51)	14 (31)	12 (01)	9 (51)	5 (01)	
Colchester	bef 43 AD	129 (75e)	77 (71)	65 (61)	57 (51)	49 (31)	38 (01)	19 (51)	12 (01)	
Coventry	1043?	337 (75e)	335 (71)	306 (61)	258 (51)	178 (31)	70 (01)	36 (51)	16 (01)	13 (1749)
Crewe and Nantwich	1086?	98 (75e)	63 (71)	63 (61)	61 (51)	53 (31)	50 (01)	9 (51)	3 (01)	
Darlington	11th c?	98 (75e)	86 (71)	84 (61)	85 (51)	72 (31)	45 (01)	11 (51)	5 (01)	
Derby	51 AD?	215 (75e)	220 (71)	132 (61)	141 (51)	142 (31)	106 (01)	41 (51)	11 (01)	
Doncaster	70 AD?	285 (75e)	83 (71)	86 (61)	82 (51)	63 (31)	29 (01)	12 (51)	7 (01)	
Dudley	8th c?	301 (75e)	186 (71)	63 (61)	64 (51)	60 (31)	49 (01)	38 (51)	10 (01)	
Dundee	12th c?	195 (75e)	182 (71)	183 (61)	177 (51)	176 (31)	161 (01)	79 (51)	26 (01)	5 (1746)
Durham	11th c?	86 (75e)	25 (71)	21 (61)	19 (51)	16 (31)	15 (01)	13 (51)	8 (01)	
Edinburgh	617?	470 (75e)	454 (71)	468 (61)	467 (51)	439 (31)	317 (01)	161 (51)	67 (01)	57 (1755) 36 (1705e)
Elmbridge[1]	bef 1st c	112 (75e)	115 (71)	107 (61)	89 (51)	42 (31)	24 (01)	5s (51)	3s (01)	
Erewash	1086?	101 (75e)	34 (71)	35 (61)	34 (51)	33 (31)	25 (01)	6s (51)	2s (01)	
Exeter	200 BC?	94 (75e)	96 (71)	80 (61)	76 (51)	66 (31)	53 (01)	33 (51)	17 (01)	
Fareham	1086?	86 (75e)	80 (71)	58 (61)	43 (51)	12 (31)	8 (01)	3 (51)	3 (01)	
Gateshead	1080?	221 (75e)	94 (71)	103 (61)	115 (51)	122 (31)	110 (01)	26 (51)	9s (01)	
Gedling[2]	1086?	102 (75e)	78 (71)	66 (61)	55 (51)	36 (31)	19 (01)	2 (51)	0.8 (01)	
Gillingham	15th c?	93 (75e)	87 (71)	73 (61)	71 (51)	62 (31)	43 (01)	8s (51)	4s (01)	
Glasgow	1202?	881 (75e)	897 (71)	1055 (61)	1090 (51)	1088 (31)	762 (01)	345 (51)	81 (01)	24 (1755)
Gloucester	bef 49 AD	91 (75e)	90 (71)	70 (61)	67 (51)	53 (31)	48 (01)	18 (51)	8 (01)	13 (1708)
Gosport	10th c?	85 (75e)	76 (71)	62 (61)	58 (51)	38 (31)	29 (01)	7 (51)	8 (11)	
Gravesham	bef 1st c	96 (75e)	54 (71)	51 (61)	45 (51)	35 (31)	27 (01)	17 (51)	2 (01)	

[1] Inc Esher, Walton-on-Thames, Weybridge.
[2] Inc Arnold, Carlton.

Figure 47. Center of London, showing Post Office Tower (left background), Westminster Abbey (center), and the Houses of Parliament on the bank of the Thames (right). Although it was the world's largest city throughout most of the nineteenth century, London now ranks third in Europe and eleventh in the world, suburbs included. (Credit: British Information Services.)

EUROPE

City	Date Settled	Latest	ca 1970	ca 1960	ca 1950	ca 1930	ca 1900	ca 1850	ca 1800	18th c
Grimsby	8th c?	94 (75e)	96 (71)	97 (61)	95 (51)	92 (31)	63 (01)	9 (51)	2s (01)	
Guildford	9th c?	120 (75e)	57 (71)	54 (61)	48 (51)	31 (31)	16 (01)	7 (51)	3 (01)	
Halton[1]	915?	111 (75e)	93 (71)	78 (61)	73 (51)	59 (31)	45 (01)	8 (51)	2s (01)	
Harrogate	74 AD?	133 (75e)	62 (71)	56 (61)	50 (51)	40 (31)	28 (01)	4 (51)	1s (01)	
Hartlepool	649?	97 (75e)	97 (71)	18 (61)	17 (51)	21 (31)	23 (01)	10 (51)	1s (01)	
Havant	bef 1066	116 (75e)	109 (71)	75 (61)	35 (51)	21 (31)	4 (01)	2 (51)	2 (01)	
Hove	13th c?	89 (75e)	73 (71)	73 (61)	70 (51)	55 (31)	37 (01)	4s (51)	0.1s (01)	
Hull	13th c	280 (75e)	286 (71)	303 (61)	299 (51)	314 (31)	240 (01)	85 (51)	22 (01)	
Inverclyde	6th c?	106 (75e)	70 (71)	75 (61)	77 (51)	79 (31)	68 (01)	37 (51)	17 (01)	
Ipswich	991?	123 (75e)	123 (71)	117 (61)	107 (51)	88 (31)	67 (01)	33 (51)	11 (01)	
Kirklees[2]	7th c?	375 (75e)	182 (71)	184 (61)	182 (51)	167 (31)	123 (01)	36 (51)	7 (01)	
Knowsley[3]	1176?	191 (75e)	127 (71)	115 (61)	56 (51)	5 (31)	5 (01)	3s (51)	2s (01)	
Lancaster	1st c?	125 (75e)	50 (71)	48 (61)	52 (51)	43 (31)	40 (01)	15 (51)	9 (01)	
Langbaurgh[4]	1086?	151 (75e)		68 (61)	61 (51)	51 (31)	19 (01)	1s (51)	0.7s (01)	
Leeds	bef 1080	749 (75e)	496 (71)	511 (61)	505 (51)	483 (31)	429 (01)	172 (51)	53 (01)	16 (1771)
Leicester	bef 43 AD	291 (75e)	284 (71)	273 (61)	285 (51)	258 (31)	212 (01)	61 (51)	17 (01)	6 (1712e)
Lichfield	7th c	88 (75e)	23 (71)	14 (61)	11 (51)	9 (31)	8 (01)	7 (51)	5 (01)	
Lincoln	bef 70 AD	73 (75e)	74 (71)	77 (61)	70 (51)	66 (31)	49 (01)	18 (51)	7 (01)	
Liverpool	1190?	549 (75e)	610 (71)	747 (61)	789 (51)	856 (31)	685 (01)	376 (51)	78 (01)	34 (1773), 5 (1700e)
*London	43 AD?	2553 (75e), 7111s (75e)	2772 (71), 7452s (71)	3200 (61), 8172s (61)	3348 (51), 8348s (51)	4397 (31), 8216s (31)	4537 (01), 6581s (01)	2362 (51)	865 (01)	676 (1750e), 674 (1701e), 272 (1622e)
Londonderry	546	84 (75e)	52 (71)	54 (61)	50 (51)	45 (26)	40 (01)	20 (51)	14 (31)	
Luton	1st c?	166 (75e)	161 (71)	132 (61)	110 (51)	69 (31)	36 (01)	11 (51)	3s (01)	

Population (thousands)

[1] Inc Runcorn, Widnes.
[2] Inc Dewsbury, Huddersfield.
[3] Inc Huyton-with-Roby, Kirkby.
[4] Inc Eston, Redcar.

EUROPE

City	Date Settled	Population (thousands)								
		Latest	ca 1970	ca 1960	ca 1950	ca 1930	ca 1900	ca 1850	ca 1800	18th c
Macclesfield	1086?	148 (75e)	44 (71)	38 (61)	36 (51)	35 (31)	35 (01)	39 (51)	9 (01)	
Maidstone	1st c?	125 (75e)	71 (71)	60 (61)	54 (51)	42 (31)	34 (01)	21 (51)	8 (01)	
Manchester	80 AD	506 (75e)	544 (71)	662 (61)	703 (51)	766 (31)	544 (01)	303 (51)	70 (01)	22 (1773) 9 (1727e)
Medway[1]	1st c?	145 (75e)	113 (71)	99 (61)	88 (51)	74 (31)	68 (01)	43 (51)	18 (01)	
Middlesbrough	1086?	153 (75e)		157 (61)	147 (51)	139 (31)	91 (01)	7 (51)	0.2s (01)	
Motherwell	1208?	160 (75e)	74 (71)	73 (61)	68 (51)	65 (31)	51 (01)	4 (51)	3s (01)	
Newcastle-under-Lyme	12th c	121 (75e)	77 (71)	76 (61)	70 (51)	23 (31)	20 (01)	11 (51)	5 (01)	
Newcastle upon Tyne	1080	296 (75e)	222 (71)	270 (61)	292 (51)	286 (31)	215 (01)	88 (51)	28 (01)	12 (1665e)
Newport	1126?	133 (75e)	112 (71)	108 (61)	106 (51)	89 (31)	67 (01)	19 (51)	1 (01)	
Northampton	6th c?	140 (75e)	127 (71)	105 (61)	104 (51)	92 (31)	87 (01)	27 (51)	7 (01)	
North Bedfordshire	571?	130 (75e)	73 (71)	64 (61)	53 (51)	43 (31)	35 (01)	12 (51)	4 (01)	
North Tyneside[2]	7th c?	206 (75e)	164 (71)	167 (61)	144 (51)	124 (31)	72 (01)	29 (51)	8s (01)	
Norwich	570?	122 (75e)	122 (71)	120 (61)	121 (51)	126 (31)	112 (01)	68 (51)	37 (01)	36s (1752) 29s (1693)
Nottingham	9th c	287 (75e)	301 (71)	312 (61)	306 (51)	276 (31)	240 (01)	57 (51)	29 (01)	11 (1739)
Nuneaton	1086?	113 (75e)	67 (71)	57 (61)	54 (51)	46 (31)	25 (01)	5 (51)	5s (01)	
Ogwr[3]	1116?	128 (75e)	69 (71)	69 (61)	70 (51)	69 (31)	43 (01)	16s (51)	8s (01)	
Oldham	11th c?	228 (75e)	106 (71)	115 (61)	123 (51)	140 (31)	137 (01)	53 (51)	12 (01)	
Oxford	912?	117 (75e)	109 (71)	106 (61)	99 (51)	81 (31)	49 (01)	28 (51)	12 (01)	
Peterborough	bef 655	115 (75e)	70 (71)	62 (61)	53 (51)	44 (31)	31 (01)	9 (51)	3 (01)	
Plymouth	11th c?	258 (75e)	239 (71)	204 (61)	208 (51)	213 (31)	108 (01)	52 (51)	16 (01)	
Poole	1224?	113 (75e)	107 (71)	92 (61)	83 (51)	57 (31)	19 (01)	9 (51)	5 (01)	

[1] Inc Chatham, Rochester.
[2] Inc Longbenton, Tynemouth, Wallsend.
[3] Inc Bridgend, Maesteg, Ogmore and Garw, Porthcawl.

EUROPE

City	Date Settled	Latest	ca 1970	ca 1960	ca 1950	ca 1930	ca 1900	ca 1850	ca 1800	18th c
Portsmouth	12th c	201 (75e)	197 (71)	215 (61)	234 (51)	252 (31)	188 (01)	72 (51)	32 (01)	
Preston	1094?	132 (75e)	98 (71)	113 (61)	121 (51)	119 (31)	113 (01)	70 (51)	12 (01)	
Reading	871	133 (75e)	133 (71)	120 (61)	114 (51)	97 (31)	72 (01)	21 (51)	10 (01)	
Reigate and Banstead	967?	114 (75e)	101 (71)	96 (61)	76 (51)	31 (31)	26 (01)	5 (51)	0.9 (01)	
Renfrew[1]	560?	209 (75e)	155 (71)	146 (61)	140 (51)	126 (31)	109 (01)	47 (51)	17 (01)	4 (1753)
Rhondda	951?	86 (75e)	89 (71)	100 (61)	111 (51)	141 (31)	114 (01)	2s (51)	0.5s (01)	
Rochdale	1st c?	211 (75e)	91 (71)	86 (61)	88 (51)	90 (31)	83 (01)	29 (51)	6 (01)	
Rotherham	1086?	249 (75e)	85 (71)	85 (61)	82 (51)	75 (31)	54 (01)	6 (51)	3s (01)	
Rugby	1086?	86 (75e)	59 (71)	52 (61)	45 (51)	24 (31)	17 (01)	6 (51)	1 (01)	
Rushcliffe	14th c?	89 (75e)	28 (71)	27 (61)	23 (51)	18 (31)	7 (01)	0.3s(51)	0.2s (01)	
Saint Albans	15 BC?	123 (75e)	52 (71)	50 (61)	44 (51)	29 (31)	16 (01)	7 (51)	3 (01)	
Saint Helens	12th c?	195 (75e)	104 (71)	109 (61)	113 (51)	107 (31)	84 (01)	15 (51)	3s (01)	
Salford	1086?	266 (75e)	131 (71)	155 (61)	178 (51)	223 (31)	221 (01)	64 (51)	14 (01)	5 (1773)
Salisbury	1220	105 (75e)	35 (71)	35 (61)	33 (51)	26 (31)	17 (01)	12 (51)	8 (01)	
Sandwell[2]	1086?	315 (75e)	331 (71)	266 (61)	267 (51)	242 (31)	180 (01)	8 (51)	1 (01)	
Scarborough	bef 1066	98 (75e)	44 (71)	43 (61)	45 (51)	42 (31)	38 (01)	13 (51)	6 (01)	
Sefton[3]	7th c?	305 (75e)	215 (71)	224 (61)	217 (51)	175 (31)	115 (01)	5 (51)	3s (01)	
Sheffield	1086?	560 (75e)	520 (71)	494 (61)	513 (51)	518 (31)	381 (01)	135 (51)	31 (01)	13 (1755)
Slough	12th c?	103 (75e)	87 (71)	81 (61)	66 (51)	34 (31)	11 (01)	1 (51)	2s (01)	
Solihull	13th c?	200 (75e)	107 (71)	96 (61)	68 (51)	25 (31)	8 (01)	3 (51)	2s (01)	
Southampton	755?	215 (75e)	215 (71)	205 (61)	178 (51)	176 (31)	105 (01)	35 (51)	8 (01)	
Southend-on-Sea	1121	159 (75e)	163 (71)	165 (61)	152 (51)	120 (31)	29 (01)	1 (51)		
South Ribble[4]	1st c?	90 (75e)	50 (71)	38 (61)	30 (51)	24 (31)	18 (01)	11s (51)	6s (01)	
South Tyneside	1245	172 (75e)	101 (71)	110 (61)	107 (51)	113 (31)	97 (01)	29 (51)	8 (01)	

Population (thousands)

[1] Inc Barrhead, Johnstone, Paisley, Renfrew.
[2] Inc Warley, West Bromwich.
[3] Inc Bootle, Crosby, Southport.
[4] Inc Leyland, Walton-le-Dale.

EUROPE

City	Date Settled	Latest	ca 1970	ca 1960	ca 1950	ca 1930	Population (thousands) ca 1900	ca 1850	ca 1800	18th c
Spelthorne[1]	1086?	97 (75e)	97 (71)	83 (61)	63 (51)	34 (31)	12 (01)	2 (51)	3s (01)	
Stafford	bef 913	114 (75e)	55 (71)	48 (61)	40 (51)	29 (31)	21 (01)	12 (51)	4 (01)	
Stockport	12th c?	293 (75e)	140 (71)	143 (61)	142 (51)	125 (31)	79 (01)	54 (51)	15 (01)	
Stockton-on-Tees	12th c?	164 (75e)		81 (61)	74 (51)	68 (31)	51 (01)	10 (51)	5s (01)	
Stoke-on-Trent	1086?	256 (75e)	265 (71)	265 (61)	275 (51)	277 (31)	268 (01)	84 (51)	23s (01)	
Stratford-on-Avon	693?	99s (75e)	19 (71)	17 (61)	15 (51)	12 (31)	8 (01)	3 (51)	2 (01)	
Sunderland	674	298 (75e)	217 (71)	190 (61)	182 (51)	186 (31)	146 (01)	64 (51)	12 (01)	
Swansea	1099?	190 (75e)	173 (71)	167 (61)	161 (51)	165 (31)	95 (01)	31 (51)	6 (01)	
Taff-Ely	1750?	90 (75e)	34 (71)	35 (61)	39 (51)	43 (31)	32 (01)	12s (91)		
Tameside[2]	1086?	223 (75e)	147 (71)	135 (61)	132 (51)	126 (31)	120 (01)	62 (51)	18s (01)	
Thamesdown	1086?	142 (75e)	91 (71)	92 (61)	69 (51)	62 (31)	45 (01)	5s (51)	1s (01)	
Thurrock	1149?	128 (75e)	125 (71)	114 (61)	82 (51)	35 (31)	19 (01)	2s (51)	0.7s (01)	
Torbay	12th c?	110 (75e)	109 (71)	54 (61)	53 (51)	46 (31)	34 (01)	8 (51)	0.8s (01)	
Torfaen[3]	12th c?	90 (75e)	69 (71)	62 (61)	56 (51)	7 (31)	6 (01)	4 (51)	1s (01)	
Trafford[4]	1212?	227 (75e)	151 (71)	152 (61)	145 (51)	106 (31)	59 (01)	4 (51)	4s (01)	
Tunbridge Wells	1606	95 (75e)	45 (71)	40 (61)	38 (51)	35 (31)	33 (01)	11 (51)	4s (01)	
Vale of Glamorgan[5]	12th c?	107 (75e)	66 (71)	63 (61)	60 (51)	57 (31)	41 (01)	0.2s(51)	0.1s (01)	
Wakefield	7th c?	305 (75e)	60 (71)	61 (61)	60 (51)	59 (31)	41 (01)	22 (51)	8 (01)	
Walsall	996?	271 (75e)	185 (71)	118 (61)	115 (51)	103 (31)	86 (01)	26 (51)	5 (01)	
Warrington	79 AD?	165 (75e)	68 (71)	76 (61)	81 (51)	79 (31)	64 (01)	23 (51)	11 (01)	
Wigan	1st c?	310 (75e)	81 (71)	79 (61)	85 (51)	85 (31)	61 (01)	32 (51)	11 (01)	
Winchester	50 BC?	89 (75e)	31 (71)	29 (61)	26 (51)	23 (31)	21 (01)	14 (51)	6 (01)	

[1] Inc Staines, Sunbury-on-Thames.
[2] Inc Ashton-under-Lyne, Denton, Hyde, Stalybridge.
[3] Inc Cwmbran, Pontypool.
[4] Inc Altrincham, Sale, Stretford.
[5] Inc Barry, Penarth.

EUROPE

City	Date Settled	Population (thousands)								
		Latest	ca 1970	ca 1960	ca 1950	ca 1930	ca 1900	ca 1850	ca 1800	18th c
Windsor and Maidenhead	1st c?	127 (75e)	75 (71)	62 (61)	50 (51)	38 (31)	27 (01)	14 (51)	4 (01)	
Wirral[1]	10th c?	348 (75e)	323 (71)	320 (61)	309 (51)	273 (31)	175 (01)	24 (51)	0.8s (01)	
Wolverhampton	985?	269 (75e)	269 (71)	151 (61)	163 (51)	133 (31)	94 (01)	50 (51)	13 (01)	
Worcester	bef 1st c	74 (75e)	73 (71)	66 (61)	62 (51)	51 (31)	47 (01)	28 (51)	11 (01)	
Worthing	1st c?	93 (75e)	88 (71)	80 (61)	69 (51)	46 (31)	20 (01)	5 (51)	1s (01)	
Wrekin[2]	7th c?	110 (75e)	43 (71)	24 (61)	19 (51)	15 (31)	14 (01)	5 (51)	8 (01)	
Wrexham Maelor	bef 180	108 (75e)	39 (71)	35 (61)	31 (51)	19 (31)	15 (01)	7 (51)	1s (01)	
Wyre[3]	1086?	99 (75e)	56 (71)	49 (61)	43 (51)	33 (31)	15 (01)	3 (51)	0.6s (01)	
York	71 AD?	103 (75e)	105 (71)	104 (61)	105 (51)	94 (31)	78 (01)	36 (51)	16 (01)	
USSR										
Archangel	1553	383 (76e)	343 (70)	258 (59)	251 (39)	73 (26)	21 (97)	15 (56)	11 (11)	
Armavir	1848	158 (76e)	145 (70)	111 (59)	84 (39)	75 (26)	18 (97)			
Astrakhan	13th c	458 (76e)	410 (70)	305 (59)	259 (39)	177 (26)	113 (97)	35 (56)	38 (11)	
Baku	5th c?	943 (76e)	852 (70)	643 (59)	544 (39)	453 (26)	112 (97)	8 (56)		
		1406s (76e)	1266s (70)	968s (59)	733s(39)					
Belgorod	1237?	219 (76e)	151 (70)	72 (59)	34 (39)	31 (26)	27 (97)	13 (56)	8 (11)	
Berezniki	1883	172 (76e)	146 (70)	106 (59)	51 (39)	11 (26)				
Bobruysk	16th c	185 (76e)	138 (70)	98 (59)	84 (39)	51 (26)	34 (97)	17 (56)	2 (11)	
Brest	1017?	162 (76e)	122 (70)	74 (59)	41 (39)	51 (31)	47 (97)	17 (56)	4 (11)	
Bryansk	1146	375 (76e)	318 (70)	207 (59)	174 (39)	27 (26)	25 (97)	11 (56)	5 (11)	
Cheboksary	1371?	278 (76e)	216 (70)	104 (59)	31 (39)	9 (26)	5 (97)	5 (56)	4 (11)	
Cherepovets	14th c	238 (76e)	188 (70)	92 (59)	32 (39)	22 (26)	7 (97)	3 (56)	1 (11)	
Cherkassy	14th c	221 (76e)	158 (70)	85 (59)	52 (39)	40 (26)	30 (97)	12 (56)	5 (11)	
Chernigov	7th c?	225 (76e)	159 (70)	90 (59)	69 (39)	35 (26)	28 (97)	4 (56)	5 (11)	

[1] Inc Bebington, Birkenhead, Wallasey, Wirral.
[2] Inc Telford (for Dawley), Wellington.
[3] Inc Fleetwood, Thornton Cleveleys.

EUROPE

Population (thousands)

City	Date Settled	Latest	ca 1970	ca 1960	ca 1950	ca 1930	ca 1900	ca 1850	ca 1800	18th c
Chernovtsy	1407?	209 (76e)	187 (70)	152 (59)	106 (39)	112 (30)	70 (00)	21 (51)	5 (00?e)	
Dneprodzerzhinsk	1750?	248 (76e)	227 (70)	194 (59)	148 (39)	34 (26)	17 (97)	3 (61)		
Dnepropetrovsk	1783	976 (76e)	862 (70)	661 (59)	528 (39)	233 (26)	113 (97)	13 (56)	9 (11)	
Donetsk	1869	967 (76e)	879 (70)	708 (59)	474 (39)	106 (26)	28 (97)			
Dzerzhinsk	bef 1917	245 (76e)	221 (70)	164 (59)	103 (39)	9 (26)	1 (20)			
Engels	1747	159 (76e)	130 (70)	91 (59)	69 (39)	34 (26)	22 (97)	13 (67?e)		
Gomel	1142?	349 (76e)	272 (70)	168 (59)	139 (39)	86 (26)	37 (97)	10 (56)		
Gorkiy	1221	1305 (76e)	1170 (70)	941 (59)	644 (39)	185 (26)	90 (97)	36 (56)	14 (11)	
Gorlovka	1867	342 (76e)	335 (70)	308 (59)	189 (39)	23 (26)	2 (97)			
Grodno	bef 1128	176 (76e)	132 (70)	73 (59)	49 (39)	50 (31)	47 (97)	15 (56)	11 (11)	
Groznyy	1818	381 (76e)	341 (70)	250 (59)	172 (39)	71 (26)	16 (97)	3 (67)		
Ivanovo	16th c	458 (76e)	420 (70)	335 (59)	285 (39)	111 (26)	54 (97)	6 (61)		
Izhevsk	1760	522 (76e)	422 (70)	285 (59)	176 (39)	63 (26)	41 (97)	21 (59)		
Kalinin	1135	395 (76e)	345 (70)	261 (59)	216 (39)	108 (26)	54 (97)	13 (56)	17 (11)	55 (1755)
Kaliningrad	1255	345 (76e)	297 (70) / 306s (70)	204 (59)	372 (39)	316 (33)	189 (00)	75 (49)	55 (02)	39 (1723)
Kaluga	1389?	255 (76e)	211 (70)	134 (59)	89 (39)	52 (26)	50 (97)	31 (56)	23 (11)	
Kaunas	1030	352 (76e)	305 (70)	219 (59)	152 (39)	92 (23)	71 (97)	20 (56)	2 (11)	
Kazan	1437	958 (76e)	869 (70)	667 (59)	406 (39)	179 (26)	130 (97)	56 (56)	54 (11)	
Kerch	6th c BC	152 (76e)	128 (70)	98 (59)	104 (39)	35 (26)	33 (97)	13 (56)	1 (11)	
Kharkov	1656	1385 (76e)	1223 (70)	953 (59)	840 (39)	417 (26)	174 (97)	31 (56)	10 (11)	
Kherson	1778	315 (76e)	261 (70)	158 (59)	97 (39)	59 (26)	59 (97)	34 (56)	9 (11)	
Khmelnitskiy	1493?	161 (76e)	113 (70)	62 (59)	37 (39)	32 (26)	23 (97)	6 (56)	2 (11)	
Kiev	430?	2013 (76e)	1632 (70)	1110 (59)	851 (39)	514 (26)	248 (97)	62 (56)	23 (11)	
Kirov	1174	376 (76e)	333 (70)	252 (59)	144 (39)	62 (26)	25 (97)	15 (56)	4 (11)	
Kirovabad	12th c?	211 (76e)	190 (70)	136 (59)	99 (39)	57 (26)	34 (97)	11 (56)		29 (1766)
Kirovograd	1754	224 (76e)	189 (70)	132 (59)	103 (39)	66 (26)	61 (97)	13 (56)	5 (11)	
Kishinev	1420?	471 (76e)	356 (70)	216 (59)	112 (39)	115 (30)	108 (97)	63 (56)	7 (12e)	
Klaypeda	1252	169 (76e)	140 (70)	90 (59)	47 (39)	37 (31)	21 (05)	11 (49)	5 (02)	
Kostroma	1152	247 (76e)	223 (70)	172 (59)	121 (39)	74 (26)	41 (97)	15 (56)	10 (11)	

Figure 48. Panorama of central Moscow, now the largest city in Europe, with the Kremlin towers in the foreground. (Credit: Embassy of the USSR, Washington.)

EUROPE

Population (thousands)

City	Date Settled	Latest	ca 1970	ca 1960	ca 1950	ca 1930	ca 1900	ca 1850	ca 1800	18th c
Kramatorsk	1897	167 (76e)	150 (70)	115 (59)	94 (39)	12 (26)	66 (97)	9 (56)	4 (25)	
Krasnodar	1794	543 (76e)	464 (70)	313 (59)	193 (39)	163 (26)	63 (97)	20 (56)	8 (11)	
Kremenchug	1571	202 (76e)	148 (70)	87 (59)	90 (39)	59 (26)	15 (97)	3 (67?e)		
Krivoy Rog	17th c	634 (76e)	573 (70)	401 (59)	192 (39)	31 (26)	76 (97)	41 (56)	23 (11)	
Kursk	9th c	363 (76e)	284 (70)	205 (59)	120 (39)	99 (26)	32 (97)	8 (67)		
Kutaisi	6th c BC	177 (76e)	161 (70)	128 (59)	78 (39)	48 (26)	90 (97)	24 (56)	4 (11)	
Kuybyshev	1586	1186 (76e)	1045 (70)	806 (59)	390 (39)	176 (26)	90 (97)	24 (56)		
Leninakan	773	188 (76e)	165 (70)	108 (59)	68 (39)	42 (26)	31 (97)	12 (56)		
Leningrad	1703	3911 (76e)	3513 (70)	2985 (59)	3103 (39)	1614 (26)	1265 (97)	491 (56)	336 (11)	150 (1760?e)
		4372s(76e)	3950s(70)	3321s(59)	3385s(39)				271 (05)	70 (1725e)
Lipetsk	13th c	363 (76e)	289 (70)	157 (59)	67 (39)	21 (26)	21 (97)	11 (56)	5 (11)	
Lvov	1250?	629 (76e)	553 (70)	411 (59)	340 (39)	316 (31)	160 (00)	68 (51)	39 (95)	
Lyubertsy	1910	154 (76e)	139 (70)	95 (59)	48 (39)	6 (26)				
Makeyevka	1899	437 (76e)	392 (70)	371 (59)	254 (39)	51 (26)				
Makhachkala	1844	231 (76e)	186 (70)	119 (59)	87 (39)	32 (26)	10 (97)	4 (67)		
Melitopol	18th c	155 (76e)	137 (70)	95 (59)	76 (39)	25 (26)	15 (97)	4 (56)		
Minsk	1067?	1175 (76e)	907 (70)	509 (59)	237 (39)	132 (26)	91 (97)	26 (56)	11 (11)	
		1189s(76e)	917s(70)							
Mogilev	1267	264 (76e)	202 (70)	122 (59)	99 (39)	50 (26)	43 (97)	23 (56)	6 (11)	
*Moscow	1147?	7563 (76e)	6942 (70)	6009 (59)	4537 (39)	2026 (26)	1039 (97)	369 (56)	270 (11)	153 (1770)
		7734s(76e)	7077s(70)	6044s(59)	4542s(39)					150 (1700e)
Murmansk	1915	369 (76e)	309 (70)	222 (59)	119 (39)	9 (26)				
Naberezhnyye Chelny	bef 1898	225 (76e)	38 (70)	16 (59)	9 (39)	5 (32)	3 (00?e)			
Nalchik	1818	195 (76e)	146 (70)	88 (59)	48 (39)	13 (26)	5 (97)	1 (67?e)		
Nikolayev	1788	436 (76e)	331 (70)	235 (59)	174 (39)	105 (26)	92 (97)	44 (56)	4 (11)	
Novgorod	859?	172 (76e)	128 (70)	61 (59)	40 (39)	33 (26)	26 (97)	13 (56)	6 (11)	
Novocherkassk	1805	183 (76e)	162 (70)	123 (59)	81 (39)	62 (26)	52 (97)	18 (56)	6 (11)	
Novorossiysk	1838	150 (76e)	133 (70)	93 (59)	95 (39)	68 (26)	17 (97)			
Odessa	14th c	1023 (76e)	892 (70)	664 (59)	599 (39)	421 (26)	404 (97)	101 (56)	11 (11)	

EUROPE

City	Date Settled	Latest	ca 1970	ca 1960	ca 1950	ca 1930	ca 1900	ca 1850	ca 1800	18th c
					Population (thousands)					
Ordzhonikidze	1784	276 (76e)	236 (70)	164 (59)	131 (39)	78 (26)	44 (97)	6 (63)		
Orel	1566	282 (76e)	232 (70)	150 (59)	111 (39)	78 (26)	70 (97)	35 (56)	25 (11)	
Orenburg	1743	435 (76e)	344 (70)	267 (59)	172 (39)	123 (26)	72 (97)	14 (56)	5 (11)	
Orsk	1735	243 (76e)	225 (70)	176 (59)	66 (39)	14 (26)	14 (97)	2 (63)	2 (00?e)	
Penza	1666	436 (76e)	374 (70)	255 (59)	160 (39)	92 (26)	60 (97)	24 (56)	15 (11)	
Perm	1568	957 (76e)	850 (70)	629 (59)	306 (39)	85 (26)	45 (97)	9 (56)	3 (11)	
Petrozavodsk	1703	216 (76e)	184 (70)	135 (59)	70 (39)	27 (26)	13 (97)	10 (56)	5 (11)	
Podolsk	bef 1781	191 (76e)	169 (70)	129 (59)	72 (39)	20 (26)	4 (97)	4 (56)	0.9 (11)	
Poltava	1174?	270 (76e)	220 (70)	143 (59)	128 (39)	92 (26)	54 (97)	21 (56)	10 (11)	
Pskov	903?	155 (76e)	127 (70)	81 (59)	60 (39)	44 (26)	30 (97)	16 (56)	9 (11)	3 (1626)
Riga	1201?	806 (76e)	732 (70)	580 (59)	348 (39)	378 (30)	282 (97)	70 (56)	32 (11)	14 (1710e)
Rostov-na-Donu	1761	907 (76e)	789 (70)	600 (59)	510 (39)	308 (26)	119 (97)	13 (56)	4 (11)	
Rovno	1282?	162 (76e)	116 (70)	56 (59)	43 (39)	42 (31)	25 (97)	5 (56)	3 (11)	
Ryazan	1095?	432 (76e)	350 (70)	214 (59)	95 (39)	51 (26)	46 (97)	21 (56)	8 (11)	
Rybinsk	1137?	236 (76e)	218 (70)	182 (59)	144 (39)	55 (26)	25 (97)	9 (56)	3 (11)	
Saransk	1641	241 (76e)	191 (70)	91 (59)	41 (39)	15 (26)	15 (97)	5 (56)	9 (11)	
Saratov	1590	848 (76e)	757 (70)	579 (59)	372 (39)	215 (26)	137 (97)	62 (56)	27 (11)	
Sevastopol	16th c	290 (76e)	229 (70)	144 (59)	114 (39)	75 (26)	54 (97)	6 (56)	2 (82)	
Severodvinsk	1918?	180 (76e)	145 (70)	79 (59)	21 (39)					
Shakhty	1839	222 (76e)	205 (70)	196 (59)	135 (39)	33 (26)	16 (97)	4 (63)	0.8 (11)	
Simferopol	16th c?	286 (76e)	249 (70)	186 (59)	143 (39)	88 (26)	49 (97)	26 (56)	2 (11)	
Smolensk	865	258 (76e)	211 (70)	147 (59)	157 (39)	79 (26)	47 (97)	9 (56)	12 (11)	
Sochi	1896	251 (76e)	224 (70)	127 (59)	71 (39)	10 (26)	0.4 (97)			
Stavropol	1777	239 (76e)	198 (70)	141 (59)	85 (39)	59 (26)	42 (97)	17 (56)		
Sterlitamak	1766	210 (76e)	185 (70)	112 (59)	39 (39)	25 (26)	16 (97)	6 (56)	2 (11)	
Sumgait	bef 1882	168 (76e)	124 (70)	51 (59)	6 (39)					
Sumy	1658	199 (76e)	159 (70)	98 (59)	64 (39)	44 (26)	28 (97)	12 (56)	9 (11)	
Syktyvkar	16th c	157 (76e)	125 (70)	69 (59)	24 (39)	5 (26)	4 (97)	3 (56)	2 (11)	
Syzran	1683	185 (76e)	173 (70)	148 (59)	83 (39)	50 (26)	32 (97)	18 (56)	7 (11)	
Taganrog	1698	282 (76e)	254 (70)	202 (59)	189 (39)	86 (26)	51 (97)	19 (56)	7 (11)	

EUROPE

Population (thousands)

City	Date Settled	Latest	ca 1970	ca 1960	ca 1950	ca 1930	ca 1900	ca 1850	ca 1800	18th c
Tallin	1154	408 (76e)	363 (70)	282 (59)	160 (39)	125 (27)	65 (97)	20 (56)	18 (11)	
Tambov	1636	262 (76)	230 (70)	172 (59)	106 (39)	76 (26)	48 (97)	22 (56)	17 (11)	
Tartu	1030	99 (76e)	90 (70)	74 (59)	57 (39)	60 (27)	41 (97)	13 (56)	6 (11)	
Tbilisi	4th c?	1030 (76)	889 (70)	703 (59)	519 (39)	294 (26)	160 (97)	38 (56)	22 (97e)	20 (1700e)
Tolyatti	1738	463 (76)	251 (70)	72 (59)	10 (39e)	6 (26)	6 (97)	4 (56)	2 (11)	
Tula	1146?	506 (76)	462 (70)	351 (59)	285 (39)	153 (26)	115 (97)	51 (56)	52 (11)	3 (1654)
Ufa	1586	923 (76)	771 (70)	547 (59)	258 (39)	99 (26)	49 (97)	13 (56)	9 (11)	
Ulyanovsk	1648	436 (76)	351 (70)	206 (59)	98 (39)	72 (26)	42 (97)	27 (56)	13 (11)	
Vilnyus	10th c	447 (76)	372 (70)	236 (59)	215 (39)	196 (31)	155 (97)	46 (56)	56 (11) / 25 (00)	21 (1770)
Vinnitsa	14th c	288 (76e)	212 (70)	122 (59)	93 (39)	58 (26)	31 (97)	9 (56)	3 (11)	
Vitebsk	1021?	279 (76e)	231 (70)	148 (59)	167 (39)	99 (26)	66 (97)	21 (56)	17 (11)	
Vladimir	1108?	278 (76e)	234 (70)	154 (59)	67 (39)	40 (26)	28 (97)	13 (56)	6 (11)	
Volgograd	1589	918 (76)	818 (70)	591 (59)	445 (39)	148 (26)	55 (97)	7 (56)	4 (11)	
Vologda	1147?	219 (76)	178 (70)	139 (59)	95 (39)	58 (26)	28 (97)	14 (56)	10 (11)	
Volzhskiy	1951	195 (76)	142 (70)	67 (59)						
Voronezh	1586	764 (76)	660 (70)	447 (59)	344 (39)	120 (26)	81 (97)	38 (56)	22 (11)	
Voroshilovgrad	18th c	439 (76e)	383 (70)	275 (59)	215 (39)	72 (26)	20 (97)	7 (61)	0.8 (82)	
Yalta	2nd c	76 (76e)	62 (70)	44 (59)	33 (39)	29 (26)	13 (97)	1 (63)		
Yaroslavl	1024	577 (76)	517 (70)	407 (59)	309 (39)	114 (26)	72 (97)	27 (56)	24 (11)	
Yerevan	607?	928 (76)	767 (70)	493 (59)	204 (39)	65 (26)	29 (97)	13 (56)	15 (00?e)	
Yoshkar-Ola	1578	210 (76)	166 (70)	89 (59)	27 (39)	4 (26)	2 (97)	1 (67?e)		
Zaporozhye	1770	760 (76)	658 (70)	449 (59)	289 (39)	56 (26)	19 (97)	3 (56)	1 (11)	
Zhdanov	1779	467 (76)	417 (70)	284 (59)	222 (39)	41 (26)	31 (97)	7 (56)	3 (11)	
Zhitomir	1240?	229 (76e)	161 (70)	106 (59)	95 (39)	77 (26)	66 (97)	31 (56)	8 (11)	
YUGOSLAVIA										
Banja Luka	1295?	91 (71)	91 (71)	51 (61)	33 (48)	22 (31)	15 (95)	15 (50?e)	15 (00?e)	
*Belgrade	3rd c BC	870 (75e) / 775s (71)	746 (71) / 775s (71)	598 (61)	368 (48)	266 (31)	70 (00)	15 (50)	25 (89e)	

EUROPE

Population (thousands)

City	Date Settled	Latest	ca 1970	ca 1960	ca 1950	ca 1930	ca 1900	ca 1850	ca 1800	18th c
Dubrovnik	7th c	31 (71)	31 (71)	23 (61)	16 (48)	19 (31)	13 (00)	5 (51)	7 (08)	
Ljubljana	34 BC	174 (71)	174 (71)	157 (61)	115 (48)	60 (31)	37 (00)	17 (51)	11 (00?e)	
		213s (71)	213s (71)							
Maribor	1147?	97 (71)	97 (71)	85 (61)	65 (48)	33 (31)	25 (00)	7 (51)	5 (00?e)	
Nis	140?	128 (71)	128 (71)	85 (61)	49 (48)	35 (31)	25 (00)	4 (50?e)	4 (00?e)	
Novi Sad	1687	141 (71)	141 (71)	111 (61)	78 (48)	64 (31)	29 (00)	10 (51)	13 (08)	
		163s (71)	163s (71)							
Osijek	8 AD?	95 (71)	95 (71)	73 (61)	49 (48)	40 (31)	25 (00)	13 (51)	9 (00?e)	8 (1757)
Rijeka	28 AD	132 (71)	132 (71)	101 (61)	73 (48)	53 (31)	38 (00)	11 (51)	9 (10)	
Sarajevo	1262	244 (71)	244 (71)	199 (61)	114 (48)	78 (31)	42 (95)	50 (50e)	40 (00?e)	
		271s (71)	271s (71)							
Skopje	2nd c?	313 (71)	313 (71)	172 (61)	92 (48)	65 (31)	20 (00?e)	10 (50?e)	8 (00?e)	
Split	305	153 (71)	153 (71)	93 (61)	50 (48)	35 (31)	27 (00)	11 (51)	7 (00?)	
Subotica	1391?	89 (71)	89 (71)	75 (61)	113 (48)	100 (31)	82 (01)	48 (51)	28 (08?e)	
Zagreb	7th c?	566 (71)	566 (71)	457 (61)	280 (48)	186 (31)	58 (00)	14 (51)	13 (00e)	

NORTH AMERICA

Population (thousands)

City	Date Settled	Latest	ca 1970	ca 1960	ca 1950	ca 1930	ca 1900	ca 1850	ca 1800	18th c
BAHAMAS										
*Nassau	1729	102s (70)	102s (70)	81s (63)	46s (53)	20s (31)	13s (01)	8s (51)	6s (00?)	
BARBADOS										
*Bridgetown	1628	9 (70)	9 (70)	11 (60)	13 (46)	13 (21)	21 (91)	20 (51)	17 (00?e)	
		115s (70)	115s (70)	94s (60)	69s (46)					
BELIZE										
Belize	1638?	41 (72e)	39 (70)	33 (60)	22 (46)	17 (31)	9 (01)	4 (55e)	1 (00?e)	

NORTH AMERICA

City	Date Settled	Population (thousands)								
		Latest	ca 1970	ca 1960	ca 1950	ca 1930	ca 1900	ca 1850	ca 1800	18th c
BERMUDA										
*Hamilton	1790	3 (70)	3 (70)	3 (60)	2 (50)	3 (31)	2 (01)			
		14s(70)	14s(70)	14s(60)						
CANADA										
Brampton	1830	103 (76)	41 (71)	18 (61)	8 (51)	6 (31)	3 (01)	2 (71)		
Brantford	1784	67 (76)	64 (71)	55 (61)	37 (51)	30 (31)	17 (01)	4 (52)		
		83s(76)	80s(71)							
Burlington	1810	98 (76)	87 (71)	47 (61)	6 (51)	3 (31)	1 (01)			
Calgary	1875	470 (76)	403 (71)	250 (61)	129 (51)	84 (31)	4 (01)	0.4 (84)		
		470s(76)	403s(71)	279s(61)	139s(51)					
Charlottetown	1768	17 (76)	19 (71)	18 (61)	16 (51)	13 (31)	12 (01)	5 (48)	2 (28)	
		25s(76)								
Chicoutimi	1676	58 (76)	34 (71)	32 (61)	23 (51)	12 (31)	4 (01)	1 (71)		
		129s(76)	134s(71)	105s(61)						
Edmonton	1819?	461 (76)	438 (71)	281 (61)	160 (51)	79 (31)	3 (01)			
		554s(76)	496s(71)	338s(61)	173s(51)					
Halifax	1749	118 (76)	122 (71)	93 (61)	86 (51)	59 (31)	41 (01)	26 (52)	5 (91)	4 (1752)
		268s(76)	223s(71)	184s(61)	134s(51)					
Hamilton	1813	312 (76)	309 (71)	274 (61)	208 (51)	156 (31)	53 (01)	14 (52)	3 (36)	
		529s(76)	499s(71)	395s(61)	260s(51)					
Jonquiere	1847	61 (76)	28 (71)	29 (61)	22 (51)	9 (31)	2 (11)			
Kingston	1783	56 (76)	59 (71)	54 (61)	33 (51)	23 (31)	18 (01)	12 (52)	0.3 (94)	
		91s(76)	86s(71)							
Kitchener	1806	132 (76)	112 (71)	74 (61)	45 (51)	31 (31)	10 (01)	2 (61)		
		272s(76)	227s(71)	155s(61)	63s(51)					
Laval	1699	246 (76)	228 (71)	125s(61)	38s(51)	16s (31)	10s (01)	10s (52)		
London	1826	240 (76)	223 (71)	170 (61)	95 (51)	71 (31)	38 (01)	7 (52)		
		270s(76)	286s(71)	181s(61)	122s(51)					
Longueuil	1657	122 (76)	98 (71)	24 (61)	11 (51)	5 (31)	3 (01)	1 (52)		

NORTH AMERICA

City	Date Settled	Population (thousands)								
		Latest	ca 1970	ca 1960	ca 1950	ca 1930	ca 1900	ca 1850	ca 1800	18th c
Mississauga	bef 1852	250 (76)	156 (71)	63s (61)	29s (51)	10s (31)	5s (01)	8s (52)		
Montreal	1642	1081 (76)	1214 (71)	1191 (61)	1022 (51)	819 (31)	268 (01)	58 (52)	16 (16)	8 (1740)
		2802s (76)	2743s (71)	2110s (61)	1395s (51)	1010s (31)				3 (1710)
Montreal-Nord	bef 1915	97 (76)	89 (71)	48 (61)	14 (51)	5 (31)	1 (21)			
Niagara Falls	1776	69 (76)	67 (71)	22 (61)	23 (51)	19 (31)	4 (01)	2 (71)		
Oshawa	1791	107 (76)	92 (71)	62 (61)	42 (51)	23 (31)	4 (01)	1 (52)		
		135s (76)	120s (71)	81s (61)	52s (51)					
*Ottawa	1826	304 (76)	302 (71)	268 (61)	202 (51)	127 (31)	60 (01)	8 (52)	1 (30)	
		693s (76)	603s (71)	430s (61)	282s (51)	166s (31)				
Quebec	1608	177 (76)	186 (71)	172 (61)	164 (51)	131 (31)	69 (01)	42 (52)	12 (00e)	9 (1759e)
		542s (76)	481s (71)	358s (61)	275s (51)	165s (31)				7 (1720e)
Regina	1882	150 (76)	139 (71)	112 (61)	71 (51)	53 (31)	2 (01)			
		151s (76)	141s (71)							
Saint Catharines	1792	123 (76)	110 (71)	84 (61)	38 (51)	25 (31)	10 (01)	4 (52)		
		302s (76)	303s (71)	96s (61)	67s (51)					
Saint John	1783?	86 (76)	89 (71)	55 (61)	51 (51)	48 (31)	41 (01)	23 (52)	9 (21e)	
		113s (76)	107s (71)	96s (61)	78s (51)					
Saint John's	1583?	87 (76)	88 (71)	64 (61)	53 (51)	40 (35)	30 (01)	21 (52)	3 (02)	
		143s (76)	132s (71)	91s (61)	68s (51)					
Saskatoon	1883	134 (76)	126 (71)	96 (61)	53 (51)	43 (31)	0.1 (01)			
		134s (76)	126s (71)							
Sault Sainte Marie	1814	81 (76)	80 (71)	43 (61)	32 (51)	23 (31)	7 (01)	0.9 (71)		
		82s (76)	81s (71)							
Sherbrooke	1794	77 (76)	81 (71)	67 (61)	51 (51)	29 (31)	12 (01)	3 (52)		
		105s (76)	85s (71)	70s (61)	56s (51)					
Sudbury	1887	98 (76)	91 (71)	80 (61)	42 (51)	19 (31)	2 (01)			
		157s (76)	155s (71)	111s (61)	71s (51)					
Sydney	1784	31 (76)	33 (71)	34 (61)	31 (51)	23 (31)	10 (01)	2 (61)		
		89s (76)	91s (71)	106s (61)	104s (51)					

NORTH AMERICA

Population (thousands)

City	Date Settled	Latest	ca 1970	ca 1960	ca 1950	ca 1930	ca 1900	ca 1850	ca 1800	18th c
Thunder Bay	1678	111 (76) 119s (76)	108 (71) 112s (71)	90 (61) 93s (61)	66 (51) 71s(51)	46 (31)	7 (01)	2 (81)		
Toronto	1793	633 (76) 2803s (76)	713 (71) 2628s (71)	672 (61) 1824s (61)	676 (51) 1117s(51)	631 (31) 665s (31)	208 (01)	31 (52)	1 (17)	
Trois-Rivieres	1634	53 (76) 99s (76)	56 (71) 98s (71)	53 (61) 87s (61)	46 (51) 68s(51)	35 (31)	10 (01)	5 (52)	2 (00?e)	
Vancouver	1870?	410 (76) 1166s (76)	426 (71) 1082s (71)	385 (61) 790s (61)	345 (51) 531s(51)	247 (31) 273s (31)	26 (01)			
Victoria	1851?	63 (76) 218s (76)	62 (71) 196s (71)	55 (61) 154s (61)	51 (51) 104s(51)	39 (31)	21 (01)	0.2 (54)		
Windsor	1745?	197 (76) 248s (76)	203 (71) 259s (71)	114 (61) 193s (61)	120 (51) 158s(51)	63 (31) 106s (31)	12 (01)	0.1 (52)		
Winnipeg	1812	561 (76) 578s (76)	246 (71) 540s (71)	265 (61) 476s (61)	236 (51) 354s(51)	219 (31) 239s (31)	42 (01)	0.2 (71)		
COSTA RICA										
*San Jose	1736	215 (73) 406s (73)	215 (73) 406s (73)	167 (63) 257s (63)	87 (50) 140s(50)	51 (27) 63s (27)	25 (03)	6 (44)	8 (23)	
CUBA										
Camaguey	1528	222 (76e)	198 (70)	162 (62e)	110 (53)	62 (31)	25 (99)	19 (46)	49 (27)	
Cienfuegos	1738?	85 (70)	85 (70)	70 (62e)	58 (53)	50 (31)	30 (99)	4 (46)		
Guantanamo	1819	149 (76e)	129 (70)	82 (62e)	65 (53)	28 (31)	7 (99)	2 (61)		
*Havana	1519	1861 (76e)	1751 (70)	978 (62e) 1463s (62e)	785 (53) 1218s(53)	521 (31)	236 (99)	107 (46)	94 (27) 51 (91)	25 (1700?e)
Holguin	1754	152 (76e)	132 (70)	80 (62e)	58 (53)	24 (31)	6 (99)	3 (46)	6 (27?)	
Matanzas	1693	85 (70)	85 (70)	80 (62e)	64 (53)	50 (31)	36 (99)	18 (46)	11 (27)	
Santa Clara	1689	147 (76e)	130 (70)	106 (62e)	77 (53)	38 (31)	14 (99)	6 (46)	9 (27)	
Santiago de Cuba	1514	316 (76e)	278 (70)	220 (62e)	163 (53)	102 (31)	43 (99)	24 (46)	27 (27)	

NORTH AMERICA

City	Date Settled	Latest	ca 1970	ca 1960	ca 1950	ca 1930	ca 1900	ca 1850	ca 1800	18th c
DOMINICAN REPUBLIC										
Santiago de los Caballeros	1504	209 (75e)	155 (70)	86 (60)	57 (50)	34 (35)	12 (00?e)	12 (50?e)	12 (00?e)	
*Santo Domingo	1496	923 (75e)	669 (70)	370 (60)	182 (50)	71 (35) 94s (35)	22 (00e)	12 (50?e)	12 (00?e) 22s (00?e)	
EL SALVADOR										
San Miguel	1530	59 (71) 118s (71)	59 (71) 118s (71)	40 (61) 82s (61)	27 (50) 57s(50)	18 (30) 40s (30)	25 (01)	6 (50?e)		
*San Salvador	1525	388 (73e) 339s (71)	337 (71) 339s (71)	256 (61) 256s(61)	162 (50) 171s(50)	89 (30) 96s (30)	60 (01)	25 (50e)	12 (07)	
Santa Ana	1576?	96 (71) 159s (71)	96 (71) 159s (71)	73 (61) 121s(61)	52 (50) 97s(50)	41 (30) 76s (30)	48 (01)	10 (50?e)		
GUATEMALA										
*Guatemala	1776	701 (73)	701 (73)	573 (64)	284 (50)	164 (40)	62 (93)	37 (50?e)	24 (95)	
Mixco		115 (73)	115 (73)	8 (64)	4 (50)	4 (40)	4 (93)	3 (50?e)		
HAITI										
*Port-au-Prince	1749	459 (75e)	306 (71) 494s (71)	240 (60e)	134 (50)	80 (29)	60 (00?e)	25 (50?e)	15 (90e)	
HONDURAS										
San Pedro Sula	1536	148 (74)	148 (74)	59 (61) 95s(61)	21 (50) 54s(50)	13 (30) 24s (30)	7 (01)			
*Tegucigalpa	1578	271 (74)	271 (74)	134 (61) 165s(61)	72 (50) 100s(50)	17 (30) 34s (30)	24 (01)	8 (50?e)		
JAMAICA										
*Kingston	1692	170 (73e) 626s (75e)	112 (70) 476s (70)	123 (60) 377s (60)	110 (43) 202s(43)	64 (21)	49 (91)	33 (44)	26 (88e)	

NORTH AMERICA

City	Date Settled	Population (thousands)								
		Latest	ca 1970	ca 1960	ca 1950	ca 1930	ca 1900	ca 1850	ca 1800	18th c
MARTINIQUE										
*Fort-de-France	1672	99s (74)	94 (67)	78 (61)	40 (54)	48 (31)	22 (01)	13 (67?)	10 (00?e)	
			97s (67)	85s (61)	61s(54)					
MEXICO										
Acapulco	1550	402 (76e)	174 (70)	49 (60)	28 (50)	7 (30)	5 (00)	4 (50e)	5 (93)	
Aguascalientes	1575	230 (76e)	181 (70)	127 (60)	93 (50)	62 (30)	35 (00)	7 (50?e)	13 (93)	
Celaya	1570	80 (70)	80 (70)	59 (60)	34 (50)	24 (30)	26 (00)	7 (50?e)		
Chihuahua	1639	366 (76e)	257 (70)	150 (60)	87 (50)	46 (30)	30 (00)	14 (50e)	12 (03e)	
Ciudad Juarez	1662	545 (76e)	407 (70)	262 (60)	123 (50)	40 (30)	8 (00)	2 (50?e)		
Ciudad Madero	bef 1910	129 (76e)	91 (70)	54 (60)	41 (50)	22 (30)	15 (21)			
Ciudad Obregon	1907	161 (76e)	114 (70)	68 (60)	31 (50)	8 (30)	0.2(21)			
Ciudad Victoria	1750	117 (76e)	84 (70)	51 (60)	32 (50)	18 (30)	10 (00)			
Coatzacoalcos	1580	105 (76e)	70 (70)	37 (60)	20 (50)	8 (30)	3 (00)			
Cordoba	1617	109 (76e)	78 (70)	47 (60)	33 (50)	16 (30)	8 (00)	6 (50?e)		
Cuernavaca	bef 1521	313 (76e)	134 (70)	37 (60)	31 (50)	9 (30)	10 (00)	10 (50?e)		
Culiacan	1533	263 (76e)	168 (70)	85 (60)	49 (50)	18 (30)	10 (00)	11 (58)	11 (03e)	
Durango	1563	200 (76e)	151 (70)	97 (60)	59 (50)	36 (30)	31 (00)	22 (50?e)	11 (90)	
Guadalajara	1542	1641 (76e)	1194 (70)	737 (60)	377 (50)	180 (30)	101 (00)	56 (50e)	24 (92)	
		2076s (76e)	1487s(70)							
Guanajuato	bef 1400	37 (70)	37 (70)	28 (60)	23 (50)	18 (30)	41 (00)	49 (50?)	41 (03e)	
Hermosillo	1750	264 (76e)	177 (70)	96 (60)	44 (50)	20 (30)	11 (00)			
Irapuato	1547	145 (76e)	117 (70)	84 (60)	49 (50)	29 (30)	20 (00)	17 (50?e)	8 (93)	
Jalapa	1313?	183 (76e)	122 (70)	66 (60)	51 (50)	37 (30)	20 (00)	6 (60?e)	6 (91)	
Leon	1576	526 (76e)	365 (70)	210 (60)	123 (50)	69 (30)	63 (00)	20 (50?e)		
Matamoros	1748	179 (76e)	138 (70)	92 (60)	46 (50)	10 (30)	6 (00)			
Mazatlan	bef 1541	162 (76e)	120 (70)	76 (60)	41 (50)	29 (30)	18 (00)	11 (50?e)		
Merida	bef 1528	245 (76e)	212 (70)	171 (60)	143 (50)	95 (30)	44 (00)	23 (50e)	28 (93)	
Mexicali	1901	346 (76e)	267 (70)	175 (60)	65 (50)	15 (30)	7 (21)			

Figure 49. In this unusual view of Mexico, now the largest city in North America and the second largest urbanized area in the world, the Plaza of the Three Cultures combines Aztec ruins, a Spanish colonial church, and modern housing developments. (Credit: Mexican National Tourist Council.)

NORTH AMERICA

Population (thousands)

City	Date Settled	Latest	ca 1970	ca 1960	ca 1950	ca 1930	ca 1900	ca 1850	ca 1800	18th c
*Mexico	1325	8628 (76e) 11,943s (76e)	2903 (70) 6874s (70) 1777s (70)	2832 (60) 4871s (60)	2235 (50) 3050s(50)	1029 (30)	345 (00)	210 (62)	113 (90)	100 (1700?e)
Minatitlan	1822	106 (76e)	68 (70)	35 (60)	22 (50)	12 (30)	1 (00)			
Monclova	1699	116 (76e)	78 (70)	43 (60)	19 (50)	7 (30)	7 (00)			
Monterrey	1579?	1090 (76e) 1725s (76e)	858 (70)	597 (60)	333 (50)	133 (30)	62 (00)	14 (50e)	11 (03e)	
Morelia	1541	219 (76e)	161 (70)	101 (60)	63 (50)	40 (30)	37 (00)	18 (50?e)	18 (03e)	
Netzahualcoyotl	16th c?	580 (70)	580 (70)	2 (60)	0.6(50)					
Nuevo Laredo	1755	204 (76e)	149 (70)	93 (60)	58 (50)	22 (30)	7 (00)			
Oaxaca	1486	123 (76e)	100 (70)	72 (60)	47 (50)	33 (30)	35 (00)	25 (50)	18 (92)	
Orizaba	1457	112 (76e)	93 (70)	70 (60)	56 (50)	43 (30)	33 (00)	16 (68e)	9 (93)	
Pachuca de Soto	1534	84 (70)	84 (70)	65 (60)	59 (50)	43 (30)	37 (00)	5 (50?e)		
Poza Rica	1939?	170 (76e)	120 (70)	20 (60)	15 (50)	4 (40)				
Puebla	1532	499 (76e)	402 (70)	289 (60)	211 (50)	115 (30)	94 (00)	72 (48)	57 (93)	50 (1746e)
Queretaro	1440	158 (76e)	113 (70)	68 (60)	49 (50)	33 (30)	33 (00)	30 (50?e)	20 (92)	70 (1678e)
Reynosa	bef 1870	206 (76e)	137 (70)	74 (60)	34 (50)	5 (30)	2 (00)			
Saltillo	1575	222 (76e)	161 (70)	99 (60)	70 (50)	45 (30)	24 (00)	20 (50?e)	6 (03e)	
San Luis Potosi	1576	292 (76e)	230 (70)	160 (60)	126 (50)	74 (30)	61 (00)	40 (50e)	9 (93)	
Tampico	1823	231 (76e) 359s(76e)	180 (70)	123 (60)	94 (50)	68 (30)	16 (00)	7 (50?e)		
Taxco	1529	27 (70)	27 (70)	15 (60)	10 (50)	4 (30)	4 (00)	1 (30?e)		
Tepic	bef 1524	120 (76e)	88 (70)	54 (60)	25 (50)	15 (30)	15 (00)	10 (50?e)		
Tijuana	1830?	412 (76e) 536s(76e)	277 (70)	152 (60)	60 (50)	8 (30)	0.2(00)			
Toluca	1120?	148 (76e)	114 (70)	77 (60)	53 (50)	41 (30)	26 (00)	12 (50?e)	7 (93)	
Torreon	1893	257 (76e) 373s(76e)	223 (70)	180 (60)	129 (50)	66 (30)	14 (00)			
Uruapan del Progreso	1536	122 (76e)	83 (70)	46 (60)	31 (50)	17 (30)	10 (00)			

NORTH AMERICA

City	Date Settled	Latest	ca 1970	ca 1960	ca 1950	ca 1930	ca 1900	ca 1850	ca 1800	18th c
				Population (thousands)						
Veracruz	1599	277 (76e)	214 (70)	145 (60)	101 (50)	68 (30)	29 (00)	8 (50e)	16 (03e)	
Villahermosa	1596	152 (76e)	100 (70)	52 (60)	34 (50)	15 (30)	11 (00)	4 (50?e)		
NETHERLANDS ANTILLES										
*Willemstad	1527	50 (70e)	50 (70e)	44 (60) / 94s(60)	41 (48e)	19 (31)	14 (05)	8 (50?e)		
NICARAGUA										
*Managua	bef 1521	313 (74e)	396 (71)	235 (63)	109 (50)	28 (20)	30 (05e)	12 (50?e)		
PANAMA										
Colon	1850	95 (75e)	68 (70)	60 (60)	52 (50)	30 (30)	18 (11)	18 (60)		
*Panama	1673	416 (76e)	349 (70)	260 (60)	128 (50)	74 (30)	38 (11)		17s (22)	
PUERTO RICO										
Bayamon	1750	148 (70)	148 (70)	15 (60)	20 (50)	13 (30)	2 (99)			
Carolina	bef 1851	94 (70)	94 (70)	3 (60)	5 (50)	4 (30)	2 (99)			
Ponce	1680	128 (70) / 128s (70)	128 (70) / 128s(70)	114 (60) / 114s(60)	99 (50)	53 (30)	28 (99)	30 (60)		
*San Juan	1521	472 (72e) / 820s (70)	453 (70) / 820s (70)	432 (60) / 542s(60)	225 (50)	115 (30)	32 (99)	18 (60)	39 (00?e)	
TRINIDAD AND TOBAGO										
*Port of Spain	bef 1595	60 (73e)	68 (70)	94 (60)	93 (46)	70 (31)	54 (01)	18 (51)	10 (00?e)	
USA										
Abilene	1881	96 (75e)	90 (70) / 91s(70)	90 (60) / 92s(60)	46 (50)	23 (30)	3 (00)	3 (90)		
Akron	1807	252 (75e)	275 (70) / 543s(70)	290 (60) / 458s(60)	275 (50) / 367s(50)	255 (30)	43 (00)	3 (50)	2 (40)	

NORTH AMERICA

Population (thousands)

City	Date Settled	Latest	ca 1970	ca 1960	ca 1950	ca 1930	ca 1900	ca 1850	ca 1800	18th c
Alameda	1850?	72 (75e)	71 (70)	64 (60)	64 (50)	35 (30)	16 (00)	0.5 (60)		
Albany (Georgia)	1836	73 (75e)	73 (70)	56 (60)	31 (50)	15 (30)	5 (00)	2 (60)		
			77s (70)	58s (60)						
Albany (New York)	1624	110 (75e)	116 (70)	130 (60)	135 (50)	127 (30)	94 (00)	51 (50)	5 (00)	1 (1714)
			487s (70)	455s (60)	292s(50)					
Albuquerque	1706	279 (75e)	244 (70)	201 (60)	97 (50)	27 (30)	6 (00)	6 (50e)	5 (99e)	1 (1749e)
			297s (70)	241s (60)						
Alexandria (Louisiana)	1805	49 (75e)	42 (70)	40 (60)	35 (50)	23 (30)	6 (00)	0.7 (50)		
			78s (70)							
Alexandria (Virginia)	1713?	105 (75e)	111 (70)	91 (60)	62 (50)	24 (30)	15 (00)	9 (50)	5 (00)	
Allentown	1762	107 (75e)	110 (70)	108 (60)	107 (50)	93 (30)	35 (00)	4 (50)	1 (00)	
			364s (70)	256s (60)	226s(50)					
Alton	1783	36 (75e)	40 (70)	43 (60)	33 (50)	30 (30)	14 (00)	4 (50)	2 (40)	
			96s (70)							
Altoona	1849	60 (75e)	63 (70)	69 (60)	77 (50)	82 (30)	39 (00)	4 (60)		
			82s (70)	83s (60)	87s(50)					
Amarillo	1887	139 (75e)	127 (70)	138 (60)	74 (50)	43 (30)	1 (00)			
			127s (70)	138s (60)	74s(50)					
Anaheim	1857	194 (75e)	166 (70)	104 (60)	15 (50)	11 (30)	1 (00)	0.9 (70)		
Anchorage	1914	161 (75e)	48 (70)	44 (60)	11 (50)	2 (30)	2 (20)			
			111s (70)							
Anderson	1823	69 (75e)	71 (70)	49 (60)	47 (50)	40 (30)	20 (00)	0.4 (50)		
			81s (70)							
Annapolis	1649	32 (75e)	30 (70)	23 (60)	10 (50)	10 (30)	8 (00)	3 (50)	2 (20)	1 (1747) 0.4(1714)
Ann Arbor	1824	104 (75e)	100 (70)	67 (60)	48 (50)	27 (30)	15 (00)	5 (50)		
			179s (70)	115s (60)						
Appleton	1835	59 (75e)	57 (70)	48 (60)	34 (50)	25 (30)	15 (00)	2 (60)		
			130s (70)							

NORTH AMERICA

City	Date Settled	Population (thousands)								
		Latest	ca 1970	ca 1960	ca 1950	ca 1930	ca 1900	ca 1850	ca 1800	18th c
Arden-Arcade	bef 1890	82 (70)	82 (70)	73 (60)						
Arlington (Texas)	1876	111 (75e)	90 (70)	45 (60)	8 (50)	4 (30)	1 (00)	0.7 (90)		
Arlington (Virginia)	1700?	152 (75e)	174 (70)	163 (60)	135 (50)	27 (30)	3s (00)			
Asheville	1794	60 (75e)	58 (70)	60 (60)	53 (50)	50 (30)	15 (00)	0.5 (50)	0.04(00)	
			72s (70)	69s (60)	58s(50)					
Athens	1801	49 (75e)	44 (70)	31 (60)	28 (50)	18 (30)	10 (00)	2 (50)		
Atlanta	1837	436 (75e)	497 (70)	487 (60)	331 (50)	270 (30)	90 (00)	3 (50)		
			1173s(70)	768s(60)	508s(50)					
Atlantic City	1790?	44 (75e)	48 (70)	60 (60)	62 (50)	66 (30)	28 (00)	0.7 (60)		
			134s(70)	125s(60)	105s(50)					
Augusta	1735	54 (75e)	60 (70)	71 (60)	72 (50)	60 (30)	39 (00)	12 (52)	1 (91e)	1 (1764e)
			149s(70)	124s(60)	88s(50)					
Aurora (Colorado)	1890	118 (75e)	75 (70)	49 (60)	11 (50)	2 (30)	0.2(00)			
Aurora (Illinois)	bef 1834	77 (75e)	74 (70)	64 (60)	51 (50)	47 (30)	24 (00)	6 (60)		
			233s(70)	86s(60)						
Austin	1839	309 (76)	252 (70)	187 (60)	132 (50)	53 (30)	22 (00)	0.6 (50)		
			264s(70)	187s(60)	136s(50)					
Bakersfield	1868	77 (75e)	70 (70)	57 (60)	35 (50)	26 (30)	5 (00)	0.8 (80)		
			176s(70)	142s(60)						
Baltimore	1730	852 (75e)	906 (70)	939 (60)	950 (50)	805 (30)	509 (00)	169 (50)	27 (00)	0.2(1752e)
			1580s(70)	1419s(60)	1162s(50)					
Bangor	1769	32 (75)	33 (70)	39 (60)	32 (50)	29 (30)	22 (00)	14 (50)	0.3 (00)	
Baton Rouge	1719	294 (75e)	166 (70)	152 (60)	126 (50)	31 (30)	11 (00)	4 (50)	4 (10)	
			249s(70)	193s(60)	139s(50)					
Battle Creek	1831	43 (75e)	39 (70)	44 (60)	49 (50)	44 (30)	19 (00)	1 (50)		
			78s(70)							

NORTH AMERICA

Population (thousands)

City	Date Settled	Latest	ca 1970	ca 1960	ca 1950	ca 1930	ca 1900	ca 1850	ca 1800	18th c
Bay City	1831	47 (75e)	49 (70) / 78s (70)	54 (60) / 73s (60)	53 (50)	47 (30)	28 (00)	2 (60)		
Bayonne	1656	74 (75e)	73 (70)	74 (60)	77 (50)	89 (30)	33 (00)	4 (70)		
Beaumont	1835	114 (75e)	118 (70) / 118s (70)	119 (60) / 119s (60)	94 (50) / 94s (50)	58 (30)	9 (00)	0.1 (50?)		
Berkeley	1853	110 (75e)	117 (70)	111 (60)	114 (50)	82 (30)	13 (00)	5 (90)		
Bethlehem	1741	74 (75e)	73 (70)	75 (60)	66 (50)	58 (30)	7 (00)	2 (50)	0.5 (00)	1 (1756)
Beverly Hills	1906	35 (75e)	33 (70)	31 (60)	29 (50)	17 (30)	0.7 (20)			
Billings	1882	69 (75e)	62 (70) / 71s (70)	53 (60) / 61s (60)	32 (50)	16 (30)	3 (00)	0.8 (90)		
Biloxi	1719	46 (75)	48 (70) / 122s (70)	44 (60)	37 (50)	15 (30)	5 (00)	1 (70)	0.4 (01e)	
Binghamton	1787	61 (75e)	64 (70) / 167s (70)	76 (60) / 158s (60)	81 (50) / 144s (50)	77 (30)	40 (00)	5 (50)	0.3 (12e)	
Birmingham	1871	276 (75e)	301 (70) / 558s (70)	341 (60) / 521s (60)	326 (50) / 445s (50)	260 (30)	38 (00)	3 (80)		
Bismarck	1872	38 (75)	35 (70)	28 (60)	19 (50)	11 (30)	3 (00)	2 (80)		
Bloomington (Illinois)	1822	41 (75)	40 (70) / 69s (70)	36 (60)	34 (50)	31 (30)	23 (00)	2 (50)		
Bloomington (Indiana)	1815	49 (75e)	43 (70)	31 (60)	28 (50)	18 (30)	6 (00)	1 (50)		
Bloomington (Minnesota)	1843	79 (75)	82 (70)	50 (60)	10 (50)	3s (30)	1s (00)	0.4s (50)		
Boise City	1863	100 (75e)	75 (70) / 85s (70)	34 (60)	34 (50)	22 (30)	6 (00)	1 (70)		
Boston	1630	637 (75e)	641 (70) / 2653s (70)	697 (60) / 2413s (60)	801 (50) / 2233s (50)	781 (30)	561 (00)	137 (50)	25 (00)	16 (1752) / 11 (1722)
Boulder	1859	79 (75e)	67 (70) / 69s (70)	38 (60)	20 (50)	11 (30)	6 (00)	0.3 (70)		

NORTH AMERICA

Population (thousands)

City	Date Settled	Latest	ca 1970	ca 1960	ca 1950	ca 1930	ca 1900	ca 1850	ca 1800	18th c
Bridgeport	1639	143 (75e)	157 (70); 413s(70)	157 (60); 367s(60)	159 (50); 237s(50)	147 (30)	71 (00)	6 (50)	0.6 (10)	
Bristol	1727	59 (75e)	55 (70); 72s(70)	45 (60)	36 (50)	28 (30)	6 (00)	3 (50)	3 (00)	
Brockton	1700	96 (75e)	89 (70); 149s(70)	73 (60); 111s(60)	63 (50); 92s(50)	64 (30)	40 (00)	4 (50)	2 (30)	
Brownsville	1846	72 (75e)	53 (70); 53s(70)	48 (60)	36 (50)	22 (30)	6 (00)	3 (60)		
Buffalo	1803	407 (75e)	463 (70); 1087s(70)	533 (60); 1054s(60)	580 (50); 798s(50)	573 (30)	352 (00)	42 (50)	2 (10)	
Burbank	1887	86 (75e)	89 (70)	90 (60)	79 (50)	17 (30)	0.3(00e)	6 (50)		
Burlington	1773	37 (75e)	39 (70)	36 (60)	33 (50)	25 (30)	19 (00)	6 (50)	0.8 (00)	
Butte	1864	23 (75e)	23 (70)	28 (60)	33 (50)	40 (30)	30 (00)	3 (80)		
Cambridge	1630	102 (75e)	100 (70)	108 (60)	121 (50)	114 (30)	92 (00)	15 (50)	2 (00)	2 (1765)
Camden	1681	90 (76)	103 (70)	117 (60)	125 (50)	119 (30)	76 (00)	9 (50)	1 (28)	
Canton	1805	102 (75e)	110 (70); 244s(70)	114 (60); 214s(60)	117 (50); 174s(50)	105 (30)	31 (00)	3 (50)	0.8 (10)	
Casper	1888	41 (75e)	39 (70)	39 (60)	24 (50)	17 (30)	0.9(00)	0.5 (90)		
Cedar Rapids	1838	109 (75)	111 (70); 132s(70)	92 (60); 105s(60)	72 (50); 78s(50)	56 (30)	26 (00)	2 (60)		
Champaign	1854	58 (75e)	57 (70); 100s(70)	50 (60); 78s(60)	40 (50)	20 (30)	9 (00)	2 (60)		
Charleston (South Carolina)	1670	57 (75e)	67 (70); 228s(70)	66 (60); 160s(60)	70 (50); 120s(50)	62 (30)	56 (00)	43 (50)	19 (00)	11 (1770)
Charleston (West Virginia)	1788	67 (75e)	72 (70); 158s(70)	86 (60); 169s(60)	74 (50); 131s(50)	60 (30)	11 (00)	1 (50)	0.6 (00)	3 (1705e)
Charlotte	1748	281 (75e)	241 (70); 280s(70)	202 (60); 210s(60)	134 (50); 141s(50)	83 (30)	18 (00)	1 (50)	0.1 (00)	
Charlottesville	1762	42 (75e)	39 (70)	29 (60)	26 (50)	15 (30)	6 (00)	3 (70)		

NORTH AMERICA

City	Date Settled	Population (thousands)								
		Latest	ca 1970	ca 1960	ca 1950	ca 1930	ca 1900	ca 1850	ca 1800	18th c
Chattanooga	1815	162 (75e)	119 (70)	130 (60)	131 (50)	120 (30)	30 (00)	3 (53)		
Cheektowaga	1809	121 (75e)	94 (70)	71 (60)	39 (50)					
			224s (70)	205s (60)	168s (50)	15s (30)	4s (00)	3s (50)		
Chesapeake	bef 1892	104 (75e)	90 (70)	22 (60)	10 (50)	8 (30)	8 (20)			
Chester	1644	49 (75e)	56 (70)	64 (60)	66 (50)	59 (30)	34 (00)	2 (50)	0.7 (20)	0.5 (1708e)
Cheyenne	1867	47 (75e)	41 (70)	44 (60)	32 (50)	17 (30)	14 (00)	1 (70)		
Chicago	1803	3099 (75e)	3369 (70)	3550 (60)	3621 (50)	3376 (30)	1699 (00)	30 (50)	4 (40)	
			6717s (70)	5962s (60)	4921s (50)					
Chicopee	1652	58 (75e)	67 (70)	62 (60)	49 (50)	44 (30)	19 (00)	8 (50)		
Cincinnati	1789	413 (75e)	453 (70)	503 (60)	504 (50)	451 (30)	326 (00)	115 (50)	0.7 (00)	
			1111s (70)	994s (60)	813s (50)					
Clarksville	1784	52 (75e)	32 (70)	22 (60)	16 (50)	9 (30)	9 (00)	3 (70)		
Clearwater	1841	67 (75e)	52 (70)	35 (60)	16 (50)	8 (30)	0.3 (00)			
Cleveland	1796	639 (75e)	751 (70)	876 (60)	915 (50)	900 (30)	382 (00)	17 (50)	0.5 (10)	
			1960s (70)	1785s (60)	1384s (50)					
Clifton	1685	79 (75e)	82 (70)	82 (60)	65 (50)	47 (30)	26 (20)			
Colorado Springs	1859	180 (75e)	135 (70)	70 (60)	45 (50)	33 (30)	21 (00)	4 (80)		
			205s (70)	100s (60)						
Columbia (Missouri)	1819	63 (75e)	59 (70)	37 (60)	32 (50)	15 (30)	6 (00)	0.7 (50)		
			59s (70)							
Columbia (South Carolina)	1786	112 (75e)	114 (70)	97 (60)	87 (50)	52 (30)	21 (00)	6 (50)	2 (16)	
			242s (70)	163s (60)	121s (50)					
Columbus (Georgia)	1828	159 (75e)	155 (70)	117 (60)	80 (50)	43 (30)	18 (00)	6 (50)	3 (40)	
			210s (70)	158s (60)	118s (50)					
Columbus (Ohio)	1797	536 (75e)	540 (70)	471 (60)	376 (50)	291 (30)	126 (00)	18 (50)	2 (30)	
			790s (70)	617s (60)	438s (50)					
Compton	1867	75 (75e)	79 (70)	72 (60)	48 (50)	13 (30)	0.9 (10)			
Concord (California)	1868	95 (75e)	85 (70)	36 (60)	7 (50)	1 (30)	0.7 (10)			

NORTH AMERICA

City	Date Settled	Latest	ca 1970	ca 1960	ca 1950	ca 1930	ca 1900	ca 1850	ca 1800	18th c
Concord (New Hampshire)	1726	29 (75e)	30 (70)	29 (60)	28 (50)	25 (30)	20 (00)	9 (50)	2 (00)	0.8(1767)
Corpus Christi	1839	215 (75e)	205 (70) 213s(70)	168 (60) 177s(60)	108 (50) 123s(50)	28 (30)	5 (00)	0.5 (50)		
Council Bluffs	1827	59 (75e)	60 (70)	56 (60)	45 (50)	42 (30)	26 (00)	2 (60)		
Covington	1815	44 (75e)	53 (70)	60 (60)	64 (50)	65 (30)	43 (00)	9 (50)	0.7 (30)	
Cranston	1638	74 (75e)	74 (70)	67 (60)	55 (50)	43 (30)	13 (00)	4 (50)	2 (00)	1 (1755)
Dallas	1842	813 (75e)	844 (70) 1339s(70)	680 (60) 932s(60)	434 (50) 539s(50)	260 (30)	43 (00)	0.2 (50)		
Danbury	1684	55 (75e)	51 (70) 67s(70)	23 (60)	22 (50)	22 (30)	17 (00)	6 (50)	3 (00)	2 (1756)
Davenport	1836	100 (75)	98 (70) 266s(70)	89 (60) 227s(60)	75 (50) 195s(50)	61 (30)	35 (00)	2 (50)		
Dayton	1796	206 (75e)	244 (70) 686s(70)	262 (60) 502s(60)	244 (50) 347s(50)	201 (30)	85 (00)	11 (50)	0.4 (10)	
Daytona Beach	1870	48 (75e)	45 (70) 115s(70)	37 (60)	30 (50)	17 (30)	0.3(10)			
Dearborn	1795?	99 (75e)	104 (70)	112 (60)	95 (50)	50 (30)	0.8(00)	1 (50)		
Dearborn Heights	1796?	79 (75e)	80 (70)	61s(60)	20s(50)	1s (30)				
Decatur	1829	90 (75e)	90 (70) 100s(70)	78 (60) 90s(60)	66 (50) 74s(50)	58 (30)	21 (00)	4 (60)		
Denver	1858	485 (75e)	515 (70) 1047s(70)	494 (60) 804s(60)	416 (50) 499s(50)	288 (30)	134 (00)	5 (60)		
Des Moines	1843	194 (75e)	201 (70) 256s(70)	209 (60) 241s(60)	178 (50) 200s(50)	143 (30)	62 (00)	1 (50)		
Detroit	1701	1335 (75e)	1514 (70) 3974s(70)	1670 (60) 3538s(60)	1850 (50) 2659s(50)	1569 (30)	286 (00)	21 (50)	0.8 (10)	0.8(1765)
Downey	1873	86 (75e)	88 (70)	83 (60)	9 (40)		1 (00e)			
Dubuque	1837	62 (75)	62 (70) 66s(70)	57 (60) 59s(60)	50 (50)	42 (30)	36 (00)	3 (50)		

NORTH AMERICA

City	Date Settled	Population (thousands)								
		Latest	ca 1970	ca 1960	ca 1950	ca 1930	ca 1900	ca 1850	ca 1800	18th c
Duluth	1852	94 (75e)	101 (70)	107 (60)	105 (50)	101 (30)	53 (00)	0.1 (60)		
			138s(70)	145s(60)	143s(50)					
Dundalk	bef 1894	85 (70)	85 (70)	82 (60)	39 (50)	2 (30?)				
Durham	1852?	101 (75e)	95 (70)	78 (60)	71 (50)	52 (30)	7 (00)	0.3 (70)		
			101s(70)	85s(60)	73s(50)					
East Los Angeles	1874?	105 (70)	105 (70)	104 (60)	42 (40)					
East Orange	1678	73 (75e)	75 (70)	77 (60)	79 (50)	68 (30)	22 (00)	4s (70)		
East Saint Louis	1797	58 (75e)	70 (70)	82 (60)	82 (50)	74 (30)	30 (00)	6 (70)		
Elgin	1835	61 (76)	56 (70)	49 (60)	44 (50)	36 (30)	22 (00)	3 (60)		
Elizabeth	1664	104 (75e)	113 (70)	108 (60)	113 (50)	115 (30)	52 (00)	6 (50)	3 (10)	
Elmira	1788	37 (75e)	40 (70)	47 (60)	50 (50)	47 (30)	36 (00)	8 (50)		
			74s(70)							
El Paso	1827	386 (75e)	322 (70)	277 (60)	130 (50)	102 (30)	16 (00)	0.4 (60)		
			337s(70)	277s(60)	137s(50)					
Elyria	1817	52 (75e)	53 (70)	44 (60)	30 (50)	26 (30)	9 (00)	1 (50)		
Erie	1795	128 (75e)	129 (70)	138 (60)	131 (50)	116 (30)	53 (00)	6 (50)	0.1 (00)	
			175s(70)	177s(60)	152s(50)					
Euclid	1798	63 (75e)	72 (70)	63 (60)	41 (50)	13 (30)	2 (10)	0.1 (50?e)		
Eugene	1851	92 (75e)	78 (70)	51 (60)	36 (50)	19 (30)	3 (00)	0.9 (70)		
			141s(70)	96s(60)						
Evanston	1826	77 (75e)	80 (70)	79 (60)	74 (50)	63 (30)	19 (00)	0.8 (60)		
Evansville	1812	134 (75e)	139 (70)	142 (60)	129 (50)	102 (30)	59 (00)	3 (50)		
			142s(70)	144s(60)	138s(50)					
Everett	1862	48 (75e)	54 (70)	40 (60)	34 (50)	31 (30)	8 (00)	3 (91e)		
Fairbanks	1901	30 (75e)	15 (70)	13 (60)	6 (50)	2 (30)	4 (10)			
Fairfield	1859	50 (75e)	44 (70)	15 (60)	3 (50)	1 (30)	0.8(10)			
Fall River	1656	100 (75e)	97 (70)	100 (60)	112 (50)	115 (30)	105 (00)	12 (50)	1 (10)	
			139s(70)	124s(60)	118s(50)					
Fargo	1871	56 (75)	53 (70)	47 (60)	38 (50)	29 (30)	10 (00)	3 (80)		
			85s(70)	73s(60)						

NORTH AMERICA

City	Date Settled	Latest	ca 1970	ca 1960	ca 1950	ca 1930	ca 1900	ca 1850	ca 1800	18th c
				Population (thousands)						
Fayetteville	1729?	66 (75e)	54 (70) 161s(70)	47 (60)	35 (50)	13 (30)	5 (00)	5 (50)	2 (00)	
Fitchburg	1730?	39 (75e)	43 (70) 78s(70)	43 (60) 72s(60)	43 (50)	41 (30)	32 (00)	5 (50)	1 (00)	0.3(1765)
Flint	1819	174 (75e)	193 (70) 330s(70)	197 (60) 278s(60)	163 (50) 198s(50)	156 (30)	13 (00)	2 (50)		
Fort Lauderdale	1838	153 (75e)	140 (70) 614s(70)	84 (60) 320s(60)	36 (50)	9 (30)	0.1(00)			
Fort Smith	1817	68 (77)	63 (70) 76s(70)	53 (60) 62s(60)	48 (50) 56s(50)	31 (30)	12 (00)	1 (50)		
Fort Wayne	bef 1685	185 (75e)	178 (70) 225s(70)	162 (60) 180s(60)	134 (50) 140s(50)	115 (30)	45 (00)	4 (50)	2 (40)	
Fort Worth	1843	358 (75e)	393 (70) 677s(70)	356 (60) 503s(60)	279 (50) 316s(50)	163 (30)	27 (00)	7 (80)		
Fremont	1797	118 (75e)	101 (70)	44 (60)	5 (50)	6 (40e)	2 (00e)			
Fresno	1872	177 (75e)	166 (70) 263s(70)	134 (60) 213s(60)	92 (50) 131s(50)	53 (30)	12 (00)	1 (80)		
Fullerton	1887	94 (75e)	86 (70)	56 (60)	14 (50)	11 (30)	2 (10)			
Gadsden	1836?	50 (75e)	54 (70) 68s(70)	58 (60) 69s(60)	56 (50)	24 (30)	4 (00)	2 (80)		
Gainesville	1830	72 (75e)	65 (70) 69s(70)	30 (60)	27 (50)	10 (30)	4 (00)	0.3 (60)		
Galveston	1816	60 (75e)	62 (70) 62s(70)	67 (60) 118s(60)	67 (50) 72s(50)	53 (30)	38 (00)	4 (50)		
Garden Grove	1876	118 (75e)	121 (70)	84 (60)	4 (50)	2 (40?e)	0.1(00e)			
Garland	1886	111 (75e)	81 (70)	39 (60)	11 (50)	2 (30)	0.8(00)	0.5 (90)		
Gary	1906	168 (75e)	175 (70)	178 (60)	134 (50)	100 (30)	17 (10)			
Gastonia	bef 1877	49 (75e)	47 (70) 95s(70)	37 (60)	23 (50)	17 (30)	5 (00)	0.2 (80)		
Glendale	1886	132 (75e)	133 (70)	119 (60)	96 (50)	63 (30)	3 (10)			

NORTH AMERICA

Population (thousands)

City	Date Settled	Latest	ca 1970	ca 1960	ca 1950	ca 1930	ca 1900	ca 1850	ca 1800	18th c
Grand Forks	1871	42 (75e)	39 (70)	34 (60)	27 (50)	17 (30)	8 (00)	2 (80)		
Grand Rapids	1824?	188 (75e)	198 (70)	177 (60)	177 (50)	169 (30)	88 (00)	3 (50)		
			353s(70)	294s(60)	227s(50)					
Great Falls	1883	61 (75e)	60 (70)	55 (60)	39 (50)	29 (30)	15 (00)	4 (90)		
			71s(70)	58s(60)						
Green Bay	bef 1669	88 (76)	88 (70)	63 (60)	53 (50)	37 (30)	19 (00)	2 (50)	0.5 (20e)	
			129s(70)	97s(60)						
Greensboro	1749	156 (75e)	144 (70)	120 (60)	74 (50)	54 (30)	10 (00)	0.5 (70)	0.4 (29)	
			152s(70)	123s(60)	83s(50)					
Greenville	1797	58 (76)	61 (70)	66 (60)	58 (50)	29 (30)	12 (00)	1 (50)	0.2 (10)	
			157s(70)	127s(60)	63s(50)					
Gulfport	1887	43 (75e)	41 (70)	30 (60)	23 (50)	13 (30)	1 (00)	0.7 (80)		
Hamilton	1791	66 (75e)	68 (70)	72 (60)	58 (50)	52 (30)	24 (00)	1 (53?e)	0.3 (10)	
			91s(70)	90s(60)						
Hammond	1851	105 (75e)	108 (70)	112 (60)	88 (50)	65 (30)	12 (00)			
Hampton	1610	125 (75e)	121 (70)	89 (60)	6 (50)	6 (30)	3 (00)	1 (50)		
Harrisburg	1785	58 (75e)	68 (70)	80 (60)	90 (50)	80 (30)	50 (00)	8 (50)	1 (00)	
			241s(70)	210s(60)	170s(50)					
Hartford	1635	138 (75e)	158 (70)	162 (60)	177 (50)	164 (30)	80 (00)	14 (50)	5 (00)	3 (1756)
			465s(70)	382s(60)	301s(50)					
Hayward	1854	93 (75e)	93 (70)	73 (60)	14 (50)	6 (30)	2 (00)	0.5 (70)		
Helena	1864	26 (75e)	23 (70)	20 (60)	18 (50)	12 (30)	11 (00)	3 (70)		
Hialeah	1921	118 (75e)	102 (70)	67 (60)	20 (50)	3 (30)				
High Point	1853	61 (75e)	63 (70)	62 (60)	40 (50)	37 (30)	4 (00)	1 (80)		
			94s(70)	67s(60)						
Hollywood	1921	119 (75e)	107 (70)	35 (60)	14 (50)	3 (30)				
Holyoke	1745	46 (75e)	50 (70)	53 (60)	55 (50)	57 (30)	46 (00)	3 (50)		
Hot Springs	1807	38 (75e)	36 (70)	28 (60)	29 (50)	20 (30)	10 (00)	0.2 (60)		
Houston	1836	1327 (75e)	1233 (70)	938 (60)	596 (50)	292 (30)	45 (00)	2 (50)		
			1678s(70)	1140s(60)	701s(50)					

NORTH AMERICA

City	Date Settled	Population (thousands)								
		Latest	ca 1970	ca 1960	ca 1950	ca 1930	ca 1900	ca 1850	ca 1800	18th c
Huntington	1871	69 (75e)	74 (70)	84 (60)	86 (50)	76 (30)	12 (00)	3 (80)		
Huntington Beach	bef 1904	150 (75e)	168s (70)	166s (60)	156s(50)	4 (30)	0.8(10)			
Huntsville	1807	136 (75e)	138 (70)	72 (60)	16 (50)	12 (30)	8 (00)	3 (50)		
			147s(70)	75s(60)						
Idaho Falls	1865	37 (75)	36 (70)	33 (60)	19 (50)	9 (30)	1 (00)			
Independence	1825	111 (75e)	112 (70)	62 (60)	37 (50)	15 (30)	7 (00)	3 (60)		
Indianapolis	1820	715 (75e)	745 (70)	476 (60)	427 (50)	364 (30)	169 (00)	8 (50)	3 (40)	
			820s (70)	639s(60)	502s(50)					
Inglewood	1887	87 (75e)	90 (70)	63 (60)	46 (50)	19 (30)	2 (10)	3 (60)		
Irving	1902	104 (75e)	97 (70)	46 (60)	3 (50)	0.7(30)	0.4(20)			
Jackson (Michigan)	1829	44 (75e)	45 (70)	51 (60)	51 (50)	55 (30)	25 (00)	2 (50)		
			79s(70)	71s(60)						
Jackson (Mississippi)	bef 1800	167 (75e)	154 (70)	144 (60)	98 (50)	48 (30)	8 (00)	3 (60)		
			190s(70)	147s(60)	100s(50)					
Jacksonville	1816	535 (75e)	529 (70)	201 (60)	205 (50)	130 (30)	28 (00)	1 (50)		
			530s(70)	373s(60)	243s(50)					
Jefferson City	1821	35 (75e)	32 (70)	28 (60)	25 (50)	22 (30)	10 (00)	3 (60)	1 (40)	
Jersey City	1629?	244 (75e)	261 (70)	276 (60)	299 (50)	317 (30)	206 (00)	7 (50)	1 (20)	
Johnstown	1800	40 (75e)	42 (70)	54 (60)	63 (50)	67 (30)	36 (00)	1 (50)	0.9 (40)	
			96s(70)	96s(60)	93s(50)					
Joliet	1831	74 (75)	79 (70)	67 (60)	52 (50)	43 (30)	29 (00)	3 (50)		
			154s(70)	117s(60)						
Kalamazoo	1829	80 (75e)	86 (70)	82 (60)	58 (50)	55 (30)	24 (00)	3 (50)		
			152s(70)	116s(60)	83s(50)					
Kansas City (Kansas)	1843	168 (75e)	168 (70)	122 (60)	130 (50)	122 (30)	51 (00)	3 (80)		
Kansas City (Missouri)	1821	473 (75e)	507 (70)	476 (60)	457 (50)	400 (30)	164 (00)	4 (60)		
			1102s(70)	921s(60)	698s(50)					

NORTH AMERICA

Population (thousands)

City	Date Settled	Latest	ca 1970	ca 1960	ca 1950	ca 1930	ca 1900	ca 1850	ca 1800	18th c
Kenosha	1835	81 (75e)	79 (70) 84s(70)	68 (60) 73s(60)	54 (50)	50 (30)	12 (00)	3 (50)	0.3 (40)	
Key West	1822	25 (75e)	29 (70)	34 (60)	26 (50)	13 (30)	17 (00)	2 (50)	0.7 (40)	
Killeen	1882	49 (75e)	36 (70) 74s(70)	23 (60)	7 (50)	1 (30)	0.8(00)	0.3 (90)		
Knoxville	1786	183 (75e)	175 (70) 191s(70)	112 (60) 173s(60)	125 (50) 148s(50)	106 (30)	33 (00)	2 (50)	0.4 (00)	
Kokomo	1842	52 (75e)	44 (70)	47 (60)	39 (50)	33 (30)	11 (00)	1 (60)		
La Crosse	1841	49 (75e)	51 (70) 63s(70)	48 (60)	48 (50)	40 (30)	29 (00)	4 (60)		
Lafayette (Indiana)	1825	49 (75e)	45 (70) 79s(70)	42 (60)	36 (50)	26 (30)	18 (00)	6 (50)		
Lafayette (Louisiana)	bef 1804	75 (75e)	69 (70) 79s(70)	40 (60)	34 (50)	15 (30)	3 (00)	2 (90)		
Lake Charles	1803?	76 (75e)	78 (70) 88s(70)	63 (60) 89s(60)	41 (50)	16 (30)	7 (00)	0.4 (60)		
Lakeland	1881	50 (75e)	42 (70)	41 (60)	31 (50)	19 (30)	1 (00)	0.6 (90)		
Lakewood (California)	1934	82 (75e)	83 (70)	67 (60)	2 (40)					
Lakewood (Colorado)	1872	120 (75e)	93 (70)	19 (60)	4 (50)	1 (30)	0.5(00)			
Lakewood (Ohio)	1808	65 (75e)	70 (70)	66 (60)	68 (50)	71 (30)	3 (00)	1s (50)		
Lancaster	1709?	57 (75e)	58 (70) 117s(70)	61 (60) 94s(60)	64 (50) 76s(50)	60 (30)	41 (00)	12 (50)	4 (00)	2 (1754e)
Lansing	1843	127 (75e)	132 (70) 230s(70)	108 (60) 169s(60)	92 (50) 134s(50)	78 (30)	16 (00)	1 (50)		
Laredo	1755	77 (75e)	69 (70) 70s(70)	61 (60) 61s(60)	52 (50)	33 (30)	13 (00)	1 (60)	1 (23)	
Las Cruces	1848	40 (75e)	38 (70)	29 (60)	12 (50)	6 (30)	4 (10)	1s (70)		

NORTH AMERICA

Population (thousands)

City	Date Settled	Latest	ca 1970	ca 1960	ca 1950	ca 1930	ca 1900	ca 1850	ca 1800	18th c
Las Vegas	1905	146 (75e)	126 (70) / 237s (70)	64 (60) / 89s (60)	25 (50)	5 (30)	2 (20)			
Lawrence (Kansas)	1854	51 (75e)	46 (70)	33 (60)	23 (50)	14 (30)	11 (00)	2 (60)		
Lawrence (Massachusetts)	1655	67 (75e)	67 (70) / 200s (70)	71 (60) / 167s (60)	81 (50) / 112s (50)	85 (30)	63 (00)	8 (50)		
Lawton	1901	76 (75e)	74 (70) / 96s (70)	62 (60) / 62s (60)	35 (50)	12 (30)	6 (07)			
Lewiston	1770	41 (75e)	42 (70) / 65s (70)	41 (60) / 65s (60)	41 (50)	35 (30)	24 (00)	4 (50)	0.9 (00)	
Lexington	1779	186 (75e)	108 (70) / 160s (70)	63 (60) / 112s (60)	56 (50)	46 (30)	26 (00)	9 (50)	2 (00)	
Lima	1831	51 (75e)	54 (70) / 70s (70)	51 (60) / 63s (60)	50 (50)	42 (30)	22 (00)	0.8 (50)		
Lincoln	1864	163 (75e)	150 (70) / 153s (70)	129 (60) / 138s (60)	99 (50) / 100s (50)	76 (30)	40 (00)	13 (80)		
Little Rock	1820	140 (74)	132 (70) / 223s (70)	108 (60) / 185s (60)	102 (50) / 154s (50)	82 (30)	38 (00)	2 (50)		
Livonia	1832	115 (75e)	110 (70)	67 (60)	18 (50)	3s (30)	1s (00)			
Long Beach	1881	336 (75e)	359 (70)	344 (60)	251 (50)	142 (30)	2 (00)			
Longview	1865	52 (75e)	46 (70)	40 (60)	25 (50)	5 (30)	4 (00)	2 (80)		
Lorain	1807	85 (75e)	78 (70) / 192s (70)	69 (60) / 143s (60)	51 (50)	45 (30)	16 (00)	2 (80)		
Los Angeles	1781	2727 (75e)	2810 (70) / 8345s (70)	2479 (60) / 6489s (60)	1970 (50) / 3997s (50)	1238 (30)	102 (00)	2 (50)	0.3 (00)	
Louisville	1778	336 (75e)	362 (70) / 740s (70)	391 (60) / 607s (60)	369 (50) / 473s (50)	308 (30)	205 (00)	43 (50)	0.4 (00)	
Lowell	1653	91 (75e)	94 (70) / 183s (70)	92 (60) / 119s (60)	97 (50) / 107s (50)	100 (30)	95 (00)	33 (50)	6 (30)	

NORTH AMERICA

Population (thousands)

City	Date Settled	Latest	ca 1970	ca 1960	ca 1950	ca 1930	ca 1900	ca 1850	ca 1800	18th c
Lubbock	1891	164 (75e)	149 (70) 150s(70)	129 (60) 129s(60)	72 (50)	21 (30)	2 (10)			
Lynchburg	1787?	63 (75e)	54 (70) 71s(70)	55 (60) 59s(60)	48 (50)	41 (30)	19 (00)	8 (50)	5 (30)	
Lynn	1629	79 (75e)	90 (70)	94 (60)	100 (50)	102 (30)	69 (00)	14 (50)	3 (00)	
Macon	1806	121 (75e)	122 (70) 128s(70)	70 (60) 114s(60)	70 (50) 93s(50)	54 (30)	23 (00)	6 (50)	3 (30)	2 (1765)
Madison	1837	169 (74)	172 (70)	127 (60)	96 (50)	58 (30)	19 (00)	2 (50)		
Manchester	1722	83 (75e)	88 (70) 204s(70)	88 (60) 158s(60)	83 (50) 110s(50)	77 (30)	57 (00)	14 (50)	0.6 (00)	0.2(1767)
Mansfield	1808?	57 (75e)	55 (70) 78s(70)	47 (60)	44 (50)	34 (30)	18 (00)	4 (50)		
McAllen	1904	49 (75e)	38 (70) 91s(70)	33 (60)	20 (50)	9 (30)	5 (20)			
Melbourne	1878	40 (75e)	40 (70) 179s(70)	12 (60)	4 (50)	3 (30)	0.1(00)	0.1 (90)		
Memphis	1819	661 (75e)	624 (70) 664s(70)	498 (60) 545s(60)	396 (50) 406s(50)	253 (30)	102 (00)	9 (50)	0.7 (30)	
Meriden	1661	58 (75e)	56 (70) 98s(70)	52 (60) 52s(60)	44 (50)	38 (30)	24 (00)	4 (50)	1 (10)	
Meridian	1831	46 (74)	45 (70)	49 (60)	42 (50)	32 (30)	14 (00)	3 (70)		
Mesa	1878	101 (75)	63 (70)	34 (60)	17 (50)	4 (30)	0.7(00)			
Metairie	bef 1892	136 (70)	136 (70)	73 (60)	3 (40)					
Miami	1870	365 (75e)	335 (70) 1220s(70)	292 (60) 853s(60)	249 (50) 459s(50)	111 (30)	2 (00)			
Miami Beach	1913	94 (75e)	87 (70)	63 (60)	46 (50)	6 (30)	0.6(20)			
Middletown	1802	48 (75e)	49 (70)	42 (60)	34 (50)	30 (30)	9 (00)	1 (50)		
Midland	1885	63 (75e)	59 (70) 60s(70)	63 (60) 63s(60)	22 (50)	5 (30)	2 (10)			

NORTH AMERICA

City	Date Settled	Latest	Population (thousands)							
			ca 1970	ca 1960	ca 1950	ca 1930	ca 1900	ca 1850	ca 1800	18th c
Milwaukee	1818	669 (75)	717 (70)	741 (60)	637 (50)	578 (30)	285 (00)	20 (50)	2 (40)	
			1252s (70)	1150s (60)	829s (50)					
Minneapolis	1847	378 (75e)	434 (70)	483 (60)	522 (50)	464 (30)	203 (00)	3 (60)		
			1704s (70)	1377s (60)	985s (50)					
Mobile	1711	196 (75e)	190 (70)	195 (60)	129 (50)	68 (30)	38 (00)	21 (50)	1 (13)	0.3 (1785)
			258s (70)	260s (60)	183s (50)					
Modesto	1870	84 (75e)	62 (70)	37 (60)	17 (50)	14 (30)	2 (00)	2 (80)		
Monroe	1785	61 (75e)	56 (70)	52 (60)	39 (50)	26 (30)	5 (00)	0.4 (50)		
			91s (70)	81s (60)						
Montgomery	1817	153 (75e)	133 (70)	134 (60)	107 (50)	66 (30)	30 (00)	5 (50)	2 (40)	
			139s (70)	143s (60)	109s (50)					
Mount Vernon	1664	68 (75e)	73 (70)	76 (60)	72 (50)	61 (30)	21 (00)	3 (70)		
Muncie	1827	78 (75e)	69 (70)	69 (60)	58 (50)	47 (30)	21 (00)	0.7 (50)		
			90s (70)	78s (60)						
Muskegon	1812	44 (75e)	45 (70)	46 (60)	48 (50)	41 (30)	21 (00)	0.5 (50)	0.9 (00)	0.5 (1767)
			106s (70)	95s (60)	85s (50)					
Nashua	1656	61 (75e)	56 (70)	39 (60)	35 (50)	31 (30)	24 (00)	6 (50)	0.3 (00)	
			61s (70)							
Nashville	1780	423 (75e)	448 (70)	171 (60)	174 (50)	154 (30)	81 (00)	10 (50)		
			448s (70)	347s (60)	259s (50)					
Natchez	1716	21 (75e)	20 (70)	24 (60)	23 (50)	13 (30)	12 (00)	4 (50)	2 (10)	0.7 (1727)
Newark	1666	340 (75e)	382 (70)	405 (60)	439 (50)	442 (30)	246 (00)	39 (50)	7 (10)	0.8 (1760e)
New Bedford	1640	100 (75e)	102 (70)	102 (60)	109 (50)	113 (30)	62 (00)	16 (50)	4 (00)	
			134s (70)	127s (60)	125s (50)					
New Britain	1686	79 (75e)	83 (70)	82 (60)	74 (50)	68 (30)	26 (00)	3 (50)	0.9 (00)	
			131s (70)	100s (60)	123s (50)					
New Haven	1638	127 (75e)	138 (70)	152 (60)	164 (50)	163 (30)	108 (00)	20 (50)	4 (00)	5 (1756)
			348s (70)	279s (60)	245s (50)					

Figure 50. Skyscrapers of midtown New York, the largest urbanized area in North America and in the world, as seen from the East River. The United Nations Secretariat Building is in the foreground. In the background, from left to right, are the Empire State, Chrysler, and Pan Am Buildings. (Credit: New York Convention and Visitors Bureau, Inc.)

NORTH AMERICA

City	Date Settled	Latest	ca 1970	ca 1960	ca 1950	ca 1930	ca 1900	ca 1850	ca 1800	18th c
New Orleans	1718	560 (75e)	593 (70)	628 (60)	570 (50)	459 (30)	287 (00)	116 (50)	8 (97)	3 (1769)
			962s(70)	845s(60)	660s(50)					
Newport	1639	29 (75e)	35 (70)	47 (60)	38 (50)	28 (30)	22 (00)	10 (50)	7 (00)	5 (1749)
										2 (1708)
Newport News	1621	139 (75e)	138 (70)	114 (60)	42 (50)	34 (30)	20 (00)	0.8 (70)		
			268s(70)	209s(60)						
New Rochelle	1688	72 (75e)	75 (70)	77 (60)	60 (50)	54 (30)	15 (00)	0.3 (70)	1s (10)	0.3(1710)
Newton	1639	89 (75e)	91 (70)	92 (60)	82 (50)	65 (30)	34 (00)	5 (50)	1 (00)	1 (1765)
New York	1624	7482 (75e)	7896 (70)	7782 (60)	7892 (50)	6930 (30)	3437 (00)	516 (50)	60 (00)	10 (1756)
			16,208s(70)	14,115s(60)	12,296s(50)					
Niagara Falls	1807	81 (75e)	86 (70)	102 (60)	91 (50)	75 (30)	19 (00)	3 (70)	0.5 (20)	
Norfolk	1682	287 (75e)	308 (70)	305 (60)	214 (50)	130 (30)	47 (00)	14 (50)	7 (00)	5 (1698)
			668s(70)	508s(60)	385s(50)					
North Charleston	17th c	59 (75e)	20s (70)	22s(60)	13s(50)	3s (30)	2s (00)	5s (50)	5s (00)	
Norwalk (California)	1874	87 (75e)	92 (70)	89 (60)	4 (40)		0.5(13e)			
Norwalk (Connecticut)	1649	77 (75e)	79 (70)	68 (60)	49 (50)	36 (30)	6 (00)	5 (50)	5 (00)	3 (1756)
			107s(70)	82s(60)	55s(50)					
Norwich	1659	41 (75e)	42 (70)	39 (60)	23 (50)	23 (30)	17 (00)	6 (50)	3 (00)	6 (1756)
			140s(70)	844s(60)						
Oakland	1848	331 (75e)	362 (70)	368 (60)	385 (50)	284 (30)	67 (00)	2 (60)		
Oak Park	1833	60 (75e)	63 (70)	61 (60)	64 (50)	64 (30)	19 (10)			
Odessa	1881	84 (75e)	78 (70)	80 (60)	29 (50)	2 (30)				
Ogden	1845?	69 (75e)	69 (70)	70 (60)	57 (50)	40 (30)	16 (00)	1 (60)		
			150s(70)	122s(60)						
Oklahoma City	1889	366 (75e)	369 (70)	324 (60)	244 (50)	185 (30)	10 (00)	4 (90)		
			583s(70)	429s(60)	275s(50)					

NORTH AMERICA

Population (thousands)

City	Date Settled	Latest	ca 1970	ca 1960	ca 1950	ca 1930	ca 1900	ca 1850	ca 1800	18th c
Omaha	1854	371 (75e)	347 (70) 492s(70)	302 (60) 392s(60)	251 (50) 310s(50)	214 (30)	103 (00)	2 (60)		
Orange	1868	82 (75e)	77 (70)	26 (60)	10 (50)	8 (30)	1 (00)	0.7 (80)		
Orlando	1837	113 (75e)	99 (70) 305s(70)	88 (60) 201s(60)	52 (50) 73s(50)	27 (30)	2 (00)			
Oshkosh	1836	50 (75e)	53 (70) 55s(70)	45 (60)	41 (50)	40 (30)	28 (00)	2 (53)		
Owensboro	1800?	51 (75e)	50 (70) 53s(70)	42 (60)	34 (50)	23 (30)	13 (00)	1 (50)	0.2 (30)	
Oxnard	1898	87 (75e)	71 (70) 245s(70)	40 (60)	22 (50)	6 (30)	3 (10)			
Palm Beach	1872	11 (75e)	9 (70)	6 (60)	4 (50)	2 (30)	0.3(95)			
Palm Springs	1876	27 (75e)	21 (70)	13 (60)	8 (50)	3 (40)	0.05(00?e)			
Palo Alto	1876	52 (75e)	56 (70)	52 (60)	25 (50)	14 (30)	2 (00)			
Parma	1826?	99 (75e)	100 (70)	83 (60)	29 (50)	14 (30)	1 (00)	1 (50)		
Pasadena (California)	1874	108 (75e)	113 (70)	116 (60)	105 (50)	76 (30)	9 (00)	0.4 (80)		
Pasadena (Texas)	1895	95 (75e)	89 (70)	59 (60)	22 (50)	2 (30)				
Passaic	1678	50 (75e)	55 (70)	54 (60)	58 (50)	63 (30)	28 (00)	7 (80)		
Paterson	1679	136 (75e)	145 (70)	144 (60)	139 (50)	139 (30)	105 (00)	11 (50)	0.3 (10)	
Pawtucket	1671	72 (75e)	77 (70)	81 (60)	81 (50)	77 (30)	39 (00)	4 (50)	1 (30)	
Pensacola	1698	64 (75e)	60 (70) 167s(70)	57 (60) 128s(60)	43 (50)	32 (30)	18 (00)	2 (50)	0.6 (02)	
Peoria	1691?	126 (75e)	127 (70) 247s(70)	103 (60) 181s(60)	112 (50) 155s(50)	105 (30)	56 (00)	5 (50)	1 (40)	
Petersburg	1646	45 (75e)	36 (70) 101s(70)	37 (60)	35 (50)	29 (30)	22 (00)	14 (50)	4 (00)	
Philadelphia	1677?	1816 (75e)	1950 (70) 4022s(70)	2003 (60) 3635s(60)	2072 (50) 2922s(50)	1951 (30)	1294 (00)	121 (50) 340s (50)	41 (00) 69s (00)	24 (1760) 4 (1700e)

NORTH AMERICA

Population (thousands)

City	Date Settled	Latest	ca 1970	ca 1960	ca 1950	ca 1930	ca 1900	ca 1850	ca 1800	18th c
Phoenix	1867	669 (75)	582 (70) / 863s(70)	439 (60) / 552s(60)	107 (50) / 216s(50)	48 (30)	6 (00)	2 (80)		
Pine Bluff	1819	55 (75e)	57 (70) / 61s(70)	44 (60)	37 (50)	21 (30)	11 (00)	0.5 (50)		
Pittsburgh	1758?	459 (75e)	520 (70) / 1846s(70)	604 (60) / 1804s(60)	677 (50) / 1533s(50)	670 (30)	322 (00)	47 (50)	2 (00)	0.4(1765)
Pittsfield	1752	55 (75e)	57 (70) / 63s(70)	58 (60) / 62s(60)	53 (50)	50 (30)	22 (00)	6 (50)	2 (00)	
Plainfield	1684	44 (75e)	47 (70)	45 (60)	42 (50)	34 (30)	15 (00)	5 (70)		
Pocatello	1882	41 (75e)	40 (70)	29 (60)	26 (50)	16 (30)	4 (00)			
Pomona	1875	82 (75e)	87¹ (70)	67 (60)	35 (50)	21 (30)	6 (00)	4 (90)		
Pontiac	1818	76 (75e)	85 (70)	82 (60) / 187s(60)	74 (50)	65 (30)	10 (00)	2 (50)	2 (40)	
Port Arthur	1895	54 (75e)	57 (70) / 116s(70)	67 (60) / 116s(60)	58 (50) / 82s(50)	51 (30)	0.9(00)			
Portland (Maine)	1716	60 (75e)	65 (70) / 107s(70)	73 (60) / 112s(60)	78 (50) / 113s(50)	71 (30)	50 (00)	21 (50)	4 (00)	0.7(1753e)
Portland (Oregon)	1845	357 (75e)	381 (70) / 823s(70)	373 (60) / 652s(60)	374 (50) / 513s(50)	302 (30)	90 (00)	0.8 (50)		
Portsmouth	1752	109 (75e)	111 (70)	115 (60)	80 (50)	46 (30)	17 (00)	9 (50)	2 (90)	
Poughkeepsie	1687	32 (75e)	32 (70) / 103s(70)	38 (60)	41 (50)	40 (30)	24 (00)	14s (50)	3s (00)	
Providence	1636	168 (75e)	179 (70) / 795s(70)	207 (60) / 660s(60)	249 (50) / 583s(50)	253 (30)	176 (00)	42 (50)	8 (00)	3 (1749) / 1 (1708)
Provo	1849	56 (75e)	53 (70) / 104s(70)	36 (60) / 61s(60)	29 (50)	15 (30)	6 (00)	2 (60)		
Pueblo	1842	105 (75e)	97 (70) / 103s(70)	91 (60) / 103s(60)	64 (50) / 73s(50)	50 (30)	28 (00)	3 (80)		

¹ Urbanized area population for 1970 included in Los Angeles urbanized area.

NORTH AMERICA

City	Date Settled	Latest	ca 1970	ca 1960	ca 1950	ca 1930	ca 1900	ca 1850	ca 1800	18th c
Quincy	1625	91 (75e)	88 (70)	87 (60)	84 (50)	72 (30)	24 (00)	5 (50)	1 (00)	
Racine	1834	95 (75e)	95 (70)	89 (60)	71 (50)	68 (30)	29 (00)	5 (50)		
Raleigh	bef 1771	134 (75e)	117s (70) 154s (70)	96s (60) 94s (60)	77s(50) 69s(50)	37 (30)	14 (00)	5 (50)	0.7 (00)	
Rapid City	1876	48 (75e)	44 (70)	42 (60)	25 (50)	10 (30)	1 (00)	0.3 (80)		
Reading	1733	82 (75e)	88 (70) 168s (70)	98 (60) 160s (60)	109 (50) 155s(50)	111 (30)	79 (00)	16 (50)	2 (00)	
Reno	1868	78 (75e)	73 (70) 100s (70)	51 (60) 70s (60)	32 (50)	19 (30)	4 (00)	1 (70)		
Richmond (California)	1899	70 (75e)	79 (70)	72 (60)	100 (50)	20 (30)	7 (10)			
Richmond (Virginia)	1737	233 (75e)	249 (70) 416s (70)	220 (60) 333s (60)	230 (50) 258s(50)	183 (30)	85 (00)	28 (50)	6 (00)	
Riverside	1870	151 (75e)	140 (70) 585s (70)	84¹ (60)	47¹(50)	30 (30)	8 (00)	5 (90)		
Roanoke	1834	101 (75e)	92 (70) 157s (70)	97 (60) 125s (60)	92 (50) 107s(50)	69 (30)	21 (00)	0.7 (80)		
Rochester (Minnesota)	1854	56 (75e)	54 (70) 57s (70)	41 (60)	30 (50)	21 (30)	7 (00)	1 (60)		
Rochester (New York)	1812	267 (75e)	296 (70) 601s (70)	319 (60) 493s (60)	332 (50) 409s(50)	328 (30)	163 (00)	36 (50)	9 (30)	
Rockford	1834	145 (75e)	147 (70) 206s (70)	127 (60) 172s (60)	93 (50) 122s(50)	86 (30)	31 (00)	2 (50)		
Rock Island	1816	49 (75e)	50 (70)	52 (60)	49 (50)	38 (30)	19 (00)	2 (50)		
Rome	1786	49 (75e)	50 (70)	52 (60)	42 (50)	32 (30)	15 (00)	4 (60)		
Royal Oak	1822	79 (75e)	86 (70)	81 (60)	47 (50)	23 (30)	0.5(00)		1 (00?e)	

¹ Urbanized area population for 1960 and 1950 included in San Bernardino urbanized area.

NORTH AMERICA

Population (thousands)

City	Date Settled	Latest	ca 1970	ca 1960	ca 1950	ca 1930	ca 1900	ca 1850	ca 1800	18th c
Sacramento	1839	261 (75e)	257 (70)	192 (60)	138 (50)	94 (30)	29 (00)	7 (50)		
			637s(70)	452s(60)	212s(50)					
Saginaw	1816	86 (75e)	92 (70)	98 (60)	93 (50)	81 (30)	42 (00)	1 (50)		
			148s(70)	129s(60)	106s(50)					
Saint Augustine	1565	13 (75e)	12 (70)	15 (60)	14 (50)	12 (30)	4 (00)	2 (50)	5 (00e)	3 (1773)
Saint Clair Shores	1818?	86 (75e)	88 (70)	77 (60)	20 (50)	7 (30)				
Saint Joseph	1826	78 (75e)	73 (70)	80 (60)	79 (50)	81 (30)	103 (00)	9 (60)	0.9 (46)	
			77s(70)	81s(60)	82s(50)					
Saint Louis	1764	525 (75e)	622 (70)	750 (60)	857 (50)	822 (30)	575 (00)	78 (50)	2 (10)	0.9(1769)
			1883s(70)	1668s(60)	1400s(50)					
Saint Paul	1838	280 (75e)	310 (70)	313 (60)	311 (50)	272 (30)	163 (00)	1 (50)		
			495s(70)	325s(60)	115s(50)					
Saint Petersburg	185f	236 (76)	216 (70)	181 (60)	97 (50)	40 (30)	2 (00)	0.3 (90)		
Salem (Massachusetts)	1626	39 (75e)	41 (70)	39 (60)	42 (50)	43 (30)	36 (00)	20 (50)	9 (00)	4 (1765)
Salem (Oregon)	1841	78 (75e)	69 (70)	49 (60)	43 (50)	26 (30)	4 (00)	1 (70)		
			94s(70)							
Salinas	1856	70 (75e)	59 (70)	29 (60)	14 (50)	10 (30)	3 (00)	0.6 (70)		
			62s(70)							
Salt Lake City	1847	170 (75e)	176 (70)	189 (60)	182 (50)	140 (30)	54 (00)	10 (53)		
			479s(70)	349s(60)	227s(50)					
San Angelo	1882	66 (75e)	64 (70)	59 (60)	52 (50)	25 (30)	10 (10)			
			64s(70)	59s(60)						
San Antonio	1718	773 (75e)	654 (70)	588 (60)	408 (50)	232 (30)	53 (00)	3 (50)	2 (10?e)	
			773s(70)	642s(60)	450s(50)					
San Bernardino	1851	102 (75)	105¹ (70)	92 (60)	63 (50)	37 (30)	6 (00)	2 (80)		
				378s(60)	136s(50)					

¹Urbanized area population for 1970 included in Riverside urbanized area.

NORTH AMERICA

Population (thousands)

City	Date Settled	Latest	ca 1970	ca 1960	ca 1950	ca 1930	ca 1900	ca 1850	ca 1800	18th c
San Diego	1769	774 (75e)	697 (70) / 1198s(70)	573 (60) / 836s(60)	334 (50) / 433s(50)	148 (30)	18 (00)	3 (52)	2 (02)	
San Francisco	1776	665 (75e)	716 (70) / 2988s(70)	740 (60) / 2395s(60)	775 (50) / 2022s(50)	634 (30)	343 (00)	35 (52)	0.8 (03e)	
San Jose	1777	556 (75e)	446 (70) / 1025s(70)	204 (60) / 603s(60)	95 (50) / 176s(50)	58 (30)	21 (00)	3 (52)	0.2 (00e)	
San Mateo	1851	78 (75e)	79 (70)	70 (60)	42 (50)	13 (30)	2 (00)			
Santa Ana	1869	177 (75e)	157 (70)	100 (60)	46 (50)	30 (30)	5 (00)	0.7 (80)		
Santa Barbara	1782	72 (75e)	70 (70) / 130s(70)	59 (60) / 73s(60)	45 (50)	34 (30)	7 (00)	3 (80)	1 (03e)	
Santa Clara	1777	83 (75e)	88 (70)	59 (60)	12 (50)	6 (30)	4 (00)	2 (80)		
Santa Cruz	1791	37 (75e)	32 (70) / 74s (70)	26 (60)	22 (50)	14 (30)	6 (00)	0.9 (60)		
Santa Fe	1609	45 (75e)	41 (70)	33 (60)	28 (50)	11 (30)	6 (00)	5 (50e)	5 (99e)	2 (1749e)
Santa Monica	1875	92 (75e)	88 (70)	83 (60)	72 (50)	37 (30)	3 (00)	0.4 (80)		
Santa Rosa	1833	65 (75e)	50 (70) / 75s(70)	31 (60)	18 (50)	11 (30)	7 (00)	0.4 (60)		
Sarasota	1885	47 (75e)	40 (70) / 167s(70)	34 (60)	19 (50)	8 (30)	0.8(10)			
Savannah	1733	110 (75e)	118 (70) / 164s(70)	149 (60) / 170s(60)	120 (50) / 128s(50)	85 (30)	54 (00)	15 (50)	5 (00)	0.6(1748e)
Schenectady	1662	75 (75e)	78 (70)	82 (60)	92 (50)	96 (30)	32 (00)	9 (50)	5 (00)	0.2(1698)
Scottsdale	1882	78 (75)	68 (70)	10 (60)	2 (50)	3s (30)	1s (20)			
Scranton	1771?	96 (75e)	104 (70) / 204s(70)	111 (60) / 211s(60)	126 (50) / 236s(50)	143 (30)	102 (00)	2 (50)		
Seaside	1900	37 (75e)	36 (70) / 93s(70)	19 (60)	10 (50)	2 (40)				
Seattle	1851	487 (75e)	531 (70) / 1238s(70)	557 (60) / 864s(60)	468 (50) / 622s(50)	366 (30)	81 (00)	1 (70)		
Sheboygan	1843	49 (75e)	48 (70)	46 (60)	42 (50)	39 (30)	23 (00)	4 (60)		

NORTH AMERICA

City	Date Settled	Population (thousands)								
		Latest	ca 1970	ca 1960	ca 1950	ca 1930	ca 1900	ca 1850	ca 1800	18th c
Shreveport	1837	186 (75e)	182 (70)	164 (60)	127 (50)	77 (30)	16 (00)	2 (50)		
			235s(70)	209s(60)	150s(50)					
Silver Spring	1842?	77 (70)	77 (70)	66 (60)	65 (50e)	7 (30?e)		0.05(70?e)		
Simi Valley	1850	70 (75e)	60 (70)	2 (60)	0.3(50?e)	0.2(30?e)	0.06(00?e)			
Sioux City	1848	86 (75e)	86 (70)	89 (60)	84 (50)	79 (30)	33 (00)	0.8 (60)		
			96s(70)	98s(60)	90s(50)					
Sioux Falls	1865	74 (75e)	72 (70)	65 (60)	53 (50)	33 (30)	10 (00)	2 (80)		
			75s(70)	67s(60)						
Skokie	1834	68 (75e)	69 (70)	59 (60)	15 (50)	5 (30)	0.5(00)			
Somerville	1630	81 (75e)	89 (70)	95 (60)	102 (50)	104 (30)	62 (00)	4 (50)		
South Bend	1820	117 (75e)	126 (70)	132 (60)	116 (50)	104 (30)	36 (00)	2 (50)		
			289s(70)	219s(60)	168s(50)					
Spartanburg	bef 1785	47 (75e)	45 (70)	44 (60)	37 (50)	29 (30)	11 (00)	1 (50)		
			74s(70)							
Spokane	1872	174 (75e)	171 (70)	182 (60)	162 (50)	116 (30)	37 (00)	0.3 (80)		
			230s(70)	227s(60)	176s(50)					
Springfield (Illinois)	1818	87 (75e)	92 (70)	83 (60)	82 (50)	72 (30)	34 (00)	5 (50)	3 (40)	
			121s(70)	111s(60)	97s(50)					
Springfield (Massachusetts)	1636	171 (75e)	164 (70)	174 (60)	162 (50)	150 (30)	62 (00)	12 (50)	2 (00)	3 (1765)
			514s(70)	459s(60)	357s(50)					
Springfield (Missouri)	1830	132 (75e)	120 (70)	96 (60)	67 (50)	58 (30)	23 (00)	0.4 (50)		
			121s(70)	97s(60)	76s(50)					
Springfield (Ohio)	1799	77 (75e)	82 (70)	83 (60)	79 (50)	69 (30)	38 (00)	5 (50)	2 (20)	
			94s(70)	90s(60)	82s(50)					
Stamford	1641	105 (75e)	109 (70)	93 (60)	74 (50)	46 (30)	16 (00)	5 (50)	4 (00)	
			185s(70)	167s(60)	118s(50)					
Sterling Heights	bef 1835	93 (76)	61 (70)	15s(60)	7s(50)	2s (30)	2s (00)	0.9s(50)		
Steubenville	1786	28 (75e)	31 (70)	32 (60)	36 (50)	35 (30)	14 (00)	6 (50)	3 (30)	3 (1756)
			85s(70)	82s(60)						

NORTH AMERICA

City	Date Settled	Latest	ca 1970	ca 1960	ca 1950	ca 1930	ca 1900	ca 1850	ca 1800	18th c
					Population (thousands)					
Stockton	1844	118 (75)	110 (70)	86 (60)	71 (50)	48 (30)	18 (00)	3 (52)		
			162s(70)	142s(60)	113s(50)					
Sunnyvale	bef 1901	102 (75e)	95 (70)	53 (60)	10 (50)	3 (30)	2 (20)			
Syracuse	1797	183 (75e)	197 (70)	216 (60)	221 (50)	209 (30)	108 (00)	22 (50)	3 (30)	
			376s(70)	333s(60)	265s(50)					
Tacoma	1864?	151 (75e)	155 (70)	148 (60)	144 (50)	107 (30)	38 (00)	1 (80)		
			333s(70)	215s(60)	168s(50)					
Tallahassee	bef 1539	84 (75e)	73 (70)	48 (60)	27 (50)	11 (30)	3 (00)	2 (60)	0.9 (30)	
			79s(70)							
Tampa	1823	280 (75e)	278 (70)	275 (60)	125 (50)	101 (30)	16 (00)	1 (50)		
			369s(70)	302s(60)	179s(50)					
Taylor	1847	77 (76)	70 (70)	45 (60)	8 (50)	2s (30)	1s (00)	0.3s(50)		
Tempe	1870	94 (75)	64 (70)	25 (60)	8 (50)	2 (30)	0.9(00)			
Terre Haute	1811	64 (75e)	70 (70)	72 (60)	64 (50)	63 (30)	37 (00)	4 (50)		
			81s(70)	81s(60)	78s(50)					
Texas City	1893	41 (75e)	39 (70)	32 (60)	17 (50)	4 (30)	3 (20)			
			84s(70)							
Thousand Oaks	1881?	56 (75e)	36 (70)	3 (60)	1 (50)	0.2(30?)				
Toledo	1817	368 (75e)	384 (70)	318 (60)	304 (50)	291 (30)	132 (00)	4 (50)	1 (40)	
			488s(70)	438s(60)	364s(50)					
Topeka	1854	119 (75e)	125 (70)	119 (60)	79 (50)	64 (30)	34 (00)	0.8 (60)		
			132s(70)	119s(60)	89s(50)					
Torrance	1911	140 (75e)	135 (70)	101 (60)	22 (50)	7 (30)				
Towson	1768	78 (70)	78 (70)	19 (60)	11 (40)	2 (30?e)	1 (00?e)	1 (70?e)		
Trenton	1679	101 (75e)	105 (70)	114 (60)	128 (50)	123 (30)	73 (00)	6 (50)	3 (10)	
			274s(70)	242s(60)	189s(50)					
Troy	1786	60 (75e)	63 (70)	67 (60)	72 (50)	73 (30)	61 (00)	29 (50)	5 (00)	
Tucson	1776	299 (75)	263 (70)	213 (60)	45 (50)	33 (30)	8 (00)	3 (70)		
			294s(70)	227s(60)						

NORTH AMERICA

City	Date Settled	Population (thousands)								
		Latest	ca 1970	ca 1960	ca 1950	ca 1930	ca 1900	ca 1850	ca 1800	18th c
Tulsa	1832?	332 (75e)	330 (70)	262 (60)	183 (50)	141 (30)	1 (00)	0.3 (57)		
			369s(70)	299s(60)	206s(50)					
Tuscaloosa	1816	69 (75e)	66 (70)	63 (60)	46 (50)	21 (30)	5 (00)	4 (60)	2 (40)	
			86s(70)	77s(60)						
Tyler	1846	61 (75e)	58 (70)	51 (60)	39 (50)	17 (30)	8 (00)	1 (60)		
			60s(70)	52s(60)						
Upper Darby	1683	92 (75e)	96 (70)	93 (60)	85 (50)	47 (30)	4 (00)	2 (50)	0.9 (00)	
Utica	1786?	82 (75e)	92 (70)	100 (60)	102 (50)	102 (30)	56 (00)	18 (50)	3 (10)	
			180s(70)	188s(60)	117s(50)					
Vallejo	1850	71 (75e)	72 (70)	61 (60)	26 (50)	16 (30)	8 (00)	6 (80)		
Ventura	1782	63 (75e)	58 (70)	29 (60)	17 (50)	12 (30)	2 (00)	1 (80)		
Vineland	1861	54 (75e)	47 (70)	38 (60)	8 (50)	8 (30)	4 (00)	3 (87)		
			74s(70)							
Virginia Beach	1887	214 (75e)	172 (70)	8 (60)	5 (50)	2 (30)	0.2(00e)			
Waco	1849	98 (75e)	95 (70)	98 (60)	85 (50)	53 (30)	21 (00)	3 (70)		
			119s(70)	116s(60)	93s(50)					
Waltham	1634	56 (75e)	62 (70)	55 (60)	47 (50)	39 (30)	23 (00)	4 (50)	0.9 (00)	0.7(1765)
Warren (Michigan)	1837?	173 (75e)	179 (70)	89 (60)	0.7(50)	0.5(30)	0.3(00)	0.7 (50)		
Warren (Ohio)	1798	60 (75e)	63 (70)	60 (60)	50 (50)	41 (30)	9 (00)	2 (60)	2 (40)	2 (1748)
Warwick	1642	86 (75e)	84 (70)	69 (60)	43 (50)	23 (30)	21 (00)	8 (50)	3 (00)	0.5(1708)
*Washington	1665?	693 (76e)	757 (70)	764 (60)	802 (50)	487 (30)	279 (00)	40 (50)	3 (00)	
			2481s(70)	1808s(60)	1287s(50)					
Waterbury	1674	107 (75e)	108 (70)	107 (60)	104 (50)	100 (30)	46 (00)	5 (50)	3 (00)	2 (1756)
			157s(70)	142s(60)	132s(50)					
Waterloo	1845	73 (75)	76 (70)	72 (60)	65 (50)	46 (30)	13 (00)	1 (60)		
			113s(70)	103s(60)	84s(50)					
Waukegan	1835	65 (75)	65 (70)	56 (60)	39 (50)	33 (30)	9 (00)	3 (60)		
West Allis	1827	69 (75e)	72 (70)	68 (60)	43 (50)	35 (30)	7 (10)			

NORTH AMERICA

Population (thousands)

City	Date Settled	Latest	ca 1970	ca 1960	ca 1950	ca 1930	ca 1900	ca 1850	ca 1800	18th c
Westland	1839	93 (75e)	87 (70)	56s(60)	17s(50)	10s (30)	2s (00)	2s (50)		
West Palm Beach	1880	61 (75e)	57 (70)	56 (60)	43 (50)	27 (30)	0.6(00)			
			288s(70)	173s(60)						
Wheeling	1769	44 (75e)	48 (70)	53 (60)	59 (50)	62 (30)	39 (00)	11 (50)	0.5 (00e)	
			93s(70)	101s(60)	107s(50)					
White Plains	1683	48 (75e)	50 (70)	50 (60)	43 (50)	36 (30)	8 (00)	1 (50)	0.6 (00)	
Whittier	1887	72 (75e)	73 (70)	34 (60)	23 (50)	15 (30)	2 (00)	0.6 (90)		
Wichita	1864	265 (75e)	277 (70)	255 (60)	168 (50)	111 (30)	25 (00)	5 (80)		
			302s(70)	292s(60)	194s(50)					
Wichita Falls	1876	95 (75e)	96 (70)	102 (60)	68 (50)	44 (30)	2 (00)	2 (90)		
			96s(70)	102s(60)						
Wilkes-Barre	1769	57 (75e)	59 (70)	64 (60)	77 (50)	87 (30)	52 (00)	3 (50)	0.8 (00)	
			223s(70)	234s(60)	272s(50)					
Williamsburg	1633	11 (75e)	9 (70)	7 (60)	7 (50)	4 (30)	2 (00)	0.9 (50)	1 (82)	
Wilmington (Delaware)	1638	76 (75e)	80 (70)	96 (60)	110 (50)	107 (30)	77 (00)	14 (50)	4 (10)	0.6(1740e)
			371s(70)	284s(60)	187s(50)					
Wilmington (North Carolina)	1732	54 (75e)	46 (70)	44 (60)	45 (50)	32 (30)	21 (00)	7 (50)	2 (00)	
			58s(70)							
Winston-Salem	1766	141 (75e)	135 (70)	111 (60)	88 (50)	75 (30)	14 (00)	1 (53)	0.2 (00)	
			145s(70)	128s(60)	92s(50)					
Woodbridge	1665	96 (75e)	99 (70)	79 (60)	36 (50)	25 (30)	8 (00)	5 (50)	4 (10)	
Worcester	1713	172 (75e)	177 (70)	187 (60)	203 (50)	195 (30)	118 (00)	17 (50)	2 (00)	1 (1765)
			247s(70)	225s(60)	219s(50)					
Yakima	1858	49 (75e)	46 (70)	43 (60)	38 (50)	22 (30)	3 (00)	2 (90)		
			65s(70)							
Yonkers	bef 1639	193 (75e)	204 (70)	191 (60)	153 (50)	135 (30)	48 (00)	4 (50)	1 (00)	
York	1741	49 (75e)	50 (70)	55 (60)	60 (50)	55 (30)	34 (00)	7 (50)	3 (00)	
Youngstown	1797	132 (75e)	141 (70)	167 (60)	168 (50)	170 (30)	45 (00)	3 (50)	1 (10)	2 (1777e)
			397s(70)	373s(60)	298s(50)					

NORTH AMERICA

City	Date Settled	Latest	ca 1970	ca 1960	ca 1950	ca 1930	ca 1900	ca 1850	ca 1800	18th c
						Population (thousands)				
VIRGIN ISLANDS (USA)										
*Charlotte Amalie	1666	12 (70)	12 (70)	13 (60)	11 (50)	7 (30)	9 (01)	11 (50)	11 (35)	7 (46)

OCEANIA

City	Date Settled	Latest	ca 1970	ca 1960	ca 1950	ca 1930	ca 1900	ca 1850	ca 1800	18th c
						Population (thousands)				
AUSTRALIA										
Adelaide	1836	857s (76)	809s (71)	588s (61)	382s (47)	313s (33)	163s (01)	18 (61)	7 (46)	
Brisbane	1824	958s (76)	818s (71)	622s (61)	402s (47)	300s (33)	119s (01)	6 (61)		
*Canberra	1824?	208s (76)	156s (71)	56s (61)	15s (47)	7 (33)				
Geelong	1837	122s (76)	115s (71)	88s (61)	45s (47)	39s (33)	23s (01)	17 (61)		
Hobart	1804	132s (76)	130s (71)	116s (61)	77s (47)	60s (33)	34s (01)	19 (61)	4 (41)	
Melbourne	1835	2479s (76)	2394s (71)	1912s (61)	1226s (47)	992s (33)	502s (01)	140s (61); 39s (51)		
Newcastle	1804	363s (76)	250s (71)	209s (61)	127s (47)	104s (33)	55s (01)	4s (61); 1 (51)		
Perth	1829	806s (76)	642s (71)	420s (61)	273s (47)	207s (33)	36 (01)	3 (61)		
Sydney	1788	3021s (76)	2725s (71)	2183s (61)	1484s (47)	1235s (33)	488s (01)	93s (61); 54s (51)	3 (00)	
Wollongong	1815	211s (76)[1]	186s (71)[1]	132s (61)[1]	18 (47)	11 (33)	4 (01)	1 (61)		
FIJI										
*Suva	bef 1877	64 (76); 80s (66)	54 (66); 80s (66)	37 (56); 53s (56)	11 (46); 24s (46)	13 (31e)				

[1] For Greater Wollongong, established in 1947; comparable populations: 63s (1947), 43s (1933), 29s (1911).

Figure 51. Aerial view of Sydney, the largest city and the largest urbanized area in Oceania, with the famous Sydney Harbour steel arch bridge in the background. (Credit: Australian Information Service.)

OCEANIA

City	Date Settled	Population (thousands)								
		Latest	ca 1970	ca 1960	ca 1950	ca 1930	ca 1900	ca 1850	ca 1800	18th c
FRENCH POLYNESIA										
*Papeete	1840?	25 (71)	25 (71)	20 (62)	12 (46)	7 (31)	4 (97)	3 (65)		
		65s (71)	65s (71)	36s (62)						
(HAWAII), USA										
Hilo	bef 1778	26 (70)	26 (70)	26 (60)	27 (50)	19 (30)	7 (10)	4 (72)		
*Honolulu	bef 1794	344s (75e)	325 (70)	294 (60)	248 (50)	138 (30)	39 (00)	14 (66)		
		442s (70)	442s (70)	351s(60)						
NEW CALEDONIA										
*Noumea	1854	42 (69)	42 (69)	35 (63)	10 (46)	10 (26)	7 (98)			
		56s (76)	47s (69)	22 (56)						
NEW ZEALAND										
Auckland	1841	151 (76)	152 (71)	144 (61)	127 (51)	88 (26)	34 (01)	5 (47)		
		743s (76)	650s (71)	448s (61)	329s(51)	193s (26)	67s (01)			
Christchurch	1850	172 (76)	166 (71)	152 (61)	124 (51)	83 (26)	18 (01)	1 (58)		
		295s (76)	276s (71)	221s(61)	174s(51)	119s (26)	57s (01)			
Dunedin	1845	83 (76)	82 (71)	73 (61)	69 (51)	68 (26)	25 (01)	0.4 (49)		
		113s (76)	111s (71)	105s(61)	95s(51)	85s (26)	52s (01)			
Hamilton	1864	88 (76)	75 (71)	42 (61)	30 (51)	14 (26)	1 (01)	0.7 (74)		
		95s (76)	81s (71)	51s(61)	33s(51)	17s (26)				
Manukau	bef 1853	139 (76)	104 (71)	28s(61)	16s(51)	7s (26)	12s (01)	9s (78)		
*Wellington	1840	140 (76)	137 (71)	124 (61)	120 (51)	99 (26)	44 (01)	3 (48)		
		327s (76)	308s (71)	151s(61)	133s(51)	122s (26)	49s (01)			
PAPUA NEW GUINEA										
*Port Moresby	1883	113 (76)	77 (71)	42 (66)	1 (49–50e)	2 (30?e)				
				14 (56)						

OCEANIA

City	Date Settled	Population (thousands)								
		Latest	ca 1970	ca 1960	ca 1950	ca 1930	ca 1900	ca 1850	ca 1800	18th c
SAMOA										
*Apia	bef 1850	32s (76)	30s (71)	22s (61)	12s (51)	5 (36)	1 (00?e)			

SOUTH AMERICA

City	Date Settled	Population (thousands)								
		Latest	ca 1970	ca 1960	ca 1950	ca 1930	ca 1900	ca 1850	ca 1800	18th c
ARGENTINA										
Almirante Brown	1874	242 (70)	242 (70)	137 (60)	36 (47)	11 (14)	4 (95)			
Avellaneda	1818	338 (70)	338 (70)	327 (60)	274 (47)	46 (14)	10 (95)	6 (69)		
Bahia Blanca	1828	209 (75e)	161 (70)	121 (60)	113 (47)	44 (14)	9 (95)	1 (69)		
		182s (70)	182s (70)	127s (60)						
*Buenos Aires	1580	2975 (75e)	2972 (70)	2967 (60)	2981 (47)	1561 (14)	663 (95)	178 (69)	40 (97e)	11 (1744)
		9245s (75e)	8436s (70)	6739s (60)	4722s (47)	2034s (14)	781s (95)	92 (55)	11 (13)	4 (1665)
Cordoba	1573	903 (75e)	782 (70)	586 (60)	370 (47)	105 (14)	48 (95)	29 (69)	4 (97e)	
		791s (70)	791s (70)	592s (60)						
Corrientes	1588	162 (75e)	137 (70)	98 (60)	57 (47)	29 (14)	16 (95)	11 (69)		
General San Martin	1856	361 (70)	361 (70)	279 (60)	270 (47)	12 (14)	3 (95)	1 (69)		
General Sarmiento	1862	314 (70)	314 (70)	167 (60)	11 (47)	6 (14)	2 (95)			
La Matanza	1784	658 (70)	658 (70)	402 (60)	89 (47)	2 (14)	2 (95)	1 (69)	2 (15e)	
Lanus	1800	450 (70)	450 (70)	375 (60)	244 (47)	33 (14)				
La Plata	1864	391 (70)	391 (70)	337 (60)	207 (47)	90 (14)	45 (95)	0.6 (69)		
		521s (75e)	479s (70)	404s (60)						
Lomas de Zamora	1862	411 (70)	411 (70)	272 (60)	126 (47)	22 (14)	9 (95)			
Mar del Plata		361 (75e)	302 (70)	211 (60)	115 (47)	28 (14)	5 (95)			

Figure 52. At the center of Buenos Aires, the largest urbanized area in South America. Avenida Nueve de Julio, looking toward the Plaza de la Republica with its 221-ft (67-m) high Obelisk, erected in 1936 to commemorate the 400th anniversary of the first founding of the city. The same anniversary of the permanent settlement of Buenos Aires will be observed in 1980. (Credit: Organization of American States.)

SOUTH AMERICA

Population (thousands)

City	Date Settled	Latest	ca 1970	ca 1960	ca 1950	ca 1930	ca 1900	ca 1850	ca 1800	18th c
Mendoza	1561	119 (70)	119 (70)	109 (60)	97 (47)	59 (14)	29 (95)	8 (69)	5 (12)	
		606s(75e)	471s(70)	331s(60)						
Merlo	1730	185 (70)	185 (70)	100 (60)	8 (47)	4 (14)	2 (95)			
Moron	1600	486 (70)	486 (70)	342 (60)	110 (47)	11 (14)	4 (95)	1 (69)	1 (01e)	
Parana	1730	139 (75e)	128 (70)	108 (60)	84 (47)	36 (14)	24 (95)	10 (69)		
Quilmes	1670	355 (70)	355 (70)	245 (60)	115 (47)	19 (14)	4 (95)	2 (69)	0.8 (01e)	
Rosario	1725	821 (75e)	750 (70)	627 (60)	468 (47)	223 (14)	94 (95)	23 (69)	5 (16)	
		807s(70)	807s(70)	669s(60)				10 (58)		
Salta	1582	216 (75e)	176 (70)	117 (60)	67 (47)	28 (14)	17 (95)	12 (69)	5 (01e)	
								8 (54)		
San Isidro	1719	250 (70)	250 (70)	188 (60)	90 (47)	9 (14)	5 (95)	0.9 (69)	2 (01e)	
San Juan	1562	113 (70)	113 (70)	107 (60)	82 (47)	17 (14)	10 (95)	8 (69)	6 (1777)	
		265s(75e)	218s(70)	147s(60)						
Santa Fe	1573	265 (75e)	245 (70)	209 (60)	169 (47)	60 (14)	22 (95)	11 (69)	7 (17)	
								6 (58)	4 (97e)	
Tres de Febrero	bef 1851	313 (70)	313 (70)	263 (60)	15 (47e)	6 (14)				
Tucuman	1685	350 (75e)	322 (70)	272 (60)	194 (47)	91 (14)	34 (95)	17 (69)	4 (12)	
		366s(70)	366s(70)	297s(60)				17 (45)		
Vicente Lopez	1878	285 (70)	285 (70)	248 (60)	150 (47)	12 (14)				
BOLIVIA										
Cochabamba	1574	205 (76)	153 (70e)	110 (60e)	81 (50)	36 (29e)	22 (00)	41 (58)	22 (88)	
*La Paz	bef 1548	655 (76)	538 (70e)	412 (60e)	321 (50)	147 (29e)	53 (00)	43 (45)	21 (96)	
Potosi	1545	77 (76)	69 (70e)	56 (60e)	46 (50)	34 (29e)	21 (00)	23 (58)	23 (79)	13 (1675e)
Santa Cruz	1595	257 (76)	116 (70e)	67 (60e)	43 (50)	30 (29e)	16 (00)	10 (58)	6 (00?e)	
*Sucre	bef 1538	62 (76)	51 (70e)	56 (60e)	40 (50)	35 (29e)	21 (00)	24 (58)	18 (26e)	160 (1611e)
BRAZIL										
Aracaju	1855	226s(75e)	179 (70)	113 (60)	68 (50)	37 (20)	16 (90)	6 (72)		
			184s(70)	116s(60)	78s(50)	37s (20)	21s (00)	10s (72)		

SOUTH AMERICA

City	Date Settled	Latest	Population (thousands)							
			ca 1970	ca 1960	ca 1950	ca 1930	ca 1900	ca 1850	ca 1800	18th c
Belem	1616		565 (70)	360 (60)	225 (50)	145 (20)		12 (33)	12 (00?e)	4 (1730e)
		772s (75e)	633s (70)	402s (60)	255s(50)		97s (00)	62s (72)		
Belo Horizonte	1701		1107 (70)	643 (60)	339 (50)			8 (64e)		
		1557s (75e)	1235s (70)	693s (60)	353s(50)	236s (20)	13s (00)			
*Brasilia	1957		282 (70)	90 (60)						
		763s (75e)	537s (70)	142s (60)						
Campina Grande	1697		163 (70)	116 (60)	72 (50)	56 (20)	21 (90)	15 (72)		
		236s (75e)	195s (70)	207s (60)	173s(50)	56s (20)	38s (00)	15s (72)		
Campinas	1773		328 (70)	180 (60)	99 (50)	42 (20)	20 (90)	6 (50?e)	7 (22)	
		473s (75e)	376s (70)	219s (60)	137s(50)	71s (20)	68s (00)	31s (72)		
Campos	1634		153 (70)	91 (60)	62 (50)		23 (90)	20 (72)		
		337s (75e)	319s (70)	292s (60)	238s(50)	116s (20)	91s (00)	89s (72)		
Curitiba	1654		484 (70)	345 (60)	138 (50)	48 (20)	23 (90)	13 (72)		
		766s (75e)	609s (70)	361s (60)	181s(50)	176s (20)	50s (00)	13s (72)		
Duque de Caxias	1886?		257 (70)	173 (60)	74 (50)	57 (20)				
		537s (75e)	431s (70)	244s (60)	92s(50)	79s (20)				
Feira de Santana	bef 1696		127 (70)	62 (60)	27 (50)	14 (20)	11 (90)	8 (72)		
		227s (75e)	187s (70)	142s (60)	107s(50)	78s (20)	63s (00)	52s (72)		
Florianopolis	1700		116 (70)	74 (60)	48 (50)	20 (20)	11 (90)	9 (72)		
		168s (75e)	138s (70)	99s (60)	68s(50)	41s (20)	32s (00)	26s (72)		
Fortaleza	1609		520 (70)	355 (60)	205 (50)	68 (20)		21 (72)		
		1110s (75e)	858s (70)	515s (60)	270s(50)	79s (20)	48s (00)	42s (72)	12s (10)	
Goiania	1935		362 (70)	133 (60)	40 (50)	15 (40)				
		518s (75e)	381s (70)	154s (60)	52s(50)					
Guarulhos	1560		222 (70)	78 (60)	16 (50)	6 (20)		3 (72)		
		311s (75e)	237s (70)	101s (60)	35s(50)	6s (20)	3s (00)			
Jaboatao	1648		52 (70)	34 (60)	34 (50)	14 (20)	9 (90)	12 (72)		
		259s (75e)	201s (70)	105s (60)	57s(50)	48s (20)	23s (00)			
Joao Pessoa	1585		197 (70)	136 (60)	96 (50)	39 (20)				
		288s (75e)	222s (70)	167s (60)	119s(50)	53s (20)	29s (00)	25s (72)		

SOUTH AMERICA

Population (thousands)

City	Date Settled	Latest	ca 1970	ca 1960	ca 1950	ca 1930	ca 1900	ca 1850	ca 1800	18th c
Juiz de Fora	1850?		219 (70)	125 (60)	85 (50)	51 (20)	23 (90)	19 (72)		
Jundiai	bef 1655	284s (75e)	146 (70), 239s (70)	80 (60), 182s (60)	39 (50), 127s (50)	25 (20), 118s (20)	12 (90), 91s (00)	5 (59e), 38s (72)		
Londrina	1931	205s (75e)	156 (70), 228s (70)	72 (60), 119s (60)	33 (50), 69s (50)	11 (40)				
Maceio	1815	284s (75e)	243 (70), 264s (70)	153 (60), 170s (60)	99 (50), 121s (50)	39 (20), 74s (20)	36s (00)	18 (72), 28s (72)		
Manaus	1669	324s (75e)	284 (70), 312s (70)	154 (60), 175s (60)	90 (50), 108s (50)	44 (20), 76s (20)	23 (90), 45s (00)	9 (72), 29s (72)		
Natal	1669	389s (75e)	251 (70), 264s (70)	154 (60), 163s (60)	95 (50), 98s (50)	31 (20), 31s (20)	16s (00)	20s (72)	1 (00?e)	
Niteroi	1565	347s (75e)	292 (70), 324s (70)	229 (60), 245s (60)	171 (50), 186s (50)	25 (20), 86s (20)	65s (00)	21 (72), 48s (72)		
Nova Iguacu	1567?	932s (75e)	331 (70), 727s (70)	135 (60), 359s (60)	59 (50), 146s (50)	3 (20?)				
Olinda	1535	251s (75e)	187 (70), 196s (70)	101 (60), 110s (60)	38 (50), 62s (50)	16 (20), 52s (20)	31s (00)	8 (45e), 13s (72)		2 (1730e)
Osasco	bef 1908	377s (75e)	283 (70), 283s (70)	116 (60)[1], 116s (60)[1]	43 (50)[1]	15 (40)[1]				
Pelotas	1780	232s (75e)	150 (70), 208s (70)	121 (60), 178s (60)	78 (50), 128s (50)	48 (20), 82s (20)	40s (00)	15 (72), 21s (72)		
Petropolis	1814	217s (75e)	116 (70), 189s (70)	94 (60), 150s (60)	61 (50), 108s (50)	38 (20), 68s (20)	19s (00)	7 (72), 7s (72)		
Porto Alegre	1740	1044s (75e)	870 (70), 886s (70)	618 (60), 641s (60)	375 (50), 394s (50)	154 (20), 179s (20)	74s (00)	25 (72), 44s (72)	4 (03)	
Recife	1535?	1250s (75e)	1061 (70), 1061s (70)	789 (60), 797s (60)	512 (50), 525s (50)	239s (20)	113s (00)	74 (45), 117s (72)	25 (10e)	7 (1730e)

[1] Included in Sao Paulo.

Figure 53. Aerial view of the skyline of Sao Paulo, the largest city in South America. (Credit: Varig Brazilian Airlines.)

SOUTH AMERICA

Population (thousands)

City	Date Settled	Latest	ca 1970	ca 1960	ca 1950	ca 1930	ca 1900	ca 1850	ca 1800	18th c
Ribeirao Preto	1856		191 (70)	116 (60)	63 (50)	62 (20)	12 (90)	6 (72)		
		259s (75e)	213s (70)	147s (60)	92s (50)	69s (20)	59s (00)	6s (72)		
Rio de Janeiro	1565		4252 (70)	3223 (60)	2303 (50)			275s (72)	43 (99)	29 (1749e)
		4858s (75e)	4252s (70)	3307s (60)	2377s (50)	1158s (20)	811s (06)	266s (49)		10 (1730e)
Salvador	1549		1007 (70)	631 (60)	389 (50)				46 (05)	34 (1757)
		1237s (75e)	1007s (70)	656s (60)	417s (50)	283s (20)	206s (00)	129s (72)	100s (03e)	22 (1706)
Santo Andre	1551		415 (70)	230 (60)	97 (50)	7 (20)				
		515s (75e)	419s (70)	245s (60)	107s (50)					
Santos	1536		341 (70)	262 (60)	198 (50)	103 (20)	13 (90)	9 (72)	6 (00?e)	2 (1730e)
		396s (75e)	346s (70)	266s (60)	204s (50)	103s (20)	50s (00)	9s (72)		
Sao Bernardo do Campo	1552		188 (70)	62 (60)	20 (50)	6 (20)	7 (90)	3 (72)		
		267s (75e)	202s (70)	82s (60)	29s (50)		10s (00)			
Sao Goncalo	1647		161 (70)	64 (60)	21 (50)		9 (90)			
		534s (75e)	430s (70)	248s (60)	127s (50)	25s (20)	19s (00)			
Sao Joao de Meriti	1645		164 (70)	103 (60)	44 (50)	31 (20)	3 (90)	3 (72)		
		366s (75e)	302s (70)	192s (60)	76s (50)	47s (20)				
Sao Luis	1612		168 (70)	125 (60)	80 (50)			32s (72)	12 (00?e)	2 (1730e)
		330s (75e)	265s (70)	160s (60)	104s (50)	53s (20)	29s (00)			
Sao Paulo	1554		5925 (70)	3165 (60)	2017 (50)			31s (72)	9 (36)	2 (1765)
		7199s (75e)	5925s (70)	3825s (60)	2198s (50)	579s (20)	240s (00)		22s (36)	
Sorocaba	1654		166 (70)	109 (60)	69 (50)	40 (20)	17 (90)	14 (72)		
		208s (75e)	168s (70)	138s (60)	94s (50)	43s (20)	19s (00)	14s (72)		
Teresina	1851		181 (70)	100 (60)	51 (50)	57 (20)		22s (72)		
		290s (75e)	220s (70)	145s (60)	91s (50)	57s (20)	45s (00)			
Vitoria	1535		122 (70)	83 (60)	50 (50)	19 (20)	7 (90)	4 (72)		
		164s (75e)	133s (70)	85s (60)	51s (50)	22s (20)	12s (00)	16s (72)		
CHILE										
Antofagasta	1866	150 (75e)	138 (70)	88 (60)	62 (52)	58 (30)	14 (95)	6 (75)		
Concepcion	1550	170 (75e)	196 (70)	148 (60)	120 (52)	78 (30)	40 (95)	14 (65)	10 (00?e)	
		500s (75e)								

SOUTH AMERICA

City	Date Settled	Latest	ca 1970	ca 1960	ca 1950	ca 1930	ca 1900	ca 1850	ca 1800	18th c
					Population (thousands)					
*Santiago	1541	3186s(75e)	2662s(70)	1907s (60)	1350s(52)	696s (30)	256 (95)	115 (65)	46 (00?e)	5 (1657)
Valparaíso	1544?	249 (75e)	293 (70)	253 (60)	219 (52)	193 (30)	122 (95)	70 (65)	6 (00?e)	
		592s(75e)						52 (54)		
Viña del Mar	1586?	229 (75e)	153 (70)	115 (60)	85 (52)	49 (30)	11 (95)			
COLOMBIA										
Barranquilla	1629	719 (76e)	661 (73)	493 (64)	276 (51)	150 (38)	40 (05)	6 (51)		
*Bogotá	bef 1538	3102 (76e)	2696 (73)	1563 (64)	639 (51)	326 (38)	100 (05)	30 (51)	24 (00)	20 (1723)
Bucaramanga	1622	324 (76e)	292 (73)	217 (64)	103 (51)	42 (38)	20 (05)	10 (51)		
Cali	1536	1003 (76e)	898 (73)	618 (64)	241 (51)	88 (38)	31 (05)	12 (51)	6 (97)	
Cartagena	1533	340 (76e)	293 (73)	218 (64)	111 (51)	73 (38)	10 (05)	10 (51)	24 (00?e)	
Cúcuta	1733	293 (76e)	220 (73)	147 (64)	70 (51)	37 (38)	15 (05)	6 (51)	4 (93)	
Ibagué	1551	223 (76e)	176 (73)	125 (64)	54 (51)	27 (38)	25 (05)	7 (51)	2 (00?e)	
Manizales	1849	251 (76e)	200 (73)	190 (64)	89 (51)	51 (38)	22 (05)	3 (51)		
Medellín	1616	1195 (76e)	1056 (73)	718 (64)	328 (51)	144 (38)	55 (05)	14 (51)	6 (26e)	
Pereira	1863	228 (76e)	174 (73)	147 (64)	76 (51)	31 (38)	0.6(05)	0.6 (70)		
ECUADOR										
Guayaquil	1537	823 (74)	823 (74)	511 (62)	259 (50)	92 (19)	45 (90)	28 (50?e)	14 (05)	15 (1734e)
*Quito	1000?	600 (74)	600 (74)	355 (62)	219 (50)	81 (22)	52 (06)	36 (57)	28 (80)	
FRENCH GUIANA										
*Cayenne	1664	30s(74)	20 (67)	18 (62)	11 (46)	14 (26)	13 (01)	5 (36)	1 (00?e)	
			25s(67)							
GUYANA										
*Georgetown	1625	66 (70)	66 (70)	73 (60)	74 (46)	63 (31)	49 (91)	26 (51)	8 (00?e)	
		167s(70)	167s(70)	148s(60)	94s(46)		55s (91)			
PARAGUAY										
*Asunción	1537	442 (76e)	389 (72)	289 (62)	203 (50)	77 (28)	52 (00)	10 (50?e)	7 (93)	

SOUTH AMERICA

Population (thousands)

City	Date Settled	Latest	ca 1970	ca 1960	ca 1950	ca 1930	ca 1900	ca 1850	ca 1800	18th c
PERU										
Arequipa	bef 1425	302 (72)	302 (72)	135 (61)	97 (50e)	77 (40)	35 (96e)	29 (76)	28 (04)	
Callao	1537	297 (72)	297 (72)	214 (61)	88 (50e)	82 (40)	29 (98)	34 (76)	5 (00?e)	
Chiclayo	1720	188 (72)	188 (72)	89 (61)	40 (50e)	32 (40)	13 (89)	12 (76)		
Cuzco	11th c	121 (72)	121 (72)	80 (61)	56 (50e)	41 (40)	30 (96e)	18 (76) 45 (58)	32 (94)	
*Lima	bef 1532	3303s (72)	3303s (72)	1553s (61)	835s(50e)	541s (40)	130 (03)	101 (76) 94 (57)	53 (91)	37 (1700)
Trujillo	1534	240 (72)	240 (72)	100 (61)	48 (50e)	37 (40)	6 (06e)	8 (76)	6 (00?e)	26 (1614)
SURINAM										
*Paramaribo	1540?	102 (71)	102 (71)	111 (64)	74 (50)	48 (31)	32 (01)	18 (60)	20 (00?e)	
URUGUAY										
*Montevideo	1726	1230 (75)	1230 (75)	1159 (63) 1203s(63)	784 (49e)	482 (31)	268 (00)	34 (52)	14 (03)	
VENEZUELA										
Barquisimeto	1552	331 (71)	331 (71)	199 (61)	105 (50)	36 (26)	27 (91)	26 (73)	15 (07)	
*Caracas	1567	2487s(75e)	2184s(71)	1336s(61)	694s(50)	168s (26)	98s (91)	49 (73) 68s (73)	31 (02)	19 (1759)
El Valle	1567	186 (71)	186 (71)	111 (61)	38 (50)	6 (26)	5 (91?)	4 (73)		
Maracaibo	1571	652 (71)	652 (71)	422 (61)	236 (50)	84 (26)	35 (91)	26 (73)	22 (01e)	12 (1700e)
Maracay	17th c	255 (71)	255 (71)	135 (61)	65 (50)	13 (26)	6 (91)	5 (73)	8 (00?e)	
Petare	1704	228 (71)	228 (71)	78 (61)	21 (50)	10 (26)	9 (91)	6 (73)		
San Felix de Guayana	1590	144 (71)	144 (71)	37 (61)	4 (50)	1s (26)	1s (91)	0.7s(73)		
Valencia	1555	367 (71)	367 (71)	164 (61)	89 (50)	45 (26)	54 (91)	29 (73)	7 (00)	

Table 11b. Location, Elevation, and Climatic Data
AFRICA

City	Latitude and Longitude		Elev (ft)	Period of Record	Temperature (°F)							Annual Precipitation (in)
					Avg	Avg Warmest Month	Avg Coolest Month	Absolute High	Absolute Low	Avg Annual High	Avg Annual Low	
ALGERIA												
*Algiers	36.47N,	3.03E	194	1931–60	63.1	77.2 Aug	50.5 Jan	107	32	102	38	27.20
Annaba	36.54N,	7.46E	10	1935–49	64.4	77.9 Aug	52.9 Jan	115	32			29.53
Constantine	36.22N,	6.37E	1906	1935–49	59.2	77.2 Aug	44.2 Jan	114	23	104	29	17.56
Oran	35.42N,	0.38W	295	1931–60	63.0	77.2 Aug	50.4 Jan	110	31	100	40	15.51
ANGOLA												
*Luanda	8.48S,	13.14E	140	1931–60	75.6	80.4 Mar	68.4 Jul	98	58	91	60	14.45
BENIN												
Cotonou	6.21N,	2.26E	43	1951–60	80.4	83.5 Mar	77.0 Aug	98	65	93	67	52.72
*Porto-Novo	6.29N,	2.37E	66	1941–70	81.3	84.0 Mar	77.5 Aug					52.32
BOTSWANA												
*Gaborone	24.40S,	25.54E	3329	28 yr bef 1956	67.5	77.7 Mar	53.2 Jul	111	20			19.96
BURUNDI												
*Bujumbura	3.23S,	29.22E	2569	1951–60	73.8	75.2 Sep	72.5 Jul	94	53			32.72
CAMEROON												
Douala	4.03N,	9.42E	43	1951–60	79.5	81.7 Feb	76.3 Aug	96	65	91	68	161.77
*Yaounde	3.52N,	11.31E	2494	1951–60	74.1	76.1 Feb	71.8 Jul	96	56	92	59	60.91

AFRICA

City	Latitude and Longitude	Elev (ft)	Period of Record	Temperature (°F) Avg	Avg Warmest Month	Avg Coolest Month	Absolute High	Absolute Low	Avg Annual High	Avg Annual Low	Annual Precipitation (in)
(CANARY ISLANDS), SPAIN											
*Las Palmas	28.06N, 15.24W	43	1882–1920, 1943–55	68.4	74.5 Aug	64.0 Jan	99	46	89	51	22.95
Santa Cruz de Tenerife	28.27N, 16.14W	13	1931–60	69.4	76.5 Jul	63.3 Jan	109	47			9.88
CAPE VERDE											
*Praia	14.55N, 23.31W	92	1941–60	75.9	79.9 Sep	72.0 Feb	94	56	90	63	10.28
CENTRAL AFRICAN EMPIRE											
*Bangui	4.23N, 18.35E	1250	1941–60	78.8	81.3 Mar	77.2 Jul	108	45	100	58	61.42
CHAD											
*Ndjamena	12.07N, 15.03E	984	1951–60	82.4	90.9 Apr	74.3 Jan	114	47	113	50	25.43
CONGO											
*Brazzaville	4.16S, 15.17E	1683	1941–60	76.8	79.5 Apr	70.5 Jul	98	51	96	56	53.98
DJIBOUTI											
*Djibouti	11.36N, 43.09E	20	1912–14, 1939–60	86.0	95.9 Jul	78.3 Jan	117	63	113	68	5.12
EGYPT											
Alexandria	31.12N, 29.54E	13	1901–34	68.0	78.6 Aug	56.5 Jan	111	37	103	43	7.40
Aswan	24.05N, 32.53E	433	1913–60	80.4	93.0 Jul	60.3 Jan	124	35	119	41	0.04
Asyut	27.11N, 31.11E	217	1900–34	71.1	84.9 Jul	52.9 Jan	121	32	112	36	0.20
*Cairo	30.03N, 31.15E	79	1909–34	69.1	81.5 Aug	53.8 Jan	117	31	108	39	1.14
Damanhur	31.02N, 30.28E	20	1929–34	68.9	79.3 Aug	56.1 Jan	109	36			4.06

AFRICA

City	Latitude and Longitude	Elev (ft)	Period of Record	Temperature (°F) Avg	Avg Warmest Month	Avg Coolest Month	Absolute High	Absolute Low	Avg Annual High	Avg Annual Low	Annual Precipitation (in)
Giza	30.01N, 31.13E	82	1909–34	67.5	79.9 Jul	52.0 Jan	117	25	111	37	1.10
Ismailia	30.35N, 32.16E	33	1886–1954	70.9	84.4 Aug	55.9 Jan	120	31	117	34	11.22
Luxor	25.41N, 32.39E	256	1941–47	76.3	90.7 Jul	56.3 Jan	119	33			0.02
Mahalla al Kubra	30.58N, 31.10E		1907–29[1]	66.4	79.7 Jul	50.9 Jan	117	32			2.40
Mansurah	31.03N, 31.23E	23	1927–34	70.5	82.4 Aug	56.5 Feb	116	32			2.13
Port Said	31.16N, 32.18E	72	1941–60	70.2	81.3 Aug	57.6 Jan	113	32	100	43	2.95
Shubra al Khaymah	30.06N, 31.15E		(see Cairo)								
Suez	29.58N, 32.33E	13	1910–34	71.6	83.8 Aug	57.2 Jan	113	35	108	41	0.87
Tanta	30.47N, 31.00E	46	1927–34	67.8	79.9 Jul	53.2 Jan	116	33			1.65
Zagazig	30.35N, 31.31E	36	1913–18, 1925–34	67.6	79.9 Jul	52.7 Jan	114	31			1.18
ETHIOPIA											
*Addis Ababa	9.02N, 38.47E	7900	27 yr bef 1960	60.4	64.2 Mar	59.0 Jul	94	32	85	35	49.06
Asmara	15.19N, 38.55E	7789	1922–39, 1945–52	62.5	65.5 May	59.0 Jan	88	31	84	36	19.37
Harar	9.19N, 42.09E	6089	1908–18	67.5	69.4 Apr	65.5 Aug	90	45			35.04
GABON											
*Libreville	0.23N, 9.27E	10	1951–60	78.8	80.8 Apr	75.4 Jul	99	62	94	65	122.83
GAMBIA											
*Banjul	13.27N, 16.34W	7	34 yr bef 1961	77.9	80.8 Oct	73.6 Jan	106	45	104	49	48.19
GHANA											
*Accra	5.31N, 0.12W	213	1941–60	79.7	82.0 Mar	75.7 Aug	100	59	93	65	30.98
Kumasi	6.43N, 1.37W	961	1941–60	77.9	80.4 Mar	74.5 Aug	100	51	97	54	57.68

[1]Climatic data for Qurashiyah, nr Mahalla al Kubra.

AFRICA

City	Latitude and Longitude	Elev (ft)	Period of Record	Temperature (°F)							Annual Precipitation (in)
				Avg	Avg Warmest Month	Avg Coolest Month	Absolute High	Absolute Low	Avg Annual High	Avg Annual Low	
GUINEA											
*Conakry	9.31N, 13.43W	151	29 yr bef 1961	80.1	82.0 Apr	77.0 Aug	96	63	94	65	170.91
GUINEA-BISSAU											
*Bissau	11.51N, 15.35W	66	1941–60	79.3	81.3 May	75.9 Jan					78.19
IVORY COAST											
*Abidjan	5.19N, 4.02W	23	1941–60	79.7	82.0 Mar	75.7 Aug	97	56	93	64	84.41
Bouake	7.41N, 5.02W	1194	16 yr bef 1961	79.9	83.1 Apr	77.0 Aug	104	57	102	60	47.64
KENYA											
Mombasa	4.03S, 39.40E	52	45 yr bef 1955	79.0	82.4 Mar	75.9 Jul	96	61	91	67	46.85
*Nairobi	1.17S, 36.49E	5453	1910–20, 1929–60	64.4	65.7 Mar	58.8 Jul	87	41	83	45	37.17
LIBERIA											
*Monrovia	6.19N, 10.49W	75	1953–56	78.4	80.6 Mar	75.9 Jul	93	55	91	63	202.01
LIBYA											
Bengasi	32.07N, 20.04E	82	46 yr bef 1947	68	78.5 Aug	56.5 Jan	109	37	104	41	10.47
*Tripoli	32.54N, 13.11E	72	31 yr bef 1941	67.5	79.9 Aug	54.9 Jan	114	33	106	39	15.94
MADAGASCAR											
*Antananarivo	18.55S, 47.31E	4531	44 yr bef 1954	65.5	70 Dec	58 Jul	95	34	90	38	53.4
(MADEIRA ISLANDS), PORTUGAL											
*Funchal	32.38N, 16.54W	82	1931–60	65.3	71.4 Aug	60.1 Feb	103	40	88	47	24.25

AFRICA

City	Latitude and Longitude		Elev (ft)	Period of Record	Temperature (°F)							Annual Precipitation (in)
					Avg	Avg Warmest Month	Avg Coolest Month	Absolute High	Absolute Low	Avg Annual High	Avg Annual Low	
MALAWI												
Blantyre	15.48S,	35.01E	3501	1931-60[1]	70.2	75.4 Nov	62.8 Jul	95	41	93	47	45.47
*Lilongwe	13.59S,	33.47E	3501	1958-61	68.0	73.4 Nov	59.9 Jul	94	31			35.98
MALI												
*Bamako	12.39N,	8.00W	1089	1941-60	82.6	91.4 Apr	77.5 Jan	117	47	111	52	43.27
Timbuktu	16.46N,	3.01W	978	1951-60	84.7	93.7 May	72.7 Jan	119	41	117	44	8.86
MAURITANIA												
*Nouakchott	18.06N,	15.57W	16	1934-36, 1941-60	78.3	86.0 Sep	69.8 Jan	115	40	112	47	5.90
MAURITIUS												
*Port Louis	20.10S,	57.30E	181	1921-50	73.4	78.8 Jan	68.0 Jul	95	50	91	54	50.98
MOROCCO												
Casablanca	33.36N,	7.37W	164	1924-60	63.5	73.4 Aug	54.3 Jan	110	31	99	36	15.91
Fez	34.03N,	4.59W	1368	1921-55	64.4	81.0 Aug	49.5 Jan	119	24	112	28	21.18
Marrakesh	31.37N,	8.00W	1509	1924-60	67.5	83.8 Aug	52.9 Jan	120	27	115	31	12.83
Meknes	33.54N,	5.33W	1831	30 yr bef 1974	63.0	77.9 Aug	49.5 Jan					22.72
Oujda	34.40N,	1.54W	1526	30 yr bef 1974	62.6	79.0 Aug	48.9 Jan					12.83
*Rabat-Sale	34.02N,	6.50W	213	1921-55	64.0	73.4 Aug	54.5 Jan	118	32	105	37	19.80
Tangier	35.47N,	5.48W	246	30 yr bef 1974	64.2	75.4 Aug	54.3 Jan	106	28	94	36	29.45
MOZAMBIQUE												
*Maputo	25.58S,	32.33E	194	1921-50	72.0	77.9 Feb	64.8 Jul	114	45	105	48	29.68

[1] Temperature data for Zomba, nr Blantyre.

AFRICA

City	Latitude and Longitude	Elev (ft)	Period of Record	Avg	Avg Warmest Month	Avg Coolest Month	Absolute High	Absolute Low	Avg Annual High	Avg Annual Low	Annual Precipitation (in)
NAMIBIA											
*Windhoek	22.35S, 17.05E	5428	55 yr bef 1960	66.6	74.1 Jan	55.8 Jul	97	25	95	33	14.49
NIGER											
*Niamey	13.30N, 2.08E	728	1945–60	84.0	92.5 Apr	76.3 Jan	114	47	113	50	25.04
NIGERIA											
Abeokuta	7.10N, 3.20E	220	1909–13, 1916–29 (see Ibadan)	79.0	80.8 Mar	76.6 Aug					46.6
Ado-Ekiti	7.38N, 5.13E	1650									
Enugu	6.27N, 7.28E	466	1951–60	80.1	83.5 Mar	77.7 Aug	99	55	97	62	65.39
Ibadan	7.23N, 3.53E	651	24 yr bef 1955	80.1	83.8 Mar	75.9 Aug	102	50	99	57	49.68
Ilesha	7.38N, 4.45E	1100?	(see Oshogbo)								
Ilorin	8.45N, 4.32E	1079	1916–29, 1951–60	79.2	83.5 Mar	76.1 Aug	104	48	101	53	52.01
Iwo	7.38N, 4.11E		(see Ibadan)								
Kaduna	10.31N, 7.26E	2113	1919–54	77.0	83.5 Apr	73.9 Jan	105	46	101	52	50.31
Kano	12.00N, 8.30E	1533	1921–50	79.6	88.1 Apr	70.7 Jan	114	43	107	48	32.80
*Lagos	6.26N, 3.23E	9	67 yr bef 1961	80.4	82.9 Mar	77.2 Aug	104	60	95	67	69.33
Mushin	6.32N, 3.22E		(see Lagos)								
Ogbomosho	8.05N, 4.11E	1200	14 yr bef 1926	77.0	81.3 Feb	73.4 Aug					76.3
Onitsha	6.10N, 6.47E	300	1943–54[1]	79.0	81.5 Mar	75.9 Jul	98	55			
Oshogbo	7.50N, 4.35E	990	1937	78.6	84.7 Mar	74.3 Aug			100	50	44.2
Port Harcourt	4.43N, 7.05E	64	1951–60	78.8	80.6 Mar	76.6 Jul	99	60			94.3
Zaria	11.07N, 7.44E	2100	24 yr bef 1952	(see Kaduna for temperature data)							44.4
REUNION											
*Saint-Denis	20.52S, 55.28E	36	1941–60	75.0	79.7 Jan	70.3 Jul					55.87

[1]Temperature data for Benin, nr Onitsha.

AFRICA

City	Latitude and Longitude		Elev (ft)	Period of Record	Temperature (°F) Avg	Avg Warmest Month	Avg Coolest Month	Absolute High	Absolute Low	Avg Annual High	Avg Annual Low	Annual Precipitation (in)
RHODESIA												
Bulawayo	20.09S,	28.36E	4405	1931–60	65.7	71.8 Oct	56.1 Jun	99	28	96	35	23.19
*Salisbury	17.50S,	31.03E	4831	1931–60	64.8	70.3 Oct	56.5 Jun	95	32	91	35	33.98
RWANDA												
*Kigali	1.57S,	30.04E	5053	1953–57, 1959	68.7	70.0 Aug	67.6 May	88	49			39.57
SENEGAL												
*Dakar	14.40N,	17.26W	79	1947–70	75.9	81.5 Sep	69.4 Feb	109	53	100	58	22.76
SIERRA LEONE												
*Freetown	8.29N,	13.13W	86	1881–1960	79.9	81.9 Mar	77.4 Aug	98	62	95	64	143.27
SOMALIA												
*Mogadishu	2.02N,	45.21E	33	1931–60	81.0	84.6 Apr	78.6 Aug	97	59	92	67	16.30
SOUTH AFRICA												
Bloemfontein	29.08S,	26.13E	4678	1880–1940	60.7	73.0 Jan	46.1 Jul	113	12	96	21	20.77
*Cape Town	33.55S,	18.27E	40	1881–1940	63.2	71.1 Feb	55.5 Jul	105	28	97	32	24.65
Durban	29.53S,	31.02E	22	1921–50	70.3	76.3 Feb	63.4 Jul	107	39	95	42	40.20
Germiston	26.15S,	28.10E	5450	1921–50	60.6	67.5 Jan	50.4 Jul	93	19	89	24	29.81
Johannesburg	26.11S,	28.03E	5750	1893–1940	60.8	68.6 Jan	49.3 Jul	96	22	86	27	32.60
Pietermaritzburg	29.37S,	30.23E	2128	1916–40	66.0	73.5 Feb	56.1 Jul	106	26	98	31	36.54
Port Elizabeth	33.58S,	25.37E	176	1881–1940	64.2	70.7 Feb	58.3 Jul	107	31	99	33	22.73
*Pretoria	25.45S,	28.12E	4375	1920–40	63.7	72.3 Jan	51.4 Jul	101	20	92	27	28.28
Vereeniging	26.41S,	27.56E	4725	1903–40	61.5	71.2 Jan	47.7 Jul	100	12	92	18	26.49

AFRICA

City	Latitude and Longitude	Elev (ft)	Period of Record	Avg	Avg Warmest Month	Avg Coolest Month	Absolute High	Absolute Low	Avg Annual High	Avg Annual Low	Annual Precipitation (in)
SUDAN											
*Khartoum	15.36N, 32.32E	1257	1931–60	83.7	91.9 Jun	72.5 Jan	118	41	114	47	6.34
Omdurman	15.38N, 32.30E		(see Khartoum)								
TANZANIA											
*Dar es Salaam	6.48S, 39.17E	47	41 yr bef 1960	77.9	82.4 Feb	74.5 Jul	96	59	92	62	40.98
Zanzibar	6.10S, 39.11E	61	1921–50	79.5	82.6 Feb	76.5 Aug	102	67	98	68	55.54
TOGO											
*Lome	6.08N, 1.13E	66	1951–60	79.5	82.2 Mar	76.3 Aug	98	60			35.16
TUNISIA											
Sfax	34.44N, 10.46E	69	1901–50	65.8	79.0 Aug	52.5 Jan	111	25	98	36	7.76
*Tunis	36.48N, 10.12E	217	60 yr bef 1951	64.5	80 Aug	50.5 Jan	118	30	109	35	17.48
UGANDA											
*Kampala	0.19N, 32.35E	3910	1931–50	70.0	72.1 Jan	68.0 Aug	97	53	91	57	45.17
UPPER VOLTA											
*Ouagadougou	12.22N, 1.31W	997	25 yr bef 1960	83.5	91.4 Apr	77.2 Jan	118	47			34.61
ZAIRE											
Kananga	5.54S, 22.25E	2215	1940–59	76.5	77.2 Mar	74.5 Jul	94	55	94	58	62.76
*Kinshasa	4.18S, 15.18E	951	1940–57	77.5	80.1 Apr	72.0 Jul	100	53	96	59	53.50
Kisangani	0.30N, 25.12E	1362	1927–41, 1951–59	77.5	79.2 Apr	76.1 Jul	97	59	96	65	67.05
Lubumbashi	11.40S, 27.28E	4035	1919–59	68.9	74.7 Oct	61.5 Jul	99	33	97	37	48.90
Mbuji-Mayi	6.09S, 23.36E										

AFRICA

City	Latitude and Longitude	Elev (ft)	Period of Record	Temperature (°F)							Annual Precipitation (in)
				Avg	Avg Warmest Month	Avg Coolest Month	Absolute High	Absolute Low	Avg Annual High	Avg Annual Low	
ZAMBIA											
Kitwe	12.49S, 28.13E	4429	(see Ndola)								
*Lusaka	15.25S, 28.17E	4196	1941–60	69.1	76.5 Oct	61.5 Jun	100	39	95	42	31.22
Ndola	12.58S, 28.38E	4137	10 yr bef 1959	67.5	73.9 Oct	58.5 Jun	97	28	93	32	45.12

ASIA

City	Latitude and Longitude	Elev (ft)	Period of Record	Temperature (°F)							Annual Precipitation (in)
				Avg	Avg Warmest Month	Avg Coolest Month	Absolute High	Absolute Low	Avg Annual High	Avg Annual Low	
AFGHANISTAN											
Herat	34.22N, 62.09E	3025	1958–63	60.8	84.9 Jul	39.2 Jan	117	10	115	16	9.09
*Kabul	34.30N, 69.13E	5903	40 yr bef 1945	57.2	78.6 Jul	30.9 Jan	112	–7	100	2	12.20
Kandahar	31.27N, 65.43E	3462	1958–63	65.3	88.7 Jul	41.5 Jan	111	14	109	19	8.86
BAHRAIN											
*Manama	26.13N, 50.35E	3	1901–14, 1927–60	78.6	92.5 Aug	62.2 Jan	113	41			3.23
BANGLADESH											
Chittagong	22.20N, 91.50E	46	1931–60	78.3	82.9 May	67.8 Jan	102	45	95	48	112.52
*Dacca	23.43N, 90.25E	27	1950–59	78.2	84.8 May	66.1 Jan					73.35
Khulna	22.48N, 89.33E	16	20 yr bef 1963	78.9	84.9 May	66.8 Jan	104	45			67.85
Narayanganj	23.37N, 90.30E	26	1931–60	79.3	84.6 Sep	67.6 Jan	105	44	99	49	79.21
BRUNEI											
*Bandar Seri Begawan	4.53N, 114.56E	10	5 yr bef 1967	81.5	82 Apr	80.5 Jan	99	70			131.0

ASIA

City	Latitude and Longitude	Elev (ft)	Period of Record	Temperature (°F) Avg	Avg Warmest Month	Avg Coolest Month	Absolute High	Absolute Low	Avg Annual High	Avg Annual Low	Annual Precipitation (in)
BURMA											
Bassein	16.47N, 94.44E	33	1891–1938, 1951–58	80.6	86.0 Apr	75.2 Jan	105	50			108.98
Henzada	17.38N, 95.28E										
Mandalay	22.00N, 96.05E	75	1891–1961	81.9	89.6 Apr	70.3 Jan	114	44	108	50	33.98
Moulmein	16.30N, 97.38E	72	1891–1922, 1952–60	80.6	85.1 Apr	77.4 Jan	102	54			189.76
Myingyan	21.28N, 95.23E		(see Mandalay)								
Pegu	17.20N, 96.29E		(see Rangoon)								
*Rangoon	16.47N, 96.10E	20	1881–1940, 1949–60	81.3	86.7 Apr	77.2 Jan	107	55	102	59	104.29
CAMBODIA											
*Phnom-Penh	11.33N, 104.55E	36	42 yr bef 1961	82.2	85.5 Apr	78.4 Dec	105	57			51.97
CHINA											
Amoy	24.27N, 118.05E	131	1915, 1917–42	71.2	84.2 Aug	56.5 Feb	100	39	95	42	46.69
Anking	30.31N, 117.02E	128	1934–37, 1942	63.0	85.8 Jul	37.9 Jan	103	18			40.87
Anshan	41.07N, 122.57E	115	1925–50	47.7	77.5 Jul	13.3 Jan	100	−21			27.99
Antung	40.08N, 124.24E	20	1924–40, 1949–52	47.3	76.3 Aug	14.7 Jan	100	−25			40.28
Canton	23.07N, 113.15E	59	1912–37, 1943–52	71.4	82.9 Jul	56.5 Jan	100	31	97	36	67.72
Changchih	36.11N, 113.06E	2999	1941–44, 1954–55	48.7	73.8 Jul	19.6 Jan	98	−20			20.00
Changchow (Fukien)	24.31N, 117.40E		(see Amoy)								
Changchow (Kiangsu)	31.47N, 119.58E		(see Wusih)								
Changchun	43.52N, 125.21E	709	1909–42, 1947–52	40.5	74.3 Jul	1.8 Jan	103	−33	94	−23	24.88
Changsha	28.12N, 112.58E	157	1924–38, 1946–50	63.1	83.8 Aug	41.0 Jan	109	17	102	21	52.13
Chefoo	37.32N, 121.24E	151	1924–42	54.7	78.4 Jul	28.6 Jan	104	5	97	13	24.13
Chengchow	34.45N, 113.40E	358	1924–37	61.0	85.3 Jul	32.5 Jan	110	1			32.36
Chengtu	30.40N, 104.04E	1634	1932–53	62.6	79.7 Jul	43.2 Jan	104	25	93	32	45.12
Chinchow	41.07N, 121.06E		1936–56	47.8	77.2 Jul	13.8 Jan	101	−15			

ASIA

City	Latitude and Longitude	Elev (ft)	Period of Record	Temperature (°F)							Annual Precipitation (in)
				Avg	Avg Warmest Month	Avg Coolest Month	Absolute High	Absolute Low	Avg Annual High	Avg Annual Low	
Chinkiang	32.13N, 119.26E	39	1881–1915	60.4	82.9 Aug	37.6 Jan	104	10			40.36
Chinwangtao	39.56N, 119.37E	10	1909–16, 1930–36	49.5	75.9 Aug	21.0 Jan	96	–8			26.26
Chuchow	27.50N, 113.09E		(see Changsha)								
Chungking	29.34N, 106.35E	659	1924–38, 1940–53	65.5	83.8 Aug	46.6 Jan	111	27	105	34	42.87
Foochow	26.05N, 119.18E	289	1924–44, 1951–52	67.6	83.3 Jul	51.3 Feb	104	27			57.09
Foshan	23.02N, 113.07E	50	(see Canton)								
Fushun	41.52N, 123.53E		(see Mukden)								
Fusin	42.06N, 121.46E										
Hangchow	30.15N, 120.10E	16	1904–47, 1950–52	61.3	82.9 Jul	39.7 Jan	108	13			58.66
Hantan	36.35N, 114.29E		1939–43[1]	58.5	82.8 Jul	30.0 Jan	111	6			20.91
Harbin	45.45N, 126.39E	476	1909–42, 1949–52	37.9	73.9 Jul	–7.1 Jan	102	–42	92	–31	22.72
Hengyang	26.54N, 112.36E	308	12 yr bef 1953	64.0	84.4 Jul	42.1 Jan	106	21	105	23	58.19
Hofei	31.51N, 117.17E	85	1953–55	59.9	82.9 Aug	32.9 Jan	100	–5			32.68
Hoihow	20.03N, 110.19E	46	1924–42	76.1	84.7 Jun	64.6 Jan	102	37			60.35
Hokang	47.05N, 130.20E		(see Kiamusze)								
Huhehot	40.47N, 111.37E	3484	1930–43, 1946–50	42.4	72.0 Jul	7.3 Jan	100	–33			14.41
Hwainan	32.40N, 117.00E		(see Pengpu)								
Hwangshih	30.13N, 115.06E		(see Wuhan)								
Ichang	30.42N, 111.17E	436	1924–38	63.5	84.0 Jul	40.3 Jan	111	20	106	26	44.57
Ichun	47.42N, 128.54E										
Ipin	28.46N, 104.34E	938	11 yr bef 1953	65.7	82.2 Aug	47.1 Jan	108	29			46.66
Kaifeng	34.51N, 114.21E	246	1931–37, 1940–43	58.8	83.1 Jul	30.9 Jan	109	5			24.41
Kalgan	40.50N, 114.56E	2494	1937–43, 1947–52	46.9	73.6 Jul	16.3 Jan	101	–14			14.45
Kashgar	39.29N, 75.58E	4629	49 yr bef 1941	54.7	80.1 Jul	22.5 Jan	106	–15	102	0	3.07
Kiamusze	46.50N, 130.21E	266	1938–42, 1949–53	38.1	73.2 Jul	–4.4 Jan	98	–39			24.76

[1] Climatic data for Anyang, nr Hantan.

ASIA

City	Latitude and Longitude	Elev (ft)	Period of Record	Temperature (°F)							Annual Precipitation (in)
				Avg	Avg Warmest Month	Avg Coolest Month	Absolute High	Absolute Low	Avg Annual High	Avg Annual Low	
Kingtehchen	29.16N, 117.11E										
Kirin	43.51N, 126.33E	689	1911–13	39.9	72.9 Jul	1.6 Jan	99	−37			26.38
Kisi	45.18N, 130.58E	719	1949–52	38.8	72.3 Jul	0.1 Jan	100	−31			20.98
Kokiu	23.23N, 103.09E	5709	1938–59[1]	66.7	73.8 Jul	56.1 Jan	98	24	96	36	35.83
Kunming	25.04N, 102.41E	6211	39 yr bef 1953	61.0	70.2 Jul	49.3 Jan	91	22	87	28	39.53
Kweilin	25.17N, 110.17E	548	1935–43, 1949–50	66.9	83.1 Jul	48.6 Jan	103	23			77.40
Kweiyang	26.35N, 106.43E	3514	1920–53	60.1	76.5 Jul	41.0 Jan	103	15	95	23	50.16
Lanchow	36.03N, 103.41E	4948	1932–52	49.1	73.0 Jul	20.3 Jan	100	−10	98	0	13.31
Lhasa	29.39N, 91.06E	12,002	1935–38, 1941–49	47.8	62.6 Jun	31.5 Jan	89	3	82	5	57.52
Liaoyang	41.17N, 123.11E	85	(see Mukden)								
Liaoyuan	42.55N, 125.09E		(see Szeping)								
Liuchow	24.19N, 109.24E	322	1938–46	57.6							57.99
Loyang	34.41N, 112.28E	453	1932–37, 1950–52		82.0 Jul	30.4 Jan	111	−4			18.90
Luchow	28.53N, 105.23E	1000	(see Ipin)								
Lushun-Talien	38.55N, 121.39E	315	1905–40, 1950–52	50.5	76.1 Aug	22.6 Jan	97	−4	91	2	22.05
Malipo	23.09N, 104.44E										
Mukden	41.48N, 123.27E	138	1905–42, 1947–52	45.1	76.8 Jul	9.0 Jan	103	−28	98	−18	27.99
Mutankiang	44.35N, 129.36E	761	1909–43, 1949–53	37.0	72.0 Jul	−4.2 Jan	99	−49			20.51
Nanchang	28.41N, 115.53E	161	1929–33, 1946–52	63.3	85.1 Aug	41.9 Jan	103	21			69.69
Nanchung	30.48N, 106.04E	978	1940–46	65.8	84.4 Jul	46.4 Jan	109	29			36.97
Nanking	32.03N, 118.47E	200	1905–37, 1947–50	59.5	81.9 Jul	36.0 Jan	109	7	101	13	38.94
Nanning	22.49N, 108.19E	246	1922–39, 1941–50	72.0	83.3 Jul	56.5 Jan	102	34			52.05
Nantung	32.02N, 120.53E	361	1917–37, 1949–50	58.5	80.4 Jul	35.2 Jan	108	9			41.02
Neikiang	29.35N, 105.03E	1165	1936–50	64.9	81.7 Jul	47.1 Jan	105	27			36.89
Ningpo	29.53N, 121.33E	15	1881–1915	61.9	82.4 Aug	41.7 Jan	103	14			53.54

[1]Climatic data for Mengtzu, nr Kokiu.

ASIA

City	Latitude and Longitude	Elev (ft)	Period of Record	Temperature (°F)								Annual Precipitation (in)
				Avg	Avg Warmest Month	Avg Coolest Month	Absolute		Avg Annual			
							High	Low	High	Low		
Paoki	34.22N, 107.07E	2041	1952-55	55.4	77.4 Jul	31.3 Jan	103	2				22.24
Paoting	38.52N, 115.29E	72	1913-37, 1950	53.8	80.4 Jul	22.1 Jan	109	-12				17.48
Paotow	40.36N, 110.03E	3425	1935-37, 1949-53	43.5	72.9 Jul	10.8 Jan	101	-27				11.97
*Peking	39.56N, 116.24E	171	78 yr bef 1953	53.2	79.0 Jul	23.5 Jan	109	-9	99	3		24.53
Pengpu	32.57N, 117.21E	66	1926-37	63.1	85.3 Jul	34.3 Jan	105	-3				28.11
Penki	41.20N, 123.45E		(see Mukden)									
Shanghai	31.14N, 121.28E	15	1873-1953	59.5	81.0 Aug	38.1 Jan	105	10	102	16		45.00
Shangkiu	34.23N, 115.37E		1953-55[1]	58.5	80.8 Jul	30.6 Jan	105	-1				25.12
Shaohing	30.00N, 120.35E		(see Hangchow)									
Shaoyang	27.15N, 111.28E	817	1936-44	63.5	81.9 Jul	42.1 Jan	101	24				51.38
Shihkiachwang	38.03N, 114.29E	269	1939-43, 1949-52	55.8	80.4 Jul	25.5 Jan	109	-16				17.09
Shiukwan	24.48N, 113.35E	285	1951-55	68.5	84.0 Jul	50.4 Jan	108	24				59.17
Siakwan	25.34N, 100.14E	5118	1952-55[2]	61.7	69.8 Jun	49.8 Jan	87	30				38.23
Sian	34.16N, 108.54E	1352	1922-25, 1931-52	57.2	82.2 Jul	31.1 Jan	115	-13	107	9		22.68
Siangtan	27.51N, 112.54E	640	(see Changsha)									
Sinhailien	34.36N, 119.13E	13	1951-53	57.9	81.1 Aug	32.5 Jan	102	6				32.20
Sining	36.37N, 101.46E	7363	1936-49	44.4	64.9 Jul	20.5 Jan	90	-10				14.84
Sinsiang	35.19N, 113.52E		(see Chengchow)									
Soochow	31.18N, 120.37E	52	1936-37	61.2	81.9 Jul	40.3 Jan						
Suchow	34.16N, 117.11E	112	1926-43, 1949-52	58.1	81.9 Jul	31.6 Jan	110	-1				27.76
Swatow	23.22N, 116.40E	14	1880-1941	71.1	84.0 Aug	57.0 Jan	101	31	98	40		58.62
Szeping	43.10N, 124.20E	535	1934-44, 1949-52	42.6	75.4 Jul	3.7 Jan	100	-38				27.91
Taichow	32.29N, 119.55E		(see Chinkiang)									
Taiyuan	37.52N, 112.33E	2566	1916-48, 1950-52	50.0	77.0 Jul	18.9 Jan	107	-21	100	-8		15.55

[1] Climatic data for Pohsien, nr Shangkiu.
[2] Climatic data for Paoshan, nr Siakwan.

ASIA

City	Latitude and Longitude	Elev (ft)	Period of Record	Temperature (°F)							Annual Precipitation (in)
				Avg	Avg Warmest Month	Avg Coolest Month	Absolute High	Absolute Low	Avg Annual High	Avg Annual Low	
Tangshan	39.38N, 118.11E	187	1922-37	54.3	80.6 Jul	24.3 Jan	104	-9			13.54
Tatung	40.05N, 113.18E	3442	1925-37, 1939-43	44.1	72.9 Jul	11.1 Jan	100	-21			13.89
Tientsin	39.08N, 117.12E	11	1904-42, 1944-52	54.1	80.2 Jul	24.6 Jan	109	-5	103	3	20.75
Tsamkong	21.12N, 110.23E	85	1921-40	73.9	83.7 Jul	61.3 Jan	100	38			54.45
Tsiaotso	35.15N, 113.13E		(see Chengchow)								
Tsinan	36.40N, 117.00E	180	1919-52	58.6	83.1 Jul	29.8 Jan	109	-3	103	6	24.84
Tsingtao	36.04N, 120.19E	253	1898-1952	53.8	77.2 Aug	30.0 Jan	97	2			25.47
Tsitsihar	47.22N, 123.57E	482	1930-42, 1949-53	36.9	73.4 Jul	-4.9 Jan	99	-39			14.92
Tzekung	29.24N, 104.47E		(see Neikiang)								
Tzepo	36.29N, 117.50E	590	(see Tsinan)								
Urumtsi	43.48N, 87.35E	2996	22 yr bef 1948	43.0	76.5 Jul	3.2 Jan	112	-43			9.17
Weifang	36.43N, 119.06E	207	1929-37	58.1	82.4 Jul	28.2 Jan	109	1			24.80
Wenchow	28.01N, 120.39E	16	1924-41	64.4	83.8 Jul	46.0 Jan	105	27			67.28
Wuhan	30.32N, 114.18E	75	1905-40, 1951-53	62.2	84.0 Jul	38.8 Jan	108	9	100	23	47.32
Wuhu	31.21N, 118.22E	43	1880-1915	61.3	83.1 Aug	39.2 Jan	106	11			46.26
Wusih	31.35N, 120.18E	23	1936-37	61.2	81.0 Jul	39.7 Jan					
Wutungkiao	29.21N, 103.48E		1936-50[1]	64.0	79.5 Jul	46.9 Jan	105	27			53.82
Yangchow	32.24N, 119.26E	656	(see Chinkiang)								
Yangchuan	37.54N, 113.36E	2181	(see Taiyuan)								
Yenan	36.38N, 109.27E	3400									
Yinchwan	38.28N, 106.19E	3645	1951-53, 1955	39.0	75.4 Jul	15.6 Jan	103	-23			7.99
Yingkow	40.40N, 122.17E	11	1905-43, 1949-53	47.5	77.0 Jul	14.2 Jan	98	-24	93	-13	26.42
CYPRUS											
*Nicosia	35.11N, 33.23E	508	1921-50	66.6	83.6 Jul	50.0 Jan	116	23	105	31	14.76

[1]Climatic data for Loshan, nr Wutungkiao.

ASIA

City	Latitude and Longitude	Elev (ft)	Period of Record	Temperature (°F)							Annual Precipitation (in)
				Avg	Avg Warmest Month	Avg Coolest Month	Absolute High	Absolute Low	Avg Annual High	Avg Annual Low	
GAZA STRIP											
*Gaza	31.30N, 34.28E	157	1921–34	67.5	78.1 Aug	55.4 Jan	112	30	104	38	13.70
HONG KONG											
*Hong Kong	22.17N, 114.09E	109	1884–1941, 1947–48	72.0	82.0 Jul	59.2 Feb	97	32	93	43	84.92
INDIA											
Agra	27.11N, 78.01E	553	1931–60	78.1	95.0 Jun	58.6 Jan	120	28	114	36	26.18
Ahmadabad	23.02N, 72.37E	180	1931–60	81.1	92.3 May	68.5 Jan	118	36	111	47	30.83
Ajmer	26.27N, 74.38E	1593	1931–60	76.5	92.1 May	58.5 Jan	114	27	110	35	20.47
Aligarh	27.53N, 78.05E	615	29 yr bef 1961	77.5	93.6 Jun	58.3 Jan	115	33	114	38	24.72
Allahabad	25.27N, 81.51E	322	1931–60	79.0	94.5 May	61.5 Jan	120	34	115	40	37.20
Ambala	30.21N, 76.50E	892	1931–60	76.3	93.0 Jun	56.8 Jan	118	30	114	36	30.20
Amravati	21.29N, 78.11E	1214	1931–60	81.0	95.0 May	71.2 Dec	116	41	113	50	34.49
Amritsar	31.35N, 74.53E	768	1951–60	73.8	91.0 Jun	52.7 Jan	116	27	113	32	22.17
Asansol	23.41N, 86.59E	414	1931–60	79.5	90.9 May	66.0 Jan	117	41	113	46	57.83
Bangalore	12.59N, 77.35E	3021	1931–60	74.5	81.1 Apr	68.9 Dec	102	46	97	53	35.00
Bareilly	28.21N, 79.25E	568	1931–60	77.4	91.8 Jun	59.5 Jan	116	30	113	40	42.36
Belgaum	15.52N, 74.30E	2470	1931–60	75.2	81.7 Apr	70.9 Dec	105	44	100	49	51.30
Bhagalpur	25.15N, 87.00E	161	1951–60	79.3	89.6 May	64.2 Jan	112	44			42.64
Bhatpara	22.52N, 88.24E		(see South Dum Dum)								
Bhavnagar	21.46N, 72.09E	55	1931–60	81.0	91.0 May	66.7 Jan	116	33	112	44	24.41
Bhilainagar-Durg	21.13N, 81.26E	967	1901–50	(see Raipur for temperature data)							50.28
Bhopal	23.16N, 77.24E	1716	1931–60	77.0	92.3 May	64.4 Jan	114	33	111	40	49.61
Bikaner	28.01N, 73.18E	734	23 yr bef 1961	78.6	95.9 Jun	56.5 Jan	121	27	115	32	12.05
Bombay	18.58N, 72.50E	37	1931–60	81.1	86.2 May	75.6 Jan	105	53	98	60	71.06
Calcutta	22.32N, 88.22E	21	1931–60	80.4	88.0 May	68.4 Jan	111	44	106	49	63.98

ASIA

City	Latitude and Longitude	Elev (ft)	Period of Record	Temperature (°F)							Annual Precipitation (in)
				Avg	Avg Warmest Month	Avg Coolest Month	Absolute High	Absolute Low	Avg Annual High	Avg Annual Low	
Calicut	11.15N, 75.46E	26	1931–60	81.1	84.7 Apr	78.3 Jul	99	61	94	66	125.12
Chandigarh	30.43N, 76.47E	1300	(see Ambala)								
Cochin	9.58N, 76.14E	10	1931–60	80.8	83.7 Apr	78.6 Jul	100	61	92	69	119.96
Coimbatore	11.00N, 76.58E	1341	1931–60	79.5	84.2 Apr	75.6 Dec	104	53	99	60	24.17
Cuttack	20.30N, 85.50E	89	1931–60	81.9	91.0 May	71.4 Dec	118	46	112	52	60.94
Dehra Dun	30.19N, 78.02E	2238	1931–60	71.2	84.9 Jun	54.7 Jan	111	30	106	36	84.58
Delhi	28.40N, 77.13E	770	1901–50	(see New Delhi for temperature data)							23.43
Dhanbad	23.48N, 86.27E	843	27 yr bef 1961	78.3	90.1 May	65.1 Jan	115	41	112	46	53.35
Durgapur	23.30N, 87.20E		(see Asansol)								
Gauhati	26.11N, 91.44E	180	1931–60	76.3	83.3 Aug	61.5 Jan	105	41			64.33
Gaya	24.47N, 85.00E	381	1931–60	79.2	93.7 May	62.8 Jan	117	38	114	43	46.65
Gorakhpur	26.45N, 83.22E	254	1931–60	78.3	90.3 May	61.5 Jan	119	35	108	43	49.13
Guntur	16.18N, 80.27E		1931–60[1]	82.0	90.0 May	74.5 Jan	118	57	110	61	35.31
Gwalior	26.13N, 78.10E	681	21 yr bef 1961	78.1	95.9 Jun	60.1 Jan	119	30	116	34	32.44
Howrah	22.35N, 88.20E		48 yr bef 1951	(see Calcutta for temperature data)							60.47
Hubli-Dharwar	15.21N, 75.10E	2297	1931–60[2]	78.6	85.8 Apr	73.0 Dec	107	52	104	55	26.34
Hyderabad	17.23N, 78.28E	1788	1931–60	78.4	90.3 May	69.1 Dec	112	43	108	48	30.39
Indore	22.43N, 75.50E	1823	1931–60	75.9	90.1 May	64.0 Jan	114	27	108	38	36.57
Jabalpur	23.10N, 79.57E	1289	1931–60	77.4	93.0 May	64.2 Dec	116	32	113	38	56.34
Jaipur	26.55N, 75.49E	1431	1931–60	77.0	91.8 Jun	59.2 Jan	118	28	113	36	23.54
Jamnagar	22.28N, 70.04E	60	1931–60	78.8	88.7 Jun	65.3 Jan	112	35	106	43	18.35
Jamshedpur	22.48N, 86.11E	423	1931–60	80.6	92.1 May	66.6 Dec	117	39	113	44	56.65
Jhansi	25.26N, 78.35E	848	1931–60	79.2	96.3 May	61.9 Jan	118	33	115	40	37.17
Jodhpur	26.17N, 73.02E	736	1931–60	80.1	93.9 May	62.6 Jan	120	28	115	38	14.41

[1] Temperature data for Masulipatnam, nr Guntur.
[2] Temperature data for Gadag, nr Hubli-Dharwar.

ASIA

City	Latitude and Longitude	Elev (ft)	Period of Record	Temperature (°F)							Annual Precipitation (in)
				Avg	Avg Warmest Month	Avg Coolest Month	Absolute High	Absolute Low	Avg Annual High	Avg Annual Low	
Jullundur	31.19N, 75.34E		1901–50	(see Amritsar for temperature data)							25.87
Kanpur	26.28N, 80.21E	413	1931–60	78.3	93.9 May	60.3 Jan	117	33	115	39	31.69
Kolhapur	16.42N, 74.13E	1880	1951–60	77.4	84.7 Apr	74.1 Dec	107	48	105	51	39.09
Kota	25.11N, 75.50E	843	1931–60	80.6	97.0 May	63.5 Jan	118	35	115	43	30.94
Lucknow	26.51N, 80.55E	364	1931–60	78.4	92.8 May	61.0 Jan	119	34	114	40	40.00
Ludhiana	30.54N, 75.51E	812	1931–60	76.3	93.4 Jun	55.4 Jan	119	29	116	35	25.51
Madras	13.05N, 80.17E	51	1931–60	83.5	90.9 May	76.1 Jan	113	57	107	63	49.92
Madurai	9.56N, 78.07E	437	1931–60	84.0	89.4 May	77.9 Jan	108	60	105	65	35.20
Malegaon	20.33N, 74.32E	1432	1931–60	78.8	90.5 May	69.4 Dec	116	31	110	41	21.34
Mangalore	12.52N, 74.53E	72	1931–60	80.8	84.6 Apr	79.7 Jan	100	60	95	65	133.78
Mathura	27.30N, 77.41E		1901–50	(see Agra for temperature data)							21.42
Meerut	28.59N, 77.42E	733	1951–60	76.5	92.1 Jun	57.6 Jan	115	33	112	38	31.93
Moradabad	28.50N, 78.47E	197	1901–50	(see Bareilly for temperature data)							38.66
Mysore	12.18N, 76.39E	2518	1931–60	75.9	81.9 Apr	71.1 Dec	103	51	98	55	31.89
Nagpur	21.09N, 79.06E	1017	1951–60	80.4	96.1 May	68.7 Dec	118	39	115	44	48.90
Nasik	19.59N, 73.48E	1961	1901–50	(see Malegaon for temperature data)							27.44
*New Delhi	28.36N, 77.12E	714	23 yr bef 1961	77.4	93.7 Jun	57.7 Jan	118	31	113	37	25.98
Patna	25.35N, 85.15E	173	1931–60	79.2	90.3 May	63.1 Jan	115	36	111	44	45.90
Poona	18.32N, 73.52E	1834	1931–60	77.2	85.8 May	69.8 Dec	110	35	109	43	26.02
Raipur	21.14N, 81.38E	971	1931–60	80.4	95.9 May	68.4 Dec	117	39	114	48	53.50
Rajahmundry	16.59N, 81.47E		1931–60[1]	81.9	90.1 May	73.6 Dec	117	57	110	60	41.61
Rajkot	22.18N, 70.47E	453	1931–60	80.1	90.7 May	66.9 Jan	118	31	110	41	23.39
Ranchi	23.21N, 85.20E	2149	1931–60	74.7	87.6 May	62.1 Jan	110	37	107	42	59.57
Saharanpur	29.58N, 77.33E	896	1931–60[2]	74.5	90.0 Jun	55.9 Jan	116	28	111	36	36.85

[1] Temperature data for Kakinada, nr Rajahmundry.
[2] Temperature data for Roorkee, nr Saharanpur.

ASIA

City	Latitude and Longitude	Elev (ft)	Period of Record	Temperature (°F)							Annual Precipitation (in)
				Avg	Avg Warmest Month	Avg Coolest Month	Absolute High	Absolute Low	Avg Annual High	Avg Annual Low	
Salem	11.39N, 78.10E	913	1931–60	82.4	88.0 May	76.6 Dec	109	52	104	60	37.60
Sangli	16.52N, 74.34E		1931–60[1]	77.9	86.0 May	71.8 Dec	108	41	110	50	22.36
Sholapur	17.41N, 75.55E	1570	1931–60	80.8	91.2 May	72.5 Dec	114	40	110	50	26.69
Simla	31.06N, 77.10E	7225	1931–60	56.5	68.4 Jun	41.4 Jan	87	13	83	24	60.16
South Dum Dum	22.37N, 88.25E	33	21 yr bef 1961	79.5	87.4 May	66.7 Jan	109	41	106	47	66.42
South Suburban	22.30N, 88.19E		(see Calcutta)								
Surat	21.10N, 72.50E	39	1931–60	81.9	88.5 May	73.6 Jan	114	40	109	50	42.17
Thana	19.12N, 72.58E		1901–50	(see Bombay for temperature data)							97.52
Tiruchirapalli	10.49N, 78.41E	255	1931–60	83.8	89.1 May	77.5 Dec	111	57	105	63	34.25
Tirunelveli	8.44N, 77.42E	213	1951–60[2]	84.7	89.1 May	79.5 Dec	108	65	104	68	31.34
Trivandrum	8.29N, 76.55E	210	20 yr bef 1961	80.8	83.7 Apr	79.0 Jul	100	61	95	67	71.34
Tuticorin	8.47N, 78.08E	6	1951–60[2]	84.7	89.1 May	79.5 Dec	108	65	104	68	23.70
Ujjain	23.11N, 75.46E	1680	1951–60[3]	77.9	90.9 May	65.5 Jan	113	41	109	44	35.24
Ulhasnagar-Kalyan	19.15N, 73.08E		1901–50	(see Bombay for temperature data)							92.72
Vadodara	22.18N, 73.12E	115	28 yr bef 1961	80.6	91.8 May	68.7 Jan	116	30	113	41	39.61
Varanasi	25.20N, 83.00E	250	1931–60	78.8	93.6 May	61.5 Jan	117	35	113	40	42.36
Vellore	12.56N, 79.08E	702	1931–60	82.2	90.5 May	74.1 Dec	112	53	112	53	41.54
Vijayawada	16.31N, 80.37E	80	1931–60[4]	82.0	90.0 May	74.5 Jan	118	57	110	61	37.76
Visakhapatnam	17.42N, 83.18E	10	1931–60	81.0	87.6 May	72.7 Jan	112	55	103	59	37.56
Warangal	18.18N, 79.35E	883	1931–60	81.9	93.9 May	72.5 Dec	116	47	112	53	36.38

[1] Temperature data for Miraj, nr Sangli.
[2] Temperature data for Palayamcottai, nr Tirunelveli and Tuticorin.
[3] Temperature data for Ratlam, nr Ujjain.
[4] Temperature data for Masulipatnam, nr Vijayawada.

ASIA

City	Latitude and Longitude	Elev (ft)	Period of Record	Temperature (°F)							Annual Precipitation (in)
				Avg	Avg Warmest Month	Avg Coolest Month	Absolute High	Absolute Low	Avg Annual High	Avg Annual Low	
INDONESIA											
Ambon	3.43S, 128.12E	14	22 yr bef 1937	78.8	80.8 Feb	76.1 Jul	96	66	93	69	138.98
Banjarmasin	3.20S, 114.36E	66	1956–65	80.2	81.1 Apr	79.3 Jan					108.46
Bandung	6.54S, 107.36E	2520	1913–38	72.7	73.2 Oct	71.8 Jul	94	52	90	55	85.00
Bogor	6.35S, 106.47E	820	1933–37	77.9	78.3 May	77.0 Jan					166.34
Cirebon	6.44S, 108.34E	13	1937	81.1	83.3 Nov	79.9 Dec			95	68	88.90
Denpasar	8.39S, 115.13E	131	1956–65	82.0	83.1 Jan	80.1 Aug					68.39
*Jakarta	6.10S, 106.49E	23	1931–60	80.4	81.3 Sep	79.2 Jan	98	66	92	69	69.29
Jambi	1.36S, 103.37E	33	1956–65	79.3	80.1 May	78.4 Jan					122.01
Kediri	7.49S, 112.01E	203	1879–1941	(see Surabaya for temperature data)							70.63
Kupang	10.10S, 123.35E	499	1951–60	81.0	83.7 Oct	79.0 Jul	101	60	98	63	55.75
Malang	7.59S, 112.38E	1460	1929–38	74.8	76.1 Oct	72.5 Jul	96	50	93	54	78.46
Medan	3.35N, 98.40E	82	1914–32, 1956–65	78.8	80.2 May	77.2 Jan	96	60	95	66	80.83
Menado	1.29N, 124.51E	13	19 yr bef 1934	79.2	79.9 Oct	78.1 Jan	96	63			106.38
Padang	0.57S, 100.21E	3	1913–36	79.5	80.4 May	78.8 Oct	94	68			175.28
Palembang	3.00S, 104.46E	26	1956–65	79.9	82.0 Apr	78.6 Jan					93.74
Pontianak	0.02S, 109.20E	10	1916–36	80.2	81.0 May	79.3 Dec	96	68			131.69
Semarang	6.57S, 110.25E	7	1956–65	80.4	83.8 Nov	79.0 Jul	99	63	97	64	80.04
Surabaya	7.15S, 112.45E	23	1920–25, 1927–38	80.2	81.9 Nov	78.4 Jul	97	62	94	65	59.76
Surakarta	7.35S, 110.50E	341	1933–37[1]	78.6	80.1 Oct	77.0 Jul			96	63	85.24
Tanjungkarang-Telukbetung	5.27S, 105.16E	33	1879–1941	(see Jakarta for temperature data)							83.70
Ujung Pandang	5.07S, 119.24E	7	1956–65	79.5	81.1 Apr	78.3 Dec	95	58	91	64	109.21
Yogyakarta	7.48S, 110.22E	377	1933–37[1]	78.6	80.1 Oct	77.0 Jul			96	63	78.82

[1] Temperature data for Klaten, nr Surakarta and Yogyakarta.

ASIA

City	Latitude and Longitude	Elev (ft)	Period of Record	Temperature (°F)								Annual Precipitation (in)
				Avg	Avg Warmest Month	Avg Coolest Month	Absolute		Avg Annual		
							High	Low	High	Low	
IRAN											
Abadan	30.20N, 48.16E	10	1951–60	77.2	97.0 Jul	55.2 Jan	124	24	119	32	5.67
Ahwaz	31.19N, 48.42E	66	11 yr bef 1954	77.0	96.8 Jul	59.0 Dec	129	19			7.48
Hamadan	34.48N, 48.30E	5824	13 yr bef 1954	53.6	77.0 Aug	33.8 Jan	102	−29	101	5	15.91
Isfahan	32.40N, 51.38E	5217	38 yr bef 1947	59.7	82.2 Jul	36.5 Jan	110	−3	104	12	4.61
Kermanshah	34.19N, 47.04E	4337	1936–51	55.5	77.5 Jul	34 Jan	112	−17	105	5	16.3
Meshed	36.18N, 59.36E	3232	33 yr bef 1948	56.1	77.0 Jul	34.5 Jan	110	−18	102	3	8.23
Qom	34.39N, 50.54E	3045	(see Tehran)								
Rasht	37.16N, 49.36E	−23	17 yr bef 1931[1]	60.4	77.7 Jul	43.0 Jan	99	−2			65.59
Shiraz	29.36N, 52.32E	5049	1951–60	63.0	83.3 Jul	43.7 Jan	108	7			14.80
Tabriz	38.05N, 46.18E	4469	1898–1914, 1942–60	53.1	77.0 Jul	29.3 Jan	107	−14			10.35
*Tehran	35.40N, 51.26E	3908	1894–1960	61.5	84.9 Jul	36.7 Jan	109	−5	104	15	9.02
IRAQ											
*Baghdad	33.21N, 44.25E	111	1938–62	72.1	93.7 Jul	50.2 Jan	123	18	117	25	6.14
Basra	30.30N, 47.47E	8	1937–62	75.7	93.2 Jul	54.0 Jan	123	24	116	31	6.34
Karbala	32.36N, 44.02E	85	(see Baghdad)								
Kirkuk	35.28N, 44.28E	1086	1938–61	71.1	94.1 Jul	47.7 Jan					15.00
Mosul	36.20N, 43.08E	730	1927–62	67.3	90.9 Jul	45.0 Jan	124	12	117	23	14.76
Najaf	31.59N, 44.20E	180	(see Baghdad)								
ISRAEL											
Haifa	32.49N, 35.00E	16	21 yr bef 1951	71	83 Aug	57 Jan	112	27	103	41	26.1
*Jerusalem	31.47N, 35.13E	2658	1921–60	63.3	75.7 Aug	48.0 Jan	108	24	101	31	23.11
Tel Aviv-Jaffa	32.05N, 34.46E	10	1951–60	68.0	80.4 Aug	56.5 Jan			102	34	21.06

[1] Climatic data for Bandar-e Pahlavi, nr Rasht.

ASIA

City	Latitude and Longitude	Elev (ft)	Period of Record	Temperature (°F)							Annual Precipitation (in)
				Avg	Avg Warmest Month	Avg Coolest Month	Absolute High	Low	Avg Annual High	Low	
JAPAN											
Akashi	34.38N, 134.59E	20	(see Kobe)								
Akita	39.43N, 140.07E	30	1941–70	51.6	75.7 Aug	30.7 Jan	98	-12	93	6	71.14
Amagasaki	34.43N, 135.25E		(see Osaka)								
Aomori	40.49N, 140.45E	13	1956–70	49.3	72.7 Aug	28.4 Jan	97	-12			56.07
Asahikawa	43.46N, 142.22E	364	1941–70	43.2	69.6 Aug	16.7 Jan	97	-42			45.63
Beppu	33.17N, 131.30E		(see Oita)								
Chiba	35.36N, 140.07E	79	(see Tokyo)								
Fuji	35.09N, 138.39E		(see Shizuoka)								
Fujisawa	35.21N, 139.29E		(see Yokosuka)								
Fukui	36.04N, 136.13E	30	1948–70	56.8	79.2 Aug	36.3 Jan	101	5			97.32
Fukuoka	33.35N, 130.24E	7	1941–70	60.3	81.0 Aug	41.5 Jan	98	17			67.13
Fukushima	37.45N, 140.28E	220	1941–70	54.1	77.2 Aug	33.6 Jan	102	-1			45.00
Fukuyama	34.29N, 133.22E	5	1942–70	58.3	80.8 Aug	38.7 Jan	101	18			47.52
Funabashi	35.42N, 139.59E		(see Tokyo)								
Gifu	35.25N, 136.45E	43	1941–70	58.5	80.6 Aug	37.9 Jan	101	6			74.96
Hachinohe	40.30N, 141.29E	89	1941–70	49.3	72.1 Aug	28.8 Jan	98	4			43.15
Hachioji	35.39N, 139.20E		(see Tokyo)								
Hakodate	41.45N, 140.43E	108	1941–70	46.8	70.7 Aug	25.0 Jan	92	-7	86	2	45.00
Hamamatsu	34.42N, 137.44E	95	1941–70	59.9	79.3 Aug	41.2 Jan	99	21			75.00
Higashiosaka	34.39N, 135.35E		(see Osaka)								
Himeji	34.49N, 134.42E	125	1948–70	58.1	80.2 Aug	38.1 Jan	97	14			54.14
Hirakata	34.48N, 135.38E		(see Osaka)								
Hiratsuka	35.19N, 139.21E		(see Yokohama)								
Hiroshima	34.24N, 132.27E	95	1941–70	58.6	80.4 Aug	39.4 Jan	101	17	95	22	64.73
Hitachi	36.36N, 140.39E	171	(see Mito)								
Ibaraki	34.49N, 135.34E		(see Osaka)								

ASIA

City	Latitude and Longitude	Elev (ft)	Period of Record	Temperature (°F)							Annual Precipitation (in)
				Avg	Avg Warmest Month	Avg Coolest Month	Absolute High	Low	Avg Annual High	Low	
Ichihara	35.31N, 140.05E		(see Tokyo)								
Ichikawa	35.44N, 139.55E	75	(see Tokyo)								
Ichinomiya	35.18N, 136.48E		(see Nagoya)								
Ise	34.29N, 136.42E	7	1941–70[1]	58.6	80.1 Aug	39.7 Jan	101	18			67.48
Iwaki	36.57N, 140.54E	10	? yr bef 1961	54.7	75.0 Aug	36.9 Jan					55.04
Kagoshima	31.36N, 130.33E	16	1941–70	62.6	81.1 Aug	44.1 Jan	99	20	94	24	95.79
Kamakura	35.19N, 139.33E		(see Yokosuka)								
Kanazawa	36.34N, 136.39E	89	1941–70	56.7	79.2 Aug	36.7 Jan	101	15			104.80
Kashiwa	35.52N, 139.59E		(see Tokyo)								
Kasugai	35.14N, 136.58E		(see Nagoya)								
Kawagoe	35.55N, 139.29E		(see Tokyo)								
Kawaguchi	35.48N, 139.43E		(see Tokyo)								
Kawasaki	35.32N, 139.43E		(see Yokohama)								
Kitakyushu	33.52N, 130.50E	23	(see Shimonoseki)								
Kobe	34.41N, 135.10E	190	1941–70	59.9	81.1 Aug	40.1 Jan	100	20			53.82
Kochi	33.33N, 133.33E	2	1941–70	61.0	80.2 Aug	41.4 Jan	101	18			104.09
Kofu	35.39N, 138.35E	892	1941–70	56.5	78.3 Aug	34.9 Jan	102	–3			44.45
Koriyama	37.24N, 140.23E	837	(see Fukushima)								
Koshigaya	35.54N, 139.48E		(see Tokyo)								
Kumamoto	32.48N, 130.43E	125	1941–70	60.6	81.1 Aug	40.5 Jan	102	15			76.34
Kurashiki	34.35N, 133.46E	10	(see Okayama)								
Kure	34.14N, 132.34E	89	1957–70	60.6	82.0 Aug	41.2 Jan	100	19			57.33
Kurume	33.19N, 130.31E	43	(see Fukuoka)								
Kushiro	42.58N, 144.23E	108	1941–70	41.9	64.2 Aug	20.1 Jan	87	–19	80	–10	43.78
Kyoto	35.00N, 135.45E	135	1941–70	58.6	81.1 Aug	38.3 Jan	101	11			64.48

[1]Climatic data for Tsu, nr Ise.

ASIA

City	Latitude and Longitude	Elev (ft)	Period of Record	Temperature (°F)							Annual Precipitation (in)
				Avg	Avg Warmest Month	Avg Coolest Month	Absolute High	Low	Avg Annual High	Low	
Machida	35.33N, 139.28E	318	(see Tokyo)								
Maebashi	36.23N, 139.04E	367	1941–70	56.5	77.9 Aug	36.7 Jan	100	11			48.27
Matsudo	35.47N, 139.54E	82	(see Tokyo)								
Matsumoto	36.14N, 137.58E	2001	1941–70	51.8	75.4 Aug	29.8 Jan	101	-13			41.65
Matsuyama	33.50N, 132.45E	105	1941–70	59.7	80.4 Aug	41.0 Jan	99	17			54.37
Mito	36.22N, 140.28E	95	1941–70	55.4	76.6 Aug	36.0 Jan	98	9			54.22
Miyazaki	31.54N, 131.26E	23	1941–70	62.2	80.4 Aug	44.1 Jan	100	18			102.13
Morioka	39.42N, 141.09E	505	1941–70	49.5	73.8 Aug	27.3 Jan	99	-5			50.35
Nagano	36.39N, 138.11E	1371	1941–70	52.3	76.6 Aug	29.3 Jan	101	1			39.92
Nagasaki	32.48N, 129.55E	89	1951–70	61.9	81.7 Aug	43.2 Jan	100	22	93	25	77.80
Nagoya	35.10N, 136.55E	167	1941–70	58.5	80.4 Aug	37.8 Jan	103	13			60.63
Naha	26.13N, 127.40E	89	1941–70	72.1	82.8 Jul	60.8 Jan	96	20	93	45	83.39
Nara	34.41N, 135.50E	345	1954–70	57.7	79.3 Aug	37.8 Jan	99	19			55.99
Neyagawa	34.46N, 135.38E		(see Osaka)								
Niigata	37.55N, 139.03E	7	1941–70	55.4	78.6 Aug	35.2 Jan	102	9	95	21	72.83
Nikko	36.45N, 139.37E	2001	1944–70	43.7	64.9 Aug	23.9 Jan	92	-3	88	11	89.41
Nishinomiya	34.43N, 135.20E		(see Osaka)								
Numazu	35.06N, 138.52E	23	1897–1926	59.5	78.6 Aug	41.7 Jan	98	17			82.44
Oita	33.14N, 131.36E	15	1941–70	59.5	79.5 Aug	41.4 Jan	99	16			66.49
Okayama	34.39N, 133.55E	10	1950–70	58.1	80.6 Aug	37.9 Jan	99	16			47.92
Okazaki	34.57N, 137.10E		(see Nagoya)								
Omiya	35.54N, 139.38E		(see Tokyo)								
Osaka	34.40N, 135.30E	23	1941–67	60.1	82.4 Aug	40.1 Jan	101	18	96	25	54.72
Otaru	43.13N, 141.00E	79	1943–70	46.8	71.1 Aug	25.0 Jan	94	0			49.02
Otsu	35.00N, 135.52E		(see Kyoto)								
Sagamihara	35.33N, 139.22E		(see Yokohama)								
Sakai	34.35N, 135.28E		(see Osaka)								

ASIA

City	Latitude and Longitude	Elev (ft)	Period of Record	Temperature (°F) Avg	Avg Warmest Month	Avg Coolest Month	Absolute High	Absolute Low	Avg Annual High	Avg Annual Low	Annual Precipitation (in)
Sapporo	43.03N, 141.21E	56	1941–70	46.0	71.1 Aug	22.8 Jan	96	–19	90	–7	44.92
Sasebo	33.10N, 129.43E	43	1952–70	61.3	81.5 Aug	42.4 Jan	101	23			78.90
Sendai	38.15N, 140.53E	125	1941–70	52.9	75.2 Aug	33.1 Jan	98	11			49.02
Shimizu	35.01N, 138.29E	98	(see Shizuoka)								
Shimonoseki	33.57N, 130.57E	151	1941–70	59.7	80.1 Aug	41.9 Jan	99	20			67.87
Shizuoka	34.58N, 138.23E	43	1941–70	60.3	79.5 Aug	42.3 Jan	101	20			92.72
Suita	34.45N, 135.32E		(see Osaka)								
Takamatsu	34.20N, 134.03E	30	1942–70	58.8	80.2 Aug	39.9 Jan	98	18			46.66
Takasaki	36.20N, 139.01E	312	(see Maebashi)								
Takatsuki	34.51N, 135.37E		(see Osaka)								
Tokorozawa	35.47N, 139.28E		(see Tokyo)								
Tokushima	34.04N, 134.34E	4	1941–70	59.9	80.4 Aug	41.2 Jan	99	21			66.66
*Tokyo	35.42N, 139.46E	13	1941–70	59.0	80.1 Aug	39.4 Jan	101	15	93	21	59.18
Toyama	36.41N, 137.13E	30	1941–70	55.9	78.6 Aug	35.4 Jan	99	11			94.02
Toyohashi	34.46N, 137.23E	98	(see Hamamatsu)								
Toyonaka	34.47N, 135.28E		(see Osaka)								
Toyota	35.05N, 137.09E	246	(see Nagoya)								
Urawa	35.51N, 139.39E	66	(see Tokyo)								
Utsunomiya	36.33N, 139.52E	394	1941–70	54.9	76.8 Aug	33.8 Jan	99	5			57.64
Wakayama	34.13N, 135.11E	46	1941–70	60.4	81.1 Aug	41.4 Jan	101	21			57.29
Yamagata	38.15N, 140.15E	495	1941–70	51.8	76.1 Aug	29.8 Jan	105	–4			47.64
Yao	34.37N, 135.36E	10	(see Osaka)								
Yokkaichi	34.58N, 136.37E		(see Nagoya)								
Yokohama	35.27N, 139.39E	128	1941–70	58.6	79.0 Aug	39.9 Jan	99	17			64.25
Yokosuka	35.18N, 139.40E	39	1894–1929	58.1	77.4 Aug	40.5 Jan	98	20			76.34

ASIA

City	Latitude and Longitude	Elev (ft)	Period of Record	Temperature (°F)							Annual Precipitation (in)
				Avg	Avg Warmest Month	Avg Coolest Month	Absolute High	Absolute Low	Avg Annual High	Avg Annual Low	
JORDAN											
*Amman	31.57N, 35.56E	2513	1931–60	63.5	78.1 Aug	46.8 Jan	109	21	103	29	10.75
Zarqa	32.04N, 36.05E	2001	1923–65	(see Amman for temperature data)							5.04
KASHMIR-JAMMU											
*Jammu	32.44N, 74.52E	1201	1931–60	75.7	93.2 Jun	55.9 Jan	117	33	113	40	43.94
*Srinagar	34.05N, 74.50E	5205	1931–60	55.9	76.3 Jul	33.8 Jan	101	−4	97	18	25.94
KOREA: NORTH											
Chongjin	41.46N, 129.49E	164	1905–34	45.5	71.1 Aug	21.6 Jan					22.20
Hamhung	39.54N, 127.32E	108	1910–37, 1941–54	49.1	74.1 Aug	21.7 Jan	101	−14			35.67
Hungnam	39.30N, 127.14E		(see Hamhung)								
Kaesong	37.58N, 126.33E	197	1917–37	50.7	77.4 Aug	21.2 Jan					50.98
Kimchaek	40.41N, 129.12E	13	1905–36	46.2	71.4 Aug	21.2 Jan	99	−12			31.81
*Pyongyang	39.01N, 125.45E	95	1907–54	48.9	75.9 Aug	17.6 Jan	98	−19			37.01
Sinuiju	40.06N, 124.24E	20	1931–54	47.7	75.4 Aug	15.8 Jan	98	−18			54.49
Wonsan	39.10N, 127.26E	120	1905–54	50.5	73.9 Aug	25.5 Jan	103	−7	96	2	52.8
KOREA: SOUTH											
Chongju	36.38N, 127.30E	194	1914–34	52.5	78.4 Aug	24.4 Jan	98	−18			46.93
Chonju	35.49N, 127.09E	167	1919–34	53.8	78.6 Aug	28.6 Jan	99	4			49.72
Inchon	37.28N, 126.38E	226	1931–60	52.0	77.2 Aug	24.8 Jan	98	−6			42.87
Kwangju	35.09N, 126.55E	115	1913–37	54.7	79.5 Aug	30.0 Jan	101	−3			47.91
Masan	35.11N, 128.34E	89	1916–25	57.0	80.1 Aug	34.3 Jan	101	6			55.39
Mokpo	34.47N, 126.23E	105	1931–60	56.1	79.0 Aug	33.8 Jan	99	6			44.33
Pusan	35.06N, 129.03E	226	1934–60	56.8	77.7 Aug	35.2 Jan	96	7	91	14	54.37
*Seoul	37.34N, 127.00E	279	1931–60	52.0	77.7 Aug	23.2 Jan	99	−12	96	−4	49.57

ASIA

City	Latitude and Longitude	Elev (ft)	Period of Record	Temperature (°F)							Annual Precipitation (in)
				Avg	Avg Warmest Month	Avg Coolest Month	Absolute High	Absolute Low	Avg Annual High	Avg Annual Low	
Songnam	37.25N, 127.00E	121	(see Seoul)								
Suwon	37.16N, 127.01E		(see Seoul)								
Taegu	35.52N, 128.36E	190	1921–50	54.5	79.2 Aug	28.8 Jan	103	–4			37.64
Taejon	36.20N, 127.26E	187	1917–37	53.4	78.8 Aug	24.3 Jan					52.72
Ulsan	35.33N, 129.19E	102	1919–34	56.7	80.1 Aug	34.3 Jan	102	3			47.60
KUWAIT											
*Kuwait	29.20N, 47.59E	16	1908–53	77.0	99.0 Jul	55.0 Jan	121	32	115	38	5.04
LAOS											
*Vientiane	17.58N, 102.36E	558	1951–60	78.6	83.8 Apr	70.5 Dec	104	39	101	47	67.48
LEBANON											
*Beirut	33.53N, 35.30E	79	1881–1960	70.3	83.5 Aug	56.8 Jan	107	30	95	40	35.04
Tripoli	34.26N, 35.51E	26	1951–60	66.4	78.8 Aug	53.6 Jan					29.33
MACAO											
*Macao	22.12N, 113.32E	66	1931–60	72.5	83.3 Jul	59.2 Jan	101	29			72.68
MALAYSIA											
George Town	5.25N, 100.20E	17	49 yr bef 1939	82	83 Mar	80.5 Sep	98	65	95	69	107.7
Ipoh	4.35N, 101.05E	126	? yr bef 1963	81.7	82.6 Jun	80.8 Dec	99	63			101.60
*Kuala Lumpur	3.10N, 101.42E	127	1930–41, 1948–54	81.5	82.5 Mar	80.5 Dec	98	64	96	67	96.1
Malacca	2.12N, 102.15E	23	1882–1953	80	81 Apr	80 Oct	99	61	94	68	86.8
MONGOLIA											
*Ulan-Bator	47.55N, 106.53E	4295	1936–60	23.9	64.9 Jul	–20.6 Jan	101	–56	91	–41	10.00

ASIA

City	Latitude and Longitude	Elev (ft)	Period of Record	Avg	Avg Warmest Month	Avg Coolest Month	Absolute High	Absolute Low	Avg Annual High	Avg Annual Low	Annual Precipitation (in)
NEPAL											
*Kathmandu	27.42N, 85.20E	4388	1891–1961	65.3	75.7 Jul	49.8 Jan	99	27	94	30	54.21
PAKISTAN											
Faisalabad	31.25N, 73.05E	605	1916–40	75.6	93.9 Jun	53.4 Jan	117	28	115	33	12.05
Gujranwala	32.09N, 74.11E	738	1963–66	73.0	91.7 Jun	52.7 Jan	118	29	114	30	21.53
Hyderabad	25.22N, 68.22E	98	1931–60	81.5	93.7 Jun	63.0 Jan	122	30			6.18
*Islamabad	33.42N, 73.10E	1800	(see Rawalpindi)								
Karachi	24.52N, 67.03E	20	1931–60	78.4	86.7 Jun	66.0 Jan	118	32	107	45	8.03
Lahore	31.35N, 74.18E	702	1931–60	75.7	93.0 Jun	54.0 Jan	120	28	116	32	19.37
Multan	30.11N, 71.29E	413	1931–60	78.6	97.0 Jun	55.6 Jan	123	27	118	33	6.57
Peshawar	34.01N, 71.33E	1177	1931–60	81.0	91.6 Jun	51.3 Jan	122	26	116	31	14.29
Rawalpindi	33.36N, 73.04E	1676	1881–1960	70.0	90.1 Jun	50.2 Jan	117	25	114	29	36.42
Sargodha	32.05N, 72.40E	614	(see Faisalabad)								
Sialkot	32.30N, 74.31E	830	? yr bef 1921	74.8	92.9 Jun	54.4 Jan					31.83
PHILIPPINES											
Bacolod	10.41N, 122.56E	23	6 yr bef 1919	80.2	82.9 May	78.6 Jan	101	59	96	64	85.55
Caloocan	14.38N, 121.03E		(see Manila)								
Cebu	10.18N, 123.54E	138	52 yr bef 1961	81.3	83.1 May	79.3 Jan	96	65	93	69	63.74
Davao	7.04N, 125.36E	63	30 yr bef 1961	80.6	81.9 Apr	79.3 Jan	99	61	96	66	75.91
Iloilo	10.42N, 122.34E	46	52 yr bef 1961	80.6	82.9 May	78.4 Jan	99	64	96	69	88.43
Makati	14.34N, 121.02E		(see Manila)								
*Manila	14.36N, 120.59E	51	71 yr bef 1961	80.1	83.5 May	77.0 Jan	101	58	98	61	81.46
Pasay	14.33N, 121.00E		(see Manila)								
Pasig	14.35N, 121.05E		(see Manila)								
Quezon City	14.38N, 121.00E	232	1952–55	80.1	84.4 May	75.4 Dec	101	56			91.62
Zamboanga	6.54N, 122.04E	15	51 yr bef 1961	80.1	80.8 Apr	79.3 Jan	96	60	92	67	44.25

Temperature (°F) spans: Avg Warmest Month, Avg Coolest Month, Absolute (High, Low), Avg Annual (High, Low).

ASIA

City	Latitude and Longitude	Elev (ft)	Period of Record	Temperature (°F) Avg	Avg Warmest Month	Avg Coolest Month	Absolute High	Absolute Low	Avg Annual High	Avg Annual Low	Annual Precipitation (in)
QATAR											
*Doha	25.17N, 51.32E		1951–60[1]	81.5	98.1 Aug	62.1 Jan					2.5
SAUDI ARABIA											
*Jidda	21.30N, 39.12E	20	1951–60	83.3	90.3 Jun	74.1 Jan	117	49	110	52	2.09
Mecca	21.27N, 39.49E	853									
Medina	24.28N, 39.36E	1949	1957–60	81.1	94.8 Aug	63.5 Jan					2.44
*Riyadh	24.38N, 46.43E	1938	1941–45, 1951–60	76.8	92.5 Jul	57.6 Jan	113	19	112	29	3.82
Taif	21.16N, 40.24E	5348									
SINGAPORE											
*Singapore	1.17N, 103.51E	33	1903–39	80.6	81.5 May	79.0 Jan	97	66	93	69	95.08
SRI LANKA											
*Colombo	6.56N, 79.51E	24	1931–60	80.4	82.3 May	79.0 Dec	96	59	92	65	94.3
Kandy	6.18N, 80.38E	1572	1931–60	75.8	78.8 Apr	73.6 Jan			91	55	79.7
SYRIA											
Aleppo	36.10N, 37.10E	1243	1952–61	63.3	84.0 Aug	43.2 Jan	117	9	108	17	12.09
*Damascus	33.30N, 36.18E	2320	1952–61	63.9	81.9 Aug	45.7 Jan	113	21	105	25	8.98
Hama	35.09N, 36.44E	919	1956–60	64.4	84.0 Aug	45.3 Jan					11.77
Homs	34.44N, 36.43E	1667	(see Hama)								
TAIWAN											
Chiayi	23.29N, 120.27E	102	(see Tainan)								

[1] Temperature data for Dhahran, Saudi Arabia, nr Doha.

ASIA

City	Latitude and Longitude	Elev (ft)	Period of Record	Temperature (°F)							Annual Precipitation (in)
				Avg	Avg Warmest Month	Avg Coolest Month	Absolute High	Absolute Low	Avg Annual High	Avg Annual Low	
Hsinchu	24.48N, 120.58E	108	1938–52	71.4	82.4 Jul	58.8 Feb	99	37	97	43	69.41
Kaohsiung	22.38N, 120.17E	95	1932–52	75.7	82.2 Jul	66.0 Jan	99	45	95	48	75.71
Keelung	25.08N, 121.44E	10	1903–52	71.1	82.4 Jul	60.1 Jan	100	37	97	46	119.80
Panchiao	25.01N, 121.27E		(see Taipei)								
Pingtung	22.40N, 120.29E		(see Kaohsiung)								
Sanchung	25.04N, 121.29E		(see Taipei)								
Taichung	24.09N, 120.41E	253	1897–1952	72.3	82.0 Jul	60.4 Jan	102	30	95	39	70.24
Tainan	23.00N, 120.12E	43	1897–1952	73.9	82.2 Jul	62.8 Jan	100	36	95	43	72.40
*Taipei	25.03N, 121.30E	26	1897–1952	71.2	82.8 Jul	59.0 Feb	102	32	97	41	82.68
THAILAND											
*Bangkok	13.45N, 100.31E	26	1931–60	82.6	86.4 Apr	77.9 Dec	106	52	103	57	58.74
TURKEY											
Adana	37.01N, 35.18E	66	30 yr bef 1961	65.5	82.4 Aug	48.4 Jan	114	19	104	25	24.06
*Ankara	39.56N, 32.52E	2959	30 yr bef 1961	53.1	73.9 Jul	31.6 Jan	104	-13	98	2	14.17
Antioch	36.14N, 36.07E	305	21 yr bef 1961	64.6	81.9 Aug	46.4 Jan	109	6			44.92
Bursa	40.11N, 29.04E	328	30 yr bef 1961	57.9	75.6 Jul	41.7 Jan	109	-3	100	16	28.54
Diyarbakir	37.55N, 40.14E	2133	42 yr bef 1975	60.6	87.8 Jul	35.2 Jan	115	-12			19.53
Eskisehir	39.46N, 30.32E	2576	30 yr bef 1961	51.4	70.5 Jul	31.6 Jan	102	-15			14.49
Gaziantep	37.05N, 37.22E	2756	21 yr bef 1961	57.9	80.8 Jul	36.7 Jan	109	0			21.65
Izmir	38.25N, 27.09E	82	30 yr bef 1961	63.5	81.7 Jul	47.5 Jan	109	12	103	24	27.28
Kayseri	38.43N, 35.30E	3514	27 yr bef 1961	51.4	73.0 Jul	28.9 Jan	105	-26			14.41
Konya	37.52N, 32.31E	3366	30 yr bef 1961	52.7	73.6 Jul	31.6 Jan	100	-19	96	-1	12.40
Samsun	41.17N, 36.20E	144	30 yr bef 1961	57.7	73.9 Aug	44.2 Feb	102	17	93	24	28.78
Trebizond	41.00N, 39.43E	121	1927–60	57.9	73.6 Aug	44.2 Feb	101	19	92	26	31.46

ASIA

City	Latitude and Longitude	Elev (ft)	Period of Record	Temperature (°F)							Annual Precipitation (in)
				Avg	Avg Warmest Month	Avg Coolest Month	Absolute High	Absolute Low	Avg Annual High	Avg Annual Low	
UNITED ARAB EMIRATES											
*Abu Zaby	24.28N, 54.22E		(see Dubayy)								
Dubayy	25.18N, 55.18E	49	5 yr bef 1967	79	92 Jul	64 Jan	117	46			2.4
USSR											
Aktyubinsk	50.17N, 57.10E	719	46 yr bef 1961	38.5	72.1 Jul	3.9 Jan	109	-54	100	-33	10.20
Alma-Ata	43.15N, 76.57E	2779	1915-60	47.7	73.9 Jul	18.7 Jan	108	-36	99	-15	22.05
Andizhan	40.45N, 72.22E	1565	1931-57	56.3	81.1 Jul	26.6 Jan	111	-20	104	-2	8.27
Angarsk	52.34N, 103.54E	1368	1951-60	29.7	64.0 Jul	-8.1 Jan	99	-60	91	-44	
Ashkhabad	37.57N, 58.23E	745	1931-60	61.2	88.2 Jul	35.8 Jan	118	-15	108	3	8.27
Barnaul	53.22N, 83.45E	518	1881-1960	34.0	67.5 Jul	0.1 Jan	100	-62	93	-45	18.90
Biysk	52.34N, 85.15E	745	1934-60	32.9	66.0 Jul	-0.8 Jan	102	-63	93	-45	18.70
Blagoveshchensk	50.16N, 127.32E	427	1881-90, 1892-1960	32.0	70.5 Jul	-11.7 Jan	106	-49	91	-36	20.59
Bratsk	56.21N, 101.55E	1243	1957-62	28.0	64.8 Jul	-8.7 Jan	99	-71	91	-49	11.85
Bukhara	39.48N, 64.25E	738	1951-60	57.0	81.5 Jul	30.9 Jan	117	-15	106	1	5.31
Chelyabinsk	55.10N, 61.24E	758	? yr bef 1960	35.2	65.7 Jul	3.7 Jan	102	-49			15.47
Chimkent	42.18N, 69.36E	1782	1920-60	53.4	79.3 Jul	26.6 Jan	111	-29	104	-11	19.02
Chita	52.03N, 113.30E	2202	1890-1919, 1924-56	27.1	65.8 Jul	-15.9 Jan	106	-65	93	-45	13.70
Dushanbe	38.33N, 68.48E	2697	1937-57	57.6	80.8 Jul	33.1 Jan	111	-18	104	7	23.78
Dzhambul	42.54N, 71.22E	2106	1928-60	48.2	73.9 Jul	21.2 Jan	111	-42	100	-22	11.30
Frunze	42.54N, 74.36E	2480	1927-60	49.6	75.4 Jul	21.9 Jan	108	-36	100	-15	15.04
Irkutsk	52.16N, 104.20E	1536	1881-1960	30.0	63.7 Jul	-5.6 Jan	97	-58	90	-42	15.87
Kamensk-Uralskiy	56.25N, 61.54E		(see Sverdlovsk)								
Karaganda	49.50N, 73.10E	1804	1933-60	36.1	68.5 Jul	4.8 Jan	104	-56	95	-35	12.32
Kemerovo	55.20N, 86.05E	505	1933-60	31.3	65.1 Jul	-2.6 Jan	100	-67	90	-44	
Khabarovsk	48.30N, 135.06E	285	66 yr bef 1961	34.5	70.0 Jul	-8.1 Jan	104	-45	90	-31	19.2
Kokand	40.30N, 70.57E	1339	1925-60	56.3	81.5 Jul	27.9 Jan	111	-17	104	3	

ASIA

City	Latitude and Longitude	Elev (ft)	Period of Record	Temperature (°F)							Annual Precipitation (in)
				Avg	Avg Warmest Month	Avg Coolest Month	Absolute High	Low	Avg Annual High	Low	
Komsomolsk-na-Amure	50.35N, 137.02E	66	1932-33, 1935-60	30.7	67.8 Jul	-14.1 Jan	102	-58	88	-44	16.50
Kopeysk	55.07N, 61.37E		(see Chelyabinsk)								
Krasnoyarsk	56.01N, 92.50E	637	1931-60	32.9	65.3 Jul	1.0 Jan	103	-56			15.16
Kurgan	55.26N, 65.18E	259	? yr bef 1948[1]	32.4	65.7 Jul	-0.9 Jan					
Kustanay	53.10N, 63.35E	558	1950-63	34.9	68.4 Jul	0.1 Jan	108	-60	97	-36	
Magnitogorsk	53.27N, 59.04E	1263									
Namangan	41.00N, 71.40E	1555	1924-60	56.1	81.7 Jul	25.9 Jan	111	-20	104	0	7.40
Nizhniy Tagil	57.55N, 59.57E	732	1881-1912, 1915	33.6	64.0 Jul	1.8 Jan					19.17
Norilsk	69.20N, 88.06E	217	1906, 1908-22, 1924[2]	12.4	54.9 Jul	-22.4 Jan					8.39
Novokuznetsk	53.45N, 87.06E	771	1931-37, 1944-54	33.3	65.3 Jul	0.0 Jan	100	-62	91	-42	
Novosibirsk	55.02N, 82.55E	446	1930-57	31.8	65.7 Jul	-2.2 Jan	100	-58	93	-45	15.12
Omsk	55.00N, 73.24E	410	1930-60	32.0	64.9 Jul	-2.6 Jan	104	-56	93	-40	13.43
Osh	40.32N, 72.48E	3324	1881-1950	52.2	75.9 Jul	26.1 Jan	102	-15			14.17
Pavlodar	52.18N, 76.57E	387	1906-11, 1922-60	35.4	70.2 Jul	-0.2 Jan	108	-53	97	-38	10.00
Petropavlovsk	54.52N, 69.06E	440	62 yr bef 1961	32.9	65.8 Jul	-1.7 Jan	106	-63	93	-36	14.02
Petropavlovsk-Kamchatskiy	53.01N, 158.39E	105	48 yr bef 1961	35.4	56.3 Aug	16.7 Feb	88	-29	77	-8	27.32
Prokopyevsk	53.53N, 86.45E		1925-60[3]	32.7	64.9 Jul	0.1 Jan	100	-58	91	-42	
Rubtsovsk	51.30N, 81.15E	715	1924-60	34.9	68.5 Jul	0.0 Jan	106	-56	97	-42	
Samarkand	39.40N, 66.58E	2382	1932-60	55.2	77.9 Jul	31.5 Jan	113	-22	102	0	12.60
Semipalatinsk	50.28N, 80.13E	663	64 yr bef 1949	37.8	72.0 Jul	2.8 Jan	108	-56	100	-40	10.51
Sverdlovsk	56.51N, 60.36E	925	1931-60	34.9	64.0 Jul	5.7 Jan	99	-45	87	-34	18.19
Tashkent	41.20N, 69.18E	1565	1881-1960	55.9	80.4 Jul	30.4 Jan	111	-22	106	-2	14.65
Temirtau	50.05N, 72.56E		(see Karaganda)								

[1] Climatic data for Staro-Sidorovo, nr Kurgan.
[2] Temperature data for Dudinka, nr Norilsk.
[3] Temperature data for Kiselevsk, nr Prokopyevsk.

ASIA

City	Latitude and Longitude	Elev (ft)	Period of Record	Temperature (°F) Avg	Avg Warmest Month	Avg Coolest Month	Absolute High	Absolute Low	Avg Annual High	Avg Annual Low	Annual Precipitation (in)
Tomsk	56.30N, 84.58E	456	1881–1960	30.9	64.6 Jul	-2.6 Jan	97	-67	90	-47	21.10
Tselinograd	51.10N, 71.30E	1139	73 yr bef 1961	34.5	68.4 Jul	0.7 Jan	108	-62	97	-36	12.01
Tyumen	57.09N, 65.26E	249	1884–1942	34.3	65.5 Jul	2.1 Jan	104	-58	93	-35	15.94
Ulan-Ude	51.50N, 107.37E	1686	66 yr bef 1961	28.9	66.9 Jul	-13.7 Jan	104	-60	95	-44	7.95
Uralsk	51.14N, 51.22E	115	1893–1917, 1923–60	39.9	72.7 Jul	6.4 Jan	108	-45	100	-29	10.79
Ust-Kamenogorsk	49.58N, 82.40E	932	1921–22, 1926–48	37.4	70.2 Jul	2.8 Jan	106	-54	99	-49	14.21
Vladivostok	43.08N, 131.54E	364	1917–60	39.2	68.0 Aug	6.1 Jan	97	-24	86	-15	29.29
Yakutsk	62.00N, 129.40E	338	1931–60	13.8	67.1 Jul	-44.9 Jan	97	-84	89	-72	8.39
Yuzhno-Sakhalinsk	46.57N, 142.44E	72	1946–60	35.8	63.1 Aug	7.2 Jan	93	-38	86	-26	
Zlatoust	55.10N, 59.40E	1503	? yr bef 1959	33.1	61.3 Jul	4.3 Jan	93	-51	85	-38	21.54
VIETNAM											
Da Nang	16.04N, 108.13E	20	1931–44, 1947–62	77.9	84.4 Jun	70.3 Jan	90	52			81.61
Haiphong	20.52N, 106.41E	381	1904–44, 1957–60	73.6	82.6 Jun	62.1 Jan	107	44	99	48	69.21
*Hanoi	21.02N, 105.51E	56	52 yr bef 1961	74.1	83.8 Jun	62.1 Jan	109	42	102	46	66.06
Ho Chi Minh City	10.45N, 106.40E	30	1897–1944, 1947–60	81.7	85.5 Apr	78.8 Dec	104	57	98	62	76.26
Hue	16.28N, 107.36E	49	1931–44, 1947–62	77.2	84.7 Jun	68.0 Jan	104	48	104	54	118.78
Nha Trang	12.15N, 109.11E	20	1898–1936, 1941–60	79.5	82.8 Jun	75.0 Jan	103	58	98	65	43.11
Qui Nhon	13.46N, 109.14E	16	21 yr bef 1940	80	87 Aug	72 Dec	108	59	101	63	63.78
YEMEN: NORTH											
*Sana	15.23N, 44.12E	7260	? yr bef 1936	63.5	71.1 Jun	57.3 Jan					19.68

ASIA

City	Latitude and Longitude	Elev (ft)	Period of Record	Temperature (°F) Avg	Avg Warmest Month	Avg Coolest Month	Absolute High	Absolute Low	Avg Annual High	Avg Annual Low	Annual Precipitation (in)
YEMEN: SOUTH											
*Aden	12.46N, 45.01E	12	1926–60	84.0	91.2 Jun	77.7 Jan	109	61			1.61

EUROPE

City	Latitude and Longitude	Elev (ft)	Period of Record	Temperature (°F) Avg	Avg Warmest Month	Avg Coolest Month	Absolute High	Absolute Low	Avg Annual High	Avg Annual Low	Annual Precipitation (in)
ALBANIA											
*Tirane	41.20N, 19.50E	416	1931–60	60.8	77.0 Jul	45.1 Jan	105	13			46.81
AUSTRIA											
Graz	47.04N, 15.27E	1237	1931–60	46.9	66.2 Jul	25.2 Jan	99	–11	88	5	33.07
Innsbruck	47.16N, 11.24E	1909	1931–60	47.5	64.6 Jul	27.0 Jan	97	–16	92	0	35.87
Linz	48.18N, 14.18E	853	1881–1930	47.8	65.8 Jul	30.2 Jan	100	–17	92	3	38.50
Salzburg	47.48N, 13.02E	1427	1931–60	46.6	64.0 Jul	27.5 Jan	97	–23	91	–1	50.31
*Vienna	48.12N, 16.22E	663	1931–60	49.6	67.8 Jul	29.5 Jan	98	–14	90	7	25.98
BELGIUM											
Anderlecht	50.50N, 4.18E	98	(see Brussels)								
Antwerp	51.13N, 4.25E	16	1901–30	50.2	64.4 Jul	37.6 Jan					28.07
Bruges	51.13N, 3.14E	39	(see Ghent)								
*Brussels	50.50N, 4.20E	328	1901–30	49.8	63.1 Jul	37.8 Jan	98	0	90	13	32.87

EUROPE

City	Latitude and Longitude	Elev (ft)	Period of Record	Temperature (°F) Avg	Avg Warmest Month	Avg Coolest Month	Absolute High	Absolute Low	Avg Annual High	Avg Annual Low	Annual Precipitation (in)
Charleroi	50.25N, 4.26E	341	1901-30[1]	49.8	63.0 Jul	38.3 Jan					30.83
Ghent	51.03N, 3.43E	23	1901-30	50.4	64.6 Jul	37.9 Jan					32.28
La Louviere	50.28N, 4.11E	377	1901-30[1]	49.8	63.0 Jul	38.3 Jan					30.83
Liege	50.38N, 5.34E		(see Aachen, West Germany)								
Louvain	50.53N, 4.42E	80	1901-30[2]	48.9	62.2 Jul	36.9 Jan	98	0	90	13	32.87
Schaerbeek	50.51N, 4.23E		(see Brussels)								
BULGARIA											
Burgas	42.30N, 27.28E	16	1900-18, 1926-39	54.7	73.8 Jul	35.2 Jan	106	-3	98	12	21.77
Pleven	43.25N, 24.37E	381	1894-1939	52.3	74.1 Jul	28.6 Jan	110	-19	98	-1	23.97
Plovdiv	42.09N, 24.45E	525	30 yr bef 1951	54.5	74.5 Jul	31.5 Jan	104	-25	100	1	20.1
Ruse	43.50N, 25.57E	151	1894-1939	51.1	72.3 Jul	27.1 Jan					22.68
Sliven	42.40N, 26.19E	853	1894-1939	54.1	74.1 Jul	34.0 Jan	103	-4	96	9	24.92
*Sofia	42.41N, 23.19E	1850	1931-60	50.7	70.3 Jul	28.9 Jan	99	-17	96	-2	24.49
Stara Zagora	42.25N, 25.38E	768	1900-18, 1928-39	55.2	75.7 Jul	34.0 Jan					23.39
Varna	43.13N, 27.55E	157	1892-1919, 1924-39	53.6	72.7 Jul	33.8 Jan	107	-12	96	8	19.76
CZECHOSLOVAKIA											
Bratislava	48.09N, 17.07E	538	1901-50	49.8	68.4 Jul	29.1 Jan	101	-22	93	3	26.38
Brno	49.11N, 16.37E	745	1901-50	47.1	65.1 Jul	28.2 Jan	98	-23	91	1	21.53

[1] Climatic data for Paturages, nr Charleroi and La Louviere.
[2] Climatic data for Uccle, nr Louvain.

EUROPE

City	Latitude and Longitude	Elev (ft)	Period of Record	Temperature (°F)							Annual Precipitation (in)
				Avg	Avg Warmest Month	Avg Coolest Month	Absolute High	Low	Avg Annual High	Low	
Havirov	49.47N, 18.27E		(see Ostrava)								
Karlovy Vary	50.14N, 12.53E	1263	1901–50	45.1	62.4 Jul	28.2 Jan	97	–18	90	–1	25.94
Kosice	48.44N, 21.15E	676	1901–50	47.1	66.4 Jul	25.9 Jan	99	–23	92	–3	26.10
Olomouc	49.36N, 17.16E	722	1901–50	47.1	65.3 Jul	27.1 Jan	96	–31	92	–7	24.09
Ostrava	49.51N, 18.18E	712	1901–50	47.5	65.7 Jul	28.4 Jan	97	–26	92	–7	30.28
Plzen	49.45N, 13.25E	1161	1901–50	46.0	64.0 Jul	28.4 Jan	98	–21	92	–2	20.39
*Prague	50.04N, 14.27E	633	1901–50	48.2	66.2 Jul	30.4 Jan	101	–20	93	1	19.17
DENMARK											
Alborg	57.03N, 9.56E	10	1931–60	45.7	61.5 Jul	30.6 Feb					22.68
Arhus	56.09N, 10.13E	161	1934–54	46	62 Jul	31 Jan	87	–12	82	10	26.6
*Copenhagen	55.40N, 12.35E	16	1931–60	47.3	64.0 Jul	31.8 Feb	91	–3	83	11	23.70
Frederiksberg	55.41N, 12.32E		(see Copenhagen)								
Odense	55.24N, 10.23E	49	1926–54	46.5	61.5 Jul	31.5 Feb	90	–11	83	7	23.8
FINLAND											
Espoo	60.13N, 24.40E	46	(see Helsinki)								
*Helsinki	60.10N, 24.58E	39	1931–60	41.7	64.0 Jul	21.2 Feb	89	–23	82	–11	25.47
Lahti	60.58N, 25.40E	338	(see Tampere)								
Oulu	65.01N, 25.28E	33	1881–1935	35.6	61.2 Jul	14.0 Feb					19.76
Tampere	61.30N, 23.45E	312	1931–60	38.8	62.2 Jul	17.6 Feb	91	–32	85	–16	19.96
Turku	60.27N, 22.17E	52	1931–60	40.3	62.8 Jul	20.1 Feb	97	–29	85	–14	23.15
Vantaa	60.18N, 24.51E		(see Helsinki)								

EUROPE

City	Latitude and Longitude		Elev (ft)	Period of Record	Temperature (°F)								Annual Precipitation (in)
					Avg	Avg Warmest Month	Avg Coolest Month	Absolute		Avg Annual			
								High	Low	High	Low		
FRANCE													
Aix-en-Provence	43.32N,	5.26E	705	1921–50	57.6	73.2 Jul	43.2 Jan						26.18
Ajaccio	41.55N,	8.44E	125	1931–60	58.3	71.2 Aug	46.6 Jan	103	23	96	29		26.46
Amiens	49.54N,	2.18E	118	1891–1914, 1923–34[1]	49.3	63.3 Jul	36.5 Jan						25.63
Angers	47.28N,	0.33W	154	1931–60	52.5	65.8 Jul	39.7 Jan						23.62
Angouleme	45.39N,	0.09E	315	1931–60	54.0	67.3 Jul	40.5 Jan						32.56
Annecy	45.54N,	6.07E	1490	1891–1930	48.9	66.2 Jul	30.9 Jan						50.04
Argenteuil	48.57N,	2.15E	115	(see Paris)									
Avignon	43.57N,	4.49E	180	1891–1914, 1921–34	56.7	73.0 Jul	40.8 Jan						26.46
Bayonne	43.29N,	1.29W	10	1931–60	(see Biarritz for temperature data)								47.56
Besancon	47.15N,	6.02E	1207	1931–60	50.5	65.8 Jul	33.9 Jan						43.31
Bethune	50.32N,	2.38E	98	1891–1914, 1923–34[1]	49.3	63.3 Jul	36.5 Jan						24.37
Biarritz	43.29N,	1.34W	226	1931–60	56.5	67.8 Aug	45.7 Jan			96	24		58.07
Bordeaux	44.50N,	0.34W	23	1931–60	54.5	67.3 Aug	41.7 Jan	102	9	96	19		35.43
Boulogne-Billancourt	48.50N,	2.15E	98	(see Paris)									
Boulogne-sur-Mer	50.43N,	1.37E	240	1931–60	50.7	62.8 Aug	39.2 Jan						23.58
Brest	48.24N,	4.29W	322	1931–60	51.4	61.0 Aug	42.8 Feb	95	7	86	24		44.45
Bruay-en-Artois	50.29N,	2.33E	131	1891–1914, 1923–34[1]	49.3	63.3 Jul	36.5 Jan						
Caen	49.11N,	0.21W	85	1901–30	50.9	62.8 Jul	40.3 Jan	99	6	91	10		27.16
Calais	50.57N,	1.50E	7	1921–50	49.8	62.2 Aug	39.0 Jan	94	4	88	17		25.59
Cannes	43.33N,	7.01E	10	1921–50	57.9	71.6 Jul	45.1 Jan						36.22
Clermont-Ferrand	45.47N,	3.05E	1335	1931–60	51.8	66.6 Jul	36.9 Jan	107	–10	97	9		22.17
Denain	50.20N,	3.23E	131	1891–1914, 1923–34[1]	49.3	63.3 Jul	36.5 Jan						
Dijon	47.19N,	5.01E	1109	1931–60	51.1	67.1 Jul	34.5 Jan	107	–8	94	11		29.09

[1] Temperature data for Arras, nr Amiens, Bethune, Bruay-en-Artois and Denain.

EUROPE

City	Latitude and Longitude	Elev (ft)	Period of Record	Temperature (°F) Avg	Avg Warmest Month	Avg Coolest Month	Absolute High	Absolute Low	Avg Annual High	Avg Annual Low	Annual Precipitation (in)
Douai	50.22N, 3.04E	79	(see Lille)								25.24
Dunkerque	51.03N, 2.22E	26	1931–60	50.7	63.3 Aug	38.8 Jan	99	−6	94	10	38.78
Grenoble	45.10N, 5.43E	810	1931–60	51.8	68.2 Jul	34.7 Jan					28.62
La Rochelle	46.10N, 1.09W	26	1931–60	54.9	67.6 Aug	42.4 Jan	102	12	95	19	26.57
Le Havre	49.30N, 0.08E	16	1931–60	51.3	63.1 Aug	40.1 Jan					26.38
Le Mans	48.00N, 0.12E	253	1931–60	52.0	65.8 Jul	38.8 Jan					
Lens	50.26N, 2.50E	131	(see Lille)								
Lille	50.38N, 3.04E	79	1931–60	49.8	63.3 Aug	36.5 Jan	96	0	91	12	25.08
Limoges	45.51N, 1.15E	942	1931–60	51.1	64.6 Jul	37.9 Jan	102	0	97	11	36.77
Lorient	47.45N, 3.22W	138	1931–60	52.3	63.0 Aug	42.3 Jan					37.80
Lyons	45.45N, 4.51E	738	1931–60	52.7	69.3 Jul	36.0 Jan	105	−13	95	11	32.01
Marseilles	43.18N, 5.24E	532	1931–60	58.5	73.0 Jul	44.6 Jan	101	9	93	23	24.88
Maubeuge	50.17N, 3.58E	459	1931–60	(see Lille for temperature data)							30.51
Metz	49.08N, 6.10E	860	1931–60	49.8	65.5 Jul	34.2 Jan					26.53
Montbeliard	47.31N, 6.48E	1207	1931–60	(see Besancon for temperature data)							41.61
Montpellier	43.36N, 3.53E	144	1931–60	56.8	72.1 Jul	42.4 Jan					27.91
Montreuil	48.52N, 2.26E	394	(see Paris)								
Mulhouse	47.45N, 7.20E	787	1931–60	49.6	66.0 Jul	32.7 Jan					27.56
Nancy	48.41N, 6.12E	915	1931–60	49.3	64.8 Jul	33.6 Jan	96	−4	89	7	28.03
Nanterre	48.54N, 2.12E	528	(see Paris)								
Nantes	47.13N, 1.33W	62	1931–60	53.2	66.0 Jul	41.0 Jan	102	5	94	17	30.79
Nice	43.42N, 7.15E	151	1931–60	59.0	72.3 Jul	46.9 Jan	98	24	92	31	33.94
Nimes	43.50N, 4.21E	374	1931–60	57.7	74.3 Jul	42.6 Jan					29.25
Orleans	47.55N, 1.54E	381	1931–60	51.1	65.5 Jul	37.0 Jan	105	4	93	15	23.62
*Paris	48.52N, 2.20E	197	1931–60	52.7	67.1 Jul	38.3 Jan	104	4	93	16	24.37

EUROPE

City	Latitude and Longitude	Elev (ft)	Period of Record	Avg	Avg Warmest Month	Avg Coolest Month	Absolute High	Absolute Low	Avg Annual High	Avg Annual Low	Annual Precipitation (in)
Pau	43.18N, 0.22W	679	1931–60	54.3	67.1 Jul	41.9 Jan	108	12	95	25	44.72
Perpignan	42.41N, 2.53E	197	1931–60	59.7	74.8 Jul	45.9 Jan	102	8	91	16	25.16
Poitiers	46.35N, 0.20E	387	1931–60	52.3	66.2 Jul	38.8 Jan	105	–3	91	16	25.98
Rennes	48.05N, 1.41W	177	1931–60	52.3	64.6 Jul	40.6 Jan	101	–4	93	9	26.34
Rheims	49.15N, 4.02E	282	1931–60	50.4	64.9 Jul	35.6 Jan					23.54
Roubaix	50.42N, 3.10E	115	(see Lille)								
Rouen	49.26N, 1.05E	72	1931–60	50.5	63.7 Jul	37.9 Jan					26.77
Saint-Denis	48.56N, 2.22E	98	1931–60	(see Paris for temperature data)							22.72
Saint-Etienne	45.26N, 4.24E	1772	1931–60	50.7	66.0 Jul	35.8 Jan					27.28
Saint-Nazaire	47.17N, 2.12W	26	1931–60	(see Nantes for temperature data)							30.51
Strasbourg	48.35N, 7.45E	932	1931–60	50.0	66.4 Jul	33.1 Jan	99	–8	93	9	23.90
Thionville	49.22N, 6.10E	656	(see Saarbrucken, West Germany)								
Toulon	43.07N, 5.56E	92	1931–60	59.7	73.0 Jul	47.7 Jan	96	14	92	29	27.80
Toulouse	43.36N, 1.26E	456	1931–60	54.9	69.4 Jul	40.3 Jan	111	1	99	16	25.94
Tourcoing	50.43N, 3.09E	148	(see Lille)								
Tours	47.23N, 0.41E	180	1931–60	52.5	66.6 Jul	38.7 Jan					27.13
Troyes	48.18N, 4.05E	361	1881–1917, 1921–34[1]	48.9	64.4 Jul	34.0 Jan					23.35
Valence	44.56N, 4.54E	420	1931–60	(see Grenoble for temperature data)							33.27
Valenciennes	50.21N, 3.32E	98	(see Lille)								
Versailles	48.48N, 2.08E	604	1931–60	(see Paris for temperature data)							24.49
Villeurbanne	45.46N, 4.53E	581	(see Lyons)								

Temperature (°F)

GERMANY: EAST
*Berlin (see under Germany: West)

[1] Temperature data for Chaumont, nr Troyes.

EUROPE

City	Latitude and Longitude	Elev (ft)	Period of Record	Temperature (°F) Avg	Avg Warmest Month	Avg Coolest Month	Absolute High	Absolute Low	Avg Annual High	Avg Annual Low	Annual Precipitation (in)
Brandenburg	52.25N, 12.33E	102	1955–56, 1958–59, 1962	(see Potsdam for temperature data)							21.85
Cottbus	51.46N, 14.20E	236	1901–50	47.8	65.5 Jul	30.9 Jan					23.07
Dessau	51.50N, 12.15E	200	1893–1930	47.8	64.4 Jul	32.0 Jan					
Dresden	51.03N, 13.45E	371	1901–50	47.1	64.0 Jul	30.6 Jan	100	−18	91	5	25.94
Erfurt	50.59N, 11.02E	656	1901–50	46.0	62.2 Jul	30.4 Jan					19.68
Gera	50.52N, 12.05E	673	1881–1930[1]	46.0	62.2 Jul	30.4 Jan					
Gorlitz	51.10N, 15.00E	689	1901–50	46.2	64.5 Jul	29.7 Jan					27.83
Gotha	50.57N, 10.43E	1011	(see Erfurt)								
Halle	51.30N, 12.00E	328	(see Leipzig)								
Jena	50.56N, 11.35E	476	1901–72	47.8	63.7 Jul	26.8 Jan					16.30
Karl-Marx-Stadt	50.50N, 12.55E	1014	1901–50	44.6	61.2 Jul	28.6 Jan					25.67
Leipzig	51.18N, 12.20E	387	1901–50	46.9	64.4 Jul	30.9 Jan	97	−15	90	4	22.05
Magdeburg	52.10N, 11.40E	164	1901–50	47.3	63.9 Jul	31.5 Jan					19.92
Potsdam	52.24N, 13.04E	105	1901–50	47.3	64.6 Jul	30.7 Jan					23.03
Rostock	54.05N, 12.08E	43	1901–50	47.3	63.5 Jul	32.7 Jan					22.17
Schwerin	53.38N, 11.23E	131	1901–50	47.1	63.5 Jul	31.8 Jan					24.68
Zwickau	50.44N, 12.30E	876	1881–1930[1]	46.0	62.2 Jul	30.4 Jan					
GERMANY: WEST											
Aachen	50.46N, 6.06E	568	1931–60	49.3	63.5 Jul	35.2 Jan	94	7			33.07
Augsburg	48.22N, 10.53E	1608	1881–1930	46.8	64.2 Jul	29.5 Jan	98	1			33.23
Baden-Baden	48.45N, 8.15E	594	1881–1930	48.7	64.4 Jul	33.4 Jan					43.19

[1]Temperature data for Crimmitschau, nr Gera and Zwickau.

EUROPE

City	Latitude and Longitude	Elev (ft)	Period of Record	Avg	Avg Warmest Month	Avg Coolest Month	Absolute High	Low	Avg Annual High	Low	Annual Precipitation (in)
Bayreuth	49.57N, 11.35E	1132	1881–1930	46.0	63.0 Jul	29.3 Jan					23.43
Bergisch Gladbach	50.59N, 7.08E	600	(see Cologne)								
Berlin	52.33N, 13.22E	112	1931–60	48.2	65.7 Jul	30.7 Jan	99	–15	91	6	23.11
Bielefeld	52.02N, 8.32E	387	1881–1930	47.7	62.4 Jul	33.3 Jan					33.03
Bochum	51.29N, 7.13E	328	1957–61	(see Essen for temperature data)							32.28
*Bonn	50.44N, 7.06E	197	1957–61	(see Cologne for temperature data)							26.50
Bottrop	51.31N, 6.55E	197	(see Essen)								
Bremen	53.05N, 8.48E	10	1931–60	48.2	63.3 Jul	33.1 Jan	94	–7	87	9	26.30
Bremerhaven	53.33N, 8.35E	23	1881–1930	47.5	62.6 Jul	33.3 Jan	93	4	84	11	31.97
Brunswick	52.16N, 10.32E	230	1893–1930	47.8	63.7 Jul	32.4 Jan	95	–5			26.61
Cologne	50.56N, 6.57E	174	1881–1930	50.4	65.1 Jul	36.3 Jan	96	–3	89	14	27.40
Darmstadt	49.52N, 8.39E	472	1881–1930	49.1	65.3 Jul	33.3 Jan	94	–2	90	4	25.04
Dortmund	51.31N, 7.27E	249	1960–63, 1965	48.4	60.8 Jun	33.6 Jan	96	–3	87	6	37.28
Duisburg	51.26N, 6.45E	85	1957–61	(see Essen for temperature data)							28.66
Dusseldorf	51.13N, 6.46E	118	1957–61	(see Solingen for temperature data)							31.22
Erlangen	49.36N, 11.01E	919	(see Nuremberg)								
Essen	51.27N, 7.01E	249	1931–60	49.3	63.5 Jul	34.7 Jan	94	11			35.31
Esslingen	48.45N, 9.18E	755	(see Stuttgart)								
Flensburg	54.47N, 9.26E	66	1931–60	46.6	61.9 Jul	32.4 Jan					31.65
Frankfurt am Main	50.07N, 8.41E	322	1931–60	50.4	66.9 Jul	33.4 Jan	100	–7	91	9	23.78
Freiburg im Breisgau	48.00N, 7.51E	912	1931–60	50.5	66.9 Jul	34.2 Jan	103	–7	91	8	34.80
Furth	49.28N, 11.00E	974	1957–61	(see Nuremberg for temperature data)							25.47
Gelsenkirchen	51.31N, 7.06E	171	1957–61	(see Essen for temperature data)							33.90
Gottingen	51.32N, 9.56E	492	1893–1930	47.3	63.0 Jul	32.0 Jan	92	–14	86	–8	23.90
Hagen	51.21N, 7.28E	348	1957–61	(see Dortmund for temperature data)							37.72
Hamburg	53.33N, 10.00E	20	1931–60	47.5	63.1 Jul	32.0 Jan	94	–4	85	10	29.13

EUROPE

City	Latitude and Longitude	Elev (ft)	Period of Record	Temperature (°F)							Annual Precipitation (in)
				Avg	Avg Warmest Month	Avg Coolest Month	Absolute High	Absolute Low	Avg Annual High	Avg Annual Low	
Hamm	51.41N, 7.48E	207	(see Dortmund)								25.08
Hannover	52.22N, 9.43E	180	1931–60	48.0	63.7 Jul	32.4 Jan	98	–13	88	7	31.46
Heidelberg	49.25N, 8.42E	374	1881–1930	50.4	66.2 Jul	34.5 Jan					28.82
Heilbronn	49.08N, 9.13E	495	1957–61	(see Heidelberg for temperature data)							
Herne	51.33N, 7.13E	194	(see Essen)								30.43
Hildesheim	52.09N, 9.58E	262	1957–61	(see Hannover for temperature data)							
Ingolstadt	48.46N, 11.26E	1227	1881–1930[1]	45.5	63.1 Jul	27.7 Jan					
Iserlohn	51.22N, 7.42E	810	1881–1930[2]	47.5	62.1 Jul	33.8 Jan					
Kaiserslautern	49.27N, 7.45E	787	1957–61	(see Saarbrucken for temperature data)							28.31
Karlsruhe	49.01N, 8.24E	377	1931–60	50.2	67.1 Jul	33.4 Jan	99	–3			29.76
Kassel	51.19N, 9.30E	548	1931–60	48.2	64.2 Jul	31.8 Jan	99	–16	90	6	23.42
Kiel	54.20N, 10.08E	46	1881–1930	45.7	61.3 Jul	32.0 Jan	92	–4	81	12	28.23
Koblenz	50.21N, 7.36E	230	1881–1930	50.2	65.3 Jul	35.2 Jan	91	3	89	9	24.29
Krefeld	51.20N, 6.34E	125	1881–1930	48.9	63.7 Jul	35.2 Jan					25.28
Leverkusen	51.01N, 6.59E	144	(see Essen)								
Lubeck	53.52N, 10.42E	43	1931–60	47.7	63.9 Jul	32.2 Jan	96	2			24.88
Ludwigshafen	49.29N, 8.27E	312	1891–1930	(see Mannheim for temperature data)							22.24
Mainz	50.00N, 8.15E	269	1881–1930[3]	48.6	62.4 Jul	35.4 Jan					23.78
Mannheim	49.29N, 8.28E	318	1881–1930	50.0	66.6 Jul	33.6 Jan	99	4			25.55
Marl	51.39N, 7.05E	171	(see Essen)								
Moers	51.27N, 6.39E	95	(see Essen)								

[1] Temperature data for Karlshuid, nr Ingolstadt.
[2] Temperature data for Arnsberg, nr Iserlohn.
[3] Temperature data for Alzey, nr Mainz.

EUROPE

City	Latitude and Longitude	Elev (ft)	Period of Record	Avg	Avg Warmest Month	Avg Coolest Month	Absolute High	Absolute Low	Avg Annual High	Avg Annual Low	Annual Precipitation (in)
Monchengladbach	51.12N, 6.26E	197	(see Solingen)								
Mulheim an der Ruhr	51.26N, 6.53E	131	(see Essen)								
Munich	48.09N, 11.35E	1706	1931–60	46.2	63.5 Jul	28.2 Jan	93	−14	86	−1	34.88
Munster	51.58N, 7.38E	197	1931–60	49.1	63.9 Jul	34.3 Jan	96	−17	91	8	30.59
Neuss	51.12N, 6.42E	131	(see Solingen)								
Nuremberg	49.27N, 11.05E	1014	1931–60	47.3	64.8 Jul	29.5 Jan	100	−18	91	1	23.31
Oberhausen	51.28N, 6.51E	131	1957–61	(see Essen for temperature data)							33.50
Offenbach am Main	50.06N, 8.46E	322	(see Frankfurt am Main)								
Oldenburg in Oldenburg	53.10N, 8.12E	16	1893–1930	47.1	62.4 Jul	33.3 Jan					31.38
Osnabruck	52.16N, 8.03E	210	1893–1930	47.8	62.8 Jul	34.0 Jan	93	0	86	4	34.92
Paderborn	51.43N, 8.46E	361	(see Bielefeld)								
Pforzheim	48.53N, 8.42E	896	1957–61	(see Stuttgart for temperature data)							30.16
Recklinghausen	51.37N, 7.12E	358	(see Dortmund)								
Regensburg	49.01N, 12.06E	1112	1881–1930	45.9	63.7 Jul	27.7 Jan	94	−9	91	−1	23.27
Remscheid	51.11N, 7.12E	1198	(see Solingen)								
Reutlingen	48.29N, 9.13E	1237	1881–1930[1]	47.1	64.0 Jul	30.0 Jan					
Saarbrucken	49.14N, 7.00E	623	1931–60	49.3	64.8 Jul	33.6 Jan	92	−1	89	5	30.94
Salzgitter	52.05N, 10.20E	469	(see Brunswick)								
Siegen	50.52N, 8.02E	919	1881–1930	45.5	59.7 Jul	31.8 Jan					
Solingen	51.11N, 7.05E	725	1881–1930	47.5	61.5 Jul	34.2 Jan	97	12	91	8	43.62
Stuttgart	48.46N, 9.11E	804	1931–60	49.8	66.2 Jul	33.4 Jan	102	−13	91	8	26.06
Trier	49.45N, 6.38E	410	1931–60	48.9	64.2 Jul	33.3 Jan	92	2	89	7	28.31
Ulm	48.24N, 10.00E	1568	1931–60	46.8	63.9 Jul	28.8 Jan					27.64
Velbert	51.20N, 7.03E	807	(see Wuppertal)								
Wiesbaden	50.05N, 8.15E	381	1957–61	(see Frankfurt am Main for temperature data)							25.75

[1]Temperature data for Tubingen, nr Reutlingen.

EUROPE

City	Latitude and Longitude	Elev (ft)	Period of Record	Temperature (°F) Avg	Avg Warmest Month	Avg Coolest Month	Absolute High	Absolute Low	Avg Annual High	Avg Annual Low	Annual Precipitation (in)
Wilhelmshaven	53.31N, 8.08E	13	1893–1930	46.9	61.5 Jul	33.4 Jan					27.28
Witten	51.26N, 7.20E	269	(see Essen)								
Wolfsburg	52.26N, 10.48E	184	(see Brunswick)								
Wuppertal	51.16N, 7.11E	525	1881–1930	48.6	63.1 Jul	34.7 Jan					45.16
Wurzburg	49.48N, 9.56E	594	1931–60	48.4	65.1 Jul	30.7 Jan	94	–6	90	2	22.05
GIBRALTAR											
*Gibraltar	36.06N, 5.21W	11	1951–60	64.8	75.4 Aug	55.0 Jan	101	33	96	38	32.09
GREECE											
*Athens	37.58N, 23.43E	453	1931–60	64.0	81.7 Jul	48.7 Jan	109	20	100	29	15.83
Iraklion	35.20N, 25.09E	98	1931–60	66.0	79.7 Aug	54.0 Jan	114	32	99	38	16.77
Patras	38.15N, 21.44E	15	1931–60	63.7	79.5 Aug	49.5 Jan	110	25	100	32	29.49
Peristeri	38.01N, 23.41E		(see Athens)								
Piraeus	37.57N, 23.38E	30	(see Athens)								
Salonika	40.38N, 22.56E	200	1931–60	61.0	81.1 Jul	41.9 Jan	107	15	101	23	17.68
HUNGARY											
*Budapest	47.30N, 19.05E	377	1931–60	52.2	72.0 Jul	30.0 Jan	103	–10	95	8	24.80
Debrecen	47.32N, 21.38E	404	1931–60	50.5	71.2 Jul	27.1 Jan	102	–22	95	–3	22.99
Gyor	47.41N, 17.38E	387	1965, 1967–71	50.0	69.1 Jul	28.8 Jan	97	–11	93	–2	20.98
Miskolc	48.06N, 20.47E	394	1931–60	49.5	70.0 Jul	26.1 Jan					23.62
Pecs	46.05N, 18.14E	830	1931–60	52.7	72.7 Jul	30.7 Jan					26.26
Szeged	46.15N, 20.10E	269	1931–60	52.7	73.4 Jul	29.5 Jan	102	–20	96	2	21.97
Szekesfehervar	47.12N, 18.25E	364	(see Budapest)								

EUROPE

City	Latitude and Longitude	Elev (ft)	Period of Record	Temperature (°F)							Annual Precipitation (in)
				Avg	Avg Warmest Month	Avg Coolest Month	Absolute High	Absolute Low	Avg Annual High	Avg Annual Low	
ICELAND											
*Reykjavik	64.09N, 21.57W	92	1931–60	41.0	52.2 Jul	31.3 Jan	74	4	67	10	31.69
IRELAND											
Cork	51.54N, 8.28W	56	27 yr bef 1951	50.5	60.5 Jul	42.5 Dec	85	15	79	23	41.3
*Dublin	53.20N, 6.15W	51	1921–50	50.0	60.4 Jul	41.9 Jan	86	8	79	19	26.78
Limerick	52.40N, 8.37W	59	1941–60[1]	50.4	60.1 Jul	41.0 Jan	87	12	81	19	36.73
ITALY											
Alessandria	44.54N, 8.37E	312	25 yr bef 1926	54.3	74.8 Jul	32.9 Jan					21.89
Ancona	43.38N, 13.30E	52	1946–55	59	76.5 Jul	45 Jan	98	20	96	27	25.35
Arezzo	43.25N, 11.53E	971	1881–1925	56.7	74.3 Jul	40.3 Jan					32.60
Bari	41.08N, 16.51E	16	1931–48, 1951–55	61.5	76.6 Jul	47.5 Jan	111	18			20.71
Bergamo	45.41N, 9.43E	817	1921–50	(see Brescia for temperature data)							48.94
Bologna	44.29N, 11.20E	180	1931–55	57.4	77.9 Jul	36.1 Jan	103	9			23.66
Bolzano	46.31N, 11.22E	860	1946–55	53	71 Jul	32 Jan	100	5			27.05
Brescia	45.33N, 10.15E	489	31 yr bef 1926	55.8	74.8 Jul	36.0 Jan					34.72
Cagliari	39.13N, 9.07E	13	1931–42, 1945–55	63.9	78.3 Aug	50.9 Jan	104	25			16.97
Catania	37.30N, 15.06E	33	1892–1930	64.4	79.7 Aug	50.5 Jan	112	26			27.40
Catanzaro	38.54N, 16.35E	1050	1921–50								39.72
Como	45.47N, 9.05E	659	1889–1906	54.5	73.4 Jul	36.1 Jan					50.20
Cosenza	39.18N, 16.15E	778	1921–50								40.98
Cremona	45.07N, 10.02E	148	1921–50	(see Piacenza for temperature data)							34.13
Ferrara	44.50N, 11.35E	30	1921–43, 1946–50	(see Bologna for temperature data)							22.68
Florence	43.46N, 11.15E	164	1931–55	58.5	76.5 Jul	41.4 Jan	105	11			33.07

[1]Temperature data for Shannon Airport, nr Limerick.

EUROPE

City	Latitude and Longitude	Elev (ft)	Period of Record	Temperature (°F)							Annual Precipitation (in)
				Avg	Avg Warmest Month	Avg Coolest Month	Absolute High	Absolute Low	Avg Annual High	Avg Annual Low	
Foggia	41.27N, 15.34E	249	1946–55	61	79 Aug	45 Jan	114	14			17.87
Forli	44.13N, 12.03E	112	1921–43, 1946–50	(see Bologna for temperature data)							29.72
Genoa	44.25N, 8.57E	62	1931–55	61.0	76.1 Jul	46.2 Jan	100	18	90	28	45.12
La Spezia	44.07N, 9.50E	10	28 yr bef 1951	(see Genoa for temperature data)							54.13
Leghorn	43.33N, 10.19E	10	1901–30	59.9	75.6 Jul	45.3 Jan	100	20			32.32
Lucca	43.50N, 10.29E	56	1921–27, 1931–50	(see Florence for temperature data)							48.66
Mantua	45.09N, 10.48E	59	26 yr bef 1926	56.5	77.0 Jul	35.8 Jan					23.74
Messina	38.11N, 15.34E	10	1901–30	64.2	78.6 Aug	52.5 Feb	105	28			36.93
Milan	45.28N, 9.12E	397	1931–43, 1946–55	56.7	77.0 Jul	35.6 Jan	101	1			35.91
Modena	44.40N, 10.55E	112	1921–50	(see Bologna for temperature data)							24.84
Monza	45.35N, 9.16E	532	1921–50	(see Milan for temperature data)							44.53
Naples	40.50N, 14.15E	33	1931–55	62.8	77.9 Jul	48.2 Jan	100	24	93	30	33.46
Novara	45.28N, 8.38E	522	1889–1906	55.4	75.0 Jul	36.1 Jan					36.14
Padua	45.25N, 11.53E	39	1921–50	(see Verona for temperature data)							33.31
Palermo	38.07N, 13.22E	46	1931–55	64.9	78.1 Aug	53.1 Jan	113	29			28.78
Parma	44.48N, 10.20E	187	1931–46	(see Bologna for temperature data)							30.20
Pavia	45.10N, 9.10E	253	1921–50	(see Piacenza for temperature data)							34.45
Perugia	43.08N, 12.22E	1618	1946–55	56.5	73.5 Jul	40.5 Jan	100	9	95	21	34.41
Pescara	42.28N, 14.13E	13	1946–55	58.5	73.5 Jul	44 Jan	105	21			27.24
Piacenza	45.01N, 9.40E	200	30 yr bef 1926	54.1	74.1 Jul	32.7 Jan	99	–6			31.26
Pisa	43.43N, 10.23E	16	1901–30	57.9	73.0 Aug	43.2 Jan	100	13			38.03
Pistoia	43.55N, 10.54E	213	15 yr bef 1923	58.1	75.4 Jul	42.3 Jan					47.80
Prato	43.53N, 11.06E	207	1921–50	(see Florence for temperature data)							38.90
Ravenna	44.25N, 12.12E	13	1924–43, 1946–50	(see Bologna for temperature data)							27.56
Reggio di Calabria	38.06N, 15.39E	49	1921–50	(see Messina for temperature data)							23.47
Reggio nell'Emilia	44.43N, 10.36E	190	1921–50	(see Bologna for temperature data)							27.72
Rimini	44.04N, 12.34E	23	1921–43	(see Florence for temperature data)							29.37

EUROPE

City	Latitude and Longitude	Elev (ft)	Period of Record	Temperature (°F) Avg	Avg Warmest Month	Avg Coolest Month	Absolute High	Absolute Low	Avg Annual High	Avg Annual Low	Annual Precipitation (in)
*Rome	41.54N, 12.29E	66	1931–55	61.2	77.7 Jul	45.9 Jan	108	15			29.92
Salerno	40.41N, 14.47E	13	1921–50	(see Naples for temperature data)							51.14
Sassari	40.43N, 8.34E	738	1883–1930	59.9	74.5 Aug	46.6 Jan					23.47
Sesto San Giovanni	45.32N, 9.14E	466	(see Milan)								
Siena	43.19N, 11.21E	1056	1921–22, 1927–50	(see Perugia for temperature data)							33.98
Syracuse	37.04N, 15.18E	56	32 yr bef 1926	64.2	79.0 Aug	51.8 Jan					25.24
Taranto	40.28N, 17.14E	49	1901–30	62.2	78.3 Aug	47.8 Jan	108	24			17.52
Terni	42.34N, 12.37E	427	1882–1900, 1906[1]	57.9	75.7 Jul	41.9 Jan					33.03
Torre del Greco	40.47N, 14.22E	125	1921–42	(see Naples for temperature data)							33.27
Trent	46.04N, 11.08E	637	1866–1906	53.1	72.5 Jul	32.2 Jan					36.02
Treviso	45.40N, 12.15E	49	29 yr bef 1926	56.7	75.9 Jul	37.6 Jan					36.93
Trieste	45.40N, 13.46E	7	1951–60	57.4	74.8 Jul	41.2 Jan	99	6			38.07
Turin	45.03N, 7.40E	784	1898–1915, 1926–34	53.2	73.2 Jul	33.3 Jan	100	6	96	15	30.67
Udine	46.03N, 13.14E	371	1923–58	54.5	72.1 Jul	36.7 Jan	99	0	94	20	53.82
Venice	45.27N, 12.21E	3	30 yr bef 1926	56.5	75.6 Jul	37.4 Jan	96	14			29.80
Verona	45.27N, 11.00E	194	1946–55	55.5	74 Jul	36.5 Jan	100	0			25.35
Vicenza	45.33N, 11.33E	128	1921–43, 1946–50	(see Verona for temperature data)							40.83
LUXEMBOURG											
*Luxembourg	49.36N, 6.07E	984	1931–60	47.8	63.3 Jul	32.5 Jan	99	–10	87	9	29.13
MALTA											
*Valletta	35.54N, 14.32E	233	90 yr bef 1948	66	79 Aug	55 Jan	105	34	97	42	20.3

[1]Climatic data for Teramo, nr Terni.

EUROPE

City	Latitude and Longitude		Elev (ft)	Period of Record	Temperature (°F)							Annual Precipitation (in)
					Avg	Avg Warmest Month	Avg Coolest Month	Absolute High	Low	Avg Annual High	Low	
MONACO												
*Monaco	43.44N,	7.25E	180	1921–50	61.5	74.8 Aug	50.4 Jan	93	27	86	36	31.34
NETHERLANDS												
*Amsterdam	52.23N,	4.54E	5	1921–50	50	64 Jul	37 Jan	95	3	86	17	25.6
Apeldoorn	52.13N,	5.57E		(see Utrecht)								
Arnhem	52.00N,	5.53E		(see Utrecht)								
Breda	51.35N,	4.46E	23	1931–60[1]	51.1	65.7 Jul	36.3 Jan					29.17
Delft	52.00N,	4.22E	10	1931–60[2]	50.9	65.1 Aug	36.7 Jan					29.57
Dordrecht	51.48N,	4.40E	13	1931–60[1]	51.1	65.7 Jul	36.3 Jan					29.17
Eindhoven	51.26N,	5.30E	66	1931–60[3]	51.3	66.9 Jul	35.6 Jan					27.32
Enschede	52.13N,	6.55E	95	(see Munster, West Germany)								
Groningen	53.13N,	6.34E	16	1921–50	48.5	63.5 Jul	34.5 Jan	98	−5	90	10	29.4
Haarlem	52.20N,	4.36E	10	(see Amsterdam)								
*Hague	52.04N,	4.19E	12	1931–60[2]	50.9	65.1 Aug	36.7 Jan					29.57
Heerlen	50.53N,	5.59E	377	(see Aachen, West Germany)								
Hilversum	52.14N,	5.10E	16	(see Utrecht)								
Leiden	52.10N,	4.30E	10	1931–60[2]	50.9	65.1 Aug	36.7 Jan					29.57
Maastricht	50.51N,	5.42E	161	1900–36	(see Aachen, West Germany, for temperature data)							25.94
Nijmegen	51.50N,	5.52E	82	1931–60[3]	51.3	66.9 Jul	35.6 Jan					27.32
Rotterdam	51.55N,	4.29E	12	1921–50	50	64 Jul	37 Jan	100	2	90	15	28.6

[1] Climatic data for Oudenbosch, nr Breda and Dordrecht.
[2] Climatic data for Naaldwijk, nr Delft, Hague, and Leiden.
[3] Climatic data for Gemert, nr Eindhoven and Nijmegen.

EUROPE

City	Latitude and Longitude		Elev (ft)	Period of Record	Temperature (°F)							Annual Precipitation (in)
					Avg	Avg Warmest Month	Avg Coolest Month	Absolute		Avg Annual		
								High	Low	High	Low	
's Hertogenbosch	51.41N,	5.19E	26	1931–60[1]	51.3	66.9 Jul	35.6 Jan					27.32
Tilburg	51.34N,	5.05E	59	1931–60[1]	51.3	66.9 Jul	35.6 Jan					27.32
Utrecht	52.07N,	5.05E	10	1931–60	50.7	65.7 Jul	35.6 Jan	98	–13			30.12
Velsen	52.27N,	4.39E	16	(see Amsterdam)								
Zaanstad	52.27N,	4.49E	10	(see Amsterdam)								
NORWAY												
Bergen	60.23N,	5.20E	144	1931–60	46.0	59.0 Jul	34.3 Feb	89	3	78	14	77.09
*Oslo	59.55N,	10.45E	315	1931–60	42.6	63.1 Jul	23.5 Jan	93	–21	83	–2	29.13
Stavanger	58.58N,	5.45E	26	1931–60	45.3	58.5 Jul	32.7 Feb	85	12	80	17	39.96
Trondheim	63.25N,	10.25E	417	1931–60	40.8	57.9 Jul	25.9 Jan	95	–22	82	1	33.74
POLAND												
Bialystok	53.09N,	23.10E	456	1931–60	44.4	64.9 Jul	23.2 Jan					21.42
Bielsko-Biala	49.50N,	19.00E	1014	1931–60	45.9	63.5 Jul	26.4 Jan					39.80
Bydgoszcz	53.16N,	17.33E	230	1931–60	46.2	65.7 Jul	27.0 Jan	98	–24			20.83
Bytom	50.22N,	18.53E	935	(see Katowice)								
Chorzow	50.19N,	18.56E	925	(see Katowice)								
Czestochowa	50.49N,	19.07E	856	1931–60	46.0	65.1 Jul	25.9 Jan					25.98
Elblag	54.10N,	19.25E	23	(see Gdansk)								
Gdansk	54.20N,	18.40E	43	1931–60	45.1	63.1 Jul	27.7 Jan	96	–18	87	1	23.23
Gdynia	54.31N,	18.30E	7	(see Gdansk)								
Gliwice	50.20N,	18.40E	728	(see Katowice)								
Gorzow Wielkopolski	52.44N,	15.14E	213	1931–60	47.1	65.3 Jul	28.8 Jan	98	–21			21.06
Jastrzebie Zdroj	49.57N,	18.35E	745	(see Katowice)								

[1]Climatic data for Gemert, nr 's Hertogenbosch and Tilburg.

EUROPE

City	Latitude and Longitude	Elev (ft)	Period of Record	Temperature (°F) Avg	Avg Warmest Month	Avg Coolest Month	Absolute High	Absolute Low	Avg Annual High	Avg Annual Low	Annual Precipitation (in)
Kalisz	51.45N, 18.05E	459	30 yr bef 1936	46.8	65.1 Jul	28.8 Jan	99	-25			21.30
Katowice	50.15N, 18.59E	932	1931-60	46.0	64.4 Jul	26.1 Jan					26.81
Kielce	50.51N, 20.39E	879	1931-60	45.3	64.8 Jul	24.6 Jan					25.55
Krakow	50.05N, 19.55E	778	1931-60	47.5	66.7 Jul	26.8 Jan	96	-28	90	-2	27.09
Lodz	51.46N, 19.25E	614	1931-60	45.7	64.9 Jul	25.7 Jan	97	-24			23.07
Lublin	51.20N, 22.30E	561	1931-60	45.5	65.5 Jul	24.4 Jan					22.20
Olsztyn	53.47N, 20.29E	377	1931-60	44.4	63.9 Jul	24.8 Jan					23.23
Opole	50.40N, 17.57E	515	1931-60	47.1	65.7 Jul	27.5 Jan	100	-26			26.26
Plock	52.33N, 19.42E	207	1891-1930	(see Warsaw for temperature data)							18.62
Poznan	52.25N, 16.58E	282	1931-60	46.8	65.7 Jul	27.3 Jan	96	-20	93	1	20.67
Radom	51.26N, 21.10E	584	1881-1930	46.4	65.7 Jul	27.3 Jan					23.23
Ruda Slaska	50.18N, 18.51E	938	(see Katowice)								
Rybnik	50.07N, 18.32E	787	(see Katowice)								
Rzeszow	50.03N, 22.00E	656	1931-60	46.0	65.8 Jul	24.3 Jan					24.96
Sosnowiec	50.16N, 19.07E	820	(see Katowice)								
Szczecin	53.26N, 14.34E	3	1931-60	46.9	64.6 Jul	29.5 Jan	97	-16	88	5	22.36
Tarnow	50.01N, 20.59E	686	1881-1930	47.8	66.4 Jul	28.8 Jan					29.09
Torun	53.01N, 18.35E	226	1881-1930	46.0	65.1 Jul	28.0 Jan					19.49
Tychy	50.08N, 18.59E	919	(see Katowice)								
Walbrzych	50.48N, 16.19E	1476	1881-1930[1]	45.1	62.4 Jul	27.7 Jan					21.34
*Warsaw	52.15N, 21.00E	348	1931-60	46.6	66.6 Jul	25.7 Jan	98	-27	90	-3	
Wloclawek	52.39N, 19.05E	203	(see Torun)								
Wodzislaw Slaski	50.00N, 18.28E	804	(see Katowice)								
Wroclaw	51.07N, 17.02E	394	1931-60	47.3	65.8 Jul	28.4 Jan	98	-26	90	1	22.99
Zabrze	50.18N, 18.47E	820	(see Katowice)								

[1]Temperature data for Klodzko, nr Walbrzych.

EUROPE

City	Latitude and Longitude		Elev (ft)	Period of Record	Avg	Avg Warmest Month	Avg Coolest Month	Absolute High	Absolute Low	Avg Annual High	Avg Annual Low	Annual Precipitation (in)
PORTUGAL												
Coimbra	40.12N,	8.25W	463	1931–60	60.6	72.0 Aug	49.5 Jan	114	25	104	29	37.87
*Lisbon	38.43N,	9.08W	312	1931–60	61.9	72.5 Aug	51.4 Jan	103	29	98	32	27.87
Ponta Delgada	37.44N,	25.40W	115	1931–60	63.3	71.6 Aug	57.6 Feb	87	37	82	45	37.72
Porto	41.09N,	8.37W	328	1931–60	57.9	67.6 Aug	48.2 Jan	104	25	97	28	45.28
ROMANIA												
Arad	46.11N,	21.19E	331	1896–1955	51.4	70.5 Jul	30.0 Jan	105	−22	95	2	22.72
Bacau	46.34N,	26.54E	597	1896–1955	48.6	69.4 Jul	24.3 Jan	102	−26	89	−8	21.42
Baia Mare	47.40N,	23.35E	637	1896–1955	48.9	67.8 Jul	27.7 Jan	103	−22	89	−2	38.43
Braila	45.16N,	27.59E	49	1896–1955	52.0	73.6 Jul	27.9 Jan	105	−16	96	0	17.32
Brasov	45.38N,	25.35E	1941	1896–1955	46.0	64.0 Jul	25.0 Jan	99	−21	89	−5	29.41
*Bucharest	44.26N,	26.06E	269	1896–1955	51.6	73.2 Jul	27.0 Jan	106	−22	99	−1	22.83
Buzau	45.09N,	26.50E	394	1896–1955	50.9	72.5 Jul	27.7 Jan	103	−20	96	−2	20.16
Cluj	46.46N,	23.36E	1027	1896–1955	46.8	66.0 Jul	24.1 Jan	98	−30	90	−7	24.13
Constanta	44.11N,	28.39E	105	1896–1955	52.2	72.0 Jul	31.5 Jan	101	−13	90	5	14.92
Craiova	44.19N,	23.48E	345	1896–1955	51.4	72.9 Jul	27.5 Jan	107	−23	99	−4	20.59
Galati	45.27N,	28.03E	98	1896–1955	50.9	72.7 Jul	26.4 Jan	102	−19	95	−2	16.77
Hunedoara	45.45N,	22.54E	797	1896–1955	49.5	68.4 Jul	27.0 Jan					24.61
Iasi	47.10N,	27.36E	335	1896–1955	49.3	70.3 Jul	25.5 Jan	104	−23	96	−8	20.39
Oradea	47.04N,	21.56E	449	1896–1955	50.9	70.2 Jul	29.3 Jan	103	−20	94	2	25.00
Pitesti	44.51N,	24.52E	1004	1896–1955	49.6	69.4 Jul	27.7 Jan	103	−17	96	−4	27.56
Ploiesti	44.57N,	26.01E	505	1896–1955	51.1	71.6 Jul	28.2 Jan	103	−22	96	−2	23.15
Resita	45.18N,	21.55E	742	1896–1955[1]	51.1	69.8 Jul	30.6 Jan					32.28
Satu Mare	47.48N,	22.53E	417	1896–1955	49.5	68.2 Jul	27.7 Jan					26.30

[1] Temperature data for Caransebes, nr Resita.

EUROPE

City	Latitude and Longitude	Elev (ft)	Period of Record	Temperature (°F)							Annual Precipitation (in)
				Avg	Avg Warmest Month	Avg Coolest Month	Absolute High	Absolute Low	Avg Annual High	Avg Annual Low	
Sibiu	45.48N, 24.09E	1365	1896–1955	48.0	67.3 Jul	25.2 Jan	99	−25	91	−8	26.06
Timisoara	45.45N, 21.13E	295	1896–1955	51.6	70.9 Jul	29.8 Jan	104	−32	95	0	24.84
Tirgu Mures	46.33N, 24.34E	1014	1896–1955	47.7	66.9 Jul	24.3 Jan	102	−27	90	−8	25.04
SPAIN											
Albacete	38.59N, 1.51W	2251	1931–60	55.9	75.4 Jul	39.6 Jan	105	−8	100	14	13.82
Alicante	38.21N, 0.29W	10	1939–60	64.4	79.0 Aug	51.8 Jan	106	24	98	31	13.35
Almeria	36.50N, 2.27W	56	1934–60	64.2	77.5 Aug	53.1 Jan	108	32	99	40	9.09
Badajoz	38.53N, 6.58W	604	1931–60	62.1	78.4 Jul	47.5 Jan	113	23	107	27	18.66
Badalona	41.27N, 2.15E	75	(see Barcelona)								
Baracaldo	43.18N, 2.59W	128	(see San Sebastian)								
Barcelona	41.23N, 2.11E	39	1931–60	61.5	75.9 Jul	48.9 Jan	98	20	92	31	23.39
Bilbao	43.15N, 2.58W	10	26 yr bef 1927	(see San Sebastian for temperature data)							41.93
Burgos	42.21N, 3.42W	2808	1931–47, 1950–60	51.1	66.0 Jul	36.5 Jan	99	0			22.17
Cadiz	36.32N, 6.18W	16	27 yr bef 1961	64.4	76.8 Aug	52.5 Jan	106	28			20.59
Cartagena	37.36N, 0.59W	20	1881–1901	63.0	76.3 Aug	51.3 Jan					12.60
Castellon de la Plana	39.59N, 0.02W	95	1931–60	63.0	76.5 Aug	50.7 Jan	103	19			16.34
Cordova	37.53N, 4.46W	328	1931–60	64.2	82.2 Jul	48.4 Jan	112	21			26.14
Elche	38.15N, 0.42W	266	(see Alicante)								
Gijon	43.32N, 5.40W	30	1931–60	57.0	67.3 Jul	48.7 Jan	95	23			40.83
Granada	37.13N, 3.41W	2261	25 yr bef 1961	59.9	78.3 Jul	44.6 Jan	109	13	100	25	20.83
Hospitalet	41.22N, 2.08E	26	(see Barcelona)								
Huelva	37.16N, 6.57W	13	1931–60	64.4	77.4 Aug	52.0 Jan	110	22			18.31
Jerez de la Frontera	36.41N, 6.08W	161	(see Cadiz)								
La Coruna	43.22N, 8.23W	79	1931–60	57.0	66.0 Aug	49.6 Feb	95	27	85	33	38.15
Leon	42.36N, 5.34W	2697	1938–60	51.8	67.5 Jul	37.0 Jan	100	1			20.75

EUROPE

City	Latitude and Longitude	Elev (ft)	Period of Record	Avg	Avg Warmest Month	Avg Coolest Month	Absolute High	Absolute Low	Avg Annual High	Avg Annual Low	Annual Precipitation (in)
Lerida	41.37N, 0.37E	728	(see Tarragona)								
Logrono	42.28N, 2.27W	1260	1911–14, 1925	54.5	70.3 Jul	40.1 Jan	102	14	97	22	17.24
*Madrid	40.24N, 3.41W	2100	1931–60	57.0	75.6 Jul	40.8 Jan	105	32	100	37	18.46
Malaga	36.43N, 4.25W	33	1931–60	65.3	78.1 Aug	54.5 Jan	113	14	104	28	11.85
Murcia	37.59N, 1.07W	141	1931–60	64.4	79.3 Aug	51.3 Jan					
Oviedo	43.22N, 5.50W	705	45 yr bef 1933	54.1	64.8 Aug	44.8 Jan					36.89
Palma	39.34N, 2.39E	108	1931–60	62.2	76.1 Aug	50.2 Jan	102	25	95	33	17.76
Pamplona	42.49N, 1.38W	1476	1931–60	54.1	68.4 Jul	40.3 Jan	105	5	99	19	42.80
Sabadell	41.33N, 2.06E	617	(see Barcelona)								
Salamanca	40.58N, 5.39W	2553	1931–60	53.4	70.7 Jul	38.7 Jan	103	10	100	16	16.54
San Sebastian	43.19N, 1.59W	26	1931–60	55.6	66.2 Aug	46.0 Jan	100	10			59.29
Santa Coloma de Gramanet	41.27N, 2.13E	59	(see Barcelona)								
Santander	43.28N, 3.48W	13	1931–60	57.0	66.7 Aug	48.6 Feb	104	25	90	32	47.13
Saragossa	41.38N, 0.53W	656	1931–60	58.5	74.7 Aug	43.0 Jan	108	5	100	22	13.42
Seville	37.23N, 5.59W	20	1931–60	65.8	82.2 Jul	50.9 Jan	117	26	110	31	21.06
Tarragona	41.07N, 1.15E	226	1931–50	60.4	73.8 Aug	48.0 Jan	94	21			18.98
Tarrasa	41.34N, 2.01E	909	(see Barcelona)								
Toledo	39.52N, 4.01W	1729	1931–60	59.0	79.0 Jul	42.6 Jan	108	15			14.80
Valencia	39.28N, 0.22W	10	1938–60	62.6	76.1 Aug	50.5 Jan	107	19	96	31	16.50
Valladolid	41.39N, 4.43W	2270	1943–60	53.8	70.3 Jul	37.9 Jan	103	9	99	18	14.25
Vigo	42.14N, 8.43W	66	1931–60	59.0	67.8 Aug	50.4 Jan	102	27			50.20
Vitoria	42.51N, 2.40W	1719	1931–60	53.1	66.9 Aug	40.3 Jan	103	0			33.23
SWEDEN											
Boras	57.43N, 12.55E	427	1931–60	43.3	61.7 Jul	26.6 Feb					35.35
Eskilstuna	59.22N, 16.30E	26	(see Vasteras)								

EUROPE

City	Latitude and Longitude	Elev (ft)	Period of Record	Temperature (°F)							Annual Precipitation (in)
				Avg	Avg Warmest Month	Avg Coolest Month	Absolute High	Absolute Low	Avg Annual High	Avg Annual Low	
Goteborg	57.43N, 11.58E	102	1931-60	45.7	62.6 Jul	29.8 Feb	90	-15	83	6	26.38
Helsingborg	56.03N, 12.42E	131	(see Copenhagen, Denmark)								
Jonkoping	57.47N, 14.11E	325	1931-60	43.2	61.3 Jul	26.8 Feb	92	-28	84	-2	21.10
Linkoping	58.25N, 15.37E	210	1931-60	44.2	63.9 Jul	26.6 Feb	94	-26			20.79
Malmo	55.36N, 13.00E	10	1901-30	45.9	61.9 Jul	32.5 Jan					22.91
Norrkoping	58.36N, 16.11E	33	(see Linkoping)								
Orebro	59.17N, 15.13E	118	1931-60	42.4	62.2 Jul	24.8 Jan	93	-22	83	3	25.24
*Stockholm	59.20N, 18.03E	144	1931-60	43.9	64.0 Jul	26.4 Feb	97	-26			21.85
Sundsvall	62.23N, 17.18E	13	1931-60	39.0	60.4 Jul	19.6 Jan					21.50
Uppsala	59.52N, 17.38E	79	1931-60	42.4	63.1 Jul	24.3 Feb	99	-28	86	-8	21.81
Vasteras	59.37N, 16.33E	59	1931-60	42.6	63.0 Jul	24.6 Jan	97	-24			20.00
SWITZERLAND											
Basel	47.34N, 7.36E	1040	1901-60	48.6	65.1 Jul	32.4 Jan	102	-11	95	7	31.26
*Bern	46.57N, 7.28E	1877	1901-60	46.9	63.7 Jul	29.8 Jan	96	-9	88	8	39.37
Biel	47.10N, 7.15E	1457	1901-40	(see Bern for temperature data)							43.70
Geneva	46.12N, 6.10E	1411	1901-60	48.6	64.9 Jul	32.4 Jan	101	-1	91	13	36.73
Lausanne	46.32N, 6.39E	1831	1901-60	48.9	65.3 Jul	32.7 Jan					41.89
Luzern	47.03N, 8.17E	1634	1901-60	47.3	64.0 Jul	30.4 Jan					45.43
Saint-Gall	47.28N, 9.24E	2178	1901-60	45.3	61.5 Jul	29.1 Jan					51.18
Winterthur	47.30N, 8.45E	1608	(see Zurich)								
Zurich	47.22N, 8.31E	1539	1901-60	46.8	63.0 Jul	30.2 Jan	99	-11	88	9	44.41
TURKEY											
Edirne	41.40N, 26.34E	157	30 yr bef 1961	56.1	76.3 Jul	35.6 Jan	107	-8	99	8	23.98
Istanbul	41.01N, 28.58E	131	30 yr bef 1961	57.0	74.1 Aug	41.9 Jan	103	3	96	23	26.14

EUROPE

City	Latitude and Longitude		Elev (ft)	Period of Record	Temperature (°F)							Annual Precipitation (in)
					Avg	Avg Warmest Month	Avg Coolest Month	Absolute High	Low	Avg Annual High	Low	
UK												
Aberdeen	57.08N,	2.06W	79	1921–47	47.0	57.6 Jul	38.7 Jan	83	12	76	20	32.93
Barnsley	53.33N,	1.29W	459	1916–50	(see Kirklees for temperature data)							26.85
Basildon	51.34N,	0.25E	95	(see London)								
Bath	51.23N,	2.22W	67	1926–50	50.6	62.8 Jul	40.6 Jan					30.93
Belfast	54.36N,	5.56W	66	24 yr bef 1963	48.8	59.8 Jul	40.0 Jan					38.26
Beverley	53.50N,	0.25W	33	1916–50	(see Hull for temperature data)							25.63
Birmingham	52.29N,	1.53W	425	1921–50	49.3	62.1 Jul	38.6 Jan	92	11	85	21	29.87
Blackburn	53.45N,	2.29W	610	1921–50	47.4	59.4 Jul	37.4 Feb					43.9
Blackpool	53.50N,	3.03W	65	1926–49	49.3	60.7 Jul	39.6 Jan					35.11
Bolton	53.35N,	2.26W	342	1921–50	48.5	60.3 Jul	38.5 Jan					46.85
Bournemouth	50.43N,	1.54W	139	1926–50	50.9	62.7 Jul	41.3 Feb					31.2
Bradford	53.48N,	1.45W	439	1921–50	47.9	60.0 Jul	38.0 Jan					34.15
Brighton	50.49N,	0.08W	32	1926–40, 1948–50	51.2	62.9 Aug	41.1 Feb					31.52
Bristol	51.27N,	2.35W	386	1916–50	(see Bath for temperature data)							32.20
Burnley	53.48N,	2.14W	458	1921–50	47.5	59.3 Jul	37.7 Jan					42.9
Bury	53.36N,	2.18W	220	1916–50	(see Bolton for temperature data)							41.73
Calderdale	53.44N,	1.52W	795	1916–50	(see Kirklees for temperature data)							38.56
Cambridge	52.12N,	0.07E	41	1921–50	49.7	62.6 Jul	38.9 Jan	96	1	86	16	21.72
Canterbury	51.17N,	1.05E	135	1921–44	50.5	63.3 Jul	39.6 Jan					27.09
Cardiff	51.29N,	3.10W	203	1921–50	50.2	61.4 Jul	40.6 Jan	91	2	83	21	42.10
Carlisle	54.53N,	2.56W	39	1921–50[1]	47.6	58.9 Jul	38.3 Jan					31.30
Charnwood	52.46N,	1.12W	269	1916–50	(see Leicester for temperature data)							28.15
Cheltenham	51.54N,	2.05W	214	1926–50	50.1	62.6 Jul	39.7 Jan					26.30
Chester	53.12N,	2.55W	66	1921–50[2]	50.0	61.5 Jul	40.5 Jan					30.83

[1]Temperature data for Nithsdale, nr Carlisle.
[2]Temperature data for Alyn and Deeside, nr Chester.

EUROPE

City	Latitude and Longitude		Elev (ft)	Period of Record	Temperature (°F)								Annual Precipitation (in)
					Avg Warmest Month		Avg Coolest Month		Absolute		Avg Annual		
					Avg	Month	Month		High	Low	High	Low	
Chesterfield	53.15N,	1.25W	226	1916–50	(see Sheffield for temperature data)								29.76
Chichester	50.50N,	0.47W	40	1916–50	(see Portsmouth for temperature data)								29.7
Colchester	51.53N,	0.54E	89	1921–50[1]	49.9	63.0 Jul	38.6 Jan						22.60
Coventry	52.25N,	1.30W	241	1921–50	49.2	62.1 Jul	38.4 Jan						26.54
Crewe and Nantwich	53.06N,	2.26W	181	1921–50[2]	47.5	59.8 Jul	37.5 Jan						29.1
Darlington	54.32N,	1.34W	174	1916–50	(see Durham for temperature data)								25.75
Derby	52.55N,	1.28W	211	1916–50	(see Nottingham for temperature data)								28.80
Doncaster	53.32N,	1.07W	41	1916–50	(see Sheffield for temperature data)								23.62
Dudley	52.30N,	2.05W	745	1916–50	(see Birmingham for temperature data)								28.68
Dundee	56.27N,	2.58W	147	1921–50	47.2	59.0 Jul	37.5 Jan						31.14
Durham	54.46N,	1.34W	336	1921–50	47.5	59.5 Jul	37.7 Jan						26.30
Edinburgh	55.57N,	3.12W	441	1921–50	47.4	58.5 Jul	38.7 Jan		83	15	78	22	27.53
Elmbridge	51.21N,	0.22W	33	1916–50	(see Guildford for temperature data)								23.78
Erewash	52.58N,	1.18W	135	1916–50	(see Nottingham for temperature data)								25.7
Exeter	50.43N,	3.31W	110	1921–50[3]	50.5	61.0 Jul	42.2 Jan						31.90
Fareham	50.51N,	1.11W	39	1916–50	(see Portsmouth for temperature data)								32.09
Gateshead	54.58N,	1.35W	510	1921–50	47.0	58.7 Jul	37.2 Jan						27.6
Gedling	52.58N,	1.05W		(see Nottingham)									
Gillingham	51.24N,	0.33E	10	1925–50[4]	49.3	62.0 Jul	38.8 Jan						24.21
Glasgow	55.51N,	4.16W	180	1921–50	47.0	58.0 Jul	37.9 Jan						38.30
Gloucester	51.53N,	2.14W	33	1916–50	(see Cheltenham for temperature data)								25.71
Gosport	50.48N,	1.08W	30	(see Portsmouth)									
Gravesham	51.26N,	0.22E	12	(see London)									

[1] Temperature data for Braintree, nr Colchester.
[2] Temperature data for East Staffordshire, nr Crewe and Nantwich.
[3] Temperature data for East Devon, nr Exeter.
[4] Temperature data for Tonbridge and Malling, nr Gillingham.

EUROPE

City	Latitude and Longitude	Elev (ft)	Period of Record	Temperature (°F)							Annual Precipitation (in)
				Avg	Avg Warmest Month	Avg Coolest Month	Absolute High	Absolute Low	Avg Annual High	Avg Annual Low	
Grimsby	53.35N, 0.05W	14	1921–50	49.4	61.1 Jul	39.4 Jan					25.0
Guildford	51.13N, 0.34W	184	1921–50	50.2	62.9 Jul	39.6 Jan					28.0
Halton	53.21N, 2.44W	30	1916–50	(see Liverpool for temperature data)							31.5
Harrogate	53.59N, 1.32W	478	1936–50	47.9	59.7 Jul	36.4 Jan					31.18
Hartlepool	54.41N, 1.13W	30	1916–50	(see Durham for temperature data)							24.5
Havant	50.51N, 0.59W	25	1916–50	(see Portsmouth for temperature data)							28.33
Hove	50.50N, 0.11W	174	(see Brighton)								
Hull	53.45N, 0.20W	8	1921–50	49.7	62.3 Jul	39.4 Jan					25.39
Inverclyde	55.56N, 4.45W	199	1921–50	48.4	59.0 Jul	39.6 Jan					62.52
Ipswich	52.04N, 1.10E	190	1921–50[1]	50.3	62.9 Jul	39.5 Jan					24.4
Kirklees	53.39N, 1.47W	762	1925–50	47.5	59.6 Jul	37.6 Jan					37.62
Knowsley	53.24N, 2.51W	49	1916–50	(see Liverpool for temperature data)							34.49
Lancaster	54.04N, 2.50W	23	1921–50	49.2	60.8 Jul	39.3 Jan					38.39
Langbaurgh	54.37N, 1.04W	26	1916–50	(see Durham for temperature data)							23.94
Leeds	53.48N, 1.32W	307	1916–50	(see Bradford for temperature data)							28.20
Leicester	52.38N, 1.07W	237	1921–50	49.0	61.5 Jul	38.3 Jan					26.39
Lichfield	52.41N, 1.49W	259	1916–50	(see Birmingham for temperature data)							26.9
Lincoln	53.14N, 0.32W	25	1916–50	(see Nottingham for temperature data)							23.74
Liverpool	53.23N, 3.00W	198	1921–50	49.3	60.4 Jul	40.0 Jan	87	15	82	24	35.06
*London	51.31N, 0.05W	149	1921–50	50.6	63.7 Jul	39.8 Jan	99	9	90	20	24.00
Londonderry	55.00N, 7.20W	325	1921–35[2]	48.6	57.1 Jul	42.2 Feb					46.1
Luton	51.53N, 0.25W	381	1921–50	48.9	61.7 Jul	38.4 Jan					25.10
Macclesfield	53.15N, 2.07W	500	1921–50	48.0	60.3 Jul	37.7 Jan					36.34

[1] Temperature data for Suffolk Coastal, nr Ipswich.
[2] Temperature data for Malin Head, Ireland, nr Londonderry.

EUROPE

City	Latitude and Longitude	Elev (ft)	Period of Record	Temperature (°F)							Annual Precipitation (in)
				Avg	Avg Warmest Month	Avg Coolest Month	Absolute High	Absolute Low	Avg Annual High	Avg Annual Low	
Maidstone	51.16N, 0.31E	36	1925-50[1]	49.3	62.0 Jul	38.8 Jan					26.73
Manchester	53.28N, 2.14W	125	1921-50	49.8	61.6 Jul	40.0 Jan					33.79
Medway	51.23N, 0.31E	125	1925-50[1]	49.3	62.0 Jul	38.8 Jan					26.6
Middlesbrough	54.35N, 1.14W	30	1916-50	(see Durham for temperature data)							24.33
Motherwell	55.47N, 4.00W	175	1916-50	(see Glasgow for temperature data)							35.4
Newcastle-under-Lyme	53.00N, 2.14W	587	1921-50[2]	47.5	59.8 Jul	37.5 Jan					31.57
Newcastle upon Tyne	54.58N, 1.37W	255	1916-50	(see Durham for temperature data)							27.69
Newport	51.35N, 3.00W	265	1921-50	50.3	62.8 Jul	40.6 Jan					43.56
Northampton	52.14N, 0.54W	258	1916-50	(see Coventry for temperature data)							24.65
North Bedfordshire	52.08N, 0.27W	118	1916-50	(see Cambridge for temperature data)							20.83
North Tyneside	55.03N, 1.28W	92	1921-50	48.5	59.1 Jul	39.9 Jan					26.57
Norwich	52.38N, 1.18E	110	1921-47	49.7	62.8 Jul	38.6 Jan					25.58
Nottingham	52.51N, 1.08W	192	1921-50	49.5	62.2 Jul	39.1 Jan					24.19
Nuneaton	52.31N, 1.28W	269	1916-50	(see Coventry for temperature data)							25.98
Ogwr	51.30N, 3.35W	551	1916-50	(see Cardiff for temperature data)							72.17
Oldham	53.33N, 2.07W	550	1916-50	(see Macclesfield for temperature data)							43.5
Oxford	51.45N, 1.15W	208	1921-50	50.1	62.8 Jul	39.5 Jan	95	3	87	18	25.66
Peterborough	52.35N, 0.15W	9	1916-50	(see Leicester for temperature data)							21.38
Plymouth	50.23N, 4.06W	117	1921-50	51.8	61.6 Jul	43.1 Feb	88	16	80	25	37.76
Poole	50.43N, 1.59W	18	1916-50	(see Bournemouth for temperature data)							31.34
Portsmouth	50.48N, 1.06W	7	1926-50	51.9	63.7 Jul	41.7 Feb					27.78
Preston	53.46N, 2.42W	110	1916-50	(see South Ribble for temperature data)							39.47
Reading	51.28N, 0.59W	152	1921-50	50.6	63.2 Jul	39.8 Jan					25.72
Reigate and Banstead	51.14N, 0.13W	331	1916-50	(see Guildford for temperature data)							31.26

[1] Temperature data for Tonbridge and Malling, nr Maidstone and Medway.
[2] Climatic data for East Staffordshire, nr Newcastle-under-Lyme.

EUROPE

City	Latitude and Longitude		Elev (ft)	Period of Record	Temperature (°F)							Annual Precipitation (in)
					Avg			Absolute		Avg Annual		
					Avg	Avg Warmest Month	Avg Coolest Month	High	Low	High	Low	
Renfrew	55.50N,	4.26W	106	1921–50	48.6	60.0 Jul	39.2 Jan	86	0	80	15	44.22
Rhondda	51.40N,	3.30W	1220	1916–50	(see Newport for temperature data)							93.87
Rochdale	53.38N,	2.09W	360	1916–50	(see Bolton for temperature data)							44.2
Rotherham	53.26N,	1.21W	282	1916–50	(see Sheffield for temperature data)							24.72
Rugby	52.22N,	1.15W	390	1921–47	48.6	61.3 Jul	38.1 Jan					25.51
Rushcliffe	52.55N,	1.07W	79	1916–50	(see Nottingham for temperature data)							24.3
Saint Albans	51.45N,	0.20W	272	1924–50	48.9	61.7 Jul	38.1 Jan					25.2
Saint Helens	53.28N,	2.44W	160	1916–50	(see Liverpool for temperature data)							34.3
Salford	53.30N,	2.16W	70	1916–50	(see Manchester for temperature data)							35.0
Salisbury	51.04N,	1.47W	249	1916–50	(see Southampton for temperature data)							31.02
Sandwell	52.30N,	2.00W	440	1916–50	(see Birmingham for temperature data)							28.51
Scarborough	54.17N,	0.26W	118	1926–50	49.6	60.9 Jul	39.9 Jan					25.6
Sefton	53.30N,	2.58W	35	1921–50	49.3	60.7 Jul	39.6 Jan					32.20
Sheffield	53.23N,	1.28W	428	1921–50	49.2	61.5 Jul	39.2 Jan					31.86
Slough	51.31N,	0.36W	70	1916–50	(see Reading for temperature data)							25.75
Solihull	52.25N,	1.45W	420	1916–50	(see Birmingham for temperature data)							29.7
Southampton	50.54N,	1.23W	65	1921–50	51.0	62.8 Jul	40.8 Jan	93	12	85	20	31.64
Southend-on-Sea	51.33N,	0.43E	90	1926–40, 1946–50	51.3	64.6 Aug	39.7 Jan					21.15
South Ribble	53.41N,	2.42W	125	1921–45	48.4	59.8 Jul	38.7 Jan					
South Tyneside	55.00N,	1.25W	40	1916–50	(see North Tyneside for temperature data)							26.6
Spelthorne	51.26N,	0.31W	55	1916–50	(see Guildford for temperature data)							24.21
Stafford	52.49N,	2.06W	247	1921–50[1]	47.5	59.8 Jul	37.5 Jan					29.8
Stockport	53.25N,	2.10W	135	1916–50	(see Bolton for temperature data)							31.1
Stockton-on-Tees	54.35N,	1.25W	62	1916–50	(see Durham for temperature data)							23.62
Stoke-on-Trent	53.00N,	2.11W	390	1921–50[1]	47.5	59.8 Jul	37.5 Jan					30.9

[1]Temperature data for East Staffordshire, nr Stafford and Stoke-on-Trent.

EUROPE

City	Latitude and Longitude	Elev (ft)	Period of Record	Temperature (°F) Avg	Avg Warmest Month	Avg Coolest Month	Absolute High	Low	Avg Annual High	Low	Annual Precipitation (in)
Stratford-on-Avon	52.11N, 1.42W	161	1916–50	(see Worcester for temperature data)							23.98
Sunderland	54.55N, 1.22W	216	1916–50	(see Durham for temperature data)							25.11
Swansea	51.37N, 3.56W	32	1926–50	51.5	62.2 Jul	42.1 Feb					42.16
Taff-Ely	51.36N, 3.20W	331	1916–50	(see Cardiff for temperature data)							62.24
Tameside	53.30N, 2.06W	394	1916–50	(see Manchester for temperature data)							39.2
Thamesdown	51.34N, 1.47W	476	1921–50[1]	48.4	60.4 Jul	38.4 Jan					29.0
Thurrock	51.29N, 0.20E	9	1916–50	(see London for temperature data)							20.4
Torbay	50.28N, 3.30W	27	1926–50	51.9	62.2 Jul	43.2 Feb					34.98
Torfaen	51.42N, 3.02W	541	1916–50	(see Newport for temperature data)							51.65
Trafford	53.27N, 2.19W	59	1916–50	(see Manchester for temperature data)							33.58
Tunbridge Wells	51.08N, 0.17E	351	1926–50	49.3	62.0 Jul	38.5 Jan					30.3
Vale of Glamorgan	51.29N, 3.09W	98	1916–50	(see Cardiff for temperature data)							36.50
Wakefield	53.42N, 1.29W	115	1921–50	49.2	61.8 Jul	38.7 Jan					26.4
Walsall	52.35N, 1.58W	455	1916–50	(see Birmingham for temperature data)							28.3
Warrington	53.23N, 2.36W	43	1916–50	(see Liverpool for temperature data)							33.7
Wigan	53.32N, 2.37W	128	1916–50	(see Bolton for temperature data)							39.8
Winchester	51.01N, 1.19W	120	1916–50	(see Southampton for temperature data)							31.97
Windsor and Maidenhead	51.31N, 0.42W	81	1916–50	(see Reading for temperature data)							24.76
Wirral	53.24N, 3.02W	198	1925–35	50.0	62.2 Jul	40.7 Feb					29.01
Wolverhampton	52.36N, 2.08W	430	1916–50	(see Birmingham for temperature data)							28.45
Worcester	52.12N, 2.12W	94	1926–50	49.5	62.1 Jul	39.3 Jan					25.31
Worthing	50.48N, 0.23W	25	1926–50	51.0	62.6 Aug	40.9 Jan					27.48
Wrekin	52.40N, 2.28W	341	1931–50[2]	49.4	61.5 Jul	38.4 Jan					26.81

[1] Temperature data for Kennet, nr Thamesdown.
[2] Temperature data for Shrewsbury and Atcham, nr Wrekin.

EUROPE

City	Latitude and Longitude		Elev (ft)	Period of Record	Avg	Avg Warmest Month	Avg Coolest Month	Absolute High	Absolute Low	Avg Annual High	Avg Annual Low	Annual Precipitation (in)
Wrexham Maelor	53.03N,	3.00W	269	1921-50[1]	50.0	61.5 Jul	40.5 Jan					30.28
Wyre	53.55N,	3.00W	30	1916-50	(see Blackpool for temperature data)							35.0
York	53.57N,	1.05W	57	1921-50	49.3	61.9 Jul	38.9 Jan	90	7	85	19	24.70
USSR												
Archangel	64.34N,	40.32E	10	1881-1960	33.4	60.1 Jul	9.5 Jan	93	-49	86	-29	19.45
Armavir	45.00N,	41.08E	518	1932-60	49.8	72.9 Jul	25.9 Jan	108	-29	99	-8	
Astrakhan	46.21N,	48.03E	-72	1881-1960	48.9	77.5 Jul	19.8 Jan	104	-29	97	-13	6.89
Baku	40.23N,	49.51E	7	1891-1917, 1921-60	57.9	78.3 Jul	38.8 Jan	104	9	97	23	9.37
Belgorod	50.36N,	36.34E	604	1925-41, 1943-60	43.3	68.4 Jul	18.3 Jan	106	-35	93	-17	18.23
Berezniki	59.24N,	56.46E	410	1891-1954	33.3	62.6 Jul	3.6 Jan		-53			18.86
Bobruysk	53.09N,	29.14E		(see Mogilev)								
Brest	52.06N,	23.42E	472	1951-60	45.1	66.0 Jul	24.1 Feb	100	-44	90	-22	21.57
Bryansk	53.15N,	34.22E	528	1936-41, 1947-60	40.8	65.1 Jul	16.7 Jan	100	-47	91	-27	21.93
Cheboksary	56.09N,	47.15E	607	1927-60	37.2	65.5 Jul	8.6 Jan	97	-56	86	-27	
Cherepovets	59.08N,	37.54E	413	1937-60	36.7	63.1 Jul	11.7 Jan					
Cherkassy	49.26N,	32.04E	308									
Chernigov	51.30N,	31.18E	443	1881-1960	43.7			102	-29			21.22
Chernovtsy	48.18N,	25.56E	1148	1881-1960	46.0			100	-26			24.57
Dneprodzerzhinsk	48.30N,	34.37E		(see Dnepropetrovsk)								
Dnepropetrovsk	48.27N,	34.59E	322	1891-1911	48.5	71 Jul	20.5 Jan	101	-25	95	-11	19.4
Donetsk	48.00N,	37.48E	659	(see Makeyevka)								
Dzerzhinsk	56.15N,	43.24E		(see Gorkiy)								
Engels	51.30N,	46.07E		(see Saratov)								
Gomel	52.25N,	31.00E	453	1928-40								24.17

[1] Temperature data for Alyn and Deeside, nr Wrexham Maelor.

EUROPE

City	Latitude and Longitude	Elev (ft)	Period of Record	Avg	Avg Warmest Month	Avg Coolest Month	Absolute High	Absolute Low	Avg Annual High	Avg Annual Low	Annual Precipitation (in)
Gorkiy	56.20N, 44.00E	532	1922–60	37.6	64.6 Jul	10.4 Jan	99	–42	90	–24	23.15
Gorlovka	48.18N, 38.03E		(see Makeyevka)								
Grodno	53.41N, 23.50E	423	1894–1904, 1914	44.1	64.2 Jul	23.9 Jan					
Groznyy	43.20N, 45.42E	404	1938–60	50.2	74.8 Jul	25.5 Jan	106	–27	100	–11	19.25
Ivanovo	57.00N, 40.59E	420	1891–1918, 1922–48	37.9	65.3 Jul	11.1 Jan	100	–51	90	–29	23.35
Izhevsk	56.51N, 53.14E	479	1933–60	35.8	65.7 Jul	6.4 Jan	99	–51	91	–33	
Kalinin	56.52N, 35.55E	446	? yr bef 1948	39.0	64.2 Jul	14.0 Jan					24.29
Kaliningrad	54.43N, 20.30E	20	1921–50	45.0	63.9 Jul	27.3 Feb	97	–24	90	–3	29.25
Kaluga	54.31N, 36.16E	663	20 yr bef 1912	38.5	64 Jul	12 Jan	94	–36	89	–23	22.9
Kaunas	54.54N, 23.54E	246	1931–40, 1942–60	42.8	64.6 Jul	22.8 Jan	96	–23	89	–13	24.61
Kazan	55.45N, 49.08E	276	1881–1960	38.5	69.0 Jul	9.0 Jan	102	–47	93	–27	17.20
Kerch	45.21N, 36.28E	13	1881–1915	51.6	74.1 Jul	29.7 Jan	93	–13	89	6	14.76
Kharkov	50.00N, 36.15E	410	1931–37, 1941, 1944–60	44.8	70.0 Jul	19.2 Jan	98	–31	92	–19	20.43
Kherson	46.38N, 32.36E	59	? yr bef 1948	49.9	73.9 Jul	25.9 Jan	102	–26	96	–6	13.94
Khmelnitskiy	49.25N, 27.00E		(see Vinnitsa)								
Kiev	50.26N, 30.31E	440	1931–60	45.3	68.7 Jul	21.0 Jan	102	–26	93	–12	24.21
Kirov	58.33N, 49.42E	545	1881–1960	34.7	64.0 Jul	6.4 Jan	99	–49	90	–31	19.13
Kirovabad	40.41N, 46.22E	1024	47 yr bef 1961	55.8	77.7 Jul	34.0 Jan	104	0	99	14	9.76
Kirovograd	48.30N, 32.18E	427	1881–1915	46.0	69.6 Jul	21.6 Jan	104	–31			18.66
Kishinev	47.00N, 28.50E	295	1951–60	49.3	72.3 Jul	27.3 Feb					18.54
Klaypeda	55.43N, 21.07E	26	1851–1900	44.2	63.3 Jul	26.8 Jan					
Kostroma	57.46N, 40.55E	456	1925–60	36.9	63.7 Jul	10.8 Jan	99	–51	90	–27	21.97
Kramatorsk	48.43N, 37.32E										
Krasnodar	45.02N, 39.00E	95	1896–1942, 1944–60	51.4	73.8 Jul	28.8 Jan	108	–33	99	–9	25.12
Kremenchug	49.04N, 33.25E										
Krivoy Rog	47.55N, 33.21E	325	1883, 1885–96	49.1	75.6 Jul	21.6 Jan	103	–20	99	–7	16.38

EUROPE

City	Latitude and Longitude	Elev (ft)	Period of Record	Avg	Avg Warmest Month	Avg Coolest Month	Absolute High	Absolute Low	Avg Annual High	Avg Annual Low	Annual Precipitation (in)
Kursk	51.42N, 36.12E	738	67 yr bef 1961	41.7	66.7 Jul	16.5 Jan	99	−36	90	−17	23.50
Kutaisi	42.15N, 42.40E	374									
Kuybyshev	53.12N, 50.09E	446	1935–60	38.8	69.3 Jul	7.2 Jan	102	−45	95	−26	15.33
Leninakan	40.48N, 43.50E	5105	1926–60	42.4	66.7 Aug	13.1 Jan	97	−42	91	−20	18.31
Leningrad	59.55N, 30.15E	7	1881–1960	39.7	64.0 Jul	17.8 Feb	91	−33	86	−15	22.36
Lipetsk	52.37N, 39.35E	541	58 yr bef 1961	41.2	68.4 Jul	13.5 Jan	102	−36	93	−22	19.61
Lvov	49.50N, 24.00E	984	1911–30, 1941–50	46.0	65.5 Jul	25.0 Jan	97	−29	92	−4	23.86
Lyubertsy	55.41N, 37.53E		(see Moscow)								
Makeyevka	48.02N, 37.58E	623	? yr bef 1948	45.0	71.6 Jul	18.5 Jan					17.28
Makhachkala	42.58N, 47.30E	−69	1882–1960	53.2	76.5 Jul	31.3 Jan	99	−15	95	5	16.93
Melitopol	46.50N, 35.22E	46	27 yr bef 1916	49.3	74.7 Jul	24.3 Jan					14.92
Minsk	53.54N, 27.34E	732	1931–41, 1944–60	41.5	64.6 Jul	18.5 Jan	92	−27	86	−19	23.86
Mogilev	53.54N, 30.21E	499	1881–1930	41.7	64.4 Jul	19.2 Jan	96	−30			25.12
*Moscow	55.45N, 37.35E	548	1931–60	39.9	66.2 Jul	14.2 Jan	100	−43	89	−19	22.64
Murmansk	68.58N, 33.05E	187	1931–60	32.4	56.1 Jul	11.5 Feb					14.80
Naberezhnyye Chelny	55.42N, 52.19E	249	1948–55	36.3	66.7 Jul	6.4 Jan					
Nalchik	43.29N, 43.37E	1447	1944–60	47.5	70.0 Jul	24.6 Jan	100	−24			23.86
Nikolayev	46.58N, 32.00E	98	? yr bef 1960	49.5	73.6 Jul	24.8 Jan	102	−22	97	−3	15.28
Novgorod	58.31N, 31.17E	79	70 yr bef 1961	39.0	63.1 Jul	16.5 Jan	93	−49	86	−24	22.44
Novocherkassk	47.25N, 40.06E	341	1922–36	47.5	73.8 Jul	21.0 Jan		−26		−8	
Novorossiysk	44.43N, 37.47E	121	1881–1942, 1944–60	54.9	74.7 Aug	36.7 Jan	102	−11	95	5	26.06
Odessa	46.28N, 30.44E	138	1921–50	49.1	73.2 Jul	26.2 Jan	99	−18	91	−1	14.72
Ordzhonikidze	43.00N, 44.40E	2192	1934–35, 1938–60	46.2	67.5 Jul	23.0 Jan	102	−29	91	−9	32.95
Orel	52.55N, 36.05E	666	1935–41, 1947–60	40.3	65.8 Jul	15.4 Jan	100	−38	91	−22	21.46
Orenburg	51.45N, 55.06E	358	1886–1960	39.0	71.4 Jul	5.4 Jan	108	−44	99	−31	15.12
Orsk	51.12N, 58.34E	673	1925–60	37.4	70.3 Jul	2.5 Jan	108	−47	99	−35	
Penza	53.13N, 45.00E	764	1887–1919, 1923–60	39.0	67.6 Jul	10.2 Jan	100	−45	93	−24	19.21

EUROPE

City	Latitude and Longitude	Elev (ft)	Period of Record	Avg	Avg Warmest Month	Avg Coolest Month	Absolute High	Absolute Low	Avg Annual High	Avg Annual Low	Annual Precipitation (in)
Perm	58.00N, 56.15E	535	1921-50	35.2	64.6 Jul	5.7 Jan	98	-49	88	-37	24.06
Petrozavodsk	61.49N, 34.20E	361	1949-63	36.0	60.6 Jul	12.9 Jan	93	-40	82	-24	20.94
Podolsk	55.26N, 37.33E		(see Moscow)								
Poltava	49.35N, 34.34E	525	1886-1915	44.4	69.1 Jul	18.9 Jan	100	-35			19.09
Pskov	57.50N, 28.20E	138	60 yr bef 1961	40.3	63.7 Jul	18.5 Jan	97	-42	86	-20	
Riga	56.57N, 24.06E	30	1881-1960	43.2	64.4 Jul	23.9 Jan	95	-24	86	-6	24.96
Rostov-na-Donu	47.14N, 39.42E	217	1886-1960	47.7	73.2 Jul	21.7 Jan	104	-27	97	-11	18.66
Rovno	50.37N, 26.15E	768	1951-60	44.6	66.0 Jul	23.0 Feb					22.17
Ryazan	54.38N, 39.44E	512	1892-1915	39.9	67.5 Jul	12.4 Jan					19.49
Rybinsk	58.03N, 38.50E	341									
Saransk	54.11N, 45.11E	230	1927-42, 1944-60	38.7	66.7 Jul	10.2 Jan	100	-47	91	-38	
Saratov	51.34N, 46.02E	492	1936-55	41.5	71.8 Jul	10.6 Jan	106	-42	97	-22	15.16
Sevastopol	44.36N, 33.32E	23	? yr bef 1960	54.0	73.9 Jul	35.6 Jan	97	-4	93	6	14.21
Severodvinsk	64.34N, 39.50E	3	1945-60	33.4	58.6 Jul	10.8 Jan	91	-47	84	-29	
Shakhty	47.42N, 40.13E	384	1937-60	46.2	72.5 Jul	19.6 Jan	104	-29	97	-13	
Simferopol	44.57N, 34.06E	669	1931-60	50.2	70.9 Jul	31.3 Jan	104	-20			20.79
Smolensk	54.47N, 32.03E	764	1951-60	39.2	63.1 Jul	15.3 Feb	89	-28	86	-18	23.62
Sochi	43.35N, 39.45E	39	1916-51	56.1	73.0 Aug	40.8 Jan	102	5	91	19	53.39
Stavropol	45.03N, 41.58E	1552	1940-60	48.4	71.4 Jul	25.3 Jan	104	-33	95	-4	26.10
Sterlitamak	53.37N, 55.58E	423	? yr bef 1950	36.7	67.8 Jul	4.3 Jan					
Sumgait	40.36N, 49.38E	-66	1935-36, 1938-60	56.5	76.8 Aug	37.6 Jan	108	5	100	18	7.28
Sumy	50.54N, 34.48E	449	1881-1960	42.8			100	-33			19.96
Syktyvkar	61.40N, 50.48E	427	1888-1960	32.7	61.9 Jul	4.8 Jan	95	-60	88	-40	16.93
Syzran	53.11N, 48.27E	187	54 yr bef 1961	39.9	70.3 Jul	8.6 Jan	106	-47	97	-27	13.62
Taganrog	47.12N, 38.56E	46	1945-60	48.4	74.3 Jul	22.3 Jan	100	-27	95	-11	16.42
Tallin	59.25N, 24.45E	39	1948-60	41.0	61.9 Jul	22.1 Feb	91	-26	85	-8	22.56
Tambov	52.43N, 41.27E	459	1927-60	40.6	68.4 Jul	12.6 Jan	104	-38	93	-24	19.09

EUROPE

City	Latitude and Longitude	Elev (ft)	Period of Record	Temperature (°F) Avg	Avg Warmest Month	Avg Coolest Month	Absolute High	Absolute Low	Avg Annual High	Avg Annual Low	Annual Precipitation (in)
Tartu	58.23N, 26.43E	217	1881–1960	40.6	63.1 Jul	20.1 Feb	95	–31	86	–13	22.0
Tbilisi	41.42N, 44.45E	1322	1931–60	55.2	76.3 Jul	34.3 Jan	101	0	93	12	20.00
Tolyatti	53.31N, 49.20E	151	1938–42	39.9	69.6 Jul	9.5 Jan	104	–49	97	–26	
Tula	54.12N, 37.37E	541	? yr bef 1948	39.7	66.0 Jul	13.5 Jan					22.01
Ufa	54.44N, 55.56E	568	? yr bef 1960	36.7	66.9 Jul	5.7 Jan	99	–42	90	–31	22.83
Ulyanovsk	54.20N, 48.24E	558	1937–54	37.8	67.3 Jul	7.2 Jan	104	–54	93	–31	16.57
Vilnyus	54.41N, 25.19E	486	1921–50	43.3	64.6 Jul	21.6 Jan					25.94
Vinnitsa	49.14N, 28.29E	791	1881–1960	44.1			100	–33			21.42
Vitebsk	55.12N, 30.11E	545	? yr bef 1948[1]	39.8	62.2 Jul	17.4 Jan					26.85
Vladimir	56.10N, 40.25E	551	? yr bef 1948	37.9	64.9 Jul	10.9 Jan					20.08
Volgograd	48.45N, 44.25E	75	44 yr bef 1943	45.7	75.7 Jul	14.9 Jan	109	–33	100	–17	12.52
Vologda	59.13N, 39.54E	430	1884–1918, 1920–60	36.3	63.0 Jul	11.1 Jan	95	–54	88	–31	20.16
Volzhskiy	48.49N, 44.44E		(see Volgograd)								
Voronezh	51.38N, 39.12E	482	36 yr bef 1961	41.7	67.8 Jul	15.3 Jan	106	–40	93	–20	20.43
Voroshilovgrad	48.34N, 39.20E	194	? yr bef 1948	45.7	72.0 Jul	19.4 Jan					18.46
Yalta	44.30N, 34.10E	13	? yr bef 1959	55.6	75.6 Jul	38.7 Jan	99	5			21.46
Yaroslavl	57.37N, 39.52E	322	(see Kostroma)								
Yerevan	40.11N, 44.30E	2986	1935–60	52.9	77.2 Jul	24.8 Jan	106	–24	100	0	12.32
Yoshkar-Ola	56.40N, 47.55E	328	1940–60	36.1	64.8 Jul	7.3 Jan	100	–53	91	–36	
Zaporozhye	47.49N, 35.11E	161	1951–60	48.6	73.6 Jul	24.6 Feb					17.76
Zhdanov	47.06N, 37.33E	10	? yr bef 1963	47.1	72.7 Jul	22.3 Jan	102	–24	92	–7	16.73
Zhitomir	50.15N, 28.40E	597	1881–1960	44.2			100	–31			22.44
YUGOSLAVIA											
Banja Luka	44.46N, 17.10E	535	1955–64	50.7	68.9 Jul	31.6 Jan					40.20
*Belgrade	44.50N, 20.30E	433	1931–60	53.2	72.7 Jul	31.6 Jan	107	–14	99	4	27.60

[1]Climatic data for Novoye Korolevo, nr Vitebsk.

EUROPE

City	Latitude and Longitude	Elev (ft)	Period of Record	Temperature (°F)							Annual Precipitation (in)
				Avg	Avg Warmest Month	Avg Coolest Month	Absolute High	Absolute Low	Avg Annual High	Avg Annual Low	
Dubrovnik	42.39N, 18.07E	161	1955–64	61.5	76.5 Aug	48.2 Jan					54.76
Ljubljana	46.02N, 14.30E	981	1955–64	49.5	69.1 Jul	29.1 Jan					54.61
Maribor	46.33N, 15.39E	902	1955–64	48.6	66.6 Jul	28.2 Jan					41.10
Nis	43.19N, 21.54E	663	1955–64	52.9	71.4 Aug	32.9 Jan	108	−9	100	4	21.85
Novi Sad	45.15N, 19.50E	433	1955–64	52.5	71.2 Jul	30.4 Jan					24.41
Osijek	45.33N, 18.42E	292	1955–64	51.3	70.2 Jul	28.6 Jan					27.80
Rijeka	45.21N, 14.24E	341	1955–64	56.8	73.0 Jul	41.5 Jan					58.31
Sarajevo	43.50N, 18.25E	2067	1931–60	49.6	67.5 Aug	29.5 Jan	104	−14	96	4	35.00
Skopje	42.00N, 21.29E	787	1931–60	54.3	74.8 Jul	34.0 Jan	105	−11	101	3	21.50
Split	43.31N, 16.26E	400	1931–60	61.0	78.1 Jul	46.0 Jan	100	17	96	24	32.13
Subotica	46.06N, 19.40E	361	1955–64[1]	51.6	71.2 Jul	28.6 Jan					22.36
Zagreb	45.48N, 16.00E	515	1931–60	52.9	71.6 Jul	32.4 Jan	94	−4	91	10	34.02

NORTH AMERICA

City	Latitude and Longitude	Elev (ft)	Period of Record	Temperature (°F)							Annual Precipitation (in)
				Avg	Avg Warmest Month	Avg Coolest Month	Absolute High	Absolute Low	Avg Annual High	Avg Annual Low	
BAHAMAS *Nassau	25.05N, 77.21W	18	1921–50	76.7	82.3 Aug	70.8 Feb	94	41	92	53	42.65
BARBADOS *Bridgetown	13.06N, 59.37W	181	1912–48	78.5	80.5 Jun	76 Feb	95	61	90	64	50.2

[1] Temperature data for Palic, nr Subotica.

NORTH AMERICA

City	Latitude and Longitude	Elev (ft)	Period of Record	Temperature (°F) Avg	Avg Warmest Month	Avg Coolest Month	Absolute High	Absolute Low	Avg Annual High	Avg Annual Low	Annual Precipitation (in)
BELIZE											
Belize	17.30N, 88.12W	16	1941–50, 1953–60	79.3	82.5 Aug	74.7 Jan	97	49	92	55	69.46
BERMUDA											
*Hamilton	32.17N, 64.46W	158	1921–50	70.7	80.5 Aug	62.0 Feb	99	40	91	46	57.20
CANADA											
Brampton	43.41N, 79.46W	722	15–19 yr bef 1971	45.1	68.7 Jul	20.9 Jan	99	−34			31.43
Brantford	43.08N, 80.16W	706	20–24 yr bef 1971	46.5	70.1 Jul	22.5 Jan	105	−30			30.31
Burlington	43.19N, 79.47W	281	15–19 yr bef 1971	46.5	69.9 Jul	22.0 Jan	101	−16			29.89
Calgary	51.03N, 114.05W	3428	1941–70	38.2	61.7 Jul	12.3 Jan	97	−49	91	−33	17.21
Charlottetown	46.14N, 63.08W	8	1941–70	42.9	66.3 Jul	19.8 Feb	98	−23			41.69
Chicoutimi	48.26N, 71.06W	19	1941–70	38.0	65.8 Jul	5.6 Jan	104	−49			35.45
Edmonton	53.33N, 113.28W	2188	1941–70	37.1	63.5 Jul	5.5 Jan	99	−57	89	−41	17.58
Halifax	44.39N, 63.36W	57	1941–70	45.6	65.4 Aug	26.1 Feb	99	−21	89	−8	51.92
Hamilton	43.15N, 79.51W	306	14–19 yr bef 1971	48.3	72.4 Jul	24.9 Jan	106	−23			31.22
Jonquiere	48.25N, 71.15W	484	(see Chicoutimi)								
Kingston	44.14N, 76.29W	264	25–29 yr bef 1971	45.0	70.0 Jul	18.0 Jan	97	−25			35.43
Kitchener	43.27N, 80.29W	1100	1941–70	45.2	69.1 Jul	20.2 Jan	101	−29			34.90
Laval	45.33N, 73.43W	130	(see Montreal)								
London	42.59N, 81.14W	809	1941–70	45.5	68.9 Jul	21.2 Jan	106	−27	95	−14	36.40
Longueuil	45.32N, 73.31W	55	(see Montreal)								
Mississauga	43.34N, 79.37W	391	(see Toronto)								
Montreal	45.30N, 73.35W	104	1941–70	45.0	70.9 Jul	16.0 Jan	97	−29	90	−19	39.33
Montreal-Nord	45.36N, 73.38W	69	(see Montreal)								
Niagara Falls	43.06N, 79.04W	584	20–24 yr bef 1971	48.9	72.1 Jul	25.4 Jan	101	−13			34.43
Oshawa	43.54N, 78.51W	325	(see Toronto)								

NORTH AMERICA

City	Latitude and Longitude	Elev (ft)	Period of Record	Temperature (°F)							Annual Precipitation (in)
				Avg	Avg Warmest Month	Avg Coolest Month	Absolute High	Absolute Low	Avg Annual High	Avg Annual Low	
*Ottawa	45.25N, 75.42W	284	1941–70	42.4	68.9 Jul	12.3 Jan	102	−38	93	−24	33.08
Quebec	46.49N, 71.13W	130	15–19 yr bef 1971	41.5	68.2 Jul	12.9 Jan	97	−34	89	−23	45.80
Regina	50.25N, 104.39W	1885	1941–70	35.7	66.0 Jul	0.9 Jan	111	−58	96	−42	15.66
Saint Catharines	43.10N, 79.15W	347	20–24 yr bef 1971	48.7	71.6 Jul	25.8 Jan	104	−12			30.95
Saint John	45.16N, 66.03W	42	25–29 yr bef 1971	42.6	62.3 Aug	20.3 Jan	94	−28	83	−13	51.40
Saint John's	47.33N, 52.43W	200	25–29 yr bef 1971	40.8	59.8 Aug	24.4 Feb	93	−21	83	−5	59.50
Saskatoon	52.07N, 106.38W	1574	1941–70	34.9	65.8 Jul	−1.7 Jan	104	−55	95	−41	13.88
Sault Sainte Marie	46.30N, 84.20W	634	10–14 yr bef 1971	39.3	62.7 Jul	23.6 Jan	97	−38			38.90
Sherbrooke	45.24N, 71.54W	521	1941–70	42.6	68.1 Jul	14.7 Jan	98	−42			38.29
Sudbury	46.30N, 81.00W	857	15–19 yr bef 1971	40.7	67.5 Jul	10.1 Jan	99	−47			29.83
Sydney	46.09N, 60.11W	42	1941–70	42.8	64.3 Jul	22.1 Feb	98	−25	87	−11	52.78
Thunder Bay	48.25N, 89.14W	615	25–29 yr bef 1971	36.3	63.5 Jul	5.4 Jan	104	−42	89	−30	29.07
Toronto	43.39N, 79.23W	273	1941–70	48.0	71.3 Jul	24.1 Jan	105	−27	93	−11	31.10
Trois-Rivieres	46.21N, 72.34W	51	1941–70	40.8	67.8 Jul	10.4 Jan	100	−38			39.28
Vancouver	49.17N, 123.07W	38	1941–70	49.7	63.4 Jul	36.3 Jan	92	0	86	13	42.05
Victoria	48.25N, 123.22W	55	1941–70	49.3	61.5 Jul	37.3 Jan	97	−2	86	20	33.72
Windsor	42.18N, 83.01W	602	1941–70	48.6	72.1 Jul	24.3 Jan	101	−27			32.91
Winnipeg	49.53N, 97.09W	757	1941–70	36.2	67.5 Jul	−0.9 Jan	108	−54	94	−38	21.06
COSTA RICA											
*San Jose	9.56N, 84.05W	3845	1941–60	68.5	70.5 May	66.2 Jan	92	49	88	51	76.26
CUBA											
Camaguey	21.23N, 77.55W	397	1952–60	77.0	81.3 Aug	70.9 Jan			95	49	52.83
Cienfuegos	22.09N, 80.27W	98	26 yr bef 1920	76.3	80.4 Aug	70.0 Jan	95	45	94	50	38.31
Guantanamo	20.08N, 75.12W	75	13 yr bef 1925	77.2	80.2 Jul	73.6 Jan					23.11
*Havana	23.08N, 82.22W	161	1921–50	76.8	81.3 Aug	72.0 Jan	96	50	94	54	44.76

NORTH AMERICA

City	Latitude and Longitude	Elev (ft)	Period of Record	Temperature (°F)							Annual Precipitation (in)
				Avg	Avg Warmest Month	Avg Coolest Month	Absolute High	Absolute Low	Avg Annual High	Avg Annual Low	
Holguin	20.53N, 76.15W		1952-60[1]	80.1	84.4 Aug	75.2 Jan			93	61	43.94
Matanzas	23.03N, 81.35W		16 yr bef 1925	75.7	81.3 Aug	69.3 Feb			93	53	51.51
Santa Clara	22.24N, 79.58W	377	1900-19[2]	73.2	77.7 Aug	67.5 Jan					55.04
Santiago de Cuba	20.01N, 75.49W	115	16 yr bef 1925	78.8	82.2 Aug	75.0 Jan	99	54			43.79
DOMINICAN REPUBLIC											
Santiago de los Caballeros	19.27N, 70.42W	728	1951-60[3]	78.8	82.2 Sep	73.4 Jan					37.68
*Santo Domingo	18.28N, 69.54W	46	1921-49	78	80.5 Aug	75 Jan	98	59	95	61	55.8
EL SALVADOR											
San Miguel	13.28N, 88.12W	345	11 yr bef 1972	80.1	83.7 Apr	78.3 Dec	113	53	108	60	68.00
*San Salvador	13.42N, 89.12W	2290	1931-60	73.0	75.6 Apr	71.6 Nov	105	45	99	51	70.55
Santa Ana	13.59N, 89.34W	2116	13 yr bef 1972	72.9	75.6 Apr	70.7 Dec	99	46			79.61
GUATEMALA											
*Guatemala	14.38N, 90.31W	4928	1931-60	64.6	67.5 May	61.5 Jan	90	41	87	45	50.43
Mixco	14.38N, 90.36W	5551	(see Guatemala)								
HAITI											
*Port-au-Prince	18.32N, 72.20W	135	1921-50	79.9	82.9 Jul	77.2 Jan	101	60	98	63	51.50
HONDURAS											
San Pedro Sula	15.27N, 88.02W	233	1952-60	79.7	84.0 May	74.1 Dec					52.44
*Tegucigalpa	14.06N, 87.13W	3304	1951-60	70.3	73.4 Apr	66.4 Jan	93	44			37.52

[1] Climatic data for Gibara, nr Holguin.
[2] Climatic data for Camajuani, nr Santa Clara.
[3] Temperature data for San Francisco de Macoris, nr Santiago de los Caballeros.

NORTH AMERICA

City	Latitude and Longitude	Elev (ft)	Period of Record	Temperature (°F)							Annual Precipitation (in)
				Avg	Avg Warmest Month	Avg Coolest Month	Absolute High	Absolute Low	Avg Annual High	Avg Annual Low	
JAMAICA											
*Kingston	18.00N, 76.48W	110	33 yr bef 1944	79.5	81.5 Jul	76.5 Jan	97	57	94	62	31.5
MARTINIQUE											
*Fort-de-France	14.36N, 61.05W	13	1932–60	77.5	79.3 Sep	75.1 Feb	96	56	91	62	73.19
MEXICO											
Acapulco	16.50N, 99.55W	13	1921–60	81.7	83.7 Jul	79.0 Jan	96	60			55.16
Aguascalientes	21.53N, 102.18W	6195	17 yr bef 1940	64.0	72.1 May	55.4 Jan	98	24			26.22
Celaya	20.31N, 100.49W	5932	1921–35	68.7	77.7 May	58.6 Jan	107	24			26.06
Chihuahua	28.38N, 106.05W	4692	1941–60	65.1	81.5 Jun	50.0 Dec	101	11			12.09
Ciudad Juarez	31.44N, 106.29W	3734	1921–35	62.1	80.6 Jul	41.7 Jan	106	3			7.36
Ciudad Madero	22.16N, 97.50W	377	(see Tampico)								
Ciudad Obregon	27.29N, 109.56W	131	1921–35	79.7	92.7 Aug	64.8 Jan	118	30			8.31
Ciudad Victoria	23.44N, 99.08W	1053	1921–35	72.9	82.2 Aug	59.9 Dec					35.55
Coatzacoalcos	18.09N, 94.25W	7	1921–35	77.0	81.1 May	71.2 Jan	106	50			113.54
Cordoba	18.53N, 96.56W	3032	1921–35	68.4	73.6 May	61.9 Jan	102	36			89.29
Cuernavaca	18.55N, 99.15W	5059	1921–35	68.5	73.8 May	65.1 Jan	91	44			40.91
Culiacan	24.48N, 107.24W	276	24 yr bef 1940	76.5	84.7 Jun	67.3 Jan	106	38			21.30
Durango	24.02N, 104.40W	6198	16 yr bef 1940	63.3	72.9 Jun	53.6 Dec	95	23			17.83
Guadalajara	20.40N, 103.20W	5141	1951–60	66.7	73.0 May	59.4 Jan	97	25			34.84
Guanajuato	21.01N, 101.15W	6726	1921–35, 1951–60	63.9	70.0 May	58.5 Dec	93	32			27.01
Hermosillo	29.04N, 110.58W	778	1921–35	75.6	89.6 Jul	60.1 Dec	113	36			12.60
Irapuato	20.41N, 101.28W	5656	1921–35	66.7	73.6 May	58.3 Jan	101	24			29.13
Jalapa	19.32N, 96.55W	4682	1921–35	64.0	68.4 May	58.1 Jan	94	36			62.56
Leon	21.07N, 101.40W	6185	1931–60	66.7	73.6 May	59.4 Jan	98	27			24.61
Matamoros	25.53N, 97.30W	26	1921–35	73.9	84.7 Aug	61.2 Jan	103	24			29.41

NORTH AMERICA

City	Latitude and Longitude	Elev (ft)	Period of Record	Temperature (°F)			Absolute		Avg Annual		Annual Precipitation (in)
				Avg	Avg Warmest Month	Avg Coolest Month	High	Low	High	Low	
Mazatlan	23.13N, 106.25W	256	1921-60	75.4	82.4 Jul	67.5 Feb	93	52	90	55	31.69
Merida	20.58N, 89.37W	30	1921-60	78.4	82.2 May	73.4 Jan	106	51	99	56	36.61
Mexicali	32.40N, 115.29W	0	1921-35	72.1	91.2 Jul	53.2 Jan	118	25			2.99
*Mexico	19.24N, 99.09W	7546	1931-60	59.4	63.3 May	54.1 Jan	93	24	85	33	27.99
Minatitlan	17.59N, 94.31W	210	1921-35	79.2	84.0 May	73.8 Jan	107	50			113.23
Monclova	26.54N, 101.25W	1923	1921-35	70.0	82.6 Aug	54.7 Jan	108	18			15.16
Monterrey	25.40N, 100.19W	1765	1931-60	72.1	82.4 Jul	59.7 Jan	107	22	102	31	25.79
Morelia	19.42N, 101.07W	6368	1921-60	63.5	69.3 May	56.3 Jan	88	35			30.04
Netzahualcoyotl	19.36N, 99.00W	7474	1921-44	59.9	64.6 Jun	53.4 Jan	93	17			22.32
Nuevo Laredo	27.30N, 99.31W	561	1921-35	76.3	90.5 Jul	57.4 Jan	113	19			16.02
Oaxaca	17.03N, 96.43W	5086	1941-50	70.3	74.5 May	65.8 Dec	100	36			25.24
Orizaba	18.51N, 97.06W	4213	1921-35	65.1	69.8 May	59.2 Jan	99	35			83.27
Pachuca de Soto	20.07N, 98.44W	7960	1921-35	57.6	62.1 May	53.2 Jan					16.50
Poza Rica	20.33N, 97.27W	197	1921-35[1]	76.3	82.9 Jun	67.1 Jan	105	39			54.41
Puebla	19.03N, 98.12W	7094	1931-60	62.1	66.0 Apr	56.5 Jan	87	29			32.99
Queretaro	20.36N, 100.23W	5528	1921-35	64.0	68.5 Jun	57.0 Jan	97	23			20.39
Reynosa	26.07N, 98.18W	125	(see Matamoros)								
Saltillo	25.25N, 101.00W	5246	33 yr bef 1961	63.9	72.4 Jun	53.4 Jan	100	15			15.41
San Luis Potosi	22.09N, 100.59W	6158	1921-35	63.7	70.7 May	55.2 Jan	99	27	95	32	14.21
Tampico	22.13N, 97.51W	39	1921-60	75.6	82.9 Aug	66.0 Jan	103	27			42.64
Taxco	18.33N, 99.36W	5840	1921-35	70.3	76.5 May	66.6 Jan	98	46			57.28
Tepic	21.30N, 104.54W	3002	1921-35	69.6	74.7 Jun	63.0 Jan	102	35			47.09
Tijuana	32.32N, 117.01W	95	1921-35	61.7	70.5 Aug	54.7 Jan	101	26			12.40
Toluca	19.17N, 99.40W	8793	1921-35	51.3	58.8 May	49.8 Jan	80	27			31.14
Torreon	25.33N, 103.26W	3708	1951-60	72.5	83.8 Jun	58.3 Dec					6.50

[1] Climatic data for Tuxpan, nr Poza Rica.

NORTH AMERICA

City	Latitude and Longitude	Elev (ft)	Period of Record	Temperature (°F)			Absolute		Avg Annual		Annual Precipitation (in)
				Avg	Avg Warmest Month	Avg Coolest Month	High	Low	High	Low	
Uruapan del Progreso	19.25N, 101.58W	5361	1921–35	67.1	72.3 May	60.8 Jan	98	41			66.26
Veracruz	19.12N, 96.08W	10	1921–60	77.2	82.0 Aug	70.0 Jan	96	49	94	55	65.83
Villahermosa	17.59N, 92.55W	33	1921–35	78.8	83.1 Aug	72.0 Jan	106	54			75.51
NETHERLANDS ANTILLES											
*Willemstad	12.06N, 68.57W	75	1937–46	81.9	84.4 Sep	79.3 Jan	96	63	94	69	17.87
NICARAGUA											
*Managua	12.09N, 86.17W	184	1953–60	80.4	82.9 May	78.3 Jan					47.52
PANAMA											
Colon	9.22N, 79.54W	36	1908–40	80.8	81.9 May	79.9 Dec					130.75
*Panama	8.57N, 79.32W	118	1951–60[1]	80.4	82.9 Apr	79.0 Oct	97	63	94	67	74.92
PUERTO RICO											
Bayamon	18.24N, 66.09W	54	(see San Juan)								
Carolina	18.23N, 65.57W	39	(see San Juan)								
Ponce	18.01N, 66.37W	53	1941–70	79.5	81.8 Aug	76.6 Jan					35.12
*San Juan	18.28N, 66.07W	57	1941–70	78.4	80.7 Sep	75.5 Jan	96	60	92	65	59.49
TRINIDAD AND TOBAGO											
*Port of Spain	10.39N, 61.31W	67	1881–1940	78.1	79.9 May	75.9 Feb	101	52	95	62	63.11
USA											
Abilene	32.28N, 99.43W	1719	1941–70	64.5	83.9 Jul	43.7 Jan	111	–9	102	13	23.59

[1]Climatic data for Balboa Heights, Canal Zone, nr Panama.

NORTH AMERICA

City	Latitude and Longitude	Elev (ft)	Period of Record	Temperature (°F) Avg	Avg Warmest Month	Avg Coolest Month	Absolute High	Absolute Low	Avg Annual High	Avg Annual Low	Annual Precipitation (in)
Akron	41.05N, 81.31W	874	1941–70	49.6	71.7 Jul	26.3 Jan	104	–21			35.13
Alameda	37.46N, 122.15W	25	(see Oakland)								
Albany (Georgia)	31.34N, 84.09W	184	1941–70	66.9	81.8 Jul	50.7 Jan	106	–2			48.84
Albany (New York)	42.39N, 73.45W	20	1941–70	49.1	73.1 Jul	23.8 Jan	104	–28	96	–12	34.38
Albuquerque	35.05N, 106.39W	4950	1941–70	56.8	78.7 Jul	35.2 Jan	104	–17			7.77
Alexandria (Louisiana)	31.18N, 92.27W	79	1941–70	64.9	80.9 Jul	47.7 Jan					54.06
Alexandria (Virginia)	38.48N, 77.03W	47	(see Washington)								
Allentown	40.36N, 75.28W	255	1941–70	51.0	74.1 Jul	27.8 Jan	105	–12			42.49
Alton	38.53N, 90.10W	488	(see Saint Louis)								
Altoona	40.31N, 78.24W	1180	1931–60	49.6	70.9 Jul	28.5 Jan	99	–20			43.83
Amarillo	35.12N, 101.50W	3676	1941–70	57.4	78.7 Jul	36.0 Jan	108	–16	102	0	20.28
Anaheim	33.50N, 117.55W	165	(see Santa Ana, USA)								
Anchorage	61.13N, 149.53W	118	1941–70	35.0	57.9 Jul	11.8 Jan	92	–38	81	–32	14.74
Anderson	40.10N, 85.41W	874	1941–70	52.3	74.3 Jul	28.2 Jan	105	–20			37.43
Annapolis	38.59N, 76.30W	20	1941–70	55.6	76.7 Jul	34.5 Jan	106	–6			39.03
Ann Arbor	42.17N, 83.45W	880	1941–70	49.5	72.7 Jul	24.9 Jan	100	–13			30.54
Appleton	44.16N, 88.25W	723	1941–70	45.3	71.4 Jul	16.9 Jan	107	–32			30.23
Arden-Arcade	38.36N, 121.25W	60	(see Sacramento)								
Arlington (Texas)	32.44N, 97.07W	616	(see Fort Worth)								
Arlington (Virginia)	38.53N, 77.07W	200	(see Washington)								
Asheville	35.36N, 82.33W	1985	1941–70	55.7	73.5 Jul	37.9 Jan	99	–7			45.18
Athens	33.57N, 83.23W	771	1941–70	61.7	78.4 Jul	44.5 Jan	108	–3			49.98
Atlanta	33.45N, 84.24W	1050	1941–70	60.8	78.0 Jul	42.4 Jan	103	–9	96	15	48.34
Atlantic City	39.22N, 74.26W	10	1941–70	54.3	74.3 Jul	34.8 Jan	106	–9			41.11
Augusta	33.28N, 81.58W	143	1941–70	63.4	80.4 Jul	45.8 Jan	106	3			42.63
Aurora (Colorado)	39.44N, 104.52W	5342	(see Denver)								
Aurora (Illinois)	41.45N, 88.19W	638	1941–70	48.9	72.9 Jul	22.2 Jan	111	–25			34.05

NORTH AMERICA

City	Latitude and Longitude	Elev (ft)	Period of Record	Temperature (°F)							Annual Precipitation (in)
				Avg	Avg Warmest Month	Avg Coolest Month	Absolute High	Absolute Low	Avg Annual High	Avg Annual Low	
Austin	30.16N, 97.45W	505	1941–70	68.1	84.7 Aug	49.7 Jan	109	–2	103	21	32.49
Bakersfield	35.22N, 119.01W	420	1941–70	64.9	83.9 Jul	47.5 Jan	118	13			5.72
Baltimore	39.17N, 76.37W	20	1941–70	57.7	79.4 Jul	36.1 Jan	107	–7	98	6	41.31
Bangor	44.48N, 68.46W	20	1941–70	43.8	68.1 Jul	18.2 Jan					40.25
Baton Rouge	30.27N, 91.11W	57	1941–70	67.4	82.0 Jul	51.0 Jan	110	2			54.05
Battle Creek	42.19N, 85.11W	885	1941–70	48.2	71.8 Jul	23.3 Jan	104	–24			34.14
Bay City	43.36N, 83.53W	595	1941–70	48.1	72.0 Jul	23.1 Jan	109	–25			27.18
Bayonne	40.40N, 74.07W	40	(see Newark)								
Beaumont	30.05N, 94.06W	20	1931–60	69.3	84.4 Jul	54.3 Jan	106	12			56.49
Berkeley	37.52N, 122.16W	40	1941–70	57.1	63.6 Sep	49.3 Jan	106	25			23.20
Bethlehem	40.37N, 75.23W	237	1931–60	53.2	75.5 Jul	31.1 Jan	105	–16			41.05
Beverly Hills	34.04N, 118.25W	292	(see Los Angeles)								
Billings	45.47N, 108.30W	3117	1941–70	46.3	71.8 Jul	21.9 Jan	112	–49			14.15
Biloxi	30.24N, 88.53W	22	1941–70	68.2	82.1 Jul	52.4 Jan	104	1			61.86
Binghamton	42.06N, 75.55W	865	1941–70	46.0	69.1 Jul	22.0 Jan	103	–28			37.35
Birmingham	33.31N, 86.49W	600	1941–70	62.4	79.9 Jul	44.2 Jan	107	–10	99	12	53.23
Bismarck	46.48N, 100.47W	1670	1941–70	41.4	70.8 Jul	8.2 Jan	114	–45	100	–33	16.16
Bloomington (Illinois)	40.29N, 89.00W	830	1941–70	52.5	75.7 Jul	26.2 Jan	114	–24			36.83
Bloomington (Indiana)	39.10N, 86.32W	752	1941–70	54.7	76.2 Jul	31.2 Jan	110	–20			43.20
Bloomington (Minnesota)	44.49N, 93.16W	830	(see Minneapolis)								
Boise City	43.37N, 116.12W	2704	1941–70	50.9	74.5 Jul	29.0 Jan	113	–28	104	–3	11.50
Boston	42.21N, 71.03W	21	1941–70	51.3	73.3 Jul	29.2 Jan	104	–18			42.52
Boulder	40.01N, 105.17W	5430	1941–70	52.0	73.9 Jul	33.0 Jan	104	–33			18.91
Bridgeport	41.11N, 73.11W	10	1941–70	51.9	73.8 Jul	30.2 Jan	103	–20			38.61
Bristol	41.40N, 72.57W	240	(see Hartford)								
Brockton	42.05N, 71.01W	130	1941–70[1]	49.3	71.4 Jul	27.5 Jan	104	–19			41.84

[1] Temperature data for Taunton, nr Brockton.

NORTH AMERICA

City	Latitude and Longitude	Elev (ft)	Period of Record	Temperature (°F) Avg	Avg Warmest Month	Avg Coolest Month	Absolute High	Low	Avg Annual High	Low	Annual Precipitation (in)
Brownsville	25.54N, 97.30W	57	1941–70	73.8	84.4 Jul	60.3 Jan	104	12	99	28	25.09
Buffalo	42.53N, 78.52W	585	1941–70	47.1	70.1 Jul	23.7 Jan	99	−21			36.11
Burbank	34.11N, 118.19W	560	1941–70	63.5	74.6 Aug	53.7 Jan	111	21			14.89
Burlington	44.29N, 73.13W	110	1941–70	44.4	69.8 Jul	16.8 Jan	101	−30			32.54
Butte	46.00N, 112.32W	5755	1941–70	38.4	62.2 Jul	15.1 Jan	100	−52			12.17
Cambridge	42.22N, 71.06W	20	(see Boston)								
Camden	39.57N, 75.07W	30	(see Philadelphia)								
Canton	40.48N, 81.23W	1030	(see Akron)								
Casper	42.51N, 106.19W	5123	1941–70	45.4	71.0 Jul	23.2 Jan	104	−40			11.22
Cedar Rapids	41.58N, 91.40W	730	1941–70	49.0	73.9 Jul	20.4 Jan	109	−25			35.60
Champaign	40.07N, 88.15W	740	1941–70[1]	52.3	75.3 Jul	26.9 Jan	109	−25			37.42
Charleston (South Carolina)	32.47N, 79.56W	9	1941–70	66.2	81.4 Jul	50.0 Jan	104	7	98	22	48.92
Charleston (West Virginia)	38.21N, 81.38W	601	1941–70	55.7	75.8 Jul	34.4 Jan	108	−17	99	1	40.28
Charlotte	35.13N, 80.51W	720	1941–70	60.5	78.5 Jul	42.1 Jan	104	−5			42.72
Charlottesville	38.02N, 78.30W	480	1941–70	57.0	77.1 Jul	36.0 Jan	107	−9			44.34
Chattanooga	35.03N, 85.19W	675	1941–70	59.8	78.8 Jul	40.2 Jan	106	−10			51.92
Cheektowaga	42.54N, 78.45W	630	(see Buffalo)								
Chesapeake	36.49N, 76.17W	20	(see Norfolk)								
Chester	39.51N, 75.22W	23	(see Philadelphia)								
Cheyenne	41.08N, 104.49W	6062	1941–70	45.9	69.1 Jul	26.6 Jan	100	−38	92	−18	14.65
Chicago	41.52N, 87.38W	595	1941–70	50.6	74.7 Jul	24.3 Jan	105	−23	95	−10	34.44
Chicopee	42.09N, 72.37W	92	[see Springfield (Massachusetts)]								
Cincinnati	39.06N, 84.31W	550	1941–70	54.9	76.2 Jul	32.1 Jan	109	−25			40.03
Clarksville	36.32N, 87.21W	444	1941–70	59.1	79.2 Jul	37.9 Jan	112	−14			47.50
Clearwater	27.58N, 82.48W	29	1941–70	72.5	82.2 Aug	61.0 Jan					53.00

[1] Climatic data for Urbana, nr Champaign.

NORTH AMERICA

City	Latitude and Longitude	Elev (ft)	Period of Record	Temperature (°F) Avg	Avg Warmest Month	Avg Coolest Month	Absolute High	Absolute Low	Avg Annual High	Avg Annual Low	Annual Precipitation (in)
Cleveland	41.30N, 81.42W	660	1941–70	49.7	71.4 Jul	26.9 Jan	103	–19			34.99
Clifton	40.53N, 74.09W	70	(see Paterson)								
Colorado Springs	38.50N, 104.49W	6012	1941–70	48.4	70.7 Jul	28.6 Jan	100	–27			15.73
Columbia (Missouri)	38.57N, 92.20W	748	1941–70	54.4	77.3 Jul	29.3 Jan	113	–26			36.96
Columbia (South Carolina)	34.00N, 81.02W	190	1941–70	64.4	81.1 Jul	47.0 Jan	107	–2	100	16	44.75
Columbus (Georgia)	32.28N, 84.59W	265	1941–70	64.3	80.6 Jul	46.9 Jan	106	–3			50.94
Columbus (Ohio)	39.58N, 83.00W	780	1941–70	51.5	73.6 Jul	28.4 Jan	106	–20	96	–5	37.01
Compton	33.54N, 118.13W	66	(see Torrance)								
Concord (California)	37.58N, 122.02W	65	(see Berkeley)								
Concord (New Hampshire)	43.12N, 71.32W	288	1941–70	45.6	69.7 Jul	20.6 Jan	102	–37			36.17
Corpus Christi	27.48N, 97.24W	35	1941–70	71.9	85.1 Aug	56.3 Jan	105	11			28.53
Council Bluffs	41.16N, 95.52W	984	(see Omaha)								
Covington	39.05N, 84.31W	513	1941–70	54.0	75.6 Jul	31.1 Jan					39.04
Cranston	41.47N, 71.26W	60	(see Providence)								
Dallas	32.47N, 96.48W	435	1941–70	66.2	85.8 Aug	45.4 Jan	111	–3	102	13	35.94
Danbury	41.24N, 73.28W	375	1931–60	49.1	71.2 Jul	27.1 Jan	100	–16			45.44
Davenport	41.31N, 90.35W	590	1931–60	51.3	76.9 Jul	23.8 Jan	111	–27			33.88
Dayton	39.46N, 84.12W	745	1941–70	52.0	74.6 Jul	28.1 Jan	108	–28			37.60
Daytona Beach	29.13N, 81.01W	7	1941–70	70.5	81.1 Aug	58.4 Jan	102	18			50.22
Dearborn	42.19N, 83.10W	604	(see Detroit)								
Dearborn Heights	42.19N, 83.17W		1941–70[1]	49.1	72.3 Jul	24.6 Jan	98	–13			31.69
Decatur	39.51N, 88.57W	682	1941–70	53.5	76.5 Jul	28.3 Jan	113	–25			38.39
Denver	39.45N, 104.59W	5280	1941–70	51.9	74.2 Jul	32.9 Jan	105	–30			12.95
Des Moines	41.35N, 93.37W	805	1941–70	49.0	75.1 Jul	19.4 Jan	110	–30	100	–14	30.85
Detroit	42.20N, 83.03W	585	1941–70	49.9	73.3 Jul	25.5 Jan	105	–24	94	–5	30.96

[1] Climatic data for Detroit Metropolitan Airport, nr Dearborn Heights.

NORTH AMERICA

City	Latitude and Longitude	Elev (ft)	Period of Record	Temperature (°F) Avg	Avg Warmest Month	Avg Coolest Month	Absolute High	Absolute Low	Avg Annual High	Avg Annual Low	Annual Precipitation (in)
Downey	33.56N, 118.07W	118	1941–70	(see Los Angeles for temperature data)							13.74
Dubuque	42.30N, 90.40W	612	1941–70	48.7	74.0 Jul	20.0 Jan	110	−32			30.70
Duluth	46.47N, 92.06W	610	1941–70	38.6	65.6 Jul	8.5 Jan	106	−41	91	−28	30.18
Dundalk	39.16N, 76.31W	20	(see Baltimore)								
Durham	36.00N, 78.55W	405	1941–70	59.5	77.9 Jul	40.9 Jan	105	−6			44.21
East Los Angeles	34.01N, 118.09W	280	(see Los Angeles)								
East Orange	40.46N, 74.13W	170	(see Newark)								
East Saint Louis	38.37N, 90.09W	418	(see Saint Louis)								
Elgin	42.02N, 88.17W	717	1941–70	[see Aurora (Illinois) for temperature data]							34.37
Elizabeth	40.40N, 74.13W	21	1941–70	53.8	75.5 Jul	31.6 Jan	105	−16			46.64
Elmira	42.06N, 76.48W	860	1941–70	48.1	70.8 Jul	25.0 Jan	107	−24			33.59
El Paso	31.46N, 106.29W	3762	1941–70	63.4	82.3 Jul	43.6 Jan	109	−8	103	14	7.77
Elyria	41.22N, 82.07W	730	1941–70[1]	50.0	72.0 Jul	26.9 Jan					33.90
Erie	42.07N, 80.05W	685	1941–70	47.1	68.7 Jul	25.1 Jan	99	−16			38.20
Euclid	41.34N, 81.32W	648	(see Cleveland)								
Eugene	44.03N, 123.05W	422	1941–70	52.6	66.9 Jul	39.4 Jan	105	−4			42.56
Evanston	42.03N, 87.41W	601	(see Chicago)								
Evansville	37.58N, 87.34W	385	1941–70	56.0	77.8 Jul	32.6 Jan	108	−23			41.88
Everett	47.59N, 122.12W	30	1941–70	50.5	62.7 Jul	38.2 Jan	98	1			35.39
Fairbanks	64.51N, 147.43W	448	1941–70	25.7	60.7 Jul	−11.9 Jan	99	−66			11.22
Fairfield	38.15N, 122.02W	12	1941–70[2]	60.4	75.7 Jul	45.0 Jan					25.56
Fall River	41.42N, 71.09W	40	1941–70	50.7	72.8 Jul	28.9 Jan	100	−18			43.04
Fargo	46.52N, 96.47W	900	1941–70	40.8	70.7 Jul	5.9 Jan	114	−48			19.62
Fayetteville	35.03N, 78.53W	100	1941–70	61.3	79.6 Jul	43.0 Jan	108	0			46.69

[1] Climatic data for Oberlin, nr Elyria.
[2] Climatic data for Vacaville, nr Fairfield.

NORTH AMERICA

City	Latitude and Longitude	Elev (ft)	Period of Record	Temperature (°F)							Annual Precipitation (in)
				Avg	Avg Warmest Month	Avg Coolest Month	Absolute High	Absolute Low	Avg Annual High	Avg Annual Low	
Fitchburg	42.35N, 71.48W	458	1941–70	48.7	72.3 Jul	24.7 Jan	103	−21			43.59
Flint	43.01N, 83.42W	715	1941–70	46.8	69.7 Jul	22.3 Jan	108	−28			29.77
Fort Lauderdale	26.07N, 80.08W	7	1941–70	75.2	82.7 Aug	66.8 Jan	100	28			60.08
Fort Smith	35.23N, 94.25W	423	1941–70	61.3	82.2 Jul	39.0 Jan	113	−15			42.27
Fort Wayne	41.04N, 85.08W	790	1941–70	49.9	73.0 Jul	25.3 Jan	106	−24			35.80
Fort Worth	32.45N, 97.20W	670	1941–70	65.5	84.9 Aug	44.8 Jan	112	−8			32.30
Fremont	37.32N, 121.57W	50	(see Santa Clara, USA)								
Fresno	36.44N, 119.47W	285	1941–70	62.3	80.6 Jul	45.3 Jan	115	17	109	26	10.24
Fullerton	33.52N, 117.55W	161	1941–70	(see Santa Ana, USA, for temperature data)							13.36
Gadsden	34.01N, 86.01W	555	1941–70	61.3	79.2 Jul	42.8 Jan	105	2			53.24
Gainesville	29.40N, 82.20W	185	1941–70	69.9	81.2 Aug	57.0 Jan	104	6			54.59
Galveston	29.18N, 94.48W	5	1941–70	69.8	83.3 Aug	53.9 Jan	101	8			42.20
Garden Grove	33.46N, 117.55W	93	(see Santa Ana, USA)								
Garland	32.54N, 96.38W	541	(see Dallas)								
Gary	41.36N, 87.20W	590	1941–70	50.3	74.0 Jul	24.7 Jan					34.87
Gastonia	35.16N, 81.11W	825	1941–70	61.2	79.1 Jul	42.9 Jan	107	4			46.68
Glendale	34.08N, 118.15W	573	1941–70	(see Burbank for temperature data)							16.84
Grand Forks	47.55N, 97.03W	830	1941–70	39.6	69.7 Jul	4.1 Jan	109	−44			21.27
Grand Rapids	42.58N, 85.40W	610	1941–70	47.8	71.5 Jul	23.2 Jan	108	−24			32.39
Great Falls	47.30N, 111.17W	3330	1941–70	44.9	69.3 Jul	20.5 Jan	107	−49			14.99
Green Bay	44.31N, 88.01W	590	1941–70	43.7	69.2 Jul	15.4 Jan	104	−36			27.01
Greensboro	36.04N, 79.47W	839	1941–70	58.1	77.2 Jul	38.7 Jan	104	−7			41.36
Greenville	34.51N, 82.24W	966	1941–70	61.0	78.7 Jul	42.6 Jan	103	4			49.54
Gulfport	30.22N, 89.05W	19	(see Biloxi)								
Hamilton	39.24N, 84.34W	600	1941–70	54.0	75.3 Jul	31.5 Jan	111	−16			38.86
Hammond	41.38N, 87.30W	590	(see Gary)								
Hampton	37.01N, 76.20W	3	1941–70	59.3	78.6 Jul	40.0 Jan	103	4			42.31

NORTH AMERICA

City	Latitude and Longitude	Elev (ft)	Period of Record	Temperature (°F)			Absolute		Avg Annual		Annual Precipitation (in)
				Avg	Avg Warmest Month	Avg Coolest Month	High	Low	High	Low	
Harrisburg	40.16N, 76.53W	365	1941-70	53.4	76.1 Jul	30.1 Jan	107	-14			36.47
Hartford	41.46N, 72.41W	40	1941-70	49.1	72.7 Jul	24.8 Jan	102	-26			43.37
Hayward	37.40N, 122.05W	116	(see Oakland)								
Helena	46.36N, 112.02W	4124	1941-70	43.2	67.9 Jul	18.1 Jan	105	-42			11.38
Hialeah	25.50N, 80.17W	8	1941-70	74.3	81.9 Aug	65.8 Jan					63.14
High Point	35.57N, 80.00W	940	1941-70	59.5	77.4 Jul	40.9 Jan					42.91
Hollywood	26.01N, 80.09W	10	(see Fort Lauderdale)								
Holyoke	42.12N, 72.36W	152	1941-70	63.5	[see Springfield (Massachusetts) for temperature data]							42.11
Hot Springs	34.30N, 93.03W	599	1941-70	63.5	82.3 Jul	43.1 Jan	115	-12			55.10
Houston	29.45N, 95.22W	40	1941-70	68.9	83.4 Aug	52.1 Jan	108	5	100	22	48.19
Huntington	38.25N, 82.27W	565	1941-70	55.2	75.3 Jul	34.3 Jan	108	-15			38.88
Huntington Beach	33.40N, 118.05W	35	(see Long Beach)								
Huntsville	34.44N, 86.35W	636	1941-70	60.8	79.5 Jul	40.9 Jan	101	-4			52.16
Idaho Falls	43.29N, 112.02W	4709	1941-70	44.1	68.9 Jul	18.9 Jan	103	-43			8.89
Independence	39.06N, 94.25W	1051	[see Kansas City (Missouri)]								
Indianapolis	39.46N, 86.10W	710	1941-70	52.3	75.0 Jul	27.9 Jan	107	-25	96	-8	38.74
Inglewood	33.58N, 118.21W	140	(see Los Angeles)								
Irving	32.49N, 96.56W	470	(see Dallas)								
Jackson (Michigan)	42.15N, 84.24W	940	1941-70	48.1	71.7 Jul	23.0 Jan	105	-18			30.35
Jackson (Mississippi)	32.18N, 90.11W	298	1941-70	65.0	81.7 Jul	47.1 Jan	107	-5			49.19
Jacksonville	30.20N, 81.40W	20	1941-70	68.4	81.0 Jul	54.6 Jan	105	10	97	25	54.47
Jefferson City	38.34N, 92.10W	557	1941-70	55.2	77.4 Jul	31.0 Jan					39.01
Jersey City	40.44N, 74.04W	20	1941-70	52.9	74.8 Jul	31.0 Jan	106	-14			41.40
Johnstown	40.20N, 78.55W	1185	1941-70	51.2	72.7 Jul	29.0 Jan					44.30
Joliet	41.32N, 88.05W	607	1941-70	50.5	74.2 Jul	23.9 Jan	101	-20			35.57
Kalamazoo	42.17N, 85.35W	755	1941-70	49.7	72.9 Jul	24.8 Jan	109	-17			33.99
Kansas City (Kansas)	39.07N, 94.38W	750	[see Kansas City (Missouri)]								

NORTH AMERICA

City	Latitude and Longitude	Elev (ft)	Period of Record	Temperature (°F)							Annual Precipitation (in)
				Avg	Avg Warmest Month	Avg Coolest Month	Absolute High	Absolute Low	Avg Annual High	Avg Annual Low	
Kansas City (Missouri)	39.05N, 94.35W	750	1941–70	56.1	79.1 Jul	30.5 Jan	113	–22	99	–5	36.48
Kenosha	42.36N, 87.50W	610	(see Racine)								
Key West	24.33N, 81.48W	7	1941–70	78.2	84.7 Aug	70.7 Jan	100	41			39.99
Killeen	31.07N, 97.43W	833	1931–60[1]	67.1	84.6 Jul	47.9 Jan	112	–4			33.94
Knoxville	35.58N, 83.55W	890	1941–70	59.5	78.2 Jul	39.6 Jan	104	–16			46.14
Kokomo	40.29N, 86.08W	828	1941–70	52.5	75.5 Jul	27.6 Jan	110	–22			37.86
La Crosse	43.48N, 91.15W	649	1941–70	46.4	72.8 Jul	16.1 Jan	108	–43			29.08
Lafayette (Indiana)	40.25N, 86.54W	550	1941–70[2]	50.7	73.6 Jul	26.0 Jan	111	–18			37.91
Lafayette (Louisiana)	30.13N, 92.01W	40	1941–70	67.9	81.9 Jul	51.9 Jan	107	6			57.01
Lake Charles	30.14N, 93.13W	16	1941–70	68.3	82.4 Jul	52.3 Jan	106	3			55.47
Lakeland	28.03N, 81.57W	206	1941–70	72.1	81.9 Aug	60.8 Jan	101	20			49.43
Lakewood (California)	33.51N, 118.08W	50	(see Long Beach)								
Lakewood (Colorado)	39.44N, 105.05W	5355	1941–70	51.1	73.1 Jul	31.9 Jan					14.95
Lakewood (Ohio)	41.29N, 81.48W	685	(see Cleveland)								
Lancaster	40.02N, 76.18W	355	1941–70	52.7	74.3 Jul	30.5 Jan	104	–18			40.48
Lansing	42.44N, 84.33W	830	1941–70	47.5	70.9 Jul	22.6 Jan	102	–33			30.39
Laredo	27.30N, 99.30W	440	1941–70	73.9	87.9 Jul	56.5 Jan	115	5			17.87
Las Cruces	32.19N, 106.47W	3895	1931–60	60.0	79.5 Jul	40.9 Jan					8.01
Las Vegas	36.10N, 115.09W	2030	1941–70	65.8	89.6 Jul	44.2 Jan	117	8	110	18	3.76
Lawrence (Kansas)	38.58N, 95.14W	840	1941–70	56.1	79.1 Jul	30.3 Jan	114	–26			37.43
Lawrence (Massachusetts)	42.42N, 71.10W	65	1941–70	49.5	72.6 Jul	25.8 Jan	106	–25			39.83
Lawton	34.37N, 98.25W	1111	1941–70	62.3	83.3 Jul	39.7 Jan	115	–11			30.18
Lewiston	44.06N, 70.13W	196	1941–70	45.7	70.0 Jul	20.4 Jan	102	–28			43.20
Lexington	38.03N, 84.30W	955	1941–70	55.2	76.2 Jul	32.9 Jan	108	–21			44.49
Lima	40.45N, 84.06W	865	1941–70	51.2	73.4 Jul	27.3 Jan	102	–17			35.54

[1] Climatic data for Temple, nr Killeen.
[2] Climatic data for Frankfort, nr Lafayette.

NORTH AMERICA

City	Latitude and Longitude	Elev (ft)	Period of Record	Temperature (°F)							Annual Precipitation (in)
				Avg	Avg Warmest Month	Avg Coolest Month	Absolute High	Absolute Low	Avg Annual High	Avg Annual Low	
Lincoln	40.49N, 96.42W	1150	1941–70	51.0	77.3 Jul	22.2 Jan	115	−29	102	−15	26.66
Little Rock	34.45N, 92.17W	300	1941–70	61.0	81.4 Jul	39.5 Jan	110	−13	90	10	48.52
Livonia	42.23N, 83.22W	663	(see Detroit)								
Long Beach	33.46N, 118.11W	35	1941–70	63.3	73.3 Aug	54.2 Jan	111	25			10.25
Longview	32.29N, 94.44W	339	1941–70	65.5	83.5 Jul	46.2 Jan	113	−7			47.18
Lorain	41.28N, 82.11W	610	1941–70[1]	50.0	72.0 Jul	26.9 Jan					33.90
Los Angeles	34.03N, 118.14W	340	1941–70	64.8	74.1 Aug	56.7 Jan	110	28	100	35	14.05
Louisville	38.15N, 85.46W	450	1941–70	55.6	76.9 Jul	33.3 Jan	107	−20	98	0	43.11
Lowell	42.38N, 71.19W	100	1941–70	50.2	73.7 Jul	26.4 Jan	103	−29			41.25
Lubbock	33.35N, 101.51W	3241	1941–70	59.7	79.7 Jul	39.1 Jan	107	−16			18.41
Lynchburg	37.25N, 79.09W	517	1941–70	56.3	75.8 Jul	36.6 Jan	106	−8			38.27
Lynn	42.28N, 70.57W	34	(see Boston)								
Macon	32.50N, 83.38W	335	1941–70	65.1	81.4 Jul	47.8 Jan	106	3	93	−18	44.46
Madison	43.04N, 89.23W	860	1941–70	44.9	70.1 Jul	16.8 Jan	107	−37			30.25
Manchester	42.59N, 71.28W	175	1941–70	46.5	70.0 Jul	22.2 Jan	100	−29			39.65
Mansfield	40.45N, 82.31W	1154	1941–70	48.4	69.9 Jul	25.5 Jan	96	−20			36.90
McAllen	26.12N, 98.14W	122	1941–70	73.7	85.2 Aug	59.4 Jan					20.41
Melbourne	28.05N, 80.36W	22	1941–70	72.2	81.5 Aug	61.8 Jan	100	22			50.79
Memphis	35.09N, 90.03W	275	1941–70	61.6	81.6 Jul	40.5 Jan	106	−13			49.10
Meriden	41.32N, 72.47W	150	(see Hartford)								
Meridian	32.22N, 88.42W	341	1941–70	64.5	81.2 Jul	46.9 Jan	105	−7			49.58
Mesa	33.25N, 111.50W	1161	1941–70	68.9	89.2 Jul	50.3 Jan	119	15			7.52
Metairie	29.58N, 90.09W	5	(see New Orleans)								
Miami	25.47N, 80.12W	10	1941–70	75.5	82.9 Aug	67.2 Jan	100	27	92	39	59.80
Miami Beach	25.47N, 80.08W	6	1941–70	76.1	83.1 Aug	68.5 Jan	98	30			46.54

[1]Climatic data for Oberlin, nr Lorain.

NORTH AMERICA

City	Latitude and Longitude	Elev (ft)	Period of Record	Temperature (°F)							Annual Precipitation (in)
				Avg	Avg Warmest Month	Avg Coolest Month	Absolute High	Absolute Low	Avg Annual High	Avg Annual Low	
Middletown	39.31N, 84.24W	666	1941–70	(see Hamilton, USA, for temperature data)							39.27
Midland	32.00N, 102.05W	2779	1941–70	63.9	82.3 Jul	43.6 Jan	109	–11			13.51
Milwaukee	43.02N, 87.54W	635	1941–70	45.7	69.9 Jul	19.4 Jan	105	–25			29.07
Minneapolis	44.59N, 93.16W	815	1941–70	43.5	71.2 Jul	11.8 Jan	108	–41	99	–21	25.94
Mobile	30.42N, 88.03W	5	1941–70	67.4	81.6 Jul	51.2 Jan	104	–1	98	23	66.98
Modesto	37.39N, 121.00W	88	1941–70	60.6	76.3 Jul	45.1 Jan	111	15			11.87
Monroe	32.30N, 92.07W	77	1941–70	65.1	82.3 Jul	46.7 Jan					49.62
Montgomery	32.23N, 86.19W	160	1941–70	64.8	81.0 Jul	47.5 Jan	107	–5	99	17	49.86
Mount Vernon	40.55N, 73.49W	100	1941–70[1]	52.2	74.1 Jul	29.8 Jan					46.39
Muncie	40.11N, 85.23W	950	1941–70	(see Anderson for temperature data)							38.25
Muskegon	43.14N, 86.16W	625	1941–70	47.3	70.0 Jul	24.0 Jan	99	–30			31.53
Nashua	42.45N, 71.28W	152	1941–70	46.7	70.1 Jul	22.8 Jan	105	–35			41.04
Nashville	36.10N, 86.47W	450	1941–70	59.4	79.6 Jul	38.3 Jan	107	–15	98	4	46.00
Natchez	31.34N, 91.23W	202	1941–70	66.4	81.9 Jul	49.4 Jan					54.38
Newark	40.44N, 74.10W	55	1941–70	53.9	76.4 Jul	31.4 Jan	105	–14			41.45
New Bedford	41.38N, 70.56W	15	1941–70	51.9	72.7 Jul	31.4 Jan	102	–12			39.77
New Britain	41.40N, 72.47W	200	(see Hartford)								
New Haven	41.18N, 72.55W	40	1941–70	50.6	72.3 Jul	28.9 Jan	101	–15			41.81
New Orleans	29.57N, 90.04W	5	1941–70	67.5	81.9 Jul	52.9 Jan	102	7	96	26	56.77
Newport	41.29N, 71.19W	10	1941–70[2]	50.1	69.5 Jul	31.3 Feb	95	–10			40.51
Newport News	36.59N, 76.25W	22	(see Hampton)								
New Rochelle	40.54N, 73.47W	72	1941–70[1]	52.2	74.1 Jul	29.8 Jan					46.39
Newton	42.21N, 71.12W	33	(see Boston)								

[1] Climatic data for Scarsdale, nr Mount Vernon and New Rochelle.
[2] Climatic data for Block Island, nr Newport.

NORTH AMERICA

City	Latitude and Longitude	Elev (ft)	Period of Record	Temperature (°F)							Annual Precipitation (in)
				Avg	Avg Warmest Month	Avg Coolest Month	Absolute High	Absolute Low	Avg Annual High	Avg Annual Low	
New York	40.45N, 74.00W	55	1941–70	54.5	76.6 Jul	32.2 Jan	106	–15	98	2	40.19
Niagara Falls	43.06N, 79.03W	570	20–24 yr bef 1971[1]	48.9	72.1 Jul	25.4 Jan	101	–13			34.43
Norfolk	36.51N, 76.17W	10	1941–70	59.3	78.3 Jul	40.5 Jan	105	2	97	14	44.68
North Charleston	32.53N, 80.00W	25	1941–70	64.7	80.2 Jul	48.6 Jan	104	7	98	22	52.12
Norwalk (California)	33.54N, 118.05W	93	(see Los Angeles)								
Norwalk (Connecticut)	41.07N, 73.22W	60	1941–70	50.4	72.4 Jul	28.3 Jan					45.30
Norwich	41.31N, 72.05W	35	(see New Haven)								
Oakland	37.48N, 122.16W	25	1941–70	57.4	64.5 Sep	48.6 Jan	107	23			18.69
Oak Park	41.53N, 87.46W	630	(see Chicago)								
Odessa	31.51N, 102.22W	2890	(see Midland)								
Ogden	41.14N, 111.58W	4299	1941–70	51.4	76.9 Jul	27.8 Jan	106	–26			20.11
Oklahoma City	35.28N, 97.31W	1195	1941–70	61.1	82.6 Jul	37.9 Jan	113	–17	104	5	32.23
Omaha	41.16N, 95.56W	1040	1941–70	51.5	77.2 Jul	22.6 Jan	114	–32	100	–13	30.18
Orange	33.47N, 117.51W	176	(see Santa Ana, USA)								
Orlando	28.33N, 81.23W	70	1941–70	72.5	82.7 Aug	60.4 Jan	103	20			51.21
Oshkosh	44.01N, 88.33W	761	1941–70	45.5	71.3 Jul	17.0 Jan	107	–34			28.36
Owensboro	37.46N, 87.07W	765	1941–70	57.3	78.2 Jul	34.9 Jan	107	–21			44.28
Oxnard	34.12N, 119.10W	50	1941–70	59.5	65.6 Aug	53.8 Jan	104	26			14.25
Palm Beach	26.43N, 80.02W	10	(see West Palm Beach)								
Palm Springs	33.50N, 116.33W	430	1941–70	71.6	90.8 Jul	54.3 Jan					5.33
Palo Alto	37.26N, 122.10W	63	1941–70	57.5	66.0 Jul	47.3 Jan					15.37
Parma	41.23N, 81.43W	846	(see Cleveland)								
Pasadena (California)	34.09N, 118.09W	830	1941–70	63.6	74.5 Aug	54.2 Jan	113	21			18.90
Pasadena (Texas)	29.43N, 95.13W	34	(see Houston)								
Passaic	40.51N, 74.07W	70	(see Paterson)								

[1] Climatic data for Niagara Falls, Canada, nr Niagara Falls, USA.

NORTH AMERICA

City	Latitude and Longitude	Elev (ft)	Period of Record	Temperature (°F) Avg	Avg Warmest Month	Avg Coolest Month	Absolute High	Absolute Low	Avg Annual High	Avg Annual Low	Annual Precipitation (in)
Paterson	40.55N, 74.10W	100	1941–70	53.2	75.9 Jul	30.3 Jan	106	–16			47.93
Pawtucket	41.53N, 71.23W	25	(see Providence)								
Pensacola	30.25N, 87.13W	15	1941–70	68.0	81.8 Jul	52.1 Jan	103	7	95	23	64.22
Peoria	40.42N, 89.36W	470	1941–70	50.8	75.1 Jul	23.8 Jan	113	–27			35.06
Petersburg	37.14N, 77.24W	87	1941–70[1]	60.0	79.1 Jul	40.5 Jan	106	–10			42.12
Philadelphia	39.57N, 75.09W	100	1941–70	54.6	76.8 Jul	32.3 Jan	106	–11	96	5	39.93
Phoenix	33.27N, 112.04W	1090	1941–70	70.3	91.2 Jul	51.2 Jan	118	16	114	29	7.05
Pine Bluff	34.13N, 92.01W	221	1941–70	64.2	83.0 Jul	44.2 Jan	112	–6			50.72
Pittsburgh	40.26N, 80.00W	745	1941–70	53.0	74.6 Jul	30.6 Jan	103	–20	95	–1	36.22
Pittsfield	42.27N, 73.15W	1015	1931–60	44.9	67.8 Jul	21.8 Jan	95	–25			44.42
Plainfield	40.37N, 74.25W	100	1941–70	52.9	74.8 Jul	30.7 Jan	106	–17			46.97
Pocatello	42.52N, 112.27W	4464	1941–70	46.7	71.5 Jul	23.2 Jan	105	–31			10.80
Pomona	34.04N, 117.45W	861	1941–70	62.0	74.5 Aug	50.8 Jan	117	22			16.45
Pontiac	42.38N, 83.17W	932	1941–70	48.5	71.9 Jul	23.5 Jan	104	–22			30.06
Port Arthur	29.54N, 93.56W	10	1941–70	68.5	83.1 Aug	52.0 Jan	107	11			55.07
Portland (Maine)	43.40N, 70.15W	25	1941–70	45.0	68.0 Jul	21.5 Jan	103	–39			40.80
Portland (Oregon)	45.31N, 122.41W	77	1941–70	52.6	67.1 Jul	38.1 Jan	107	–3	95	16	37.61
Portsmouth	36.50N, 76.18W	10	(see Norfolk)								
Poughkeepsie	41.42N, 73.56W	175	1931–60	51.0	74.6 Jul	27.3 Jan	104	–21			40.21
Providence	41.50N, 71.25W	80	1941–70	50.0	72.1 Jul	28.4 Jan	102	–17			42.75
Provo	40.14N, 111.39W	4549	64 yr bef 1971	(see Salt Lake City for temperature data)							14.22
Pueblo	38.16N, 104.37W	4695	1941–70	52.8	76.4 Jul	30.1 Jan	105	–31	99	–14	11.91
Quincy	42.15N, 71.00W	42	(see Boston)								
Racine	42.44N, 87.47W	630	1941–70	48.0	71.8 Jul	22.3 Jan	107	–24			33.34
Raleigh	35.47N, 78.38W	365	1941–70	60.5	79.0 Jul	41.5 Jan	105	–2			45.91

[1] Climatic data for Hopewell, nr Petersburg.

NORTH AMERICA

City	Latitude and Longitude	Elev (ft)	Period of Record	Temperature (°F)							Annual Precipitation (in)
				Avg	Avg Warmest Month	Avg Coolest Month	Absolute High	Absolute Low	Avg Annual High	Avg Annual Low	
Rapid City	44.05N, 103.14W	3231	1941–70	48.1	73.2 Jul	24.5 Jan	109	–34			18.62
Reading	40.20N, 75.56W	265	1941–70	54.3	76.5 Jul	31.9 Jan	105	–14			40.85
Reno	39.31N, 119.49W	4491	1941–70	49.4	69.3 Jul	31.9 Jan	106	–19			7.20
Richmond (California)	37.56N, 122.21W	12	(see Berkeley)								
Richmond (Virginia)	37.32N, 77.26W	160	1941–70	57.8	77.9 Jul	37.5 Jan	107	–12	98	10	42.59
Riverside	33.59N, 117.22W	851	1941–70	63.3	76.1 Aug	52.1 Jan	118	19			9.92
Roanoke	37.16N, 79.57W	905	1941–70	55.9	75.2 Jul	36.4 Jan	105	–12			39.03
Rochester (Minnesota)	44.01N, 92.28W	988	1941–70	43.6	70.1 Jul	12.9 Jan	108	–42			27.47
Rochester (New York)	43.10N, 77.36W	515	1941–70	47.9	71.2 Jul	24.0 Jan	102	–22			31.33
Rockford	42.17N, 89.06W	715	1941–70	48.1	72.8 Jul	20.2 Jan	112	–25			36.72
Rock Island	41.29N, 90.34W	563	1941–70[1]	49.8	74.5 Jul	21.5 Jan	106	–26			35.80
Rome	43.13N, 75.27W	440	(see Syracuse)								
Royal Oak	42.30N, 83.09W	693	(see Detroit)								
Sacramento	38.35N, 121.30W	30	1941–70	60.3	75.2 Jul	45.1 Jan	115	17	104	28	17.22
Saginaw	43.26N, 83.56W	595	1941–70	46.7	70.6 Jul	21.4 Jan	111	–23			30.07
Saint Augustine	29.53N, 81.19W	7	1941–70	69.7	80.9 Jul	56.8 Jan					55.23
Saint Clair Shores	42.30N, 82.53W	586	(see Detroit)								
Saint Joseph	39.46N, 94.51W	850	1941–70	53.7	78.2 Jul	26.2 Jan	110	–24			35.65
Saint Louis	38.38N, 90.12W	455	1941–70	57.1	79.9 Jul	32.2 Jan	115	–23	99	–8	38.59
Saint Paul	44.57N, 93.06W	780	(see Minneapolis)								
Saint Petersburg	27.46N, 82.38W	20	1941–70	73.5	82.9 Aug	62.2 Jan	98	28			54.60
Salem (Massachusetts)	42.31N, 70.53W	13	(see Boston)								
Salem (Oregon)	44.56N, 123.02W	163	1941–70	52.3	66.6 Jul	38.8 Jan	108	–10			41.08
Salinas	36.40N, 121.39W	44	1941–70	57.1	63.6 Sep	50.0 Jan	110	18			13.81
Salt Lake City	40.45N, 111.53W	4266	1941–70	51.0	76.7 Jul	28.0 Jan	107	–30	102	–1	15.17

[1] Climatic data for Moline, nr Rock Island.

NORTH AMERICA

City	Latitude and Longitude	Elev (ft)	Period of Record	Temperature (°F) Avg	Avg Warmest Month	Avg Coolest Month	Absolute High	Absolute Low	Avg Annual High	Avg Annual Low	Annual Precipitation (in)
San Angelo	31.28N, 100.26W	1847	1941–70	66.2	84.7 Jul	46.4 Jan	111	1			17.53
San Antonio	29.26N, 98.29W	650	1941–70	68.8	84.7 Jul	50.7 Jan	107	0			27.54
San Bernardino	34.06N, 117.17W	1080	1941–70	64.0	78.2 Jul	52.0 Jan	116	17	91	36	16.11
San Diego	32.43N, 117.09W	20	1941–70	62.9	71.4 Aug	55.2 Jan	111	25	90	37	9.45
San Francisco	37.47N, 122.25W	65	1941–70	56.9	64.1 Sep	48.3 Jan	106	20			19.53
San Jose	37.20N, 121.53W	90	1941–70	59.6	68.4 Jul	49.5 Jan	104	22	99	26	13.65
San Mateo	37.34N, 122.19W	22	1941–70	58.3	65.5 Sep	49.5 Jan					19.51
Santa Ana	33.46N, 117.52W	133	1941–70	63.0	72.7 Aug	54.2 Jan	112	22			12.92
Santa Barbara	34.25N, 119.42W	100	1941–70	60.3	67.5 Aug	53.2 Jan	115	20			17.41
Santa Clara	37.21N, 121.57W	94	1941–70	59.1	68.1 Jul	48.6 Jan	111	19			14.15
Santa Cruz	36.58N, 122.01W	15	1941–70	56.6	63.4 Sep	48.8 Jan	108	20			31.37
Santa Fe	35.41N, 105.56W	6950	1941–70	49.6	70.4 Jul	29.8 Jan	97	–13	87	–1	13.83
Santa Monica	34.01N, 118.29W	100	1941–70	61.0	67.5 Aug	55.5 Jan					14.13
Santa Rosa	38.26N, 122.43W	150	1941–70	57.4	67.0 Aug	46.1 Jan	112	15			30.54
Sarasota	27.20N, 82.32W	18	(see Saint Petersburg)								
Savannah	32.05N, 81.06W	20	1941–70	65.9	81.1 Jul	49.9 Jan	105	8			51.15
Schenectady	42.49N, 73.57W	220	1941–70	48.0	72.9 Jul	21.8 Jan	102	–28			34.32
Scottsdale	33.29N, 111.56W	1250	1941–70	69.8	89.7 Jul	51.6 Jan	119	15			8.06
Scranton	41.25N, 75.40W	725	1941–70	49.9	72.5 Jul	26.8 Jan	103	–19	96	–3	36.57
Seaside	36.37N, 121.50W	20	(see Salinas)								
Seattle	47.37N, 122.20W	10	1941–70	53.2	66.1 Jul	41.2 Jan	100	0	90	21	32.59
Sheboygan	43.46N, 87.45W	589	1941–70	46.3	70.4 Jul	20.5 Jan	107	–25			29.89
Shreveport	32.31N, 93.45W	204	1941–70	65.9	83.2 Jul	47.2 Jan	110	–5			44.72
Silver Spring	38.59N, 77.02W	340	(see Washington)								
Simi Valley	34.16N, 118.47W	690	1941–70[1]	63.6	74.6 Aug	54.2 Jan					16.12

[1]Climatic data for San Fernando, nr Simi Valley.

NORTH AMERICA

City	Latitude and Longitude	Elev (ft)	Period of Record	Temperature (°F)			Absolute		Avg Annual		Annual Precipitation (in)
				Avg	Avg Warmest Month	Avg Coolest Month	High	Low	High	Low	
Sioux City	42.30N, 96.24W	1110	1941–70	48.4	75.3 Jul	18.0 Jan	111	−35	99	−19	25.74
Sioux Falls	43.33N, 96.44W	1395	1941–70	45.4	73.3 Jul	14.2 Jan	110	−42			24.72
Skokie	42.03N, 87.45W	625	(see Chicago)								
Somerville	42.23N, 71.06W	41	(see Boston)								
South Bend	41.41N, 86.15W	710	1941–70	49.1	72.3 Jul	24.0 Jan	109	−22			36.20
Spartanburg	34.56N, 81.57W	875	1941–70[1]	60.6	78.3 Jul	42.3 Jan	104	−6			47.54
Spokane	47.40N, 117.26W	1890	1941–70	47.3	69.7 Jul	25.4 Jan	108	−30	99	−7	17.42
Springfield (Illinois)	39.48N, 89.39W	610	1941–70	52.7	76.1 Jul	26.7 Jan	112	−24			35.02
Springfield (Massachusetts)	42.06N, 72.36W	85	1941–70	50.9	73.6 Jul	27.1 Jan	104	−18			42.57
Springfield (Missouri)	37.13N, 93.18W	1300	1941–70	56.1	77.8 Jul	32.9 Jan	113	−29	96	−5	39.70
Springfield (Ohio)	39.56N, 83.48W	980	[see Columbus (Ohio)]								
Stamford	41.03N, 73.32W	35	(see Bridgeport)								
Sterling Heights	42.33N, 83.02W	645	1941–70[2]	48.4	71.9 Jul	24.1 Jan					27.69
Steubenville	40.22N, 80.37W	660	1941–70	52.0	73.3 Jul	29.6 Jan					39.65
Stockton	37.57N, 121.17W	20	1941–70	60.7	76.7 Jul	44.6 Jan	113	19			14.17
Sunnyvale	37.23N, 122.02W	130	(see San Jose, USA)								
Syracuse	43.03N, 76.09W	400	1941–70	48.1	71.5 Jul	21.8 Jan	102	−26			36.41
Tacoma	47.15N, 122.26W	110	1941–70	52.1	64.8 Jul	39.9 Jan	99	7	90	18	36.88
Tallahassee	30.27N, 84.17W	216	1941–70	67.7	81.1 Jul	52.6 Jan	104	−2			61.58
Tampa	27.57N, 82.27W	15	1941–70	72.2	82.2 Aug	60.4 Jan	98	18	95	31	49.38
Taylor	42.13N, 83.16W	612	1941–70[3]	49.1	72.3 Jul	24.6 Jan	98	−13			31.69
Tempe	33.25N, 111.56W	1150	1941–70	68.6	88.9 Jul	50.5 Jan	118	15			7.63
Terre Haute	39.28N, 87.24W	496	1941–70	53.3	75.9 Jul	28.7 Jan	110	−18			38.41

[1] Climatic data for Greenville-Spartanburg Airport, nr Spartanburg.
[2] Climatic data for Mount Clemens, nr Sterling Heights.
[3] Climatic data for Detroit Metropolitan Airport, nr Taylor.

NORTH AMERICA

City	Latitude and Longitude	Elev (ft)	Period of Record	Temperature (°F) Avg	Avg Warmest Month	Avg Coolest Month	Absolute High	Absolute Low	Avg Annual High	Avg Annual Low	Annual Precipitation (in)
Texas City	29.24N, 94.54W	12	(see Galveston)								
Thousand Oaks	34.10N, 118.50W	800	1941–70[1]	63.6	74.6 Aug	54.2 Jan					16.12
Toledo	41.39N, 83.33W	585	1941–70	51.1	74.9 Jul	26.4 Jan	105	–17			31.92
Topeka	39.03N, 95.40W	930	1941–70	54.3	78.2 Jul	28.0 Jan	114	–25			34.66
Torrance	33.49N, 118.18W	75	1941–70	61.5	69.3 Aug	54.2 Jan	111	24			12.21
Towson	39.24N, 76.36W	465	47 yrs bef 1971	54.6	76.2 Jul	35.7 Jan					42.52
Trenton	40.13N, 74.46W	35	1941–70	54.0	75.9 Jul	32.1 Jan	106	–14			40.17
Troy	42.44N, 73.41W	34	[see Albany (New York)]								
Tucson	32.13N, 110.58W	2390	1941–70	68.2	86.5 Jul	51.1 Jan	112	6			11.05
Tulsa	36.09N, 96.00W	804	1941–70	60.2	82.1 Jul	36.6 Jan	115	–16			36.90
Tuscaloosa	33.12N, 87.34W	172	1941–70	63.6	81.2 Jul	45.0 Jan	108	5			49.91
Tyler	32.21N, 95.18W	558	1941–70[2]	65.5	83.5 Jul	46.2 Jan	113	–7			47.18
Upper Darby	39.58N, 75.16W	100	(see Philadelphia)								
Utica	43.06N, 75.14W	415	1941–70[3]	45.4	69.3 Jul	19.5 Jan	100	–26			41.30
Vallejo	38.07N, 122.15W	10	(see San Francisco)								
Ventura	34.17N, 119.18W	48	1941–70	(see Oxnard for temperature data)							13.56
Vineland	39.29N, 75.02W	110	1941–70[4]	54.4	76.1 Jul	32.7 Jan	104	–13			44.40
Virginia Beach	36.51N, 75.58W	13	(see Norfolk)								
Waco	31.33N, 97.08W	405	1941–70	67.1	85.7 Aug	47.0 Jan	111	–5			31.26
Waltham	42.23N, 71.14W	48	(see Boston)								
Warren (Michigan)	42.30N, 83.02W	619	(see Detroit)								
Warren (Ohio)	41.14N, 80.48W	904	1941–70	50.2	71.8 Jul	27.5 Jan	105	–25			34.94

[1] Climatic data for San Fernando, nr Thousand Oaks.
[2] Climatic data for Longview, nr Tyler.
[3] Climatic data for Little Falls, nr Utica.
[4] Climatic data for Hammonton, nr Vineland.

NORTH AMERICA

City	Latitude and Longitude	Elev (ft)	Period of Record	Temperature (°F)							Annual Precipitation (in)
				Avg	Avg Warmest Month	Avg Coolest Month	Absolute High	Absolute Low	Avg Annual High	Avg Annual Low	
Warwick	41.42N, 71.27W	174	(see Providence)								
*Washington	38.54N, 77.01W	25	1941–70	57.3	78.7 Jul	35.6 Jan	106	–15	98	4	38.89
Waterbury	41.33N, 73.03W	260	(see Hartford)								
Waterloo	42.30N, 92.20W	850	1941–70	46.4	72.6 Jul	16.3 Jan	112	–34			33.74
Waukegan	42.21N, 87.50W	669	1941–70	48.4	71.9 Jul	22.5 Jan	108	–24			32.63
West Allis	43.00N, 88.00W	723	(see Milwaukee)								
Westland	42.18N, 83.23W	650	1941–70[1]	49.1	72.3 Jul	24.6 Jan	98	–13			31.69
West Palm Beach	26.43N, 80.03W	16	1941–70	74.5	82.3 Aug	65.5 Jan	101	30			62.06
Wheeling	40.04N, 80.43W	650	1941–70	52.4	73.9 Jul	30.4 Jan	105	–15			37.06
White Plains	41.02N, 73.46W	467	1941–70[2]	51.3	73.9 Jul	28.1 Jan					44.38
Whittier	33.58N, 118.02W	250	1941–70	(see Los Angeles for temperature data)							13.95
Wichita	37.41N, 97.20W	1290	1941–70	56.6	80.7 Jul	31.3 Jan	114	–22			30.58
Wichita Falls	33.55N, 98.29W	945	1941–70	64.1	85.8 Jul	41.5 Jan	113	–12			27.22
Wilkes-Barre	41.15N, 75.53W	640	1941–70	49.4	72.2 Jul	26.0 Jan	103	–19			37.71
Williamsburg	37.16N, 76.42W	81	59 yr bef 1971	58.6	77.4 Jul	40.5 Jan					46.20
Wilmington (Delaware)	39.45N, 75.33W	135	1941–70	53.6	75.0 Jul	31.8 Jan	107	–15			42.99
Wilmington (North Carolina)	34.14N, 77.55W	32	1941–70	63.7	80.4 Jul	46.4 Jan	104	5			53.59
Winston-Salem	36.06N, 80.15W	860	1931–60	59.1	78.0 Jul	40.6 Jan	100	1			44.13
Woodbridge	40.34N, 74.17W	16	1941–70[3]	53.2	74.9 Jul	31.2 Jan	98	–12			43.23
Worcester	42.16N, 71.48W	475	1941–70	47.1	70.1 Jul	23.6 Jan	102	–24			45.24
Yakima	46.36N, 120.31W	1075	1941–70	49.8	70.7 Jul	27.5 Jan	111	–25			8.00
Yonkers	40.56N, 73.54W	10	(see New York)								
York	39.58N, 76.44W	370	1941–70	53.6	75.3 Jul	31.3 Jan	107	–27			40.02

[1] Climatic data for Detroit Metropolitan Airport, nr Westland.
[2] Climatic data for Bedford Hills, nr White Plains.
[3] Climatic data for New Brunswick, nr Woodbridge.

NORTH AMERICA

City	Latitude and Longitude	Elev (ft)	Period of Record	Temperature (°F)							Annual Precipitation (in)
				Avg	Avg Warmest Month	Avg Coolest Month	Absolute High	Absolute Low	Avg Annual High	Avg Annual Low	
Youngstown	41.06N, 80.39W	840	1941–70	48.7	70.7 Jul	25.7 Jan	100	–18			37.99
VIRGIN ISLANDS (USA)											
*Charlotte Amalie	18.21N, 64.56W	15	1941–70	79.9	82.5 Aug	76.9 Feb					42.99

OCEANIA

City	Latitude and Longitude	Elev (ft)	Period of Record	Temperature (°F)							Annual Precipitation (in)
				Avg	Avg Warmest Month	Avg Coolest Month	Absolute High	Absolute Low	Avg Annual High	Avg Annual Low	
AUSTRALIA											
Adelaide	34.56S, 138.36E	140	109 yr bef 1967	62.8	73.5 Jan	51.9 Jul	118	32	110	36	20.77
Brisbane	27.30S, 153.01E	134	79 yr bef 1967	68.8	76.9 Jan	58.7 Jul	110	36	100	39	43.17
*Canberra	35.20S, 149.10E	1906	38 yr bef 1967	56.1	69.1 Jan	42.8 Jul	107	18	99	21	25.34
Geelong	38.09S, 144.21E	57	1911–40	58.0	67.1 Feb	49.2 Jul	111	27			20.85
Hobart	42.55S, 147.20E	177	1921–50	54.2	61.5 Feb	46.6 Jul	105	28	96	31	24.05
Melbourne	37.50S, 145.00E	114	110 yr bef 1967	58.5	67.6 Jan	48.9 Jul	114	27	105	31	25.82
Newcastle	32.55S, 151.45E	106	1911–40	64.4	72.1 Jan	54.5 Jul	112	37			42.17
Perth	31.56S, 115.50E	210	69 yr bef 1967	64.5	74.6 Feb	55.4 Jul	112	34	105	38	34.68
Sydney	33.53S, 151.12E	138	107 yr bef 1967	63.3	71.6 Jan	53.3 Jul	114	36	100	39	45.28
Wollongong	34.25S, 150.54E	150	1911–40	63.0	70.8 Feb	54.3 Jul	115	34			44.04
FIJI											
*Suva	18.08S, 178.25E	30	1921–50	77.1	80.5 Feb	73.6 Jul	98	55	93	60	124.41
FRENCH POLYNESIA											
*Papeete	17.32S, 149.34W	7	1941–60	79.2	81.5 Mar	76.6 Jul	95	61	93	64	72.60

OCEANIA

City	Latitude and Longitude	Elev (ft)	Period of Record	Temperature (°F)							Annual Precipitation (in)
				Avg	Avg Warmest Month	Avg Coolest Month	Absolute High	Low	Avg Annual High	Low	
(HAWAII), USA											
Hilo	19.43N, 155.05W	60	1941–70	73.4	75.9 Aug	71.0 Feb	94	51			133.57
*Honolulu	21.19N, 157.52W	21	1941–70	76.6	80.7 Aug	72.3 Jan	93	52	86	59	22.90
NEW CALEDONIA											
*Noumea	22.16S, 166.27E	10	1941–60	73.6	79.5 Feb	67.8 Jul	99	52	94	56	42.64
NEW ZEALAND											
Auckland	36.51S, 174.45E	160	1931–60	59.5	66.5 Jan	51.5 Jul	90	33	81	37	48.9
Christchurch	43.35S, 172.36E	22	1931–60	52.6	61.5 Jan	42.5 Jul	96	21	88	25	26.3
Dunedin	45.52S, 170.30E	5	1931–60	51.6	59.5 Jan	43.5 Jul	94	23	84	29	31.0
Hamilton	37.47S, 175.17E	131	38 yr bef 1951	55.1	63.3 Feb	46.5 Jul			85	20	45.95
Manukau	36.58S, 174.48E	40	(see Auckland)								
*Wellington	41.28S, 174.51E	415	1931–60	54.3	62.5 Jan	47.5 Jul	88	29	80	32	47.5
PAPUA NEW GUINEA											
*Port Moresby	9.29S, 147.08E	92	1941–60	80.4	81.9 Dec	78.4 Jul	98	64	95	69	40.87
SAMOA											
*Apia	13.50S, 171.44W	7	1921–50	79.5	80.1 Jan	78.4 Jul	98	63	90	66	111.77

SOUTH AMERICA

City	Latitude and Longitude	Elev (ft)	Period of Record	Temperature (°F)							Annual Precipitation (in)
				Avg	Avg Warmest Month	Avg Coolest Month	Absolute High	Low	Avg Annual High	Low	
ARGENTINA											
Almirante Brown	34.48S, 58.23W	82	(see Buenos Aires)								

SOUTH AMERICA

City	Latitude and Longitude	Elev (ft)	Period of Record	Temperature (°F)							Annual Precipitation (in)
				Avg	Avg Warmest Month	Avg Coolest Month	Absolute High	Low	Avg Annual High	Low	
Avellaneda	34.39S, 58.23W	23	(see Buenos Aires)								
Bahia Blanca	38.43S, 62.17W	66	1901–50	59.9	74.7 Jan	46.2 Jul	109	18	103	23	21.26
*Buenos Aires	34.36S, 58.27W	82	1901–50	61.7	74.3 Jan	50.0 Jul	110	22	99	27	38.62
Cordoba	31.24S, 64.11W	1450	1901–50	63.0	75.2 Jan	50.5 Jul	114	13	105	21	27.16
Corrientes	27.28S, 58.50W	180	1901–50	71.1	81.7 Jan	60.8 Jul	112	30	103	35	47.40
General San Martin	34.34S, 58.32W	69	(see Buenos Aires)								
General Sarmiento	34.33S, 58.43W	157	1941–50	61.3	73.8 Jan	49.3 Jul	105	18			36.97
La Matanza	34.40S, 58.33W	89	(see Buenos Aires)								
Lanus	34.43S, 58.24W	33	(see Buenos Aires)								
La Plata	34.55S, 57.57W	59	1941–50	61.0	72.3 Jan	49.3 Jul	92	23			39.21
Lomas de Zamora	34.46S, 58.24W	69	(see Buenos Aires)								
Mar del Plata	38.00S, 57.33W	79	1901–50	56.7	67.5 Jan	46.6 Jul	106	20			30.83
Mendoza	32.53S, 68.49W	2523	1901–50	60.1	74.3 Jan	45.3 Jun	109	15	102	23	7.76
Merlo	34.40S, 58.45W	56	(see Buenos Aires)								
Moron	34.39S, 58.37W	75	(see Buenos Aires)								
Parana	31.44S, 60.32W	243	1941–50	64.4	77.2 Jan	52.5 Jul	113	22	103	30	37.44
Quilmes	34.44S, 58.16W	62	(see Buenos Aires)								
Rosario	32.57S, 60.40W	85	1901–50	64.2	76.5 Jan	51.6 Jun	111	13	103	24	36.42
Salta	24.47S, 65.25W	3878	1901–50	63.3	71.4 Dec	52.3 Jul	103	14			27.20
San Isidro	34.27S, 58.30W	75	(see Buenos Aires)								
San Juan	31.32S, 68.31W	2110	1901–50	63.0	78.1 Jan	47.1 Jul	115	17			3.66
Santa Fe	31.38S, 60.42W	121	1941–50[1]	64.6	76.1 Jan	52.3 Jul	111	19			34.41
Tres de Febrero	34.36S, 58.33W	85	(see Buenos Aires)								
Tucuman	26.49S, 65.13W	1385	1901–50	66.6	77.0 Jan	54.1 Jul	118	19			37.56
Vicente Lopez	34.32S, 58.28W	62	1941–50	62.1	73.2 Jan	50.4 Jul	93	19			35.28

[1]Temperature data for Angel Gallardo, nr Santa Fe.

SOUTH AMERICA

City	Latitude and Longitude		Elev (ft)	Period of Record	Temperature (°F)							Annual Precipitation (in)
					Avg	Avg Warmest Month	Avg Coolest Month	Absolute High	Absolute Low	Avg Annual High	Avg Annual Low	
BOLIVIA												
Cochabamba	17.24S,	66.09W	8390	1941–60	62.8	66.9 Nov	56.7 Jun					20.31
*La Paz	16.30S,	68.09W	11,910	1921–50	50.0	53.4 Nov	45.5 Jul	80	26	75	28	21.93
Potosi	19.35S,	65.45W	13,045	1969	50.5	54.9 Nov	45.0 Jun			75	27	6.69
Santa Cruz	17.48S,	63.10W	1365	1943–60	75.0	79.0 Jan	68.7 Jun					48.39
*Sucre	19.02S,	65.17W	9331	5 yr bef 1953	54	58 Nov	49 Jul	88	25	87	31	27.8
BRAZIL												
Aracaju	10.55S,	37.04W	16	1931–60	77.9	80.4 Mar	74.7 Aug	97	60	90	65	49.33
Belem	1.27S,	48.29W	46	1893–1910	78.6	79.9 Nov	77.4 Feb	98	64	93	67	86.10
Belo Horizonte	19.55S,	43.56W	2723	1954–62	69.8	73.0 Feb	64.6 Jun	97	36	94	49	56.15
*Brasilia	15.47S,	47.55W	3809	1961–65	69.4	72.7 Sep	64.0 Jun	94	43	91	46	64.02
Campina Grande	7.13S,	35.53W	1804	1931–60	73.2	75.9 Feb	69.1 Jul	95	58			27.09
Campinas	22.54S,	47.05W	2274	1889–1943	68.0	72.7 Jan	61.3 Jul	98	32			56.42
Campos	21.45S,	41.18W	46	1912–19	72.7	77.9 Feb	67.8 Jul	102	45			42.64
Curitiba	25.25S,	49.15W	3117	1921–50	61.2	67.6 Jan	54.0 Jul	99	16	90	32	57.17
Duque de Caxias	22.47S,	43.18W	13	(see Rio de Janeiro)								
Feira de Santana	12.15S,	38.57W	820									
Florianopolis	27.35S,	48.34W	7	14 yr bef 1965	69.3	76.6 Jan	61.7 Jul	97	34	91	44	46.69
Fortaleza	3.43S,	38.30W	82	1931–60	80.1	81.0 Jan	78.8 Jul	95	64	89	69	49.45
Goiania	16.40S,	49.16W	2493	1958–65	73.2	77.2 Sep	67.6 Jun	102	36	98	42	55.16
Guarulhos	23.29S,	46.31W	2493	(see Sao Paulo)								
Jaboatao	8.07S,	35.01W	148	(see Recife)								
Joao Pessoa	7.07S,	34.53W	148	1931–60	77.9	80.2 Feb	74.7 Jul	94	61			65.67
Juiz de Fora	21.45S,	43.20W	2218	1893–1922	67.1	73.9 Feb	59.9 Jul	98	33			57.91
Jundiai	23.11S,	46.52W	2461	1904–20, 1928	66.7	71.4 Feb	60.8 Jun					58.19
Londrina	23.18S,	51.09W	1969	1932–56	69.1	74.8 Jan	62.2 Jun	104	32			56.69

SOUTH AMERICA

City	Latitude and Longitude	Elev (ft)	Period of Record	Temperature (°F)							Annual Precipitation (in)
				Avg	Avg Warmest Month	Avg Coolest Month	Absolute High	Absolute Low	Avg Annual High	Avg Annual Low	
Maceio	9.40S, 35.43W	16	1931–60	77.7	80.1 Feb	74.7 Jul	94	63	90	66	65.12
Manaus	3.08S, 60.01W	197	1910–29, 1931–40	80.4	82.2 Sep	79.3 Jan	101	64	97	67	71.3
Natal	5.47S, 35.13W	10	1931–60	79.2	81.1 Feb	75.9 Jul	98	59	89	65	60.91
Niteroi	22.53S, 43.07W	8	1954–63	75.2	80.8 Jan	69.3 Jun	104	48	102	53	44.61
Nova Iguacu	22.45S, 43.27W	85	(see Rio de Janeiro)								
Olinda	8.01S, 34.51W	98	1941–60	78.4	81.0 Feb	75.0 Aug	92	64	89	66	68.23
Osasco	23.32S, 46.46W	2366	(see Sao Paulo)								
Pelotas	31.46S, 52.20W	23	1912–19	63.9	73.6 Feb	54.1 Jun	102	28			52.05
Petropolis	22.31S, 43.10W	2667	1913–19	64.8	70.2 Jan	59.5 Jul	92	33			83.54
Porto Alegre	30.04S, 51.11W	13	22 yr bef 1924	67	77.5 Feb	57.5 Jun	105	25	101	32	49.1
Recife	8.03S, 34.54W	180	1931–60	78.6	80.8 Feb	75.6 Jul	94	63	90	67	69.41
Ribeirao Preto	21.10S, 47.48W	1700	1901–17	70.2	75.0 Feb	62.6 Aug	104	29	96	56	55.12
Rio de Janeiro	22.54S, 43.14W	102	1921–50	73.2	78.4 Feb	68.4 Jul	102	50	96	56	40.20
Salvador	12.59S, 38.31W	28	1931–60	76.6	79.3 Feb	73.2 Aug	96	62	91	65	72.13
Santo Andre	23.40S, 46.31W	2438	(see Sao Paulo)								
Santos	23.57S, 46.20W	13	1888–1941	71.4	77.4 Feb	65.5 Jul	107	41	101	49	87.87
Sao Bernardo do Campo	23.42S, 46.33W	2507	(see Sao Paulo)								
Sao Goncalo	22.51S, 43.04W	43	(see Rio de Janeiro)								
Sao Joao de Meriti	22.48S, 43.22W	233	(see Rio de Janeiro)								
Sao Luis	2.31S, 44.16W	13	13 yr bef 1963	79.7	81.0 Oct	79.0 Apr	96	66	92	70	74.05
Sao Paulo	23.32S, 46.37W	2690	1921–50	64.6	70.0 Feb	58.1 Jul	95	35	92	40	54.41
Sorocaba	23.29S, 47.27W	1805	20 yr bef 1942	68.9	74.5 Feb	61.9 Jul					42.01
Teresina	5.05S, 42.49W	213	1931–60	81.3	85.1 Oct	79.2 Mar	105	58	102	63	51.06
Vitoria	20.19S, 40.21W	10	1928–32, 1934–58	75.2	81.0 Feb	71.8 Jul	100	50	95	58	51.02
CHILE											
Antofagasta	23.39S, 70.24W	400	1951–60	61.5	68.5 Feb	56.1 Jul	85	41	81	46	0.02

SOUTH AMERICA

City	Latitude and Longitude	Elev (ft)	Period of Record	Temperature (°F)							Annual Precipitation (in)
				Avg	Avg Warmest Month	Avg Coolest Month	Absolute High	Absolute Low	Avg Annual High	Avg Annual Low	
Concepcion	36.50S, 73.03W	43	? yr bef 1950	56.3	62.6 Jan	50.0 Jul			90	34	51.57
*Santiago	33.27S, 70.40W	1706	1921–50	57.4	68.7 Jan	46.2 Jul	99	24	92	27	13.90
Valparaiso	33.02S, 71.38W	135	22 yr bef 1933	57.7	63.1 Jan	52.5 Jul	94	36	86	39	20.94
Vina del Mar	33.02S, 71.34W		(see Valparaiso)								
COLOMBIA											
Barranquilla	10.59N, 74.48W	16	1942–53, 1957–61	82.0			108	55	99	65	29.29
*Bogota	4.36N, 74.05W	8675	1941–60	57.2	58.8 Apr	56.8 Aug	77	32	75	35	35.51
Bucaramanga	7.08N, 73.09W	3035	1946–48, 1950–61	73.9			102	58	90	62	49.92
Cali	3.27N, 76.31W	3432	1942–49, 1952–59	74.8			95	57	90	62	44.09
Cartagena	10.25N, 75.32W	0	1942–50, 1952–61	80.8			107	56	94	68	33.07
Cucuta	7.54N, 72.31W	965	1942–54, 1958–61	81.1			102	52	94	69	17.56
Ibague	4.27N, 75.14W	4098	1941–47, 1949–59, 1961	71.4			95	45	91	55	80.83
Manizales	5.04N, 75.37W	7021	1942, 1944–52	64.6			84	45	80	51	80.91
Medellin	6.15N, 75.35W	5056	1942–55, 1957–61	70.7			102	45	91	50	51.57
Pereira	4.49N, 75.43W	4672	1942, 1944, 1946–53	71.2			97	46	89	54	80.87
ECUADOR											
Guayaquil	2.10S, 79.50W	10	1931–60	76.8	79.5 Mar	73.8 Aug	98	57	96	61	43.31
*Quito	0.13S, 78.30W	9249	1931–60	55.4	55.8 Sep	55.0 Nov	86	32	81	35	48.54
FRENCH GUIANA											
*Cayenne	4.56N, 52.20W	30	1951–60	78.1	79.3 Oct	77.5 Jan	97	65	94	69	147.40
GUYANA											
*Georgetown	6.49N, 58.10W	7	1887–1940	80.6	82.0 Sep	79.3 Jan	93	68	90	70	87.44

SOUTH AMERICA

City	Latitude and Longitude	Elev (ft)	Period of Record	Temperature (°F) Avg	Avg Warmest Month	Avg Coolest Month	Absolute High	Absolute Low	Avg Annual High	Avg Annual Low	Annual Precipitation (in)
PARAGUAY											
*Asuncion	25.16S, 57.40W	210	1921-50	74.7	84.0 Jan	63.7 Jul	110	29	105	35	49.84
PERU											
Arequipa	16.24S, 71.33W	7559	1896-1911, 1921-25	58.3	58.8 Sep	57.5 Jul	82	30	79	32	4.31
Callao	12.04S, 77.09W	20	9 yr bef 1956[1]	67.3	73.6 Feb	61.7 Aug					0.47
Chiclayo	6.46S, 79.51W	89	1963-72	69.3			93	54	91	55	1.63
Cuzco	13.31S, 71.59W	11,152	1931-43	55	58 Nov	50.5 Jul	84	16	82	23	32.0
*Lima	12.03S, 77.03W	505	1931-60	65.1	72.3 Feb	59.2 Aug	93	49	89	51	1.22
Trujillo	8.07S, 79.02W	108	1963-72	66.4			85	50	82	53	0.54
SURINAM											
*Paramaribo	5.50N, 55.13W	12	1931-60	81.1	83.3 Oct	79.5 Jan	99	62	96	67	88.90
URUGUAY											
*Montevideo	34.53S, 56.11W	72	1901-60	61.3	72.5 Jan	50.9 Jul	109	25	92	36	39.92
VENEZUELA											
Barquisimeto	10.04N, 69.19W	1857	1938-47, 1951-60	75.2	76.3 Oct	73.9 Jan	97	59	94	61	20.43
*Caracas	10.30N, 66.55W	3025	1941-60	69.4	71.1 May	66.2 Jan	92	45	87	49	32.88
El Valle	10.27N, 66.55W	2904	1940-44	69.3	71.1 May	66.2 Jan	92	43	89	45	31.38
Maracaibo	10.40N, 71.37W	20	1938-46, 1951-60	82.4	84.3 Aug	80.5 Jan	102	66	99	69	18.70
Maracay	10.15N, 67.36W	1460	1938-47, 1951-60	76.5	79.2 Apr	74.3 Jan	99	47	97	54	35.79
Petare	10.29N, 66.49W	2769	1971	67.8	70.0 May	65.3 Jan					30.43
San Felix de Guayana	8.17N, 62.44W										
Valencia	10.11N, 68.00W	1568	1937-46	76.3	78.1 Apr	75.2 Jan	101	61	94	61	45.31

[1] Climatic data for La Punta, nr Callao.

Table 11c. Supplemental Information
AFRICA

Algeria

*Algiers: ^{b}alt Alger; off Jazair. ^{c}B of Algiers of Mediterranean Sea. ^{d}Universite d'Alger (1859) 17 – 1530; Universite des Sciences et de la Technologie d'Alger (1974) 4 – 480. ^{e}Bibliotheque Nationale (1835) 650; Bibliotheque de l'Universite . . . (1879) 600.

Annaba: ^{b}for Bone. ^{c}Mediterranean Sea. ^{d}Universite de Annaba (1975) ? – ? ^{e}Bibliotheque Municipale (?) 20.

Constantine: ^{b}off Qusantina. ^{c}Rummel R. ^{d}Universite de Constantine (1961) 8 – 1023. ^{e}Bibliotheque Municipale (1862) 35.

Oran: ^{b}off Ouahran. ^{c}Mediterranean Sea. ^{d}Universite d'Oran (1961) 7 – 557. ^{e}Bibliotheque Municipale (1861) 32.

Angola

*Luanda: ^{b}for Loanda, Sao Paulo de Loanda. ^{c}Atlantic O. ^{d}Universidade de Luanda (1962) 3 – 293. ^{e}Biblioteca Nacional de Angola (1938) 25; Biblioteca Municipal (?) 18.

Benin

Cotonou: ^{b}for Kotonu. ^{c}Bight of Benin of G of Guinea. ^{d}Universite Nationale du Benin (1962) 3 – 126. ^{e}Bibliotheque de l'Universite . . . (1962?) 17; Bibliotheque du Service des Travaux Publiques (?) ?

*Porto-Novo: ^{b}none. ^{c}Porto-Novo Lagoon. ^{d}none. ^{e}Bibliotheque Nationale (1961) 7.

Botswana

*Gaborone: ^{b}for Gaberones. ^{c}nr Notwani R. ^{d}University of Botswana, Lesotho and Swaziland, Gaborone campus (1972) 0.2 – 24. ^{e}Botswana National Library Service (1967) 80.

Burundi

*Bujumbura: ^{b}for Usumbura. ^{c}Tanganyika L. ^{d}Universite Officielle de Bujumbura (1958) 0.6 – 122. ^{e}Bibliotheque de l'Universite . . . (1960) 40; Bibliotheque Publique (?) 26.

Cameroon

Douala: ^{b}alt Duala. ^{c}Wouri R. ^{d}none. ^{e}Bibliotheque du Centre IFAN du Cameroun (?) 5.

*Yaounde: ^{b}alt Yaunde. ^{c}none. ^{d}Universite de Yaounde (1962) 8 – 390. ^{e}Bibliotheque de l'Universite . . . (1962) 65; Bibliotheque du Centre Culturel Francais (1960) 20; Bibliotheque Nationale du Cameroun (?) 10.

(Canary Islands), Spain

*Las Palmas: ^{a}Gran Canaria I. ^{b}alt Las Palmas de Gran Canaria, Palmas. ^{c}Atlantic O. ^{d}none. ^{e}Biblioteca del Museo Canario (1879) 40; Biblioteca Publica Provincial (?) 28.

Santa Cruz de Tenerife: ^{a}Tenerife I. ^{b}none. ^{c}Atlantic O. ^{d}Universidad de La Laguna (at La Laguna, nr Santa Cruz de Tenerife) (1701) 15 – 648. ^{e}Biblioteca de la Universidad . . . (at La Laguna, nr Santa Cruz de Tenerife) (1701) 61; Biblioteca Municipal (?) ?

Cape Verde

*Praia: ^{a}Sao Tiago I. ^{b}none. ^{c}Atlantic O. ^{d}none.

Central African Empire

*Bangui: ^{b}none. ^{c}Ubangi R. ^{d}Universite Jean-Bedel Bokassa (1970) 0.4 – 84. ^{e}Bibliotheque Nationale (bef 1966) ?

Chad

*Ndjamena: ^{b}for Fort-Lamy. ^{c}Shari R; Logone R. ^{d}Universite du Tchad (1971) 0.6 – 57. ^{e}Bibliotheque de l'Universite ... (1971) 3; Bibliotheque du Centre de Documentation Pedagogique (1962) 2; Bibliotheque de l'Ecole Nationale d'Administration (1963?) ?

Congo

*Brazzaville: ^{b}none. ^{c}Congo R. ^{d}Universite Marien-Ngouabi (1959) 4 – 230. ^{e}Bibliotheque de l'Universite ... (1959) 70; Bibliotheque Publique (1950) 15.

Djibouti

*Djibouti: ^{b}alt Jibuti. ^{c}G of Aden. ^{d}none.

Egypt

Alexandria: ^{b}off Iskandariyah. ^{c}Mediterranean Sea; Mareotis (off Maryut) L. ^{d}University of Alexandria (1942) 75 – 2955. ^{e}University ... Library (1942) 1000; Municipal Library (1892) 90.

Aswan: ^{b}alt Assuan. ^{c}Nile R. ^{d}none.

Asyut: ^{b}alt Assiut; for Siut. ^{c}Nile R. ^{d}University of Asyut (1949) 13 – 380. ^{e}University ... Library (1957) 60.

*Cairo: ^{b}off Qahirah. ^{c}Nile R. ^{d}Ain Shams University (1950) 69 – 1262; Al-Azhar University (970) 32 – 1354; American University in Cairo (1919) 2 – 195. ^{e}Egyptian National Library (alt Darul-Kutub) (1870) 1500; Ain Shams University Library (1839) 410; Bibliotheque de l'Institut d'Egypte (1798) 180.

Damanhur: ^{b}none. ^{c}nr Rosetta (off Rashid) R (Nile R distributary). ^{d}none. ^{e}Municipal Library (?) 13.

Giza: ^{b}alt Gizeh; inc Imbabah since 1960; off Jizah. ^{c}Nile R. ^{d}University of Cairo (1908) 62 – 3402. ^{e}University ... Library (1828) 1000. ^{f}Imbabah: 136 (1960), 1 (1947?).

Ismailia: ^{b}off Ismailiyah. ^{c}Timsah L. ^{d}none.

Luxor: ^{b}off Uqsur. ^{c}Nile R. ^{d}none.

Mahalla al Kubra: ^{b}for Mehallet el Kebir. ^{c}nr Damietta (off Dumyat) R (Nile R distributary). ^{d}none.

Mansurah: ^{b}alt Mansura. ^{c}Damietta (off Dumyat) R (Nile R distributary). ^{d}Mansurah University (1960) 26 – 500. ^{e}Municipal Library (?) 22.

Port Said: ^{b}off Bur Said. ^{c}Mediterranean Sea; Manzala Lagoon. ^{d}none.

Shubra al Khaymah: ^{b}alt Shubra el-Kheima. ^{c}Nile R. ^{d}none.

Suez: ^{b}off Suways. ^{c}G of Suez of Red Sea. ^{d}none.

Tanta: ^{b}none. ^{c}nr Rosetta (off Rashid) R (Nile R distributary). ^{d}Tanta University (1963) 27 – 1151. ^{e}Municipal Library (?) 23.

Zagazig: ^{b}off Zaqaziq. ^{c}none. ^{d}Zagazig University (1974) ? – ? ^{e} Sharqiyah Provincial Council Library (?) 15.

Ethiopia

*Addis Ababa: ^{b}none. ^{c}nr Akaki R. ^{d}University of Addis Ababa (1950) 5 – 650. ^{e}University ... Library (1961) 320; National Library of Ethiopia (1944) 100.

Asmara: ^{a}*Eritrea. ^{b}none. ^{c}nr Anseba R. ^{d}University of Asmara (1958) 1 – 107. ^{e}University ... Library (1958?) 32; Public Library (1955) 5.

Harar: ^{b}none. ^{c}Errer R. ^{d}none. ^{e}National Military Academy Library (1950) 6.

Gabon

*Libreville: ^{b}none. ^{c}G of Guinea; Gabon (for Gabun) R. ^{d}Universite Nationale du Gabon (1970) 0.6 – 135. ^{e}Bibliotheque du Centre d'Information (1960) 10.

Gambia

*Banjul: ^{b}for Bathurst. ^{c}Atlantic O; Gambia R. ^{d}none. ^{e}British Council Library (1946) 20; National Library (1946) 20.

Ghana

*Accra: ^{b}for Akkra. ^{c}G of Guinea. ^{d}University of Ghana (at Legon, nr Accra) (1927) 4 – 550. ^{e}Ghana Library Board (1950) 800; Balme Library (alt University . . . Library) (at Legon, nr Accra) (1948) 278.

Kumasi: ^{b}for Coomassie. ^{c}none. ^{d}University of Science and Technology (1951) 2 – 260. ^{e}University . . . Library (1951) 85; Ashanti Regional Library of the Ghana Library Board (?) 40.

Guinea

*Conakry: ^{b}for Konakry. ^{c}Atlantic O. ^{d}none. ^{e}Bibliotheque de l'Institut Polytechnique de Conakry (1963) 18; Bibliotheque Nationale (1960) 10.

Guinea-Bissau

*Bissau: ^{a}Bissau I. ^{b}none. ^{c}Atlantic O; Geba R. ^{d}none. ^{e}Biblioteca Nacional (1970) 25; Biblioteca do Museu da Guine-Bissau (1947) 14; Biblioteca do Centro de Estudos da Guine-Bissau (1945) 10.

Ivory Coast

*Abidjan: ^{b}none. ^{c}Ebrie Lagoon. ^{d}Universite Nationale de Cote d'Ivoire (1959) 8 – 403. ^{e}Bibliotheque de l'Universite . . . (1963) 50; Bibliotheque Municipale (1952) 50; Bibliotheque Nationale (1968) 16.

Bouake: ^{b}none. ^{c}none. ^{d}none.

Kenya

Mombasa: ^{b}alt Mvita. ^{c}Indian O. ^{d}none. ^{e}Seif Bin Salim Public Library (1903) 19.

*Nairobi: ^{b}none. ^{c}Nairobi R; Mathari R. ^{d}University of Nairobi (1954) 5 – 600. ^{e}University . . . Library (1956) 170; McMillan Memorial Library (1931) 150; Kenya National Library Service (1967) 87.

Liberia

*Monrovia: ^{b}none. ^{c}Atlantic O. ^{d}University of Liberia (1851) 2 – 190. ^{e}University . . . Library (1862) 80; Cuttingham College and Divinity School Library (1889) 60; Government Public Library (1959) 20.

Libya

Bengasi: ^{a}*Cyrenaica. ^{b}alt Benghazi; off Banghazi. ^{c}Mediterranean Sea. ^{d}University of Garyounis (1955) 9 – 354. ^{e}University . . . Library (1955) 127; Public Library (1955) 11.

*Tripoli: ^{a}*Tripolitania. ^{b}off Tarabulus. ^{c}Mediterranean Sea. ^{d}Alfateh University (1955) 4 – 582. ^{e}Public Library (1917) 35.

Madagascar

*Antananarivo: ^{b}for Tananarive. ^{c}Ikopa R; Anosy L. ^{d}Universite de Madagascar (1896) 7 – 250. ^{e}Bibliotheque Nationale (1961) 150; Bibliotheque Universitaire (1960) 120.

(Madeira Islands), Portugal

*Funchal: ^{a}Madeira I. ^{b}none. ^{c}Atlantic O. ^{d}none. ^{e}Biblioteca Municipal (1838) 36.

Malawi

Blantyre: ^{b}inc Limbe since 1949. ^{c}Mudi R. ^{d}University of Malawi, Blantyre campus (alt Malawi Polytechnic) (1964) 0.3 – 83. ^{e}Malawi National Library Service (1968) 90. ^{f}Limbe: 9 (1949e).

*Lilongwe: ^{b}none. ^{c}Lilongwe R. ^{d}none. ^{e}British Council Library (1963) 8.

Mali

*Bamako: ^{b}none. ^{c}Niger R. ^{d}none. ^{e}Centre Francais de Documentation (1962) 13; Bibliotheque‑ Nationale (1913) 8; Bibliotheque Municipale (1949) ?

Timbuktu: ^{b}off Tombouctou. ^{c}nr Niger R. ^{d}none. ^{e}Centre de Documentation Arabe (1966) ?

Mauritania

*Nouakchott: ^{b}none. ^{c}nr Atlantic O. ^{d}none. ^{e}Bibliotheque Nationale (1965) 10; Bibliotheque Publique Centrale (?) ?

Mauritius

*Port Louis: ^{b}none. ^{c}Indian O. ^{d}University of Mauritius (at Reduit, nr Port Louis) (1965) 1 – 138. ^{e}Municipal Library (1951) 60; Mauritius Institute Public Library (1902) 51; University . . . Library (at Reduit, nr Port Louis) (1965?) 45.

Morocco

Casablanca: ^{b}off Dar al Baida. ^{c}Atlantic O. ^{d}none. ^{e}Bibliotheque Municipale (1917) 240.

Fez: ^{b}alt Fes; off Fas. ^{c}Fez (alt Fes; off Fas) R. ^{d}Universite Qarawiyine (859) 0.9 – 607. ^{e}Bibliotheque du Centre Culturel Francais (1960) 21; Bibliotheque Municipale (1950) 7.

Marrakesh: ^{b}alt Marrakech; for Morocco; off Marrakush. ^{c}Issil R. ^{d}Universite Ben Youssef de Marrakech (14th c) 1 – ?. ^{e}Bibliotheque de l'Universite . . . (14th c ?) 18; Bibliotheque du Centre Culturel Francais (1960) 15; Bibliotheque Municipale (1923) 13.

Meknes: ^{b}for Mequinez; off Miknasa. ^{c}Boufekrane R. ^{d}none. ^{e}Bibliotheque de la Province de Meknes (?) 53; Bibliotheque Generale (?) 14.

Oujda: ^{b}none. ^{c}nr Isly R. ^{d}none. ^{e}Bibliotheque du Centre Culturel Ibn Khaldoun (?) 30; Bibliotheque de Dar el Hikma (1964) 14; Bibliotheque Municipale (?) 5.

*Rabat-Sale: ^{b}inc Rabat, Sale; Sale: for Sallee; off Sla. ^{c}Atlantic O; Bou Regreg R. ^{d}Universite Mohammed V (1912) 20 – 641. ^{e}Bibliotheque Generale et Archives (1920) 230; Bibliotheque de l'Universite Mohammed V (1957) ? ^{f}Rabat: 368 (1971), 227 (1960), 156 (1952), 53 (1931), 31 (1900e), 28 (1850e), 17 (1800e); Sale: 156 (1971), 76 (1960), 47 (1952), 26 (1931), 30 (1900e), 10 (1850e).

Tangier: ^{b}alt Tanger; off Tanjah. ^{c}Strait of Gibraltar. ^{d}none. ^{e}Biblioteca Publica Espanola Antonio de Nebrija (1941) 36; Bibliotheque du Centre Culturel Francais (?) 27.

Mozambique

*Maputo: ^{b}for Lourenco Marques. ^{c}Delagoa B of Indian O. ^{d}Universidade Eduardo Mondlane (1962) 1 – 250. ^{e}Biblioteca Nacional de Mocambique (1961) 110; Biblioteca da Universidade . . . (1963) 75.

Namibia

*Windhoek: ^{b}for Aigams, Windhuk. ^{c}none. ^{d}none. ^{e}Public Library (1924) 52.

Niger

*Niamey: ^{b}none. ^{c}Niger R. ^{d}Universite de Niamey (1971) 0.8 – 200. ^{e}Bibliotheque de l'Ecole Nationale d'Administration du Niger (1963) 11.

Nigeria

Abeokuta: ^{b}none. ^{c}Ogun R. ^{d}none.

Ado-Ekiti: ^{b}none. ^{c}nr Ogbesse R. ^{d}none. ^{e}Ado-Ekiti District Council Library (1966) 0.5.

Enugu: ^{b}none. ^{c}Asata R; Aria R. ^{d}none. ^{e}East Central State Central Library (1956) 134.

Ibadan: ^{b}none. ^{c}Ogunpa R. ^{d}University of Ibadan (1948) 7 – 669; University of Ife, Ibadan campus (1961) 7 – 891; ^{e}University . . . Library (1948) 278; Western State Library (1956) 94.

Ilesha: ^{b}none. ^{c}nr Shasha R. ^{d}none.

Ilorin: [b]none. [c]Awun R. [d]University of Ilorin (1976) 0.2 – 56. [e]Kwara State Library (?) ?

Iwo: [b]none. [c]nr Oba R. [d]none.

Kaduna: [b]none. [c]Kaduna R. [d]none. [e]North Central State Library (1952) 40; British Council Library (1961) 20.

Kano: [b]none. [c]Jakara R. [d]Bayero University (1960) 1 – 126. [e]Kano State Library (1968) 40.

*Lagos: [b]none. [c]Bight of Benin of G of Guinea; Lagos (for Cradu) Lagoon. [d]University of Lagos (1962) 5 – 486. [e]University . . . Library (1962) 126; National Library of Nigeria (1964) 100; City Library (1946) 69.

Mushin: [b]none. [c]nr Ogun R. [d]none.

Ogbomosho: [b]none. [c]nr Oba R. [d]none.

Onitsha: [b]none. [c]Niger R. [d]none. [e]Divisional Library of East Central State Library Board (?) ?

Oshogbo: [b]none. [c]Oshun R. [d]none.

Port Harcourt: [b]none. [c]Bonny R (Niger R distributary). [d]none. [e]College of Science and Technology Library (1971) 20; State Library (?) ?

Zaria: [b]none. [c]Galma R. [d]Ahmadu Bello University (1946) 14 – 1173. [e]Kashim Ibrahim Library of the . . . University (1962) 112.

Reunion

*Saint-Denis: [b]none. [c]Indian O; Saint-Denis R. [d]none. [e]Bibliotheque Centrale de Pret (1956) 91; Bibliotheque Departementale (1856) 55.

Rhodesia

Bulawayo: [b]none. [c]Matsheumhlope R. [d]none. [e]Public Library (1896) 60; National Free Library of Rhodesia (1944) 50.

*Salisbury: [b]none. [c]Makabusi R. [d]University of Rhodesia (1953) 2 – 239. [e]University . . . Library (1956) 259; Library of Parliament (1899) 90; Queen Victoria Public Library (1902) 49.

Rwanda

*Kigali: [b]none. [c]nr Nyabarongo R. [d]none. [e]Bibliotheque Publique (?) 10.

Senegal

*Dakar: [b]none. [c]Atlantic O. [d]Universite de Dakar (1918) 4 – 480. [e]Bibliotheque de l'Universite . . . (1950) 201; Bibliotheque de l'Institut Fondamental d'Afrique Noire (1936) 55.

Sierra Leone

*Freetown: [b]none. [c]Atlantic O. [d]University of Sierra Leone (1827) 2 – 266. [e]Sierra Leone Library Board (1959) 398; University . . . Library (1827) 125.

Somalia

*Mogadishu: [b]alt Mogadiscio, Muqdisho; off Hamar. [c]Indian O. [d]Universita Nazionale della Somalia (1954) 3 – 446. [e]National Library (1934) 8; Biblioteca dell'Universita . . . (1954) ?

South Africa

Bloemfontein: [a]*Orange Free State. [b]none. [c]Bloem(spruit) Creek. [d]University of the Orange Free State (alt Universiteit van die Oranje-Vrystaat) (1855) 8 – 532. [e]Orange Free State Provincial Library Service (alt Oranje-Vrystaat Provinsiale Biblioteekdiens) (1948) 1300; Public Library (alt Openbare Biblioteek) (1875) 213.

*Cape Town: [a]*Cape of Good Hope. [b]alt Kaapstad. [c]Table B of Atlantic O. [d]University of Cape Town (alt Universiteit van Kaapstad) (1829) 8 – 500. [e]Cape Provincial Library Service (alt Kaapse Provinsiale Biblioteekdiens (1945) 4073; City Library (alt Stadsbiblioteek) (1952) 1100; University . . . Library (alt Universiteit . . . Biblioteek) (1905) 648; South African Public Library (alt Suid-Afrikaanse Openbare Biblioteek) (1820) 520.

Durban: [a]Natal. [b]none. [c]Natal B of Indian O. [d]University of Durban-Westville (alt Universiteit van Durban-Westville) (1960) 4 – 272; University of Natal (alt Universiteit van Natal), Durban campus (1922) ? – ? [e]Municipal Library (alt Munisipale Biblioteek) (1853) 600; University of Natal Library (alt Universiteit van Natal Biblioteek), Durban campus (1936) 236.

Germiston: [a]Transvaal. [b]none. [c]Germiston L. [d]none. [e]Carnegie Public Library (alt Carnegie Openbare Biblioteek) (1909) 88.

Johannesburg: [a]Transvaal. [b]none. [c]none. [d]University of the Witwatersrand (alt Universiteit van Witwatersrand) (1896 at Kimberley, relocated 1904) 11 – 1089; Rand Afrikaans University (alt Randse Afrikaanse Universiteit) (1966) 3 – 220. [e]Public Library (alt Openbare Biblioteek) (1890) 1404; University . . . Library (alt Universiteit . . . Biblioteek) (1922) 600.

Pietermaritzburg: [a]*Natal. [b]none. [c]Umsindusi R. [d]University of Natal (alt Universiteit van Natal) (1909) 8 – 790 [Pietermaritzburg campus: (1909) ? – ?]. [e]Natal Provincial Library Service (alt Natalse Provinsiale Biblioteekdiens) (1951) 909; University . . . Library (alt Universiteit . . . Biblioteek) (1912) 376 [Pietermaritzburg campus: (1912) 140].

Port Elizabeth: [a]Cape of Good Hope. [b]none. [c]Algoa B of Indian O. [d]University of Port Elizabeth (alt Universiteit van Port Elizabeth) (1964) 3 – 260. [e]Municipal Library (alt Munisipale Biblioteek) (1848) 300.

*Pretoria: [a]*Transvaal. [b]none. [c]Apies R. [d]University of Pretoria (alt Universiteit van Pretoria) (1908) 15 – 1005. [e]Transvaal Provinical Library Service (alt Transvaalse Provinsiale Biblioteekdiens) (1943) 3270; State Library (alt Staatsbiblioteek) (1887) 700; University . . . Library (alt Universiteit . . . Biblioteek) (1908) 491.

Vereeniging: [a]Transvaal. [b]none. [c]Vaal R. [d]none. [e]Public Library (alt Openbare Biblioteek) (1912) 23.

Sudan

*Khartoum: [b]off Khurtum. [c]Nile R; Blue Nile R. [d]University of Khartoum (1902) 8 – 788; University of Cairo, Khartoum campus (1955) 5 – 80. [e]University of Khartoum Library (1945) 209.

Omdurman: [b]off Umm Durman. [c]Blue Nile R. [d]Islamic University of Omdurman (1912) 1 – 65. [e]. . . University Library (1912) 70; Public Library (1951) 20.

Tanzania

*Dar es Salaam: [a]*Tanganyika. [b]none. [c]Indian O. [d]University of Dar es Salaam (1961) 3 – 535. [e]Tanzania Library Service (1964) 900; University . . . Library (1961) 203.

Zanzibar: [a]*Zanzibar; Zanzibar I. [b]alt Unguja. [c]Zanzibar Channel. [d]none. [e]Shree Vanik Library (?) 7; Museum Library (?) 3.

Togo

*Lome: [b]none. [c]Bight of Benin of G of Guinea. [d]Universite du Benin (1962) 2 – 121. [e]Bibliotheque Nationale (1960) 8; Bibliotheque de l'Universite . . . (1970) 5.

Tunisia

Sfax: [b]off Safaqis. [c]Mediterranean Sea. [d]none. [e]Bibliotheque Publique (?) ?

*Tunis: [b]none. [c]Tunis Lagoon. [d]Universite de Tunis (1945) 16 – 1427. [e]Bibliotheque Nationale de Tunisie (1883) 500; Bibliotheque Publique Centrale (1852) 150.

Uganda

*Kampala: [b]none. [c]nr Victoria L. [d]Makerere University (1922) 4 – 377. [e]. . . University Library (1922) 400; Public Library Board (1964) 140.

Upper Volta

*Ouagadougou: [b]for Wagadugu. [c]nr Ouadmana R. [d]Universite de Ouagadougou (1965) 1 – 87. [e]Bibliotheque de l'Universite. . . (1965?) 30.

Zaire

Kananga: [b]for Luluabourg. [c]Lulua R. [d]none. [e]Bibliotheque Publique (1896) 25.

*Kinshasa: [b]for Leopoldville. [c]Congo R. [d]Universite Nationale du Zaire (1925) 15 – 1410 [Kinshasa campus: (1925) 5 – 513]. [e]Bibliotheque Centrale de l'Universite . . . (1954) 386 [Kinshasa campus: (1954) 300]; Bibliotheque Centrale du Zaire (1949) 90.

Kisangani: [b]for Stanleyville. [c]Congo R; Tshope R. [d]Universite Nationale du Zaire, Kisangani campus (1963) 2 – 94. [e]Bibliotheque Centrale de l'Universite . . . , Kisangani campus (1963?) 26; Bibliotheque Publique (1930) 20.

Lubumbashi: [b]for Elisabethville. [c]Lubumbashi R. [d]Universite Nationale du Zaire, Lubumbashi campus (1956) 3 – 65. [e]Bibliotheque Centrale de l'Universite . . . , Lubumbashi campus (1955) 60.

Mbuji-Mayi: [b]for Bakwanga. [c]Bushimaie R. [d]none.

Zambia

Kitwe: [b]none. [c]nr Kafue R. [d]none. [e]Central Technical Library of Rhokana Corporation, Ltd (1954) 20; Hammarskjold Memorial Library (1963) 20.

*Lusaka: [b]none. [c]none. [d]University of Zambia (1965) 3 – 352. [e]University . . . Library (1965?) 480; Zambia Library Service (1962) 300; Public Library (1943) 70.

Ndola: [b]none. [c]none. [d]none. [e]Public Library (1936) 50.

ASIA

Afghanistan

Herat: [b]none. [c]Hari (Rud) R. [d]none.

*Kabul: [b]for Cabul. [c]Kabul (for Cabul) R. [d]Kabul University (1931) 10 – 1115. [e]. . . University Library (1931) 120; Public Library (1920) 65.

Kandahar: [b]alt Qandahar. [c]Arghandab R. [d]none. [e]Teachers' Training School Library (1956) ?

Bahrain

*Manama: [a]Bahrain I. [b]off Manamah. [c]Persian G. [d]none. [e]British Council Library (?) 7.

Bangladesh

Chittagong: [b]for Islamabad, Porto Grande. [c]Karnaphuli R. [d]University of Chittagong (1966) 3 – 283. [e]University . . . Central Library (1869) 57; Municipal Public Library (1904) 15.

*Dacca: [b]for Jahangirnagar. [c]Burhi Ganga R (Brahmaputra R distributary). [d]University of Dacca (1910) 58 – ?; Bangladesh University of Engineering and Technology (1961) 2 – 164; Jahangirnagar University (1970) 0.9 – 112. [e]University . . . Library (1921) 340; Central Public Library (1958) 75; National Assembly Library (1947) 45.

Khulna: [b]none. [c]Bhairab R (Ganges R distributary). [d]none. [e]Public Library (1964) 33; Brajalal College Library (?) 16.

Narayanganj: [b]none. [c]Dhaleswari R (Brahmaputra R distributary); Lakhya R (Brahmaputra R distributary). [d]none. [e]Rahmatullah Muslim Institute Library (?) 4; Municipal Public Library (?) ?

Brunei

*Bandar Seri Begawan: [b]for Borneo, Brunei, Brunei Town. [c]Brunei (for Borneo) R. [d]none. [e]Language and Literature Bureau Library (?) 36.

Burma

Bassein: [b]off Puthein. [c]Bassein (off Puthein) R. [d]none. [e]Bassein College Library (1964) 19; State Library (1963) 1.

Henzada: [b]none. [c]Irrawaddy R. [d]none. [e]Seikta Thukha Library (?) 9.

Mandalay: [b]off Mandale. [c]Irrawaddy R. [d]Arts and Science University, Mandalay (1923) 7 – 429. [e]. . . University Library (1958) 103.

Moulmein: ^{b}off Mawlamyaing. ^{c}G of Martaban of Andaman Sea; Salween R; Ataran R; Gyaing R. ^{d}none. ^{e}State Library (1955) 13; Moulmein College Library (1964) 5.

Myingyan: ^{b}none. ^{c}Irrawaddy R. ^{d}none.

Pegu: ^{b}none. ^{c}Pegu R. ^{d}none. ^{e}Thahaya Yuwa Library (?) ?; Youth League Free Library (?) ?

*Rangoon: ^{b}off Yangon. ^{c}Rangoon (off Yangon) R; Pegu R; Myitmaka R. ^{d}Arts and Science University, Rangoon (1885) 7 – 424. ^{e}Universities' Central Library (1929) 110; National Library (1952) 49.

Cambodia

*Phnom-Penh: ^{b}none. ^{c}Mekong R; Bassac R; Tonle Sap R. ^{d}Universite de Phnom-Penh (1953) 8 – 350; Universite Technique (1958) 0.9 – 234; Universite des Beaux-Arts (1965) 0.7 – 194; Universite Bouddhique (1954) 0.2 – 20. ^{e}Bibliotheque de l'Institut Bouddhique (1923) 40; Bibliotheque Nationale (1921) 33.

China

Amoy: ^{a}Fukien. ^{b}for Szeming; off Hsiamen. ^{c}Formosa (alt Taiwan) Strait; Kiulung (off Chiulung) R. ^{d}Amoy University (1921) 3 – 1000? ^{e}. . . University Library (1921) ?

Anking: ^{a}Anhwei. ^{b}for Hwaining; off Anching. ^{c}Yangtze R. ^{d}none.

Anshan: ^{a}Liaoning. ^{b}for Shaho. ^{c}nr Taitzu R. ^{d}none.

Antung: ^{a}Liaoning. ^{b}none. ^{c}Yalu R. ^{d}none.

Canton: ^{a}*Kwangtung. ^{b}alt Kwangchow; off Kuangchou. ^{c}Pearl R. ^{d}Chinan University (1958) 2 – ?; Chungshan University (?) ? – ?; South China Technical University (?) ? – ?; Sun Yat-sen University (1924) ? – ? ^{e}Chungshan Library of Kwangtung Province (1909) 980; Chinan University Library (1958) 400; Sun Yat-sen University Library (1924) ?

Changchih: ^{a}Shansi. ^{b}for Luan. ^{c}nr Chang R. ^{d}none.

Changchow: ^{a}Fukien. ^{b}for Lungki; off Changchou. ^{c}Lungki (off Lungchi) R. ^{d}none.

Changchow: ^{a}Kiangsu. ^{b}for Wutsin; off Changchou. ^{c}nr Ho L. ^{d}none.

Changchun: ^{a}*Kirin. ^{b}for Hsinking. ^{c}Itung R. ^{d}Kirin University (1958) ? – ? ^{e}Changchun Library (?) 245; . . . University Library (1958) ?

Changsha: ^{a}*Hunan. ^{b}none. ^{c}Siang R. ^{d}Hunan University (1959) ? – ? ^{e}Chungshan Library of Hunan Province (1912) 410; . . . University Library (1959) ?

Chefoo: ^{a}Shantung. ^{b}off Yentai. ^{c}G of Chihli [alt Po (Hai) G] of Yellow Sea. ^{d}none.

Chengchow: ^{a}*Honan. ^{b}for Chenghsien; off Chengchou. ^{c}nr Kialu (off Chialu) R. ^{d}Chengchow University (?) ? – ? ^{e}. . . University Library (?) ?

Chengtu: ^{a}*Szechwan. ^{b}none. ^{c}Chin R (arm of Min R). ^{d}Szechwan University (1931) 4 – 700?; Chengtu Technical University (1954) ? – ? ^{e}Szechwan Library (1912) 895; Chengtu Technical University Library (1954) ?; Szechwan University Library (1931) ?

Chinchow: ^{a}Liaoning. ^{b}off Chinchou. ^{c}Taling R. ^{d}none.

Chinkiang: ^{a}Kiangsu. ^{b}off Chenchiang. ^{c}Yangtze R. ^{d}none.

Chinwangtao: ^{a}Hopeh. ^{b}off Chinhuangtao. ^{c}G of Liaotung of Yellow Sea. ^{d}none.

Chuchow: ^{a}Hunan. ^{b}off Chuchou. ^{c}Siang R. ^{d}none.

Chungking: ^{a}Szechwan. ^{b}for Pahsien; off Chungching. ^{c}Yangtze R; Kialing R. ^{d}Chungking Technical University (?) ? – ?; Chungking University (?) ? – ? ^{e}Chungking Library (?) 1850; Chungking Technical University Library (?) ?; Chungking University Library (?) ?

Foochow: ^{a}*Fukien. ^{b}for Minhow; off Fuchou. ^{c}Min R. ^{d}Foochow University (?) ? – ? ^{e}Fukien Library (1908) 310; . . . University Library (?) ?

Foshan: ^{a}Kwangtung. ^{b}alt Fatshan; for Namhoi, Nanhai. ^{c}Foshan (alt Fatshan) R (Si R distributary). ^{d}none.

Fushun: ^{a}Liaoning. ^{b}none. ^{c}Hun (off Shen) R. ^{d}none.

Fusin: ^{a}Liaoning. ^{b}off Fouhsinhsien. ^{c}Hsi R. ^{d}none.

Hangchow: ^{a}*Chekiang. ^{b}off Hangchou. ^{c}Tsientang R; Hsi (alt West) L. ^{d}Chekiang University (1927) 6 – ?; Hangchow University (1959?) ? – ? ^{e}Chekiang Library (1872) 1050; Chekiang University Library (1927) ?; Hangchow University Library (1959) ?

Hantan: ^{a}Hopeh. ^{b}none. ^{c}Fuyang R. ^{d}none.

Harbin: ^{a}*Heilungkiang. ^{b}for Pinkiang; off Haerhpin. ^{c}Sungari R. ^{d}Harbin Polytechnic University (1920) 4 – ?; Heilungkiang University (?) ? – ? ^{e}Harbin Library (1950) 430; Harbin Polytechnic University Library (1920)?

Hengyang: ^{a}Hunan. ^{b}for Hengchow. ^{c}Siang R; Lei R; Chen R. ^{d}none.

Hofei: ^{a}*Anhwei. ^{b}for Luchow. ^{c}Chintou R. ^{d}Anhwei University (?) ? – ?; Hofei Polytechnic University (?) ? – ? ^{e}Anhwei Library (?) 330; Anhwei University Library (?) ?; Hofei Polytechnic University Library (?) ?

Hoihow: ^{a}Kwangtung; Hainan I. ^{b}off Haikou. ^{c}Hainan Strait; Nantu R. ^{d}none.

Hokang: ^{a}Heilungkiang. ^{b}off Haoli. ^{c}Alingta R. ^{d}none.

Huhehot: ^{a}*Inner Mongolia. ^{b}alt Kuku-Khoto; for Kweisui; inc Kweihwating, Suiyuan; off Huhohaote. ^{c}Tahei R. ^{d}Inner Mongolian University (1957) ? – ? ^{e}Library of Inner Mongolia (?) 125; . . . University Library (1957) ?

Hwainan: ^{a}Anhwei. ^{b}for Tienkiaan; off Huainan. ^{c}Hwai R. ^{d}none.

Hwangshih: ^{a}Hupeh. ^{b}inc Hwangshihkang, Shihhweiyao; off Huangshih. ^{c}Yangtze R. ^{d}none.

Ichang: ^{a}Hupeh. ^{b}none. ^{c}Yangtze R. ^{d}none.

Ichun: ^{a}Heilungkiang. ^{b}none. ^{c}Tangwang R. ^{d}none.

Ipin: ^{a}Szechwan. ^{b}for Suchow, Suifu. ^{c}Yangtze R; Min R. ^{d}none.

Kaifeng: ^{a}Honan. ^{b}none. ^{c}nr Yellow R. ^{d}none. ^{e}Honan Library (1909) 395.

Kalgan: ^{a}Hopeh. ^{b}for Wanchuan; off Changchiakou. ^{c}Chin R. ^{d}none.

Kashgar: ^{a}Sinkiang. ^{b}for Shufu, Sufu; off Koshih. ^{c}Kashgar (off Kashihkaerh) R. ^{d}none.

Kiamusze: ^{a}Heilungkiang. ^{b}off Chiamussu. ^{c}Sungari R. ^{d}none.

Kingtehchen: ^{a}Kiangsi. ^{b}for Fowliang; off Chingtechen. ^{c}Chang R. ^{d}none.

Kirin: ^{a}Kirin. ^{b}for Yungki; off Chilin. ^{c}Sungari R. ^{d}none.

Kisi: ^{a}Heilungkiang. ^{b}off Chihsi. ^{c}Muleng R. ^{d}none.

Kokiu: ^{a}Yunnan. ^{b}for Kokiuchang; off Kochiu. ^{c}nr Hsiao R. ^{d}none.

Kunming: ^{a}*Yunnan. ^{b}for Yunnan. ^{c}Tien L. ^{d}Yunnan University (1923) ? – ? ^{e}Yunnan Library (1909) 810; . . . University Library (1934) ?

Kweilin: ^{a}Kwangsi. ^{b}off Kueilin. ^{c}Kwei (off Kuei) R. ^{d}none. ^{e}First Library of Kwangsi (1909) 280.

Kweiyang: ^{a}*Kweichow. ^{b}for Kweichu; off Kueiyang. ^{c}Nanming R; Niulu R. ^{d}Kweichow University (1958) ? – ?; Kweiyang Technical University (1958) ? – ? ^{e}People's Library of Kweiyang (?) 280; Kweichow University Library (1958) ?; Kweiyang Technical University Library (1958) ?

Lanchow: ^{a}*Kansu. ^{b}for Kaolan; off Lanchou. ^{c}Yellow R. ^{d}Lanchow University (1946) 2 – 700? ^{e}Kansu Library (?) 380; Municipal Library (1957) 380; . . . University Library (1946?) ?

Lhasa: ^{a}*Tibet. ^{b}off Lasa. ^{c}Lasa R. ^{d}none.

Liaoyang: ^{a}Liaoning. ^{b}none. ^{c}Taitzu R. ^{d}none.

Liaoyuan: ^{a}Kirin. ^{b}for Peifeng, Sian, Tungliao. ^{c}Tungliao R. ^{d}none.

Liuchow: ^{a}Kwangsi. ^{b}for Maping; off Liuchou. ^{c}Liu R. ^{d}none.

Loyang: ^{a}Honan. ^{b}for Honan. ^{c}Lo R. ^{d}none. ^{e}Library of the Honan Institute for Agricultural Machinery (?) ?

Luchow: ^{a}Szechwan. ^{b}for Luhsien; off Luchou. ^{c}Yangtze R; To R. ^{d}none.

Lushun-Talien: ^{a}Liaoning. ^{b}alt Luta; Lushun: for Port Arthur, Ryojun; Talien: for Dairen, Dalny. ^{c}G of Chihli [alt Po (Hai) G] of Yellow Sea; Korea B of Yellow Sea. ^{d}Talien Technical University (1950) ? – ? ^{e}Lushun-Talien Library (?) 740; . . . University Library (1950) ? ^{f}Lushun: 34 (1930), 14 (1900?e); Talien: 293 (1930), 40 (1900?e).

Malipo: ^{a}Yunnan. ^{b}none. ^{c}nr Chouyangta R. ^{d}none.

Mukden: ^{a}*Liaoning. ^{b}for Fengtien; off Shenyang. ^{c}Hun (off Shen) R. ^{d}Liaoning University (?) ? – ?; Northeastern China Technical University (?) ? – ? ^{e}Liaoning Library (1948) 1300; Shenyang Library (1908) 580.

Mutankiang: ^{a}Heilungkiang. ^{b}off Mutanchiang. ^{c}Mutan R. ^{d}none.

Nanchang: ^{a}*Kiangsi. ^{b}none. ^{c}Kan R. ^{d}none. ^{e}Kiangsi Library (1921) 520.

Nanchung: ^{a}Szechwan. ^{b}for Shunking. ^{c}Kialing R. ^{d}none.

Nanking: ^{a}*Kiangsu. ^{b}for Kiangning; off Nanching. ^{c}Yangtze R. ^{d}Nanking Technical University (?)

? – ?; Nanking University (1902) ? – ? ^{e}Nanking Library (1908) 2400; Second Library of Kiangsu (1914?) 210; Nanking Technical University Library (?) ?; Nanking University Library (1902) ?

Nanning: ^{a}*Kwangsi. ^{b}for Yungning. ^{c}Yu R. ^{d}none. ^{e}Kwangsi Agricultural Institute Library (?) ?; Kwangsi Medical College Library (?) ?

Nantung: ^{a}Kiangsu. ^{b}for Tungchow. ^{c}Yangtze R. ^{d}none. ^{e}Nantung Medical College Library (?) ?

Neikiang: ^{a}Szechwan. ^{b}off Neichiang. ^{c}To R. ^{d}none.

Ningpo: ^{a}Chekiang. ^{b}for Ninghsien. ^{c}Yung (alt Ningpo) R; Fenghua R; Yao R. ^{d}none.

Paoki: ^{a}Shensi. ^{b}off Paochi. ^{c}Wei R. ^{d}none.

Paoting: ^{a}Hopeh. ^{b}for Tsingyuan. ^{c}Fu R. ^{d}none. ^{e}Hopeh Library (?) 195.

Paotow: ^{a}Inner Mongolia. ^{b}off Paotou. ^{c}Yellow R. ^{d}none.

*Peking: ^{a}(independent city). ^{b}for Peiping; off Peiching. ^{c}Pei L; Chung L; Nan L. ^{d}Tsinghua University (1911) 12 – 2600; Peking University (1898) 10 – 2000?; China University of Science and Technology (1958) 6 – ?; People's University of China (1950) 1 – ? ^{e}National Library of Peking (1909) 4600; Central Library of the Chinese Academy of Sciences (1951) 4000; Peking University Library (1902) 3000; National Capital Library (?) 520; People's University of China Library (1950) ?; Tsinghua University Library (1911) ?

Pengpu: ^{a}Anhwei. ^{b}off Pangfou. ^{c}Hwai R. ^{d}none.

Penki: ^{a}Liaoning. ^{b}for Penkihu; off Penchi. ^{c}Taitzu R. ^{d}none.

Shanghai: ^{a}(independent city). ^{b}none. ^{c}East China Sea; Yangtze R; Whangpoo (off Huangpu) R; Wusung R. ^{d}Shanghai University (1895) 7 – 1400?; Tung Chi University (1907) 6 – 550?; Futan University (1905) 3 – ?; Chiao Tung University (1896) ? – ? ^{e}Shanghai Library (1952) 6500; Municipal People's Library (1912) 920; Chiao Tung University Library (1896) ?; Futan University Library (1922) 1200; Shanghai University Library (1895) ?

Shangkiu: ^{a}Honan. ^{b}for Kweiteh; off Shangchiuhsien. ^{c}Pao R. ^{d}none.

Shaohing: ^{a}Chekiang. ^{b}off Shaohsing. ^{c}nr Hangchow B of East China Sea. ^{d}none.

Shaoyang: ^{a}Hunan. ^{b}for Paoking. ^{c}Tzu R. ^{d}none.

Shihkiachwang: ^{a}*Hopeh. ^{b}for Shihmen; off Shihchiachuang. ^{c}nr Sha R. ^{d}none.

Shiukwan: ^{a}Kwangtung. ^{b}for Chukiang, Kukong, Shaochow, Shiuchow; off Shaokuan. ^{c}Pei R; Wu R; Cheng R. ^{d}none.

Siakwan: ^{a}Yunnan. ^{b}off Hsiakuan. ^{c}Yangpi R; Erh L. ^{d}none.

Sian: ^{a}*Shensi. ^{b}for Changan, Siking, Singan; off Hsian. ^{c}nr Wei R. ^{d}Northwestern University (1937) 3 – 300?; Shensi University of Science and Technology (1960) ? – ?; Sian University (1884) ? – ? ^{e}Shensi Library (1909) 380; Northwestern University Library (1937) ?; Shensi University ... Library (1960) ?

Siangtan: ^{a}Hunan. ^{b}off Hsiangtan. ^{c}Siang R. ^{d}none.

Sinhailien: ^{a}Kiangsu. ^{b}inc Lienyunkang, Sinpu, Tunghai (for Haichow); off Hsinhailien. ^{c}Yellow Sea; Yen R. ^{d}none. ^{f}Lienyunkang: 77 (1946e); Tunghai: 48 (1948e).

Sining: ^{a}*Tsinghai. ^{b}off Hsining. ^{c}Huang (alt Hsining, Sining) R. ^{d}none. ^{e}Tsinghai Library (?) 130.

Sinsiang: ^{a}Honan. ^{b}off Hsinhsiang. ^{c}Wei R. ^{d}none.

Soochow: ^{a}Kiangsu. ^{b}for Wuhsien; off Suchou. ^{c}nr Yangcheng L; nr Tushu L. ^{d}none.

Suchow: ^{a}Kiangsu. ^{b}for Tungshan; off Hsuchou. ^{c}nr Wei Shan L. ^{d}none.

Swatow: ^{a}Kwangtung. ^{b}off Shantou. ^{c}South China Sea; Han R. ^{d}none.

Szeping: ^{a}Kirin. ^{b}for Szepingkai; off Ssuping. ^{c}Tiaotzu R. ^{d}none.

Taichow: ^{a}Kiangsu. ^{b}for Taihsien; off Taichou. ^{c}Tung R; Hsi R. ^{d}none.

Taiyuan: ^{a}*Shansi. ^{b}for Yangku. ^{c}Fen R. ^{d}none.

Tangshan: ^{a}Hopeh. ^{b}none. ^{c}Tung R. ^{d}none.

Tatung: ^{a}Shansi. ^{b}none. ^{c}Yu R. ^{d}none.

Tientsin: ^{a}(independent city). ^{b}off Tienching. ^{c}Hai R; Chinchung R; Hsinkai R; Tzuya R; Weiching R; Yun R. ^{d}Nankai University (1914) 3 – 460; Hopeh University (1960?) ? – ?; Tientsin University (1895) ? – ? ^{e}Municipal Library (1908) 880; Hopeh University Library (1960) ?; Nankai University Library (1919)?

Tsamkong: [a]Kwangtung. [b]for Fort-Bayard, Haikang, Leichow, Siying; off Chanchiang. [c]Kwangchow B of South China Sea. [d]none. [f]60 (1900?e).

Tsiaotso: [a]Honan. [b]off Chiaotso. [c]Yunliang R. [d]none.

Tsinan: [a]*Shantung. [b]for Licheng; off Chinan. [c]Siaoching (off Hsiaoching) R; Taming L. [d]Tsinan Technical University (?) ? – ? [e]Shantung Library (1909) 680; . . . University Library (?) ?

Tsingtao: [a]Shantung. [b]off Chingtao. [c]Yellow Sea. [d]Shantung University (1926) ? – ?; Tsingtao Technical University (?) ? – ? [e]Tsingtao Library (?) 330; Shantung University Library (1926) ?; Tsingtao Technical University Library (?) ? [f]527 (1935e).

Tsitsihar: [a]Heilungkiang. [b]for Lungkiang, Pukwei; off Chichihaerh. [c]Nonni R. [d]none.

Tzekung: [a]Szechwan. [b]inc Kungtsing, Tzeliutsing; off Tzukung. [c]Tsin (off Ching) R. [d]none. [f]Tzeliutsing: 100 (1926e).

Tzepo: [a]Shantung. [b]inc Poshan, Tzechwan; off Tzupo. [c]Chenghuang (alt Hsiaofu) R. [d]none. [f]Poshan: 50 (1920?e).

Urumtsi: [a]*Sinkiang. [b]alt Tihwa, Urumchi; off Wulumuchi. [c]Wulumuchi R. [d]Sinkiang University (1960) 2 – ? [e]Library of the Sinkiang Uighur Autonomous Region (?) 85; . . . University Library (1960) ?

Weifang: [a]Shantung. [b]for Weihsien. [c]Pailang R. [d]none.

Wenchow: [a]Chekiang. [b]for Yungkia; off Wenchou. [c]Wu (alt Ou) R. [d]none. [e]Chekiang Institute of Forestry Library (?) ?

Wuhan: [a]*Hupeh. [b]inc Hankow, Hanyang, Wuchang. [c]Yangtze R; Han (Shui) R; Tung L. [d]Wuhan University (1913) 4 – 700?; Central China Technical University (?) ? – ?; Hupeh University (?) ? – ? [e]Hupeh Library (1908) 780; Wuhan Library (?) 280; Central China Technical University Library (?) ?; Hupeh University Library (?) ?; Wuhan University Library (1913) ? [f]Hankow: 782 (1935e), 850 (1900?e), 690 (1850e); Hanyang: 137 (1934e), 100 (1900?e), 100 (1850e); Wuchang: 434 (1935e), 550 (1900?e), 206 (1850e).

Wuhu: [a]Anhwei. [b]none. [c]Yangtze R; Chingi R; Suiyang R. [d]none.

Wusih: [a]Kiangsu. [b]off Wuhsi. [c]nr Tai L. [d]none.

Wutungkiao: [a]Szechwan. [b]none. [c]Min R. [d]none.

Yangchow: [a]Kiangsu. [b]for Kiangtu; off Yangchou. [c]Nanyun R. [d]none. [e]North Kiangsu Agricultural Institute Library (?) ?

Yangchuan: [a]Shansi. [b]none. [c]Mien (Shui) R. [d]none.

Yenan: [a]Shensi. [b]for Fushih. [c]Yen R. [d]none.

Yinchwan: [a]*Ningsia. [b]for Ningsia; off Yinchuan. [c]nr Yellow R. [d]Ningsia University (bef 1962) 1 – 290. [e]. . . University Library (1962) ?

Yingkow: [a]Liaoning. [b]for Yingtze; off Yingkou. [c]G of Liaotung of Yellow Sea; Liao R. [d]none.

Cyprus

*Nicosia: [b]off Levkosia. [c]Pedias R. [d]none. [e]Library of Phaneromeni (1934) 45; Library of the Pan-Cyprian Gymnasium (1927) 35.

Gaza Strip

*Gaza: [b]alt Azzah; off Ghazzah. [c]Mediterranean Sea. [d]none.

Hong Kong

*Hong Kong: [b]alt Hsiangchiang, Hsiangkang; inc Kowloon, New Kowloon, Victoria. [c]South China Sea; Pearl R. [d]Chinese University of Hong Kong (1949) 4 – 484; University of Hong Kong (1887) 5 – 500. [e]Urban Council Library (1962) 677; . . . University Library (1949) 524; University . . . Library (1912) 513. [f]Kowloon: 716 (1971), 727 (1961), 232 (1931); New Kowloon: 1479 (1971), 853 (1961), 23 (1931); Victoria: 996 (1971), 1005 (1961), 411 (1931).

India

Agra: [a]Uttar Pradesh. [b]none. [c]Yamuna R. [d]Agra University (1927) 92 – ? [e]. . . University Library (1935) 107; Agra District Central Library (1957) 9.

Ahmadabad: [a]*Gujarat. [b]alt Ahmedabad; for Asawal. [c]Sabarmati R. [d]Gujarat University (1949) 101 – ?; National University of Gujarat (1920) 0.6 – 45. [e]National University of Gujarat Library (1920) 190; Gujarat University Library (1949) 186; Gujarat State Central Library (1933) 124.

Ajmer: [a]Rajasthan. [b]none. [c]Ana L. [d]none. [e]City Public Library (1899) 35.

Aligarh: [a]Uttar Pradesh. [b]inc Koil. [c]none. [d]Aligarh Muslim University (1877) 10 – 816. [e]. . . University Library (1921) 460; Lytton Library (1880) 45 (in 1950).

Allahabad: [a]Uttar Pradesh. [b]none. [c]Ganges R; Yamuna R. [d]University of Allahabad (1887) 21 – ? [e]University . . . Library (1916) 293; Public Library (1864) 63.

Ambala: [a]Haryana. [b]for Umballa. [c]nr Ghaggar R. [d]none. [e]Hans Raj Library of D. A. V. College (1886) 24.

Amravati: [a]Maharashtra. [b]alt Amraoti. [c]nr Pedhi R. [d]none. [e]Amravati District Central Library (1923) 40.

Amritsar: [a]Punjab. [b]none. [c]none. [d]Guru Nanak University (1969) 45 – ? [e]. . . University Library (1969) 60; Pandit Motilal Nehru City Public Library (1900) 42.

Asansol: [a]West Bengal. [b]none. [c]nr Damodar R; nr Barakar R. [d]none. [e]Burdwan District Library (1959) 5.

Bangalore: [a]*Karnataka. [b]none. [c]nr Pinikini R. [d]Bangalore University (1858) 31 – 1960; University of Agricultural Sciences (1964) 2 – ?; Indian Institute of Science (univ) (1909) 1 – ? [e]. . . Institute Library (1909) 185; . . . University Library (1865) 161; Karnataka State Central Library (1914) 77.

Bareilly: [a]Uttar Pradesh. [b]for Bareli. [c]Jooah R; Sunkra R. [d]none. [e]Bareilly College Library (1837) 33; Bareilly District Central Library (1957) 7.

Belgaum: [a]Karnataka. [b]none. [c]Markandeya R. [d]none. [e]General Library (1848) 11 (in 1950).

Bhagalpur: [a]Bihar. [b]for Boglipore, Sujanganj. [c]Ganges R. [d]Bhagalpur University (1887) 32 – 1569. [e]. . . University Library (1960) 55.

Bhatpara: [a]West Bengal. [b]none. [c]Hooghly R (Ganges R distributary). [d]none. [e]Bhatpara Literary Association and Library (?) ?

Bhavnagar: [a]Gujarat. [b]alt Bhaunagar. [c]G of Cambay of Arabian Sea. [d]none. [e]Bhavnagar District Central Library (1959) 30.

Bhilainagar-Durg: [a]Madhya Pradesh. [b]inc Bhilainagar, Durg; Bhilainagar: for Bhilai; Durg: for Drug. [c]nr Seonath R. [d]none. [f]Bhilainagar: 157 (1971), 86 (1961); Durg: 68 (1971), 47 (1961), 20 (1951), 13 (1931), 4 (1901), 4 (1881).

Bhopal: [a]*Madhya Pradesh. [b]none. [c]Bess R; Patra R; Pukhta-Pul Talao L. [d]Bhopal University (1946) 12 – 612. [e]Maulana Azad Central Library (?) 60.

Bikaner: [a]Rajasthan. [b]none. [c]none. [d]none. [e]State Divisional Library (1937) 43; Dungar College Library (1935) 14 (in 1950).

Bombay: [a]*Maharashtra. [b]none. [c]Arabian Sea. [d]University of Bombay (1832) 119 – ?; Shreemati Nathibai Damodar Thackersey Women's University (1916) 19 – 634; Indian Institute of Technology, Bombay (univ) (1958) 2 – 200. [e]Asiatic Society of Bombay Library (alt Maharashtra State Central Library) (1804) 518; University . . . Library (1869) 382.

Calcutta: [a]*West Bengal. [b]none. [c]Hooghly R (Ganges R distributary). [d]University of Calcutta (1817) 227 – ?; Rabindra Bharati University (1962) 5 – 214; Jadavpur University (1906) 4 – 474. [e]National Library (1836) 1511; University . . . Library (1857) 461.

Calicut: [a]Kerala. [b]alt Kozhikode. [c]Arabian Sea; Kallayi R. [d]University of Calicut (1968) 50 – ? [e]University . . . Library (1968?) 16; City Public Library (?) ?

Chandigarh: [a]Chandigarh; *Haryana; *Punjab. [b]none. [c]nr Ghaggar R. [d]Panjab University (1947 at Solon, relocated 1957) 75 – ? [e]. . . University Library (1947 at Solon, relocated 1957?) 390; Central State Library (?) 65.

Cochin: ^{a}Kerala. ^{b}inc Ernakulam. ^{c}Arabian Sea; Vembanad Lagoon. ^{d}University of Cochin (1971) 0.5 – ? ^{e}University . . . Library (1971?) 60; Public Library (?) ? ^{f}Ernakulam: 117 (1961), 62 (1951), 37 (1931), 22 (1901), 14 (1875).

Coimbatore: ^{a}Tamil Nadu. ^{b}none. ^{c}Noyil R. ^{d}Tamil Nadu Agricultural University (1971) 2 – ? ^{e}. . . University Library (1876) 87; Coimbatore District Central Library (1952) 20.

Cuttack: ^{a}Orissa. ^{b}none. ^{c}Mahanadi R. ^{d}none. ^{e}Kanika Library of Ravenshaw College (1919) 67.

Dehra Dun: ^{a}Uttar Pradesh. ^{b}for Dehra. ^{c}Rispana R; Bindal R. ^{d}none. ^{e}Forest Research Institute Central Library (1906) 75; Mahatma Kushiram Public Library and Reading Room (1921) 5 (in 1950).

Delhi: ^{a}*Delhi. ^{b}none. ^{c}Yamuna R. ^{d}University of Delhi (1881) 75 – ? ^{e}Public Library (1951) 674; University . . . Library (1922) 493.

Dhanbad: ^{a}Bihar. ^{b}none. ^{c}nr Damodar R. ^{d}none. ^{e}Indian School of Mines Library (1926) 23; Central Fuel Research Institute Library (1955) 19; Dhanbad District Central Library (1956) 11.

Durgapur: ^{a}West Bengal. ^{b}none. ^{c}Damodar R. ^{d}none. ^{e}Central Mechanical Engineering Research Institute Library (1958) 26.

Gauhati: ^{a}Assam. ^{b}none. ^{c}Brahmaputra R. ^{d}Gauhati University (at Jhalukbari, nr Gauhati) (1914) 55 – 2457. ^{e}. . . University Library (at Jhalukbari, nr Gauhati) (1948) 138; Gauhati District Central Library (?) 39.

Gaya: ^{a}Bihar. ^{b}none. ^{c}Phalgu R. ^{d}Magadh University (at Bodh Gaya, nr Gaya) (1944) 49 – ? ^{e}. . . University Library (at Bodh Gaya, nr Gaya) (1962?) 70; Mannulal Library (1911) 24 (in 1950); Gaya District Central Library (1855) 14.

Gorakhpur: ^{a}Uttar Pradesh. ^{b}none. ^{c}Rapti R. ^{d}University of Gorakhpur (1933) 61 – ? ^{e}University . . . Library (1957) 107; Gorakhpur District Central Library (1957) 9.

Guntur: ^{a}Andhra Pradesh. ^{b}for Guntoor. ^{c}none. ^{d}none. ^{e}Guntur District Central Library (1918) 31.

Gwalior: ^{a}Madhya Pradesh. ^{b}inc Lashkar since 1951. ^{c}Sonrekha (Nadi) Creek. ^{d}Jiwaji University (1887) 31 – ? ^{e}Government Central Library (1928) 77; . . . University Library (1964) 35. ^{f}Lashkar: 151 (1951), 79 (1931), 89 (1901), 88 (1881).

Howrah: ^{a}West Bengal. ^{b}inc Bally (alt Baly) since 1969. ^{c}Hooghly R (Ganges R distributary). ^{d}none. ^{e}Sibpur Public Library (?) 21; Bally Public Library (?) 20. ^{f}Bally: 131 (1961), 63 (1951), 30 (1931), 19 (1901), 14 (1872).

Hubli-Dharwar: ^{a}Karnataka. ^{b}inc Dharwar, Hubli. ^{c}nr Bedti R. ^{d}Karnatak University (1917) 43 – 3122. ^{e}. . . University Library (1950) 163; Hubli-Dharwar City Central Library (1948) 50. ^{f}Dharwar: 77 (1961), 66 (1951), 42 (1931), 31 (1901), 27 (1872); Hubli: 171 (1961), 130 (1951), 90 (1931), 60 (1901), 38 (1872).

Hyderabad: ^{a}*Andhra Pradesh. ^{b}for Bhagnagar, Haidarabad; inc Secunderabad since 1951. ^{c}Musi R; Hussain L. ^{d}Osmania University (1887) 69 – ?; Jawaharlal Nehru Technological University (1946) 3 – 399; Andhra Pradesh Agricultural University (1964) 2 – 245; University of Hyderabad (1974) ? – ? ^{e}Osmania University Library (1919) 276; Andhra Pradesh State Central Library (1891) 184. ^{f}Secunderabad: 225 (1951), 121 (1931), 96 (1901).

Indore: ^{a}Madhya Pradesh. ^{b}none. ^{c}Katki R. ^{d}University of Indore (1884) 18 – 817. ^{e}University . . . Library (1964) 57; Indore Christian College Library (1887) 30; General Library (1854) 21 (in 1950).

Jabalpur: ^{a}Madhya Pradesh. ^{b}for Jubbulpore. ^{c}nr Narmada R. ^{d}University of Jabalpur (1933) 18 – 683; Jawaharlal Nehru Agricultural University (1948) 2 – ? ^{e}. . . University Library (1966) 84; University . . . Library (1957) 79; Government Central Library (1956) 45.

Jaipur: ^{a}*Rajasthan. ^{b}none. ^{c}nr Dhund (Nadi) Creek. ^{d}University of Rajasthan (1873) 82 – ? ^{e}University . . . Library (1947) 195; Public Library (1866) 90.

Jamnagar: ^{a}Gujarat. ^{b}for Navanagar. ^{c}G of Cutch of Arabian Sea. ^{d}Gujarat Ayurved University (1946) 3 – ? ^{e}Jamnagar District Central Library (1956) 34; . . . University Library (1946?) 18.

Jamshedpur: ^{a}Bihar. ^{b}none. ^{c}Subarnarekha R. ^{d}none. ^{e}National Metallurgical Laboratory Library (1950) 32.

Jhansi: ^{a}Uttar Pradesh. ^{b}none. ^{c}nr Betwa R. ^{d}none. ^{e}Jhansi District Central Library (1957) 7.

Jodhpur: [a]Rajasthan. [b]none. [c]nr Umed L. [d]University of Jodhpur (1892) 10 – 571. [e]University . . . Library (1962) 118; Sumer Public Library (1915) 40.

Jullundur: [a]Punjab. [b]none. [c]nr East Bein (alt White Bein) R. [d]none. [e]Jullundur District Central Library (1955) 20; Lajpat Rai Library (1918) 12 (in 1950).

Kanpur: [a]Uttar Pradesh. [b]for Cawnpore. [c]Ganges R. [d]Kanpur University (1955) 39 – ?; Indian Institute of Technology, Kanpur (univ) (1960) 2 – 270. [e]. . . Institute Library (1960?) 146; Government Agricultural Library (1904) 62; Kanpur District Central Library (1957) 7.

Kolhapur: [a]Maharashtra. [b]none. [c]Panchaganga R. [d]Shivaji University (1962) 74 – 2568. [e]. . . University Library (1962) 89; Rajaram College Library (1880) 61; Kolhapur District Central Library (1850) 30.

Kota: [a]Rajasthan. [b]alt Kotah. [c]Chambal R. [d]none. [e]Government Girls Higher Secondary School Library (?) 4; Herbert College Library (?) ?

Lucknow: [a]*Uttar Pradesh. [b]none. [c]Gumti R. [d]University of Lucknow (1911) 23 – 600. [e]University . . . Library (1921) 280; Uttar Pradesh Legislative Library (1931) 160; Amir-ud-Daula Government Public Library (alt Uttar Pradesh State Central Library) (1910) 48.

Ludhiana: [a]Punjab. [b]none. [c]none. [d]Punjab Agricultural University (1947) 2 – 909. [e]. . . University Library (1962) 134.

Madras: [a]*Tamil Nadu. [b]none. [c]B of Bengal; Cooum R; Adyar R. [d]University of Madras (1794) 152 – ?; Indian Institute of Technology, Madras (univ) (1959) 2 – 186. [e]University . . . Library (1857) 328; Connemara Public Library (alt Tamil Nadu State Central Library) (1896) 220.

Madurai: [a]Tamil Nadu. [b]alt Madura. [c]Vaigai R. [d]Madurai University (1958) 77 – ? [e]. . . University Library (1958) 99; American College Library (1842?) 27; Madurai District Central Library (1952) 14.

Malegaon: [a]Maharashtra. [b]none. [c]Girna R. [d]none.

Mangalore: [a]Karnataka. [b]none. [c]Arabian Sea; Netravati R; Gurpur R. [d]none. [e]South Kanara District Central Library (1950) 74; Saint Aloysius College Library (1880) 49 (in 1950).

Mathura: [a]Uttar Pradesh. [b]for Muttra. [c]Yamuna R. [d]none. [e]Uttar Pradesh College of Veterinary Science and Animal Husbandry Library (?) 8.

Meerut: [a]Uttar Pradesh. [b]none. [c]nr Kali (Nadi) Creek. [d]Meerut University (1966) 23 – 1942. [e]Meerut College Library (1892) 85; Tialk (Lyall) Library and Reading Room (1886) 30; . . . University Library (1966?) 24; Meerut District Central Library (1957) 7.

Moradabad: [a]Uttar Pradesh. [b]none. [c]Ramganga R. [d]none. [e]K. G. K. College Library (1940?) 19.

Mysore: [a]Karnataka. [b]none. [c]nr Kaveri R. [d]University of Mysore (1833) 95 – ? [e]University . . . Library (1916) 373; Oriental Research Institute Library (1891) 48; City Central Library (1915) 14 (in 1950).

Nagpur: [a]Maharashtra. [b]none. [c]Nag R; Chamar R. [d]Nagpur University (1923) 92 – 4124. [e]. . . University Library (1927) 243; Nagpur District Central Library (1955) 71.

Nasik: [a]Maharashtra. [b]none. [c]Godavari R. [d]none. [e]Nasik District Central Library (1840) 35; Hansraj Pragji Thackersey College Library (1924) 27.

*New Delhi: [a]Delhi. [b]none. [c]Yamuna R. [d]Indian Institute of Technology, Delhi (univ) (1961) 2 – 164; Jawaharlal Nehru University (1969) 2 – 240; Jamia Millia Islamia (univ) (1920 at Aligarh, relocated 1925) 1 – 201; All-India Institute of Medical Sciences (univ) (1956) 0.7 – ? [e]Central Secretariat Library (1900) 280; Parliament Library (1921) 270; Indian Agricultural Research Institute Library (1905) 200.

Patna: [a]*Bihar. [b]none. [c]Ganges R. [d]Patna University (1863) 12 – 735. [e]. . . University Library (1917) 186; Shrimati Radhika Sinha Institute and Sachchidananda Sinha Library (alt Bihar State Central Library) (1924) 78.

Poona: [a]Maharashtra. [b]none. [c]Mutha Mula R; Mutha R; Mula R. [d]University of Poona (1885) 93 – ? [e]University . . . Library (1950) 180; Bai Jerbai Wadia Library of Fergusson College (1885) 130; Poona District Central Library (1947) 80.

Raipur: [a]Madhya Pradesh. [b]none. [c]Karun R. [d]Ravishankar University (1948) 24 – 1024. [e]. . . University Library (1948) 33.

Rajahmundry: ^{a}Andhra Pradesh. ^{b}none. ^{c}Godavari R. ^{d}none. ^{e}Gauthami Library (1898) 70.

Rajkot: ^{a}Gujarat. ^{b}none. ^{c}Aji R. ^{d}Saurashtra University (1965) 44 – ? ^{e}. . . University Library (1968) 45; Sir Lakhajiraj Library and Reading Room (1868) 40; Rajkot District Central Library (1956) 34.

Ranchi: ^{a}Bihar. ^{b}none. ^{c}Subarnarekha R. ^{d}Ranchi University (1899) 48 – 1571. ^{e}. . . University Library (1899?) 66; British Council Library (1962) 23; Ranchi District Central Library (1953) 13.

Saharanpur: ^{a}Uttar Pradesh. ^{b}none. ^{c}Dhamola R. ^{d}none. ^{e}City Library (?) ?

Salem: ^{a}Tamil Nadu. ^{b}none. ^{c}Tirumanimutar (alt Salem) R. ^{d}none. ^{e}Salem District Central Library (1953) 10.

Sangli: ^{a}Maharashtra. ^{b}none. ^{c}Krishna R. ^{d}none.

Sholapur: ^{a}Maharashtra. ^{b}none. ^{c}nr Ekruk L. ^{d}none. ^{e}Dayanand College Library (1940?) 33; Sholapur District Central Library (1857) 26.

Simla: ^{a}*Himachal Pradesh. ^{b}none. ^{c}nr Sutlej R. ^{d}Himachal Pradesh University (1962) 14 – 228. ^{e}Indian Institute of Advanced Study Library (1965) 50; . . . University Library (1965) 40.

South Dum Dum: ^{a}West Bengal. ^{b}none. ^{c}nr Hooghly R (Ganges R distributary). ^{d}none.

South Suburban: ^{a}West Bengal. ^{b}inc Behala. ^{c}nr Hooghly R (Ganges R distributary). ^{d}none. ^{e}Behala Library (?) ?

Surat: ^{a}Gujarat. ^{b}none. ^{c}Tapti R. ^{d}South Gujarat University (1965) 24 – 985. ^{e}. . . University Library (1968) 31; Maganlal Thakordas Balmukandas College Library (1918) 28; Andrews Library (alt Surat District Central Library) (1850) 15.

Thana: ^{a}Maharashtra. ^{b}none. ^{c}Thana (Nadi) Creek. ^{d}none.

Tiruchirapalli: ^{a}Tamil Nadu. ^{b}for Trichinopoly. ^{c}Kaveri R. ^{d}none. ^{e}National College Library (1919) 17; Tiruchirapalli District Central Library (1952) 7.

Tirunelveli: ^{a}Tamil Nadu. ^{b}for Tinnevelly. ^{c}Tambraparni R. ^{d}none. ^{e}Tirunelveli District Central Library (1952) 23.

Trivandrum: ^{a}*Kerala. ^{b}none. ^{c}Arabian Sea. ^{d}University of Kerala (1937) 116 – ? ^{e}University . . . Library (1943) 160; Kerala State Central Library (1851) 95.

Tuticorin: ^{a}Tamil Nadu. ^{b}none. ^{c}G of Mannar of Indian O. ^{d}none.

Ujjain: ^{a}Madhya Pradesh. ^{b}none. ^{c}Sipra R. ^{d}Vikram University (1896) 19 – ? ^{e}. . . University Library (1896?) 68; Madhav College Library (1896?) 26; Yuvraj General Library (1913) 10 (in 1950).

Ulhasnagar-Kalyan: ^{a}Maharashtra. ^{b}inc Kalyan, Ulhasnagar. ^{c}Ulhas R. ^{d}none. ^{f}Kalyan: 100 (1971), 73 (1961), 59 (1951), 26 (1931), 11 (1901), 13 (1872); Ulhasnagar: 168 (1971), 108 (1961), 81 (1951).

Vadodara: ^{a}Gujarat. ^{b}for Baroda. ^{c}Vishvamitri R. ^{d}Maharaja Sayajirao University of Vadodara (1881) 20 – 862. ^{e}. . . University Library (1950) 262; Vadodara District Central Library (1910) 175.

Varanasi: ^{a}Uttar Pradesh. ^{b}alt Banaras; for Benares. ^{c}Ganges R. ^{d}Sampurnanand Sanskrit University (1958) 31 – ?; Banaras Hindu University (1898) 14 – 1068. ^{e}Banaras Hindu University Library (1916) 499; Sampurnanand Sanskrit University Library (1958?) 102; Carmichael Library (1872) 60; Varanasi District Central Library (1957) 6.

Vellore: ^{a}Tamil Nadu. ^{b}none. ^{c}Palar R. ^{d}none. ^{e}Christian Medical College Library (1942?) 14.

Vijayawada: ^{a}Andhra Pradesh. ^{b}alt Vijayavada; for Bezwada. ^{c}Krishna R. ^{d}none. ^{e}Ram Mohan Free Reading Room and Library (1911) 31.

Visakhapatnam: ^{a}Andhra Pradesh. ^{b}alt Vishakhapatnam; for Vizagapatam. ^{c}B of Bengal. ^{d}Andhra University (at Waltair, nr Visakhapatnam) (1926) 33 – ? ^{e}. . . University Library (at Waltair, nr Visakhapatnam) (1927) 216.

Warangal: ^{a}Andhra Pradesh. ^{b}alt Hanamkonda. ^{c}none. ^{d}none. ^{e}Warangal District Central Library (1958) 8.

Indonesia

Ambon: ^{a}Ambon I. ^{b}alt Amboina. ^{c}Ambon B of Banda Sea. ^{d}Universitas Pattimura (1956) 1 – 229. ^{e}Perpustakaan Universitas (1956?) 12; Perpustakaan Negara (?) 4.

Banjarmasin: [a]Borneo I. [b]for Bandjermasin. [c]Barito R; Martapura R. [d]Universitas Lambung Mangkurat (1958) 2 – 766. [e]Perpustakaan Negara (?) 2; Perpustakaan Universitas (1960) ?

Bandung: [a]Java I. [b]for Bandoeng. [e](Ci)kapundung R. [d1]Universitas Pajajaran (1952) 10 – 1919; Institut Teknologi Bandung (univ) (1920) 7 – 493; Universitas Katolik Parahyangan (1955) 3 – 250. [e]Perpustakaan Pusat Institut . . . (1920) 130; Perpustakaan Universitas Pajajaran (1957) 110.

Bogor: [a]Java I. [b]for Buitenzorg. [c](Ci)liwung R. [d]Institut Pertanian Bogor (univ) (1940) 1 – 443; Universitas Bogor (1958) 0.3 – 60; Universitas Ibnu Chaldun Bogor (1958) ? – ? [e]Bibliotheca Bogoriensis (alt Lembaga Perpustakaan Biologi dan Pertanian) (1842) 357.

Cirebon: [a]Java I. [b]for Cheribon, Tjirebon. [c]Java Sea. [d]Universitas Islam Syarief Hidayatullah Cirebon (?) ? – ?

Denpasar: [a]Bali I. [b]none. [c]Badung (for Badoeng) R. [d]Universitas Udayana (1950) 2 – 487. [e]Perpustakaan Universitas (1950?) 12.

*Jakarta: [a]Java I. [b]for Batavia, Djakarta. [c]Java Sea; (Ci)liwung R. [d1]Universitas Indonesia (1920) 10 – 2018; Universitas Trisakti (1966) 6 – 691; Universitas Krishnadwipayana (1952) 2 – 128; Universitas Kristen Indonesia (1953) 2 – 426; Universitas Tarumanegara (1959) 2 – 249; Universitas Ibnu Chaldun (1956) 1 – 80; Universitas Katolik Indonesia Atma Jaya (1960) 1 – 230; Universitas 17 Agustus 1945 (1952) 1 – 213. [e]Perpustakaan Museum Pusat (1778) 350; Perpustakaan Seyarah Politik dan Sosial (?) 75.

Jambi: [a]Sumatra I. [b]for Djambi. [c]Hari (for Djambi) R. [d]Universitas Negeri Jambi (1963) 0.8 – 215. [e]Perpustakaan Universitas . . . (1963?) 16; Perpustakaan Negara (?) ?

Kediri: [a]Java I. [b]none. [c]Brantas R. [d]none.

Kupang: [a]Timor I. [b]for Koepang. [c]Savu Sea. [d]Universitas Nusa Cendana (1962) 1 – 226. [e]Perpustakaan Universitas (1962?) 7; Perpustakaan Negara (?) 2.

Malang: [a]Java I. [b]for Singasari. [c]Brantas R. [d]Universitas Brawijaya (1957) 4 – 497; Universitas Merdeka (1962) 1 – 147. [e]Perpustakaan Pusat Institut Keguruan dan Ilmu Pendidikan (?) 16.

Medan: [a]Sumatra I. [b]none. [c]Deli R. [d]Universitas Sumatera Utara (1952) 7 – 626; Universitas Islam Sumatera Utara (1952) 1 – 279; Universitas Tyut Nya' Dhien (1956) 0.6 – 149. [e]Perpustakaan Balai Penelitian Perkebunan Medan (1918) 20; Perpustakaan Universitas Islam Sumatera Utara (1952?) 8; Perpustakaan Negara (?) 4.

Menado: [a]Celebes I. [b]alt Manado. [c]Celebes Sea; Menado (alt Manado) R. [d]Universitas Sam Ratulangi (1961) 3 – 946; Universitas Sulawesi Utara (1961) ? – ? [e]Perpustakaan Universitas Sam Ratulangi (1961?) 9; Perpustakaan Negara (?) 1.

Padang: [a]Sumatra I. [b]none. [c]Indian O; Padang R. [d]Universitas Andalas (1956) 4 – 500. [e]Perpustakaan Universitas (1956?) ?

Palembang: [a]Sumatra I. [b]none. [c]Musi (for Moesi) R. [d]Universitas Sriwijaya (1953) 3 – 1355. [e]Perpustakaan Universitas (1953?) 12.

Pontianak: [a]Borneo I. [b]none. [c]Kapuas-Kecil R (Kapuas R distributary); Landak R. [d]Universitas Tanjungpura (1963) 1 – 154; Universitas Kalimantan Barat (?) ? – ?

Semarang: [a]Java I. [b]alt Samarang. [c]Java Sea; Semarang (alt Samarang) R. [d]Universitas Diponegoro (1956) 6 – 830; Universitas Islam Sultan Agung (1962) 0.9 – 265. [e]Perpustakaan Negara (?) 21; Perpustakaan Universitas Diponegoro (1960) 15.

Surabaya: [a]Java I. [b]for Soerabaja, Surabaja. [c]Surabaya (for Surabaja) Strait; Mas R. [d]Universitas Airlangga (1954) 4 – 1032; Institut Teknologi 10 Nopember Surabaya (univ) (1960) 3 – 484; Universitas Kristen Petra (1965) 0.6 – 72. [e]Perpustakaan Universitas Airlangga (1954) 55.

Surakarta: [a]Java I. [b]alt Solo; for Soerakarta. [c]Solo R; Pepe R. [d]Universitas Cokroaminoto (1955) 4 – 100. [e]Perpustakaan Konservatori Karawitan Indonesia (?) ?; Perpustakaan Yayasan Paheman Radya Pustaka (?) ?

Tanjungkarang-Telukbetung: [a]Sumatra I. [b]inc Tanjungkarang, Telukbetung; Tanjungkarang: for Tandjoengkarang; Telukbetung: for Teloekbetoeng. [c]Lampung B of Sunda Strait. [d]Universitas Lampung (1961) 1 – 199. [e]Perpustakaan Universitas (1961?) ?

[1]Universities with 1000 or more students.

Ujung Pandang: [a]Celebes I. [b]for Macassar, Makasar, Makassar. [c]Makassar Strait. [d]Universitas Hasanuddin (1949) 7 – 1427; Universitas Veteran Republik Indonesia (1959) 2 – 224; Universitas Sawerigadang (1945) 1 – 158. [e]Perpustakaan Universitas Hasanuddin (1949) 95; Perpustakaan Negara (?) 12.

Yogyakarta: [a]Java I. [b]for Djokjakarta, Jogjakarta. [c]Codeh (for Tjodeh) R. [d][1]Universitas Gajah Mada (1949) 16 – 2951; Universitas Islam Indonesia (1945) 5 – 246 [Yogyakarta campus: (1945) ? – ?]. [e]Perpustakaan Universitas Gajah Mada (1949?) 210; Perpustakaan Negara (1949) 75; Badan Wakaf Perpustakaan Islam (?) 41.

Iran

Abadan: [b]none. [c](Shatt al) Arab R. [d]none. [e]Abadan Institute of Technology Library (1939) 22.

Ahwaz: [b]alt Ahvaz. [c]Karun R. [d]Jundi Shapur University (1955) 4 – 402. [e]. . . University Library (1955) ?

Hamadan: [b]none. [c]Qareh (Su) [alt Qara (Chai)] R. [d]Bu Ali Sina University (1973) ? – ?

Isfahan: [b]alt Ispahan; off Esfahan. [c]Zayandeh R. [d]University of Isfahan (1949) 6 – 369. [e]University . . . Library (1950) 140; Municipal Library (?) 31.

Kermanshah: [b]none. [c]nr Qareh (Su) [alt Qara (Chai)] R. [d]Razi University (1974) 1 – 73. [e]Public Library (bef 1969) ?

Meshed: [b]off Mashhad. [c]nr Kashaf R. [d]University of Ferdowsi (1940) 5 – 412. [e]University . . . Library (1925?) 100; Astaneh Razavy Library (15th c?) 55; Public Library (?) 18.

Qom: [b]alt Ghom, Qum; for Kum. [c]Qom (alt Ghom, Qum; for Kum) R. [d]none.

Rasht: [b]alt Resht. [c]Siah R. [d]University of Guilan (1975) ? – ?

Shiraz: [b]none. [c]Khoshk R. [d]Pahlavi University (1945) 4 – 560. [e]. . . University Library (1934) 170; Fars National Library (?) 14.

Tabriz: [b]none. [c]Talkheh (alt Aji) R. [d]University of Azarabadegan (1946) 8 – 647. [e]University . . . Library (1947) 70; Public Library (?) 13.

*Tehran: [b]alt Teheran. [c]none. [d]University of Tehran (1934) 18 – 2078; National University of Iran (1960) 8 – 365; Arya-Mehr University of Technology (1965) 3 – 304; Farah Pahlavi University (1975) 2 – 143; Farabi University (1975) ? – ? [e]University . . . Library (1932) 321; Parliament Library (1924) 120; National Library (1935) 100; National Pahlavi Library (1965) ?

Iraq

*Baghdad: [b]alt Bagdad. [c]Tigris R. [d]University of Baghdad (1908) 19 – 1509; Al-Mustansiriya University (1963) 12 – 473; University of Technology (1974) 6 – 300. [e]Central Library of the University . . . (1958) 220; Al-Awqaf Library (1928) 100; National Library of the Iraq Museum (1934) 100; University . . . College of Medicine Library (1927) 62; National Library (1955) 52.

Basra: [b]off Basrah. [c](Shatt al) Arab R. [d]University of Basra (1964) 11 – 427. [e]Bashayan el Abbasi Library (16th c) 11; Public Library (?) ?

Karbala: [b]alt Kerbela. [c]Jadwal R. [d]none. [e]Public Library (?) ?

Kirkuk: [b]alt Kerkuk; for Zor. [c]Qada (alt Qadha) (Chai) Creek. [d]none. [e]Public Library (?) ?

Mosul: [b]off Mawsil. [c]Tigris R. [d]University of Mosul (1967) 10 – 576. [e]Central Library of the University . . . (1965) 107; Public Library (1930) 66.

Najaf: [b]alt Nedjef. [c]Najaf L. [d]none. [e]Public Library (?) ?

Israel

Haifa: [b]off Hefa. [c]B of Acre of Mediterranean Sea; Kishon R. [d]Technion–Israel Institute of Technology (univ) (1912) 8 – 1449; University of Haifa (1963) 5 – 722. [e]University . . . Library (1951) 250; Central Library of Technion (1925) 180; Pevsner Public Library (1930) 170; Borochov Library (1921) 105.

[1]Universities with 1000 or more students.

*Jerusalem: ^{b}alt Quds ash Sharif; inc Jordanian Jerusalem; off Yerushalayim. ^{c}nr Dead (Sea) L. ^{d}Hebrew University of Jerusalem (1918) 14 – 2067. ^{e}Jewish National and University Library (1884) 2000; City Library (1961) 300. ^{f}Jordanian Jerusalem: 60 (1961), 70 (1952).

Tel Aviv-Jaffa: ^{b}inc Jaffa, Tel Aviv; Jaffa: off Yafo. ^{c}Mediterranean Sea. ^{d}Tel Aviv University (1935) 12 – 2775; Bar-Ilan University (at Ramat Gan, nr Tel Aviv-Jaffa) (1953) 7 – 918; Everyman's University (1974) 5 – ? ^{e}Tel Aviv University Library (1953) 420; Municipal Library (1885) 418. ^{f}Jaffa: 51 (1931), 21 (1900e), 5 (1850e).

Japan

Akashi: ^{a}Honshu I. ^{b}none. ^{c}Akashi Channel. ^{d}none.

Akita: ^{a}Honshu I. ^{b}none. ^{c}Sea of Japan; Omono R. ^{d}Akita University (1875) 3 – 612. ^{e}Akita Prefectural Library (1899) 219; . . . University Library (1911) 170.

Amagasaki: ^{a}Honshu I. ^{b}none. ^{c}Osaka B of Inland (alt Seto) Sea; Yodo R. ^{d}none. ^{e}Municipal Library (1919) 68.

Aomori: ^{a}Honshu I. ^{b}none. ^{c}Aomori B of Tsugaru Strait. ^{d}none. ^{e}Aomori Prefectural Library (1928) 85.

Asahikawa: ^{a}Hokkaido I. ^{b}alt Asahigawa. ^{c}Ishikari R. ^{d}none. ^{e}Shimomura Bunko Library (1918) 10 (in 1946).

Beppu: ^{a}Kyushu I. ^{b}none. ^{c}Beppu B of Inland (alt Seto) Sea. ^{d}Beppu University (college) (1946) 0.5 – 60. ^{e}. . . University Library (1946?) 50; Municipal Library (1922) 6 (in 1946).

Chiba: ^{a}Honshu I. ^{b}none. ^{c}Tokyo B of Pacific O. ^{d}Chiba University (1872) 8 – 1349. ^{e}. . . University Library (1949) 496; Central Library of Chiba Prefecture (1924) 268.

Fuji: ^{a}Honshu I. ^{b}inc Yoshiwara since 1966. ^{c}Fuji R. ^{d}none. ^{f}Yoshiwara: 81 (1960), 33 (1950), 3 (1898).

Fujisawa: ^{a}Honshu I. ^{b}none. ^{c}Sagami B of Pacific O. ^{d}none.

Fukui: ^{a}Honshu I. ^{b}none. ^{c}Asuwa R. ^{d}Fukui University (1923) 2 – 300. ^{e}. . . University Library (1923?) 128; Fukui Prefectural Library (1909) 65.

Fukuoka: ^{a}Kyushu I. ^{b}inc Hakata. ^{c}Hakata B of East China Sea; Naka R. ^{d}Fukuoka University (1934) 21 – 463; Kyushu University (1903) 11 – 3116; Seinan Gakuin University (1916) 7 – 346. ^{e}Kyushu University Library (1911) 1778; Fukuoka University Library (1934?) 300; Fukuoka Prefectural Library (1918) 203.

Fukushima: ^{a}Honshu I. ^{b}none. ^{c}Abukuma R. ^{d}Fukushima University (college) (1921) 2 – 145. ^{e}Fukushima Prefectural Library (1929) 120.

Fukuyama: ^{a}Honshu I. ^{b}none. ^{c}Inland (alt Seto) Sea; Ashida R. ^{d}none. ^{e}Library of the Faculty of Fisheries and Animal Husbandry of Hiroshima University (1949) 11; Yoshikura Library (1910) ?

Funabashi: ^{a}Honshu I. ^{b}none. ^{c}Tokyo B of Pacific O. ^{d}Chiba Institute of Technology (univ) (at Narashino, nr Funabashi) (1942 at Machida, relocated 1950) 5 – 322.

Gifu: ^{a}Honshu I. ^{b}for Imaizumi. ^{c}Nagara R. ^{d}Gifu University (at Kakamigahara, nr Gifu) (1875) 4 – 1066. ^{e}. . . University Library (at Kakamigahara, nr Gifu) (1875?) 363; Gifu Prefectural Library (1909) 174.

Hachinohe: ^{a}Honshu I. ^{b}none. ^{c}Mabechi R. ^{d}none. ^{e}Municipal Library (1913) 21 (in 1946).

Hachioji: ^{a}Honshu I. ^{b}none. ^{c}Asa R; Kawaguchi R. ^{d}none. ^{e}Tokyo Metropolitan Hachioji Library (1911) 2 (in 1946).

Hakodate: ^{a}Hokkaido I. ^{b}none. ^{c}Tsugaru Strait. ^{d}none. ^{e}Municipal Library (1926) 125.

Hamamatsu: ^{a}Honshu I. ^{b}none. ^{c}Pacific O. ^{d}none. ^{e}Hamamatsu Technical College Library (1924) 35 (in 1946); Municipal Library (1920) 16 (in 1946).

Higashiosaka: ^{a}Honshu I. ^{b}for Fuse. ^{c}nr Yodo R. ^{d}Kinki University (1925) 31 – 530. ^{e}. . . University Library (1925?) 212.

Himeji: ^{a}Honshu I. ^{b}none. ^{c}Ichi R. ^{d}Himeji Institute of Technology (univ) (1944) 1 – 150. ^{e}. . . Institute Library (1944?) 44; Himeji Higher School Library (1923) 41 (in 1946); Municipal Library (1912) 20 (in 1946).

Hirakata: ^{a}Honshu I. ^{b}none. ^{c}Yodo R. ^{d}none. ^{e}Osaka College of Dental Medicine Library (1931) 5 (in 1946); Osaka Women's Higher Medical College Library (1915) 5 (in 1946).

Hiratsuka: [a]Honshu I. [b]none. [c]Sagami B of Pacific O. [d]none. [e]National Institute of Agricultural Sciences Library (1950) 30.

Hiroshima: [a]Honshu I. [b]none. [c]Hiroshima B of Inland (alt Seto) Sea. [d]Hiroshima University (1902) 10 – 1626; Hiroshima Commercial College (univ) (1952) 3 – ? [e]. . . University Library (1949) 1261; Hiroshima Prefectural Library (?) 100.

Hitachi: [a]Honshu I. [b]none. [c]Pacific O. [d]none.

Ibaraki: [a]Honshu I. [b]none. [c]nr Yodo R. [d]none.

Ichihara: [a]Honshu I. [b]none. [c]Tokyo B of Pacific O. [d]none.

Ichikawa: [a]Honshu I. [b]none. [c]Edo R. [d]none.

Ichinomiya: [a]Honshu I. [b]none. [c]nr Kiso R. [d]none. [e]Municipal Library (1915) 6 (in 1946).

Ise: [a]Honshu I. [b]for Uji-Yamada. [c]Ise B of Pacific O. [d]Kogakukan College (univ) (1962) 0.9 – ? [e]Mie Shinto Library (1928) 195.

Iwaki: [a]Honshu I. [b]for Onahama; inc Taira since 1965. [c]Pacific O; Kamata R. [d]none. [f]Taira: 71 (1965), 71 (1960), 43 (1950), 25 (1930), 11 (1898), 4 (1874).

Kagoshima: [a]Kyushu I. [b]none. [c]Kagoshima B of East China Sea. [d]Kagoshima University (1901) 6 – 1291. [e]. . . University Library (1949) 563; Kagoshima Prefectural Library (1912) 222.

Kamakura: [a]Honshu I. [b]none. [c]Sagami B of Pacific O. [d]none. [e]Nomura Research Institute of Technology and Economics Documentation Center (?) 120; Municipal Library (1911) 10 (in 1946).

Kanazawa: [a]Honshu I. [b]none. [c]Sea of Japan; Ono R. [d]Kanazawa University (1923) 6 – 732. [e]. . . University Library (1949) 717; Central Library of Ishikawa Prefecture (1912) 165; Municipal Library (1929) 157.

Kashiwa: [a]Honshu I. [b]none. [c]nr Tone R. [d]none.

Kasugai: [a]Honshu I. [b]none. [c]Shonai R. [d]Chubu Institute of Technology (univ) (1938) 5 – 204. [e]. . . Institute Library (1938?) 76.

Kawagoe: [a]Honshu I. [b]none. [c]Shingashi R. [d]none. [e]Municipal Library (1915) 35 (in 1946).

Kawaguchi: [a]Honshu I. [b]none. [c]Ara R. [d]none.

Kawasaki: [a]Honshu I. [b]none. [c]Tokyo B of Pacific O. [d]none. [e]Kanagawa Prefectural Library (1958) 148.

Kitakyushu: [a]Kyushu I. [b]inc Kokura, Moji, Tobata, Wakamatsu, Yahata (alt Yawata). [c]Kammon Strait. [d]Kyushu Institute of Technology (univ) (1907) 2 – 200. [e]Yawata Iron and Steel Works Company Library (?) 145; . . . Institute Library (1909) 114. [f]:

	1960	1950	1930	1898	1874
Kokura	286	199	88	28	7
Moji	152	124	108	25	
Tobata	109	88	52	3	
Wakamatsu	107	90	57	29	
Yahata	332	210	168	3	

Kobe: [a]Honshu I. [b]inc Hyogo (alt Hiogo) since 1878. [c]Osaka B of Inland (alt Seto) Sea; Minato R. [d]Kobe University (1903) 10 – 1367; Konan University (1918) 7 – 300; Kobe University of Commerce (1929) 2 – 130; Kobe Municipal University of Foreign Studies (1946) 0.9 – ? [e]Kobe University Library (1903) 1181; Municipal Library (1912) 278; Konan University Library (1918?) 263. [f]Hyogo: 30 (1877), 22 (1796).

Kochi: [a]Shikoku I. [b]none. [c]Tosa B of Pacific O; Kagami R. [d]Kochi University (1874) 2 – 338. [e]. . . University Library (1925) 190; Kochi Prefectural Library (1915) 142.

Kofu: [a]Honshu I. [b]none. [c]Ara R. [d]Yamanashi University (1924) 3 – 625. [e]. . . University Library (1924?) 213; Yamanashi Prefectural Library (1931) 138.

Koriyama: [a]Honshu I. [b]none. [c]Abukuma R. [d]none. [e]Municipal Library (1944) 8 (in 1946).

Koshigaya: [a]Honshu I. [b]none. [c]Moto-Ara R. [d]none.

Kumamoto: [a]Kyushu I. [b]none. [c]Shira R. [d]Kumamoto University (1874) 6 – 945. [e]. . . University Library (1949) 556.

Kurashiki: [a]Honshu I. [b]none. [c]Kurashiki R. [d]none. [e]Library of the Ohara Institute of Agricultural Biology of Okayama University (1921) 142.

Kure: [a]Honshu I. [b]none. [c]Hiroshima B of Inland (alt Seto) Sea. [d]none. [e]Municipal Library (1924) 32 (in 1946).

Kurume: [a]Kyushu I. [b]none. [c]Chikugo R. [d]Kurume University (1928) 3 – 880. [e]. . . University Library (1928?) 147; Municipal Library (1938) 30 (in 1946).

Kushiro: [a]Hokkaido I. [b]none. [c]Pacific O; Kushiro R. [d]none. [e]Municipal Popular Library (1925) 7 (in 1946).

Kyoto: [a]Honshu I. [b]alt Kioto; for Miyako. [c]Kamo R. [d1]Doshisha University (1875) 24 – 1000; Ritsumeikan University (1900) 21 – 600; Kyoto University (1897) 15 – 3114; Ryukoku University (1639) 13 – 652; Kyoto Industrial University (1965) 11 – 240; Bukkyo University (1887) 3 – 220; Doshisha Women's University (1876) 3 – 200; Kyoto University of Foreign Studies (1950) 3 – ?; Kyoto Women's University (1920) 3 – 194; University of Buddhism (1903) 3 – 220. [e]Kyoto University Library (1899) 3438; Doshisha University Library (1875) 559; Ryukoku University Library (1639) 495; Kyoto Prefectural Library (1899) 380.

Machida: [a]Honshu I. [b]none. [c]Sakai R. [d]Tamagawa University (1929) 5 – 413. [e]. . . University Library (1929) 336.

Maebashi: [a]Honshu I. [b]for Umayabashi. [c]Tone R. [d]Gumma University (1876) 4 – 573. [e]. . . University Library (1949) 222; Municipal Library (1915) 92.

Matsudo: [a]Honshu I. [b]none. [c]Edo R. [d]none. [e]Library of the Faculty of Horticulture of Chiba University (1909) 12.

Matsumoto: [a]Honshu I. [b]for Fukashi. [c]Narai R. [d]Shinshu University (1910) 6 – 2250. [e]. . . University Library (1920) 333; Municipal Library (1921) 74.

Matsuyama: [a]Shikoku I. [b]none. [c]nr Inland (alt Seto) Sea. [d]Ehime University (1896) 4 – 500; Matsuyama University of Commerce (1923) 4 – 147. [e]Ehime University Library (1949) 320; Matsuyama University of Commerce Library (1923) 120; Ehime Prefectural Library (1935) 82.

Mito: [a]Honshu I. [b]none. [c]Naka R. [d]Ibaraki University (1920) 5 – 665. [e]. . . University Library (1920?) 319; Ibaraki Prefectural Library (1903) 95.

Miyazaki: [a]Kyushu I. [b]none. [c]Oyodo R. [d]Miyazaki University (1923) 3 – 400. [e]. . . University Library (1926) 169; Miyazaki Prefectural Library (1902) 78.

Morioka: [a]Honshu I. [b]none. [c]Kitakami R. [d]Iwate University (1902) 3 – 300. [e]. . . University Library (1902) 198; Iwate Prefectural Library (1922) 132.

Nagano: [a]Honshu I. [b]for Zenkoji. [c]Sai R. [d]none. [e]Nagano Prefectural Library (1929) 174.

Nagasaki: [a]Kyushu I. [b]none. [c]East China Sea. [d]Nagasaki University (1857) 4 – 700. [e]Nagasaki Prefectural Library (1912) 780; . . . University Library (1949) 376.

Nagoya: [a]Honshu I. [b]none. [c]Ise B of Pacific O; Shonai R; Yata R. [d]Meijo University (1924) 18 – 470; Aichi Gakuin University (1876) 9 – 140; Chukyo University (1927) 9 – 110; Nagoya University (1871) 8 – 1904; Aichi Institute of Technology (univ) (1912) 5 – ?; Nanzan University (1932) 5 – 285; Nagoya Institute of Technology (univ) (1905) 4 – 580; Kinjo Gakuin University (1889) 3 – 264; Nagoya Municipal University (1931) 2 – 350. [e]Nagoya University Library (1939) 1051; Municipal Library (1920) 247; Aichi Prefectural Library (?) 170.

Naha: [a]Okinawa I. [b]alt Nawa. [c]East China Sea; Kokuba R. [d]University of the Ryukyus (1950) 4 – 615; Okinawa University (1956) 3 – ? [e]Shikiya Memorial Library (alt University . . . Library) (1950) 200; Ryukyu Islands Central Library (1950) 46.

Nara: [a]Honshu I. [b]none. [c]Saho R; Noto R. [d]Tenri University (college) (at Tenri, nr Nara) (1925) 2 – 246; Nara Women's University (1908) 1 – 257. [e]Tenri Central Library (alt Tenri University Library) (at Tenri, nr Nara) (1930) 1083; Nara Women's University Library (1909) 200; Nara Prefectural Library (1908) 164.

Neyagawa: [a]Honshu I. [b]alt Neyakawa. [c]nr Yodo R. [d]none.

Niigata: [a]Honshu I. [b]for Kambaratsu. [c]Sea of Japan; Shinano R. [d]Niigata University (1910) 6 – 1199. [e]. . . University Library (1949) 509; Niigata Prefectural Library (1915) 230.

Nikko: [a]Honshu I. [b]none. [c]Daiya R. [d]none. [e]Nikko Bunko Library (1924) 16 (in 1946).

---

[1]Universities with 3000 or more students.

Nishinomiya: [a]Honshu I. [b]none. [c]Osaka B of Inland (alt Seto) Sea. [d]Kwansei Gakuin University (1889) 14 – 582; Mukogawa Women's University (1946) 4 – 250; Kobe Jogakuin University (1875) 2 – 160. [e]Kwansei Gakuin University Library (1889) 450.

Numazu: [a]Honshu I. [b]none. [c]Suruga B of Pacific O; Kano R. [d]none. [e]Numazu Bunko Library (1898) 9 (in 1946).

Oita: [a]Kyushu I. [b]for Funai. [c]Beppu B of Inland (alt Seto) Sea; Oita R. [d]Oita University (college) (1875) 2 – 160. [e]. . . University Library (1922) 128; Oita Prefectural Library (1902) 65.

Okayama: [a]Honshu I. [b]none. [c]Asahi R. [d]Okayama University (1874) 7 – 1218. [e]. . . University Library (1949) 894.

Okazaki: [a]Honshu I. [b]none. [c]Yahagi R. [d]none. [e]Aichi Gakugei College Library (1945?) 125; Municipal Library (1922) ?

Omiya: [a]Honshu I. [b]none. [c]nr Shiba R. [d]none. [e]Municipal Library (1921) 4 (in 1946).

Osaka: [a]Honshu I. [b]none. [c]Osaka B of Inland (alt Seto) Sea; Yodo R. [d][1]Osaka University (1843) 9 – 1243; Osaka University of Economics (1932) 9 – 242; Osaka Municipal University (1928) 6 – 1176; Osaka Institute of Technology (univ) (1922) 5 – ?; Osaka University of Education (1943) 4 – 290; Osaka University of Foreign Studies (1921) 3 – 369. [e]Osaka University Library (1931) 1259; Osaka Municipal University Library (1928) 984; Osaka Prefectural Library (1903) 820.

Otaru: [a]Hokkaido I. [b]none. [c]Ishikari (alt Otaru) B of Sea of Japan. [d]Otaru University of Commerce (1910) 1 – ? [e]. . . University Library (1911) 115.

Otsu: [a]Honshu I. [b]none. [c]Biwa L. [d]none. [e]Shiga Prefectural Library (1943) 52.

Sagamihara: [a]Honshu I. [b]inc Kami-Mizo, Ono. [c]Sakai R. [d]none. [e]Azabu Veterinary College Library (1949) 22.

Sakai: [a]Honshu I. [b]none. [c]Osaka B of Inland (alt Seto) Sea; Yamato R. [d]University of Osaka Prefecture (1939) 5 – 635. [e]University . . . Library (1939?) 403; Municipal Library (1872) 62.

Sapporo: [a]Hokkaido I. [b]none. [c]Toyohira R. [d]Hokkaido University (1872) 11 – 1784; Hokkai Gakuen University (1887) 6 – 157. [e]Hokkaido University Library (1876) 1595; Hokkaido College of Education Library (1943?) 264; Hokkai Gakuen University Library (1887?) 168; Hokkaido Prefectural Library (1924) 92.

Sasebo: [a]Kyushu I. [b]none. [c]East China Sea. [d]none. [e]Municipal Library (1918) 5 (in 1946).

Sendai: [a]Honshu I [b]none. [c]Hirose R. [d]Tohoku University (1907) 12 – 2702; Tohoku Gakuin University (1886) 10 – 350. [e]Tohoku University Library (1911) 1726; Tohoku Gakuin University Library (1890) 152; Miyagi Prefectural Library (1881) 110.

Shimizu: [a]Honshu I. [b]none. [c]Suruga B of Pacific O. [d]none. [e]Nautical College Library (1943) 20 (in 1946); Municipal Library (1931) 1 (in 1946).

Shimonoseki: [a]Honshu I. [b]for Akamagaseki, Bakwan. [c]Kammon Strait. [d]none. [e]Shimonoseki College of Fisheries Library (?) 46; Municipal Library (1882) 26 (in 1946).

Shizuoka: [a]Honshu I. [b]for Fuchu, Shunpei, Sumpu. [c]Suruga B of Pacific O. [d]Shizuoka University (1875) 7 – 540. [e]. . . University Library (1949) 210; Central Library of Shizuoka Prefecture (1922) 175.

Suita: [a]Honshu I. [b]none. [c]Yodo R. [d]Kansai University (1886) 25 – 477; Osaka Gakuin University (1962) 3 – ? [e]Kansai University Library (1886) 750.

Takamatsu: [a]Shikoku I. [b]none. [c]Inland (alt Seto) Sea. [d]Kagawa University (1923) 3 – 262. [e]. . . University Library (1924) 307; Kagawa Prefectural Library (1934) ?

Takasaki: [a]Honshu I. [b]none. [c]Karasu R. [d]none. [e]Municipal Library (1910) 32 (in 1946).

Takatsuki: [a]Honshu I. [b]none. [c]Yodo R. [d]none.

Tokorozawa: [a]Honshu I. [b]none. [c]nr Iruma R. [d]none.

Tokushima: [a]Shikoku I. [b]none. [c]Kii Channel; Yoshino R. [d]Tokushima University (1922) 4 – 657; Tokushima University of Arts and Science (1966) 0.9 – ? [e]Tokushima University Library (1949) 257; Tokushima Prefectural Library (1916) 88.

[1]Universities with 1000 or more students.

*Tokyo: ^{a}Honshu I. ^{b}alt Tokio; for Edo, Yedo. ^{c}Tokyo B of Pacific O; Sumida R; Ara R. $^{d^1}$Nihon University (1889) 102 – 3505; Waseda University (1882) 48 – 2159; Chuo University (1885) 36 – 1356; Meiji University (1881) 34 – 1335; Hosei University (1880) 30 – 970; Tokai University (1943) 27 – 921; Keio University (1858) 26 – 1060; Toyo University (1887) 21 – 393; Komazawa University (1759) 19 – 360; University of Tokyo (1789?) 19 – 3649; Senshu University (1880) 18 – 465; Aoyama Gakuin University (1874) 17 – 883; Kokugakuin University (1882) 13 – 537; Rikkyo (alt Saint Paul's) University (1874) 13 – 725; Meiji Gakuin University (1877) 12 – 410; Jochi (alt Sophia) University (1913) 10 – 772. ^{e}National Diet Library (1948) 6560; University of Tokyo Library (1893) 4040; Nihon University Library (1889) 1130; Waseda University Library (1882) 1019.

Toyama: ^{a}Honshu I. ^{b}none. ^{c}Jintsu R. ^{d}Toyama University (1920) 4 – 400. ^{e}. . . University Library (1920?) 208; Toyama Prefectural Library (1940) 165.

Toyohashi: ^{a}Honshu I. ^{b}for Yoshida. ^{c}Atsumi B of Pacific O. ^{d}Aichi University (1946) 7 – 110. ^{e}. . . University Library (1946?) 160; Municipal Library (1912) 78.

Toyonaka: ^{a}Honshu I. ^{b}none. ^{c}Yodo R. ^{d}none. ^{e}Naniwa Higher School Library (1926) 27 (in 1946).

Toyota: ^{a}Honshu I. ^{b}for Koromo. ^{c}Yahagi R. ^{d}none. ^{e}Toyota Motor Company Documentation Center (1936) 62.

Urawa: ^{a}Honshu I. ^{b}none. ^{c}Ara R. ^{d}Saitama University (college) (1921) 5 – 701. ^{e}. . . University Library (1924) 130; Saitama Prefectural Library (1924) 120.

Utsunomiya: ^{a}Honshu I. ^{b}none. ^{c}Ta R. ^{d}Utsunomiya University (1922) 3 – 320. ^{e}Tochigi Prefectural Library (?) 197; . . . University Library (1922) 182.

Wakayama: ^{a}Honshu I. ^{b}none. ^{c}Kitan Strait; Kino R. ^{d}Wakayama University (1871) 2 – 191. ^{e}. . . University Library (1923) 265; Wakayama Prefectural Library (1908) 68.

Yamagata: ^{a}Honshu I. ^{b}none. ^{c}nr Mogami R. ^{d}Yamagata University (1920) 5 – 560. ^{e}. . . University Library (1949) 311; Yamagata Prefectural Library (1910) 108.

Yao: ^{a}Honshu I. ^{b}none. ^{c}nr Yamato R. ^{d}none.

Yokkaichi: ^{a}Honshu I. ^{b}none. ^{c}Ise B of Pacific O. ^{d}none. ^{e}Municipal Library (1908) 52.

Yokohama: ^{a}Honshu I. ^{b}none. ^{c}Tokyo B of Pacific O. ^{d}Kanagawa University (1929) 12 – 340; Kanto Gakuin University (1884) 7 – 330; Yokohama National University (1876) 6 – 1170; Yokohama Municipal University (1928) 3 – 410. ^{e}Yokohama National University Library (1949) 463; Kanagawa Prefectural Library (1930) 427; Municipal Library (1925) 184.

Yokosuka: ^{a}Honshu I. ^{b}none. ^{c}Tokyo B of Pacific O. ^{d}none. ^{e}Kanto Auto Works Engineering Library (?) 10; Fuji Electric Company Central Research Library (?) 7.

Jordan

*Amman: ^{b}none. ^{c}(Wadi) Amman R. ^{d}University of Jordan (1962) 5 – 281. ^{e}University . . . Library (1962) 150; Municipal Public Library (1960) 50.

Zarqa: ^{b}alt Zerka, Zerqa. ^{c}Zarqa (alt Zerka, Zerqa) R. ^{d}none.

Kashmir-Jammu

*Jammu: ^{b}none. ^{c}Tawi R. ^{d}University of Jammu (1969) 11 – ? ^{e}University . . . Library (1969) 80; Shri Ranbir Library (1879) 18 (in 1950).

*Srinagar: ^{b}for Cashmere. ^{c}Jhelum R. ^{d}University of Kashmir (1948) 18 – 957. ^{e}University . . . Library (1948) 91.

Korea: North

Chongjin: ^{b}for Seishin. ^{c}Sea of Japan. ^{d}none. ^{e}City Library (?) 23; North Hamgyong Province Library (?) ?

Hamhung: ^{b}for Kanko. ^{c}Tongsong R. ^{d}none. ^{e}South Hamgyong Province Library (?) 40; City Library (?) ?

Hungnam: ^{b}for Konan. ^{c}Sea of Japan; Tongsong R. ^{d}none.

---

[1]Universities with 10,000 or more students.

Kaesong: [b]for Kaijo, Songdo. [c]nr Han R; nr Yesong R. [d]none. [e]City Library (?) 28.

Kimchaek: [b]for Joshin, Songjin. [c]Sea of Japan; Susong R. [d]none.

*Pyongyang: [b]for Heijo. [c]Taedong R. [d]Kim Il-song Comprehensive University (1946) 16 – 900. [e]State Central Library (1964) 1520; . . . University Library (1946) 60; South Pyongan Province Library (?) ?

Sinuiju: [b]for Shingishu. [c]Yalu R. [d]none. [e]North Pyongan Province Library (?) ?

Wonsan: [b]for Gensan. [c]Sea of Japan. [d]none. [e]Kangwon Province Library (?) ?

Korea: South

Chongju: [b]alt Cheongju; for Seishu. [c]Musim R. [d]none. [e]Chung Buk National College Library (1946?) 150.

Chonju: [b]alt Jeonju; for Zenshu. [c]Chonju (alt Jeonju) R. [d]Chunpuk National University (1951) 6 – 286. [e]. . . University Library (1952) 85.

Inchon: [b]alt Incheon; for Chemulpo, Zinsen. [c]Yellow Sea. [d]Inha University (1952) 3 – 120. [e]. . . University Library (1955) 53; Public Library (1921) 18 (in 1959).

Kwangju: [b]alt Gwangju; for Koshu. [c]nr Yongsan R. [d]Chonnam National University (1952) 6 – 321; Chosun University (1946) 6 – 358. [e]Chosun University Library (1950) 158; Chonnam National University Library (1952) 58.

Masan: [b]for Masampo. [c]Chinhae B of Korea Strait. [d]none. [e]Masan College Library (?) ?

Mokpo: [b]alt Mogpo; for Moppo. [c]Yellow Sea. [d]none. [e]Commerce College Library of Chonnam National University (?) ?

Pusan: [b]alt Busan; for Fusan. [c]Korea Strait. [d]Pusan National University (1945) 8 – 373; Dong A University (1947) 6 – 400. [e]Dong A University Library (1950) 139; Pusan National University Library (1955) 95.

*Seoul: [b]alt Kyongsong; for Hanyang, Keijo; off Soul. [c]Han R. [d][1]Seoul National University (1946) 16 – 1700; Yonsei University (1885) 11 – 2331; Korea University (1905) 10 – 425; Kyung Hee University (1949) 10 – 635; Hanyang University (1939) 9 – 726; Chungang University (1919) 8 – 767; Ewha Women's University (1886) 8 – 995; Dongguk University (1906) 5 – 300; Konkuk University (1946) 5 – 472; Sung Kyun Kwan University (992) 5 – 300. [e]Seoul National University Library (1946) 957; National Central Library (1923) 450; Sung Kyun Kwan University Library (992?) 400.

Songnam: [b]alt Seongnam. [c]nr Han R. [d]none.

Suwon: [b]alt Suweon; for Suigen. [c]none. [d]none. [e]Office of Rural Development Library (1945) 31.

Taegu: [b]alt Daegu; for Taikyu. [c]Tae R. [d]Yeungnam University (1948) 10 – 264; Kyongbuk National University (1951) 7 – 527. [e]Yeungnam University Library (1948?) 250; Kyongbuk National University Library (1952) 130.

Taejon: [b]alt Daejeon; for Taiden. [c]Taejon (alt Daejeon) R. [d]Chungnam National University (1952) 5 – 300. [e]. . . University Library (1955) 49.

Ulsan: [b]for Urusan. [c]Ulsan B of Korea Strait; Ulsan R. [d]none.

Kuwait

*Kuwait: [b]alt Kuweit; off Kuwayt. [c]Persian G. [d]Kuwait University (1962) 7 – 1000. [e]. . . University Library (1966) 210; Kuwait Central Library (1936) 95.

Laos

*Vientiane: [b]none. [c]Mekong R. [d]Universite Sisavangvong (1958) 2 – 149. [e]Bibliotheque de l'Alliance Francaise (?) 10; British Council Library (?) 10; Bibliotheque Nationale (?) ?

Lebanon

*Beirut: [b]for Beyrouth; off Bayrut. [c]Mediterranean Sea. [d]Beirut Arab University (1960) 27 – 178; Universite Libanaise (1951) 15 – 733; Universite Saint-Joseph (1846) 3 – 370; American Uni-

[1]Universities with 5000 or more students.

versity of Beirut (1866) 2 − 501. [e]American University of Beirut Library (1866) 407; Bibliotheque Orientale de l'Universite Saint-Joseph (1881) 150; Bibliotheque Nationale du Liban (1921) 120.

Tripoli: [b]off Tarabulus. [c]Mediterranean Sea. [d]none.

Macao

*Macao: [a]Macao (alt Aomen, Chungshan; off Macau) I. [b]alt Aomen; off Macau. [c]South China Sea; Pearl R. [d]none. [e]Biblioteca Nacional de Macau (1929) 60.

Malaysia

George Town: [a]Peninsular Malaysia; Penang I. [b]alt Penang; off Pinang. [c]Strait of Malacca. [d]University of Science, Malaysia (1969) 3 − 320. [e]University . . . Library (1969) 150; Penang Public Library (1817) 36.

Ipoh: [a]Peninsular Malaysia. [b]none. [c]Kinta R. [d]none. [e]Public Library (1951) 100.

*Kuala Lumpur: [a]Peninsular Malaysia. [b]none. [c]Klang R. [d]University of Malaya (1905) 8 − 925; National University of Malaysia (1970) 4 − 329; University of Technology, Malaysia (1954) 2 − 189. [e]University . . . Library (1957) 558; National University of Malaysia Library (1970?) 157; Selangor State Library (1972) 140; Ministry of Agriculture Library (1907) 70; Public Library (1966) 45; National Library of Malaysia (1971) 43.

Malacca: [a]Peninsular Malaysia. [b]off Melaka. [c]Strait of Malacca; Malacca (off Melaka) R. [d]none. [e]Malacca Library (1966) 10.

Mongolia

*Ulan-Bator: [b]for Kulun, Urga; off Ulaanbaatar. [c]Tuul R. [d]Mongolian State University (1942) 3 − 426. [e]State Public Library (1921) 1100; . . . University Library (1942) 400.

Nepal

*Kathmandu: [b]alt Katmandu. [c]Baghmati R. [d]Tribhuvan University (1958) 17 − 1287. [e]. . . University Library (1959) 67; Nepal National Library (?) 20.

Pakistan

Faisalabad: [b]for Lyallpur. [c]none. [d]Pakistan Agricultural University (1909) 4 − 364. [e]. . . University Library (1909) 57.

Gujranwala: [b]none. [c]none. [d]none. [e]Islamia College Library (?) 11; Municipal Public Library (?) 6.

Hyderabad: [b]for Nerankot. [c]Indus R. [d]University of Sind (at Jam Shoro, nr Hyderabad) (1947 at Karachi, relocated 1951) 3 − 390. [e]Alama I. I. Kazi Central Library of the University . . . (at Jam Shoro, nr Hyderabad) (1947) 111; Hayat-e-Adab Library (?) 38.

*Islamabad: [b]none. [c]Soan R; Rawal L; Lohi Shir L. [d]Quaid-i-Azam University (1965) 0.9 − 144. [e]National Assembly Library (1947 at Karachi, relocated 1970?) 48; . . . University Library (1967) 14; National Library (?) ?

Karachi: [b]none. [c]Arabian Sea; Layari R. [d]University of Karachi (1950) 7 − 440. [e]University . . . Library (1950) 201; Liaquat Memorial Library (150) 100.

Lahore: [b]none. [c]Ravi R. [d]University of the Punjab (1858) 9 − 400; Pakistan University of Engineering and Technology (1923) 3 − 201. [e]University . . . Library (1882) 267; Punjab Public Library (1884) 200.

Multan: [b]none. [c]nr Chenab R. [d]none. [e]Emerson College Library (?) 18; Public Library (?) 10.

Peshawar: [b]none. [c]nr Bara R. [d]University of Peshawar (1913) 8 − 386. [e]University . . . Library (1950) 95.

Rawalpindi: [b]none. [c]Leh R. [d]none. [e]Gordon College Library (1893?) 37.

Sargodha: [b]none. [c]none. [d]none. [e]Municipal Public Library (?) 13; De Montmorency College Library (1934?) 12.

Sialkot: [b]for Sealkote. [c]Aik (Nala) Creek. [d]none. [e]Murray College Library (1889?) 26; Municipal Library (?) 5.

Philippines

Bacolod: [a]Negros I. [b]none. [c]Guimaras Strait; Lupit R. [d]University of Negros Occidental-Recoletos (1941) 7 – 222. [e]University . . . Library (1941) 32.

Caloocan: [a]Luzon I. [b]none. [c]Manila B of South China Sea. [d]Araneta University Foundation (at Malabon, nr Caloocan) (1946) 14 – 494; University of the East, Caloocan campus (?) ? – 65. [e]. . . University Library (at Malabon, nr Caloocan) (1946?) 19.

Cebu: [a]Cebu I. [b]for Zebu. [c]Bohol Strait. [d]University of the Visayas (1919) 20 – 500; Southwestern University (1946) 12 – 410; University of San Carlos (1595) 10 – 400; University of the Southern Philippines (1927) 7 – 193. [e]University of San Carlos Library (1945) 145.

Davao: [a]Mindanao I. [b]none. [c]Davao G of Pacific O; Davao R. [d]University of Mindanao (1946) 15 – 338. [e]University . . . Library (1952) 36; City Library (1952) 19.

Iloilo: [a]Panay I. [b]none. [c]Iloilo Strait; Iloilo R. [d]University of San Agustin (1904) 10 – 350; Central Philippine University (1905) 8 – 274; University of Iloilo (at Rizal, nr Iloilo) (1947) ? – ? [e]. . . University Library (1905) 64; University of San Agustin Library (1904) 58; Iloilo Provincial Library (1916) 13.

Makati: [a]Luzon I. [b]for San Pedro Macati. [c]Pasig R. [d]none. [e]Estrella Public Library (?) ?

*Manila: [a]Luzon I. [b]none. [c]Manila B of South China Sea; Pasig R. [d1]University of the East (1946) 64 – 1548 [Manila campus: (1946) ? – 1483]; Pontifical University of Santo Tomas (1611) 43 – 1606; Far Eastern University (1928) 40 – 1150; Feati University (1946) 30 – 850; Adamson University (1932) 17 – 471; Manuel L. Quezon University (1947) 14 – 520; Centro Escolar University (1907) 10 – 368; University of Manila (1913) 10 – 356; Philippine Women's University (1919) 8 – 555; Arellano University (1938) 7 – 216; National University (1900) 7 – 182; Manila Central University (1904) 6 – 202. [e]National Library (1901) 686; Pontifical University of Santo Tomas Library (1605) 234; Ateneo de Manila University Library (1859?) 170; University of the East Library (1947) 143; Far Eastern University Library (1934) 128; City Library (1946) 86.

Pasay: [a]Luzon I. [b]for Pineda, Rizal. [c]Manila B of South China Sea. [d]none. [e]Lopez Memorial Museum and Library (1960) 14; City Library (1950) 9.

Pasig: [a]Luzon I. [b]none. [c]Pasig R; Marikina R. [d]none. [e]Rizal Provincial Library (1951) 6.

Quezon City: [a]Luzon I. [b]none. [c]Marikina R; Dario R. [d]University of the Philippines (1908 at Manila, relocated 1948) 18 – 2452. [e]University . . . Library (1911 at Manila, relocated 1948) 700 [Main library: (1911 at Manila, relocated 1948) 281].

Zamboanga: [a]Mindanao I. [b]for Samboangan. [c]Basilan Strait. [d]none. [e]City Library (1924) 9.

Qatar

*Doha: [b]off Dawhah. [c]Persian G. [d]none. [e]Qatari Public Library (?) 48.

Saudi Arabia

*Jidda: [a]Hejaz. [b]alt Jedda; off Juddah. [c]Red Sea. [d]King Abdul Aziz University (1962) 7 – 330 [Jidda campus: (1967) ? – ?]. [e]. . . University Library (1967) 115 [Jidda campus: (1967) 75].

Mecca: [a]*Hejaz. [b]alt Mekka; off Makkah. [c]nr (Wadi) Shayi R. [d]King Abdul Aziz University, Mecca campus (1962) ? – ? [e]Mecca Campus Library of . . . University (1962?) 40; Abbas Kattan Library (?) 10; Library of Alharam (?) 7.

Medina: [a]Hejaz. [b]off Madinah. [c](Wadi) Buthan R. [d]Islamic University (1961) 1 – 60. [e]. . . University Library (1961) 30.

*Riyadh: [a]*Nejd. [b]off Riyad. [c](Wadi) Hanifah R. [d]Islamic University of Imam Muhammad Ibn Saud (1950) 11 – 498; University of Riyadh (1951) 7 – 700. [e]University . . . Library (1957) 122; National Library (1968) 16.

Taif: [a]Hejaz. [b]none. [c]none. [d]none.

[1]Universities with 5000 or more students.

Singapore

*Singapore: [b]off Singapura. [c]Singapore Strait; Johore Strait. [d]University of Singapore (1905) 6 – 486; Nanyang University (1953) 2 – 179. [e]National Library (1884) 703; University . . . Library (1953) 631.

Sri Lanka

*Colombo: [b]none. [c]Indian O; Colombo L. [d]University of Sri Lanka (1870) 16 – 1507 [campuses at or nr Colombo; (1870) 11 – 1112]. [e]University . . . Library (1870) 442 [campuses at or nr Colombo: (1870) 198]; Public Library (1925) 132; National Museum Library (1870) 90.

Kandy: [b]for Candy. [c]Mahaweli Ganga R; Kandy L. [d]University of Sri Lanka, Peradeniya campus (at Peradeniya, nr Kandy) (1942) 5 – 395. [e]University . . . Library, Peradeniya campus (at Peradeniya, nr Kandy) (1921) 204; Municipal Public Library (?) ?

Syria

Aleppo: [b]off Halab. [c]Quwayq (alt Kuweik) R. [d]University of Aleppo (1946) 16 – 737. [e]University . . . Library (1960) 90; National Library (1924) ?

*Damascus: [b]off Dimashq. [c]Barada R. [d]University of Damascus (1903) 42 – 745. [e]University . . . Library (1919) 150; National Library (1880) 100.

Hama: [b]off Hamah. [c]Orontes R. [d]none. [e]Library of the Cultural Center (?) ?

Homs: [b]alt Hums; for Lebda; off Hims. [c]nr Orontes R. [d]none. [e]National Library (?) ?

Taiwan

Chiayi: [b]alt Chiai, Kiayi; for Kagi. [c]Pachang R. [d]none. [e]Chiayi County Library (1946) 25.

Hsinchu: [b]alt Sinchu; for Chuchien, Shinchiku. [c]Formosa (alt Taiwan) Strait. [d]National Tsing Hua University (1911 at Peking, China; relocated 1956) 0.8 – 148. [e]. . . University Library (1956) 54.

Kaohsiung: [b]for Takao; inc Kigo. [c]Formosa (alt Taiwan) Strait; Kaohsiung (alt Ai) R. [d]none. [e]City Library (1945) 81. [f]Kigo: 7 (1911).

Keelung: [b]alt Chilung; for Kiirun. [c]East China Sea; Keelung (alt Chilung) R. [d]none. [e]City Library (1932) 20.

Panchiao: [b]alt Panchiau; for Itahashi, Taipeihsien. [c]nr Tanshui (for Tamsui) R. [d]none. [e]Taipei County Library (1948) 27.

Pingtung: [b]for Ako, Akow, Heito. [c]nr Hsiatanshui R. [d]none. [e]Pingtung County Chieh-Shou Library (?) 35.

Sanchung: [b]alt Shanchung. [c]Tanshui (for Tamsui) R. [d]none.

Taichung: [b]for Taichu. [c]Yanagi R; Midori R. [d]National Chung Hsing University (1919) 10 – 1222; Tunghai (Christian) University (1955) 5 – 427. [e]Providence College Library (1963?) 240; National Chung Hsing University Library (1961) 193; Tunghai (Christian) University Library (1955) 125; Provincial Taichung Library (1947) 109.

Tainan: [b]for Taiwan. [c]Formosa (alt Taiwan) Strait. [d]National Cheng Kung University (1927) 8 – 580. [e]. . . University Library (1927) 250; City Library (1919) 63.

*Taipei: [b]for Taihoku; inc Daitotei, Moko (alt Banka, Manka). [c]Tanshui (for Tamsui) R; Keelung (alt Chilung) R. [d]National Taiwan University (1928) 13 – 1469; Fu Jen Catholic University (1923 at Peking, China; relocated 1961) 10 – 494; Soochow University (1900 at Soochow, China; relocated 1951) 10 – 698; National Taiwan Normal University (1946) 8 – 770; National Chengchi University (1927 at Nanking, China; relocated 1954) 6 – 685. [e]National Taiwan University Library (1945) 1254; National Central Library (1954) 500; National Chengchi University Library (?) 430; Taiwan Provincial Library (1915) 382. [f]Daitotei: 50 (1901); Moko: 29 (1901), 50 (1864e).

Thailand

*Bangkok: [b]inc Thon Buri bef 1937 and since 1971; off Krungthep Mahanakhon. [c]Chao Phraya R. [d][1]Chulalongkorn University (1902) 15 – 2203; Thammasat University (1933) 11 – 686; Kasetsart

[1]Universities with 1000 or more students.

University (1904) 7 – 840; Mahidol University (1880) 5 – 1400; Sri Nakharinwirot University (1954) 5 – 304; King Mongkut's Institute of Technology (univ) (1954) 4 – 500; Silpakorn University (alt University of Fine Arts) (1943) 3 – 160. [e]National Library (1905) 724; Thammasat University Library (1933) 212. [f]Thon Buri: 628 (1970), 404 (1960), 177 (1947).

Turkey

Adana: [b]for Seyhan. [c]Seyhan R. [d]Cukurova Universitesi (1973) 0.6 – ? [e]Merkez Ramazanoglu Kutuphane (?) 9.

*Ankara: [b]for Angora. [c]Ankara R; Cubuk R. [d]Ankara Universitesi (1925) 15 – 1905; Hacettepe Universitesi (1206 at Kayseri, relocated 1967) 13 – 1747; Orta-Dogu Teknik Universitesi (1956) 10 – 974. [e]Milli Kutuphane (1946) 650; Ankara Universitesi Kutuphane (1925?) 500.

Antioch: [b]for Hatay; off Antakya. [c]Orontes R. [d]none.

Bursa: [b]alt Brusa. [c]Gok R. [d]Bursa Universitesi (1975) 1 – 179. [e]Merkez Genel Kutuphane (?) 33.

Diyarbakir: [b]alt Diyarbekir. [c]Tigris R. [d]Diyarbakir Universitesi (1966) 0.8 – 232. [e]Merkez Genel Kutuphane (?) 14.

Eskisehir: [b]alt Eskishehir. [c]Porsuk R. [d]Anadolu Universitesi (1973) 0.2 – 33. [e]Merkez Genel Kutuphane (?) 14.

Gaziantep: [b]for Aintab, Antep. [c]Kavaklik Creek; Ainleben Creek. [d]none. [e]Merkez Genel Kutuphane (?) 9.

Izmir: [b]for Smyrna. [c]G of Izmir of Aegean Sea. [d]Ege Universitesi (1955) 17 – 1500. [e]Merkez Milli Kutuphane (1912) 97.

Kayseri: [b]none. [c]Kizil (Irmak) R. [d]none. [e]Merkez Genel Kutuphane (?) 15.

Konya: [b]alt Konia. [c]none. [d]Selcuk Universitesi (1975) 0.4 – ? [e]Merkez Genel Kutuphane (1947) 36.

Samsun: [b]none. [c]Black Sea. [d]19 Mayis Universitesi (1975) 0.3 – 46. [e]Merkez Gazi Kutuphane (?) 16.

Trebizond: [b]off Trabzon. [c]Black Sea. [d]Karadeniz Teknik Universitesi (1963) 3 – 250. [e]. . . Universitesi Kutuphane (1963?) 30.

United Arab Emirates

*Abu Zaby: [a]*Abu Dhabi; Abu Zaby (alt Abu Dhabi) I. [b]alt Abu Dhabi. [c]Persian G. [d]none. [e]Public Library (?) ?

Dubayy: [a]*Dubai. [b]alt Dubai; for Dibai. [c]Persian G. [d]none. [e]Public Library (?) 15; British Council Library (1970) 6.

USSR

Aktyubinsk: [a]Kazakhstan. [b]none. [c]Ilek R. [d]none. [e]Aktyubinsk Province Library (1932) 96.

Alma-Ata: [a]*Kazakhstan. [b]for Vernyy. [c]Bolshaya Almatinka R. [d]Kazakh S. M. Kirov State University (1934) 10 – 947. [e]Central Scientific Library of the Kazakh SSR Academy of Sciences (1933) 2329; National A. S. Pushkin State Library of the Kazakh SSR (1931) 1614; Central Library of the . . . University (1934) 500.

Andizhan: [a]Uzbekistan. [b]none. [c]Andizhan(-Say) R. [d]none. [e]Andizhan Province Z. M. Babur Library (1906) 258.

Angarsk: [a]Russia in Asia. [b]none. [c]Angara R. [d]none. [e]City Library (1951) 83.

Ashkhabad: [a]*Turkmenia. [b]for Poltoratsk. [c]Geami R. [d]Turkmen A. M. Gorkiy State University (1950) 4 – ? [e]Karl Marx State Library of the Turkmen SSR (1895) 1432; . . . University Library (1950) 349; Central Scientific Library of the Turkmen SSR Academy of Sciences (1941) 205.

Barnaul: [a]Russia in Asia. [b]none. [c]Ob R; Barnaulka R. [d]none. [e]Altai Regional Library (1888) 549; Barnaul I. I. Polzunov Polytechnic Institute Library (1942) 346.

Biysk: [a]Russia in Asia. [b]alt Biisk, Bisk. [c]Biya R. [d]none. [e]Biysk Pedagogic Institute Library (1931) 184; Central City Library (1899) 111.

Blagoveshchensk: [a]Russia in Asia. [b]none. [c]Amur R; Zeya R. [d]none. [e]Amur Province Library (1937) 246; Blagoveshchensk M. I. Kalinin Pedagogic Institute Library (1930) 168.

Bratsk: [a]Russia in Asia. [b]none. [c]Angara R. [d]none. [e]City Library (1957) 24.

Bukhara: [a]Uzbekistan. [b]alt Bokhara. [c]nr Zeravshan R. [d]none. [e]Central Library of the Bukhara Sergo Ordzhonikidze Pedagogic Institute (1930) 430; Bukhara Province Ibn Sina Library (1921) 204.

Chelyabinsk: [a]Russia in Asia. [b]none. [c]Miass R. [d]none. [e]Chelyabinsk Province 50th Anniversary of the October Revolution Public Library (1898) 841; Chelyabinsk Lenin Komsomol Polytechnic Institute Library (1943) 683.

Chimkent: [a]Kazakhstan. [b]none. [c]Badam R. [d]none. [e]Kazakh Technological Institute Library (1943) 174; Chimkent Province A. S. Pushkin Library (1815) 157; Chimkent N. K. Krupskaya Pedagogic Institute Library (1937) 134.

Chita: [a]Russia in Asia. [b]none. [c]Ingoda R. [d]none. [e]Chita Province A. S. Pushkin Library (1895) 370.

Dushanbe: [a]*Tadzhikistan. [b]for Dyushambe, Stalinabad. [c]Kafirnigan R; Dushanbinka R. [d]Tadzhik V. I. Lenin State University (1948) 13 – 750. [e]National A. Firdousi State Library of the Tadzhik SSR (1933) 1322; Central Scientific Library of the Tadzhik SSR Academy of Sciences (1933) 478; . . . University Library (1948) 330.

Dzhambul: [a]Kazakhstan. [b]for Auliye-Ata. [c]Talas R. [d]none. [e]Dzhambul Province C. Valikhanov Library (1898) 131.

Frunze: [a]*Kirgizia. [b]for Pishpek. [c]Alamedin R; Alarcha R; Dzhirgozar R. [d]Kirgiz State University (1951) 12 – 550. [e]National N. G. Chernyshevskiy State Library of the Kirgiz SSR (1934) 1604; Scientific Library of the . . . University (1932) 506; Central Scientific Library of the Kirgiz SSR Academy of Sciences (1943) 315.

Irkutsk: [a]Russia in Asia. [b]none. [c]Angara R; Irkut R. [d]Irkutsk A. A. Zhdanov State University (1918) 9 – 500. [e]. . . University Library (1918) 1762; Irkutsk Polytechnic Institute Library (1930) 621; Irkutsk Province I. I. Molchanov-Sibirskiy Library (1861) 621.

Kamensk-Uralskiy: [a]Russia in Asia. [b]for Kamensk. [c]Iset R; Kamenka R. [d]none. [e]Technical Library of the Ural Aluminum Works (1936) 70.

Karaganda: [a]Kazakhstan. [b]none. [c]nr Nura R. [d]Karaganda State University (1972) ? – ? [e]Karaganda Polytechnic Institute Library (1953) 314; Karaganda Province N. V. Gogol Library (1938) 222; Karaganda Pedagogic Institute Library (1938) 217.

Kemerovo: [a]Russia in Asia. [b]for Shcheglovsk. [c]Tom R. [d]none. [e]Kemerovo Province Scientific Library (1920) 451; Kuznetsk Basin Polytechnic Institute Library (1950) 306.

Khabarovsk: [a]Russia in Asia. [b]none. [c]Amur R. [d]none. [e]Khabarovsk Region Scientific Library (1894) 980.

Kokand: [a]Uzbekistan. [b]alt Khokand. [c]Sokh R. [d]none. [e]Kokand Mukimi Pedagogic Institute Library (1930) 157.

Komsomolsk-na-Amure: [a]Russia in Asia. [b]for Permskoye. [c]Amur R. [d]none. [e]Komsomolsk-na-Amure Pedagogic Institute Library (1954) 135; Komsomolsk-na-Amure Polytechnic Institute Library (1955) 123; Central City N. Ostrovskiy Library (1934) 82.

Kopeysk: [a]Russia in Asia. [b]alt Kopeisk; for Kopi. [c]nr Miass R. [d]none.

Krasnoyarsk: [a]Russia in Asia. [b]none. [c]Yenisey R. [d]Krasnoyarsk State University (1970) ? – ? [e]Krasnoyarsk Region Scientific Library (1935) 889.

Kurgan: [a]Russia in Asia. [b]none. [c]Tobol R. [d]none. [e]Kurgan Province Library (1943) 488.

Kustanay: [a]Kazakhstan. [b]none. [c]Tobol R. [d]none. [e]Kustanay Province L. N. Tolstoy Library (1919) 211; Amangeldy Pedagogic Institute Library (1939) 196.

Magnitogorsk: [a]Russia in Asia. [b]none. [c]Ural R. [d]none. [e]Magnitogorsk G. I. Nosov Mining and Metallurgical Institute Library (1932) 297.

Namangan: [a]Uzbekistan. [b]none. [c]nr Naryn R. [d]none. [e]Namangan Province Library (1918) 142; Namangan Khamza Pedagogic Institute Library (1946) 127.

Nizhniy Tagil: [a]Russia in Asia. [b]alt Nizhni Tagil. [c]Tagil R. [d]none. [e]Scientific-Technical Library of the Ural Car-Building Works (1932) 187; Scientific-Technical Library of the Nizhniy Tagil V. I. Lenin Metallurgical Combine (1934) 178; Nizhniy Tagil Pedagogic Institute Library (1939) 174.

Norilsk: [a]Russia in Asia. [b]none. [c]nr Norilskaya R. [d]none. [e]Central Technical Library of the A. P. Zavenyagin Mining and Metallurgical Combine (1938) 508.

Novokuznetsk: [a]Russia in Asia. [b]for Stalinsk. [c]Tom R; Kondoma R; Aba R. [d]none. [e]Academician I. P. Bardin Scientific-Technical Library of the Kuznetsk Metallurgical Combine (1927) 1014; Siberian Sergo Ordzhonikidze Metallurgical Institute Library (1930) 287; Metallurgical Palace of Culture Library of the Kuznetsk Metallurgical Combine (1931) 261.

Novosibirsk: [a]Russia in Asia. [b]for Novonikolayevsk. [c]Ob R; Inya R; Yeltsovka R. [d]Novosibirsk State University (1959) 5 – 550. [e]State Public Scientific-Technical Library of the Siberian Department of the USSR Academy of Sciences (1918) 1844; Novosibirsk Province Library (1929) 843; Novosibirsk Electrical Engineering Institute Library (1953) 612.

Omsk: [a]Russia in Asia. [b]none. [c]Irtysh R; Om R. [d]none. [e]Omsk Province A. S. Pushkin Scientific Library (1899) 774; Omsk M. I. Kalinin Medical Institute Library (1920) 438.

Osh: [a]Kirgizia. [b]none. [c]Akbura R. [d]none. [e]Osh Pedagogic Institute Library (1947) 226; Osh Province Library (1949) 102.

Pavlodar: [a]Kazakhstan. [b]none. [c]Irtysh R. [d]none. [e]Pavlodar Industrial Institute Library (1960) 210; Pavlodar Province N. Ostrovskiy Library (1919) 128.

Petropavlovsk: [a]Kazakhstan. [b]none. [c]Ishim R. [d]none. [e]Petropavlosk K. D. Ushinskiy Pedagogic Institute Library (1937) 236; North Kazakhstan Province Library (1919) 155.

Petropavlovsk-Kamchatskiy: [a]Russia in Asia. [b]for Petropavlovsk. [c]Ayacha B of Pacific O. [d]none. [e]Kamchatka Province Library (1914) 185.

Prokopyevsk: [a]Russia in Asia. [b]alt Prokopevsk. [c]Aba R. [d]none. [e]Technical Library of the Kuznetsk Coal Scientific Research Institute (1944) 55.

Rubtsovsk: [a]Russia in Asia. [b]none. [c]Aley R. [d]none. [e]Scientific-Technical Library of the Altai Kalinin Tractor Works (1942) 109.

Samarkand: [a]Uzbekistan. [b]none. [c]nr Zeravshan R. [d]Samarkand Alisher Navoi State University (1927) 6 – 600. [e]Central Library of the . . . University (1927) 928; V. V. Kuybyshev Agricultural Institute Library (1929) 191; Samarkand Province A. S. Pushkin Library (1911) 188.

Semipalatinsk: [a]Kazakhstan. [b]none. [c]Irtysh R. [d]none. [e]Semipalatinsk N. K. Krupskaya Pedagogic Institute Library (1935) 190; Semipalatinsk Province N. V. Gogol Library (1883) 190.

Sverdlovsk: [a]Russia in Asia. [b]for Ekaterinburg, Yekaterinburg. [c]Iset R. [d]Ural A. M. Gorkiy State University (1920) 6 – 425. [e]Sverdlovsk V. G. Belinskiy State Public Library (1899) 1111; Ural S. M. Kirov Polytechnic Institute Library (1920) 997; Scientific Library of the . . . University (1920) 564.

Tashkent: [a]*Uzbekistan. [b]alt Tashkend. [c]Chirchik R. [d]Tashkent V. I. Lenin State University (1920) 13 – 1100. [e]Alisher Navoi State Library of the Uzbek SSR (1870) 2900; Scientific Library of the . . . University (1918) 1213; Central Library of the Uzbek SSR Academy of Sciences (1933) 936; Central Library of the Tashkent Polytechnic Institute (1929) 756.

Temirtau: [a]Kazakhstan. [b]for Samarkandskiy. [c]Nura R. [d]none. [e]Karaganda Metallurgical Combine Technological Institute Library (1963) 88.

Tomsk: [a]Russia in Asia. [b]none. [c]Tom R. [d]Tomsk V. V. Kuybyshev State University (1888) 10 – ? [e]Scientific Library of the . . . University (1888) 1382; Scientific-Technical Library of the Tomsk S. M. Kirov Polytechnic Institute (1900) 980; Tomsk Province A. S. Pushkin Library (1899) 361.

Tselinograd: [a]Kazakhstan. [b]for Akmolinsk. [c]Ishim R. [d]none. [e]Tselinograd Province Saken Seyfullin Library (1941) 388; Tselinograd Agricultural Institute Library (1958) 290.

Tyumen: [a]Russia in Asia. [b]alt Tiumen. [c]Tura R. [d]Tyumen State University (1973?) ? – ? [e]Tyumen Province Library (1875) 406.

Ulan-Ude: [a]Russia in Asia; *Buryat ASSR. [b]for Udinskoye, Verkhne-Udinsk. [c]Selenga R; Uda R. [d]none. [e]National M. Goriky Library of the Buryat ASSR (1881) 467.

Uralsk: [a]Kazakhstan. [b]for Yaitskiy Gorodok. [c]Ural R. [d]none. [e]Uralsk A. S. Pushkin Pedagogic Institute Library (1932) 396; Uralsk Province N. K. Krupskaya Library (1871) 131.

Ust-Kamenogorsk: [a]Kazakhstan. [b]for Zashchita. [c]Irtysh R; Ulba R. [d]none. [e]Ust-Kamenogorsk Road-Building Institute Library (1958) 232; Ust-Kamenogorsk Pedagogic Institute Library (1952) 194; East Kazakhstan Province A. S. Pushkin Library (1896) 162.

Vladivostok: [a]Russia in Asia. [b]none. [c]Sea of Japan. [d]Far Eastern State University (1899) 7 – 400.

[e]Central Library of the Far Eastern V. V. Kuybyshev Polytechnic Institute (1930) 433; . . . University Library (1956) 332; Maritime Region A. M. Gorkiy Library (1887) 325.

Yakutsk: [a]Russia in Asia. [b]none. [c]Lena R. [d]Yakutsk State University (1934) 6 – 400. [e]Yakutsk National A. S. Pushkin Library (1925) 783; . . . University Library (1934) 362.

Yuzhno-Sakhalinsk: [a]Russia in Asia; Sakhalin I. [b]for Toyohara, Vladimirovka. [c]Susuya R. [d]none. [e]Sakhalin Province Library (1946) 392.

Zlatoust: [a]Russia in Asia. [b]none. [c]Ay R. [d]none. [e]Scientific-Technical Library of the Zlatoust Metallurgical Works (1933) 114.

Vietnam

Da Nang: [b]for Tourane. [c]South China Sea; Da Nang R. [d]none.

Haiphong: [b]none. [c](Cua) Cam R; Kinh Thay R; Tram Bac (alt Tambac) R. [d]none.

*Hanoi: [b]for Kecho. [c]Red R; Tay L; Truc Bach L. [d]University of Hanoi (1904) 2 – 150. [e]Bibliotheque Nationale (1918) 1000.

Ho Chi Minh City: [b]for Saigon; inc Cholon. [c]Sai Gon R; (Arroyo) Chinois Creek. [d]University of Ho Chi Minh City (1917 at Hanoi, relocated 1954) 7 – 1400; Van Hanh University (1964) 4 – 160; Min Duc University (1970) 2 – 250. [e]Bibliotheque Nationale (1902) 160; University . . . Library (1955) ?

Hue: [b]none. [c]Huong (alt Parfums) R. [d]Universite de Hue (1957) 6 – 280. [e]Bibliotheque de l'Universite . . . (1957) 22.

Nha Trang: [b]for Kanh Hoa. [c]South China Sea; Cai R. [d]none. [e]Bibliotheque de l'Institut Oceanographique (?) 14.

Qui Nhon: [b]none. [c]South China Sea. [d]none.

Yemen: North

*Sana: [b]alt Sanaa. [c](Wadi) Alaf R; (Wadi) Shaub R. [d]Sana University (1970) 3 – 89. [e]Library of the Great Mosque (1925) 10; . . . University Library (1970?) ?

Yemen: South

*Aden: [b]none. [c]G of Aden. [d]Aden University (1975) ? – ? [e]Miswat Library (1951) 30.

EUROPE

Albania

*Tirane: [b]alt Tirana. [c]nr Ishm R. [d]Universiteti Shteteror i Tiranes (1957) 16 – 881. [e]Biblioteka Kombetare (1922) 550; Biblioteka Shkencore e Universitetit . . . (1957) 360.

Austria

Graz: [b]none. [c]Mur R. [d]Karl-Franzens-Universitat (1586) 12 – 1314; Technische Universitat Graz (1811) 4 – 700; Hochschule fur Musik und Darstellende Kunst in Graz (univ) (1803) 0.9 – 180. [e]Universitatsbibliothek (1577) 830; Steiermarkische Landesbibliothek am Joanneum (1811) 437.

Innsbruck: [b]none. [c]Inn R; Sill R. [d]Leopold-Franzens-Universitat (1669) 12 – 1309. [e]Universitatsbibliothek (1746) 800.

Linz: [b]none. [c]Danube R. [d]Johannes-Kepler-Universitat (1962) 4 – 566. [e]Bundesstaatliche Studienbibliothek (1774) 190; Buchereien der Stadt (?) 126.

Salzburg: [b]none. [c]Salzach R. [d]Universitat Salzburg (1622) 6 – 914; Hochschule fur Musik und Darstellende Kunst Mozarteum in Salzburg (univ) (1841) 1 – 171. [e]Universitatsbibliothek (1623) 300; . . .; Stadtbibliothek (1941) 41.

*Vienna: [b]off Wien. [c]Danube R. [d][1] Universitat Wien (1365) 28 – 3251; Wirtschafts Universitat Wien (1898) 9 – 262; Technische Universitat Wien (1815) 8 – 1336; Hochschule fur Musik und Dar-

[1]Universities with 1000 or more students.

stellende Kunst (univ) (1817) 2 – 348; Hochschule fur Bodenkultur (univ) (1872) 2 – 263; Veterinarmedizinische Universitat Wien (1767) 1 – 135. ^{e}Osterreichische Nationalbibliothek (1526) 2390; Universitatsbibliothek (1777) 1786;...; Stadtische Buchereien (1936) 535.

Belgium
Anderlecht: ^{b}none. ^{c}Senne (alt Zenne) R. ^{d}none. ^{e}Algemene Openbare Bibliotheek Davidsfonds (?) 5.
Antwerp: ^{b}off Antwerpen, Anvers. ^{c}Scheldt R. ^{d}Rijksuniversitair Centrum te Antwerpen (1852) 2 – 184; Universitair Instelling Antwerpen (1971) 1 – 188. ^{e}Stadsbibliotheek (1470) 800.
Bruges: ^{b}alt Brugge. ^{c}Rei (alt Reye, Roye) R. ^{d}none. ^{e}Bibliotheek Sint-Andriesabdij (1902) 110; Stedelijke Openbare Bibliotheek (1789) 109; Bibliotheque du Grand Seminaire (alt Bibliotheek van het Grootseminarie) (1833) 103.
*Brussels: ^{b}off Brussel, Bruxelles. ^{c}Senne (alt Zenne) R. ^{d}Universite Libre de Bruxelles (1834) 18 – 2100; Vrije Universiteit Brussel (1834) 4 – 660. ^{e}Bibliotheque Royale Albert Ier (1837) 2391; Bibliotheque de l'Universite ... (1846) 1161; Bibliotheque Fonds Quetelet (1841) 611.
Charleroi: ^{b}none. ^{c}Sambre R. ^{d}none. ^{e}Bibliotheque Communale (?) 40.
Ghent: ^{b}off Gand, Gent. ^{c}Scheldt R; Lys R. ^{d}Rijksuniversiteit te Gent (1816) 13 – 1004. ^{e}Centrale Bibliotheek van de Rijksuniversiteit ... (1797) 2054; Bibliotheque Publique Municipale (alt Openbare Stadsbibliotheek) (1945) 94.
La Louviere: ^{b}alt Louviere. ^{c}nr Sambre R. ^{d}Universite de l'Etat a Mons (at Mons, nr La Louviere) (1965) 2 – 209. ^{e}Bibliotheque Centrale de l'Universite ... (at Mons, nr La Louviere) (1797) 318; Bibliotheque Centrale du Hainaut (1958) 60.
Liege: ^{b}alt Luik. ^{c}Meuse R; Ourthe R. ^{d}Universite de Liege (1816) 9 – 347. ^{e}Bibliotheque Generale de l'Universite ... (1817) 2035; Bibliotheque Communale (1865) 247.
Louvain: ^{b}alt Leuven. ^{c}Dijle (alt Dyle) R. ^{d}Katholieke Universiteit te Leuven (1425) 18 – 864; Universite Catholique de Louvain (1425) 16 – 921. ^{e}Bibliotheque de l'Universite ... (1425) 1300; Bibliotheek der ... Universiteit (1425) 700; Stedelijke Openbare Bibliotheek (1866) 50.
Schaerbeek: ^{b}alt Schaarbeek. ^{c}Woluwe R. ^{d}none. ^{e}Bibliotheque Publique La Lecture pour Tous (1930) 19; Bibliotheque Communale de Helmet (1904) 17; Bibliotheque Communale Ouest (1904) 12.

Bulgaria
Burgas: ^{b}none. ^{c}G of Burgas of Black Sea. ^{d}none. ^{e}District Library (1890) 285.
Pleven: ^{b}for Plevna. ^{c}nr Vit R. ^{d}none. ^{e}District K. Smirnenski Library (1954) 140.
Plovdiv: ^{b}for Philippopolis. ^{c}Maritsa R. ^{d}Paisij Hilendarski University of Plovdiv (1963) 4 – 244. ^{e}National Ivan Vazov Library (1882) 768.
Ruse: ^{b}for Ruschuk. ^{c}Danube R. ^{d}none. ^{e}District L. Karavelov Library (1888) 401.
Sliven: ^{b}for Isliniye. ^{c}Novoselska R; Asenovitsa R. ^{d}none. ^{e}District Georgi Kirkov Library (1956) 90.
*Sofia: ^{b}off Sofiya. ^{c}Bogana R. ^{d}Saint Clement of Ohrid University of Sofia (1888) 13 – 1195. ^{e}National Cyril and Methodius Library (1878) 1238;... University Library (1888) 983; City Library (1928) 489.
Stara Zagora: ^{b}for Eski Zagra. ^{c}nr Syuyutliyka R. ^{d}none. ^{e}District Library (1955) 269.
Varna: ^{b}for Stalin. ^{c}Black Sea; Varna L. ^{d}none. ^{e}National City Library (1883) 410.

Czechoslovakia
Bratislava: ^{a}Slovakia. ^{b}for Pozsony, Pressburg. ^{c}Danube R. ^{d}Univerzita Komemskeho (1919) 16 – 2032; Slovenska Vysoka Skola Technicka v Bratislave (univ) (1938) 14 – 1300. ^{e}Univerzitna Kniznica (1919) 1362; Slovenska Technicka Kniznica (1938) 1100; Mestska Kniznica (1900) 200.
Brno: ^{a}Moravia. ^{b}for Brunn. ^{c}Svratka R; Svitava R. ^{d}Vysoke Uceni Technicke v Brne (univ) (1899) 11 – 1042; Univerzita Jana Evangelisty Purkyne (1919) 7 – 850; Vysoka Skola Zemedelska (univ) (1919) 4 – 385. ^{e}Statni Vedecka Knihovna (1770) 2438; Knihovna Jiriho Mahena (1921) 534.
Havirov: ^{a}Moravia. ^{b}none. ^{c}Lucina R. ^{d}none. ^{e}Vzorna Mestska Knihovna (1954) 106.

Karlovy Vary: [a]Bohemia. [b]alt Carlsbad; for Karlsbad. [c]Ohre (alt Eger) R; Tepla (alt Tepl) R. [d]none. [e]Okresni Knihovna (1948) 232.

Kosice: [a]Slovakia. [b]for Kaschau, Kassa. [c]Hornad R. [d]Univerzita Pavla Jozefa Safarika (1948) 7 – 600; Vysoka Skola Technicka v Kosiciach (univ) (1864 at Banska Bystrica, relocated 1952) 4 – ? [e]Statna Vedecka Kniznica (1657) 728; Krajska Kniznica (1657) 163.

Olomouc: [a]Moravia. [b]for Olmutz. [c]Morava (for March) R; Bystrice R. [d]Univerzita Palackeho (1573) 5 – 700. [e]Statni Vedecka Knihovna (1566) 976; Knihovni Stredisko Filosoficke Fakulty Univerzita ... (1946) 219; Vzorna Okresni Knihovna (1888) 190.

Ostrava: [a]Moravia. [b]for Mahrisch-Ostrau, Moravska Ostrava. [c]Oder R; Ostravice R. [d]none. [e]Statni Vedecka Knihovna (1951) 328; Ustredni Knihovna Vysoke Skoly Banske (1849) 296; Knihovna Mesta Ostravy (1928) 187.

Plzen: [a]Bohemia. [b]for Pilsen. [c]Berounka R; Mze R; Radbuza R; Uhlava R; Uslava R. [d]none. [e]Statni Vedecka Knihovna (1950) 463; Knihovna Mesta Plzne (1876) 267.

*Prague: [a]Bohemia. [b]for Prag; off Praha. [c]Vltava R. [d]Univerzita Karlova (1348) 21 – 2689; Ceske Vysoke Uceni Technicke v Praze (univ) (1707) 14 – 1548; Univerzita 17. Listopadu (1961) 1 – 183. [e]Statni Knihovna (1366?) 4672; Narodniho Muzea Knihovna (1818) 3000; Mestska Knihovna (1891) 1988.

Denmark

Alborg: [b]alt Aalborg. [c]Lim Fjord of Kattegat Strait. [d]Alborg Universitetscenter (1974) 3 – 338. [e]Nordjyske Landsbibliotek (1895) 500.

Arhus: [b]alt Aarhus. [c]Kattegat Strait. [d]Arhus Universitet (1928) 15 – 1786. [e]Stadsbiblioteket (1902) 1254; Kommunes Biblioteker (1934) 1000.

*Copenhagen: [a]Zealand I. [b]off Kobenhavn. [c]Oresund Strait. [d]Kobenhavns Universitet (1479) 28 – 2842; Polytekniske Laereanstalt, Danmarks Tekniske Hojskole (univ) (at Lyngby, nr Copenhagen) (1829) 3 – 649; Kongelige Veterinaer- og Landbohojskole (univ) (1856) 2 – 347; Danmarks Ingeniorakademi (univ) (at Lyngby, nr Copenhagen) (1957) 1 – 350. [e]Kongelige Bibliotek (1661) 2100; Kommunes Biblioteker (1885) 1888; Universitetsbiblioteket (1482) 1479.

Frederiksberg: [a]Zealand I. [b]none. [c]nr Oresund Strait. [d]none. [e]Kommunes Biblioteker (1887) 403.

Odense: [a]Fyn I. [b]none. [c]Odense R. [d]Odense Universitet (1964) 4 – 349. [e]Centralbibliotek (1924) 810; Universitetsbiblioteket (1965) 425.

Finland

Espoo: [b]alt Esbo. [c]G of Finland of Baltic Sea. [d]Teknillinen Korkeakoulu (univ) (1848) 6 – 411. [e]Korkeakoulun Kirjasto (1849) 250; Kaupunginkirjasto (?) 90.

*Helsinki: [b]alt Helsingfors. [c]G of Finland of Baltic Sea; Vantaan (alt Vanda) R. [d]Helsingin Yliopisto (1640 at Turku, relocated 1828) 23 – 1526. [e]... Yliopiston Kirjasto (1640 at Turku, relocated 1828) 1600; Kaupunginkirjasto (1860) 986.

Lahti: [b]none. [c]Vesi L; Jout L. [d]none. [e]Kaupunginkirjasto (1876) 186.

Oulu: [b]alt Uleaborg. [c]G of Bothnia; Oulu (alt Ule) R. [d]Oulun Yliopisto (1959) 7 – 654. [e]... Yliopiston Kirjasto (1959) 550; Kaupunginkirjasto (1866) 212.

Tampere: [b]alt Tammerfors. [c]Tammer Rapids; Nasi L; Pyha L. [d]Tampereen Yliopisto (1925) 5 – 522; Tampereen Teknillinen Korkeakoulu (univ) (1965) 2 – 144. [e]Kaupunginkirjasto (1861) 518; ... Yliopiston Kirjasto (1925) 400.

Turku: [b]alt Abo. [c]G of Bothnia; Aura R. [d]Turun Yliopisto (1920) 9 – 676; Abo Akademi (univ) (1917) 3 – 220. [e]... Yliopiston Kirjasto (1919) 1000; ... Akademis Bibliotek (1918) 871; Kaupunginkirjasto (1863) 434.

Vantaa: [b]alt Vanda. [c]Vantaan (alt Vanda) R. [d]none.

France

Aix-en-Provence: [a]Provence-Cote d'Azur. [b]alt Aix. [c]nr Arc R. [d]Universite d'Aix-Marseille (1409) 52 – 3825 [Aix-en-Provence campus: (1409) 11 – 393]. [e]Bibliotheque Interuniversitaire (1879) 845 [Aix-en-Provence campus: (1879) 346]; Bibliotheque Municipale (1810) 292.

Ajaccio: [a]Corse; Corsica I. [b]none. [c]Mediterranean Sea. [d]none. [e]Bibliotheque Municipale (1801) 60.

Amiens: [a]Picardie. [b]none. [c]Somme R. [d]Universite de Picardie (1750) 9 − 665. [e]Bibliotheque Municipale (1791) 160.

Angers: [a]Pays de la Loire. [b]none. [c]Maine R. [d]Universite d'Angers (1969) 4 − 280; Universite Catholique de l'Ouest (1875) 2 − 207. [e]Bibliotheque Lamoriciere de l'Universite Catholique de l'Ouest (1875) 200; Bibliotheque Municipale (1848) 190.

Angouleme: [a]Poitou-Charentes. [b]none. [c]Charente R. [d]none. [e]Bibliotheque Municipale (1800) 40.

Annecy: [a]Rhone-Alpes. [b]none. [c]Annecy L. [d]none. [e]Bibliotheque Municipale (1744) 80.

Argenteuil: [a]Region Parisienne. [b]none. [c]Seine R. [d]none. [e]Bibliotheque Municipale (?) 10.

Avignon: [a]Provence-Cote d'Azur. [b]none. [c]Rhone R. [d]Centre Universitaire d'Avignon (1972) 1 − 66. [e]Bibliotheque du Museum Calvet (1810) 250.

Bayonne: [a]Aquitaine. [b]none. [c]Adour R; Nive R. [d]none. [e]Bibliotheque Municipale (1850) 62.

Besancon: [a]Franche-Comte. [b]none. [c]Doubs R. [d]Universite de Besancon (1422 at Dole, relocated 1691) 11 − 690. [e]Bibliotheque Universitaire (1891) 350; Bibliotheque Municipale (1694) 250.

Bethune: [a]Nord. [b]none. [c]Lawe R. [d]none. [e]Bibliotheque Municipale (?) 13.

Biarritz: [a]Aquitaine. [b]none. [c]B of Biscay. [d]none. [e]Bibliotheque du Centre d'Etudes et de Recherches Scientifiques de Biarritz (1955) 7.

Bordeaux: [a]Aquitaine. [b]none. [c]Garonne R. [d]Universite de Bordeaux (1441) 25 − 1307. [e]Bibliotheque de l'Universite . . . (1879) 870; Bibliotheque Municipale (1736) 650.

Boulogne-Billancourt: [a]Region Parisienne. [b]for Boulogne-sur-Seine. [c]Seine R. [d]none. [e]Bibliotheque Municipale (1864) 56.

Boulogne-sur-Mer: [a]Nord. [b]alt Boulogne. [c]English Channel; Liane R. [d]none. [e]Bibliotheque Municipale (1803) 150.

Brest: [a]Bretagne. [b]none. [c]Atlantic O. [d]Universite de Bretagne Occidentale (1970) 7 − 483. [e]Bibliotheque Municipale (1853) 86.

Bruay-en-Artois: [a]Nord. [b]alt Bruay. [c]Lawe R. [d]none.

Caen: [a]Basse-Normandie. [b]none. [c]Orne R; Odon R. [d]Universite de Caen (1432) 12 − 712. [e]Bibliotheque Universitaire (1432) 600; Bibliotheque Municipale (1872) 400.

Calais: [a]Nord. [b]none. [c]Strait of Dover. [d]none. [e]Bibliotheque Municipale (1790) 50.

Cannes: [a]Provence-Cote d'Azur. [b]none. [c]Mediterranean Sea. [d]none. [e]Bibliotheque Municipale (1868) 90.

Clermont-Ferrand: [a]Auvergne. [b]none. [c]nr Sioule R. [d]Universite de Clermont-Ferrand (1810) 14 − 973. [e]Bibliotheque Municipale et Universitaire (18th c) 230.

Denain: [a]Nord. [b]none. [c]Scheldt R. [d]none.

Dijon: [a]Bourgogne. [b]none. [c]Ouche R; Suzon R. [d]Universite de Dijon (1722) 12 − 713. [e]Bibliotheque de l'Universite . . . (1880) 325; Bibliotheque Municipale (1701) 200.

Douai: [a]Nord. [b]for Douay. [c]Scarpe R. [d]none. [e]Bibliotheque Municipale (1767) 120.

Dunkerque: [a]Nord. [b]alt Dunkirk. [c]North Sea. [d]none. [e]Bibliotheque Municipale (18th c) 31.

Grenoble: [a]Rhone-Alpes. [b]none. [c]Isere R. [d]Universite de Grenoble (1339) 16 − 612. [e]Bibliotheque Municipale (1772) 650; Bibliotheque Universitaire (1879) 469.

La Rochelle: [a]Poitou-Charentes. [b]alt Rochelle. [c]B of Biscay. [d]none. [e]Bibliotheque Municipale (1750) 165.

Le Havre: [a]Haute-Normandie. [b]alt Havre. [c]English Channel; Seine R. [d]none. [e]Bibliotheque Municipale (1800) 148.

Le Mans: [a]Pays de la Loire. [b]alt Mans. [c]Maine R; Sarthe R; Huisne R. [d]Universite du Maine (1969) 3 − 253. [e]Bibliotheque Municipale (1931) 110.

Lens: [a]Nord. [b]none. [c]Deule R. [d]none.

Lille: [a]Nord. [b]none. [c]Deule R. [d]Universite de Lille (1560 at Douai, relocated 1887) 35 − 1836; Federation Universitaire et Polytechnique de Lille (1875) 5 − 546. [e]Bibliotheque Universitaire (1880 at Douai, relocated 1887) 700; Bibliotheque Municipale (1726) 385.

Limoges: [a]Limousin. [b]none. [c]Vienne R. [d]Universite de Limoges (1808) 7 − 350. [e]Bibliotheque Municipale (1804) 220.

Lorient: [a]Bretagne. [b]for Blavet. [c]Scorff R; Blavet R. [d]none. [e]Bibliotheque Municipale (19th c) 22.

Lyons: [a]Rhone-Alpes. [b]off Lyon. [c]Rhone R; Saone R. [d]Universite de Lyon (1809) 38 – ? [Lyons campus: (1809) ? – ?] ; Facultes Catholiques de Lyon (univ) (1875) 3 – 171. [e]Bibliotheque Interuniversitaire (1809?) 1036 [Lyons campus: (1809?) 861] ; Bibliotheque Municipale (1693) 780.

Marseilles: [a]Provence-Cote d'Azur. [b]off Marseille. [c]G of Lions of Mediterranean Sea. [d]Universite d'Aix-Marseille, Marseilles campus (1854) 41 – 3432. [e]Bibliotheque Interuniversitaire, Marseilles campus (1891) 499; Bibliotheque Municipale (1799) 300.

Maubeuge: [a]Nord. [b]none. [c]Sambre R. [d]none. [e]Bibliotheque Municipale (1947) 9.

Metz: [a]Lorraine. [b]none. [c]Moselle R; Seille R. [d]Universite de Metz (1970) 5 – 242. [e]Bibliotheque Municipale (1811) 200.

Montbeliard: [a]Franche-Comte. [b]none. [c]Allaine R; Luzine R. [d]none. [e]Bibliotheque Municipale (1765) 56.

Montpellier: [a]Languedoc-Roussillon. [b]none. [c]Lez R. [d]Universite de Montpellier (1181) 24 – 1001. [e]Bibliotheque Interuniversitaire (1767) 800; Bibliotheque de la Ville et du Musee Fabre (1800) 500.

Montreuil: [a]Region Parisienne. [b]alt Montreuil-sous-Bois. [c]nr Seine R. [d]none. [e]Bibliotheque Municipale (1879) 40.

Mulhouse: [a]Alsace. [b]for Mulhausen. [c]Ill R. [d]Universite du Haut-Rhin (1970) 2 – ? [e]Bibliotheque Municipale (1840) 158.

Nancy: [a]Lorraine. [b]none. [c]Meurthe R. [d]Universite de Nancy (1572) 13 – 412. [e]Bibliotheque de l'Universite . . . (1855) 705; Bibliotheque Municipale (1750) 400.

Nanterre: [a]Region Parisienne. [b]none. [c]Seine R. [d]none. [e]Bibliotheque de Documentation Internationale Contemporaine et Musee des Deux Guerres Mondiales (1918) 600.

Nantes: [a]Pays de la Loire. [b]none. [c]Loire R; Sevre Nantaise R; Erdre R. [d]Universite de Nantes (1460) 16 – 933. [e]Bibliotheque Municipale (1753) 332.

Nice: [a]Provence-Cote d'Azur. [b]for Nizza. [c]Mediterranean Sea; Paillon R. [d]Universite de Nice (1965) 16 – 707. [e]Bibliotheque Municipale (bef 1787) 210.

Nimes: [a]Languedoc-Roussillon. [b]none. [c]nr Gard R. [d]none. [e]Bibliotheque Municipale Dite Seguier (1794) 171.

Orleans: [a]Centre. [b]none. [c]Loire R. [d]Universite d'Orleans (1306) 5 – 546. [e]Bibliotheque Municipale (1714) 330.

*Paris: [a]Region Parisienne. [b]none. [c]Seine R. [d]Universite de Paris (1200) 233 – ?; Institut Catholique de Paris (univ) (1875) 10 – 835. [e]Bibliotheque Nationale (1480) 7000; Bibliotheque de l'Universite . . . (1570?) 4876, inc Bibliotheque de la Sorbonne (1762) 1800, and Bibliotheque Sainte-Genevieve (1624) 1500; Bibliotheque de l'Arsenal (1785) 1530; Bibliotheque de l'Institut de France (1796) 1500; Bibliotheque Pedagogique (1879) 1000.

Pau: [a]Aquitaine. [b]none. [c](Gave de) Pau R. [d]Universite de Pau et des Pays de l'Adour (1970) 5 – 228. [e]Bibliotheque Municipale (1789) 200.

Perpignan: [a]Languedoc-Roussillon. [b]none. [c]Tet R. [d]Centre Universitaire de Perpignan (1970) 3 – 163. [e]Bibliotheque Municipale (1349) 72.

Poitiers: [a]Poitou-Charentes. [b]none. [c]Clain R; Boivre R. [d]Universite de Poitiers (1432) 13 – ? [e]Bibliotheque Municipale (18th c) 320; Bibliotheque Universitaire (1879) 291.

Rennes: [a]Bretagne. [b]none. [c]Vilaine R; Ille R. [d]Universite de Rennes (1461 at Nantes, relocated 1735) 20 – 1038. [e]Bibliotheque de l'Universite . . . (1855) 665; Bibliotheque Municipale (1790) 268.

Rheims: [a]Champagne. [b]off Reims. [c]Vesle R. [d]Universite de Reims (1550) 11 – 669. [e]Bibliotheque Municipale Dite Carnegie (1809) 245.

Roubaix: [a]Nord. [b]none. [c]nr Scheldt R. [d]none. [e]Bibliotheque de l'Agence Univers (1934 at Lille, relocated 1952) 67; Bibliotheque Municipale (1833) 57.

Rouen: [a]Haute-Normandie. [b]none. [c]Seine R. [d]Universite de Rouen–Haute Normandie (1828) 11 – 855. [e]Bibliotheque Municipale (1791) 350.

Saint-Denis: [a]Region Parisienne. [b]none. [c]Seine R. [d]none. [e]Bibliotheque Municipale (1792) 120.

Saint-Etienne: [a]Rhone-Alpes. [b]none. [c]Furens R. [d]Universite de Saint-Etienne (1970) ? – ? [e]Bibliotheque Municipale (1842) 107.

Saint-Nazaire: [a]Pays de la Loire. [b]none. [c]B of Biscay; Loire R. [d]none. [e]Bibliotheque Municipale (1889) 61.

Strasbourg: [a]Alsace. [b]for Strassburg. [c]Ill R. [d]Universite de Strasbourg (1537) 23 – 1274. [e]Bibliotheque Nationale et Universitaire (1872) 3020; Bibliotheque Municipale (1765) 218.

Thionville: [a]Lorraine. [b]for Diedenhofen. [c]Moselle R. [d]none. [e]Bibliotheque Municipale (1842) 12.

Toulon: [a]Provence-Cote d'Azur. [b]none. [c]Mediterranean Sea. [d]Centre Universitaire de Toulon et du Var (at La Garde, nr Toulon) (1970) 2 – 107. [e]Bibliotheque Municipale (1790) 78.

Toulouse: [a]Midi-Pyrenees. [b]none. [c]Garonne R. [d]Universite de Toulouse (1229) 42 – 1463; Institut Catholique de Toulouse (univ) (1877) 1 – 171. [e]Bibliotheque Interuniversitaire (1879) 900; Bibliotheque Municipale (1782) 350.

Tourcoing: [a]Nord. [b]none. [c]nr Scheldt R. [d]none. [e]Bibliotheque Municipale (1880) 16.

Tours: [a]Centre. [b]none. [c]Loire R; Cher R. [d]Universite de Tours (1948) 12 – 562. [e]Bibliotheque Municipale (1791) 390.

Troyes: [a]Champagne. [b]none. [c]Seine R. [d]none. [e]Bibliotheque Municipale (1651) 285.

Valence: [a]Rhone-Alpes. [b]none. [c]Rhone R. [d]none. [e]Bibliotheque Municipale (1775) 90.

Valenciennes: [a]Nord. [b]none. [c]Scheldt R; Rhonelle R. [d]Centre Universitaire de Valenciennes et du Hainaut-Cambresis (1970) 2 – 189. [e]Bibliotheque Municipale (1765) 120.

Versailles: [a]Region Parisienne. [b]none. [c]nr Seine R. [d]none. [e]Bibliotheque Municipale (1803) 403.

Villeurbanne: [a]Rhone-Alpes. [b]none. [c]nr Rhone R. [d]Universite de Lyon, Villeurbanne campus (1964) ? – ? [e]Bibliotheque Interuniversitaire, Villeurbanne campus (1964) 175; Bibliotheque Municipale (1934) 100.

Germany: East

*Berlin (see under Germany: West)

Brandenburg: [b]alt Brandenburg an der Havel. [c]Havel R. [d]none. [e]Stadtbibliothek (1892) 93.

Cottbus: [b]alt Kottbus. [c]Spree R. [d]none. [e]Stadt- und Bezirksbibliothek (1925) 96.

Dessau: [b]none. [c]Mulde R. [d]none. [e]Stadtbibliothek (1897) 198.

Dresden: [b]none. [c]Elbe R; Weisseritz R. [d]Technische Universitat Dresden (1828) 13 – 2530; Hochschule fur Verkehrswesen Friedrich List (univ) (1952) 4 – 625. [e]Sachsische Landesbibliothek (1556) 1080; Universitatsbibliothek (1833) 1005; Stadt- und Bezirksbibliothek (1910) 380.

Erfurt: [b]none. [c]Gera R. [d]none. [e]Wissenschaftliche Allgemeinbibliothek (1392) 463.

Gera: [b]none. [c]White Elster (off Weisse Elster) R. [d]none. [e]Stadt- und Bezirksbibliothek (1920) 132.

Gorlitz: [b]none. [c]Neisse R. [d]none. [e]Stadtbibliothek (1779) 108; Oberlausitzische Bibliothek der Wissenschaften bei den Stadtischen Kunstsammlungen Gorlitz (1680) 100.

Gotha: [b]none. [c]nr Horsel R; nr Nesse R. [d]none. [e]Forschungsbibliothek Gotha (1647) 502.

Halle: [b]none. [c]Saale R. [d]Martin-Luther-Universitat Halle-Wittenberg [1502 at Wittenberg, united with Universitat Halle (1694) in 1817] 9 – 1924. [e]Univesitats- und Landesbibliothek Sachsen-Anhalt (1696) 3300; Stadt- und Bezirksbibliothek (1874) 194.

Jena: [b]none. [c]Saale R. [d]Friedrich-Schiller-Universitat (1548) 5 – 1320. [e]Universitatsbibliothek (1558) 2312.

Kark-Marx-Stadt: [b]for Chemnitz. [c]Chemnitz R. [d]Technische Hochschule Karl-Marx-Stadt (univ) (1836) 7 – 1269. [e]Stadt- und Bezirksbibliothek (1869) 550; Bibliothek der Technischen Hochschule . . . (1836) 393.

Leipzig: [b]none. [c]White Elster (off Weisse Elster) R; Pleisse R; Parthe R. [d]Karl-Marx-Universitat Leipzig (1409) 15 – 2500; Hochschule fur Bauwesen Leipzig (univ) (1954) 1 – 300. [e]Deutsche Bucherei (1912) 6078; Universitatsbibliothek (1543) 3019; Stadt- und Bezirksbibliothek (1914) 515.

Magdeburg: [b]none. [c]Elbe R. [d]Technische Hochschule Otto von Guericke Magdeburg (univ) (1953) 3 – 630. [e]Stadt- und Bezirksbibliothek (1525) 311.

Potsdam: [b]none. [c]Havel R; Templiner L; Jungfern L; Heiliger L. [d]none. [e]Wissenschaftliche Allgemeinbibliothek des Bezirkes Potsdam (1874) 450; Bibliothek der Deutschen Akademie fur Staats- und Rechtswissenschaft Walter Ulbricht (1949) 320.

Rostock: [b]none. [c]Warnow R. [d]Universitat Rostock (1419) 6 – 1300. [e]Universitatsbibliothek (1419) 1588; Stadt- und Bezirksbibliothek Willi Bredel (1894) 171.

Schwerin: ^{b}none. ^{c}Schweriner L. ^{d}none. ^{e}Wissenschaftliche Allgemeinbibliothek des Bezirkes Schwerin (1779) 570.

Zwickau: ^{b}none. ^{c}Zwickauer Mulde R. ^{d}none. ^{e}Ratsschulbibliothek (1500?) 80; Stadtbibliothek (1923) 69.

Germany: West

Aachen: ^{a}Nordrhein-Westfalen. ^{b}for Aix-la-Chapelle. ^{c}Wurm(bach) Creek. ^{d}Rheinisch-Westfalische Technische Hochschule (univ) (1870) 22 – 1900. ^{e}Bibliothek der . . . Technischen Hochschule (1870) 459; Stadtbibliothek (1828) 192.

Augsburg: ^{a}Bayern. ^{b}none. ^{c}Lech R. ^{d}Universitat Augsburg (1970) 4 – 297. ^{e}Universitatsbibliothek (1970?) 385; Staats- und Stadtbibliothek (1537) 364.

Baden-Baden: ^{a}Baden-Wurttemberg. ^{b}alt Baden. ^{c}Oos(bach) Creek. ^{d}none. ^{e}Stadtbucherei (1900) 44.

Bayreuth: ^{a}Bayern. ^{b}none. ^{c}Roter Main R. ^{d}Universitat Bayreuth (1972) 0.7 – 34. ^{e}Bibliothek des Fachbereichs Erziehungswissenschaft Bayreuth der Universitat Erlangen-Nurnberg (1813) 89; Stadtbibliothek (1921) 81.

Bergisch Gladbach: ^{a}Nordrhein-Westfalen. ^{b}none. ^{c}Sulz(bach) Creek. ^{d}none. ^{e}Kreisbucherei des Rheinisch-Bergischen Kreises (?) 30; Stadtbucherei (?) 25.

Berlin: ^{a}West Berlin: *West-Berlin. ^{b}inc Charlottenburg since 1920, East Berlin (off Ost-Berlin), West Berlin (off West-Berlin). ^{c}Havel R; Spree R. ^{d}East Berlin: Humboldt-Universitat zu Berlin (1809) 13 – 3000; West Berlin: Freie Universitat Berlin (1948) 32 – 3350; Technische Universitat Berlin (1799) 22 – 1550. ^{e}East Berlin: Deutsche Staatsbibliothek (1661) 5238; Universitatsbibliothek der Humboldt-Universitat zu Berlin (1831) 3598; Stadtbibliothek (1901) 987; West Berlin: Staatsbibliothek Preussischer Kulturbesitz (1661) 2560; Universitatsbibliothek der Freien Universitat Berlin (1952) 718. ^{f}Charlottenburg: 189 (1900), 9 (1849), 3 (1800), 2 (1740).

Bielefeld: ^{a}Nordrhein-Westfalen. ^{b}none. ^{c}Lutter(bach) Creek. ^{d}Universitat Bielefeld (1967) 5 – 390. ^{e}Universitatsbibliothek (1967?) 623; Stadtbibliothek (1905) 360.

Bochum: ^{a}Nordrhein-Westfalen. ^{b}none. ^{c}nr Ruhr R. ^{d}Ruhr-Universitat Bochum (1961) 23 – 1377. ^{e}Universitatsbibliothek (1965) 656; Stadtbucherei (1905) 330.

*Bonn: ^{a}Nordrhein-Westfalen. ^{b}inc Bad Godesberg (alt Godesberg) since 1968. ^{c}Rhine R. ^{d}Rheinische Friedrich-Wilhelms-Universitat Bonn (1777) 26 – 1681. ^{e}Universitatsbibliothek (1818) 1188; Abteilung Wissenschaftliche Dokumentation des Deutschen Bundestages (1949) 554; Stadtbucherei (1943) 207. ^{f}Bad Godesberg: 75 (1968e), 65 (1961), 45 (1950), 24 (1933), 9 (1900), 0.9 (1845), 0.8 (1794).

Bottrop: ^{a}Nordrhein-Westfalen. ^{b}none. ^{c}nr Emscher R. ^{d}none. ^{e}Stadtbucherei (1934) 68.

Bremen: ^{a}*Bremen. ^{b}none. ^{c}Weser R. ^{d}Universitat Bremen (1964) 5 – 320. ^{e}Universitatsbibliothek (1660) 1098; Stadtbibliothek (1902) 558.

Bremerhaven: ^{a}Bremen. ^{b}none. ^{c}Weser R. ^{d}none. ^{e}Stadtbibliothek (1873) 195.

Brunswick: ^{a}Niedersachsen. ^{b}off Braunschweig. ^{c}Oker R. ^{d}Technische Universitat Carolo-Wilhelmina (1745) 8 – 510. ^{e}Universitatsbibliothek (1748) 440; Stadtarchiv und Stadtbibliothek (1861) 234.

Cologne: ^{a}Nordrhein-Westfalen. ^{b}off Koln. ^{c}Rhine R. ^{d}Universitat zu Koln (1388) 26 – 950. ^{e}Universitats- und Stadtbibliothek (1602) 1398; Stadtbucherei (1878) 510.

Darmstadt: ^{a}Hessen. ^{b}none. ^{c}nr Gersprenz R. ^{d}Technische Universitat Darmstadt (1836) 11 – 720. ^{e}Hessische Landes- und Universitatsbibliothek (1567) 939.

Dortmund: ^{a}Nordrhein-Westfalen. ^{b}none. ^{c}Emscher R. ^{d}Universitat Dortmund (1965) 5 – 820. ^{e}Universitatsbibliothek (1965?) 379; Stadt- und Landesbibliothek (1907) 324; Stadtbucherei (1897) 293.

Duisburg: ^{a}Nordrhein-Westfalen. ^{b}inc Hamborn since 1929. ^{c}Rhine R; Ruhr R. ^{d}Gesamthochschule Duisburg (univ) (1972) 6 – 400. ^{e}Stadtbucherei (1901) 499. ^{f}Hamborn: 127 (1925), 33 (1900), 1 (1871).

Dusseldorf: ^{a}*Nordrhein-Westfalen. ^{b}none. ^{c}Rhine R; Dussel R. ^{d}Universitat Dusseldorf (1708) 7 – 670. ^{e}Universitatsbibliothek (1770) 1300; Stadtbucherei (1886) 450.

Erlangen: ^{a}Bayern. ^{b}none. ^{c}Regnitz R. ^{d}Friedrich-Alexander-Universitat Erlangen-Nurnberg (1743) 16 – 1070. ^{e}Universitatsbibliothek (1743) 1011.

Essen: [a]Nordrhein-Westfalen. [b]none. [c]Ruhr R; Emscher R; Baldeney L. [d]Universitat Essen (1972) 6 – 855. [e]Stadtbibliothek (1902) 608.

Esslingen: [a]Baden-Wurttemberg. [b]none. [c]Neckar R. [d]none. [e]Bucherei der Stadt (?) 65; Bibliothek der Padagogischen Hochschule Esslingen (?) 60.

Flensburg: [a]Schleswig-Holstein. [b]for Flensborg. [c]Flensburger (Forde) Inlet of Baltic Sea. [d]none. [e]Dansk Central Bibliotek for Sydslesvig (1920) 85; Stadtbucherei (1904) 70.

Frankfurt am Main: [a]Hessen. [b]alt Frankfort, Frankfurt. [c]Main R. [d]Johann Wolfgang Goethe-Universitat (1914) 23 – 1877. [e]Deutsche Bibliothek (1946) 1895; Stadt- und Universitats-bibliothek (1668) 1750; Senckenbergische Bibliothek (1763) 729.

Freiburg im Breisgau: [a]Baden-Wurttemberg. [b]alt Freiburg. [c]Dreisam R. [d]Albert-Ludwigs-Universitat (1457) 17 – 800. [e]Universitatsbibliothek (1457) 1145.

Furth: [a]Bayern. [b]none. [c]Regnitz R; Pegnitz R; Rednitz R. [d]none. [e]Stadtische Volksbucherei (1906) 39; Stadtbibliothek (1906) 22.

Gelsenkirchen: [a]Nordrhein-Westfalen. [b]none. [c]Emscher R. [d]none. [e]Stadtbucherei (1911) 236.

Gottingen: [a]Niedersachsen. [b]none. [c]Leine R. [d]Georg-August-Universitat (1736) 20 – 904. [e]Niedersachsische Staats- und Universitatsbibliothek (1735) 2302.

Hagen: [a]Nordrhein-Westfalen. [b]none. [c]Ennepe R; Volme R. [d]none. [e]Stadtbucherei (1899) 121.

Hamburg: [a]*Hamburg. [b]inc Altona since 1933. [c]Elbe R. [d]Universitat Hamburg (1919) 27 – 2043. [e]Stiftung Hamburger Offentliche Bucherhallen (1899) 1387; Staats- und Universitatsbibliothek (1479) 1181. [f]Altona: 242 (1933), 162 (1900), 41 (1855), 23 (1803), 18 (1769).

Hamm: [a]Nordrhein-Westfalen. [b]none. [c]Lippe R. [d]none. [e]Stadtbucherei (1895) 63.

Hannover: [a]*Niedersachsen. [b]alt Hanover. [c]Leine R. [d]Technische Universitat Hannover (1831) 13 – 600; Medizinische Hochschule Hannover (univ) (1961) 2 – 170; Tierarztliche Hochschule Hannover (univ) (1778) 1 – 246. [e]Stadtbucherei (1440) 863; Universitatsbibliothek (1831) 825; Niedersachsische Landesbibliothek (1665) 670; Stadtbibliothek (1440) 388.

Heidelberg: [a]Baden-Wurttemberg. [b]none. [c]Neckar R. [d]Ruprecht-Karl-Universitat Heidelberg (1386) 18 – 1895. [e]Universitatsbibliothek (1706) 1150.

Heilbronn: [a]Baden-Wurttemberg. [b]none. [c]Neckar R. [d]none. [e]Stadtbucherei (1903) 94.

Herne: [a]Nordrhein-Westfalen. [b]inc Wanne-Eickel since 1974; Wanne: for Bickern. [c]nr Emscher R. [d]none. [e]Stadtische Bucherei (of Herne) (1906) 132; Stadtbucherei (of Wanne-Eickel) (1902) 97. [f]Eickel: 17 (1900), 0.6 (1843), 0.5 (1818); Wanne: 24 (1900), 0.4 (1843), 0.3 (1818); Wanne-Eickel: 99 (1970), 107 (1961), 92 (1933).

Hildesheim: [a]Niedersachsen. [b]none. [c]Innerste R. [d]none. [e]Stadtarchiv und Stadtbibliothek (1888) 107.

Ingolstadt: [a]Bayern. [b]none. [c]Danube R; Schutter R. [d]none. [e]Wissenschaftliche Stadtbibliothek (?) 47.

Iserlohn: [a]Nordrhein-Westfalen. [b]none. [c]Baar(bach) Creek. [d]none. [e]Bibliothek der Staatliche Ingenieurschule fur Maschinenwesen (1852) ?

Kaiserslautern: [a]Rheinland-Pfalz. [b]none. [c]Lauter R. [d]Universitat Kaiserslautern (1970) 2 – 184. [e]Universitatsbibliothek (1970) 142; Stadtbucherei (1832) 50.

Karlsruhe: [a]Baden-Wurttemberg. [b]for Carlsruhe. [c]nr Rhine R. [d]Universitat Fridericiana (1825) 12 – 1039. [e]Badische Landesbibliothek (1500 at Pforzheim, relocated 1765) 567; Universitatsbibliothek (1840) 404.

Kassel: [a]Hessen. [b]for Cassel. [c]Fulda R. [d]Gesamthochschule Kassel (univ) (1970) 6 – 300. [e]Murhardsche Bibliothek der Stadt und Landesbibliothek (1863) 335; Stadtbucherei (1876) 118.

Kiel: [a]*Schleswig-Holstein. [b]none. [c]Baltic Sea. [d]Christian-Albrechts-Universitat Kiel (1665) 12 – 559. [e]Bibliothek des Instituts fur Weltwirtschaft an der ... Universitat (1910) 1220; Universitatsbibliothek (1665) 642; Stadtbucherei (1874) 222.

Koblenz: [a]Rheinland-Pfalz. [b]alt Coblenz. [c]Rhine R; Moselle R. [d]none. [e]Stadtbibliothek (1827) 255.

Krefeld: [a]Nordrhein-Westfalen. [b]for Crefeld. [c]Rhine R. [d]none. [e]Stadtbucherei (1900) 120.

Leverkusen: [a]Nordrhein-Westfalen. [b]inc Wiesdorf since 1930. [c]Rhine R; Wupper R. [d]none. [e]Kekule-Bibliothek (1897) 410; Stadtbucherei (1926) 77. [f]Wiesdorf: 22 (1920), 6 (1900), 1 (1861), 0.7 (1797).

Lubeck: [a]Schleswig-Holstein. [b]none. [c]Trave R; Wakenitz R. [d]none. [e]Bibliothek der Hansestadt Lubeck (1616) 650.

Ludwigshafen: [a]Rheinland-Pfalz. [b]alt Ludwigshafen am Rhein. [c]Rhine R. [d]none. [e]Stadtbucherei (1875) 197.

Mainz: [a]*Rheinland-Pfalz. [b]for Mayence. [c]Rhine R; Main R. [d]Johannes-Gutenberg-Universitat (1477) 19 – 1508. [e]Universitatsbibliothek (1946?) 844; Stadtbibliothek (1477) 422.

Mannheim: [a]Baden-Wurttemberg. [b]none. [c]Rhine R; Neckar R. [d]Universitat Mannheim (1907) 7 – 240. [e]Universitatsbibliothek (1907?) 575; Stadtbucherei (1895) 267.

Marl: [a]Nordrhein-Westfalen. [b]none. [c]nr Lippe R. [d]none. [e]Stadtbucherei (1950) 49.

Moers: [a]Nordrhein-Westfalen. [b]for Meurs. [c]Moers(bach) Creek. [d]none. [e]Stadtbucherei (1852) 30.

Monchengladbach: [a]Nordrhein-Westfalen. [b]for Munchen-Gladbach; inc Rheydt since 1974. [c]Niers R. [d]none. [e]Stadtbibliothek (of Monchengladbach) (1904) 311; Stadtbucherei (of Rheydt) (1904) 58. [f]Rheydt: 100 (1970), 94 (1961), 78 (1950), 77 (1933), 34 (1900), 8 (1849), 3 (1803).

Mulheim an der Ruhr: [a]Nordrhein-Westfalen. [b]alt Mulheim. [c]Ruhr R. [d]none. [e]Stadtbucherei (1883) 229.

Munich: [a]*Bayern. [b]off Munchen. [c]Isar R. [d]Ludwig-Maximilians-Universitat Munchen (1472) 36 – 1650; Technische Universitat Munchen (1827) 13 – 805; Ukrainische Freie Universitat (1921) 0.5 – 56. [e]Bayerische Staatsbibliothek (1558) 3700; Stadtische Bibliothek (1843) 1522; Universitatsbibliothek (1472) 1283; Bibliothek des Deutschen Museums (1903) 600.

Munster: [a]Nordrhein-Westfalen. [b]none. [c]Aa R. [d]Westfalische Wilhelms-Universitat Munster (1780) 29 – 797. [e]Universitatsbibliothek (1906) 753.

Neuss: [a]Nordrhein-Westfalen. [b]none. [c]Rhine R; Erft R. [d]none. [e]Stadtbucherei (1907) 112.

Nuremberg: [a]Bayern. [b]off Nurnberg. [c]Pegnitz R. [d]none. [e]Stadtbibliothek (1371) 586; Bibliothek des Germanischen Nationalmuseums (1852) 330.

Oberhausen: [a]Nordrhein-Westfalen. [b]none. [c]nr Emscher R. [d]none. [e]Stadtbucherei (1905) 191.

Offenbach am Main: [a]Hessen. [b]alt Offenbach. [c]Main R. [d]none. [e]Bibliothek des Deutschen Wetterdienstes (1848) 114; Stadtbucherei (1901) 74.

Oldenburg in Oldenburg: [a]Niedersachsen. [b]alt Oldenburg. [c]Hunte R. [d]Universitat Oldenburg (1970) 4 – ? [e]Landesbibliothek (1792) 297.

Osnabruck: [a]Niedersachsen. [b]none. [c]Haase R. [d]Universitat Osnabruck (1970) 3 – 300. [e]Stadtbibliothek (1902) 180.

Paderborn: [a]Nordrhein-Westfalen. [b]none. [c]Pader R. [d]Gesamthochschule Paderborn (univ) (1972) 6 – 295. [e]Erzbischofliche Akademische Bibliothek (1887) 185; Gesamthochschulbibliothek (1972?) 168.

Pforzheim: [a]Baden-Wurttemberg. [b]none. [c]Enz R; Nagold R; Wurm R. [d]none. [e]Stadtbucherei (1892) 73.

Recklinghausen: [a]Nordrhein-Westfalen. [b]none. [c]nr Emscher R. [d]none. [e]Stadtbucherei (1938) 63.

Regensburg: [a]Bayern. [b]alt Ratisbon. [c]Danube R. [d]Universitat Regensburg (1962) 9 – 750. [e]Universitatsbibliothek (1964) 1328; Furstlich Thurn und Taxissche Hofbibliothek (1773) 197; Staatliche Bibliothek (1816) 156.

Remscheid: [a]Nordrhein-Westfalen. [b]none. [c]Wupper R. [d]none. [e]Stadtbucherei (1902) 130.

Reutlingen: [a]Baden-Wurttemberg. [b]none. [c]Echaz R. [d]Eberhard-Karls-Universitat Tubingen (at Tubingen, nr Reutlingen) (1477) 18 – 588. [e]Universitatsbibliothek (at Tubingen, nr Reutlingen) (1477?) 1046; Stadtbucherei (1660) 114.

Saarbrucken: [a]*Saarland. [b]for Sarrebruck. [c]Saar (alt Sarre) R; Sulz R. [d]Universitat des Saarlandes (1947) 12 – 900. [e]Universitatsbibliothek (1950) 700; Stadtbucherei (1924) 152.

Salzgitter: [a]Niedersachsen. [b]for Watenstedt-Salzgitter. [c]Fuhse R. [d]none. [e]Stadtbucherei (1945) 64.

Siegen: [a]Nordrhein-Westfalen. [b]none. [c]Sieg R. [d]Gesamthochschule Siegen (univ) (1972) 5 – 540. [e]Gesamthochschulbibliothek (1972?) 225; Stadtbibliothek (1928) 32.

Solingen: [a]Nordrhein-Westfalen. [b]none. [c]Wupper R. [d]none. [e]Stadtbucherei (1926) 164.

Stuttgart: [a]*Baden-Wurttemberg. [b]none. [c]Neckar R. [d]Universitat Stuttgart (1829) 11 – 2610; Universitat Hohenheim (1818) 3 – 360. [e]Wurttembergische Landesbibliothek (1765) 1301; Stadtbucherei (1897) 547; Universitatsbibliothek (of the Universitat Stuttgart) (1829) 435.

Trier: [a]Rheinland-Pfalz. [b]for Treves. [c]Moselle R. [d]Universitat Trier (1970) 3 – 118. [e]Stadtbibliothek (1775) 279; Universitatsbibliothek (1970) 225.

Ulm: ^{a}Baden-Wurttemberg. ^{b}none. ^{c}Danube R. ^{d}Universitat Ulm (1967) 2 – 163. ^{e}Universitatsbibliothek (1964) 314; Stadtbibliothek (1516) 161.

Velbert: ^{a}Nordrhein-Westfalen. ^{b}none. ^{c}nr Ruhr R. ^{d}none. ^{e}Stadtbucherei (1905) 43.

Wiesbaden: ^{a}*Hessen. ^{b}none. ^{c}Rhine R. ^{d}none. ^{e}Hessische Landesbibliothek (1813) 469; Stadtbucherei (1872) 179.

Wilhelmshaven: ^{a}Niedersachsen. ^{b}none. ^{c}Jade B of North Sea. ^{d}none. ^{e}Stadtbucherei (1872) 78.

Witten: ^{a}Nordrhein-Westfalen. ^{b}none. ^{c}Ruhr R. ^{d}none. ^{e}Stadtbucherei (1911) 62.

Wolfsburg: ^{a}Niedersachsen. ^{b}none. ^{c}Aller R. ^{d}none. ^{e}Stadtbucherei (1943) 111.

Wuppertal: ^{a}Nordrhein-Westfalen. ^{b}inc Barmen, Elberfeld. ^{c}Wupper R. ^{d}Gesamthochschule Wuppertal (univ) (1972) 2 – 350. ^{e}Stadtbibliothek (1852) 462. ^{f}Barmen: 142 (1900), 36 (1849), 16 (1810); Elberfeld: 157 (1900), 39 (1849), 19 (1810).

Wurzburg: ^{a}Bayern. ^{b}none. ^{c}Main R. ^{d}Bayerische Julius-Maximilians-Universitat (1402) 13 – 937. ^{e}Universitatsbibliothek (1619) 535; Stadtbucherei (1872) 98.

Gibraltar

*Gibraltar: ^{b}none. ^{c}Strait of Gibraltar. ^{d}none. ^{e}Gibraltar Garrison Library (1793) 45.

Greece

*Athens: ^{b}off Athinai. ^{c}Kifisos R; Ilissos R. ^{d}National and Capodistrian University of Athens (1835) 28 – 1529; Polytechnic University of Athens (1836) 4 – 590. ^{e}Chamber of Deputies Library (1844) 1500; National Library of Greece (1828) 1000.

Iraklion: ^{a}Crete I. ^{b}alt Herakleion; for Candia. ^{c}Sea of Crete. ^{d}none. ^{e}Vikelaia Municipal Library (?) 35.

Patras: ^{b}off Patrai. ^{c}G of Patras of Ionian Sea. ^{d}University of Patras (1964) 2 – 186. ^{e}Municipal Library (?) ?

Peristeri: ^{b}alt Peristerion. ^{c}nr Kifisos R. ^{d}none.

Piraeus: ^{b}off Piraievs. ^{c}Saronic G of Aegean Sea. ^{d}none. ^{e}Institute of Advanced Industrial Studies Library (1938) 14; Municipal Library (1926) ?

Salonika: ^{b}alt Salonica; off Thessaloniki. ^{c}G of Salonika of Aegean Sea. ^{d}Aristotelian University of Salonika (1925) 24 – 1765. ^{e}. . . University Library (1925) 750.

Hungary

*Budapest: ^{b}inc Buda (for Ofen), Pest. ^{c}Danube R. ^{d}Budapesti Muszaki Egyetem (1782) 11 – 1659; Eotvos Lorand Tudomanyegyetem (1561) 9 – 1219; Marx Karoly Kozgazdasagtudomanyi Egyetem (1948) 4 – 328; Semmelweis Orvostudomanyi Egyetem (1769) 4 – 1155; Kerteszeti Egyetem (1853) 1 – 210; Allatorvostudomanyi Egyetem (1782) 0.5 – 124. ^{e}Fovarosi Szabo Ervin Konyvtar (1904) 2535; Orszagos Szechenyi Konyvtar (1802) 2150; Egyetemi Konyvtara (of Eotvos Lorand Tudomanyegyetem) (1635) 1257; Magyar Tudomanyos Akademia Konyvtara (1826) 960. ^{f}Buda: 50 (1851), 25 (1787), 10 (1720); Pest: 106 (1851), 22 (1787), 3 (1720).

Debrecen: ^{b}for Debreczin. ^{c}nr Berettyo R. ^{d}Kossuth Lajos Tudomanyegyetem (1538) 3 – 373; Debreceni Orvostudomanyi Egyetem (1912) 1 – 480; Debreceni Agrartudomanyi Egyetem (1868) 0.9 – 200. ^{e}Egyetemi Konyvtara (of Kossuth Lajos Tudomanyegyetem) (1916) 940; Tiszantuli Reformatus Egyhazkerulet Nagykonyvtara (1549) 551.

Gyor: ^{b}for Raab. ^{c}Mosoni Duna (alt Little Danube) R; Raba (alt Raab) R; Repce (alt Rabnitz) R. ^{d}none. ^{e}Szent-Benedek-Rend Kozponti Fokonyutara (at Gyorszentmarton, nr Gyor) (1001) 298.

Miskolc: ^{b}none. ^{c}Szvinva R. ^{d}Nehezipari Muszaki Egyetem (1763 at Banska Stiavnica, Czechoslovakia; relocated 1949) 3 – 398. ^{e}. . . Egyetem Kozponti Konyvtara (1763 at Banska Stiavnica, Czechoslovakia; relocated 1949) 358.

Pecs: ^{b}for Funfkirchen. ^{c}Pecsi (Viz) Creek. ^{d}Pecsi Tudomanyegyetem (1367) 2 – 44; Pecsi Orvostudomanyi Egyetem (1912 at Bratislava, Czechoslovakia; relocated 1921) 1 – 411. ^{e}Pecsi Tudomanyegyetem Konyvtara (1774) 300; Baranya Megyei Konyvtar (1941) 90.

Szeged: ^{b}for Szegedin. ^{c}Tisza R; Maros R. ^{d}Jozsef Attila Tudomanyegyetem Szeged (1872 at Cluj, Romania; relocated 1921) 4 – 428; Szegedi Orvostudomanyi Egyetem (1921) 2 – 506. ^{e}Jozsef

Attila Tudomanyegyetem Kozponti Konyvtara (1872 at Cluj, Romania; relocated 1921?) 584; Varosi Somogyi Konyvtar (1880) 400.

Szekesfehervar: ^{b}for Stuhlweissenburg. ^{c}nr Sarviz R. ^{d}none. ^{e}Fejer Megyei Vorosmarty Mihaly Konyvtar (1952) 60.

Iceland

*Reykjavik: ^{b}none. ^{c}Faxa B of Atlantic O. ^{d}Haskoli Islands (1911) 3 – 230. ^{e}Landsbokasafn Islands (1818) 320; Borgarbokasafn (1923) 250; Haskolabokasafn (1940) 176.

Ireland

Cork: ^{b}off Corcaigh. ^{c}Lee R; Mahon L. ^{d}University College, Cork (of the National University of Ireland) (1845) 5 – 520. ^{e}Cork County Library (1926) 336; University College Library (1845) 256; City Public Library (1790) 112.

*Dublin: ^{b}off Baile Atha Cliath. ^{c}Dublin B of Irish Sea; Liffey R. ^{d}National University of Ireland (1795) 18 – 1270 [University College, Dublin (of the . . . University) (1851) 10 – 500]; Trinity College, University of Dublin (1591) 5 – 476. ^{e}Trinity College Library (1591) 1500; Public Library (1884) 1295; University College Library (1851?) 583; National Library of Ireland (1877) 550.

Limerick: ^{b}off Luimneach. ^{c}Shannon R. ^{d}none. ^{e}Limerick County Library (1935) 178; City Public Library (1893) 47.

Italy

Alessandria: ^{a}Piemonte. ^{b}none. ^{c}Tanaro R. ^{d}none. ^{e}Biblioteca Civica (1806) 105.

Ancona: ^{a}Marche. ^{b}none. ^{c}Adriatic Sea. ^{d}Universita degli Studi di Ancona (1970) 4 – 146. ^{e}Biblioteca Comunale Luciano Benincasa (1669) 97.

Arezzo: ^{a}Toscana. ^{b}none. ^{c}nr Arno R. ^{d}none. ^{e}Biblioteca Consorziale della Citta (1603) 120.

Bari: ^{a}Puglia. ^{b}alt Bari delle Puglie. ^{c}Adriatic Sea. ^{d}Universita degli Studi di Bari (1924) 43 – 2236. ^{e}Biblioteca Nazionale Sagarriga Visconti-Volpi (1865) 225.

Bergamo: ^{a}Lombardia. ^{b}none. ^{c}Serio R; Brembo R. ^{d}none. ^{e}Biblioteca Civica Angelo Mai (1760) 550.

Bologna: ^{a}Emilia-Romagna. ^{b}none. ^{c}nr Reno R; nr Savena R. ^{d}Universita degli Studi di Bologna (1088?) 53 – 2951. ^{e}Biblioteca Comunale dell'Archiginnasio (1801) 700; Biblioteca Universitaria (1712) 666.

Bolzano: ^{a}Trentino-Alto Adige. ^{b}for Bozen. ^{c}Isarco R. ^{d}none. ^{e}Biblioteca Civica Cesare Battisti (1928) 98.

Brescia: ^{a}Lombardia. ^{b}none. ^{c}Garza R. ^{d}none. ^{e}Biblioteca Civica Queriniana (1750) 410.

Cagliari: ^{a}Sardegna; Sardinia I. ^{b}none. ^{c}G of Cagliari of Mediterranean Sea. ^{d}Universita degli Studi di Cagliari (1606) 18 – 1426. ^{e}Biblioteca Universitaria (1792) 483.

Catania: ^{a}Sicilia; Sicily I. ^{b}none. ^{c}G of Catania of Mediterranean Sea. ^{d}Universita degli Studi di Catania (1434) 27 – 1660. ^{e}Biblioteca Universitaria e Ventimilliana (1775) 266; Biblioteca Riunite Civica e A. Ursino Recupero (1693) 127.

Catanzaro: ^{a}Calabria. ^{b}none. ^{c}nr Squillace G of Ionian Sea; nr Corace R; nr Alli R. ^{d}none. ^{e}Biblioteca Comunale Filippo De Nobili (1934) 44.

Como: ^{a}Lombardia. ^{b}none. ^{c}Como L. ^{d}none. ^{e}Biblioteca Comunale (1663) 174.

Cosenza: ^{a}Calabria. ^{b}none. ^{c}Crati R; Busente R. ^{d}Universita degli Studi di Calabria (1972) 2 – 251. ^{e}Biblioteca Civica (1507) 203.

Cremona: ^{a}Lombardia. ^{b}none. ^{c}Po R. ^{d}none. ^{e}Biblioteca Governativa e Biblioteca Civica (1607) 389.

Ferrara: ^{a}Emilia-Romagna. ^{b}none. ^{c}Po di Volano R. ^{d}Universita degli Studi di Ferrara (1391) 6 – 591. ^{e}Biblioteca Comunale Ariostea (1753) 155.

Florence: ^{a}Toscana. ^{b}off Firenze. ^{c}Arno R. ^{d}Universita degli Studi di Firenze (1321) 39 – 2295. ^{e}Biblioteca Nazionale Centrale (1747) 4210; Biblioteca della Facolta di Lettre e Filosofia dell'Universita . . . (1859) 1550; Biblioteca della Facolta di Giurisprudenza dell'Universita . . . (1926) 476; Biblioteca del Gabinetto Scientifico-Letterario G. P. Vieusseux (1819) 450;

Biblioteca Marucelliana (1752) 440; . . .; Biblioteca Medicea-Laurenziana (1571) 79; Biblioteca Riccardiana (1816) 63.

Foggia: [a]Puglia. [b]none. [c]nr Celone R. [d]none. [e]Biblioteca Provinciale (1936) 135.

Forli: [a]Emilia-Romagna. [b]none. [c]Montone R. [d]none. [e]Biblioteca Comunale Aurelio Saffi (1750) 264.

Genoa: [a]Liguria. [b]off Genova. [c]G of Genoa of Ligurian Sea. [d]Universita degli Studi di Genova (1471) 30 – 2165. [e]Biblioteca Universitaria (1773) 378; Biblioteca Civica Berio (1775) 210.

La Spezia: [a]Liguria. [b]alt Spezia. [c]G of Spezia of Ligurian Sea. [d]none. [e]Biblioteca Comunale Ubaldo Mazzini (1898) 110.

Leghorn: [a]Toscana. [b]off Livorno. [c]Ligurian Sea. [d]none. [e]Biblioteca Comunale Labronica Francesco Domenico Guerrazzi (1816) 290.

Lucca: [a]Toscana. [b]none. [c]Serchio R. [d]none. [e]Biblioteca Statale (1794) 401.

Mantua: [a]Lombardia. [b]off Mantova. [c]Mincio R. [d]none. [e]Biblioteca Comunale (1780) 400.

Messina: [a]Sicilia; Sicily I. [b]none. [c]Strait of Messina. [d]Universita degli Studi di Messina (1548) 23 – 1591. [e]Biblioteca Universitaria (1548) 190; Biblioteca Painiana del Seminario Arcivescovile (1783) 180.

Milan: [a]Lombardia. [b]for Mailand; off Milano. [c]Olona R. [d]Universita degli Studi di Milano (1923) 50 – 2304; Politecnico di Milano (univ) (1862) 16 – 805; Universita Cattolica del Sacro Cuore (1920) 15 – 390; Universita Commerciale Luigi Bocconi (1902) 3 – 140. [e]Biblioteca Comunale (1890) 1150; Biblioteca Ambrosiana (1609) 850; Biblioteca Nazionale Braidense (1786) 845; Biblioteca dell'Universita Cattolica del Sacro Cuore (1921) 827.

Modena: [a]Emilia-Romagna. [b]none. [c]nr Secchia R. [d]Universita degli Studi di Modena (1175) 7 – 774. [e]Biblioteca Estense e Universitaria (14th c) 536.

Monza: [a]Lombardia. [b]none. [c]Lambro R. [d]none. [e]Biblioteca Civica (1870) 80.

Naples: [a]Campania. [b]off Napoli. [c]B of Naples of Tyrrhenian Sea. [d]Universita degli Studi di Napoli (1224) 79 – 3256. [e]Biblioteca Nazionale (1804) 1588; Biblioteca Universitaria (1615) 706.

Novara: [a]Piemonte. [b]none. [c]Agogna R; Terdoppio R. [d]none. [e]Biblioteca Riunite Civica e Negroni (1852) 173.

Padua: [a]Veneto. [b]off Padova. [c]Bacchiglione R. [d]Universita degli Studi di Padova (1222) 42 – 2379. [e]Biblioteca Universitaria (1629) 494; Biblioteca Civica (1839) 405.

Palermo: [a]Sicilia; Sicily I. [b]none. [c]G of Palermo of Tyrrhenian Sea. [d]Universita degli Studi di Palermo (1777) 39 – 1686. [e]Biblioteca Nazionale (1782) 499; Biblioteca Comunale (1760) 400.

Parma: [a]Emilia-Romagna. [b]none. [c]Parma R. [d]Universita degli Studi di Parma (1065) 16 – 1203. [e]Biblioteca Palatina (1762) 402.

Pavia: [a]Lombardia. [b]none. [c]Ticino R. [d]Universita degli Studi di Pavia (1361) 14 – 1779. [e]Biblioteca Universitaria (1772) 421; Biblioteca Civica Bonetta (1833) 73.

Perugia: [a]Umbria. [b]none. [c]nr Tiber R. [d]Universita degli Studi di Perugia (1200) 16 – 1143; Universita Italiana per Stranieri (1921) 6 – 105. [e]Biblioteca Comunale Augusta (1623) 200; Biblioteca Centrale dell'Universita . . . (1848) 180.

Pescara: [a]Abruzzi e Molise. [b]none. [c]Adriatic Sea; Pescara R. [d]Libera Universita Abruzzese degli Studi Gabriele d'Annunzio (at Chieti, nr Pescara) (1961) 4 – 152. [e]Biblioteca Provinciale Gabriele d'Annunzio (1929) 118.

Piacenza: [a]Emilia-Romagna. [b]none. [c]Po R. [d]none. [e]Biblioteca Comunale Passerini Landi (1791) 200.

Pisa: [a]Toscana. [b]none. [c]Arno R. [d]Universita degli Studi di Pisa (1343) 24 – 2062. [e]Biblioteca dell'Universita . . . (1742) 321.

Pistoia: [a]Toscana. [b]none. [c]Ombrone R. [d]none. [e]Biblioteca Comunale Forteguerriana (1696) 250.

Prato: [a]Toscana. [b]none. [c]Bisenzio R. [d]none. [e]Biblioteca Lazzariniana Comunale e Biblioteca Roncioniana (1676) 55.

Ravenna: [a]Emilia-Romagna. [b]none. [c]nr Adriatic Sea. [d]none. [e]Biblioteca Comunale Classense (1710?) 400.

Reggio di Calabria: [a]Calabria. [b]none. [c]Strait of Messina. [d]none. [e]Biblioteca Sandicci (1957) 100; Biblioteca Comunale (1819) 56.

Reggio nell'Emilia: [a]Emilia-Romagna. [b]none. [c]Crostolo R. [d]none. [e]Biblioteca Municipale (1796) 250.

Rimini: [a]Emilia-Romagna. [b]none. [c]Adriatic Sea; Marecchia R. [d]none. [e]Biblioteca Civica Gambalunga (1619) 135.

*Rome: [a]Lazio. [b]off Roma. [c]Tiber R. [d][1] Universita degli Studi di Roma (1303) 122 − 6806; Pontificia Universitas Gregoriana[2] (1551) 2 − 257; Libera Universita Internazionale degli Studi Sociali pro Deo (1945) 1 − 97; Pontificia Universitas Lateranensis[2] (1773) 1 − 121. [e]Biblioteca Nazionale Vittorio Emanuele II (1876) 2234; Biblioteca Apostolica Vaticana[2] (15th c) 1000; Biblioteca Universitaria Alessandrina (1661) 1000; Library of the Food and Agriculture Organization of the United Nations (1946) 1000; Biblioteca del Ministero dell'Agricoltura e delle Foreste (1860) 800; Biblioteca della Pontificia Universitas Gregoriana[2] (1551) 730; Biblioteca dell'Accademia Nazionale dei Lincei (1730) 430;...; Biblioteca Casanatense (1700) 314; Biblioteca Angelica (1605) 182.

Salerno: [a]Campania. [b]none. [c]G of Salerno of Tyrrhenian Sea. [d]Universita degli Studi di Salerno (1944) 18 − 398. [e]Biblioteca Provinciale (1845) 90.

Sassari: [a]Sardegna; Sardinia I. [b]none. [c]nr G of Asinara of Mediterranean Sea. [d]Universita degli Studi di Sassari (1562) 6 − 466. [e]Biblioteca Universitaria (1560?) 144; Biblioteca Comunale (1934) 70.

Sesto San Giovanni: [a]Lombardia. [b]none. [c]nr Lambro R. [d]none.

Siena: [a]Toscana. [b]none. [c]nr Arbia R. [d]Universita degli Studi di Siena (1240) 7 − 662. [e]Biblioteca Comunale degli Intronati (1758) 350.

Syracuse: [a]Sicilia; Sicily I. [b]off Siracusa. [c]Ionian Sea. [d]none. [e]Biblioteca Comunale (1857) 50.

Taranto: [a]Puglia. [b]none. [c]G of Taranto of Ionian Sea. [d]none. [e]Biblioteca Comunale Pietro Acclavio (1893) 57.

Terni: [a]Umbria. [b]none. [c]Nera R. [d]none. [e]Biblioteca Comunale (1885) 40.

Torre del Greco: [a]Campania. [b]none. [c]B of Naples of Tyrrhenian Sea. [d]none.

Trent: [a]Trentino-Alto Adige. [b]for Trient; off Trento. [c]Adige R. [d]Libera Universita degli Studi di Trento (1962) 3 − 88. [e]Biblioteca Comunale (1856) 189.

Treviso: [a]Veneto. [b]none. [c]Sile R. [d]none. [e]Biblioteca Comunale (1847) 168.

Trieste: [a]Friuli-Venezia Giulia. [b]for Triest. [c]G of Trieste of Adriatic Sea. [d]Universita degli Studi di Trieste (1877) 13 − 1164. [e]Biblioteca Civica (1793) 310.

Turin: [a]Piemonte. [b]off Torino. [c]Po R; Dora Riparia R. [d]Universita degli Studi di Torino (1404) 35 - 2281; Politecnico di Torino (univ) (1859) 11 − 564. [e]Biblioteca Nazionale Universitaria (1720) 776; Biblioteche Civiche e Raccolte Storiche (1869) 350.

Udine: [a]Friuli-Venezia Giulia. [b]none. [c]Torre R. [d]none. [e]Biblioteca Comunale Joppi (1866) 370.

Venice: [a]Veneto. [b]for Venedig; off Venezia. [c]Venice Lagoon. [d]Universita degli Studi di Venezia (1868) 6 − 365. [e]Biblioteca Nazionale Marciana (1468) 747.

Verona: [a]Veneto. [b]none. [c]Adige R. [d]none. [e]Biblioteca Civica (1792) 520.

Vicenza: [a]Veneto. [b]none. [c]Bacchiglione R; Retrone R; Astichello R. [d]none. [e]Biblioteca Civica Bertoliana (1696) 370.

Luxembourg

*Luxembourg: [b]alt Lutzelburg, Luxemburg. [c]Alzette R. [d]Universite du Travail de Luxembourg (1951) 15 − 1912; Centre Universitaire de Luxembourg (1958) 0.3 − 80. [e]Bibliotheque Nationale (1798) 650.

Malta

*Valletta: [a]Malta I. [b]alt Valetta. [c]Mediterranean Sea. [d]Royal University of Malta (at Msida, nr Valletta) (1592 at Valletta, relocated from 1967?) 1 − 166. [e]Royal Malta Library (1555) 320; ... University Library (at Msida, nr Valletta) (1769 at Valletta, relocated 1967?) 145.

Monaco

*Monaco: [b]inc Monte-Carlo. [c]Mediterranean Sea. [d]none. [e]Bibliotheque Communale (1909) 125.

---

[1]Universities with 1000 or more students.
[2]At Vatican City, nr Rome.

Netherlands

*Amsterdam: ^{b}none. ^{c}Amstel R; IJssel(meer) L. ^{d}Universiteit van Amsterdam (1632) 20 – 2106; Vrije Universiteit, Amsterdam (1880) 12 – 1024. ^{e}Universiteitsbibliotheek (of the Universiteit . . .) (1578) 2000: Openbare Leeszaal en Bibliotheek (1919) 800; Bibliotheek der . . . Universiteit (1880) 480.

Apeldoorn: ^{b}none. ^{c}IJssel R. ^{d}none. ^{e}Gemeenschappelijke Openbare Bibliotheek (1911) 72.

Arnhem: ^{b}for Arnheim. ^{c}Nederrijn R (Rhine R distributary). ^{d}none. ^{e}Stichting Arnhemse Openbare en Gelderse Wetenschappelijke Bibliotheek (1856) 300.

Breda: ^{b}none. ^{c}Mark R; Aa of Weerijs R. ^{d}none. ^{e}Katholieke Provinciale Bibliotheekcentrale voor Noord-Brabant (1915) 112; Bibliotheek Koninklijke Militaire Academie (1828) 95.

Delft: ^{b}none. ^{c}Schie R. ^{d}Technische Hogeschool te Delft (univ) (1842) 10 – 2130. ^{e}Centrale Bibliotheek der . . . Hogeschool (1842) 511; Stichting Samenwerkende Openbare Bibliotheken (1917) 65.

Dordrecht: ^{b}alt Dort. ^{c}Lower Merwede (off Beneden Merwede) R. ^{d}none. ^{e}Openbare Leeszaal en Bibliotheek (1899) 80.

Eindhoven: ^{b}none. ^{c}Dommel R; Gender R. ^{d}Technische Hogeschool te Eindhoven (univ) (1956) 4 – 599. ^{e}Openbare Leeszaal en Boekerij Sint Catharina (1916) 450; Bibliotheek der . . . Hogeschool (1956) 242.

Enschede: ^{b}none. ^{c}nr Almeloosche R. ^{d}Technische Hogeschool Twente (univ) (1961) 3 – 329. ^{e}Openbare Leeszaal en Bibliotheek (1920) 140; Bibliotheek der . . . Hogeschool (1964) 120.

Groningen: ^{b}none. ^{c}Drentse A (Riviertje) Creek; Hunze (Riviertje) Creek. ^{d}Rijksuniversiteit te Groningen (1614) 15 – 1554. ^{e}Bibliotheek der Rijksuniversiteit . . . (1615) 850; Openbare Bibliotheek (1903) 170.

Haarlem: ^{b}none. ^{c}Spaarne R. ^{d}none. ^{e}Stadsbibliotheek en Leeszaal (1596) 218.

*Hague: ^{b}alt The Hague; off 's Gravenhage. ^{c}North Sea. ^{d}none. ^{e}Koninklijke Bibliotheek (1798) 1000; Openbare Bibliotheek (1906) 650; Bibliotheek van het Vredespaleis (1913) 500.

Heerlen: ^{b}none. ^{c}Geleen R. ^{d}none. ^{e}Openbare Leeszaal en Bibliotheek (1913) 38.

Hilversum: ^{b}none. ^{c}nr IJssel(meer) L. ^{d}none. ^{e}Openbare Leeszaal en Bibliotheek (1909) 111.

Leiden: ^{b}alt Leyden. ^{c}Rhine R. ^{d}Rijksuniversiteit te Leiden (1575) 9 – 1683. ^{e}Bibliotheek der Rijksuniversiteit . . . (1575) 2200.

Maastricht: ^{b}for Maestricht. ^{c}Meuse R; Geer R. ^{d}Rijksuniversiteit Limburg (1975) 0.2 – 95. ^{e}Bibliotheek Theologische Hogeschool (1852) 300; Stadsarchief en -bibliotheek (1662) 250.

Nijmegen: ^{b}alt Nimeguen, Nimwegen. ^{c}Waal R (Rhine R distributary). ^{d}Roomsch Katholieke Universiteit (1923) 15 – 1400. ^{e}Bibliotheek van de . . . Universiteit (1923) 500; Openbare Leeszaal en Boekerij (1916) 110.

Rotterdam: ^{b}none. ^{c}Nieuwe Maas R (Rhine R distributary). ^{d}Erasmus Universiteit Rotterdam (1913) 8 – 582. ^{e}Bibliotheek en Leeszalen der Gemeente (1604) 725; Universiteitsbibliotheek (1913) 300.

's Hertogenbosch: ^{b}for Bois-le-Duc. ^{c}Dieze R; Dommel R; Aa R. ^{d}none. ^{e}Bibliotheekcentrale van de Nederlandse Capucijnen (?) 110; Openbare Bibliotheek (1915) 85.

Tilburg: ^{b}none. ^{c}Leij R. ^{d}Katholieke Hogeschool te Tilburg (univ) (1927) 3 – 202. ^{e}Gemeenschappelijke Provinciale Bibliotheekcentrale voor Noord-Brabant (1961) 470; Bibliotheek der . . . Hogeschool (1927) 250; Openbare Bibliotheek (1845) 85.

Utrecht: ^{b}none. ^{c}Rhine R. ^{d}Rijksuniversiteit te Utrecht (1636) 19 – 2337. ^{e}Bibliotheek der Rijksuniversiteit . . . (1584) 1600; Stichting Openbare Bibliotheken (1892) 440.

Velsen: ^{b}none. ^{c}North Sea. ^{d}none. ^{e}Gemeentelijke Openbare Leeszaal en Bibliotheek (1955) 53.

Zaanstad: ^{b}for Zaandam. ^{c}Zaan R. ^{d}none. ^{e}Openbare Bibliotheek (1913) 92.

Norway

Bergen: ^{b}none. ^{c}North Sea. ^{d}Universitetet i Bergen (1948) 8 – 1653. ^{e}Universitetsbiblioteket (1825) 800; Bergens Offentlige Bibliotek (1874) 440.

*Oslo: ^{b}for Christiania, Kristiania. ^{c}Oslo Fjord of Skagerrak Strait. ^{d}Universitetet i Oslo (1811) 20 – 3604. ^{e}Universitetsbiblioteket (1811) 3162; Deichmanske Bibliotek (1785) 1050.

Stavanger: [b]none. [c]Bokn Fjord of North Sea. [d]none. [e]Stavanger Bibliotek (1885) 200.

Trondheim: [b]for Nidaros, Trondhjem. [c]Trondheim Fjord of Norwegian Sea. [d]Universitetet i Trondheim (1900) 7 – 798; Norges Tekniske Hogskole (univ) (1900) 4 – 635. [e]Kongelige Norske Videnskabers Selskabs Bibliotek (1760) 450; Universitetsbiblioteket (1911) 409.

Poland

Bialystok: [b]for Belostok. [c]Suprasl (alt Biala) R. [d]Politechnika Bialostocka (univ) (?) 2 – ? [e]Wojewodzka i Miejska Biblioteka Publiczna (1919) 226; Akademia Medycznej Biblioteka (1950) 189.

Bielsko-Biala: [b]inc Biala, Bielsko; Bielsko: for Bielitz. [c]Biala R. [d]none. [e]Miejska Biblioteka Publiczna (1947) 150. [f]Biala: 23 (1931), 8 (1900), 5 (1851), 3 (1800?); Bielsko: 22 (1931), 17 (1900), 7 (1851), 3 (1800?).

Bydgoszcz: [b]for Bromberg. [c]Brda R. [d]none. [e]Miejska Biblioteka Publiczna (1903) 503.

Bytom: [b]for Beuthen; inc Rossberg since 1927. [c]Bytomka R. [d]none. [e]Miejska Biblioteka Publiczna (1946) 187. [f]Rossberg: 23 (1925), 14 (1900), 3 (1871).

Chorzow: [b]for Konigshutte, Krolewska Huta. [c]Rawa R. [d]none. [e]Miejska Biblioteka Publiczna (1945) 149.

Czestochowa: [b]for Chenstokhov. [c]Warta R; Stradomka R. [d]Politechnika Czestochowska (univ) (1949) 4 – 436. [e]Miejska Biblioteka Publiczna imienia Wladyslawa Bienganskiego (1917) 214; Biblioteka Glowna Politechniki Czestochowskiej (1950) 123.

Elblag: [b]for Elbing. [c]Elblag (for Elbing) R. [d]none. [e]Miejska Biblioteka Publiczna (1945) 182.

Gdansk: [b]for Danzig. [c]G of Danzig of Baltic Sea; Vistula R. [d]Uniwersytet Gdanski (1970) 13 – 1703; Politechnika Gdanska (univ) (1904) 9 – 1158. [e]Biblioteka Gdanska Polskiej Akademii Nauk (1596) 464; Biblioteka Glowna Politechniki Gdanskiej (1945) 390; Biblioteka Glowna Uniwersytet . . . (1970) 370; Wojewodzka i Miejska Biblioteka Publiczna (1945) 311.

Gdynia: [b]none. [c]G of Danzig of Baltic Sea. [d]none. [e]Miejska Biblioteka Publiczna (1950) 355.

Gliwice: [b]for Gleiwitz. [c]Klodnica R. [d]Politechnika Slaska imienia W. Pstrowskiego (univ) (1945) 19 – 1882. [e]Biblioteka Glowna Politechniki Slaskiej (1945) 319; Miejska Biblioteka Publiczna (1946) 173.

Gorzow Wielkopolski: [b]for Landsberg an der Warthe. [c]Warta R. [d]none. [e]Miejska Biblioteka Publiczna (1947) 118.

Jastrzebie Zdroj: [b]none. [c]nr Olza (alt Olse) R. [d]none.

Kalisz: [b]for Kalisch. [c]Prosna R. [d]none. [e]Miejska Biblioteka Publiczna (1927) 104.

Katowice: [b]for Kattowitz, Stalinogrod. [c]Rawa R. [d]Uniwersytet Slaski w Katowicach (1968) 14 – 1197. [e]Biblioteka Slaska (1922) 632; Miejska Biblioteka Publiczna (1945) 231.

Kielce: [b]for Keltsy. [c]nr Czarna Nida R. [d]none. [e]Wojewodzka i Miejska Biblioteka Publiczna (1909) 177.

Krakow: [b]alt Cracow; for Krakau. [c]Vistula R. [d]Uniwersytet Jagiellonski (1364) 16 – 1200; Politechnika w Krakowie (univ) (1835) 10 – 959. [e]Biblioteka Jagiellonska (1364) 1501; Miejska Biblioteka Publiczna (1946) 525; Biblioteka Glowna Akademii Gorniczo-Hutniczej w Krakowie (1919) 468.

Lodz: [b]none. [c]Lodka R. [d]Uniwersytet Lodzki (1945) 16 – 700; Politechnika Lodzka (univ) (1945) 12 – 1307. [e]Biblioteka Glowna Uniwersytet . . . (1945) 1187; Miejska Biblioteka Publiczna imienia L. Warynskiego (1917) 244.

Lublin: [b]for Lyublin. [c]Bystrzyca R. [d]Uniwersytet Marii Curie-Sklodowskiej (1944) 16 – 1244; Katolicki Uniwersytet Lubelski (1918) 2 – 373. [e]Biblioteka Glowna . . . Uniwersytet (1918) 488; Wojewodzka i Miejska Biblioteka Publiczna (1907) 393; Biblioteka Glowna Uniwersytet . . . (1944) 385.

Olsztyn: [b]for Allenstein. [c]Lyna R. [d]none. [e]Wojewodzka i Miejska Biblioteka Publiczna imienia Emilii Sukertowej-Biedrawiny (1946) 177; Biblioteka Glowna Wyzsza Szkola Rolnicza (1950) 170.

Opole: [b]for Oppeln. [c]Oder R. [d]none. [e]Wojewodzka i Miejska Biblioteka Publiczna (1951) 176.

Plock: [b]for Plotsk, Plozk. [c]Vistula R. [d]none. [e]Towarzystwo Naukowe Plockie Biblioteka imienia Zielinskich (1820) 120; Miejska Biblioteka Publiczna (1948) 69.

Poznan: ^{b}for Posen. ^{c}Warta R; Cybina R. ^{d}Uniwersytet imienia Adama Mickiewicza w Poznaniu (1919) 16 – 1421; Politechnika Poznanska (univ) (1915) 8 – 1074. ^{e}Biblioteka Glowna Uniwersytet ... (1902) 1945; Miejska Biblioteka Publiczna imienia Edwarda Raczynskiego (1829) 682.

Radom: ^{b}none. ^{c}Mieczna R. ^{d}none. ^{e}Miejska Biblioteka Publiczna (1922) 150.

Ruda Slaska: ^{b}for Ruda. ^{c}Bytomka R. ^{d}none. ^{e}Miejska Biblioteka Publiczna (1949) 138.

Rybnik: ^{b}for Ribnik. ^{c}Ruda R. ^{d}none. ^{e}Miejska Biblioteka Publiczna (1945) 33; Powiatowa Biblioteka Publiczna (1945) 21.

Rzeszow: ^{b}none. ^{c}Wislok R. ^{d}Politechnika Rzeszowska (univ) (1963) 5 – 350. ^{e}Wojewodzka i Miejska Biblioteka Publiczna (1945) 197; Pedagogiczna Biblioteka Wojewodzka (1945) 132; Biblioteka Glowna Politechnika ... (1963) 110.

Sosnowiec: ^{b}for Sosnovets. ^{c}Czarna Przemsza R. ^{d}none. ^{e}Miejska Biblioteka Publiczna (1922) 177.

Szczecin: ^{b}for Stettin. ^{c}Oder R. ^{d}Politechnika Szczecinska (univ) (1946) 6 – 774. ^{e}Wojewodzka i Miejska Biblioteka Publiczna (1902) 540.

Tarnow: ^{b}none. ^{c}Biala R; Dunajec R. ^{d}none. ^{e}Miejska Biblioteka Publiczna imienia Juliusza Slowackiego (1908) 148.

Torun: ^{b}for Thorn. ^{c}Vistula R. ^{d}Uniwersytet Mikolaja Kopernika w Toruniu (1945) 7 – 800. ^{e}Biblioteka Glowna Uniwersytet ... (1945) 1200; Ksiaznica Miejska imienia M. Kopernika (1923) 388.

Tychy: ^{b}for Tichau. ^{c}nr Gostynka R. ^{d}none. ^{e}Miejska Biblioteka Publiczna (1956) 87.

Walbrzych: ^{b}for Waldenburg, Waldenburg in Schlesien. ^{c}Strzegomka R. ^{d}none. ^{e}Miejska Biblioteka Publiczna (1945) 171.

*Warsaw: ^{b}off Warszawa. ^{c}Vistula R. ^{d}Uniwersytet Warszawski (1808) 26 – 2259; Politechnika Warszawska (univ) (1826) 11 – 2600. ^{e}Biblioteka Narodowa (1928) 2600; Biblioteka Glowna Uniwersytet ... (1817) 1693; Glowna Biblioteka Lekarska (1945) 1500; Biblioteka Publiczna (1907) 679.

Wloclawek: ^{b}for Vlotslavsk. ^{c}Vistula R. ^{d}none. ^{e}Miejska Biblioteka Publiczna (1945) 137.

Wodzislaw Slaski: ^{b}alt Wodzislaw; for Loslau. ^{c}nr Oder R. ^{d}none. ^{e}Powiatowa i Miejska Biblioteka Publiczna (1955) 21.

Wroclaw: ^{b}for Breslau. ^{c}Oder R; Olawa R; Sleza R. ^{d}Uniwersytet Wroclawski imienia Boleslawa Bieruta (1505) 18 – 1454; Politechnika Wroclawska (univ) (1910) 11 – 1835. ^{e}Biblioteka Glowna Uniwersytet ... (1811) 787; Biblioteka Zakladu Narodowego imienia Ossolinskich (1817 at Lvov, USSR; relocated 1946) 785.

Zabrze: ^{b}for Hindenburg, Kunzendorf. ^{c}Bytomka R. ^{d}none. ^{e}Slaska Akademia Medyczna Biblioteka (1948) 70.

Portugal

Coimbra: ^{b}none. ^{c}Mondego R. ^{d}Universidade de Coimbra (1290 at Lisbon, relocated 1537) 10 – 686. ^{e}Biblioteca Geral da Universidade ... (1716) 1050; Biblioteca Municipal (1922) 300.

*Lisbon: ^{b}off Lisboa. ^{c}Tagus R. ^{d}Universidade de Lisboa (1825) 13 – 879; Universidade Tecnica de Lisboa (1759) 13 – 1190; Universidade Nova de Lisboa (?) 0.2 – 79; Universidade Catolica Portuguesa (1971) ? – ? ^{e}Biblioteca Nacional (1796) 1300; Biblioteca da Universidade de Lisboa (1837) 500; Biblioteca da Academia das Ciencias de Lisboa (1779) 400.

Ponta Delgada: ^{a}Azores; Sao Miguel I. ^{b}none. ^{c}Atlantic O. ^{d}none. ^{e}Biblioteca Publica (1845) 130.

Porto: ^{b}alt Oporto. ^{c}Douro R. ^{d}Universidade do Porto (1762) 14 – 1120. ^{e}Biblioteca Publica Municipal (1833) 1400.

Romania

Arad: ^{b}none. ^{c}Maros R. ^{d}none. ^{e}Biblioteca Judeteana Arad (1888) 235.

Bacau: ^{b}none. ^{c}Bistrita R. ^{d}none. ^{e}Biblioteca Judeteana Bacau (1950) 150.

Baia Mare: ^{b}for Nagybanya. ^{c}nr Lapusul R. ^{d}none. ^{e}Biblioteca Judeteana Maramures (1951) 100.

Braila: ^{b}for Ibraila. ^{c}Danube R. ^{d}none. ^{e}Biblioteca Judeteana Braila (1881) 100.

Brasov: ^{b}for Brasso, Kronstadt, Orasul Stalin, Stalin. ^{c}Timis R. ^{d}Universitatea din Brasov (1948) 7 – 571. ^{e}Biblioteca Judeteana Brasov (1926) 350; Biblioteca Centrala Universitara (1949) 274.

*Bucharest: [b]off Bucuresti. [c]Dimbovita R; Colentina R. [d]Universitatea din Bucuresti (1694) 20 – 1500; Institutul Politehnic Gheorghe Gheorghiu-Dej Bucuresti (univ) (1819) 20 – 1389. [e]Biblioteca Academiei Republicii Socialiste Romania (1867) 6624; Biblioteca Centrala de Stat (1955) 5642; Biblioteca Centrala Universitara (1864) 1661.

Buzau: [b]none. [c]Buzau R. [d]none. [e]Biblioteca Judeteana Buzau (1950) 55.

Cluj: [b]for Klausenburg, Kolozsvar. [c]Somesul Mic R; Nadasul R. [d]Universitatea Babes Bolyai (1872) 11 – 900; Institutul Politehnic Cluj-Napoca (univ) (1948) 5 – 429. [e]Biblioteca Centrala Universitara Cluj (1872) 2401; Biblioteca Filialei Cluj a Academiei Republicii Socialiste Romania (1950) 500; Biblioteca Judeteana Cluj (1945) 170.

Constanta: [b]alt Constantsa; for Kustendje. [c]Black Sea. [d]none. [e]Biblioteca Judeteana Constanta (1935) 130.

Craiova: [b]none. [c]Jiu R. [d]Universitatea Craiova (1966) 10 – 570. [e]Biblioteca Judeteana Dolj (1908) 180; Biblioteca Centrala Universitara Craiova (1966) 170.

Galati: [b]for Galatz. [c]Danube R; Brates L. [d]Universitatea din Galati (1948) 5 – 380. [e]Biblioteca Judeteana V. A. Urechia (1890) 303.

Hunedoara: [b]for Eisenmarkt, Vajdahunyad. [c]nr Cerna R. [d]none. [e]Biblioteca Clubului Sindicatului Combinatului Siderurgic Hunedoara (1950) 71; Biblioteca Judeteana Hunedoara (1951) 32.

Iasi: [b]for Jassy. [c]Bahlui R. [d]Institutul Politehnic Gheorghe Asachi din Iasi (univ) (1912) 13 – 1065; Universitatea Alexandru Ioan Cuza (1860) 9 – 651. [e]Biblioteca Centrala Universitara M. Eminescu (1640) 1450; Biblioteca Institutului . . . (1937) 280; Biblioteca Judeteana Gheorghe Asachi (1950) 205; Biblioteca Filialei Iasi a Academiei Republicii Socialiste Romania (1949) 200.

Oradea: [b]alt Oradea Mare; for Grosswardein, Nagy-Varad. [c]Sebes-Koros (alt Crisul-Repede) R. [d]none. [e]Biblioteca Judeteana-Bihor (1882) 260.

Pitesti: [b]none. [c]Arges (alt Argesul) R. [d]none. [e]Biblioteca Judeteana Arges (1950) 130.

Ploiesti: [b]alt Ploesti. [c]nr Teleajen R. [d]none. [e]Biblioteca Judeteana N. Iorga Ploiesti (1921) 138.

Resita: [b]for Resiczabanya. [c]Birzava (alt Brzava) R. [d]none. [e]Biblioteca Casei de Cultura a Sindicatelor (?) 78; Biblioteca Judeteana Caras-Severin (1952) 30.

Satu Mare: [b]for Sathmar, Szatmar-Nemeti. [c]Somes R. [d]none. [e]Biblioteca Judeteana Satu Mare (1951) 45.

Sibiu: [b]for Hermannstadt, Nagyszeben. [c]Cibin R. [d]none. [e]Biblioteca Judeteana Astra Sibiu (1861) 370; Biblioteca Muzeului Brukenthal (1815) 270.

Timisoara: [b]for Temesvar. [c]Bega R. [d]Institutul Politehnic Traian Vuia Timisoara (univ) (1920) 11 – 914; Universitatea din Timisoara (1948) 5 – 432. [e]Biblioteca Judeteana Timis (1904) 300; Biblioteca Centrala Universitara (1948) 228.

Tirgu Mures: [b]alt Targu Mures; for Maros-Vasarhely. [c]Maros R. [d]none. [e]Biblioteca Judeteana Mures (1913) 300; Biblioteca Documentara Teleki-Bolyai (1557) 200.

Spain

Albacete: [a]Murcia. [b]none. [c]Don Juan (alt Balazote) R. [d]none. [e]Biblioteca Publica Provincial (?) 24.

Alicante: [a]Valencia. [b]none. [c]Mediterranean Sea. [d]none. [e]Biblioteca Publica Provincial (?) 24.

Almeria: [a]Andalucia. [b]none. [c]G of Almeria of Mediterranean Sea. [d]none. [e]Biblioteca Publica Provincial (?) 22.

Badajoz: [a]Extremadura. [b]none. [c]Guadiana R. [d]Universidad de Extremadura (at Badajoz and Caceres) (1973?) 2 – 68 [Badajoz campus: (1973?) 1 – ?]. [e]Biblioteca Publica Provincial (?) 21.

Badalona: [a]Cataluna. [b]none. [c]Mediterranean Sea. [d]none.

Baracaldo: [a]Vascongadas y Navarra. [b]alt San Vicente de Baracaldo. [c]Nervion R. [d]none.

Barcelona: [a]Cataluna. [b]none. [c]Mediterranean Sea; Besos R. [d]Universidad de Barcelona (bef 1377) 40 – 2100; Universidad Autonoma de Barcelona (1968) 25 – 1034; Universidad Politecnica de Barcelona (1851) 16 – 1120. [e]Biblioteca Central de la Diputacion Provincial de Barcelona (alt Biblioteca de Cataluna) (1914) 670; Biblioteca Universitaria y Provincial (1837) 400.

Bilbao: [a]Vascongadas y Navarra. [b]none. [c]Nervion R. [d]Universidad de Bilbao (1968) 8 – 514; Universidad de Deusto (1886) 8 – 517. [e]Biblioteca de la Universidad de Deusto (1916) 80; Biblioteca de la Universidad de Bilbao (1968?) 35; Biblioteca Municipal (?) ?; Biblioteca Provincial (?) ?

Burgos: ^{a}Castilla la Vieja. ^{b}none. ^{c}Arlanzon R. ^{d}none. ^{e}Biblioteca del Colegio Maximo de San Francisco (1880) 100; Biblioteca Publica (1871) 60.

Cadiz: ^{a}Andalucia. ^{b}none. ^{c}G of Cadiz of Atlantic O. ^{d}none. ^{e}Biblioteca Popular (?) 40.

Cartagena: ^{a}Murcia. ^{b}none. ^{c}Mediterranean Sea. ^{d}none. ^{e}Biblioteca Popular (?) 30.

Castellon de la Plana: ^{a}Valencia. ^{b}alt Castellon. ^{c}Mediterranean Sea; Seco R. ^{d}none. ^{e}Biblioteca Publica Provincial (?) 13.

Cordova: ^{a}Andalucia. ^{b}off Cordoba. ^{c}Guadalquivir R. ^{d}Universidad de Cordoba (1972) 3 – 108. ^{e}Biblioteca Publica Provincial (?) 60.

Elche: ^{a}Valencia. ^{b}none. ^{c}Vinalapo R. ^{d}none.

Gijon: ^{a}Asturias. ^{b}none. ^{c}B of Biscay. ^{d}none.

Granada: ^{a}Andalucia. ^{b}none. ^{c}Genil R. ^{d}Universidad de Granada (1531) 32 – 1460. ^{e}Biblioteca de la Universidad . . . (1532) 267; Biblioteca Publica (?) ?

Hospitalet: ^{a}Cataluna. ^{b}none. ^{c}nr Llobregat R. ^{d}none.

Huelva: ^{a}Andalucia. ^{b}none. ^{c}Odiel R; Tinto R. ^{d}none. ^{e}Biblioteca Publica Provincial (?) 18.

Jerez de la Frontera: ^{a}Andalucia. ^{b}alt Jerez; for Xeres. ^{c}Guadalete R. ^{d}none. ^{e}Biblioteca, Archivo y Coleccion Arquelogica Municipal (1873) 29.

La Coruna: ^{a}Galicia. ^{b}alt Coruna, Corunna. ^{c}Atlantic O. ^{d}none. ^{e}Biblioteca Publica Provincial (?) 40.

Leon: ^{a}Leon. ^{b}none. ^{c}Bernesga R; Torio R. ^{d}none. ^{e}Biblioteca Publica Provincial la Reina (1844) 28.

Lerida: ^{a}Cataluna. ^{b}none. ^{c}Segre R. ^{d}none. ^{e}Biblioteca Publica Provincial (?) 30.

Logrono: ^{a}Castilla la Vieja. ^{b}none. ^{c}Ebro R. ^{d}none. ^{e}Biblioteca Publica Provincial (?) 30.

*Madrid: ^{a}Castilla la Nueva. ^{b}none. ^{c}Manzanares R. ^{d}Universidad Complutense de Madrid (1498 at Alcala de Henares, relocated 1836) 97 – 4570; Universidad Politecnica de Madrid (1971) 30 – 2800; Universidad Autonoma de Madrid (1968) 9 – 793; Universidad Pontificia Comillas (1890 at Comillas, relocated from 1960) 0.9 – 120. ^{e}Biblioteca Nacional (1712) 2923; Biblioteca de la Universidad Complutense de Madrid (1498 at Alcala de Henares, relocated 1841) 665.

Malaga: ^{a}Andalucia. ^{b}none. ^{c}Mediterranean Sea. ^{d}Universidad de Malaga (1972) 6 – 483. ^{e}Biblioteca de la Universidad . . . (1972?) 30; Biblioteca Publica (1933) 27.

Murcia: ^{a}Murcia. ^{b}none. ^{c}Segura R. ^{d}Universidad de Murcia (1915) 6 – 408. ^{e}Biblioteca de la Universidad . . . (1915) 50; Biblioteca Publica (19th c) 25.

Oviedo: ^{a}Asturias. ^{b}none. ^{c}nr Nalon R; nr Nora R. ^{d}Universidad de Oviedo (1604) 23 – 1090. ^{e}Biblioteca de la Universidad . . . (1608) 170.

Palma: ^{a}Baleares; Majorca I. ^{b}alt Palma de Mallorca. ^{c}Mediterranean Sea. ^{d}none. ^{e}Biblioteca Publica Provincial (?) 80.

Pamplona: ^{a}Vascongadas y Navarra. ^{b}for Pampeluna. ^{c}Arga R. ^{d}Universidad de Navarra (1952) 8 – 755. ^{e}Biblioteca de la Universidad . . . (1952) 309.

Sabadell: ^{a}Cataluna. ^{b}none. ^{c}Ripoll R. ^{d}none. ^{e}Biblioteca del Archivo Historico (1793) 6.

Salamanca: ^{a}Leon. ^{b}none. ^{c}Tormes R. ^{d}Universidad de Salamanca (1218) 10 – 714; Universidad Pontificia de Salamanca (1134) 2 – 147. ^{e}Biblioteca de la Universidad de Salamanca (1254) 230.

San Sebastian: ^{a}Vascongadas y Navarra. ^{b}none. ^{c}B of Biscay; Urumea R. ^{d}none. ^{e}Biblioteca de los Estudios Universitarios y Tecnicos de Guipuzcoa (1956) ?

Santa Coloma de Gramanet: ^{a}Cataluna. ^{b}none. ^{c}Besos R. ^{d}none.

Santander: ^{a}Castilla la Vieja. ^{b}none. ^{c}B of Biscay. ^{d}Universidad de Santander (1972) 1 – 101. ^{e}Biblioteca de Menendez Palayo (1908) 120; Biblioteca Municipal (?) 60.

Saragossa: ^{a}Aragon. ^{b}off Zaragoza. ^{c}Ebro R; Huerva R. ^{d}Universidad de Zaragoza (1474) 23 – 941. ^{e}Biblioteca de la Universidad . . . (1542) 600; Biblioteca Publica (?) 48.

Seville: ^{a}Andalucia. ^{b}off Sevilla. ^{c}Guadalquivir R. ^{d}Universidad de Sevilla (1502) 24 – 1631. ^{e}Biblioteca de la Universidad . . . (1502) 200; Biblioteca Capitular Colombina (1450) 92.

Tarragona: ^{a}Cataluna. ^{b}none. ^{c}Mediterranean Sea; Francoli R. ^{d}none. ^{e}Biblioteca Publica Provincial (?) 65.

Tarrasa: ^{a}Cataluna. ^{b}none. ^{c}nr Ripoll R. ^{d}none. ^{e}Biblioteca de la Escuela Tecnica Superior de Ingenieros Industriales (1962) 3.

Toledo: ^{a}Castilla la Nueva. ^{b}none. ^{c}Tagus R. ^{d}none. ^{e}Biblioteca Provincial de Toledo (1775) 98.

Valencia: ^{a}Valencia. ^{b}none. ^{c}Turia R. ^{d}Universidad de Valencia (1500) 21 – 758; Universidad

Politecnica de Valencia (1968) 6 – 500. ^{e}Biblioteca de la Universidad de Valencia (1500) 300; Biblioteca Publica Provincial (?) 12.

Valladolid: ^{a}Castilla la Vieja. ^{b}none. ^{c}Pisuerga R; Esgueva R. ^{d}Universidad de Valladolid (1346) 10 – 753. ^{e}Biblioteca de la Universidad . . . (1484) 134; Biblioteca Publica Provincial (?) 10.

Vigo: ^{a}Galicia. ^{b}none. ^{c}Vigo B of Atlantic O. ^{d}none. ^{e}Biblioteca Publica Municipal (?) 8.

Vitoria: ^{a}Vascongadas y Navarra. ^{b}none. ^{c}Zapardiel R. ^{d}none. ^{e}Biblioteca Publica Provincial (?) 32.

Sweden

Boras: ^{b}none. ^{c}Viskan R. ^{d}none. ^{e}Stadsbibliotek (1860) 416.

Eskilstuna: ^{b}for Tuna. ^{c}Eskilstunaan R. ^{d}none. ^{e}Stadsbibliotek (1925) 404.

Goteborg: ^{b}alt Gothenburg. ^{c}Kattegat Strait; Gota R. ^{d}Goteborgs Universitet (1891) 22 – 1066; Chalmers Tekniska Hogskola (univ) (1829) 5 – 425. ^{e}Stadsbibliotek (1861) 1418; . . . Universitetsbibliotek (1861) 1250.

Helsingborg: ^{b}alt Halsingborg. ^{c}Oresund Strait. ^{d}none. ^{e}Stadsbibliotek (1866) 220.

Jonkoping: ^{b}none. ^{c}Vattern L; Munksjon L; Rocksjon L. ^{d}none. ^{e}Stadsbibliotek (1916) 370.

Linkoping: ^{b}none. ^{c}Stangan R. ^{d}Universitetet i Linkoping (1967) 6 – 650. ^{e}Stifts- och Landsbiblioteket i Linkoping (1926) 563.

Malmo: ^{b}none. ^{c}Oresund Strait. ^{d}Lunds Universitet (at Lund, nr Malmo) (1666) 22 – 1894. ^{e}. . . Universitetsbibliotek (at Lund, nr Malmo) (1671) 1500; Stadsbibliotek (1905) 896.

Norrkoping: ^{b}none. ^{c}Motala R. ^{d}none. ^{e}Stadsbibliotek (1913) 275.

Orebro: ^{b}none. ^{c}Svartan R. ^{d}Hogskolan i Orebro (univ) (1967) 5 – 190. ^{e}Stadsbibliotek (1862) 315.

*Stockholm: ^{b}none. ^{c}Baltic Sea; Malaren L. ^{d}Stockholms Universitet (1877) 24 – 900; Kungliga Tekniska Hogskolan (univ) (1827) 7 – 1500. ^{e}Stadsbibliotek (1927) 1642; Kungliga Biblioteket (17th c) 1000; . . . Universitetsbibliotek (1903) 700.

Sundsvall: ^{b}none. ^{c}G of Bothnia; Selangeran R. ^{d}none. ^{e}Stadsbibliotek (1894) 110.

Uppsala: ^{b}alt Upsala. ^{c}Fyris R. ^{d}Uppsala Universitet (1477) 16 – 1258; Sveriges Lantbruksuniversitet (1977?) 2 – 277. ^{e}. . . Universitetsbibliotek (1620) 2000; Stadsbibliotek (1906) 501.

Vasteras: ^{b}for Vesteras. ^{c}Svart R; Malaren L. ^{d}none. ^{e}Stifts- och Landsbiblioteket i Vasteras (1952) 535.

Switzerland

Basel: ^{b}alt Bale, Basilea. ^{c}Rhine R; Birs R; Wiese R. ^{d}Universitat Basel (1460) 5 – 500. ^{e}Offentliche Bibliothek der Universitat . . . (1460) 2131; Schweizerisches Wirtschaftsarchiv (alt Archives Economiques Suisses) (1910) 750.

*Bern: ^{b}alt Berna, Berne. ^{c}Aar R. ^{d}Universitat Bern (1528) 7 – 662. ^{e}Schweizerische Landesbibliothek (alt Bibliotheque Nationale Suisse) (1895) 1500; Stadt- und Universistatsbibliothek (1528) 1100.

Biel: ^{b}alt Bienne. ^{c}Schuss (alt Suze) R; Bieler (alt Bienne) L. ^{d}none. ^{e}Stadtbibliothek (1926) 99.

Geneva: ^{b}off Geneve, Genf, Ginevra. ^{c}Rhone R; Geneva L. ^{d}Universite de Geneve (1559) 8 – 777. ^{e}Bibliotheque Publique et Universitaire (1561) 1200; Bibliotheque des Nations Unies (alt Library of the United Nations) (1920) 800; International Labour Office Library (1920) 660.

Lausanne: ^{b}alt Losanna. ^{c}Geneva L. ^{d}Universite de Lausanne (1537) 5 – 435; Ecole Polytechnique Federale de Lausanne (univ) (1853) 2 – 151. ^{e}Bibliotheque Cantonale et Universitaire (1537) 650.

Luzern: ^{b}alt Lucerna, Lucerne. ^{c}Reuss R; Lucerne L. ^{d}none. ^{e}Zentralbibliothek (1812) 400.

Saint-Gall: ^{b}alt San Gallo, Sankt Gallen. ^{c}nr Constance L. ^{d}none. ^{e}Stadtbibliothek Vadiana (1551) 400; Stiftsbibliothek Sankt Gallen (720?) 100.

Winterthur: ^{b}none. ^{c}Toss R. ^{d}none. ^{e}Stadtbibliothek (1660) 500.

Zurich: ^{b}alt Zurigo. ^{c}Limmat R; Sihl R; Zurich(see) L. ^{d}Universitat Zurich (1523) 13 – 1236; Eidgenossische Technische Hochschule (univ) (1854) 7 – 691. ^{e}Bibliothek der . . . Hochschule (1855) 1800; Zentralbibliothek (1629) 1600; Universitatsbibliothek (1523?) 1400.

Turkey

Edirne: ^{b}for Adrianople. ^{c}Maritsa R. ^{d}none. ^{e}Merkez Selimiye Kutuphane (1575) 32.

Istanbul: ^{b}for Constantinople. ^{c}Sea of Marmara; Bosporus Strait. ^{d}Istanbul Universitesi (1453) 33 − 1659; Istanbul Teknik Universitesi (1773) 8 − 789; Bogazici Universitesi (1863) 3 − 250. ^{e}Istanbul Universitesi Kutuphane (1453?) 260; Bogazici Universitesi Kutuphane (1863) 150; Merkez Beyazit Genel Kutuphane (1882) 118.

UK (see note 1, p. 361)

Aberdeen: ^{a}Scotland. ^{b}none. ^{c}North Sea; Dee R; Don R. ^{d}University of Aberdeen (1494) 6 − 817. ^{e}University . . . Library (1495) 650; Public Library (1885) 482.

Barnsley: ^{a}England. ^{b}none. ^{c}Dearne R. ^{d}none. ^{e}Public Library (1890) 432.

Basildon: ^{a}England. ^{b}for Billericay. ^{c}nr Thames R. ^{d}none. ^{e}Basildon Branch of Essex County Library (?) ?

Bath: ^{a}England. ^{b}none. ^{c}Avon R. ^{d}University of Bath (1856) 3 − 312. ^{e}Wiltshire County Library (at West Wiltshire, nr Bath) (1923) 994; Bath Branch of Avon County Library (for Bath Public Library) (1900) 219; University . . . Library (1960) 100.

Belfast: ^{a}*Northern Ireland. ^{b}none. ^{c}Belfast (Lough) Inlet of North Channel; Lagan R. ^{d}Queen's University of Belfast (1845) 6 − 724. ^{e}Public Library (1888) 931; . . . University Library (1848) 688.

Beverley: ^{a}England. ^{b}inc Haltemprice since 1974. ^{c}Hull R. ^{d}none. ^{e}Beverley Branch of Humberside County Library (for Beverley Public Library) (1906) 44. ^{f}Haltemprice: 52 (1971), 42 (1961), 36 (1951), 17s (1931).

Birmingham: ^{a}England. ^{b}inc Sutton Coldfield since 1974. ^{c}Tame R; Rea R. ^{d}University of Birmingham (1880) 8 − 1456; University of Aston in Birmingham (1895) 4 − 455. ^{e}Public Library (1861) 2562; University of Birmingham Library (1880) 950. ^{f}Sutton Coldfield: 83 (1971), 72 (1961), 48 (1951), 30 (1931), 14 (1901), 5s (1851), 3s (1801).

Blackburn: ^{a}England. ^{b}none. ^{c}Darwen R. ^{d}none. ^{e}Blackburn Branch of Lancashire County Library (for Blackburn Public Library) (1862) 248.

Blackpool: ^{a}England. ^{b}none. ^{c}Irish Sea. ^{d}none. ^{e}Blackpool Branch of Lancashire County Library (for Blackpool Public Library) (1880) 206.

Bolton: ^{a}England. ^{b}none. ^{c}Croal R. ^{d}none. ^{e}Public Library (1853) 549.

Bournemouth: ^{a}England. ^{b}none. ^{c}English Channel. ^{d}none. ^{e}Bournemouth Branch of Dorset County Library (for Bournemouth Public Library) (1895) 287.

Bradford: ^{a}England. ^{b}inc Keighley since 1974. ^{c}Bradford (Beck) Creek. ^{d}University of Bradford (1957) 4 − 463. ^{e}Public Library (1872) 1041; University . . . Library (1966) 230. ^{f}Keighley: 55 (1971), 56 (1961), 57 (1951), 40 (1931), 42 (1901), 13 (1851), 6s (1801).

Brighton: ^{a}England. ^{b}none. ^{c}English Channel. ^{d}University of Sussex (1961) 4 − 565. ^{e}East Sussex County Library (at Lewes, nr Brighton) (1924) 1686; University . . . Library (1961) 400; Brighton Branch of East Sussex County Library (for Brighton Public Library) (1873) 350.

Bristol: ^{a}England. ^{b}none. ^{c}Avon R; Frome R. ^{d}University of Bristol (1876) 8 − 918. ^{e}Avon County Library (1974) 1609, inc Bristol Branch (for Bristol Public Library) (1876) 820; University . . . Library (1923) 630.

Burnley: ^{a}England. ^{b}none. ^{c}Calder R; Burn R. ^{d}none. ^{e}Burnley Branch of Lancashire County Library (for Burnley Public Library) (1914) 271.

Bury: ^{a}England. ^{b}inc Prestwich, Radcliffe since 1974. ^{c}Irwell R. ^{d}none. ^{e}Public Library (1901) 478. ^{f}Prestwich: 33 (1971), 34 (1961), 34 (1951), 24 (1931), 13 (1901), 4s (1851), 2s (1801); Radcliffe: 29 (1971), 27 (1961), 28 (1951), 25 (1931), 25 (1901), 5 (1851), 2s (1801).

Calderdale: ^{a}England. ^{b}for Halifax. ^{c}Hebble R. ^{d}none. ^{e}Public Library (1882) 556.

Cambridge: ^{a}England. ^{b}none. ^{c}Cam R. ^{d}University of Cambridge (1209?) 12 − 1256. ^{e}University . . . Library (1400) 3000; Cambridge Branch of Cambridgeshire County Library (for Cambridge Public Library) (1855) 245.

Canterbury: ^{a}England. ^{b}none. ^{c}Stour R. ^{d}University of Kent at Canterbury (1964) 3 − 374. ^{e}University ... Library (1964) 278; Canterbury Branch of Kent County Library (for Canterbury Public Library) (1847) 66.

Cardiff: [a]*Wales. [b]alt Caerdydd. [c]Bristol Channel; Taff R. [d]University of Wales (1822) 17 – 2240 [Cardiff campus: (1866) 7 – 931]. [e]University ... Library (1872) 1560 [Cardiff campus: (1883) 470]; South Glamorgan County Library (1923) 1112, inc Cardiff Branch (for Cardiff Public Library) (1862) 779.

Carlisle: [a]England. [b]none. [c]Eden R. [d]none. [e]Cumbria County Library (1921) 1013, inc Carlisle Branch (for Carlisle Public Library) (1893) 174.

Charnwood: [a]England. [b]for Loughborough. [c]Soar R. [d]Charnwood University of Technology (1952) 4 – 366. [e]Charnwood Branch of Leicestershire County Library (for Loughborough Public Library) (1886) 115.

Cheltenham: [a]England. [b]none. [c]Chelt R. [d]none. [e]Cheltenham Branch of Gloucestershire County Library (for Cheltenham Public Library) (1884) 190.

Chester: [a]England. [b]none. [c]Dee R. [d]none. [e]Cheshire County Library (1922) 1637, inc Chester Branch (for Chester Public Library) (1877) 206.

Chesterfield: [a]England. [b]none. [c]Rother R; Hipper R. [d]none. [e]Chesterfield Branch of Derbyshire County Library (for Chesterfield Public Library) (1879) 105.

Chichester: [a]England. [b]none. [c]Rother R. [d]none. [e]West Sussex County Library (1925) 1040.

Colchester: [a]England. [b]none. [c]Colne R. [d]University of Essex (1961) 2 – 238. [e]University ... Library (1964) 230; Colchester Branch of Essex County Library (for Colchester Public Library) (1892) 157.

Coventry: [a]England. [b]none. [c]Sherbourne R. [d]University of Warwick (1965) 3 – 375. [e]Warwickshire County Library (at Warwick, nr Coventry) (1920) 1068; Public Library (1868) 353; University ... Library (1963) 300.

Crewe and Nantwich: [a]England. [b]inc Crewe, Nantwich. [c]Weaver R. [d]none. [e]Crewe and Nantwich Branch of Cheshire County Library [for Crewe Public Library (1936) 110, and Nantwich Public Library (1889) ?]. [f]Crewe: 51 (1971), 53 (1961), 52 (1951), 46 (1931), 42 (1901), 4 (1851), 0.3s (1801); Nantwich: 12 (1971), 10 (1961), 9 (1951), 7 (1931), 8 (1901), 5 (1851), 3 (1801).

Darlington: [a]England. [b]none. [c]Skerne R. [d]none. [e]Darlington Branch of Durham County Library (for Darlington Public Library) (1885) 194.

Derby: [a]England. [b]none. [c]Derwent R. [d]none. [e]Derby Branch of Derbyshire County Library (for Derby Public Library) (1871) 342.

Doncaster: [a]England. [b]none. [c]Don R. [d]none. [e]Public Library (1869) 623.

Dudley: [a]England. [b]inc Brierley Hill since 1966, Stourbridge since 1974, Thalesowen since 1974. [c]Stour R. [d]none. [e]Public Library (1884) 698. [f]Brierley Hill: 56 (1961), 49 (1951), 45 (1931), 12 (1901), 12 (1881); Stourbridge: 54 (1971), 43 (1961), 37 (1951), 20 (1931) 16 (1901), 8 (1851), 3 (1801); Thalesowen: 54 (1971), 44 (1961), 40 (1951), 31 (1931), 4s (1901).

Dundee: [a]Scotland. [b]none. [c]Tay R. [d]University of Dundee (1881) 3 – 469; University of Saint Andrews (at Saint Andrews, nr Dundee) (1410) 3 – 303. [e]University of Saint Andrews Library (at Saint Andrews, nr Dundee) (1456) 700; Public Library (1869) 507; University of Dundee Library (1883) 300.

Durham: [a]England. [b]none. [c]Wear R. [d]University of Durham (1832) 4 – 481. [e]Durham County Library (1924) 1338; University ... Library (1833) 460.

Edinburgh: [a]*Scotland. [b]inc Leith since 1920. [c]Forth R. [d]University of Edinburgh (1583) 11 – 1462; Heriot-Watt University (1821) 3 – 248. [e]National Library of Scotland (1682) 3000; Public Library (1890) 1193; University ... Library (1580) 1150. [f]Leith: 80 (1911), 77 (1901), 31 (1851), 15s (1801).

Elmbridge: [a]England. [b]inc Esher, Walton-on-Thames, Weybridge. [c]Thames R; Mole R; Wey R. [d]none. [e]Surrey County Library (1925) 1871. [f]Esher: 64 (1971), 61 (1961), 51 (1951), 17 (1931), 9 (1901), 1s (1851), 0.8s (1801); Walton and Weybridge: 51 (1971), 46 (1961), 38 (1951); Walton-on-Thames: 18 (1931), 10 (1901), 3s (1851), 1s (1801); Weybridge: 7 (1931), 5 (1901), 1s (1851), 0.7s (1801).

Erewash: ^aEngland. ^bfor Ilkeston. ^cErewash R. ^dnone. ^eErewash Branch of Derbyshire County Library (for Ilkeston Public Library) (1901) 59.

Exeter: ^aEngland. ^bnone. ^cExe R. ^dUniversity of Exeter (1855) 4 – 462. ^eDevon County Library (1924) 1931, inc Exeter Branch (for Exeter Public Library) (1870) 360.

Fareham: ^aEngland. ^bnone. ^cPortsmouth Harbor of English Channel. ^dnone. ^eFareham Branch of Hampshire County Library (?) ?

Gateshead: ^aEngland. ^bnone. ^cTyne R. ^dnone. ^ePublic Library (1885) 390.

Gedling: ^aEngland. ^binc Arnold, Carlton. ^cTrent R. ^dnone. ^eGedling Branch of Nottinghamshire County Library [for Carlton Public Library (1888) 83, and Arnold Public Library (1906) 67]. ^fArnold: 33 (1971), 27 (1961), 21 (1951), 14 (1931), 9 (1901), 5s (1851) 3s (1801); Carlton: 45 (1971), 39 (1961), 34 (1951), 22 (1931), 10 (1901), 2 (1851), 0.8 (1801).

Gillingham: ^aEngland. ^bnone. ^cMedway R. ^dnone. ^eGillingham Branch of Kent County Library (for Gillingham Public Library) (1952) 149.

Glasgow: ^aScotland. ^bnone. ^cClyde R; Kelvin R. ^dUniversity of Glasgow (1451) 10 – 1399; University of Strathclyde (1796) 7 – 765. ^ePublic Library (1877) 2051; University of Glasgow Library (1577) 1165.

Gloucester: ^aEngland. ^bnone. ^cSevern R. ^dnone. ^eGloucestershire County Library (1917) 1147, inc Gloucester Branch (for Gloucester Public Library) (1897) 251.

Gosport: ^aEngland. ^bnone. ^cSolent; Portsmouth Harbor of English Channel. ^dnone. ^eGosport Branch of Hampshire County Library (for Gosport Public Library) (1890) 139.

Gravesham: ^aEngland. ^bfor Gravesend. ^cThames R. ^dnone. ^eGravesham Branch of Kent County Library (for Gravesend Public Library) (1894) 99.

Grimsby: ^aEngland. ^bfor Great Grimsby. ^cHumber R. ^dnone. ^eGrimsby Branch of Humberside County Library (for Grimsby Public Library) (1901) 191.

Guildford: ^aEngland. ^bnone. ^cWey R. ^dUniversity of Surrey (1891) 3 – 306. ^eUniversity ... Library (1894) 164; Guildford Branch of Surrey County Library (for Guildford Public Library) (1924) 130.

Halton: ^aEngland. ^binc Runcorn, Widnes. ^cMersey R. ^dnone. ^eHalton Branch of Cheshire County Library [for Widnes Public Library (1887) 114, and Runcorn Public Library (1882) 63]. ^fRuncorn: 36 (1971) 26 (1961), 24 (1951), 18 (1931), 16 (1901), 8 (1851), 1s (1801); Widnes: 57 (1971), 52 (1961), 49 (1951), 41 (1931), 29 (1901), 3s (1851), 1s (1801).

Harrogate: ^aEngland. ^bnone. ^cUre R. ^dnone. ^eHarrogate Branch of North Yorkshire County Library (for Harrogate Public Library) (1887) 119.

Hartlepool: ^aEngland. ^binc West Hartlepool since 1967. ^cNorth Sea. ^dnone. ^eHartlepool Branch of Cleveland County Library (for Hartlepool Public Library) (1894) 162. ^fWest Hartlepool: 77 (1961), 73 (1951), 69 (1931), 63 (1901).

Havant: ^aEngland. ^bfor Havant and Waterloo. ^cnr English Channel. ^dnone. ^eHavant Branch of Hampshire County Library (?) ?

Hove: ^aEngland. ^bnone. ^cEnglish Channel. ^dnone. ^eHove Branch of East Sussex County Library (for Hove Public Library) (1891) 148.

Hull: ^aEngland. ^boff Kingston upon Hull. ^cHumber R; Hull R. ^dUniversity of Hull (1927) 4 – 528. ^eHumberside County Library (1925) 2044, inc Hull Branch (for Hull Public Library) (1893) 709; Brynmor Jones Library of the University ... (1928) 475.

Inverclyde: ^aScotland. ^bfor Greenock. ^cClyde R. ^dnone. ^ePublic Library (1902) 143.

Ipswich: ^aEngland. ^bnone. ^cOrwell R; Gipping R. ^dnone. ^eSuffolk County Library (1925) 872, inc Ipswich Branch (for Ipswich Public Library) (1853) 269.

Kirklees: ^aEngland. ^binc Dewsbury, Huddersfield. ^cCalder R; Colne R. ^dnone. ^ePublic Library (1889) 934. ^fDewsbury: 51 (1971), 53 (1961), 53 (1951), 54 (1931), 28 (1901), 5 (1851), 5s (1801); Huddersfield: 131 (1971), 131 (1961), 129 (1951), 113 (1931), 95 (1901), 31 (1851), 7 (1801).

Knowsley: ^aEngland. ^binc Huyton-with-Roby, Kirkby. ^cAlt R. ^dnone. ^ePublic Library (1974?) 255. ^fHuyton-with-Roby: 67 (1971), 63 (1961), 56 (1951), 5 (1931), 5 (1901), 3s (1851), 2s (1801); Kirkby: 60 (1971), 52 (1961), 3s (1951), 1s (1931), 1s (1901).

Lancaster: [a]England. [b]inc Morecambe and Heysham since 1974. [c]Morecambe B of Irish Sea; Lune R. [d]University of Lancaster (1964) 4 – 446. [e]University . . . Library (1963) 350; Lancaster Branch of Lancashire County Library (for Lancaster Public Library) (1893) 109. [f]Morecambe and Heysham: 42 (1971), 40 (1961), 37 (1951), 25 (1931); Heysham: 3 (1901), 0.6s (1851), 0.4s (1801); Morecambe: 12 (1901).

Langbaurgh: [a]England. [b]for Teesside (in part); inc Eston, Redcar. [c]North Sea; Tees R. [d]none. [e]Langbaurgh Branch of Cleveland County Library (for Redcar Public Library) (1937) 62 (in 1967). [f]Eston: 37 (1961), 33 (1951), 31 (1931), 11 (1901), 0.5s (1851), 0.3s (1801); Redcar: 31 (1961), 28 (1951), 20 (1931), 8 (1901), 1s (1851), 0.4s (1801).

Leeds: [a]England. [b]none. [c]Aire R. [d]University of Leeds (1831) 9 – 1403. [e]Public Library (1870) 1516; Brotherton Library (alt University . . . Library) (1874) 1166.

Leicester: [a]England. [b]none. [c]Soar R. [d]University of Leicester (1918) 4 – 394. [e]Leicestershire County Library (1924) 1667, inc Leicester Branch (for Leicester Public Library) (1871) 420; University . . . Library (1921) 430.

Lichfield: [a]England. [b]none. [c]nr Trent R. [d]none. [e]Lichfield Branch of Staffordshire County Library (for Lichfield Public Library) (1859) 50.

Lincoln: [a]England. [b]none. [c]Witham R; Till R. [d]none. [e]Lincolnshire County Library (1924) 1128, inc Lincoln Branch (for Lincoln Public Library) (1895) 210.

Liverpool: [a]England. [b]none. [c]Mersey R. [d]University of Liverpool (1881) 7 – 944. [e]Public Library (1852) 2498; University . . . Library (1881) 865.

*London: [a]*England. [b]none. [c]Thames R; Lea R. [d]University of London (13th c) 82 – 6630. Brunel University (1957) 4 – 283; City University (1891) 2 – 323; Royal College of Arts (univ) (1837) 0.6 – 123. [e]British Library (inc British Museum Library) (1753) 12,150; Westminster Public Library (1857) 1270; Wandsworth Public Library (1885) 1039; University of London Library (1838) 1000.

Londonderry: [a]Northern Ireland. [b]alt Derry. [c]Foyle R. [d]none. [e]Londonderry Branch of Western Education and Library Board (for Londonderry Public Library) (1923) 185.

Luton: [a]England. [b]none. [c]Lea R. [d]none. [e]Luton Branch of Bedfordshire County Library (for Luton Public Library) (1910) 326.

Macclesfield: [a]England. [b]none. [c]Bollin R. [d]none. [e]Macclesfield Branch of Cheshire County Library (for Macclesfield Public Library) (1876) 99.

Maidstone: [a]England. [b]none. [c]Medway R. [d]none. [e]Kent County Library (1921) 3286, inc Maidstone Branch (for Maidstone Public Library) (1858) 108.

Manchester: [a]England. [b]none. [c]Irwell R. [d]Victoria University of Manchester (1851) 17 – 2107. [e]John Rylands University Library of Manchester (1851) 2500; Public Library (1852) 1927.

Medway: [a]England. [b]inc Chatham, Rochester. [c]Medway R; Luton R. [d]none. [e]Medway Branch of Kent County Library [for Chatham Public Library (1903) 149, and Rochester Public Library (1894) 113]. [f]Chatham: 57 (1971), 49 (1961), 44 (1951), 43 (1931), 37 (1901), 28 (1851), 11 (1801); Rochester: 56 (1971), 50 (1961), 44 (1951), 31 (1931), 31 (1901), 15 (1851), 7 (1801).

Middlesbrough: [a]England. [b]for Teesside (in part). [c]Tees R. [d]none. [e]Cleveland County Library (1974) 1101, inc Middlesbrough Branch (for Middlesbrough Public Library) (1871) 365 (in 1967). [f]Teesside: 396 (1971).

Motherwell: [a]Scotland. [b]for Motherwell and Wishaw. [c]nr Tyne R. [d]none. [e]Public Library (1906) 106. [f]Motherwell: 30 (1901), 1 (1851), 0.6s (1801); Wishaw: 21 (1901), 3 (1851), 2s (1801).

Newcastle-under-Lyme: [a]England. [b]none. [c]Lyme Brook. [d]University of Keele (1949) 3 – 311. [e]University . . . Library (1949) 359; Newcastle-under-Lyme Branch of Staffordshire County Library (for Newcastle-under-Lyme Public Library) (1891) 161.

Newcastle upon Tyne: [a]England. [b]alt Newcastle. [c]Tyne R. [d]University of Newcastle upon Tyne (1834) 7 – 993. [e]Public Library (1880) 988; University . . . Library (1871) 475.

Newport: [a]Wales. [b]alt Casnewydd-ar-Wysg. [c]Usk (alt Wysg) R. [d]none. [e]Gwent County Library (1927) 791, inc Newport Branch (for Newport Public Library) (1870) 198.

Northampton: [a]England. [b]none. [c]Nene R. [d]none. [e]Northamptonshire County Library (1926) 1057, inc Northampton Branch (for Northampton Public Library) (1876) 232.

North Bedfordshire: ^{a}England. ^{b}for Bedford. ^{c}Ouse R. ^{d}none. ^{e}Bedfordshire County Library (1925) 1087, inc North Bedfordshire Branch (for Bedford Public Library) (1937) 166.

North Tyneside: ^{a}England. ^{b}inc Longbenton, Tynemouth, Wallsend. ^{c}North Sea; Tyne R. ^{d}none. ^{e}Public Library (1869) 470. ^{f}Longbenton: 49 (1971), 47 (1961), 28 (1951), 14 (1931), 7s (1901), 2s (1851), 3s (1801); Tynemouth; 69 (1971), 70 (1961), 67 (1951), 65 (1931), 51 (1901), 29 (1851), 4s (1801); Wallsend: 46 (1971), 50 (1961), 49 (1951), 45 (1931), 21 (1901), 2s (1851), 1s (1801).

Norwich: ^{a}England. ^{b}none. ^{c}Wensum R. ^{d}University of East Anglia (1961) 3 – 321. ^{e}Norfolk County Library (1925) 1406, inc Norwich Branch (for Norwich Public Library) (1857) 397; University . . . Library (1962) 300.

Nottingham: ^{a}England. ^{b}none. ^{c}Trent R. ^{d}University of Nottingham (1881) 6 – 796. ^{e}Nottingham Branch of Nottinghamshire County Library (for Nottingham Public Library) (1868) 494; University . . . Library (1881) 480.

Nuneaton: ^{a}England. ^{b}inc Bedworth since 1974. ^{c}Anker R. ^{d}none. ^{e}Nuneaton Branch of Warwickshire County Library (for Nuneaton Public Library) (1895) 135. ^{f}Bedworth: 41 (1971), 33 (1961), 25 (1951), 12 (1931), 7s (1901), 3 (1851), 3s (1801).

Ogwr: ^{a}Wales. ^{b}inc Bridgend, Maesteg, Ogmore and Garw, Porthcawl. ^{c}Bristol Channel; Ogmore (alt Ogwr) R. ^{d}none. ^{e}Mid Glamorgan County Library (1923) 618, inc Ogwr Branch (for Bridgend Public Library) (1901) ? ^{f}Bridgend: 15 (1971), 15 (1961), 14 (1951), 10 (1931), 6 (1901), 9s (1851), 5s (1801); Maesteg: 21 (1971), 22 (1961),23 (1951), 26 (1931), 15 (1901), 7s (1851), 3s (1801); Ogmore and Garw: 19 (1971), 21 (1961), 23 (1951). 27 (1931), 20 (1901); Porthcawl: 14 (1971), 11 (1961), 10 (1951), 6 (1931), 2 (1901).

Oldham: ^{a}England. ^{b}none. ^{c}Medlock R. ^{d}none. ^{e}Public Library (1883) 390.

Oxford: ^{a}England. ^{b}none. ^{c}Thames R; Cherwell R. ^{d}University of Oxford (1200?) 11 – 1713. ^{e}Bodleian Library (alt University . . . Library) (1602) 3500; Oxfordshire County Library (1924) 864, inc Oxford Branch (for Oxford Public Library) (1854) 242.

Peterborough: ^{a}England. ^{b}none. ^{c}Nene R. ^{d}none. ^{e}Peterborough Branch of Cambridgeshire County Library (for Peterborough Public Library) (1892) 145.

Plymouth: ^{a}England. ^{b}none. ^{c}Plymouth Sound of English Channel; Tamar R; Plym R. ^{d}none. ^{e}Plymouth Branch of Devon County Library (for Plymouth Public Library) (1876) 408.

Poole: ^{a}England. ^{b}none. ^{c}English Channel. ^{d}none. ^{e}Poole Branch of Dorset County Library (for Poole Public Library) (1886) 242.

Portsmouth: ^{a}England; Portsea I. ^{b}none. ^{c}English Channel; Portsmouth Harbor of English Channel. ^{d}none. ^{e}Portsmouth Branch of Hampshire County Library (for Portsmouth Public Library) (1883) 299.

Preston: ^{a}England. ^{b}none. ^{c}Ribble R. ^{d}none. ^{e}Lancashire County Library (1925) 3191, inc Preston Branch (for Preston Public Library) (1879) 275.

Reading: ^{a}England. ^{b}none. ^{c}Thames R; Kennet R. ^{d}University of Reading (1892) 6 – 640. ^{e}Berkshire County Library (1924) 1048, inc Reading Branch (for Reading Public Library) (1883) 236; University . . . Library (1893) 500.

Reigate and Banstead: ^{a}England. ^{b}inc Banstead, Reigate. ^{c}nr Mole R. ^{d}none. ^{e}Reigate and Banstead Branch of Surrey County Library (?) ? ^{f}Banstead: 45 (1971), 42 (1961), 34 (1951), 11s (1931), 6s (1901); Reigate: 56 (1971), 54 (1961), 42 (1951), 31 (1931), 26 (1901), 5 (1851), 0.9 (1801).

Renfrew: ^{a}Scotland. ^{b}inc Barrhead, Johnstone. Paisley, Renfrew. ^{c}Clyde R; White Cart R; Black Cart R. ^{d}none. ^{e}Public Library (1871) 343. ^{f}Barrhead: 18 (1971), 14 (1961), 13 (1951), 12 (1931), 10 (1901), 6 (1851); Johnstone: 23 (1971), 18 (1961), 16 (1951), 13 (1931), 11 (1901), 6 (1851); Paisley: 95 (1971), 96 (1961), 94 (1951), 86 (1931), 79 (1901), 32 (1851), 17 (1801), 4 (1753); Renfrew: 19 (1971), 18 (1961), 17 (1951), 15 (1931), 9 (1901), 3 (1851), 2s (1801).

Rhondda: ^{a}Wales. ^{b}for Ystradyfodwg. ^{c}Rhondda Fawr R; Rhondda Fach R. ^{d}none. ^{e}Public Library (1939) 173.

Rochdale: ^{a}England. ^{b}inc Middleton since 1974. ^{c}Roch R. ^{d}none. ^{e}Public Library (1872) 458. ^{f}Middleton: 53 (1971), 57 (1961), 33 (1951), 29 (1931), 25 (1901), 6 (1851), 3s (1801).

Rotherham: ^{a}England. ^{b}none. ^{c}Don R; Rother R. ^{d}none. ^{e}Public Library (1880) 472.

Rugby: ^{a}England. ^{b}none. ^{c}Avon R. ^{d}none. ^{e}Rugby Branch of Warwickshire County Library (for Rugby Public Library) (1891) 108.

Rushcliffe: ^{a}England. ^{b}for West Bridgford. ^{c}Trent R. ^{d}none. ^{e}Nottinghamshire County Library (1924) 2314.

Saint Albans: ^{a}England. ^{b}none. ^{c}Ver R. ^{d}none. ^{e}Hertfordshire County Library (at East Hertfordshire, nr Saint Albans) (1925) 1940; Saint Albans Branch of Hertfordshire County Library (for Saint Albans Public Library) (1882) 113.

Saint Helens: ^{a}England. ^{b}none. ^{c}Sankey Brook. ^{d}none. ^{e}Public Library (1872) 299.

Salford: ^{a}England. ^{b}inc Worsley since 1974. ^{c}Irwell R. ^{d}University of Salford (1896) 4 – 481. ^{e}Public Library (1850) 533; University ... Library (1957) 180. ^{f}Worsley: 50 (1971), 40 (1961), 27 (1951), 15 (1931), 12 (1901), 10s (1851), 5s (1801).

Salisbury: ^{a}England. ^{b}off New Sarum. ^{c}Avon R; Wiley R. ^{d}none. ^{e}Salisbury Branch of Wiltshire County Library (for Salisbury Public Library) (1890) 69.

Sandwell: ^{a}England. ^{b}inc Warley (inc Oldbury, Rowley Regis, Smethwick), West Bromwich. ^{c}Tame R. ^{d}none. ^{e}Public Library (1874) 788. ^{f}Oldbury: 54 (1961), 54 (1951), 36 (1931), 25 (1901); Rowley Regis: 48 (1961), 49 (1951), 41 (1931), 35 (1901); Smethwick: 68 (1961), 76 (1951), 84 (1931), 55 (1901), 8 (1851), 1 (1801); Warley: 164 (1971); West Bromwich: 167 (1971), 96 (1961), 88 (1951), 81 (1931), 65 (1901), 35s (1851), 6s (1801).

Scarborough: ^{a}England. ^{b}none. ^{c}North Sea; Esk R. ^{d}none. ^{e}Scarborough Branch of North Yorkshire County Library (for Scarborough Public Library) (1930) 69.

Sefton: ^{a}England. ^{b}inc Bootle, Crosby, Southport. ^{c}Irish Sea; Mersey R. ^{d}none. ^{e}Public Library (1876) 628. ^{f}Bootle: 74 (1971), 83 (1961), 75 (1951), 77 (1931), 59 (1901), 4s (1851), 0.5s (1801); Crosby: 57 (1971), 59 (1961), 58 (1951), 19 (1931), 8 (1901), 3s (1851), 0.7s (1801); Southport: 84 (1971), 82 (1961), 84 (1951), 79 (1931), 48 (1901), 5 (1851), 2s (1801).

Sheffield: ^{a}England. ^{b}none. ^{c}Don R. ^{d}University of Sheffield (1828) 7 – 824. ^{e}Public Library (1856) 1199; University ... Library (1897) 600.

Slough: ^{a}England. ^{b}none. ^{c}nr Thames R. ^{d}none. ^{e}Slough Branch of Berkshire County Library (?) ?

Solihull: ^{a}England. ^{b}none. ^{c}Blythe R. ^{d}none. ^{e}Public Library (1947) 294.

Southampton: ^{a}England. ^{b}none. ^{c}English Channel; Test R; Itchen R. ^{d}University of Southampton (1862) 7 – 631. ^{e}University ... Library (1862) 500; Southampton Branch of Hampshire County Library (for Southampton Public Library) (1889) 292.

Southend-on-Sea: ^{a}England. ^{b}alt Southend. ^{c}North Sea; Thames R. ^{d}none. ^{e}Southend-on-Sea Branch of Essex County Library (for Southend-on-Sea Public Library) (1906) 316.

South Ribble: ^{a}England. ^{b}inc Leyland, Walton-le-Dale. ^{c}Ribble R. ^{d}none. ^{e}South Ribble Branch of Lancashire County Library (?) ? ^{f}Leyland: 23 (1971), 19 (1961), 15 (1951), 11 (1931), 7 (1901), 4s (1851), 2s (1801); Walton-le-Dale: 27 (1971), 19 (1961), 15 (1951), 13 (1931), 11 (1901), 7s (1851). 4s (1801).

South Tyneside: ^{a}England. ^{b}for South Shields. ^{c}North Sea; Tyne R. ^{d}none. ^{e}Public Library (1873) 263.

Spelthorne: ^{a}England. ^{b}inc Staines, Sunbury-on-Thames. ^{c}Thames R; Colne R. ^{d}none. ^{e}Spelthorne Branch of Surrey County Library (?) ? ^{f}Staines: 57 (1971), 50 (1961), 40 (1951), 21 (1931), 7 (1901), 2 (1851), 2s (1801); Sunbury-on-Thames: 40 (1971), 33 (1961), 23 (1951), 13 (1931), 5 (1901), 2s (1851), 1s (1801).

Stafford: ^{a}England. ^{b}none. ^{c}Trent R; Sow R. ^{d}none. ^{e}Staffordshire County Library (1922) 2136, inc Stafford Branch (for Stafford Public Library) (1882) 103.

Stockport: ^{a}England. ^{b}inc Cheadle and Gatley since 1974. ^{c}Mersey R; Tame R; Goyt R. ^{d}none. ^{e}Public Library (1875) 743. ^{f}Cheadle and Gatley: 61 (1971), 46 (1961), 32 (1951), 18 (1931), 11 (1901).

Stockton-on-Tees: ^{a}England. ^{b}alt Stockton; for Teesside (in part); inc Billingham, Thornaby-on-Tees since 1974. ^{c}Tees R. ^{d}none. ^{e}Stockton-on-Tees Branch of Cleveland County Library (for Stockton-on-Tees Public Library) (1877) 136 (in 1967). ^{f}Billingham: 32 (1961), 24 (1951),

19 (1931), 4s (1901), 0.7s (1851), 0.3s (1801); Thornaby-on-Tees: 23 (1961), 23 (1951), 21 (1931), 16 (1901), 2s (1851), 0.2s (1801).

Stoke-on-Trent: ^{a}England. ^{b}none. ^{c}Trent R. ^{d}none. ^{e}Stoke-on-Trent Branch of Staffordshire County Library (for Stoke-on-Trent Public Library) (1869) 395.

Stratford-on-Avon: ^{a}England. ^{b}none. ^{c}Avon R. ^{d}none. ^{e}Stratford-on-Avon Branch of Warwickshire County Library (for Stratford-on-Avon Public Library) (1905) 62.

Sunderland: ^{a}England. ^{b}none. ^{c}Wear R. ^{d}none. ^{e}Public Library (1859) 439.

Swansea: ^{a}Wales. ^{b}alt Abertawe. ^{c}Swansea B of Bristol Channel; Tawe R. ^{d}University College of Swansea of the University of Wales (1920) 3 – 470. ^{e}West Glamorgan County Library (1923) 635, inc Swansea Branch (for Swansea Public Library) (1875) 188; University College . . . Library (1920) 350.

Taff-Ely: ^{a}Wales. ^{b}for Pontypridd. ^{c}Taff R; Rhondda R. ^{d}none. ^{e}Taff-Ely Branch of Mid Glamorgan County Library (for Pontypridd Public Library) (1887) 118.

Tameside: ^{a}England. ^{b}inc Ashton-under-Lyne, Denton, Hyde, Stalybridge. ^{c}Tame R. ^{d}none. ^{e}Public Library (1881) 424. ^{f}Ashton-under-Lyne: 49 (1971), 50 (1961), 52 (1951), 52 (1931), 44 (1901), 31 (1851), 16s (1801); Denton: 38 (1971), 31 (1961), 26 (1951), 17 (1931), 15 (1901), 3s (1851), 1s (1801); Hyde: 37 (1971), 32 (1961), 31 (1951), 32 (1931), 33 (1901), 10 (1851), 1s (1801); Stalybridge: 23 (1971), 22 (1961), 23 (1951), 25 (1931), 28 (1901), 21 (1851), 1s (1801).

Thamesdown: ^{a}England. ^{b}for Swindon. ^{c}Thames R. ^{d}none. ^{e}Thamesdown Branch of Wiltshire County Library (for Swindon Public Library) (1943) 188.

Thurrock: ^{a}England. ^{b}inc Grays Thurrock, Tilbury. ^{c}Thames R. ^{d}none. ^{e}Thurrock Branch of Essex County Library (for Thurrock Public Library) (1894) 280. ^{f}Grays Thurrock: 18 (1931), 14 (1901), 2s (1851), 0.7s (1801); Tilbury: 17 (1931), 5 (1901).

Torbay: ^{a}England. ^{b}for Torquay. ^{c}English Channel. ^{d}none. ^{e}Torbay Branch of Devon County Library (for Torbay Public Library) (1907) 211.

Torfaen: ^{a}Wales. ^{b}inc Cwmbran, Pontypool. ^{c}(Afon) Lwyd R. ^{d}none. ^{e}Torfaen Branch of Gwent County Library (for Pontypool Public Library) (1901) ? ^{f}Cwmbran: 32 (1971), 22 (1961), 13 (1951); Pontypool: 37 (1971), 40 (1961), 43 (1951), 7 (1931), 6 (1901), 4 (1851), 1s (1801).

Trafford: ^{a}England. ^{b}inc Altrincham, Sale, Stretford. ^{c}Mersey R. ^{d}none. ^{e}Public Library (1891) 387. ^{f}Altrincham: 41 (1971), 41 (1961), 40 (1951), 21 (1931), 17 (1901), 4 (1851), 2s (1801); Sale: 56 (1971), 51 (1961), 43 (1951), 28 (1931), 12 (1901), 2s (1851), 0.8s (1801); Stretford: 54 (1971), 60 (1961), 62 (1951), 57 (1931), 30 (1901), 5s (1851), 1s (1801).

Tunbridge Wells: ^{a}England. ^{b}off Royal Tunbridge Wells. ^{c}nr Medway R. ^{d}none. ^{e}Tunbridge Wells Branch of Kent County Library (for Tunbridge Wells Public Library) (1921) 125.

Vale of Glamorgan: ^{a}Wales. ^{b}inc Barry, Penarth. ^{c}Bristol Channel; Severn R. ^{d}none. ^{e}Vale of Glamorgan Branch of South Glamorgan County Library [for Barry Public Library (1891) 116, and Penarth Public Library (1895) 71]. ^{f}Barry: 42 (1971), 42 (1961), 41 (1951), 39 (1931), 27 (1901), 0.07s (1851), 0.07s (1801); Penarth: 24 (1971), 21 (1961), 19 (1951), 18 (1931), 14 (1901), 0.1s (1851), 0.07s (1801).

Wakefield: ^{a}England. ^{b}none. ^{c}Aire R; Calder R. ^{d}none. ^{e}Public Library (1906) 640.

Walsall: ^{a}England. ^{b}inc Aldridge-Brownhills since 1974. ^{c}nr Tame R. ^{d}none. ^{e}Public Library (1859) 433. ^{f}Aldridge-Brownhills: 89 (1971); Aldridge: 51 (1961), 29 (1951), 14s (1931), 10s (1901); Brownhills: 26 (1961), 21 (1951), 18 (1931), 15 (1901), 4s (1851), 2s (1801).

Warrington: ^{a}England. ^{b}none. ^{c}Mersey R. ^{d}none. ^{e}Warrington Branch of Cheshire County Library (for Warrington Public Library) (1848) 147.

Wigan: ^{a}England. ^{b}inc Leigh since 1974. ^{c}Douglas R. ^{d}none. ^{e}Public Library (1878) 606. ^{f}Leigh: 46 (1971), 46 (1961), 49 (1951), 45 (1931), 40 (1901), 5 (1851), 13s (1801).

Winchester: ^{a}England. ^{b}none. ^{c}Itchen R. ^{d}none. ^{e}Hampshire County Library (1925) 2647, inc Winchester Branch (for Winchester Public Library) (1851) 95.

Windsor and Maidenhead: ^{a}England. ^{b}inc Maidenhead, New Windsor (alt Windsor). ^{c}Thames R. ^{d}none. ^{e}Windsor and Maidenhead Branch of Berkshire County Library (for Maidenhead Public Library)

(1904) 88. [f]Maidenhead: 45 (1971), 35 (1961), 27 (1951), 18 (1931), 13 (1901), 4 (1851), 0.9 (1801); New Windsor: 30 (1971), 27 (1961), 23 (1951), 20 (1931), 14 (1901), 10 (1851), 3 (1801).

Wirral: [a]England. [b]inc Bebington, Birkenhead, Wallasey, Wirral. [c]Irish Sea; Dee R; Mersey R. [d]none. [e]Public Library (1856) 710. [f]Bebington: 61 (1971), 53 (1961), 48 (1951), 27 (1931), 10 (1901), 3s (1851), 0.4s (1801); Birkenhead; 138 (1971), 142 (1961), 143 (1951), 148 (1931), 111 (1901), 24 (1851), 0.1s (1801); Wallasey: 97 (1971), 103 (1961), 101 (1951), 98 (1931), 54 (1901), 1s (1851), 0.3s (1801); Wirral: 27 (1971), 22 (1961), 17 (1951).

Wolverhampton: [a]England. [b]none. [c]nr Smestow R. [d]none. [e]Public Library (1869) 471.

Worcester: [a]England. [b]none. [c]Severn R. [d]none. [e]Hereford and Worcester County Library (1923) 1086, inc Worcester Branch (for Worcester Public Library) (1881) 161.

Worthing: [a]England. [b]none. [c]English Channel. [d]none. [e]Worthing Branch of West Sussex County Library (for Worthing Public Library) (1896) 181.

Wrekin: [a]England. [b]inc Telford (for Dawley), Wellington. [c]nr Severn R. [d]none. [e]Shropshire County Library (at Shrewsbury and Atcham, nr Wrekin) (1925) 955; Wrekin Branch of Shropshire County Library (for Wellington Public Library) (1904) ? [f]Telford: 26 (1971), 10 (1961), 8 (1951), 7 (1931), 8 (1901), 9s (1851), 4s (1801); Wellington: 17 (1971), 14 (1961), 11 (1951), 8 (1931), 6 (1901), 5 (1851), 8 (1801).

Wrexham Maelor: [a]Wales. [b]for Wrexham. [c](Afon) Clywedog R. [d]none. [e]Clwyd County Library (at Delyn, nr Wrexham Maelor) (1921) 1284; Wrexham Maelor Branch of Clwyd County Library (for Wrexham Public Library) (1879) 102.

Wyre: [a]England. [b]inc Fleetwood, Thornton Cleveleys. [c]Irish Sea; Wyre R. [d]none. [e]Wyre Branch of Lancashire County Library (for Fleetwood Public Library) (1887) 101. [f]Fleetwood: 29 (1971), 28 (1961), 28 (1951), 23 (1931), 12 (1901), 3 (1851); Thornton Cleveleys: 27 (1971), 21 (1961), 15 (1951), 10 (1931), 3 (1901).

York: [a]England. [b]none. [c]Ouse R; Foss R. [d]University of York (1963) 3 – 278. [e]J. B. Morrell Library of the University . . . (1963) 220; York Branch of North Yorkshire County Library (for York Public Library) (1893) 152.

USSR

Archangel: [a]Russia in Europe. [b]off Arkhangelsk. [c]Northern Dvina R. [d]none. [e]Archangel Province N. A. Dobrolyubov Library (1833) 471.

Armavir: [a]Russia in Europe. [b]none. [c]Kuban R; Urup R. [d]none. [e]Armavir Pedagogic Institute Library (1948) 118.

Astrakhan: [a]Russia in Europe. [b]none. [c]Volga R. [d]none. [e]Astrakhan Province N. K. Krupskaya Scientific Library (1838) 333; Astrakhan S. M. Kirov Pedagogic Institute Library (1932) 252.

Baku: [a]*Azerbaijan. [b]none. [c]Caspian (Sea) L. [d]Azerbaijan S. M. Kirov State University (1919) 11 – 700. [e]Azerbaijan National M. F. Akhundov State Library (1923) 1336; Central Library of the . . . University (1919) 1156; Central Library of the Azerbaijan M. Azizbekov Institute of Petroleum and Chemistry (1920) 625; Azerbaijan V. I. Lenin Pedagogic Institute Library (1939) 576; Central Scientific Library of the Azerbaijan SSR Academy of Sciences (1923) 494.

Belgorod: [a]Russia in Europe. [b]none. [c]Donets R. [d]none. [e]Belgorod Province Library (1955) 240.

Berezniki: [a]Russia in Europe. [b]none. [c]Kama R. [d]none. [e]Technical Library of the V. I. Lenin Horticultural Works (1940) 71.

Bobruysk: [a]White Russia. [b]alt Bobruisk. [c]Berezina R. [d]none.

Brest: [a]White Russia. [b]for Brest-Litovsk, Brzesc-Litewski. [c]Bug R. [d]none. [e]Brest Province Gorkiy Library (1940) 360; Brest A. S. Pushkin Pedagogic Institute Library (1945) 230.

Bryansk: [a]Russia in Europe. [b]alt Briansk. [c]Desna R. [d]none. [e]Bryansk Province Library (1944) 450.

Cheboksary: [a]Russia in Europe; *Chuvash ASSR. [b]none. [c]Volga R. [d]Chuvash I. N. Ulyanov State University (1967) 8 – 300. [e]Chuvash National M. Gorkiy Library (1871) 548.

Cherepovets: [a]Russia in Europe. [b]none. [c]Sheksna R. [d]none. [e]Cherepovets Pedagogic Institute Library (1875) 126.

Cherkassy: [a]Ukraine. [b]none. [c]Dnieper R. [d]none. [e]Cherkassy Province V. V. Mayakovskiy Library (1954) 301; Cherkassy 300th Anniversary of the Union of Russia and the Ukraine Pedagogic Institute Library (1930) 292.

Chernigov: [a]Ukraine. [b]none. [c]Desna R. [d]none. [e]Chernigov Province V. G. Korolenko State Library (1877) 416.

Chernovtsy: [a]Ukraine. [b]for Cernauti, Czernowitz. [c]Prut R. [d]Chernovtsy State University (1875) 9 – 450. [e]. . . University Library (1875) 984; Chernovtsy Province Library (1940) 386.

Dneprodzerzhinsk: [a]Ukraine. [b]for Kamenskoye. [c]Dnieper R. [d]none. [e]Dneprodzerzhinsk M. I. Arsenichev Industrial Institute Library (1920) 161; Palace of Culture Library (?) ?

Dnepropetrovsk: [a]Ukraine. [b]for Ekaterinoslav, Yekaterinoslav. [c]Dnieper R; Samara R. [d]Dnepropetrovsk 300th Anniversay of the Union of Russia and the Ukraine University (1918) 13 – 700. [e]Dnepropetrovsk Province October Revolution Library (1889) 617; . . . University Library (1918) 604.

Donetsk: [a]Ukraine. [b]for Stalino, Yuzovka, Yuzovo. [c]Kalmius R. [d]Donetsk State University (1965) 12 – ? [e]Donetsk Polytechnic Institute Library (1921) 806; Donetsk Province N. K. Krupskaya Scientific Library (1926) 804; . . . University Library (1937) 480.

Dzerzhinsk: [a]Russia in Europe. [b]for Chernorechye, Rastyapino. [c]Oka R. [d]none. [e]Scientific-Technical Library of the Dzerzhinsk M. I. Kalinin Chemical Works (1929) 106; N. K. Krupskaya City Library (1927) 52.

Engels: [a]Russia in Europe. [b]for Pokrovsk, Pokrovskaya Sloboda. [c]Volga R. [d]none. [e]Central City Library (1918) 228.

Gomel: [a]White Russia. [b]none. [c]Sozh R. [d]Gomel State University (1970) 5 – 250. [e]Gomel Province V. I. Lenin Library (1938) 426; Scientific Library of the . . . University (1930) 310.

Gorkiy: [a]Russia in Europe. [b]alt Gorki, Gorky; for Nizhniy Novgorod. [c]Volga R; Oka R. [d]Gorkiy N. I. Lobachevskiy State University (1918) 8 – 700. [e]Gorkiy Province V. I. Lenin Library (1930) 1305; Scientific-Technical Library of the Gorkiy A. A. Zhdanov Polytechnic Institute (1930) 531; Central Library of the . . . University (1931) 470.

Gorlovka: [a]Ukraine. [b]none. [c]nr Lugan R. [d]none. [e]Foreign-Language Pedagogic Institute Library (1949) 121.

Grodno: [a]White Russia. [b]none. [c]Neman R. [d]none. [e]Grodno Province E. F. Karskiy Library (1830) 214; Grodno Yanka Kupala Pedagogic Institute Library (1945) 208.

Groznyy: [a]Russia in Europe; *Chechen-Ingush ASSR. [b]alt Grozny. [c]Sunzha R. [d]Chechen-Ingush State University (1972) ? – ? [e]Chechen-Ingush National A. P. Chekhov Library (1905) 541; . . . University Library (1938) 343.

Ivanovo: [a]Russia in Europe. [b]for Ivanovo-Voznesensk. [c]Uvod R. [d]none. [e]Ivanovo Province Library (1919) 1105.

Izhevsk: [a]Russia in Europe; *Udmurt ASSR. [b]for Izhevskiy Zavod. [c]Izh R. [d]Udmurt State University (1973?) ? – ? [e]Udmurt National V. I. Lenin Library (1919) 727.

Kalinin: [a]Russia in Europe. [b]for Tver. [c]Volga R; Tvertsa R; Tmaka R. [d]Kalinin State University (1971) 5 – ? [e]Kalinin Province A. M. Gorkiy Library (1860) 752.

Kaliningrad: [a]Russia in Europe. [b]for Konigsberg. [c]Pregolya (for Pregel) R. [d]Kaliningrad State University (1544) 4 – 200. [e]Kaliningrad Province Library (1946) 537.

Kaluga: [a]Russia in Europe. [b]none. [c]Oka R. [d]none. [e]Kaluga Province V. G. Belinskiy Library (1944) 316.

Kaunas: [a]Lithuania. [b]alt Kovno. [c]Neman R; Viliya R. [d]Kaunas Polytechnic Institute (univ) (1920) 15 – 1326. [e]Public Library (1919) 1000; . . . Institute Library (1923) 689.

Kazan: [a]Russia in Europe; *Tatar ASSR. [b]for Kasan. [c]Kazanka R. [d]Kazan V. I. Ulyanov-Lenin State University (1804) 10 – 700. [e]N. I. Lobachevskiy Scientific Library of the . . . University (1798) 1612; National V. I. Lenin Library of the Tatar ASSR (1865) 1081.

Kerch: [a]Ukraine. [b]none. [c]Kerch Strait. [d]none. [e]Scientific-Technical Library of the Azov-Black Sea Scientific Research Institute of Fisheries and Oceanography (1921) 30.

Kharkov: [a]Ukraine. [b]none. [c]Kharkov R; Lopan R; Netetcha R; Gnilopiat R; Udi R. [d]Kharkov A. M. Gorkiy State University (1805) 7 – ? [e]V. G. Korolenko State Scientific Library (1886) 2407;

Central Scientific Library of the ... University (1805) 1611; Scientific-Technical Library of the Kharkov V. I. Lenin Polytechnic Institute (1885) 720.

Kherson: [a]Ukraine. [b]none. [c]Dnieper R. [d]none. [e]Kherson Province A. M. Gorkiy Library (1872) 358.

Khmelnitskiy: [a]Ukraine. [b]for Proskurov, Proskurow. [c]Southern Bug R. [d]none. [e]Khmelnitskiy Province N. A. Ostrovskiy Library (1901) 287.

Kiev: [a]*Ukraine. [b]off Kiyev. [c]Dnieper R. [d]Kiev T. G. Shevchenko State University (1834) 20 – 1630. [e]Central Scientific Library of the Ukrainian SSR Academy of Sciences (1919) 3594; Scientific Library of the ... University (1834) 1428; National CPSU State Library of the Ukrainian SSR (1866) 1307; Scientific-Technical Library of the Kiev 50th Anniversary of the October Revolution Polytechnic Institute (1898) 1200.

Kirov: [a]Russia in Europe. [b]for Khlynov, Viatka, Vyatka. [c]Vyatka R. [d]none. [e]Kirov Province A. I. Herzen Library (1837) 1056.

Kirovabad: [a]Azerbaijan. [b]for Elisavetpol, Gandzha, Yelisavetpol. [c]Gyandzhachay R. [d]none. [e]Kirovabad G. B. Zardabi Pedagogic Institute Library (1943) 188; Azerbaijan Agricultural Machine Construction Institute Library (1932) 180.

Kirovograd: [a]Ukraine. [b]for Elisavetgrad, Yelisavetgrad, Zinovyevsk. [c]Ingul R. [d]none. [e]Kirovograd Province N. K. Krupskaya Scientific Library (1898) 347.

Kishinev: [a]*Moldavia. [b]for Chisinau. [c]Byk R. [d]Kishinev State University (1945) 8 – 542. [e]National N. K. Krupskaya State Library of the Moldavian SSR (1832) 1099; Scientific Library of the ... University (1946) 758; Central Scientific Library of the Moldavian SSR Academy of Sciences (1947) 412.

Klaypeda: [a]Lithuania. [b]alt Klaipeda; for Memel. [c]Baltic Sea; Neman R. [d]none. [e]Klaypeda Branch of the National State Library of the Lithuanian SSR (1959) 58.

Kostroma: [a]Russia in Europe. [b]none. [c]Volga R; Kostroma R. [d]none. [e]Kostroma Province N. K. Krupskaya Library (1918) 778.

Kramatorsk: [a]Ukraine. [b]for Kramatorskaya. [c]Kazennyy Torets R. [d]none. [e]Kramatorsk Industrial Institute Library (1953) 263.

Krasnodar: [a]Russia in Europe. [b]for Ekaterinodar, Yekaterinodar. [c]Kuban R. [d]Kuban State University (1970) 10 – ? [e]Krasnodar Region A. S. Pushkin Library (1900) 770; Krasnodar Polytechnic Institute Library (1920) 585; Scientific Library of the ... University (1920) 358.

Kremenchug: [a]Ukraine. [b]alt Kremenchuk. [c]Dnieper R. [d]none.

Krivoy Rog: [a]Ukraine. [b]alt Krivoi Rog. [c]Ingulets R; Saksagan R. [d]none. [e]Krivoy Rog Ore-Mining Institute Library (1929) 393; Krivoy Rog Pedagogic Institute Library (1930) 96.

Kursk: [a]Russia in Europe. [b]none. [c]Tuskor R; Kur R. [d]none. [e]Kursk Province N. N. Aseyev Library (1935) 448.

Kutaisi: [a]Georgia. [b]for Kutais. [c]Rioni R. [d]none. [e]Scientific Library of the K. A. Tsulukidze Pedagogic Institute (1933) 244.

Kuybyshev: [a]Russia in Europe. [b]alt Kuibyshev; for Samara. [c]Volga R; Samara R. [d]Kuybyshev State University (1970) ? – ? [e]Kuybyshev Province V. I. Lenin Library (1860) 1560; Kuybyshev V. V. Kuybyshev Polytechnic Institute Library (1939) 580.

Leninakan: [a]Armenia. [b]for Aleksandropol, Gyumri. [c]Akhuryan [alt Arpa (Cayi)] R. [d]none. [e]Leninakan M. Nalbandyan Pedagogic Institute Library (1934) 79.

Leningrad: [a]Russia in Europe. [b]for Petrograd, Saint Petersburg. [c]G of Finland of Baltic Sea; Neva R. [d]Leningrad A. A. Zhdanov State University (1819) 20 – 1700. [e]M. E. Saltykov-Shchedrin State Public Library (1795) 8000; USSR Academy of Sciences Library (1714) 5526; M. Gorkiy Scientific Library of the ... University (1819) 2759; Leningrad A. I. Herzen Pedagogic Institute Library (1918) 1213.

Lipetsk: [a]Russia in Europe, [b]none. [c]Voronezh R. [d]none. [e]Lipetsk Province Scientific Library (1955) 282.

Lvov: [a]Ukraine. [b]for Lemberg, Lwow. [c]Peltev R. [d]Lvov Ivan Franko State University (1661) 13 – 700. [e]Lvov V. Stefanik State Scientific Library of the Ukrainian SSR Academy of Sciences (1940) 2824; Scientific Library of the ... University (1608) 1583; Scientific-Technical Library of the Lvov Polytechnic Institute (1844) 776; Lvov Province Yaroslav Galan Library (1940) 423.

Lyubertsy: ^{a}Russia in Europe. ^{b}none. ^{c}nr Moscow R. ^{d}none. ^{e}Scientific-Technical Library of the A. A. Skochinskiy Mining Institute (1949) 92.

Makeyevka: ^{a}Ukraine. ^{b}alt Makeevka; for Dmitriyevsk. ^{c}Gruzskaya R. ^{d}none. ^{e}Makeyevak State Library of the Scientific Research Institute of Work Safety in the Mining Industry (1927) 55.

Makhachkala: ^{a}Russia in Europe; *Daghestan ASSR. ^{b}for Petrovsk. ^{c}Caspian (Sea) L. ^{d}Daghestan V. I. Lenin State University (1957) 8 – 450. ^{e}Daghestan National A. S. Pushkin Library (1900) 382; . . . University Library (1957) 307.

Melitopol: ^{a}Ukraine. ^{b}for Novo-Aleksandrovka. ^{c}Molochnaya R. ^{d}none. ^{e}Melitopol Institute for the Mechanization of Agriculture Library (1932) 168; Melitopol Pedagogic Institute Library (1930) 86.

Minsk: ^{a}*White Russia. ^{b}none. ^{c}Svisloch R. ^{d}White Russian V. I. Lenin State University (1921) 9 – 1546. ^{e}V. I. Lenin State Library of the White Russian SSR (1922) 2800; White Russian Polytechnic Institute Library (1933) 818; Central Library of the . . . University (1921) 769; A. M. Gorkiy Government Library of the White Russian SSR Council of Ministers (1934) 663.

Mogilev: ^{a}White Russia. ^{b}none. ^{c}Dnieper R. ^{d}none. ^{e}Mogilev Province V. I. Lenin Library (1935) 435.

*Moscow: ^{a}Russia in Europe; *Russia. ^{b}off Moskva. ^{c}Moscow R; Yauza R. ^{d}Moscow M. V. Lomonosov State University (1755) 29 – 3700; Patrice Lumumba People's Friendship University (1960) 7 – 998; Moscow I. M. Sechenov Medical Institute (univ) (1755) 1 – 680. ^{e}V. I. Lenin State Library of the USSR (1828 at Leningrad, relocated 1862) 11,750; Central Library of Natural Sciences of the USSR Academy of Sciences (1934) 3784; A. M. Gorkiy Scientific Library of the Moscow M. V. Lomonosov State University (1755) 3485; Central House of the Soviet Army Library (?) 3000; Institute of Scientific Information on Social Sciences of the USSR Academy of Sciences (1918) 2872; State Public Historical Library of the Russian SFSR (1938) 2634; All-Union State Library of Foreign Literature (1921) 1562; Central Polytechnic Library of the All-Union Society for the Dissemination of Political and Scientific Knowledge (1864) 1561; Moscow Sergo Ordzhonikidze Aviation Institute Library (1933) 1387; Central Scientific Agricultural Library of the All-Union Lenin Academy of Agricultural Sciences (1930) 1234; State Public Scientific-Technical Library of the USSR (1958) 1190; Scientific-Technical Library of the Moscow Institute of Railroad Transportation Engineering (1896) 1119; Central State Scientific Medical Library (1919) 780.

Murmansk: ^{a}Russia in Europe. ^{b}none. ^{c}Barents Sea. ^{d}none. ^{e}Murmansk Province Scientific Library (1938) 412.

Naberezhnyye Chelny: ^{a}Russia in Europe; Tatar ASSR. ^{b}for Chelny. ^{c}Kama R. ^{d}none.

Nalchik: ^{a}Russia in Europe; *Kabardino-Balkar ASSR. ^{b}none. ^{c}Nalchik R. ^{d}Kabardino-Balkar State University (1957) 8 – 400. ^{e}Kabardino-Balkar N. K. Krupskaya State Scientific Library (1921) 711; . . . University Library (1932) 550.

Nikolayev: ^{a}Ukraine. ^{b}alt Nikolaev; for Vernoleninsk. ^{c}Southern Bug R; Ingul R. ^{d}none. ^{e}Nikolayev Province Aleksey Gmyrev Library (1881) 433.

Novgorod: ^{a}Russia in Europe. ^{b}none. ^{c}Volkhoy R. ^{d}none. ^{e}Novgorod Province Library (1919) 420.

Novocherkassk: ^{a}Russia in Europe. ^{b}none. ^{c}Ak(say) R. ^{d}none. ^{e}Novocherkassk Sergo Ordzhonikidze Polytechnic Institute Library (1907) 918; Central City A. S. Pushkin Library (1869) 212.

Novorossiysk: ^{a}Russia in Europe. ^{b}alt Novorossiisk. ^{c}Black Sea; Tsemes R. ^{d}none. ^{e}Central City A. M. Gorkiy Library (1943) 99.

Odessa: ^{a}Ukraine. ^{b}none. ^{c}Black Sea. ^{d}Odessa I. I. Mechnikov State University (1807) 12 – 800. ^{e}Odessa A. M. Gorkiy State Scientific Library (1830) 1811; Scientific Library of the . . . University (1817) 879; Scientific-Technical Library of the Odessa Polytechnic Institute (1918) 501; Odessa Province V. I. Lenin Scientific Library (1920) 405.

Ordzhonikidze: ^{a}Russia in Europe; *North Ossetian ASSR. ^{b}for Dzaudzhikau, Vladikavkaz. ^{c}Terek R. ^{d}North Ossetian K. L. Khetagurov State University (1970) ? – ? ^{e}North Ossetian National S. M. Kirov State Scientific Library (1895) 439; Central Library of the . . . University (1920) 341.

Orel: ^{a}Russia in Europe. ^{b}none. ^{c}Oka R; Orlik R. ^{d}none. ^{e}Orel Province N. K. Krupskaya Library (1919) 389; Orel Pedagogic Institute Library (1931) 315.

Orenburg: [a]Russia in Europe. [b]for Chkalov. [c]Ural R; Sakmara R. [d]none. [e]Orenburg Province N. K. Krupskaya Library (1896) 494.

Orsk: [a]Russia in Europe. [b]none. [c]Ural R; Or R. [d]none. [e]Orsk T. G. Shevchenko State Pedagogic Institute Library (1949) 136.

Penza: [a]Russia in Europe. [b]none. [c]Sura R; Penza R. [d]none. [e]Penza Province M. Y. Lermontov Library (1892) 448.

Perm: [a]Russia in Europe. [b]for Molotov, Yegozhikhinskiy Zavod. [c]Kama R. [d]Perm A. M. Gorkiy State University (1916) 12 – 600. [e]Perm Province M. Gorkiy Library (1831) 877; Central Library of the ... University (1916) 649; Perm Polytechnic Institute Library (1960) 547.

Petrozavodsk: [a]Russia in Europe; *Karelian ASSR. [b]for Kalininsk. [c]Onega L. [d]Petrozavodsk O. V. Kuusinen State University (1940) 7 – 450. [e]State Public Library of the Karelian ASSR (1860) 1053; Central Library of the ... University (1940) 368.

Podolsk: [a]Russia in Europe. [b]none. [c]Pakhra R. [d]none. [e]Scientific-Technical Library of the Podolsk M. I. Kalinin Mechanical Works (1932) 115.

Poltava: [a]Ukraine. [b]none. [c]Vorskla R; Kolomak R. [d]none. [e]Poltava Province I. P. Kotlyarevskiy Scientific Library (1894) 269; Poltava V. G. Korolenko Pedagogic Institute Library (1920) 209.

Pskov: [a]Russia in Europe. [b]none. [c]Velikaya R. [d]none. [e]Pskov Province Library (1944) 706.

Riga: [a]*Latvia. [b]none. [c]G of Riga of Baltic Sea; Western Dvina R. [d]Latvian Pyetr Stuchka State University (1861) 9 – 594. [e]Vilis Lacis State Library of the Latvian SSR (1919) 2268; Scientific Library of the ... University (1862) 1400; Central Library of the Latvian SSR Academy of Sciences (1524) 1277.

Rostov-na-Donu: [a]Russia in Europe. [b]alt Rostov, Rostov-on-Don. [c]Don R. [d]Rostov-na-Donu State University (1869 at Warsaw, Poland; relocated 1915) 10 – ? [e]Rostov-na-Donu Karl Marx State Scientific Library (1920) 1562; Scientific Library of the ... University (1915) 1131.

Rovno: [a]Ukraine. [b]for Rowne. [c]Ustye R. [d]none. [e]Rovno Province Library (1940) 287.

Ryazan: [a]Russia in Europe. [b]alt Riazan. [c]Oka R; Trobezh R. [d]none. [e]Ryazan Province A. M. Gorkiy Library (1858) 520.

Rybinsk: [a]Russia in Europe. [b]for Shcherbakov. [c]Volga R; Cheremukha R; Sheksna R. [d]none. [e]Central City F. Engels Library (1919) 113.

Saransk: [a]Russia in Europe; *Mordvinian ASSR. [b]none. [c]Insar R. [d]Mordvinian State University (1957) 4 – ? [e]Mordvinian National A. S. Pushkin Library (1899) 430; ... University Library (1931) 401.

Saratov: [a]Russia in Europe. [b]none. [c]Volga R. [d]Saratov N. G. Chernyshevskiy State University (1909) 10 – 700. [e]Scientific Library of the ... University (1909) 1481; Saratov Province Library (1831) 926.

Sevastopol: [a]Ukraine. [b]for Sebastopol. [c]Black Sea. [d]none. [e]Sevastopol Instrument-Making Institute Library (1960) 168; Central City L. N. Tolstoy Library (?) ?

Severodvinsk: [a]Russia in Europe. [b]for Molotovsk, Sudostroy. [c]G of Dvina of White Sea; Northern Dvina R. [d]none.

Shakhty: [a]Russia in Europe. [b]for Aleksandrovsk-Grushevskiy. [c]Grushevka R. [d]none. [e]Central City A. S. Pushkin Library (1915) 257.

Simferopol: [a]Ukraine. [b]none. [c]Salgir R. [d]Simferopol M. V. Frunze State University (1973) 5 – ? [e]Crimean Province I. Y. Franko Library (1890) 573; Central Library of the Crimean M. V. Frunze Pedagogic Institute (1918) 484.

Smolensk: [a]Russia in Europe. [b]none. [c]Dnieper R. [d]none. [e]Smolensk Province V. I. Lenin Library (1920) 463.

Sochi: [a]Russia in Europe. [b]none. [c]Black Sea; Sochi R. [d]none. [e]Central City Library (1963) 60.

Stavropol: [a]Russia in Europe. [b]for Voroshilovsk. [c]Tashla R. [d]none. [e]Stavropol Region M. Y. Lermontov Library (1853) 700.

Sterlitamak: [a]Russia in Europe; Bashkir ASSR. [b]none. [c]Belaya R; Sterlya R. [d]none. [e]Sterlitamak Pedagogic Institute Library (1944) 95.

Sumgait: [a]Azerbaijan. [b]none. [c]Sumgait R; Caspian (Sea) L. [d]none.

Sumy: [a]Ukraine. [b]none. [c]Psel R. [d]none. [e]Sumy Province Library (1939) 231.

Syktyvkar: ^{a}Russia in Europe; *Komi ASSR. ^{b}for Ust-Sysolsk. ^{c}Vychegda R; Sysola R. ^{d}Syktyvkar State University (1972) ? – ? ^{e}Komi National V. I. Lenin Library (1902) 522.

Syzran: ^{a}Russia in Europe. ^{b}none. ^{c}Volga R; Syzran R. ^{d}none.

Taganrog: ^{a}Russia in Europe. ^{b}none. ^{c}G of Taganrog of Sea of Azov. ^{d}none. ^{e}Taganrog Institute of Radio Engineering Library (1952) 362; A. P. Chekhov City Library (1876) 173.

Tallin: ^{a}*Estonia. ^{b}alt Tallinn; for Reval, Revel. ^{c}G of Finland of Baltic Sea. ^{d}Tallin Polytechnic Institute (univ) (1918) 9 – 569. ^{e}Scientific Library of the Estonian SSR Academy of Sciences (1947) 1240; F. R. Kreutzwald State Library of the Estonian SSR (1918) 1219.

Tambov: ^{a}Russia in Europe. ^{b}none. ^{c}Tsna R. ^{d}none. ^{e}Tambov Province A. S. Pushkin Library (1830) 411.

Tartu: ^{a}Estonia. ^{b}for Derpt, Dorpat, Yurev. ^{c}Ema R. ^{d}Tartu State University (1802) 6 – 620. ^{e}. . . University Library (1802) 1776.

Tbilisi: ^{a}*Georgia. ^{b}alt Tiflis. ^{c}Kura R. ^{d}Tbilisi State University (1918) 16 – 1659. ^{e}Scientific Library of the . . . University (1918) 2790; National Karl Marx State Library of the Georgian SSR (1846) 2500; Central Scientific Library of the Georgian SSR Academy of Sciences (1941) 1206; Central Library of the Georgian V. I. Lenin Polytechnic Institute (1922) 760.

Tolyatti: ^{a}Russia in Europe. ^{b}alt Togliatti; for Stavropol. ^{c}Volga R. ^{d}none. ^{e}Scientific-Technical Library of the All-Union Scientific Research Institute of Nonmetallic Construction Materials and Hydromechanics (1958) 31.

Tula: ^{a}Russia in Europe. ^{b}none. ^{c}Upa R; Tulitsa R. ^{d}none. ^{e}Tula Polytechnic Institute Library (1930) 680; Tula Province V. I. Lenin Library (1919) 481.

Ufa: ^{a}Russia in Europe; *Bashkir ASSR. ^{b}none. ^{c}Belaya R; Ufa R. ^{d}Bashkir 40th Anniversary of the October Revolution State University (1957) 6 – 215. ^{e}Bashkir National N. K. Krupskaya Library (1921) 530; . . . University Library (1908) 405.

Ulyanovsk: ^{a}Russia in Europe. ^{b}alt Ulianovsk; for Simbirsk. ^{c}Volga R; Sviyaga R. ^{d}none. ^{e}Ulyanovsk Province Library – V. I. Lenin Book Palace (1848) 653.

Vilnyus: ^{a}*Lithuania. ^{b}alt Vilna, Vilnius; for Wilno. ^{c}Viliya R; Vilnia R. ^{d}Vilnyus V. Kapsukas State University (1579) 16 – 1250. ^{e}Central Library of the Lithuanian SSR Academy of Sciences (1557) 1979; National State Library of the Lithuanian SSR (1919) 1597; Scientific Library of the . . . University (1570) 1200.

Vinnitsa: ^{a}Ukraine. ^{b}none. ^{c}Southern Bug R. ^{d}none. ^{e}Vinnitsa Province K. A. Timiryazev Library (1907) 463.

Vitebsk: ^{a}White Russia. ^{b}none. ^{c}Western Dvina R; Luchesa R. ^{d}none. ^{e}Vitebsk Province V. I. Lenin Library (1925) 338.

Vladimir: ^{a}Russia in Europe. ^{b}none. ^{c}Klyazma R; Lybed R. ^{d}none. ^{e}Vladimir Province M. Gorkiy Library (1896) 422.

Volgograd: ^{a}Russia in Europe. ^{b}for Stalingrad, Tsaritsyn. ^{c}Volga R; Tsaritsa R. ^{d}none. ^{e}Volgograd Province M. Gorkiy Library (1900) 466; Volgograd A. S. Serafimovich Pedagogic Institute Library (1945) 305.

Vologda: ^{a}Russia in Europe. ^{b}none. ^{c}Vologda R. ^{d}none. ^{e}Vologda Province I. V. Babushkin Library (1919) 404.

Volzhskiy: ^{a}Russia in Europe. ^{b}none. ^{c}Volga R. ^{d}none. ^{e}Volgograd Hydroelectric Scientific-Technical Library (1950) 168.

Voronezh: ^{a}Russia in Europe. ^{b}none. ^{c}Voronezh R. ^{d}Voronezh State University (1918) 8 – 507. ^{e}Central Library of the . . . University (1918) 821; Voronezh Province I. S. Nikitin Library (1864) 681.

Voroshilovgrad: ^{a}Ukraine. ^{b}for Lugansk. ^{c}Lugan R. ^{d}none. ^{e}Voroshilovgrad Province A. M. Gorkiy Library (1898) 479.

Yalta: ^{a}Ukraine. ^{b}none. ^{c}Black Sea. ^{d}none. ^{e}Scientific Library of the Nikita State Botanical Garden (at Nikita, nr Yalta) (1812) 53.

Yaroslavl: ^{a}Russia in Europe. ^{b}none. ^{c}Volga R; Kotorosl R. ^{d}Yaroslavl State University (1970) ? – ? ^{e}Yaroslavl Province N. A. Nekrasov Library (1902) 652; Central Library of the Yaroslavl K. D. Ushinskiy Pedagogic Institute (1918) 552.

Yerevan: [a]*Armenia. [b]alt Erevan; for Erivan. [c]Zanga R. [d]Yerevan State University (1920) 7 – 384. [e]National A. F. Myashnikyan State Library of the Armenian SSR (1832) 3163; Scientific Library of the . . . University (1921) 972; Central Library of the Armenian SSR Academy of Sciences (1935) 919.

Yoshkar-Ola: [a]Russia in Europe; *Mari ASSR. [b]alt Ioshkar-Ola; for Krasnokokshaysk, Tsarevokokshaysk. [c]Malaya Kokshaga R. [d]Mari State University (1972) ? – ? [e]National Scientific Library of the Mari ASSR (1922) 577.

Zaporozhye: [a]Ukraine. [b]alt Zaporozhe; for Aleksandrovsk. [c]Dnieper R. [d]none. [e]Zaporozhye Province A. M. Gorkiy Library (1905) 524; Zaporozhye V. Y. Chubar Machine-Building Institute Library (1920) 446.

Zhdanov: [a]Ukraine. [b]for Mariupol. [c]Sea of Azov; Kalmius R. [d]none. [e]Zhdanov Metallurgical Institute Library (1933) 234; Gorkiy Library (?) ?; Korolenko Library (?) ?; Krupskaya Library (?) ?

Zhitomir: [a]Ukraine. [b]none. [c]Teterev R; Kamenka R. [d]none. [e]Zhitomir Province October Revolution Library (1866) 371.

Yugoslavia

Banja Luka: [a]Bosnia and Herzegovina. [b]none. [c]Vrbas R. [d]Univerzitet u Banja Luci (1975) 4 – 152. [e]Narodna Biblioteka Petar Kocic (1946) 55.

*Belgrade: [a]*Serbia. [b]off Beograd. [c]Danube R; Sava R. [d]Univerzitet u Beogradu (1808) 51 – 3684; Univerzitet Umetnosti u Beogradu (1957) 1 – 239. [e]Narodna Biblioteka Socijalisticke Republike Srbije (1832) 786; Univerzitetska Biblioteka Svetozar Markovic (1844) 720; Centralna Biblioteka Srpske Akademije Nauka i Umetnosti (1842) 600; Biblioteka Grada Beograda (1929) 200.

Dubrovnik: [a]Croatia. [b]for Ragusa. [c]Adriatic Sea. [d]none. [e]Naucna Biblioteka (1936) 145.

Ljubljana: [a]*Slovenia. [b]for Laibach. [c]Ljubljanica R. [d]Univerza v Ljubljani (1595) 20 – 2216. [e]Narodna in Univerzitetna Knjiznica (1774) 1000.

Maribor: [a]Slovenia. [b]for Marburg. [c]Drava R. [d]Univerza v Mariboru (1975) 4 – 425. [e]Visokoskolska in Studijska Knjiznica (1903) 248; Mestna Knjiznica (1949) 65.

Nis: [a]Serbia. [b]alt Nish; for Nissa. [c]Nisava R. [d]Univerzitet u Nisu (1965) 21 – 825. [e]Narodna Biblioteka (1903) 95; Visa Pedagoska Skola Biblioteka (1904) 95.

Novi Sad: [a]Serbia. [b]for Neusatz, Ujvidek. [c]Danube R. [d]Univerzitet u Novom Sadu (1960) 19 – 1085. [e]Biblioteka Matice Srpske (1838 at Budapest, Hungary; relocated 1864) 550.

Osijek: [a]Croatia. [b]for Esseg, Eszek. [c]Drava R. [d]Sveuciliste u Osijeku (1975) 5 – 212. [e]Gradska Knjiznica (1946) 40.

Rijeka: [a]Croatia. [b]for Fiume. [c]Kvarner G of Adriatic Sea. [d]Sveuciliste u Rijeci (1973) 9 – 410. [e]Naucna Biblioteka (1627) 400.

Sarajevo: [a]*Bosnia and Herzegovina. [b]alt Serajevo. [c]Miljacka R. [d]Univerzitet u Sarajevu (1946) 19 – 1402. [e]Narodna Biblioteka N R Bosne i Hercegovine (1945) 575.

Skopje: [a]*Macedonia. [b]alt Skoplje; for Uskub. [c]Vardar R. [d]Univerzitet Kiril i Metodij vo Skopje (1946) 40 – 1500. [e]Narodna i Univerzitetska Biblioteka Kliment Ohridski (1944) 1000.

Split: [a]Croatia. [b]for Spalato, Spalatro. [c]Adriatic Sea. [d]Sveuciliste u Splitu (1974) 4 – 385. [e]Gradska Biblioteka (1903) 200.

Subotica: [a]Serbia. [b]for Maria-Theresiopel, Szabadka. [c]nr Koros (alt Zuti Potok) R; nr Palicsko L. [d]none. [e]Gradska Narodna Biblioteka (?) 145.

Zagreb: [a]*Croatia. [b]for Agram, Zagrab. [c]Sava R. [d]Sveuciliste u Zagrebu (1669) 31 – 3573. [e]Nacionalna i Sveucilisna Biblioteka (1607) 1069.

NORTH AMERICA

Bahamas

*Nassau: [a]New Providence I. [b]none. [c]Northeast Providence Channel. [d]none. [e]Public Library (1847) 28 (in 1938).

Barbados

*Bridgetown: [b]none. [c]Carlisle B of Atlantic O. [d]University of the West Indies, Barbados campus (1963) 1 – 135. [e]Public Library (1847) 182.

Belize

Belize: [b]none. [c]G of Honduras of Caribbean Sea; Belize R. [d]none. [e]National Library Service (?) 70; Jubilee Public Library (1935) 32.

Bermuda

*Hamilton: [a]Bermuda I. [b]none. [c]Atlantic O. [d]none. [e]Bermuda Library (1839) 122.

Canada

Brampton: [a]Ontario. [b]none. [c]Etobicoke Creek. [d]none. [e]Public Library and Art Gallery (1895) 201.

Brantford: [a]Ontario. [b]none. [c]Grand R. [d]none. [e]Public Library (1884) 118.

Burlington: [a]Ontario. [b]for Wellington Square. [c]Ontario L. [d]none. [e]Public Library (1872) 188.

Calgary: [a]Alberta. [b]none. [c]Bow R; Elbow R. [d]University of Calgary (1945) 14 – 996. [e]University . . . Library (1957) 750; Public Library (1911) 562.

Charlottetown: [a]*Prince Edward Island; Prince Edward I. [b]none. [c]Hillsborough B of Northumberland Strait; Hillsborough R; Yorke R. [d]University of Prince Edward Island (1834) 2 – 130. [e]Robertson Library of the University . . . (1917?) 181; Prince Edward Island Provincial Library (1933) 161.

Chicoutimi: [a]Quebec. [b]none. [c]Saguenay R; Chicoutimi R. [d]Universite du Quebec (alt University of Quebec), Chicoutimi campus (1969) 3 – 138. [e]Bibliotheque de l'Universite du Quebec a Chicoutimi (1969) 102; Bibliotheque Municipale (1950) 35.

Edmonton: [a]*Alberta. [b]none. [c]North Saskatchewan R. [d]University of Alberta (1906) 24 – 1550; Athabasca University (1972) 0.6 – 40. [e]University . . . Library (1909) 1685; Public Library (1913) 679.

Halifax: [a]*Nova Scotia. [b]none. [c]Atlantic O. [d]Dalhousie University (1818) 9 – 1190; Saint Mary's University (1802) 4 – 137; Mount Saint Vincent University (1914) 2 – 144; Nova Scotia Technical College (univ) (1909) 0.5 – 75; University of King's College (1789 at Windsor, Nova Scotia; relocated 1923) 0.3 – 14. [e]Dalhousie University Library (1867) 443; City Regional Library (1873) 203; Nova Scotia Provincial Library (1949) 150; Saint Mary's University Library (1841) 133.

Hamilton: [a]Ontario. [b]none. [c]Ontario L. [d]McMaster University (1887) 14 – 850. [e]Public Library (1890) 723; Mills Memorial Library of . . . University (1887) 586.

Jonquiere: [a]Quebec. [b]none. [c]Sable R. [d]none. [e]Bibliotheque Municipale (1954) 19.

Kingston: [a]Ontario. [b]none. [c]Saint Lawrence R; Ontario L. [d]Queen's University at Kingston (1841) 13 – 860; Royal Military College of Canada (univ) (1874) 0.7 – 141. [e]Douglas Library of . . . University (1841) 2234; Public Library (1834) 185.

Kitchener: [a]Ontario. [b]for Berlin. [c]Grand R. [d]University of Waterloo (at Waterloo, nr Kitchener) (1864) 17 – 850; Wilfrid Laurier University (at Waterloo, nr Kitchener) (1910) 6 – 205. [e]University . . . Library (at Waterloo, nr Kitchener) (1958) 900; Public Library (1884) 335; . . . University Library (at Waterloo, nr Kitchener) (1911) 331.

Laval: [a]Quebec; Jesus I. [b]inc Chomedey, Laval-des-Rapides, Sainte-Rose. [c]Prairies R; Mille-Iles R. [d]none. [e]Bibliotheque de l'Institut de Microbiologie et d'Hygiene de l'Universite de Montreal (1948) 26. [f]Chomedey: 30 (1961); Laval-des-Rapides: 19 (1961), 5 (1951), 3 (1931), 1 (1911); Sainte-Rose: 8 (1961), 4 (1951), 2 (1931), 1 (1901), 0.7 (1871).

London: [a]Ontario. [b]none. [c]Thames R. [d]University of Western Ontario (1878) 23 – 1275. [e]D. B. Weldon Library of the University . . . (1908) 1269; Public Library and Art Museum (1894) 469.

Longueuil: [a]Quebec. [b]inc Jacques-Cartier since 1970. [c]Saint Lawrence R. [d]none. [e]Bibliotheque du College Edouard-Montpetit (1960) 70; Bibliotheque Municipale (1967) 55. [f]Jacques-Cartier: 41 (1961), 22 (1951).

Mississauga: [a]Ontario. [b]inc Cooksville. [c]Cooksville Creek. [d]none. [e]Public Library (bef 1953) 250.

Montreal: [a]Quebec; Montreal I. [b]for Ville-Marie. [c]Saint Lawrence R; Ottawa R. [d]Universite de Montreal (1876) 32 – 1772; Concordia University (1899) 25 – 1524; McGill University (1821) 20 – 3100; Universite du Quebec (alt University of Quebec), Montreal campus (1969) 14 – 420. [e]McGill University Library (1821) 3297; Bibliotheque de la Ville (alt City Library) (1902) 1101; Bibliotheque de l'Universite de Montreal (1943) 1012.

Montreal-Nord: [a]Quebec. [b]alt Montreal North. [c]Prairies R. [d]none. [e]Bibliotheque Publique (1970) 41.

Niagara Falls: [a]Ontario. [b]none. [c]Niagara R. [d]none. [e]Public Library (1878) 131.

Oshawa: [a]Ontario. [b]none. [c]Ontario L. [d]none. [e]Public Library (1864) 255.

*Ottawa: [a]Ontario. [b]for Bytown. [c]Ottawa R; Rideau R. [d]University of Ottawa (1848) 19 – 1209; Carleton University (1942) 16 – 635. [e]Morisset Library of the University . . . (1903) 800; National Library of Canada (1953) 720; Murdoch Maxwell MacOdrum Library of Carleton University (1942) 709; Public Library (1906) 545; Library of Parliament (1867) 450.

Quebec: [a]*Quebec. [b]none. [c]Saint Lawrence R; Saint-Charles R. [d]Universite du Quebec (alt University of Quebec) (1968) 31 – 1008 (no campus at Quebec); Universite Laval (1852) 19 – 2935. [e]Bibliotheque de l'Universite Laval (1852) 702; Bibliotheque de la Legislature (1792) 540; Bibliotheque Municipale (1848) 100.

Regina: [a]*Saskatchewan. [b]none. [c]Waskana Creek. [d]University of Regina (1911) 6 – 334. [e]University . . . Library (1934) 403; Public Library (1909) 290; Saskatchewan Provincial Library (1953) 160.

Saint Catharines: [a]Ontario. [b]none. [c]Twelve Mile Creek. [d]Brock University (1962) 5 – 251. [e]Public Library (1883) 251; . . . University Library (1964) 180.

Saint John: [a]New Brunswick. [b]none. [c]B of Fundy of Atlantic O; Saint John R. [d]none. [e]Saint John Regional Library (1883) 187.

Saint John's: [a]*Newfoundland; Newfoundland I. [b]none. [c]Atlantic O. [d]Memorial University of Newfoundland (1925) 9 – 597. [e]Newfoundland Public Library Services (1935) 558; . . . University Library (1925) 470.

Saskatoon: [a]Saskatchewan. [b]none. [c]South Saskatchewan R. [d]University of Saskatchewan (1907) 14 – 1312. [e]University . . . Library (1910) 626; Wheatland Regional Library (1967) 392; Public Library (1913) 295.

Sault Sainte Marie: [a]Ontario. [b]none. [c]Saint Marys R. [d]none. [e]Public Library (1915?) 91; Algoma University College Library (1967) 40.

Sherbrooke: [a]Quebec. [b]none. [c]Saint-Francois R; Magog R. [d]Universite de Sherbrooke (1954) 8 – 833; Bishop's University (at Lennoxville, nr Sherbrooke) (1843) 1 – 70. [e]Bibliotheque de l'Universite . . . (1961) 285; John Bassett Memorial Library of . . . University (at Lennoxville, nr Sherbrooke) (1843) 115; Bibliotheque du Seminaire de Sherbrooke (1875) 65; Bibliotheque Municipale (1954) 65.

Sudbury: [a]Ontario. [b]none. [c]Junction Creek; Ramsey L. [d]Laurentian University of Sudbury (1960) 7 – 258. [e]. . . University Library (1960) 284; Public Library (1896) 187.

Sydney: [a]Nova Scotia; Cape Breton I. [b]none. [c]Atlantic O. [d]none. [e]Cape Breton Regional Library (bef 1925) 131; Sydney Campus Library of Saint Francis Xavier University (1951) 81.

Thunder Bay: [a]Ontario. [b]inc Fort William, Port Arthur. [c]Kaministikwia R; Thunder B of Superior L. [d]Lakehead University (1946) 4 – 225. [e]. . . University Library (1948) 240; Public Library (1881) 213. [f]Fort William: 45 (1961), 35 (1951), 26 (1931), 4 (1901), 0.7 (1881); Port Arthur: 45 (1961), 31 (1951), 20 (1931), 3 (1901), 1 (1881).

Toronto: [a]*Ontario. [b]for York. [c]Humber R; Ontario L. [d]University of Toronto (1827) 46 – 4969; York University (1959) 24 – 1075. [e]University . . . Library (1842) 4382; North York Public Library (1955) 1060; Toronto Public Library (1884) 864; Metropolitan Toronto Central Library (1909) 772; Scott Library of . . . University (1959) 766.

Trois-Rivieres: [a]Quebec. [b]alt Three Rivers. [c]Saint Lawrence R; Saint-Maurice R; Sainte-Marguerite R. [d]Universite du Quebec (alt University of Quebec), Trois-Rivieres campus (1969) 6 – 210.

^{e}Bibliotheque de l'Universite du Quebec a Trois-Rivieres (1963) 120; Bibliotheque Municipale (1946) 120.

Vancouver: ^{a}British Columbia. ^{b}none. ^{c}Burrard Inlet of Strait of Georgia; Fraser R. ^{d}University of British Columbia (1908) 25 – 1792; Simon Fraser University (1963) 8 – 451. ^{e}University . . . Library (1912) 1670; Public Library (1887) 760; . . . University Library (1964) 486.

Victoria: ^{a}*British Columbia; Vancouver I. ^{b}none. ^{c}Juan de Fuca Strait. ^{d}University of Victoria (1902) 7 – 560. ^{e}McPherson Library of the University . . . (1902) 740; Provincial Library (1863) 500; Greater Victoria Public Library (1864) 381.

Windsor: ^{a}Ontario. ^{b}none. ^{c}Detroit R. ^{d}University of Windsor (1857) 11 – 502. ^{e}University . . . Library (1857) 800; Public Library (1894) 564.

Winnipeg: ^{a}*Manitoba. ^{b}none. ^{c}Red R; Assiniboine R. ^{d}University of Manitoba (1877) 21 – 1294; University of Winnipeg (1871) 6 – 175. ^{e}University of Manitoba Library (1885) 1030; Public Library (1905) 407; University of Winnipeg Library (1871) 265.

Costa Rica

*San Jose: ^{b}none. ^{c}Torres R. ^{d}Universidad de Costa Rica (1814) 28 – 2662. ^{e}Biblioteca de la Universidad . . . (1946) 200; Biblioteca Nacional (1888) 175.

Cuba

Camaguey: ^{b}for Puerto Principe. ^{c}Jatibonico R; Tinima R. ^{d}Universidad Ignacio Agramonte (bef 1974) 5 – ? ^{e}Biblioteca de las Escuelas Pias (?) 10; Biblioteca Municipal (1938) ?

Cienfuegos: ^{b}none. ^{c}Cienfuegos B of Caribbean Sea. ^{d}none. ^{e}Biblioteca Pedro Modesto Hernandez del Instituto de Segunda Ensenanza (1937) 5; Biblioteca Publica Municipal Jose Marti (1931) 1.

Guantanamo: ^{b}none. ^{c}Guaso R. ^{d}none. ^{e}Biblioteca Popular Enrique Jose Varona (1938) ?

*Havana: ^{b}alt Habana; inc Marianao, San Miguel del Padron since 1969; off La Habana. ^{c}G of Mexico. ^{d}Universidad de La Habana (1728) 54 – 3066; Universidad Catolica Santo Tomas de Villanueva (1957) ? – ? ^{e}Biblioteca Nacional Jose Marti (1901) 800; Biblioteca del Instituto de Literatura y Linguistica (1793) 360; Biblioteca Central Ruben Martinez Villena de la Universidad de La Habana (1728) 208. ^{f}Marianao: 325 (1962e), 219 (1953), 71 (1931), 5 (1899), 3 (1861); San Miguel del Padron: 61 (1953), 3 (1931), 2 (1899).

Holguin: ^{b}none. ^{c}nr Maranon R. ^{d}none. ^{e}Biblioteca Jose Marti del Instituto de Segunda Ensenanza (1938) 3; Biblioteca Alex Urquiola (?) ?

Matanzas: ^{b}none. ^{c}Matanzas B of G of Mexico; San Juan R; Yumuri R. ^{d}none. ^{e}Biblioteca Publica Gener y del Monte (1835) 24.

Santa Clara: ^{b}for Villa Clara. ^{c}Monte R; Sabana R. ^{d}Universidad Central de Las Villas (1948) 8 – 600. ^{e}Biblioteca General de la Universidad . . . (1953) 160; Biblioteca Provincial Marti (1927) 9.

Santiago de Cuba: ^{b}alt Santiago. ^{c}Caribbean Sea. ^{d}Universidad de Oriente (1947) 16 – 800. ^{e}Biblioteca de la Universidad . . . (1948) 175; Biblioteca del Instituto Pedagogico (1916) 32; Biblioteca Municipal Elvira Cape de Bacardi (1899) 18.

Dominican Republic

Santiago de los Caballeros: ^{b}alt Santiago. ^{c}Yaque del Norte R. ^{d}Universidad Catolica Madre y Maestra (1962) 3 – 190. ^{e}Biblioteca de la Universidad . . . (1962) 20; Biblioteca del Ateneo Amantes de la Luz (1875) 10.

*Santo Domingo: ^{b}for Ciudad Trujillo, Trujillo. ^{c}Caribbean Sea; Ozama R. ^{d}Universidad Autonoma de Santo Domingo (1538) 29 – 1178; Universidad Nacional Pedro Henriquez Urena (1966) 8 – 500. ^{e}Biblioteca de la Universidad Autonoma de Santo Domingo (1538) 864; Biblioteca Nacional (1971) 154; Biblioteca Municipal (1914) 35.

El Salvador

San Miguel: ^{b}none. ^{c}Grande de San Miguel R. ^{d}none. ^{e}Biblioteca de la Escuela Oficial de Ciencias y Letras (bef 1942) ?

*San Salvador: [b]none. [c]Acelhuate R; Urbina R; Aseseco R. [d]Universidad de El Salvador (1841) 24 – 1972; Universidad Jose Simeon Canas (1965) 3 – 142. [e]Biblioteca Nacional (1870) 95; Biblioteca Central de la Universidad de El Salvador (1847) 95.

Santa Ana: [b]none. [c](Quebrada) Santa Lucia Creek. [d]none. [e]Biblioteca Municipal Camilo Arevalo (1896) 5.

Guatemala

*Guatemala: [b]alt Guatemala City. [c]Barranquila R; Barranca R. [d]Universidad de San Carlos de Guatemala (1676 at Antigua, relocated 1776) 24 – 1235; Universidad Rafael Landivar (1961) 3 – 250; Universidad Mariano Galvez (1966) 2 – 73; Universidad del Valle de Guatemala (1961) 0.3 – 50; Universidad Francisco Marroquin (1971) 0.3 – 60. [e]Biblioteca Nacional de Guatemala (1879) 351.

Mixco: [b]none. [c]nr Villalobos R; nr Amatitlan L. [d]none. [e]Biblioteca de Mixco (bef 1941) ?

Haiti

*Port-au-Prince: [b]none. [c]G of Gonaives of Caribbean Sea. [d]Universite d'Etat d'Haiti (1944) 2 – 211. [e]Bibliotheque de l'Institut Francais d'Haiti (1945) 32; Bibliotheque Nationale d'Haiti (1940) 25.

Honduras

San Pedro Sula: [b]none. [c]Piedras R. [d]none. [e]Biblioteca de la Sociedad la Juventud (1913) 2 (in 1938).

*Tegucigalpa: [b]none. [c]Choluteca R; Chiquito R. [d]Universidad Nacional Autonoma de Honduras (1845) 15 – 763. [e]Biblioteca Nacional de Honduras (1880) 60; Archivo Nacional de Honduras (1880) 40; Biblioteca de la Universidad . . . (1848) 29.

Jamaica

*Kingston: [b]none. [c]Caribbean Sea. [d]University of the West Indies (at Mona, nr Kingston) (1921) 7 – 694 [Mona campus: (1921) 4 – 222]. [e]Jamaica Library Service (1947) 983; University . . . Library (at Mona, nr Kingston) (1948) 378 [Mona campus: (1948) 210].

Martinique

*Fort-de-France: [b]for Fort-Royal. [c]Caribbean Sea; Madame R. [d]none. [e]Bibliotheque Schoelcher (1884) 12 (in 1938).

Mexico

Acapulco: [a]Guerrero. [b]off Acapulco de Juarez. [c]Pacific O. [d]none. [e]Biblioteca Publica Doctor Alfonso G. Alarcon (1960) 7.

Aguascalientes: [a]*Aguascalientes. [b]none. [c]nr San Pedro (alt Aguscalientes, Verde) R. [d]Universidad Autonoma de Aguascalientes (1867) 0.8 – ? [e]Biblioteca Enrique Fernandez Ledesma (1953) 5.

Celaya: [a]Guanajuato. [b]none. [c]Laja R. [d]none. [e]Biblioteca Publica Municipal (1944) 4.

Chihuahua: [a]*Chihuahua. [b]none. [c]Chuviscar R. [d]Universidad Autonoma de Chihuahua (1954) 5 – 360. [e]Biblioteca Municipal Miguel de Cervantes Saavedra (1943) 14; Biblioteca de la Universidad . . . (1955) 8.

Ciudad Juarez: [a]Chihuahua. [b]alt Juarez; for Paso del Norte. [c]Rio Grande R. [d]Universidad Autonoma de Ciudad Juarez (1968) 1 – 137. [e]Biblioteca Publica Municipal Arturo Tolentino (1945) 25.

Ciudad Madero: [a]Tamaulipas. [b]alt Madero; for Dona Cecilia, Villa de Cecilia. [c]Panuco R. [d]none. [e]Biblioteca del Instituto Tecnologico Regional (1954) 3; Biblioteca Publica de la Sociedad Cultural Amado Nervo (1920) 1.

Ciudad Obregon: [a]Sonora. [b]alt Obregon; for Cajeme. [c]nr Yaqui R. [d]none. [e]Biblioteca Central del Instituto Tecnologico de Sonora (?) 8; Biblioteca Popular (1958) 2.

Ciudad Victoria: [a]*Tamaulipas. [b]alt Victoria; for Nuevo Santander. [c]San Marcos R. [d]Universidad Autonoma de Tamaulipas (1955) 7 – 300. [e]Biblioteca Central de la Universidad . . . (1955?) 11; Biblioteca Publica del Estado de Tamaulipas Adolfo Ruiz Cortines (1962) 9.

Coatzacoalcos: [a]Veracruz. [b]for Puerto Mexico. [c]B of Campeche of G of Mexico; Coatzacoalcos R. [d]none. [e]Biblioteca Publica Municipal Salvador Diaz Miron (1961) 4.

Cordoba: [a]Veracruz. [b]none. [c]Blanco R. [d]none. [e]Biblioteca Publica Municipal (1968?) ?

Cuernavaca: [a]*Morelos. [b]none. [c]Cuernavaca R; Amatitlan R. [d]Universidad Autonoma de Morelos (1872) 1 – 235. [e]Biblioteca Central Universitaria Profesor Miguel Salinas (1887) 20.

Culiacan: [a]*Sinaloa. [b]none. [c]Culiacan R. [d]Universidad Autonoma de Sinaloa (1873) 7 – 403. [e]Biblioteca Central Licenciado Eustaquio Buelna de la Universidad . . . (1875) 30; Biblioteca Publica del Estado de Sinaloa (1940) 15.

Durango: [a]*Durango. [b]for Ciudad de Victoria, Victoria; off Victoria de Durango. [c]Tunal R. [d]Universidad Autonoma Juarez de Durango (1856) 2 – 740. [e]Biblioteca Publica del Estado de Durango (1853) 20; Biblioteca de la Universidad . . . (1957) 16.

Guadalajara: [a]*Jalisco. [b]none. [c]San Juan de Dios R; Agua Azul L. [d]Universidad de Guadalajara (1791) 21 – 3169; Universidad Autonoma de Guadalajara (1935) 12 – 802. Universidad Jaime Balmes de Occidente (1965) 0.5 – 30. [e]Biblioteca Publica del Estado de Jalisco (1861) 300.

Guanajuato: [a]*Guanajuato. [b]none. [c]Guanajuato R. [d]Universidad de Guanajuato (1732) 3 – 1153. [e]Biblioteca Armando Olivares de la Universidad . . . (1732) 40.

Hermosillo: [a]*Sonora. [b]none. [c]Sonora R. [d]Universidad Autonoma de Sonora (1938) 4 – 300; Universidad Militarizada de Mexico (?) ? – ? [e]Biblioteca Central de la Universidad de Sonora (1949) 56; Biblioteca y Museo de Sonora (1949) 40.

Irapuato: [a]Guanajuato. [b]none. [c]Irapuato R. [d]none. [e]Biblioteca de la Escuela Preparatoria de la Universidad de Guanajuato (1947) 5; Biblioteca Publica Federal Benito Juarez (1964) 2.

Jalapa: [a]*Veracruz. [b]alt Xalapa; off Jalapa Enriquez. [c](Arroyo de) Santiago Creek. [d]Universidad Veracruzana (1846) 10 – 1823 [Jalapa campus: (1846) ? – ?]. [e]Biblioteca Central de la Universidad . . . (1959) 46; Biblioteca de la Escuela Normal Veracruzana Enrique C. Rebsamen (1893) 13; Biblioteca Publica Juan Diaz Covarrubias (1948) 7.

Leon: [a]Guanajuato. [b]alt Leon de los Aldamas. [c]Turbio R. [d]none. [e]Biblioteca de la Escuela Preparatoria (1877) 33; Biblioteca del Instituto America (1954) 6; Biblioteca Publica Municipal (1924) 2.

Matamoros: [a]Tamaulipas. [b]for Nueva Santander. [c]Rio Grande R. [d]none. [e]Biblioteca Profesor Juan B. Tijerina (1958) 6.

Mazatlan: [a]Sinaloa. [b]none. [c]Pacific O. [d]none. [e]Biblioteca de la Escuela Preparatoria de la Universidad Autonoma de Sinaloa (1958) 9; Biblioteca Publica Municipal Ingeniero Manuel Bonilla (1946) 5.

Merida: [a]*Yucatan. [b]for T'ho. [c]none. [d]Universidad Autonoma de Yucatan (1624) 3 – 680. [e]Biblioteca Central de la Universidad . . . (1922) 38; Biblioteca Publica del Estado de Yucatan Manuel Cepeda Peraza (1867) 16.

Mexicali: [a]*Baja California. [b]none. [c]Nuevo R. [d]Universidad Autonoma de Baja California (1957) 5 – 1200. [e]Biblioteca Publica del Estado de Baja California (?) 3; Biblioteca Publica Municipal Esperanza Lopez Mateos (1961) 1.

*Mexico: [a]*Distrito Federal. [b]alt Mexico City; for Tenochtitlan; inc Guadalupe Hidalgo (for Gustavo A. Madero) since 1973. [c]Piedad R; Tacubaya R. [d][1] Universidad Nacional Autonoma de Mexico (1551) 120 – 15,964; Instituto Politecnico Nacional (univ) (1931) 62 – 11,000; Universidad Iberoamericana (1943) 7 – 983; Universidad Autonoma Metropolitana (1973) 6 – 1038; Universidad La Salle (1961) 3 – 719; Universidad del Valle de Mexico (1960) 2 – 250; Universidad Tecnologica de Mexico (1946) 2 – 162; Universidad Anahuac (1963) 2 – 130. [e]Biblioteca Nacional de Mexico (1844) 1000; Biblioteca Nacional de Antropologia e Historia (1831) 300; Biblioteca Publica Miguel Lerdo de Tejada (1928) 230; Biblioteca Central de la Universidad Nacional Autonoma de Mexico (1924) 200. [f]Guadalupe Hidalgo: 103 (1960), 60 (1950), 11 (1921), 6 (1900).

Minatitlan: [a]Veracruz. [b]none. [c]Coatzacoalcos R. [d]none. [e]Biblioteca Publica de la Escuela Jose Ma. Morelos y Pavon (1955) 3.

Monclova: [a]Coahuila. [b]none. [c]Monclova R. [d]none.

Monterrey: [a]*Nuevo Leon. [b]none. [c]Santa Catarina R; Silla R. [d]Universidad Autonoma de Nuevo

[1]Universities with 1000 or more students.

Leon (1857) 21 – 1380; Instituto Tecnologico y de Estudios Superiores de Monterrey (univ) (1943) 6 – 740; Universidad de Monterrey (1969?) 2 – 379; Universidad Regiomontana (1951) 2 – 337. ^{e}Biblioteca de la Universidad de Nuevo Leon (1933) 136; Biblioteca del Instituto . . . (1943) 105.

Morelia: ^{a}*Michoacan. ^{b}for Valladolid. ^{c}Grande R; Chiquito R. ^{d}Universidad Autonoma Michoacana de San Juan Nicolas de Hidalgo (1540 at Patzcuaro, relocated 1580) 11 – 1282. ^{e}Biblioteca Publica Universitaria (1874) 44.

Netzahualcoyotl: ^{a}Mexico. ^{b}none. ^{c}nr Ixtapan R. ^{d}none. ^{e}Biblioteca Manuel Pavon G. (?) 3.

Nuevo Laredo: ^{a}Tamaulipas. ^{b}none. ^{c}Rio Grande R. ^{d}none. ^{e}Biblioteca Benito Juarez (1941) 4; Biblioteca Publica Municipal (1945) 3.

Oaxaca: ^{a}*Oaxaca. ^{b}off Oaxaca de Juarez. ^{c}Jalatlaco R. ^{d}Universidad Autonoma Benito Juarez de Oaxaca (1827) 2 – 265. ^{e}Biblioteca General de la Universidad . . . (1826) 42.

Orizaba: ^{a}Veracruz. ^{b}none. ^{c}Blanco R. ^{d}Universidad Veracruzana, Orizaba campus (?) ? – ? ^{e}Biblioteca del Instituto Tecnologico Regional (1957?) 3; Biblioteca Rafael Delgado (1929) 2.

Pachuca de Soto: ^{a}*Hidalgo. ^{b}none. ^{c}Avenida de Pachuca R. ^{d}Universidad Autonoma de Hidalgo (1869) 1 – ? ^{e}Biblioteca Publica del Estado de Hidalgo (1933) 3.

Poza Rica: ^{a}Veracruz. ^{b}alt Poza Rica de Hidalgo. ^{c}Cazones R. ^{d}none. ^{e}Biblioteca Publica del Sindicado de Trabajadores Petroleros de la Republica Mexicana (1958) 6.

Puebla: ^{a}*Puebla. ^{b}off Puebla de Zaragoza. ^{c}Atoyac R; Alseseca R; San Francisco R. ^{d}Universidad Autonoma de Puebla (1578) 14 – 924; Universidad de las Americas (alt University of the Americas) (1940 at Mexico, relocated 1970?) 1 – 150. ^{e}Biblioteca Publica Jose M. Lafragua de la Universidad Autonoma de Puebla (1885) 117; Biblioteca de la Universidad de las Americas (1940 at Mexico, relocated ?) 100; Biblioteca Publica del Estado de Puebla (1648) 43.

Queretaro: ^{a}*Queretaro. ^{b}none. ^{c}Queretaro R. ^{d}Universidad Autonoma de Queretaro (1775) 1 – 300. ^{e}Biblioteca Central de la Universidad . . . (1963) 23; Biblioteca del Museo Regional (1944) 12; Biblioteca Publica Josefa Ortiz de Dominguez (1963) 6.

Reynosa: ^{a}Tamaulipas. ^{b}alt Reinosa. ^{c}Rio Grande R. ^{d}none. ^{e}Biblioteca Publica Municipal Amalia de Castillo Ledon (1955) 8.

Saltillo: ^{a}*Coahuila. ^{b}none. ^{c}(Arroyo) Barranca Creek; (Arroyo) Tortola Creek. ^{d}Universidad Autonoma de Coahuila (1867) 6 – 930. ^{e}Biblioteca Publica del Estado de Coahuila Manuel Muzquiz Blanco (1942) 12.

San Luis Potosi: ^{a}*San Luis Potosi. ^{b}none. ^{c}Santiago R. ^{d}Universidad Autonoma de San Luis Potosi (1624) 7 – 670. ^{e}Biblioteca Publica de la Universidad . . . (1877) 65.

Tampico: ^{a}Tamaulipas. ^{b}none. ^{c}Panuco R. ^{d}none. ^{e}Biblioteca Publica Municipal (1941) 10.

Taxco: ^{a}Guerrero. ^{b}off Taxco de Alarcon. ^{c}nr Amacuzac R. ^{d}none. ^{e}Biblioteca Publica Municipal Alejandro Gomez Maganda (1953) 1.

Tepic: ^{a}*Nayarit. ^{b}none. ^{c}Mololoa R. ^{d}Universidad de Nayarit (1930) 2 – 230. ^{e}Biblioteca Publica Municipal Francisco I. Madero (1925) 5.

Tijuana: ^{a}Baja California. ^{b}none. ^{c}Tijuana R. ^{d}none. ^{e}Biblioteca del Banco de Baja California (1964) 10; Biblioteca Miguel de Cervantes Saavedra (1926) 4.

Toluca: ^{a}*Mexico. ^{b}off Toluca de Lerdo. ^{c}Xicualtenco R. ^{d}Universidad Autonoma del Estado de Mexico (1828) 5 – 2000. ^{e}Biblioteca Publica Central del Estado de Mexico (1827) 45.

Torreon: ^{a}Coahuila. ^{b}none. ^{c}Nazas R. ^{d}none. ^{e}Biblioteca Municipal Jose Garcia de Letona (1945) 11.

Uruapan del Progreso: ^{a}Michoacan. ^{b}alt Uruapan. ^{c}Cupatitzio R. ^{d}none. ^{e}Biblioteca Licenciado Justo Sierra (1958) 2.

Veracruz: ^{a}Veracruz. ^{b}off Veracruz Llave. ^{c}G of Mexico. ^{d}Universidad Veracruzana, Veracruz campus (?) ? – ? ^{e}Biblioteca Publica Municipal Venustiano Carranza (1872) 15.

Villahermosa: ^{a}*Tabasco. ^{b}for San Juan Bautista. ^{c}Grijalva R. ^{d}Universidad Juarez Autonoma de Tabasco (1879) 0.8 – 171. ^{e}Biblioteca Jose Marti de la Universidad . . . (1944) 23.

Netherlands Antilles

*Willemstad: ^{a}Curacao I. ^{b}none. ^{c}Caribbean Sea. ^{d}none. ^{e}Openbare Leeszaal en Bibliotheek (1920) 67.

Nicaragua
*Managua: ^{b}none. ^{c}Managua L. ^{d}Universidad Centroamericana (1960) 2 – 196. ^{e}Biblioteca Nacional (1882) 75.

Panama
Colon: ^{a}Manzanillo I. ^{b}for Aspinwall. ^{c}Caribbean Sea. ^{d}none. ^{e}Biblioteca Mateo Iturralde (1908) 6 (in 1940).
*Panama: ^{b}alt Panama City. ^{c}B of Panama of Pacific O. ^{d}Universidad de Panama (1935) 32 – 1276; Universidad Santa Maria la Antigua (1965) 2 – 157. ^{e}Biblioteca de la Universidad de Panama (1935) 250; Canal Zone Library-Museum (at Ancon, Canal Zone, nr Panama) (1914) 249; Biblioteca Nacional (1892) 200.

Puerto Rico
Bayamon: ^{b}none. ^{c}Bayamon R. ^{d}Bayamon Central University (college) (1961) 2 – 80. ^{e}Bayamon Campus Library of Inter American University of Puerto Rico (1967) 25.
Carolina: ^{b}none. ^{c}Grande de Loiza R. ^{d}none. ^{e}Biblioteca del Seminario Episcopal del Caribe (1961) 16.
Ponce: ^{b}none. ^{c}Caribbean Sea. ^{d}Universidad Catolica de Puerto Rico (alt Catholic University of Puerto Rico) (1948) 7 – 400. ^{e}Biblioteca de la Universidad . . . (1948) 145; Biblioteca Publica (1890) 21.
*San Juan: ^{b}inc Rio Piedras since 1950. ^{c}Atlantic O. ^{d}Universidad de Puerto Rico (alt University of Puerto Rico (1903) 52 – 3337 [San Juan campus: (1903) 28 – ?]; Inter American University of Puerto Rico (1912) 22 – 1057 [San Juan campus: (1962) 6 – ?]; World University (1965) 4 – ? ^{e}Biblioteca General de la Universidad . . . (alt University . . . General Library) (1903) 498; San Juan Campus Library of Inter American University of Puerto Rico (1961) 87; Biblioteca General de Puerto Rico (1967) 65. ^{f}Rio Piedras: 132 (1950), 13 (1930), 2 (1899).

Trinidad and Tobago
*Port of Spain: ^{a}Trinidad I. ^{b}none. ^{c}G of Paria of Caribbean Sea. ^{d}University of the West Indies, Trinidad campus (at Saint Augustine, nr Port of Spain) (1960) 2 – 337. ^{e}Central Library of Trinidad and Tobago (1851) 447.

USA
Abilene: ^{a}Texas. ^{b}none. ^{c}Catclaw Creek; Kirby L; Lytle L. ^{d}Abilene Christian University (1906) 4 – 222; Hardin-Simmons University (1891) 2 – 113. ^{e}Hardin-Simmons University Library (1892) 278; Public Library (1909) 177; Brown Library of Abilene Christian University (1906) 165; Jay-Rollins Library of McMurry College (1923) 121.
Akron: ^{a}Ohio. ^{b}none. ^{c}Cuyahoga R; Little Cuyahoga R. ^{d}Kent State University (at Kent, nr Akron) (1910) 28 – 2020 [Kent campus: (1910) 21 – 1748]; University of Akron (1870) 22 – 1075 [Akron campus: (1870) 22 – ?]. ^{e}. . . University Library (at Kent, nr Akron) (1913) 1066; Akron-Summit County Public Library (1874) 833; Bierce Library of the University . . . (1872) 352.
Alameda: ^{a}California. ^{b}none. ^{c}San Francisco B of Pacific O; San Leandro B of Pacific O. ^{d}none. ^{e}Free Library (1879) 140.
Albany: ^{a}Georgia. ^{b}none. ^{c}Flint R. ^{d}none. ^{e}Dougherty Public Library (1904) 113; Margaret Rood Hazard Library of Albany State College (1903) 96.
Albany: ^{a}*New York. ^{b}for Fort Orange. ^{c}Hudson R. ^{d}State University of New York (1844) 350 – 25,500 [Albany campus: (1844) 15 – 1514]; College of Saint Rose (univ) (1920) 2 – 147. ^{e}New York State Library (1818) 5000; Albany Campus Library of the . . . University (1844) 910; Public Library (1792?) 264.
Albuquerque: ^{a}New Mexico. ^{b}none. ^{c}Rio Grande R. ^{d}University of New Mexico (1889) 22 – 1419

[Albuquerque campus: (1889) 21 – ?] ; University of Albuquerque (college) (1920) 3 – 173. [e]University of New Mexico General Library (1892) 504; Public Library (1891) 333.

Alexandria: [a]Louisiana. [b]none. [c]Red R. [d]none. [e]Rapides Parish Library (1942) 143; Alexandria Campus Library of Louisiana State University (1960) 80.

Alexandria: [a]Virginia. [b]none. [c]Potomac R. [d]none. [e]Fairfax County Public Library (at Springfield, Virginia, nr Alexandria) (1939) 1203; Alexandria Library (1794) 249; Van Noy Army Engineer Center Library (at Fort Belvoir, nr Alexandria) (1939) 106.

Allentown: [a]Pennsylvania. [b]for Northampton. [c]Lehigh R. [d]none. [e]Public Library (1912) 188; John A. W. Haas Library of Muhlenberg College (1867) 163.

Alton: [a]Illinois. [b]none. [c]Mississippi R. [d]Southern Illinois University, Edwardsville campus (at Edwardsville, nr Alton) (1957) 13 – 699. [e]Elijah P. Lovejoy Edwardsville Campus Library of . . . University (at Edwardsville, nr Alton) (1957) 512; Hayner Public Library (1891) 89.

Altoona: [a]Pennsylvania. [b]none. [c]Little Juniata R. [d]none. [e]Altoona Area Public Library (1927) 121.

Amarillo: [a]Texas. [b]none. [c]Amarillo L. [d]none. [e]Public Library (1902) 285.

Anaheim: [a]California. [b]none. [c]Santa Ana R. [d]none. [e]Public Library (1908) 326.

Anchorage: [a]Alaska. [b]none. [c]Cook Inlet of Pacific O. [d]University of Alaska, Anchorage campus (1954) 7 – 678. [e]Z. J. Loussac Public Library (1945) 154.

Anderson: [a]Indiana. [b]none. [c]West Fork of White R. [d]none. [e]Public Library (1890) 237.

Annapolis: [a]*Maryland. [b]none. [c]Severn R. [d]Saint John's College (univ) (1696) 0.6 – 95 [Annapolis campus: (1696) 0.3 – 56]. [e]Public Library of Annapolis and Anne Arundel County (bef 1923) 564; Nimitz Library of the United States Naval Academy (1845) 400; Maryland State Library (1826) 112.

Ann Arbor: [a]Michigan. [b]none. [c]Huron R. [d]University of Michigan (1817 at Detroit, relocated 1837) 46 – 5362 [Ann Arbor campus: (1837) 38 – ?] ; Eastern Michigan University (at Ypsilanti, nr Ann Arbor) (1849) 20 – 850. [e]Ann Arbor Campus Library of the University . . . (1838) 3924; . . . University Library (at Ypsilanti, nr Ann Arbor) (1849) 351; Public Library (1856) 110.

Appleton: [a]Wisconsin. [b]none. [c]Fox R. [d]Lawrence University (college) (1847) 1 – 140. [e]Seeley G. Mudd Library of . . . University (1847) 201; Public Library (1872) 137.

Arden-Arcade: [a]California. [b]none. [c]American R. [d]none. [e]Arcade Branch of Sacramento City-County Library (bef 1954) 45; Arden Branch of Sacramento City-County Library (bef 1954) 38.

Arlington: [a]Texas. [b]none. [c]West Fork of Trinity R. [d]University of Texas, Arlington campus (1895) 16 – 854. [e]Arlington Campus Library of the University . . . (1895) 442; Public Library (1922) 139.

Arlington: [a]Virginia. [b]none. [c]Potomac R. [d]George Mason University (at Fairfax, nr Arlington) (1957) 8 – 481. [e]Public Library (1937) 417; Army Library (1944) 267; Northern Virginia Community College Library (at Annandale, nr Arlington) (1965) 121; Charles Rogers Fenwick Library of . . . University (at Fairfax, nr Arlington) (1957) 117.

Asheville: [a]North Carolina. [b]none. [c]French Broad R. [d]none. [e]Pack Memorial Public Library (1879) 206.

Athens: [a]Georgia. [b]none. [c]Oconee R. [d]University of Georgia (1785) 23 – 1968. [e]University . . . Library (1831) 1523; Athens Regional Library (1888) 159.

Atlanta: [a]*Georgia. [b]none. [c]Chattahoochee R. [d]Georgia State University (1913) 20 – 971; Georgia Institute of Technology (univ) (1885) 11 – 1139 [Atlanta campus: (1885) 9 – 1049] ; Emory University (1836) 7 – 2208; Atlanta University (1865) 1 – 131; Oglethorpe University (1835 at Milledgeville, relocated 1870) 0.9 – 51. [e]Emory University Library (1836) 1150; Public Library (1867) 848; Price Gilbert Memorial Library of . . . Institute (1901) 820; William Russell Pullen Library of Georgia State University (1931) 513.

Atlantic City: [a]New Jersey. [b]none. [c]Atlantic O. [d]none. [e]Ocean City Free Public Library (at Ocean City, nr Atlantic City) (bef 1927) 53; Atlantic City Free Public Library (1900) 52.

Augusta: [a]Georgia. [b]none. [c]Savannah R. [d]Augusta College (univ) (1925) 4 – 156. [e]Augusta Regional Library (1848) 320; Reese Library of . . . College (1957) 166.

Aurora: [a]Colorado. [b]none. [c]nr South Platte R. [d]none. [e]Public Library (1929) 109.

Aurora: [a]Illinois. [b]none. [c]Fox R. [d]none. [e]Public Library (1881) 173.

Austin: [a]*Texas. [b]none. [c]Colorado R. [d]University of Texas (1881) 92 – 7709 [Austin campus: (1881) 49 – 3468]; Saint Edward's University (1884) 2 – 82. [e]Mirabeau B. Lamar Austin Campus Library of the University . . . (1883) 3726; Public Library (1925) 437; Texas State Library (1891) 338.

Bakersfield: [a]California. [b]none. [c]Kern R. [d]California State College at Bakersfield (univ) (1965) 3 – 196. [e]Kern County Library (1900) 591; . . . College Library (1970) 119.

Baltimore: [a]Maryland. [b]none. [c]Patapsco R. [d][1] Johns Hopkins University (1867) 10 – 1507; University of Maryland, Baltimore campus (1807) 10 – 1181; Morgan State University (1867) 6 – 286; University of Baltimore (1925) 6 – 265; Loyola College (univ) (1852) 5 – 203. [e]Enoch Pratt Free Library (1886) 2043; Johns Hopkins University Library (1876) 1718; Baltimore Campus Library of the University of Maryland (1813) 409.

Bangor: [a]Maine. [b]none. [c]Penobscot R. [d]University of Maine (at Orono, nr Bangor) (1862) 29 – 1647 [Orono campus: (1862) 11 – 722]. [e]Raymond H. Fogler Orono Campus Library of the University . . . (at Orono, nr Bangor) (1865) 500; Public Library (1883) 401.

Baton Rouge: [a]*Louisiana. [b]none. [c]Mississippi R. [d]Louisiana State University and Agricultural and Mechanical College (1853 at Alexandria, relocated 1869) 47 – 4671 [Baton Rouge campus: (1869) 26 – 1960]; Southern University and Agricultural and Mechanical College (1880 at New Orleans, relocated 1914) 14 – 613 [Baton Rouge campus: (1914) 10 – 457]. [e]Baton Rouge Campus Library of Louisiana State University (1860 at Alexandria, relocated 1869?) 1333; East Baton Rouge Parish Library (bef 1923) 315; Baton Rouge Campus Library of Southern University (1928) 255; Louisiana State Library (1925) 247.

Battle Creek: [a]Michigan. [b]none. [c]Kalamazoo R. [d]none. [e]Willard Library (1870) 151.

Bay City: [a]Michigan. [b]none. [c]Saginaw R. [d]none. [e]Bay County Library (1869) 253.

Bayonne: [a]New Jersey. [b]none. [c]New York B of Atlantic O; Newark B of Atlantic O; Kill van Kull. [d]none. [e]Free Public Library (1893) 131.

Beaumont: [a]Texas. [b]none. [c]Neches R; Brakes Bayou. [d]Lamar University (1923) 13 – 414. [e]Mary and John Gray Library of . . . University (1923) 246; Public Library (1926) 128.

Berkeley: [a]California. [b]none. [c]San Francisco B of Pacific O. [d]University of California (1855 at Oakland, relocated 1873) 182 – 12,500 [Berkeley campus: (1873) 35 – 2429]; John F. Kennedy University (at Orinda, nr Berkeley) (1964 at Martinez, relocated 1975?) 0.5 – 86. [e]Berkeley Campus Library of the University . . . (1868 at Oakland, relocated 1873?) 4658; Public Library (1893) 364.

Bethlehem: [a]Pennsylvania. [b]none. [c]Lehigh R. [d]Lehigh University (1865) 6 – 605. [e]Linderman Memorial Library of . . . University (1877) 581; Public Library (1901) 155; Reeves Library of Moravian College (1807) 125.

Beverly Hills: [a]California. [b]none. [c]nr Santa Monica B of Pacific O. [d]none. [e]Public Library (1929) 150.

Billings: [a]Montana. [b]none. [c]Yellowstone R. [d]Eastern Montana College (univ) (1925) 3 – 170. [e]Public Library (1901) 190; . . . College Library (1927) 103.

Biloxi: [a]Mississippi. [b]none. [c]Mississippi Sound of G of Mexico; B of Biloxi of G of Mexico. [d]none. [e]Public Library (bef 1918) 49.

Binghamton: [a]New York. [b]none. [c]Susquehanna R; Chenango R. [d]State University of New York, Binghamton campus (1946) 14 – 1085. [e]Binghamton Campus Library of . . . University (1950) 712; Public Library (1904) 324.

Birmingham: [a]Alabama. [b]none. [c]Village Creek. [d]University of Alabama in Birmingham (1966) 12 – 1191; Samford University (1841) 4 – 224. [e]Birmingham Public and Jefferson County Free Library (1886) 945; Mervyn H. Sterne Library of the University . . . (1966) 246.

Bismarck: [a]*North Dakota. [b]none. [c]Missouri R. [d]none. [e]Veterans Memorial Public Library (1917) 70; North Dakota State Library (1907) 62.

Bloomington: [a]Illinois. [b]none. [c]Sugar Creek. [d]Illinois State University (at Normal, nr Bloomington) (1857) 21 – 1250; Illinois Wesleyan University (1850) 2 – 145. [e]Milner Library of Illinois State

[1] Universities with 5000 or more students.

University (at Normal, nr Bloomington) (1857) 678; Illinois Wesleyan University Library (1850) 117; Withers Public Library (1857) 99.

Bloomington: [a]Indiana. [b]none. [c]nr Salt Creek. [d]Indiana University (1820) 77 – 4470 [Bloomington campus: (1820) 33 – 1541. [e]Bloomington Campus Library of . . . University (1824) 2763; Monroe County Public Library (1821) 140.

Bloomington: [a]Minnesota. [b]none. [c]Minnesota R. [d]none. [e]Hennepin County Library (at Edina, nr Bloomington) (1922) 973.

Boise City: [a]*Idaho. [b]alt Boise. [c]Boise R. [d]Boise State University (1932) 10 – 478. [e]Public Library (1894) 163; . . . University Library (1932) 163; Idaho State Library (1901) 116.

Boston: [a]*Massachusetts. [b]none. [c]Massachusetts B of Atlantic O; Charles R. [d][1] Northeastern University (1898) 36 – 2139; Boston University (1839 at Newbury, Vermont; relocated 1867) 24 – 2000; Boston State College (univ) (1852) 11 – 304; Suffolk University (college) (1906) 7 – 232; University of Massachusetts, Boston campus (1965) 7 – 500. [e]Public Library (1854) 3864; Boston University Library (1870) 1127; State Library of Massachusetts (1826) 867; Robert Gray Dodge Library of Northeastern University (1898) 787; Boston Athenaeum (1807) 500.

Boulder: [a]Colorado. [b]none. [c]Boulder Creek. [d]University of Colorado (1861) 34 – 3611 [Boulder campus: (1861) 22 – 1200]. [e]Boulder Campus Library of the University . . . (1876) 1315; Public Library (1882) 134.

Bridgeport: [a]Connecticut. [b]none. [c]Long Island Sound of Atlantic O; Pequennock R. [d]University of Bridgeport (1927) 7 – 548; Fairfield University (at Fairfield, nr Bridgeport) (1942) 5 – 304; Sacred Heart University (college) (1963) 2 – 139. [e]Bridgeport Public Library (1881) 477; Magnus Wahlstrom Library of the University . . . (1947) 205; Fairfield Public Library (at Fairfield, nr Bridgeport) (1903) 153.

Bristol: [a]Connecticut. [b]none. [c]Pequabuck R. [d]none. [e]Public Library (1892) 138.

Brockton: [a]Massachusetts. [b]none. [c]Salisbury Plain R. [d]Bridgewater State College (univ) (at Bridgewater, nr Brockton) (1840) 8 – 429. [e]Public Library (1867) 219; Clement C. Maxwell Library of . . . College (at Bridgewater, nr Brockton) (1840) 139.

Brownsville: [a]Texas. [b]none. [c]Rio Grande R. [d]none. [e]Texas Southmost College-Brownsville City Library (1948) 78.

Buffalo: [a]New York. [b]none. [c]Niagara R; Erie L. [d]State University of New York, campuses at or nr Buffalo (1846) 51 – 3607; Canisius College (univ) (1870) 4 – 249. [e]Buffalo and Erie County Public Library (1836) 2931; Buffalo Campus Library of . . . University (1846) 1939.

Burbank: [a]California. [b]none. [c]Los Angeles R. [d]none. [e]Public Library (1938) 248.

Burlington: [a]Vermont. [b]none. [c]Winooski R; Champlain L. [d]University of Vermont and State Agricultural College (1791) 10 – 988; Saint Michael's College (univ) (at Winooski, nr Burlington) (1903) 2 – 95. [e]Guy W. Bailey Memorial Library of the University . . . (1800) 690; Durick Library of . . . College (at Winooski, nr Burlington) (1904) 73; Fletcher Free Library (1873) 55.

Butte: [a]Montana. [b]none. [c]Clark Fork River. [d]Montana College of Mineral Science and Technology (univ) (1895) 1 – 55. [e]Free Public Library (1890) 90; . . . College Library (1900) 42.

Cambridge: [a]Massachusetts. [b]none. [c]Charles R. [d]Harvard University (1636) 20 – 5170; Massachusetts Institute of Technology (univ) (1861) 8 – 1672; Lesley College (univ) (1909) 2 – 109. [e]. . . University Library (1638) 9207; . . . Institute Library (1862) 601; Public Library (1857) 354.

Camden: [a]New Jersey. [b]none. [c]Delaware R. [d]Rutgers, the State University of New Jersey, Camden campus (1927) 5 – 166. [e]Camden Campus Library of Rutgers University (1926) 280; Camden County Free Library (at Ashland, nr Camden) (1922 at Haddonfield, relocated 1973?) 252; Camden Free Public Library (1898) 150.

Canton: [a]Ohio. [b]none. [c]Nimishillen Creek. [d]none. [e]Stark County District Library (1884) 430.

Casper: [a]Wyoming. [b]none. [c]North Platte R. [d]none. [e]Natrona County Public Library (1910) 90; Goodstein Foundation Library of Casper College (1967) 40.

---

[1]Universities with 5000 or more students.

Cedar Rapids: ^{a}Iowa. ^{b}none. ^{c}Cedar R; Cedar L. ^{d}none. ^{e}Public Library (1897) 208; Stewart Memorial Library of Coe College (1900) 132.

Champaign: ^{a}Illinois. ^{b}none. ^{c}nr Kaskaskia R; nr Embarrass (alt Embarras) R. ^{d}University of Illinois (at Urbana, nr Champaign, and Champaign) (1867) 62 – 9119 [Urbana-Champaign campus: (1867) 37 – 6785]. ^{e}Urbana-Champaign Campus Library of the University . . . (at Urbana, nr Champaign) (1868) 5227; Champaign Public Library (1876) 100; Urbana Public Library (at Urbana, nr Champaign) (1874) 96.

Charleston: ^{a}South Carolina. ^{b}none. ^{c}Atlantic O; Ashley R; Cooper R. ^{d}College of Charleston (univ) (1770) 5 – 215; Citadel (univ) (1842) 3 – 220. ^{e}Charleston County Library (1930) 342; Daniel Library of the Citadel (1842) 151; Robert Scott Small Library of the College . . . (1790) 133.

Charleston: ^{a}*West Virginia. ^{b}none. ^{c}Kanawha R; Elk R. ^{d}none. ^{e}West Virginia Library Commission Science and Culture Center (1929) 528; Kanawha County Public Library (1909) 374.

Charlotte: ^{a}North Carolina. ^{b}none. ^{c}Little Sugar Creek; Irwin Creek. ^{d}University of North Carolina, Charlotte campus (1946) 8 – 474; Johnson C. Smith University (college) (1867) 1 – 93. ^{e}Public Library of Charlotte and Mecklenburg County (1891) 627; J. Murrey Atkins Charlotte Campus Library of the University . . . (1946) 156.

Charlottesville: ^{a}Virginia. ^{b}none. ^{c}Rivanna R. ^{d}University of Virginia (1819) 27 – 1602 [Charlottesville campus: (1819) 26 – 1532]. ^{e}Alderman Charlottesville Campus Library of the University . . . (1819) 2006; Jefferson-Madison Regional Library (1921) 168.

Chattanooga: ^{a}Tennessee. ^{b}none. ^{c}Tennessee R. ^{d}University of Tennessee, Chattanooga campus (1886) 6 – 327. ^{e}Chattanooga Campus Library of the University . . . (1872) 300; Chattanooga-Hamilton County Bicentennial Library (1905) 262.

Cheektowaga: ^{a}New York. ^{b}none. ^{c}Cayuga Creek. ^{d}none. ^{e}Public Library (1938) 84.

Chesapeake: ^{a}Virginia. ^{b}inc South Norfolk. ^{c}Hampton Roads; Elizabeth R. ^{d}none. ^{e}Public Library (1963) 122. ^{f}South Norfolk: 22 (1960), 10 (1950), 8 (1930), 8 (1920).

Chester: ^{a}Pennsylvania. ^{b}none. ^{c}Delaware R. ^{d}Widener College (univ) (1821) 3 – 225; Swarthmore College (univ) (at Swarthmore, nr Chester) (1864) 1 – 160. ^{e}McCabe Library of Swarthmore College (at Swarthmore, nr Chester) (1864) 367; Wolfgram Memorial Library of Widener College (1821) 100; J. Lewis Crozer Library (1894) 50.

Cheyenne: ^{a}*Wyoming. ^{b}none. ^{c}Crow Creek. ^{d}none. ^{e}Wyoming State Library (1887) 142; Laramie County Library (1886) 106.

Chicago: ^{a}Illinois. ^{b}none. ^{c}Chicago R; Calumet (alt Grand Calumet) R; Michigan L. $^{d^1}$University of Illinois, Chicago campus (1881) 26 – 2334; Loyola University of Chicago (1869) 13 – 1077; De Paul University (1898) 11 – 566; Northeastern Illinois University (1961) 10 – 490; University of Chicago (1890) 9 – 1050; Roosevelt University (1945) 8 – 443; Chicago State University (1869) 7 – 368; Illinois Institute of Technology (univ) (1892) 7 – 735. ^{e}Public Library (1873) 5948; Joseph Regenstein Library of the University of Chicago (1891) 3622; Center for Research Library (1949) 3000; Newberry Library (1887) 1282; John Crerar Library (1895) 1155.

Chicopee: ^{a}Massachusetts. ^{b}none. ^{c}Connecticut R; Chicopee R. ^{d}none. ^{e}Public Library (1853) 129.

Cincinnati: ^{a}Ohio. ^{b}none. ^{c}Ohio R. ^{d}University of Cincinnati (1819) 39 – 3161 [Cincinnati campus: (1819) 35 – ?]; Xavier University (1831) 6 – 285; Athenaeum of Ohio (univ) (at Norwood, nr Cincinnati) (1829) 0.2 – 30. ^{e}Public Library of Cincinnati and Hamilton County (1856) 3055; Main Cincinnati Campus Library of the University . . . (1819) 1335.

Clarksville: ^{a}Tennessee. ^{b}none. ^{c}Cumberland R; Red R. ^{d}Austin Peay State University (1927) 4 – 172. ^{e}Felix G. Woodward Library of . . . University (1927) 166; Warioto Regional Library Center (1947) 76.

Clearwater: ^{a}Florida. ^{b}none. ^{c}G of Mexico; Old Tampa B of G of Mexico; Clearwater B of G of Mexico. ^{d}none. ^{e}Public Library (1911) 162.

Cleveland: ^{a}Ohio. ^{b}none. ^{c}Cuyahoga R; Erie L. ^{d}Cleveland State University (1881) 17 – 715; Case Western Reserve University (1826) 9 – 1500; John Carroll University (at University Heights, nr

^{1}Universities with 5000 or more students.

Cleveland) (1886) 4 – 265. [e]Public Library (1869) 3092; Case Western Reserve University Library (1856) 1559; Cuyahoga County Public Library (1923) 1407; Cleveland State University Library (1928) 303.

Clifton: [a]New Jersey. [b]none. [c]Passaic R. [d]Montclair State College (univ) (at Upper Montclair, nr Clifton) (1908) 16 – 687. [e]Public Library (1920) 184; Harry A. Sprague Library of . . . College (at Upper Montclair, nr Clifton) (1908) 60.

Colorado Springs: [a]Colorado. [b]none. [c]Fountain R. [d]University of Colorado, Colorado Springs campus (1955) 3 – 130, Colorado College (univ) (1874) 2 – 198. [e]United States Air Force Academy Library (1955) 395; Pikes Peak Regional Library (alt Penrose Public Library) (1885) 364; Charles Leaming Tutt Library of . . . College (1878) 268.

Columbia: [a]Missouri. [b]none. [c]nr Missouri R. [d]University of Missouri (1839) 55 – 6220 [Columbia campus: (1839) 26 – 3515]. [e]Elmer Ellis Columbia Campus Library of the University . . . (1841) 1794; State Historical Society of Missouri Library (1898) 380; Daniel Boone Regional Library (1900) 180; Hugh Stephens Library of Stephens College (1911) 105.

Columbia: [a]*South Carolina. [b]none. [c]Congaree R; Saluda R; Broad R. [d]University of South Carolina (1801) 31 – 1484 [Columbia campus: (1801) 25 – ?]; Allen University (college) (1870) 0.7 – 47. [e]Thomas Cooper Columbia Campus Library of the University . . . (1801) 1270; Richland County Public Library (1896) 248; South Carolina State Library (1943) 120.

Columbus: [a]Georgia. [b]none. [c]Chattahoochee R. [d]Columbus College (univ) (1958) 5 – 225. [e]Chattahoochee Valley Regional Library (alt W. C. Bradley Memorial Library) (bef 1904) 380.

Columbus: [a]*Ohio. [b]none. [c]Scioto R; Olentangy R. [d]Ohio State University (1870) 54 – 3637 [Columbus campus: (1870) 50 – ?]; Franklin University (college) (1902) 4 – 144; Capital University (1850) 3 – 167. [e]Columbus Campus Library of Ohio State University (1873) 3413; Public Library of Columbus and Franklin County (1872) 1168; State Library of Ohio (1817) 619.

Compton: [a]California. [b]none. [c]Los Angeles R. [d]none. [e]Compton Branch of Los Angeles County Public Library (bef 1954) 98.

Concord: [a]California. [b]none. [c]nr Sacramento R. [d]none. [e]Contra Costa County Library (at Pleasant Hill, nr Concord) (1913) 859.

Concord: [a]*New Hampshire. [b]none. [c]Merrimack R. [d]none. [e]New Hampshire State Library (1818) 709; Public Library (1855) 146.

Corpus Christi: [a]Texas. [b]none. [c]G of Mexico. [d]Texas Agricultural and Industrial University at Corpus Christi (1947) 2 – 56. [e]La Retama Public Library (1909) 310.

Council Bluffs: [a]Iowa. [b]for Kanesville. [c]Missouri R. [d]none. [e]Free Public Library (1866) 129.

Covington: [a]Kentucky. [b]none. [c]Ohio R. [d]Northern Kentucky University (at Highland Heights, nr Covington) (1873) 6 – 332. [e]Kenton County Public Library (1900) 140; Salmon P. Chase College of Law Library of Northern Kentucky University (1873?) 100.

Cranston: [a]Rhode Island. [b]none. [c]Providence R; Pawtuxet R. [d]none. [e]Public Library (bef 1927) 101.

Dallas: [a]Texas. [b]none. [c]Trinity R. [d]Southern Methodist University (1911) 10 – 861. [e]Public Library (1901) 1637; . . . University Library (1915) 1115.

Danbury: [a]Connecticut. [b]none. [c]Still R. [d]Western Connecticut State College (univ) (1903) 5 – 289. [e]Ruth A. Haas Library of . . . College (1905) 109; Public Library (1869) 98.

Davenport: [a]Iowa. [b]none. [c]Mississippi R. [d]Marycrest College (univ) (1939) 0.9 – 95. [e]Public Library (1874) 254.

Dayton: [a]Ohio. [b]none. [c]Miami (alt Great Miami) R; Stillwater R; Mad R. [d]Wright State University (1964) 15 – 592 [Dayton campus: (1964) 14 – ?]; University of Dayton (1850) 8 – 607. [e]Dayton and Montgomery County Public Library (1847) 1222; University . . . Library (1928) 452; Dayton Campus Library of . . . University (1964) 294.

Daytona Beach: [a]Florida. [b]for Daytona. [c]Atlantic O; Halifax R. [d]none. [e]Volusia County Public Library (bef 1927) 272.

Dearborn: [a]Michigan. [b]none. [c]Rouge R. [d]University of Michigan, Dearborn campus (1959) 5 – 160. [e]Henry Ford Centennial Library (1921) 239; Dearborn Campus Library of the University . . . (1959) 180.

Dearborn Heights: ^{a}Michigan. ^{b}none. ^{c}Middle Rouge R. ^{d}none. ^{e}Caroline Kennedy Library (1961) 39; John F. Kennedy, Jr, Library (1965?) 33.

Decatur: ^{a}Illinois. ^{b}none. ^{c}Sangamon R. ^{d}Millikin University (college) (1901) 2 – 217. ^{e}Public Library (1868) 225; . . . University Library (1903) 138.

Denver: ^{a}*Colorado. ^{b}none. ^{c}South Platte R. ^{d}University of Colorado, Denver campus (1883) 10 – 2281; University of Denver (1864) 8 – 640. ^{e}Public Library (1859) 1561; Penrose Library of the University of Denver (1933) 580; Colorado State Library (1862) 413.

Des Moines: ^{a}*Iowa. ^{b}none. ^{c}Des Moines R; Raccoon R. ^{d}Drake University (1881) 6 – 326. ^{e}Public Library (1866) 394; State Library Commission of Iowa (1838) 378; Cowles Library of . . . University (1908) 285.

Detroit: ^{a}Michigan. ^{b}none. ^{c}Detroit R. ^{d}Wayne State University (1868) 38 – 2142; University of Detroit (1877) 8 – 561; Marygrove College (univ) (1910) 0.9 – 72. ^{e}Public Library (1865) 2325; G. Flint Purdy Library of . . . University (1923) 1277; University . . . Library (1877) 479.

Downey: ^{a}California. ^{b}none. ^{c}San Gabriel R; Hondo R. ^{d}none. ^{e}City Library (1958) 95.

Dubuque: ^{a}Iowa. ^{b}none. ^{c}Mississippi R. ^{d}Loras College (univ) (1839) 2 – 92; University of Dubuque (1852) 1 – 66; Clarke College (univ) (1843) 0.7 – 93. ^{e}Wahlert Memorial Library of Loras College (1839) 210; Carnegie-Stout Public Library (1902) 136; Aquinas-Dubuque Theological Library (1852) 98.

Duluth: ^{a}Minnesota. ^{b}none. ^{c}Saint Louis R; Superior L. ^{d}University of Minnesota, Duluth campus (1895) 8 – 362; University of Wisconsin, Superior campus (at Superior, Wisconsin, nr Duluth) (1893) 3 – 177; College of Saint Scholastica (univ) (1902) 1 – 114. ^{e}Duluth Public Library (1869) 249; Duluth Campus Library of the University of Minnesota (1902) 213; Jim Dan Hill Superior Campus Library of the University of Wisconsin (at Superior, Wisconsin, nr Duluth) (1896) 200; Superior Public Library (at Superior, Wisconsin, nr Duluth) (1889) 111.

Dundalk: ^{a}Maryland. ^{b}none. ^{c}Patapsco R; Back R. ^{d}none. ^{e}Dundalk Branch of Baltimore County Public Library (bef 1954) 137.

Durham: ^{a}North Carolina. ^{b}none. ^{c}Ellerbe Creek. ^{d}University of North Carolina (at Chapel Hill, nr Durham) (1789) 111 – 9681 [Chapel Hill campus: (1789) 21 – 2891]; Duke University (1838 at High Point, relocated 1892) 9 – 1210; North Carolina Central University (1909) 5 – 269. ^{e}William R. Perkins Library of Duke University (1838 at High Point, relocated 1892?) 2622; Louis Round Wilson Chapel Hill Campus Library of the University . . . (at Chapel Hill, nr Durham) (1795) 2126; James E. Shepard Memorial Library of North Carolina Central University (1910) 263; Durham County Library (1897) 182.

East Los Angeles: ^{a}California. ^{b}none. ^{c}nr Hondo R. ^{d}none. ^{e}East Los Angeles College Library (1946) 74; East Los Angeles Branch of Los Angeles County Public Library (bef 1954) 51.

East Orange: ^{a}New Jersey. ^{b}none. ^{c}nr Passaic R. ^{d}Seton Hall University (at South Orange, nr East Orange) (1856) 10 – 579; Fairleigh Dickinson University, Madison campus (at Madison, nr East Orange) (1942) 5 – ?; Drew University (at Madison, nr East Orange) (1866) 2 – 191. ^{e}Rose Memorial Library of Drew University (at Madison, nr East Orange) (1867) 362; Public Library (1903) 320; McLaughlin Library of Seton Hall University (at South Orange, nr East Orange) (1856) 290.

East Saint Louis: ^{a}Illinois. ^{b}for Illinois Town. ^{c}Mississippi R. ^{d}none. ^{e}Public Library (1874) 119.

Elgin: ^{a}Illinois. ^{b}none. ^{c}Fox R. ^{d}none. ^{e}Gail Borden Public Library (1873) 139.

Elizabeth: ^{a}New Jersey. ^{b}none. ^{c}Newark B of Atlantic O; Arthur Kill. ^{d}Kean College of New Jersey (univ) (at Union, nr Elizabeth) (1855) 14 – 728. ^{e}Union Township Public Library (at Union, nr Elizabeth) (1891) 230; Nancy Thompson Library of . . . College (at Union, nr Elizabeth) (1914) 204; Elizabeth Public Library (1755) 137.

Elmira: ^{a}New York. ^{b}none. ^{c}Chemung R. ^{d}Elmira College (univ) (1853) 3 – 157. ^{e}Steele Memorial Library (1899) 358; Gannett-Tripp Learning Center of . . . College (1855) 123.

El Paso: ^{a}Texas. ^{b}for Franklin. ^{c}Rio Grande R. ^{d}University of Texas, El Paso campus (1913) 14 – 442. ^{e}Public Library (1896) 414; El Paso Campus Library of the University . . . (1913) 334.

Elyria: ^{a}Ohio. ^{b}none. ^{c}Black R. ^{d}Oberlin College (univ) (at Oberlin, nr Elyria) (1833) 3 – 265.

[e]Seeley G. Mudd Learning Center of . . . College (at Oberlin, nr Elyria) (1833) 751; Public Library (1870) 117.

Erie: [a]Pennsylvania. [b]none. [c]Erie L. [d]Gannon College (univ) (1933) 3 – 195; Pennsylvania State University, Behrend College campus (univ) (1926) 2 – 81. [e]Erie City and County Library (1899) 336.

Euclid: [a]Ohio. [b]none. [c]Erie L. [d]none. [e]Public Library (1935) 376.

Eugene: [a]Oregon. [b]none. [c]Willamette R. [d]University of Oregon (1872) 21 – 2202 [Eugene campus: (1872) 19 – 2044. [e]Eugene Campus Library of the University . . . (1881) 1303; Public Library (1895) 173.

Evanston: [a]Illinois. [b]none. [c]Michigan L. [d]Northwestern University (1851) 15 – 2748; National College of Education (univ) (1886) 4 – 144 [Evanston campus: (1886) 3 – ?]. [e]. . . University Library (1856) 2475; Public Library (1871)·273.

Evansville: [a]Indiana. [b]none. [c]Ohio R. [d]University of Evansville (1854) 5 – 283. [e]Evansville Public Library and Vanderburgh County Public Library (1875) 465; Clifford Memorial Library of the University . . . (1854?) 167.

Everett: [a]Washington. [b]none. [c]Puget Sound of Pacific O; Snohomish R. [d]none. [e]Public Library (1894) 152.

Fairbanks: [a]Alaska. [b]none. [c]Chena R. [d]University of Alaska (1917) 13 – 1787 [Fairbanks campus: (1917) 5 – 1061]. [e]Elmer E. Rasmuson Fairbanks Campus Library of the University . . . (1917) 359; Fairbanks North Star Borough Library (alt George C. Thomas Memorial Library) (1909) 37.

Fairfield: [a]California. [b]none. [c]nr Suisun Channel. [d]none. [e]Solano County Library (1914) 154.

Fall River: [a]Massachusetts. [b]none. [c]Mount Hope B of Atlantic O; Taunton R; Fall R. [d]none. [e]Public Library (1861) 254.

Fargo: [a]North Dakota. [b]none. [c]Red R. [d]North Dakota State University (1890) 8 – 573 [Fargo campus: (1890) 8 – ?]; Moorhead State University (at Moorhead, Minnesota, nr Fargo) (1887) 5 – 319. [e]North Dakota State University Library (1889) 210; Lake Agassiz Regional Library (at Moorhead, Minnesota, nr Fargo) (1961) 195; Carl B. Ylvisaker Library of Concordia College (at Moorhead, Minnesota, nr Fargo) (1891) 185; Livingston Lord Library of Moorhead State University (at Moorhead, Minnesota, nr Fargo) (1887) 177; Public Library (1900) 117.

Fayetteville: [a]North Carolina. [b]none. [c]Cape Fear R. [d]Fayetteville State University (college) (1867) 4 – 271. [e]Cumberland County Public Library (1907) 131; Charles W. Chesnutt Library of . . . University (1937) 87.

Fitchburg: [a]Masschusetts. [b]none. [c]North Nashua R. [d]Fitchburg State College (univ) (1894) 7 – 250. [e]Public Library (1859) 160; . . . College Library (1895) 125.

Flint: [a]Michigan. [b]none. [c]Flint R. [d]none. [e]Public Library (1851) 403; Genesee County Library (1942) 371.

Fort Lauderdale: [a]Florida. [b]none. [c]Atlantic O. [d]Florida Atlantic University (at Boca Raton, nr Fort Lauderdale) (1961) 7 – 419; Nova University (1964) 3 – 260. [e]S. E. Wimberly Memorial Library of Florida Atlantic University (at Boca Raton, nr Fort Lauderdale) (1963) 425; Public Library (1913) 248.

Fort Smith: [a]Arkansas. [b]none. [c]Arkansas R. [d]none. [e]Public Library (1906) 70.

Fort Wayne: [a]Indiana. [b]none. [c]Maumee R; Saint Mary's R; Saint Joseph R. [d]Indiana University-Purdue University, Fort Wayne campus (1917) 5 – 494; Saint Francis College (univ) (1890) 1 – 75. [e]Public Library of Fort Wayne and Allen County (1894) 1329.

Fort Worth: [a]Texas. [b]none. [c]Trinity R. [d]Texas Christian University (1873) 6 – 434. [e]Mary Couts Burnett Library of . . . University (1910) 835; Public Library (1901) 734.

Fremont: [a]California. [b]none. [c]Alameda Creek. [d]none. [e]Fremont Branch of Alameda County Library (1947?) 102.

Fresno: [a]California. [b]none. [c]nr San Joaquin R. [d]California State University at Fresno (1911) 17 – 966; Pacific College (univ) (1944) 0.4 – 41. [e]Fresno County Free Library (1896) 787; . . . University Library (1911) 402.

Fullerton: [a]California. [b]none. [c]nr Santa Ana R. [d]California State University at Fullerton (1957)

22 – 773. ^{e}... University Library (1959) 335; Public Library (1907) 108; William T. Boyce Library of Fullerton College (1913) 85.

Gadsden: ^{a}Alabama. ^{b}none. ^{c}Coosa R. ^{d}none. ^{e}Public Library (1906) 118.

Gainesville: ^{a}Florida. ^{b}none. ^{c}nr Newnan L. ^{d}University of Florida (1853) 29 – 3976. ^{e}University ... Library (1889) 1756; Santa Fe Regional Library (alt Gainesville Public Library) (1917) 102.

Galveston: ^{a}Texas; Galveston I. ^{b}none. ^{c}G of Mexico; Galveston B of G of Mexico; Galveston Channel. ^{d}none. ^{e}Rosenberg Library (1904) 215; Moody Medical Library (alt University of Texas Medical Branch Library) (1893) 181.

Garden Grove: ^{a}California. ^{b}none. ^{c}Santa Ana R. ^{d}none. ^{e}Garden Grove Branch of Orange County Public Library (1921?) 89.

Garland: ^{a}Texas. ^{b}none. ^{c}Duck Creek. ^{d}University of Texas, Dallas campus (at Richardson, nr Garland) (1969) 3 – 250. ^{e}Nicholson Memorial Library (alt Garland Public Library) (1933) 126; Dallas Campus Library of the University of Texas (at Richardson, nr Garland) (1964) 116; Richardson Public Library (at Richardson, nr Garland) (1959) 103.

Gary: ^{a}Indiana. ^{b}none. ^{c}Calumet (alt Grand Calumet) R; Little Calumet R; Michigan L. ^{d}Indiana University, Northwest campus (1922) 5 – 265. ^{e}Public Library (1908) 452.

Gastonia: ^{a}North Carolina. ^{b}none. ^{c}nr Catawba R. ^{d}none. ^{e}Gaston-Lincoln Regional Library (1964) 294; Gaston County Public Library (1905) 248.

Glendale: ^{a}California. ^{b}none. ^{c}Los Angeles R. ^{d}none. ^{e}Public Library (1906) 392.

Grand Forks: ^{a}North Dakota. ^{b}none. ^{c}Red R. ^{d}University of North Dakota (1883) 9 – 571 [Grand Forks campus: (1883) 9 – 520]. ^{e}Chester Fritz Grand Forks Campus Library of the University ... (1883) 279; Public City-County Library (1900) 97.

Grand Rapids: ^{a}Michigan. ^{b}none. ^{c}Grand R. ^{d}Calvin College (univ) (1876) 4 – 223; Aquinas College (univ) (1886) 2 – 122. ^{e}Public Library (1871) 612; Kent County Library (1936) 383.

Great Falls: ^{a}Montana. ^{b}none. ^{c}Missouri R; Sun R. ^{d}none. ^{e}Public Library (1889) 167.

Green Bay: ^{a}Wisconsin. ^{b}none. ^{c}Fox R; Green B of Michigan L. ^{d}University of Wisconsin, Green Bay campus (1965) 4 – 188. ^{e}Brown County Library (1889) 310; Green Bay Campus Library of the University ... (1967) 177.

Greensboro: ^{a}North Carolina. ^{b}none. ^{c}Buffalo Creek. ^{d}University of North Carolina, Greensboro campus (1891) 9 – 755; North Carolina Agricultural and Technical State University (1891) 5 – 310. ^{e}Walter Clinton Jackson Greensboro Campus Library of the University ... (1892) 597; Public Library (1902) 437.

Greenville: ^{a}South Carolina. ^{b}for Pleasantburg. ^{c}Reedy R. ^{d}Bob Jones University (1927) 4 – 304; Furman University (1825 at Edgefield, relocated 1851) 3 – 167. ^{e}Greenville County Library (1921) 339; Furman University Library (1826 at Edgefield, relocated 1851?) 221.

Gulfport: ^{a}Mississippi. ^{b}none. ^{c}Mississippi Sound of G of Mexico. ^{d}none. ^{e}Gulfport-Harrison County Library (bef 1923) 74.

Hamilton: ^{a}Ohio. ^{b}none. ^{c}Miami (alt Great Miami) R. ^{d}Miami University (at Oxford, nr Hamilton) (1809) 18 – 1051 [Oxford campus: (1809) 15 – 832]. ^{e}Edgar W. King Oxford Campus Library of ... University (at Oxford, nr Hamilton) (1824) 500; Lane Public Library (1866) 215.

Hammond: ^{a}Indiana. ^{b}none. ^{c}Calumet (alt Grand Calumet) R; Wolf L. ^{d}Purdue University, Calumet campus (1946) 7 – 325. ^{e}Public Library (1902) 289.

Hampton: ^{a}Virginia. ^{b}none. ^{c}Chesapeake B of Atlantic O; Hampton Roads; Back R. ^{d}Hampton Institute (univ) (1868) 3 – 237. ^{e}Collis P. Huntington Memorial Library of ... Institute (1903) 159; Charles H. Taylor Memorial Library (alt Hampton Public Library) (1926) 154.

Harrisburg: ^{a}*Pennsylvania. ^{b}none. ^{c}Susquehanna R. ^{d}Pennsylvania State University, Capitol campus (at Middletown, nr Harrisburg) (1966) 2 – 132. ^{e}State Library of Pennsylvania (1745 at Philadelphia, relocated 1812?) 865; Dauphin County Library (1889) 316.

Hartford: ^{a}*Connecticut. ^{b}none. ^{c}Connecticut R. ^{d}University of Hartford (at West Hartford, nr Hartford) (1877) 9 – 550; Trinity College (univ) (1823) 2 – 148; Saint Joseph College (univ) (at West Hartford, nr Hartford) (1925) 1 – 97. ^{e}Connecticut State Library (1854) 632; Trinity College Library (1823) 554; Hartford Public Library (1774) 467; Mortensen Library of the University . . . (at West Hartford, nr Hartford) (1957) 245; West Hartford Public Library (at West Hartford, nr Hartford) (1897) 220.

Hayward: [a]California. [b]none. [c]nr San Francisco B of Pacific O. [d]California State University at Hayward (1957) 13 – 675. [e]Alameda County Library (1910 at Oakland, relocated bef 1958) 507; . . . University Library (1959) 445; Chabot College Library (1961) 85; Public Library (1898) 57.

Helena: [a]*Montana. [b]none. [c]nr Missouri R; nr Helena L. [d]none. [e]Montana State Library (1946) 136; Lewis and Clark Library (1886) 75.

Hialeah: [a]Florida. [b]none. [c]nr Biscayne B of Atlantic O. [d]Biscayne College (univ) (at Opa-Locka, nr Hialeah) (1962) 2 – 83. [e]Hialeah John F. Kennedy Library (bef 1949) 126.

High Point: [a]North Carolina. [b]none. [c]nr Uwharrie (alt Uharie) R. [d]none. [e]Public Library (1926) 160; Wrenn Memorial Library of High Point College (1924) 93.

Hollywood: [a]Florida. [b]none. [c]Atlantic O. [d]Heed University (1970) 0.1 – ? [e]Hollywood Public Library (1943) 86.

Holyoke: [a]Masschusetts. [b]none. [c]Connecticut R. [d]Smith College (univ) (at Northampton, nr Holyoke) (1871) 3 – 307; Mount Holyoke College (univ) (at South Hadley, nr Holyoke) (1836) 2 – 236. [e]Smith College Library (at Northampton, nr Holyoke) (1871) 840; Forbes Library (at Northampton, nr Holyoke) (1895) 339; Williston Memorial Library of Mount Holyoke College (at South Hadley, nr Holyoke) (1837) 249; Public Library (1870) 155.

Hot Springs: [a]Arkansas. [b]none. [c]nr Ouachita R. [d]none. [e]Garland-Montgomery Regional Library (bef 1928) 76.

Houston: [a]Texas. [b]none. [c]Buffalo Bayou. [d]University of Houston (1927) 38 – 1769; Texas Southern University (1947) 7 – 366; Rice University (1891) 4 – 408; Houston Baptist University (college) (1960) 2 – 87; University of Saint Thomas (1947) 2 – 101. [e]Public Library (1848) 1737; M. D. Anderson Memorial Library of the University of Houston (1927) 911; Fondren Library of Rice University (1912) 876; Harris County Public Library (1921) 266.

Huntington: [a]West Virginia. [b]none. [c]Ohio R. [d]Marshall University (1837) 11 – 410. [e]James E. Morrow Library of . . . University (1837) 273; Cabell County Public Library (1902) 245.

Huntington Beach: [a]California. [b]none. [c]Pacific O. [d]none. [e]Huntington Beach Library (1908) 249.

Huntsville: [a]Alabama. [b]none. [c]Indian Creek. [d]Alabama Agricultural and Mechanical University (at Normal, nr Huntsville) (1875) 5 – 265; University of Alabama in Huntsville (1950) 4 – 200. [e]Redstone Scientific Information Center (1949) 200; Public Library (1817) 190; Joseph F. Drake Memorial Library of . . . University (at Normal, nr Huntsville) (1904) 157; University . . . Library (1950?) 124.

Idaho Falls: [a]Idaho. [b]none. [c]Snake R. [d]none. [e]Public Library (1909) 84.

Independence: [a]Missouri. [b]none. [c]Missouri R. [d]none. [e]Mid-Continent Public Library (1894) 976.

Indianapolis: [a]*Indiana. [b]none. [c]White R. [d]Indiana University-Purdue University, Indianapolis campus (1908) 20 – 1564; Butler University (1850) 4 – 235; Indiana Central University (1902) 3 – 168. [e]Indianapolis-Marion County Public Library (1873) 1235; Indiana State Library (1825) 708; Indianapolis Campus Library of Indiana University-Purdue University (1908) 475.

Inglewood: [a]California. [b]none. [c]nr Santa Monica B of Pacific O. [d]Northrop University (1942) 1 – 108. [e]Public Library (1917) 233.

Irving: [a]Texas. [b]none. [c]West Fork of Trinity R. [d]University of Dallas (1955) 2 – 125. [e]William A. Blakely Library of the University . . . (1956) 106; Public Library (1961) 105.

Jackson: [a]Michigan. [b]none. [c]Grand R. [d]none. [e]Jackson Public Library (1883) 124; Jackson County Library (1929) 99.

Jackson: [a]*Mississippi. [b]none. [c]Pearl R. [d]Jackson State University (1877) 8 – 359; Mississippi College (univ) (at Clinton, nr Jackson) (1826) 3 – 173. [e]Jackson Metropolitan Library (1914) 566; Mississippi Library Commission (1926) 378; Henry Thomas Sampson Library of . . . University (1916) 234.

Jacksonville: [a]Florida. [b]none. [c]Saint Johns R; Trout R. [d]University of North Florida (1965) 4 – 117; Jacksonville University (1934) 2 – 150. [e]Public Library (1905) 946.

Jefferson City: [a]*Missouri. [b]none. [c]Missouri R. [d]Lincoln University (1866) 2 – 165. [e]Missouri State

Library (1829) 190; Thomas Jefferson Library (1900) 187; Inman E. Page Library of . . . University (1866) 115.

Jersey City: [a]New Jersey. [b]none. [c]Hudson R; Hackensack R. [d]Jersey City State College (univ) (1927) 12 – 768; Stevens Institute of Technology (univ) (at Hoboken, nr Jersey City) (1870) 2 – 287. [e]Free Public Library (1891) 602; Forrest A. Irwin Library of . . . College (1927) 160.

Johnstown: [a]Pennsylvania. [b]none. [c]Conemaugh R. [d]none. [e]Cambria County Library (1870) 320.

Joliet: [a]Illinois. [b]for Juliet. [c]Des Plaines R. [d]Lewis University (at Lockport, nr Joliet) (1930) 3 – 241. [e]Lockport Township Public Library (at Lockport, nr Joliet) (1921) 116; Joliet Public Library (1875) 114; College of Saint Francis Library (1930) 85; . . . University Library (at Lockport, nr Joliet) (1952) 82.

Kalamazoo: [a]Michigan. [b]none. [c]Kalamazoo R. [d]Western Michigan University (1903) 23 – 1150. [e]Dwight B. Waldo Library of . . . University (1903) 793; Public Library (1872) 252; Upjohn Library of Kalamazoo College (1903?) 190.

Kansas City: [a]Kansas. [b]none. [c]Missouri R; Kansas R. [d]none. [e]Public Library (1890) 281.

Kansas City: [a]Missouri. [b]none. [c]Missouri R. [d]University of Missouri, Kansas City campus (1881) 12 – 1500. [e]Public Library (1873) 1156; General Kansas City Campus Library of the University . . . (1933) 423.

Kenosha: [a]Wisconsin. [b]none. [c]Michigan L. [d]University of Wisconsin, Parkside campus (1965) 5 – 165. [e]Parkside Campus Library of the University . . . (1967) 255; Gilbert M. Simmons Public Library (1897) 207.

Key West: [a]Florida; Key West I. [b]none. [c]Atlantic O; G of Mexico. [d]none. [e]Monroe County Public Library (1892) 108.

Killeen: [a]Texas. [b]none. [c]South Nolan Creek. [d]none. [e]Public Library (1959) 53.

Knoxville: [a]Tennessee. [b]none. [c]Tennessee R; Holston R. [d]University of Tennessee (1794) 49 – 3257 [Knoxville campus: (1794) 30 – 1668]. [e]James D. Hoskins Knoxville Campus Library of the University . . . (1838) 1229; Knoxville-Knox County Public Library (1879) 513.

Kokomo: [a]Indiana. [b]none. [c]Wildcat Creek. [d]none. [e]Public Library (1885) 141; Kokomo Campus Library of Indiana University (1945) 68.

La Crosse: [a]Wisconsin. [b]none. [c]Mississippi R; La Crosse R. [d]University of Wisconsin, La Crosse campus (1909) 8 – 354. [e]Murphy La Crosse Campus Library of the University . . . (1909) 262; La Crosse Public Library (1888) 178; La Crosse County Library (bef 1935) 48.

Lafayette: [a]Indiana. [b]none. [c]Wabash R. [d]Purdue University (at West Lafayette, nr Lafayette) (1869) 38 – 4988 [West Lafayette campus: (1869) 29 – 4477]. [e]West Lafayette Campus Library of . . . University (at West Lafayette, nr Lafayette) (1874) 839; Albert A. Wells Memorial Library (1883) 104.

Lafayette: [a]Louisiana. [b]none. [c]Vermilion R. [d]University of Southwestern Louisiana (1898) 12 – 522. [e]Dupre Library of the University . . . (1901) 297; Lafayette Parish Public Library (bef 1939) 117.

Lake Charles: [a]Louisiana. [b]none. [c]Calcasieu R. [d]McNeese State University (1939) 6 – 344. [e]Calcasieu Parish Public Library (bef 1903) 204; Lether E. Frazar Memorial Library of . . . University (1939) 110.

Lakeland: [a]Florida. [b]none. [c]Parker L; Hollingsworth L; Hunter L; Morton L; Wire L; Mirror L. [d]none. [e]Roux Library of Florida Southern College (1885) 153; Public Library (1926) 90.

Lakewood: [a]California. [b]none. [c]San Gabriel R. [d]none. [e]Public Library (bef 1957) 115.

Lakewood: [a]Colorado. [b]none. [c]nr South Platte R. [d]Colorado School of Mines (univ) (at Golden, nr Lakewood) (1869) 2 – 173. [e]Jefferson County Public Library (at Golden, nr Lakewood) (1953) 295; Arthur Lakes Library of . . . School (at Golden, nr Lakewood) (1874) 162.

Lakewood: [a]Ohio. [b]for East Rockport. [c]Rocky R; Erie L. [d]none. [e]Public Library (1916) 209.

Lancaster: [a]Pennsylvania. [b]none. [c]Conestoga R. [d]Millersville State College (univ) (at Millersville, nr Lancaster) (1854) 6 – 372; Franklin and Marshall College (univ) (1787) 3 – 151. [e]Lancaster County Library (1759) 293; Helen A. Ganser Library of Millersville State College (at Millersville, nr Lancaster) (1855) 263; Franklin and Marshall College Library (1787) 191.

Lansing: [a]*Michigan. [b]none. [c]Grand R; Red Cedar R. [d]Michigan State University (at East Lansing, nr Lansing) (1855) 48– 4586. [e] . . . University Library (at East Lansing, nr Lansing) (1855) 2102; Michigan State Library (1828) 1000; Lansing Public Library (1882) 204; East Lansing Public Library (at East Lansing, nr Lansing) (1923) 80.

Laredo: [a]Texas. [b]none. [c]Rio Grande R. [d]Texas Agricultural and Industrial University at Laredo (1969) 0.8 – ? [e]Harold Yeary Library of Laredo Junior College (1947) 61; Yeary Memorial Library of . . . University (1968) 45; Public Library (bef 1939) 27.

Las Cruces: [a]New Mexico. [b]none. [c]nr Rio Grande R. [d]New Mexico State University (1888) 13 – 557 [Las Cruces campus: (1888) 11 – 450]. [e] . . . University Library (1888) 464; Thomas Branigan Memorial Library (bef 1930) 67.

Las Vegas: [a]Nevada. [b]none. [c]Las Vegas Creek. [d]University of Nevada, Las Vegas campus (1951) 15 – 690. [e]James R. Dickinson Las Vegas Campus Library of the University . . . (1957) 292; Clark County Library (bef 1928) 239.

Lawrence: [a]Kansas. [b]none. [c]Kansas R. [d]University of Kansas (1855) 24 – 1975 [Lawrence campus: (1855) 22 – ?]. [e]Watson Memorial Lawrence Campus Library of the University . . . (1866) 1500; Public Library (1904) 110.

Lawrence: [a]Massachusetts. [b]none. [c]Merrimack R. [d]none. [e]Public Library (1847) 214.

Lawton: [a]Oklahoma. [b]none. [c]Cache Creek. [d]Cameron University (college) (1908) 5 – 235. [e] . . . University Library (1908) 140; Public Library (1903) 79.

Lewiston: [a]Maine. [b]none. [c]Androscoggin R. [d]none. [e]Bates College Library (1863) 180; Public Library (1902) 127.

Lexington: [a]Kentucky. [b]none. [c]Town Branch Creek. [d]University of Kentucky (1865) 40 – 3137 [Lexington campus: (1865) 22 – 2479]; Transylvania University (college) (1780) 0.7 – 59. [e]Margaret I. King Lexington Campus Library of the University . . . (1909) 568; Public Library (1795) 150; Frances Carrick Thomas Library of . . . University (1784) 93.

Lima: [a]Ohio. [b]none. [c]Ottawa R. [d]none. [e]Public Library (1901) 222.

Lincoln: [a]*Nebraska. [b]none. [c]Salt Creek; Antelope Creek. [d]University of Nebraska (1869) 39 – 3179 [Lincoln campus: (1869) 22 – 1525]; Nebraska Wesleyan University (college) (1887) 1 – 94. [e]Don L. Love Memorial Lincoln Campus Library of the University . . . (1869) 1208; City Library (1876) 414.

Little Rock: [a]*Arkansas. [b]none. [c]Arkansas R. [d]University of Arkansas, Little Rock campus (1879) 10 – 502. [e]Public Library (1910) 399; Little Rock Campus Library of the University . . . (1879) 213; Arkansas Library Commission (1935) 144.

Livonia: [a]Michigan. [b]none. [c]Middle Rouge R. [d]none. [e]Public Library (1958) 140.

Long Beach: [a]California. [b]none. [c]San Pedro B of Pacific O; Los Angeles R. [d]California State University at Long Beach (1949) 34 – 1809. [e]Public Library (1896) 652; . . . University Library (1949) 592.

Longview: [a]Texas. [b]none. [c]nr Sabine R. [d]none. [e]LeTourneau College Library (1945) 90; Nicholson Memorial Public Library (1932) 53.

Lorain: [a]Ohio. [b]for Charleston. [c]Black R; Erie L. [d]none. [e]Public Library (1900) 232.

Los Angeles: [a]California. [b]inc Hollywood. [c]San Pedro B of Pacific O; Los Angeles R. [d][1] University of California, Los Angeles campus (1881) 60 – 3134; California State University at Northridge (1956) 29 – 1385; University of Southern California (1880) 28 – 2038; California State University at Los Angeles (1947) 26 – 1510; Pepperdine University (at Malibu, nr Los Angeles) (1937 at Los Angeles, relocated 1973?) 12 – 716; Loyola Marymount University (1865) 6 – 335. [e]Los Angeles Public Library (1872) 4533; Los Angeles County Public Library (1912) 4058; Los Angeles Campus Library of the University of California (1919) 3519; Edward L. Doheny Memorial Library of the University of Southern California (1880) 1671.

Louisville: [a]Kentucky. [b]none. [c]Ohio R. [d]University of Louisville (1798) 15 – 1718; Indiana University, Southeast campus (at New Albany, Indiana, nr Louisville) (1941) 4 – 154; Bellarmine Col-

[1]Universities with 5000 or more students.

lege (univ) (1950) 2 – 88; Spalding College (univ) (1920) 1 – 92. [e]Free Public Library (1816) 1041; University . . . Library (1798) 697.

Lowell: [a]Massachusetts. [b]none. [c]Merrimack R; Concord R. [d]University of Lowell (1894) 12 – 480. [e]City Library (1845) 341; University . . . Library (1894) 103.

Lubbock: [a]Texas. [b]none. [c]Double Mountain Fork of Brazos R. [d]Texas Tech University (1923) 23 – 1399. [e]. . . University Library (1925) 906; Lubbock City-County Library (1923) 204.

Lynchburg: [a]Virginia. [b]none. [c]James R. [d]Lynchburg College (univ) (1903) 2 – 126. [e]Lipscomb Library of Randolph-Macon Woman's College (1896) 111; Public Library (1966) 80; Lynchburg College Library (1903) 79; Jones Memorial Library (1907) 69.

Lynn: [a]Massachusetts. [b]none. [c]Massachusetts B of Atlantic O. [d]none. [e]Public Library (1815) 258.

Macon: [a]Georgia. [b]none. [c]Ocmulgee R. [d]Mercer University (1833 at Penfield, relocated 1871) 4 – 185 [Macon campus: (1871) 2 – ?]. [e]Middle Georgia Regional Library (bef 1880) 250; Stetson Memorial Macon Campus Library of . . . University (1833 at Penfield, relocated 1871?) 137.

Madison: [a]*Wisconsin. [b]none. [c]Yahara R; Mendota L; Monona L; Waubesa L. [d]University of Wisconsin (1849) 143 – 9202 [Madison campus: (1849) 39 – 3103]. [e]Memorial Madison Campus Library of the University . . . (1850) 2973; State Historical Society of Wisconsin Library (1853) 1283; Public Library (1875) 445.

Manchester: [a]New Hampshire. [b]for Derryfield. [c]Merrimack R. [d]New Hampshire College (univ) (1932) 4 – 124; Notre Dame College (univ) (1950) 0.5 – 48. [e]City Library (1854) 206; Geisel Library of Saint Anselm's College (1929) 93.

Mansfield: [a]Ohio. [b]none. [c]Rocky Fork of Mohican R. [d]none. [e]Public Library (1887) 243.

McAllen: [a]Texas. [b]none. [c]nr Rio Grande R. [d]Pan American University (at Edinburg, nr McAllen) (1927) 8 – 321. [e]. . . University Library (at Edinburg, nr McAllen) (1927) 121; McAllen Memorial Library (1932) 90.

Melbourne: [a]Florida. [b]none. [c]Indian River Lagoon. [d]Florida Institute of Technology (univ) (1958) 3 – 229. [e]. . . Institute Library (1958) 72; Melbourne Public Library (1918) 72; Eau Gallie Public Library (1955) 52.

Memphis: [a]Tennessee. [b]none. [c]Mississippi R; Wolf R. [d]Memphis State University (1912) 22 – 1250. [e]Memphis-Shelby County Public Library (1893) 1183; . . . University Library (1914) 636.

Meriden: [a]Connecticut. [b]none. [c]Quinnipiac R. [d]Wesleyan University (at Middletown, nr Meriden) (1829) 2 – 313. [e]Olin Memorial Library of . . . University (at Middletown, nr Meriden) (1832) 759; Meriden Public Library (1898) 142; Russell Public Library (at Middletown, nr Meriden) (1875) 84.

Meridian: [a]Mississippi. [b]none. [c]Sowashee Creek. [d]none. [e]Public Library (1913) 107.

Mesa: [a]Arizona. [b]none. [c]Salt R. [d]none. [e]Public Library (1925) 91; Mesa Community College Library (1963) 46.

Metairie: [a]Louisiana. [b]none. [c]Pontchartrain L. [d]none. [e]Jefferson Parish Library (1949) 448.

Miami: [a]Florida. [b]none. [c]Biscayne B of Atlantic O. [d]University of Miami (at Coral Gables, nr Miami) (1925) 15 – 1674; Florida International University (1972) 8 – 221; Barry College (univ) (at Miami Shores, nr Miami) (1940) 2 – 120. [e]Otto G. Richter Library of the University . . . (at Coral Gables, nr Miami) (1926) 1118; Miami-Dade Public Library (1900) 869.

Miami Beach: [a]Florida. [b]none. [c]Atlantic O; Biscayne B of Atlantic O. [d]none. [e]Public Library (1927) 167.

Middletown: [a]Ohio. [b]none. [c]Miami (alt Great Miami) R. [d]none. [e]Public Library (1911) 139.

Midland: [a]Texas. [b]none. [c]Midland Draw. [d]none. [e]Midland County Public Library (1903) 112.

Milwaukee: [a]Wisconsin. [b]none. [c]Milwaukee R; Menomonee R; Michigan L. [d]University of Wisconsin, Milwaukee campus (1955) 25 – 1362; Marquette University (1857) 10 – 900; Milwaukee School of Engineering (univ) (1903) 2 – 154; Cardinal Stritch College (univ) (1934) 1 – 101. [e]Public Library (1847) 2389; Milwaukee Campus Library of the University . . . (1956) 625; . . . University Memorial Library (1881) 454.

Minneapolis: [a]Minnesota. [b]none. [c]Mississippi R. [d]University of Minnesota (1851) 79 – 5981

[Minneapolis campus: (1851) 67 – 5420]. [e]O. Meredith Wilson Minneapolis Campus Library of the University . . . (1851) 3047; Public Library (1860) 1311.

Mobile: [a]Alabama. [b]none. [c]Mobile B of G of Mexico; Mobile R. [d]University of South Alabama (1963) 7 – 461. [e]Public Library (bef 1902) 341; University . . . Library (1964) 165.

Modesto: [a]California. [b]none. [c]Tuolumne R. [d]none. [e]Stanislaus County Free Library (1907) 413.

Monroe: [a]Louisiana. [b]none. [c]Ouachita R. [d]Northeast Louisiana University (1931) 10 – 400. [e]Sandel Library of . . . University (1931) 395; Ouachita Parish Public Library (1916) 179.

Montgomery: [a]*Alabama. [b]none. [c]Alabama R. [d]Alabama State University (1873) 4 – 186; Auburn University, Montgomery campus (1968) 4 – 93; Troy State University, Montgomery campus (1966) 2 – ? [e]Alabama Public Library Service (1939) 349; Montgomery Campus Library of Auburn University (1969) 254; Montgomery City-County Public Library (1899) 165; George W. Trenholm Library of Alabama State University (1921) 141.

Mount Vernon: [a]New York. [b]none. [c]Bronx R. [d]none. [e]Public Library (1896) 319.

Muncie: [a]Indiana. [b]none. [c]West Fork of White R. [d]Ball State University (1898) 18 – 1351. [e]. . . University Library (1918) 683; Muncie-Center Township Public Library (1874) 274.

Muskegon: [a]Michigan. [b]none. [c]Muskegon R; Michigan L; Muskegon L. [d]none. [e]Hackley Public Library (1889) 164; Muskegon County Library (1938) 125.

Nashua: [a]New Hampshire. [b]for Dunstable. [c]Merrimack R; Nashua R. [d]Rivier College (univ) (1933) 2 – 71. [e]Public Library (1840) 175; Regina Library of . . . College (1933) 73.

Nashville: [a]*Tennessee. [b]none. [c]Cumberland R. [d]Vanderbilt University (1872) 7 – 1616; Tennessee State University (1909) 6 – 301; University of Tennessee, Nashville campus (1947) 5 – 299; George Peabody College for Teachers (univ) (1785) 2 – 162; Fisk University (1865) 1 – 125. [e]Joint University Library (1936) 1302; Public Library of Nashville and Davidson County (1904) 474; Tennessee State Library and Archives (1854) 250.

Natchez: [a]Mississippi. [b]none. [c]Mississippi R. [d]none. [e]Judge George W. Armstrong Library (1883) 66.

Newark: [a]New Jersey. [b]none. [c]Newark B of Atlantic O; Passaic R. [d]Rutgers, the State University of New Jersey, Newark campus (1892) 10 – ?; New Jersey Institute of Technology (univ) (1881) 6 – 297. [e]Public Library (1888) 1250.

New Bedford: [a]Massachusetts. [b]none. [c]Buzzards B of Atlantic O; Acushnet R. [d]Southeastern Massachusetts University (at North Dartmouth, nr New Bedford) (1895) 5 – 330. [e]Free Public Library (1852) 346; . . . University Library (at North Dartmouth, nr New Bedford) (1960) 140.

New Britain: [a]Connecticut. [b]none. [c]nr Quinnipiac R; nr Mattabessett R. [d]Central Connecticut State College (univ) (1849) 13 – 645. [e]Elihu Burritt Library of . . . College (1850) 201; Public Library (1853) 194.

New Haven: [a]Connecticut. [b]none. [c]Long Island Sound of Atlantic O; Quinnipiac R; Mill R; West R. [d]Southern Connecticut State College (univ) (1893) 13 – 611; Yale University (1701) 10 – 2613; University of New Haven (at West Haven, nr New Haven) (1920) 6 – 460; Quinnipiac College (univ) (at Hamden, nr New Haven) (1911) 3 – 225. [e]. . . University Library (1701) 6519; Free Public Library (1887) 499.

New Orleans: [a]Louisiana. [b]none. [c]Mississippi R; Pontchartrain L; Borgne L. [d]University of New Orleans (1956) 14 – 691; Tulane University of Louisiana (1834) 9 – 1004; Loyola University in New Orleans (1849) 5 – 322; Xavier University of Louisiana (1915) 2 – 158; Dillard University (college) (1869) 1 – 93. [e]Howard-Tilton Memorial Library of Tulane University of Louisiana (1834) 1218; Public Library (1843) 704.

Newport: [a]Rhode Island. [b]none. [c]Atlantic O; Narragansett B of Atlantic O. [d]University of Rhode Island (at Kingston, nr Newport) (1888) 17 – 895; Salve Regina–The Newport College (univ) (1934) 2 – 98. [e]University . . . Library (at Kingston, nr Newport) (1892) 340; United States Naval War College Library (1885) 136; Redwood Library and Athenaeum (1747) 134; . . . College Library (1947) 58; Public Library (1868) 53.

Newport News: [a]Virginia. [b]none. [c]Hampton Roads; James R. [d]none. [e]Public Library (1908) 198.

New Rochelle: [a]New York. [b]none. [c]Long Island Sound of Atlantic O. [d]Iona College (univ) (1940) 5 – 280; College of New Rochelle (univ) (1904) 3 – 230. [e]Public Library (1894) 156; Ryan Library of . . . College (1940) 150; Gill Library of the College . . . (1904) 100.

Newton: [a]Massachusetts. [b]none. [c]Charles R. [d]Boston College (univ) (1863) 14 – 945; Wellesley College (univ) (at Wellesley, nr Newton) (1870) 2 – 254. [e]Bapst Library of Boston College (1863) 909; Margaret Clapp Library of Wellesley College (at Wellesley, nr Newton) (1876) 499; Newton Free Library (1870) 320; Wellesley Free Library (at Wellesley, nr Newton) (1883) 147.

New York: [a]New York. [b]alt New York City; for New Amsterdam; inc Brooklyn since 1898. [c]Long Island Sound of Atlantic O; New York B of Atlantic O; Raritan B of Atlantic O; Jamaica B of Atlantic O; Hudson R; East R; Bronx R; Arthur Kill; Harlem R; Kill van Kull. [d1] City University of New York (1847) 251 – 18,121; New York University (1831) 30 – 5500; Long Island University (at Greenvale, nr New York) (1886 at New York, relocated 1948) 24 – 1119 [Greenvale campus: (1948) 15 – 633] [New York campus: (1886) 8 – 415]; Columbia University (1754) 23 – 4616; New York Institute of Technology (univ) (at Old Westbury, nr New York) (1955 at New York, relocated 1964) 20 – 600 [Old Westbury campus: (1964) 18 – ?] [New York campus: (1955) 2 – ?]; Saint John's University (1870) 15 – 707; Fordham University (1841) 14 – 799; Pace University (1906) 14 – 789 [New York campus: (1906) 9 – 445]; Hofstra University (at Hempstead, nr New York) (1935) 11 – 611; Adelphi University (at Garden City, nr New York) (1863) 10 – 580. [e]New York Public Library (1848) 8898; Columbia University Library (1761) 4662; Brooklyn Public Library (1897) 3520; Queens Borough Public Library (1896) 3013; Elmer Holmes Bobst Library of New York University (1835) 2179; Duane Library of Fordham University (1841) 952; Morris Raphael Cohen City College Library of the City University of New York (1849) 863; . . . ; Pierpont Morgan Library (1924) 80. [f]Brooklyn: 806 (1890), 97 (1850), 4 (1800), 0.7 (1738).

Niagara Falls: [a]New York. [b]none. [c]Niagara R. [d]Niagara University (at Niagara, nr Niagara Falls) (1856) 4 – 270. [e]Public Library (1838) 173; . . . University Library (at Niagara, nr Niagara Falls) (1856) 147.

Norfolk: [a]Virginia. [b]none. [c]Chesapeake B of Atlantic O; Hampton Roads; Elizabeth R. [d]Old Dominion University (1919) 13 – 668; Norfolk State College (univ) (1935) 7 – 405. [e]Public Library (1872) 553; . . . University Library (1959) 375.

North Charleston: [a]South Carolina. [b]none. [c]Cooper R. [d]none. [e]Cooper River Branch of Charleston County Library (?) 33.

Norwalk: [a]California. [b]none. [c]nr San Gabriel R. [d]Biola College (univ) (at La Mirada, nr Norwalk) (1907) 3 – 182. [e]Norwalk Branch of Los Angeles County Public Library (1913) 163; Rose Memorial Library of . . . College (at La Mirada, nr Norwalk) (1908) 120.

Norwalk: [a]Connecticut. [b]none. [c]Long Island Sound of Atlantic O; Norwalk R. [d]none. [e]Public Library (1879) 129.

Norwich: [a]Connecticut. [b]none. [c]Thames R; Shetucket R; Yantic R. [d]Connecticut College (univ) (at New London, nr Norwich) (1911) 2 – 184. [e]. . . College Library (at New London, nr Norwich) (1915) 301; United States Coast Guard Academy Library (at New London, nr Norwich) (1876) 103; Otis Library (1850) 98; Public Library of New London (at New London, nr Norwich) (1891) 62.

Oakland: [a]California. [b]none. [c]San Francisco B of Pacific O. [d]Mills College (univ) (1852) 1 – 107; Saint Mary's College of California (univ) (at Moraga, nr Oakland) (1863) 1 – 105; Holy Names College (univ) 1868) 0.7 – 108. [e]Public Library (1868) 737.

Oak Park: [a]Illinois. [b]none. [c]nr Des Plaines R. [d]none. [e]Public Library (1903) 171.

Odessa: [a]Texas. [b]none. [c]nr Johnson Draw. [d]University of Texas, Permian Basin campus (1970) 1 – 85. [e]Permian Basin Campus Library of the University . . . (1973) 194; Ector County Library (1938) 130.

Ogden: [a]Utah. [b]none. [c]Weber R; Ogden R. [d]none. [e]Weber County Library (1903) 246; Weber State College Library (1888) 231.

Oklahoma City: [a]*Oklahoma. [b]none. [c]North Canadian R; Hefner L. [d]Central State University (at Edmond, nr Oklahoma City) (1890) 13 – 390; Oklahoma City University (1904) 3 – 205; Bethany Nazarene College (univ) (at Bethany, nr Oklahoma City) (1909) 1 – 70. [e]Oklahoma

[1]Universities with 10,000 or more students.

County Library (1901) 600; Central State University Library (at Edmond, nr Oklahoma City) (1890) 350; Oklahoma Department of Libraries (1890) 220.

Omaha: ^{a}Nebraska. ^{b}none. ^{c}Missouri R. ^{d}University of Nebraska, Omaha campus (1908) 16 – 1654; Creighton University (1878) 5 – 781. ^{e}Public Library (1872) 479; Omaha Campus Library of the University . . . (1902) 334; Alumni Library of . . . University (1878) 217.

Orange: ^{a}California. ^{b}none. ^{c}Santa Ana R. ^{d}Chapman College (univ) (1861 at Woodland, relocated 1954) 6 – 426; West Coast University, Orange County campus (1963) 0.4 – ? ^{e}Orange County Public Library (1921) 936; Orange Public Library (1894) 235; Thurmond Clarke Memorial Library of . . . College (1923 at Woodland, relocated 1954?) 129.

Orlando: ^{a}Florida. ^{b}none. ^{c}Clear L; Holden L; Ivanhoe L; Sue L; Concord L; Lancaster L. ^{d}Florida Technological University (1963) 11 – 384; Rollins College (univ) (at Winter Park, nr Orlando) (1885) 4 – 286. ^{e}Public Library (1923) 483; Mills Memorial Library of . . . College (at Winter Park, nr Orlando) (1885) 164; . . . University Library (1966) 150.

Oshkosh: ^{a}Wisconsin. ^{b}none. ^{c}Fox R; Winnebago L. ^{d}University of Wisconsin, Oshkosh campus (1871) 10 – 520. ^{e}Forrest R. Polk Oshkosh Campus Library of the University . . . (1871) 399; Public Library (1896) 224.

Owensboro: ^{a}Kentucky. ^{b}for Rossborough. ^{c}Ohio R. ^{d}none. ^{e}Owensboro-Daviess County Public Library (1909) 90; Kentucky Wesleyan College Library (1866) 64; Brescia College Library (1950) 46.

Oxnard: ^{a}California. ^{b}none. ^{c}nr Santa Clara R. ^{d}Saint John's College (univ) (at Camarillo, nr Oxnard) (1927) 0.2 – 28. ^{e}Public Library (1906) 89; . . . College Library (at Camarillo, nr Oxnard) (1940) 79.

Palm Beach: ^{a}Florida. ^{b}none. ^{c}Atlantic O; Worth Lagoon. ^{d}none. ^{e}Society of the Four Arts Library (bef 1953) 27.

Palm Springs: ^{a}California. ^{b}none. ^{c}none. ^{d}none. ^{e}Public Library (1940) 74.

Palo Alto: ^{a}California. ^{b}none. ^{c}San Francisco B of Pacific O. ^{d}Stanford University (at Stanford, nr Palo Alto) (1885) 13 – 2563. ^{e}. . . University Library (at Stanford, nr Palo Alto) (1892) 4092; City Library (1897) 177.

Parma: ^{a}Ohio. ^{b}none. ^{c}nr Cuyahoga R. ^{d}Baldwin-Wallace College (univ) (at Berea, nr Parma) (1845) 3 – 189. ^{e}Parma Regional Library of Cuyahoga County Public Library (bef 1964) 222; Ritter Library of . . . College (at Berea, nr Parma) (1893) 152.

Pasadena: ^{a}California. ^{b}none. ^{c}Arroyo Seco Creek. ^{d}California Institute of Technology (univ) (1891) 2 – 523; Pacific Oaks College (univ) (1945) 0.3 – 44. ^{e}Public Library (1884) 443; Robert A. Millikan Memorial Library of . . . Institute (1891) 160.

Pasadena: ^{a}Texas. ^{b}none. ^{c}Buffalo Bayou. ^{d}none. ^{e}Public Library (1953) 108; Lee Davis Library of San Jacinto College (1961) 89.

Passaic: ^{a}New Jersey. ^{b}none. ^{c}Passaic R. ^{d}Fairleigh Dickinson University (at Rutherford, nr Passaic) (1941) 19 – 1999 [Rutherford campus: (1941) 5 – ?] [Teaneck campus (at Teaneck, nr Passaic): (1954) 9 – ?]. ^{e}Weiner Teaneck Campus Library of . . . University (at Teaneck, nr Passaic) (1954) 162; Teaneck Public Library (at Teaneck, nr Passaic) (1921) 155; Messler Rutherford Campus Library of . . . University (at Rutherford, nr Passaic) (1941) 141; Passaic Public Library (alt Julius Forstmann Library) (1887) 141; Rutherford Public Library (at Rutherford, nr Passaic) (1894) 90.

Paterson: ^{a}New Jersey. ^{b}none. ^{c}Passaic R. ^{d}William Paterson College (univ) (at Wayne, nr Paterson) (1855) 14 – 610. ^{e}Paterson Free Public Library (alt Danforth Memorial Library) (1885) 291; Sarah Byrd Askew Library of . . . College (at Wayne, nr Paterson) (1924) 243; Wayne Public Library (at Wayne, nr Paterson) (1922) 181.

Pawtucket: ^{a}Rhode Island. ^{b}none. ^{c}Blackstone (alt Seekonk) R. ^{d}none. ^{e}Public Library (alt Deborah Cook Sayles Memorial Library) (1852) 122.

Pensacola: ^{a}Florida. ^{b}none. ^{c}Pensacola B of G of Mexico. ^{d}University of West Florida (1955) 5 – 252. ^{e}West Florida Regional Library (1937) 203; John C. Pace Library of the University . . . (1966) 177.

Peoria: [a]Illinois. [b]none. [c]Illinois R; Peoria L. [d]Bradley University (1896) 5 – 419. [e]Public Library (1855) 406; Cullom-Davis Library of . . . University (1897) 290.

Petersburg: [a]Virginia. [b]none. [c]Appomattox R. [d]Virginia State College (univ) (1882) 6 – 242. [e]Johnston Memorial Library of . . . College (1882) 169; Chesterfield County Free Public Library (at Chester, Virginia, nr Petersburg) (1965) 117; Public Library (1924) 56.

Philadelphia: [a]Pennsylvania. [b]none. [c]Delaware R; Schuylkill R. [d][1] Temple University (1884) 35 – 3057; American College (univ) (at Bryn Mawr, nr Philadelphia) (1927) 21 – ?; University of Pennsylvania (1740) 20 – 4722; Villanova University (at Villanova, nr Philadelphia) (1842) 10 – 572; Drexel University (1891) 9 – 611; La Salle College (univ) (1863) 6 – 375; Saint Joseph's College (univ) (1851) 6 – 357. [e]Free Library (1891) 2974; University . . . Library (1750) 2700; Samuel Paley Library of Temple University (1892) 1247; Canady Library of Bryn Mawr College (at Bryn Mawr, nr Philadelphia) (1885) 418; Falvey Memorial Library of Villanova University (at Villanova, nr Philadelphia) (1848) 400.

Phoenix: [a]*Arizona. [b]none. [c]Salt R. [d]none. [e]Arizona State Library (1864 at Prescott, relocated 1889?) 1060; Public Library (1908) 983.

Pine Bluff: [a]Arkansas. [b]none. [c]Arkansas R. [d]none. [e]Public Library of Pine Bluff and Jefferson County (1913) 100; Watson Memorial Pine Bluff Campus Library of the University of Arkansas (1938) 63.

Pittsburgh: [a]Pennsylvania. [b]for Pittsburg; inc Allegheny since 1906. [c]Ohio R; Allegheny R; Monongahela R. [d]University of Pittsburgh (1787) 35 – 4029 [Pittsburgh campus: (1787) 30 – 3843; Duquesne University (1878) 8 – 621; Carnegie-Mellon University (1900) 5 – 660. [e]Carnegie Library (1895) 2086; Hillman Library of the University . . . (1873) 1824; Hunt Library of Carnegie-Mellon University (1920) 287; Duquesne University Library (1928) 273. [f]Allegheny: 130 (1900), 21 (1850), 3 (1830).

Pittsfield: [a]Massachusetts. [b]none. [c]Housatonic R. [d]none. [e]Berkshire Athenaeum (alt Pittsfield Public Library) (1871) 146.

Plainfield: [a]New Jersey. [b]for Milltown. [c]Green Brook. [d]none. [e]Public Library (1881) 171.

Pocatello: [a]Idaho. [b]none. [c]Portneuf R. [d]Idaho State University (1901) 7 – 385. [e]. . . University Library (1902) 188; Public Library (1906) 75.

Pomona: [a]California. [b]none. [c]Chino Creek. [d]California State Polytechnic University at Pomona (1938) 13 – 660; Claremont Colleges (univ) (at Claremont, nr Pomona) (1887) 5 – 484; Azusa Pacific College (univ) (at Azusa, nr Pomona) (1899) 2 – 76; La Verne College (univ) (at La Verne, nr Pomona) (1891) 2 – 80. [e]Claremont Colleges Library (at Claremont, nr Pomona) (1952) 859; . . . University Library (1938) 253; Public Library (1883) 211.

Pontiac: [a]Michigan. [b]none. [c]Clinton R. [d]Oakland University (at Rochester, Michigan, nr Pontiac) (1957) 11 – 508. [e]Kresge Library of . . . University (at Rochester, Michigan, nr Pontiac) (1959) 220; Public Library (1882) 118.

Port Arthur: [a]Texas. [b]none. [c]Sabine Lake. [d]none. [e]Gates Memorial Library (1918) 92.

Portland: [a]Maine. [b]none. [c]Casco B of Atlantic O; Presumpscot R. [d]University of Maine, Portland-Gorham campus (1933) 9 – 615. [e]Portland Campus Library of the University . . . (1878) 294; Public Library (alt Baxter Library) (1867) 231.

Portland: [a]Oregon. [b]none. [c]Willamette R. [d]Portland State University (1946) 15 – 742; Lewis and Clark College (univ) (1867) 3 – 199; University of Portland (1901) 2 – 130; Reed College (univ) (1908) 1 – 115; Warner Pacific College (univ) (1935 at Spokane, Washington; relocated 1940) 0.6 – 40. [e]Multnomah County Library (alt Library Association of Portland) (1864) 1076; Branford Price Millar Library of . . . University (1946) 396.

Portsmouth: [a]Virginia. [b]none. [c]Hampton Roads; Elizabeth R. [d]none. [e]Public Library (1914) 154.

Poughkeepsie: [a]New York. [b]none. [c]Hudson R. [d]Marist College (univ) (1930) 2 – 118; Vassar College (univ) (1861) 2 – 223. [e]Vassar College Library (1865) 483; Adriance Memorial Library (1840) 108.

---

[1] Universities with 5000 or more students.

Providence: [a]*Rhode Island. [b]none. [c]Providence R; Seekonk R. [d]Rhode Island College (univ) (1854) 8 - 364; Brown University (1764) 7 – 1350; Providence College (univ) (1917) 6 – 328. [e]. . . University Library (1767) 1527; Public Library (1878) 680.

Provo: [a]Utah. [b]none. [c]Provo R. [d]Brigham Young University (1875) 28 – 1987 [Provo campus: (1875) 27 – ?]. [e]Harold B. Lee Library of . . . University (1876) 1145; City Public Library (1904) 68.

Pueblo: [a]Colorado. [b]none. [c]Arkansas R. [d]University of Southern Colorado (1933) 7 – 327. [e]University . . . Library (1933) 165; Pueblo Regional Library (1873) 125.

Quincy: [a]Massachusetts. [b]none. [c]Massachusetts B of Atlantic O; Neponset R; Fore R. [d]none. [e]Thomas Crane Public Library (1871) 223.

Racine: [a]Wisconsin. [b]none. [c]Root R; Michigan L. [d]none. [e]Public Library (1897) 250.

Raleigh: [a]*North Carolina. [b]none. [c]Crabtree Creek. [d]North Carolina State University (1887) 18 – 1947; Shaw University (college) (1865) 2 – 149. [e]D. H. Hill Library of North Carolina State University (1889) 429; Wake County Public Library (1900) 249; North Carolina State Library (1812) 199.

Rapid City: [a]South Dakota. [b]none. [c]Rapid Creek. [d]South Dakota School of Mines and Technology (univ) (1885) 2 – 146. [e]Public Library (1903) 90; Devereaux Library of . . . School (1885) 58.

Reading: [a]Pennsylvania. [b]none. [c]Schuylkill R. [d]none. [e]Public Library (1763) 261.

Reno: [a]Nevada. [b]none. [c]Truckee R. [d]University of Nevada (1864 at Elko, relocated 1886) 30 – 1189 [Reno campus: (1886) 8 – 445]. [e]Noble H. Getchell Reno Campus Library of the University . . . (1886) 513; Washoe County Library (1904) 249.

Richmond: [a]California. [b]none. [c]San Francisco B of Pacific O. [d]Dominican College of San Rafael (univ) (at San Rafael, nr Richmond) (1890) 0.9 – 111. [e]Marin County Free Library (at San Rafael, nr Richmond) (1927) 277; Public Library (1907) 214.

Richmond: [a]*Virginia. [b]none. [c]James R. [d]Virginia Commonwealth University (1838) 18 – 2151; University of Richmond (college) (1830) 4 – 292; Virginia Union University (college) (1865) 1 – 123. [e]Virginia State Library (1823) 526; Public Library (1891) 522; Virginia Commonwealth University Library (1913) 229; Boatwright Memorial Library of the University . . . (1832) 223; County of Henrico Public Library (1966) 200.

Riverside: [a]California. [b]none. [c]Santa Ana R. [d]University of California, Riverside campus (1907) 7 – 766. [e]Riverside City and County Public Library (1876) 863; Riverside Campus Library of the University . . . (1954) 840.

Roanoke: [a]Virginia. [b]for Big Lick. [c]Roanoke R. [d]Hollins College (univ) (at Hollins, nr Roanoke) (1842) 1 – 99. [e]Roanoke Public Library (1921) 301; Roanoke County Public Library (1945) 140; Fishburn Library of . . . College (at Hollins, nr Roanoke) (1855) 127.

Rochester: [a]Minnesota. [b]none. [c]South Branch of Zumbro R. [d]none. [e]Public Library (1895) 129.

Rochester: [a]New York. [b]none. [c]Genesee R; Ontario L. [d]Rochester Institute of Technology (univ) (1829) 12 – 1028; University of Rochester (1850) 8 – 2348; Nazareth College of Rochester (univ) (1924) 3 – 142. [e]Rush Rhees Library of the University . . . (1850) 986; Public Library (1886) 830.

Rockford: [a]Illinois. [b]none. [c]Rock R. [d]Rockford College (univ) (1847) 1 – 100. [e]Public Library (1873) 283.

Rock Island: [a]Illinois. [b]none. [c]Mississippi R; Rock R. [d]Augustana College (univ) (1860) 2 – 135. [e]Denkman Memorial Library of . . . College (1860) 181; Rock Island Public Library (1872) 151; Moline Public Library (at Moline, nr Rock Island) (1873) 136.

Rome: [a]New York. [b]none. [c]Mohawk R. [d]none. [e]Jervis Public Library (1894) 100.

Royal Oak: [a]Michigan. [b]none. [c]nr Rouge R. [d]none. [e]Public Library (1915) 122.

Sacramento: [a]*California. [b]none. [c]Sacramento R; American R. [d]California State University at Sacramento (1947) 22 – 1040; University of California, Davis campus (at Davis, nr Sacramento) (1906) 19 – 1269. [e]California State Library (1850) 2844; Sacramento City-County Library (1857) 946; General Davis Campus Library of the University . . . (at Davis, nr Sacramento) (1908) 741; . . . University Library (1947) 474.

Saginaw: [a]Michigan. [b]none. [c]Saginaw R. [d]none. [e]Saginaw Public Library (alt Hoyt Public Library) (1855) 301.

Saint Augustine: ^{a}Florida. ^{b}none. ^{c}Matanzas B of Atlantic O; San Sebastian R. ^{d}none. ^{e}Louise Wise Lewis Library of Flagler College (1968) 33; Free Public Library (1874) 22.

Saint Clair Shores: ^{a}Michigan. ^{b}none. ^{c}Saint Clair L. ^{d}none. ^{e}Public Library (1935) 112.

Saint Joseph: ^{a}Missouri. ^{b}none. ^{c}Missouri R. ^{d}none. ^{e}Public Library (1890) 182.

Saint Louis: ^{a}Missouri. ^{b}none. ^{c}Mississippi R. $^{d^1}$ University of Missouri, Saint Louis campus (1960) 12 – 535; Washington University (1853) 11 – 2658; Saint Louis University (1818) 10 – 1642 [Saint Louis campus: (1818) 10 – ?]. ^{e}John M. Olin Library of Washington University (1853) 1545; Saint Louis Public Library (1865) 1367; Saint Louis County Library (1946) 1266; Pius XII Memorial Library of Saint Louis University (1818) 457.

Saint Paul: ^{a}*Minnesota. ^{b}none. ^{c}Mississippi R. ^{d}College of Saint Thomas (univ) (1885) 3 – 150; Macalester College (univ) (1853) 2 – 166; Hamline University (college) (1854 at Red Wing, relocated 1880) 1 – 121; Metropolitan State University (college) (1971) 1 – ? ^{e}Public Library (1863) 717; Minnesota Historical Society Library (1849) 305.

Saint Petersburg: ^{a}Florida. ^{b}none. ^{c}Tampa B of G of Mexico; Boca Ciega B of G of Mexico. ^{d}none. ^{e}Public Library (1909) 352.

Salem: ^{a}Massachusetts. ^{b}none. ^{c}Atlantic O. ^{d}Salem State College (univ) (1854) 8 – 312. ^{e}James Duncan Phillips Library of Essex Institute (1821) 600; Public Library (1888) 156; . . . College Library (1854) 151.

Salem: ^{a}*Oregon. ^{b}none. ^{c}Willamette R. ^{d}Oregon College of Education (univ) (at Monmouth, nr Salem) (1856) 3 – 250; Willamette University (college) (1842) 2 – 154. ^{e}Oregon State Library (1848) 330; Public Library (1904) 153; . . . College Library (at Monmouth, nr Salem) (1882) 140; . . . University Library (1844) 93.

Salinas: ^{a}California. ^{b}none. ^{c}Salinas R. ^{d}none. ^{e}Monterey County Library (1912) 285; Public Library (1900) 167.

Salt Lake City: ^{a}*Utah. ^{b}for Great Salt Lake City. ^{c}Jordan R. ^{d}University of Utah (1850) 24 – 1600. ^{e}Marriott Library of the University . . . (1850) 1130; Salt Lake County Library (1938) 527; Utah State Library Commission (1957) 483; Salt Lake City Public Library (1898) 463.

San Angelo: ^{a}Texas. ^{b}none. ^{c}Concho R. ^{d}Angelo State University (1928) 5 – 181. ^{e}Porter Henderson Library of . . . University (1928) 190; Tom Green County Library (1923) 144.

San Antonio: ^{a}Texas. ^{b}none. ^{c}San Antonio R; San Pedro R; Acequia R. ^{d}University of Texas, San Antonio campus (1959) 5 – 561; Trinity University (1869) 4 – 225; Saint Mary's University of San Antonio (college) (1852) 3 – 211; Our Lady of the Lake University of San Antonio (1896) 2 – 138; Incarnate Word College (univ) (1881) 1 – 114. ^{e}Public Library (1903) 850.

San Bernardino: ^{a}California. ^{b}none. ^{c}Santa Ana R. ^{d}California State College at San Bernardino (univ) (1960) 4 – 205; Loma Linda University (at Loma Linda, nr San Bernardino) (1905) 4 – 1416; University of Redlands (at Redlands, nr San Bernardino) (1907) 3 – 205. ^{e}San Bernardino County Library (1914) 629; University . . . Library (at Redlands, nr San Bernardino) (1909) 231; San Bernardino Public Library (1891) 200; . . . College Library (1963) 188.

San Diego: ^{a}California. ^{b}inc La Jolla. ^{c}San Diego B of Pacific O; San Diego R. ^{d}San Diego State University (1897) 35 – 1996; University of California, San Diego campus (1903) 14 – 946; United States International University (1952) 3 – 295; University of San Diego (1949) 3 – 150; Point Loma College (univ) (1902) 2 – 90; National University (1971) 1 – ? ^{e}San Diego Public Library (1882) 1289; San Diego Campus Library of the University of California (1913) 1103; San Diego County Library (1907) 598; San Diego State University Library (1898) 469.

San Francisco: ^{a}California. ^{b}none. ^{c}Pacific O; San Francisco B of Pacific O; Merced L. $^{d^1}$ San Francisco State University (1899) 25 – 1500; Golden Gate University (1881) 9 – 500; University of San Francisco (1855) 6 – 469; University of California, San Francisco campus (1864) 5 – 1414. ^{e}Public Library (1878) 1536; San Francisco State University Library (1899) 430; San Francisco Campus Library of the University of California (1864) 410.

San Jose: ^{a}California. ^{b}none. ^{c}Guadalupe R; Coyote R. ^{d}San Jose State University (1857 at San Fran-

---

[1] Universities with 5000 or more students.

cisco, relocated 1871) 32 – 1600. ^{e}Santa Clara County Free Library (1912) 915; Public Library (1872) 848; ... University Library (1872) 627.

San Mateo: ^{a}California. ^{b}none. ^{c}San Francisco B of Pacific O. ^{d}College of Notre Dame (univ) (at Belmont, nr San Mateo) (1851) 1 – 91. ^{e}San Mateo County Library (at Belmont, nr San Mateo) (1915) 557; San Mateo Public Library (1899) 262.

Santa Ana: ^{a}California. ^{b}none. ^{c}Santa Ana R. ^{d}University of California, Irvine campus (at Irvine, nr Santa Ana) (1961) 10 – 724. ^{e}Irvine Campus Library of the University ... (at Irvine, nr Santa Ana) (1965) 711; Public Library (1878) 314.

Santa Barbara: ^{a}California. ^{b}none. ^{c}Santa Barbara Channel. ^{d}University of California, Santa Barbara campus (1891) 18 – 947. ^{e}Santa Barbara Campus Library of the University ... (1909) 1100; Public Library (1882) 350.

Santa Clara: ^{a}California. ^{b}none. ^{c}Guadalupe R. ^{d}University of Santa Clara (1851) 7 – 312. ^{e}Michel Orradre Library of the University ... (1851) 235; City Library (1904) 217.

Santa Cruz: ^{a}California. ^{b}none. ^{c}Monterey B of Pacific O; San Lorenzo R. ^{d}University of California, Santa Cruz campus (1961) 12 – 421. ^{e}McHenry Santa Cruz Campus Library of the University ... (1965) 487; Public Library (1868) 271.

Santa Fe: ^{a}*New Mexico. ^{b}none. ^{c}Santa Fe R. ^{d}Saint John's College, Santa Fe campus (univ) (1964) 0.3 – 39. ^{e}New Mexico State Library (1929) 199; Public Library (1896) 90; New Mexico State Supreme Court Law Library (1853) 88; Fogelson Library of the College of Santa Fe (1874) 70.

Santa Monica: ^{a}California. ^{b}none. ^{c}Santa Monica B of Pacific O. ^{d}none. ^{e}Public Library (1890) 300.

Santa Rosa: ^{a}California. ^{b}none. ^{c}nr Russian R. ^{d}California State College of Sonoma (univ) (at Rohnert Park, nr Santa Rosa) (1960) 9 – 385. ^{e}Sonoma County Library (1869) 404; ... College Library (at Rohnert Park, nr Santa Rosa) (1961) 242.

Sarasota: ^{a}Florida. ^{b}none. ^{c}Sarasota B of G of Mexico. ^{d}University of Sarasota (1974) 0.1 – ? ^{e}Public Library (1907) 100; New College of the University of South Florida Library (1962) 93.

Savannah: ^{a}Georgia. ^{b}none. ^{c}Savannah R. ^{d}Armstrong State College (univ) (1935) 4 – 165; Savannah State College (univ) (1890) 3 – 136. ^{e}Savannah Public and Chatham-Effingham-Liberty Regional Library (1809) 331.

Schenectady: ^{a}New York. ^{b}none. ^{c}Mohawk R. ^{d}Union College (univ) (1795) 3 – 220. ^{e}Schenectady County Public Library (1895) 311; Schaffer Library of ... College (1795) 207.

Scottsdale: ^{a}Arizona. ^{b}none. ^{c}nr Salt R. ^{d}none. ^{e}Public Library (1955) 128.

Scranton: ^{a}Pennsylvania. ^{b}none. ^{c}Lackawanna R; Scranton L. ^{d}University of Scranton (1888) 4 – 210; Marywood College (univ) (1915) 3 – 214. ^{e}Alumni Memorial Library of the University ... (1926) 129; ... College Library (1915) 118; Public Library (alt Albright Memorial Library) (1893) 111.

Seaside: ^{a}California. ^{b}none. ^{c}Monterey B of Pacific O. ^{d}Monterey Institute of Foreign Studies (univ) (at Monterey, nr Seaside) (1955) 0.4 – 60. ^{e}Dudley Knox Library of the United States Naval Postgraduate School (at Monterey, nr Seaside) (1946) 260; Monterey Public Library (at Monterey, nr Seaside) (1849) 115; Pacific Grove Public Library (at Pacific Grove, nr Seaside) (1908) 65; Seaside Branch of Monterey County Library (bef 1954) 23.

Seattle: ^{a}Washington. ^{b}none. ^{c}Puget Sound of Pacific O; Washington L; Union L; Green L. ^{d}University of Washington (1861) 36 – 3775; Seattle University (1891) 3 – 208; Seattle Pacific College (univ) (1891) 2 – 170. ^{e}University ... Library (1862) 1953; Public Library (1873) 1507; King County Library (1943) 1108.

Sheboygan: ^{a}Wisconsin. ^{b}none. ^{c}Sheboygan R; Michigan L. ^{d}none. ^{e}Mead Public Library (1897) 187.

Shreveport: ^{a}Louisiana. ^{b}none. ^{c}Red R; Cross L. ^{d}none. ^{e}Shreve Memorial Library (1923) 255.

Silver Spring: ^{a}Maryland. ^{b}none. ^{c}Sligo Creek. ^{d}University of Maryland (at College Park, nr Silver Spring) (1807 at Baltimore, relocated 1856) 59 – 4718 [College Park campus: (1856) 48 – 3454]. ^{e}National Agricultural Library (at Beltsville, nr Silver Spring) (1862 at Washington, relocated 1969) 1548; College Park Campus Library of the University ... (at College Park, nr Silver Spring) (1891?) 1465; Montgomery County Public Library (at Rockville, nr Silver Spring) (1951) 1264; Prince George's County Memorial Library (at Hyattsville, nr Silver Spring) (1946)

1096; National Library of Medicine (at Bethesda, nr Silver Spring) (1836 at Washington, relocated 1962) 422; Atmospheric Sciences Library (1872) 175.

Simi Valley: ^{a}California. ^{b}for Simi. ^{c}Arroyo Simi Creek. ^{d}none. ^{e}Simi Valley Branch of Ventura County Library (bef 1954) 62.

Sioux City: ^{a}Iowa. ^{b}none. ^{c}Missouri R; Big Sioux R; Floyd R. ^{d}Morningside College (univ) (1889) 2 – 101. ^{e}Public Library (1870) 179; Petersmeyer Library of . . . College (1889) 113.

Sioux Falls: ^{a}South Dakota. ^{b}none. ^{c}Big Sioux R. ^{d}Augustana College (univ) (1860 at Chicago, relocated 1918) 2 – 178. ^{e}Mikkelsen Library of . . . College (1860 at Chicago, relocated 1918?) 146; Public Library (1886) 126.

Skokie: ^{a}Illinois. ^{b}for Niles Center. ^{c}nr Chicago R. ^{d}none. ^{e}Public Library (1941) 271.

Somerville: ^{a}Massachusetts. ^{b}none. ^{c}Mystic R. ^{d}Tufts University (at Medford, nr Somerville) (1852) 6 – 1850. ^{e}Nils Yngve Wessell Library of . . . University (at Medford, nr Somerville) (1854) 351; Medford Public Library (at Medford, nr Somerville) (1825) 220; Somerville Public Library (1873) 183.

South Bend: ^{a}Indiana. ^{b}none. ^{c}Saint Joseph R. ^{d}University of Notre Dame (1842) 9 – 1091; Indiana University, South Bend campus (1922) 6 – 266. ^{e}University . . . Library (1873) 1221; Public Library (1888) 304.

Spartanburg: ^{a}South Carolina. ^{b}none. ^{c}Fairforest Creek. ^{d}Converse College (univ) (1889) 0.8 – 83. ^{e}Spartanburg County Public Library (1892) 190; Sandor Teszler Library of Wofford College (1854) 140; Gwathmey Library of Converse College (1889) 103.

Spokane: ^{a}Washington. ^{b}for Spokane Falls. ^{c}Spokane R. ^{d}Eastern Washington State College (univ) (at Cheney, nr Spokane) (1890) 6 – 374; Gonzaga University (1887) 3 – 250; Whitworth College (univ) (1890) 2 – 152. ^{e}Spokane Public Library (1891) 457; John F. Kennedy Memorial Library of Eastern Washington State College (at Cheney, nr Spokane) (1890) 203; Spokane County Library (1943) 203; Crosby Library of . . . University (1887) 159.

Springfield: ^{a}*Illinois. ^{b}none. ^{c}Spring Creek; Sugar Creek; Springfield L. ^{d}Sangamon State University (1969) 4 – 202. ^{e}Illinois State Library (1839) 708; Lincoln Library (alt Springfield Public Library) (1867) 274; Norris L. Brookens Library of . . . University (1970) 159.

Springfield: ^{a}Massachusetts. ^{b}none. ^{c}Connecticut R; Chicopee R. ^{d}Western New England College (univ) (1919) 4 – 234; Westfield State College (univ) (at Westfield, nr Springfield) (1839) 4 – 155; Springfield College (univ) (1885) 3 – 183; American International College (univ) (1885) 2 – 124. ^{e}City Library (1857) 626.

Springfield: ^{a}Missouri. ^{b}none. ^{c}nr James R. ^{d}Southwest Missouri State University (1906) 12 – 644; Drury College (univ) (1873) 3 – 156. ^{e}. . . University Library (1907) 420; Springfield-Greene County Library (1903) 251.

Springfield: ^{a}Ohio. ^{b}none. ^{c}Mad R. ^{d}Antioch College (univ) (at Yellow Springs, nr Springfield) (1852) 4 – 432; Wittenberg University (1845) 3 – 207; Union of Experimenting Colleges and Universities (univ) (at Yellow Springs, nr Springfield) (1964) 0.8 – ? ^{e}Warder Public Library (1872) 280; Thomas Library of . . . University (1845) 221; Olive Kettering Memorial Library of . . . College (at Yellow Springs, nr Springfield) (1852) 187.

Stamford: ^{a}Connecticut. ^{b}none. ^{c}Long Island Sound of Atlantic O; Rippowam R. ^{d}none. ^{e}Ferguson Library (alt Stamford's Public Library) (1880) 315.

Sterling Heights: ^{a}Michigan. ^{b}none. ^{c}Clinton R. ^{d}none. ^{e}Macomb County Library (at Mount Clemens, nr Sterling Heights) (1946) 176; Mount Clemens Public Library (at Mount Clemens, nr Sterling Heights) (1865) 89; Sterling Heights Public Library (1971) 29.

Steubenville: ^{a}Ohio. ^{b}none. ^{c}Ohio R. ^{d}none. ^{e}Public Library of Steubenville and Jefferson County (1902) 143; Starvaggi Memorial Library of the College of Steubenville (1946) 101.

Stockton: ^{a}California. ^{b}none. ^{c}San Joaquin R; Calaveras R. ^{d}University of the Pacific (1851 at San Jose, relocated 1924) 6 – 486. ^{e}Stockton-San Joaquin County Public Library (1880) 406; University . . . Library (1852 at San Jose, relocated 1924?) 275.

Sunnyvale: ^{a}California. ^{b}none. ^{c}nr San Francisco B of Pacific O. ^{d}none. ^{e}Public Library (bef 1923) 195.

Syracuse: ^{a}New York. ^{b}none. ^{c}Onondaga Creek; Onondaga Lake. ^{d}Syracuse University (1870) 23 – 1934 [Syracuse campus: (1870) 21 – 1785]; State University of New York, Syracuse campus (1834) 9 – 745. ^{e}Syracuse University Library (1871) 1505; Public Library (1852) 396.

Tacoma: ^{a}Washington. ^{b}none. ^{c}Puget Sound of Pacific O; Puyallup R. ^{d}University of Puget Sound (1888) 5 – 253; Pacific Lutheran University (1890) 3 – 242. ^{e}Public Library (1886) 544.

Tallahassee: ^{a}*Florida. ^{b}none. ^{c}nr Ochlockonee R; nr Jackson L. ^{d}Florida State University (1851) 22 – 1653; Florida Agricultural and Mechanical University (1887) 5 – 421. ^{e}Robert Manning Strozier Library of Florida State University (1853) 1126; Florida State Library (bef 1855) 284; Coleman Memorial Library of Florida Agricultural and Mechanical University (1909) 248; Leon County Public Library (1906) 135.

Tampa: ^{a}Florida. ^{b}none. ^{c}Tampa B of G of Mexico; Old Tampa B of G of Mexico; Hillsborough B of G of Mexico; Hillsborough R. ^{d}University of South Florida (1956) 23 – 1236; University of Tampa (1930) 2 – 144. ^{e}Tampa-Hillsborough County Public Library (1917) 533; University of South Florida Library (1960) 468.

Taylor: ^{a}Michigan. ^{b}none. ^{c}South Branch of Ecorse R. ^{d}none. ^{e}Alexander Papp Public Library (bef 1954) 42; Richard J. Trolley Public Library (1968) 24.

Tempe: ^{a}Arizona. ^{b}none. ^{c}Salt R. ^{d}Arizona State University (1885) 37 – 2204. ^{e}. . . University Library (1891) 955; Public Library (1935) 70.

Terre Haute: ^{a}Indiana. ^{b}none. ^{c}Wabash R. ^{d}Indiana State University (1865) 14 – 1258 [Terre Haute campus: (1865) 11 – 1160]; Rose-Hulman Institute of Technology (univ) (1874) 1 – 72. ^{e}Cunningham Memorial Library of . . . University (1870) 650; Vigo County Public Library (1882) 254.

Texas City: ^{a}Texas. ^{b}none. ^{c}Galveston B of G of Mexico; Dickinson Bayou; Moses L. ^{d}none. ^{e}Moore Memorial Public Library (1928) 54; College of the Mainland Library (1967) 34.

Thousand Oaks: ^{a}California. ^{b}none. ^{c}Conejo Creek. ^{d}California Lutheran College (univ) (1959) 2 – 195. ^{e}. . . College Library (1961) 75; Conejo Branch of Ventura County Library (bef 1954) 72.

Toledo: ^{a}Ohio. ^{b}none. ^{c}Maumee R; Erie L. ^{d}University of Toledo (1872) 17 – 1060. ^{e}Toledo-Lucas County Public Library (1838) 1129; University . . . Library (1917) 884.

Topeka: ^{a}*Kansas. ^{b}none. ^{c}Kansas R. ^{d}Washburn University of Topeka (1865) 6 – 313. ^{e}Kansas State Library (1854) 327; Public Library (1882) 282; . . . University Library (1865) 121; Kansas State Historical Society Library (1875) 105.

Torrance: ^{a}California. ^{b}none. ^{c}Pacific O. ^{d}California State College at Dominguez Hills (univ) (at Carson, nr Torrance) (1960) 8 – 330. ^{e}Public Library (1935?) 283; . . . College Library (at Carson, nr Torrance) (1963) 168.

Towson: ^{a}Maryland. ^{b}none. ^{c}Herring Run. ^{d}Towson State University (1865) 14 – 545. ^{e}Baltimore County Public Library (1948) 1269; Albert S. Cook Library of Towson State University (1866) 237; Julia Rogers Library of Goucher College (1885) 186.

Trenton: ^{a}*New Jersey. ^{b}none. ^{c}Delaware R. ^{d}Trenton State College (univ) (1855) 12 – 546; Princeton University (at Princeton, nr Trenton) (1746) 6 – 1357; Rider College (univ) (at Lawrenceville, nr Trenton) (1865 at Trenton, relocated 1956) 6 – 310. ^{e}. . . University Library (at Princeton, nr Trenton) (1746) 2812; New Jersey State Library (1796) 585; Speer Library of Princeton Theological Seminary (at Princeton, nr Trenton) (1812) 325; Roscoe L. West Library of Trenton State College (1855) 320; Franklin F. Moore Library of Rider College (at Lawrenceville, nr Trenton) (1934 at Trenton, relocated 1956?) 275; Free Public Library (1750) 245.

Troy: ^{a}New York. ^{b}none. ^{c}Hudson R. ^{d}Rensselaer Polytechnic Institute (univ) (1824) 5 – 646; Russell Sage College (univ) (1916) 4 – 259 [Troy campus: (1916) 3 – ?]. ^{e}. . . Institute Library (1824) 160; Janes Wheelock Clark Library of . . . College (1916?) 135; Public Library (1835) 104.

Tucson: ^{a}Arizona. ^{b}none. ^{c}Santa Cruz R. ^{d}University of Arizona (1885) 29 – 2690. ^{e}University . . . Library (1891) 964; Public Library (1879) 534.

Tulsa: ^{a}Oklahoma. ^{b}none. ^{c}Arkansas R. ^{d}University of Tulsa (1894 at Muskogee, relocated 1907)

7 – 362; Oral Roberts University (1963) 3 – 180. [e]Tulsa City-County Library (1912) 648; McFarlin Library of the University . . . (1894 at Muskogee, relocated 1907?) 533.

Tuscaloosa: [a]Alabama. [b]none. [c]Black Warrior R. [d]University of Alabama (1820) 17 – 1177. [e]Amelia Gayle Gorgas Library of the University . . . (1831) 1163; Geological Survey of Alabama Library (1873) 100; William H. Sheppard Library of Stillman College (1930) 67; Friedman Library (1921) 108.

Tyler: [a]Texas. [b]none. [c]nr Neches R. [d]Texas Eastern University (1971) 1 – 57. [e]D. R. Glass Library of Texas College (1894) 72; Carnegie Public Library (1904) 56.

Upper Darby: [a]Pennsylvania. [b]none. [c]Cobbs Creek. [d]none. [e]Upper Darby Township and Sellers Free Public Library (1932) 59.

Utica: [a]New York. [b]none. [c]Mohawk R. [d]State University of New York, Utica campus (1966) 10 – 303. [e]Public Library (1842) 157.

Vallejo: [a]California. [b]none. [c]San Pablo B of Pacific O; Napa R. [d]none. [e]Public Library (alt John F. Kennedy Library) (1884) 145.

Ventura: [a]California. [b]off San Buenaventura. [c]Santa Barbara Channel. [d]none. [e]Ventura County Library (1915) 474.

Vineland: [a]New Jersey. [b]none. [c]nr Maurice R. [d]none. [e]Public Library (1901) 62; Cumberland County College Library (1966) 46.

Virginia Beach: [a]Virginia. [b]none. [c]Atlantic O; Chesapeake B of Atlantic O. [d]none. [e]Department of Public Libraries and Information (1959) 166.

Waco: [a]Texas. [b]none. [c]Brazos R. [d]Baylor University (1845 at Independence, Texas; relocated 1887) 9 – 444. [e]Moody Memorial Library of . . . University (1901) 623; Waco-McLennan County Library (1899) 218.

Waltham: [a]Massachusetts. [b]none. [c]Charles R. [d]Bentley College (univ) (1917) 5 – 181; Brandeis University (1948) 4 – 371. [e]Goldfarb Library of . . . University (1948) 450; Public Library (1865) 146; Baker-Vanguard Library of . . . College (1959) 79.

Warren: [a]Michigan. [b]none. [c]nr Clinton R. [d]none. [e]Warren Public Library (bef 1940) 292.

Warren: [a]Ohio. [b]none. [c]Mahoning R. [d]none. [e]Public Library (1848) 161.

Warwick: [a]Rhode Island. [b]none. [c]Narragansett B of Atlantic O; Greenwich B of Atlantic O; Providence R; Pawtuxet R. [d]none. [e]Public Library (1965) 136.

*Washington: [a]*District of Columbia. [b]none. [c]Potomac R; Anacostia R. [d][1] George Washington University (1821) 22 – 2376; University of the District of Columbia (1851) 16 – 1000; American University (1893) 14 – 1063; Georgetown University (1789) 11 – 2200; Howard University (1867) 9 – 1716; Catholic University of America (1887) 7 – 767. [e]Library of Congress (1800) 17,889; Public Library of the District of Columbia (1898) 1989; John K. Mullen of Denver Memorial Library of Catholic University of America (1889) 959; Howard University Library (1867) 908; Smithsonian Institution Library (1846) 900; Joseph Mark Lauinger Library of Georgetown University (1789) 867; Natural Resources Library of the Department of the Interior (1949) 850; . . . ; Folger Shakespeare Library (1932) 215.

Waterbury: [a]Connecticut. [b]none. [c]Naugatuck R; Mad R. [d]none. [e]Silas Bronson Library (1869) 172.

Waterloo: [a]Iowa. [b]none. [c]Cedar R. [d]University of Northern Iowa (at Cedar Falls, nr Waterloo) (1876) 10 – 575. [e]University . . . Library (at Cedar Falls, nr Waterloo) (1876) 358; Waterloo Public Library (1896) 165; Cedar Falls Public Library (at Cedar Falls, nr Waterloo) (1865) 75.

Waukegan: [a]Illinois. [b]none. [c]Michigan L. [d]none. [e]Public Library (1898) 173.

West Allis: [a]Wisconsin. [b]none. [c]nr Milwaukee R. [d]none. [e]Public Library (1898) 162.

Westland: [a]Michigan. [b]none. [c]Lower Rouge R. [d]none. [e]Wayne County Library (at Wayne, nr Westland) (1920) 790; Wayne-Westland Public Library (at Wayne, nr Westland) (1924) 43.

West Palm Beach: [a]Florida. [b]none. [c]Worth Lagoon. [d]none. [e]Palm Beach County Public Library (1967) 86; West Palm Beach Public Library (1895) 59.

---

[1] Universities with 5000 or more students.

Wheeling: [a]West Virginia. [b]none. [c]Ohio R. [d]none. [e]Ohio County Public Library (1859) 99; Bishop Hodges Learning Center of Wheeling College (1955) 87.

White Plains: [a]New York. [b]none. [c]nr Bronx R. [d]Pace University, Pleasantville campus (at Pleasantville, nr White Plains) (1963) 4 – 258; Manhattanville College (univ) (at Purchase, nr White Plains) (1841) 2 – 153; Nyack College (univ) (at Nyack, nr White Plains) (1882) 0.7 – 66. [e]Manhattanville College Library (at Purchase, nr White Plains) (1841) 217; Purchase Campus Library of the State University of New York (at Purchase, nr White Plains) (1967) 167; White Plains Public Library (1906) 162; Pleasantville Campus Library of Pace University (at Pleasantville, nr White Plains) (1963) 107; Mount Pleasant Public Library (at Pleasantville, nr White Plains) (1893) 99.

Whittier: [a]California. [b]none. [c]San Gabriel R. [d]Whittier College (univ) (1901) 2 – 137. [e]Public Library (1900) 235; . . . College Library (1901) 125.

Wichita: [a]Kansas. [b]none. [c]Arkansas R; Little Arkansas R. [d]Wichita State University (1892) 16 – 892; Friends University (college) (1898) 0.9 – 65. [e]Ablah Library of Wichita State University (1895) 521; Public Library (1876) 354.

Wichita Falls: [a]Texas. [b]none. [c]Wichita R. [d]Midwestern State University (1922) 5 – 216. [e]George Moffett Library of . . . University (1924) 171; Kemp Public Library (1917) 100.

Wilkes-Barre: [a]Pennsylvania. [b]none. [c]Susquehanna R. [d]Wilkes College (univ) (1933) 3 – 200. [e]Eugene Shedden Farley Library of Wilkes College (1933) 135; D. Leonard Corgan Library of King's College (1946) 128; Osterhout Free Library (1889) 124.

Williamsburg: [a]Virginia. [b]none. [c]nr James R; nr York R. [d]College of William and Mary (univ) (1693) 10 – 646 [Williamsburg campus: (1693) 6 – 447]. [e]Earl Gregg Swem Library of the College . . . (1693) 550; Research Department Library of Colonial Williamsburg (bef 1954) 31; Public Library (1910) 21.

Wilmington: [a]Delaware. [b]none. [c]Delaware R; Christina R. [d]University of Delaware (at Newark, Delaware, nr Wilmington) (1743 at New London, Pennsylvania; relocated 1765) 19 – 1266. [e]Hugh M. Morris Library of the University . . . (at Newark, Delaware, nr Wilmington) (1834) 950; Wilmington Institute Library (1788) 300.

Wilmington: [a]North Carolina. [b]none. [c]Cape Fear R. [d]none. [e]Public Library (1906) 132; William Madison Randall Wilmington Campus Library of the University of North Carolina (1947) 122.

Winston-Salem: [a]North Carolina. [b]inc Salem, Winston. [c]Salem Creek. [d]Wake Forest University (1833 at Wake Forest, relocated 1956) 4 – 728; Winston-Salem State University (college) (1892) 2 – 135. [e]Z. Smith Reynolds Library of Wake Forest University (1879 at Wake Forest, relocated 1956?) 385; Forsyth County Public Library (1903) 300.

Woodbridge: [a]New Jersey. [b]none. [c]Arthur Kill. [d]Rutgers, the State University of New Jersey (at New Brunswick, nr Woodbridge) (1766) 46 – 5161 [New Brunswick campus: (1766) 31 – ?]. [e]New Brunswick Campus Library of Rutgers University (at New Brunswick, nr Woodbridge) (1766) 1346; Woodbridge Free Public Library (1879) 298; New Brunswick Free Public Library (at New Brunswick, nr Woodbridge) (1883) 100.

Worcester: [a]Massachusetts. [b]none. [c]Blackstone R; Quinsigamond L; Indian L. [d]Worcester State College (univ) (1874) 6 – 209; Clark University (1887) 3 – 227; College of the Holy Cross (univ) (1843) 3 – 194; Worcester Polytechnic Institute (univ) (1865) 3 – 205; Assumption College (univ) (1904) 2 – 131; Ana Maria College (univ) (at Paxton, nr Worcester) (1946 at Marlborough, relocated 1952) 1 – 68. [e]Public Library (1859) 827; American Antiquarian Society Library (1812) 700; Robert Hutchings Goddard Library of . . . University (1889) 336; Dinand Library of the College . . . (1927) 330.

Yakima: [a]Washington. [b]none. [c]Yakima R. [d]none. [e]Yakima Valley Regional Library (1907) 237.

Yonkers: [a]New York. [b]none. [c]Hudson R; Bronx R. [d]Sarah Lawrence College (univ) (at Bronxville, nr Yonkers) (1926) 1 – 149. [e]Public Library (1893) 361; . . . College Library (at Bronxville, nr Yonkers) (1928) 161.

York: [a]Pennsylvania. [b]none. [c]Codorus Creek. [d]none. [e]Martin Memorial Library (1885) 126; York College of Pennsylvania Library (1968) 90.

Youngstown: [a]Ohio. [b]none. [c]Mahoning R. [d]Youngstown State University (1908) 16 – 778. [e]Reuben

McMillan Free Library Association (alt Public Library of Youngstown and Mahoning County) (1878) 668; . . . University Library (1931) 283.

Virgin Islands (USA)
*Charlotte Amalie: [a]Saint Thomas I. [b]for Saint Thomas. [c]Caribbean Sea. [d]College of the Virgin Islands (univ) (1962) 2 – 117. [e]Virgin Islands Bureau of Libraries and Museums (1920) 81; Ralph M. Paiewonsky Library of the College . . . (1963) 52.

OCEANIA

Australia
Adelaide: [a]*South Australia. [b]none. [c]Saint Vincent G of Indian O; Torrens R; Sturt R. [d]University of Adelaide (1874) 10 – 678; Flinders University of South Australia (1963) 4 – 310. [e]University . . . Library (1876) 684; State Library of South Australia (1884) 675.
Brisbane: [a]*Queensland. [b]none. [c]Brisbane R. [d]University of Queensland (1909) 18 – 1248; Griffith University (1971) 0.8 – 91. [e]University . . . Library (1911) 820; State Library of Queensland (1896) 341.
*Canberra: [a]*Australian Capital Territory. [b]none. [c]Molonglo R. [d]Australian National University (1929) 6 – 936. [e]National Library of Australia (1902 at Melbourne, relocated 1927) 1400; . . . University Library (1948) 742.
Geelong: [a]Victoria. [b]none. [c]Port Phillip B of Bass Strait; Barwon R. [d]Deakin University (1974) ? – ?[1] [e]Regional Library Service (?) 65.
Hobart: [a]*Tasmania. [b]none. [c]Derwent R. [d]University of Tasmania (1890) 3 – 280. [e]State Library of Tasmania (1849) 867; University . . . Library (1889) 280.
Melbourne: [a]*Victoria. [b]none. [c]Port Phillip B of Bass Strait; Yarra R. [d]University of Melbourne (1853) 16 – 1100; Monash University (1958) 13 – 956; La Trobe University (1964) 9 – 471. [e]State Library of Victoria (1856) 1080; University . . . Library (1855) 768; Monash University Library (1961) 682.
Newcastle: [a]New South Wales. [b]none. [c]Pacific O; Hunter R. [d]University of Newcastle (1951) 5 – 326. [e]Public Library (1948) 291; Auchmuty Library of the University . . . (1951) 265.
Perth: [a]*Western Australia. [b]none. [c]Swan R. [d]University of Western Australia (1911) 10 – 636; Murdoch University (1973) 2 – 110. [e]Library Service of Western Australia (1887) 2541 [at Perth: 1263]; University . . . Library (1913) 500.
Sydney: [a]*New South Wales. [b]none. [c]Pacific O; Parramatta R. [d]University of New South Wales (1949) 18 – 987 [Sydney campus: (1949) 18 – ?]; University of Sydney (1850) 17 – 1194; Macquarie University (1964) 10 – 650. [e]Fisher Library (alt University of Sydney Library) (1852) 1923; Library of New South Wales (1826) 1350; University of New South Wales Library (1949) 680; Macquarie University Library (1965) 359; Public Library (1909) 260.
Wollongong: [a]New South Wales. [b]off Greater Wollongong. [c]Pacific O; Nepean R. [d]University of Wollongong (1961) 2 – 136. [e]Public Library (?) 290.

Fiji
*Suva: [a]Viti Levu I. [b]none. [c]Pacific O. [d]University of the South Pacific (1968) 2 – 117. [e]University . . . Library (1968) 120; City Library (1909) 40.

French Polynesia
*Papeete: [a]Tahiti I. [b]none. [c]Pacific O. [d]none.

[1]Classes scheduled to begin in 1978.

(Hawaii), USA

Hilo: ^{a}Hawaii I. ^{b}none. ^{c}Hilo B of Pacific O; Wailuku R. ^{d}none. ^{e}Hawaii Public Library (1899) 193; Hilo Campus Library of the University of Hawaii (1947) 85.

*Honolulu: ^{a}Oahu I. ^{b}none. ^{c}Pacific O. ^{d}University of Hawaii (1907) 43 – 2327 [campuses at or nr Honolulu: (1907) 37 – 2106]. ^{e}Hawaii State Library (1879?) 2211, inc Oahu Public Library (1912?) 750; Manoa Campus Library of the University . . . (1907) 1413.

New Caledonia

*Noumea: ^{b}for Port-de-France. ^{c}Pacific O. ^{d}none. ^{e}Bibliotheque Bernheim (1905) 33; Bibliotheque de la Commission du Pacifique Sud (alt South Pacific Commission Library) (?) 25.

New Zealand

Auckland: ^{a}North I. ^{b}none. ^{c}Tasman Sea; Hauraki G of Pacific O. ^{d}University of Auckland (1882) 11 – 742. ^{e}Public Library (1880) 728; University . . . Library (1882) 680.

Christchurch: ^{a}South I. ^{b}none. ^{c}Avon R. ^{d}University of Canterbury (1873) 7 – 371. ^{e}University . . . Library (1873) 430; Canterbury Public Library (1859) 240.

Dunedin: ^{a}South I. ^{b}none. ^{c}Pacific O. ^{d}University of Otago (1869) 6 – 500. ^{e}University . . . Library (1870) 601; Public Library (1908) 286.

Hamilton: ^{a}North I. ^{b}none. ^{c}Waikato R. ^{d}University of Waikato (1964) 3 – 168. ^{e}Public Library (1960) 62.

Manukau: ^{a}North I. ^{b}none. ^{c}Tasman Sea. ^{d}none.

*Wellington: ^{a}North I. ^{b}none. ^{c}Cook Strait. ^{d}Victoria University of Wellington (1897) 7 – 450. ^{e}National Library of New Zealand (1856) 4418 [at Wellington: 1017]; . . . University Library (1897) 410; Public Library (1841) 397.

Papua New Guinea

*Port Moresby: ^{a}New Guinea I. ^{b}none. ^{c}G of Papua of Coral Sea. ^{d}University of Papua New Guinea (1965) 2 – 257. ^{e}University . . . Library (1965?) 150.

Samoa

*Apia: ^{a}Upolu I. ^{b}none. ^{c}Pacific O. ^{d}none. ^{e}Nelson Memorial Public Library (1959) 36.

SOUTH AMERICA

Argentina

Almirante Brown: ^{a}Buenos Aires. ^{b}inc Adrogue. ^{c}(Arroyo) San Francisco Creek. ^{d}none.

Avellaneda: ^{a}Buenos Aires. ^{b}for Barracas al Sud. ^{c}Plata R; Riachuelo R. ^{d}none. ^{e}Biblioteca Obreros y Empleados Municipales (?) 30.

Bahia Blanca: ^{a}Buenos Aires. ^{b}none. ^{c}Naposta Grande R. ^{d}Universidad Nacional del Sur (1948) 9 – 1136. ^{e}Biblioteca Popular de la Asociacion Bernardino Rivadavia (1882) 100; Biblioteca Central de la Universidad . . . (1948) 72.

*Buenos Aires: ^{a}*Distrito Federal. ^{b}for Buenos Ayres. ^{c}Plata R; Riachuelo R. $^{d^1}$Universidad de Buenos Aires (1821) 174 – 9101; Universidad Tecnologica Nacional (1953) 32 – 2139; Pontificia Universidad Catolica Santa Maria de los Buenos Aires (1958) 9 – 1419; Universidad Argentina de la Empresa (1962) 8 – 688; Universidad de Belgrano (1964) 7 – 571; Universidad del Salvador (1944) 4 – 895; Universidad del Museo Social Argentino (1961) 2 – 438; Universidad Argentina John F. Kennedy (1961) 1 – 211. ^{e}Bibliotecas de la Universidad de Buenos Aires (1853) 2033; Biblioteca Nacional (1810) 700; Centro de Documentacion Internacional (1959) 400.

---

[1] Universities with 1000 or more students.

Cordoba: [a]*Cordoba. [b]none. [c]Primero R. [d]Universidad Nacional de Cordoba (1613) 55 – 4212; Universidad Catolica de Cordoba (1956) 3 – 502. [e]Biblioteca Mayor de la Universidad Nacional de Cordoba (1614) 120; Biblioteca de la Universidad Catolica de Cordoba (1956) 80.

Corrientes: [a]*Corrientes. [b]none. [c]Parana R. [d]Universidad Nacional del Nordeste (1956) 20 – 1652. [e]Biblioteca Central de la Universidad . . . (1957) 32.

General San Martin: [a]Buenos Aires. [b]alt San Martin. [c]nr Plata R. [d]none. [e]Biblioteca Popular (1904) 12.

General Sarmiento: [a]Buenos Aires. [b]alt Sarmiento; for San Miguel. [c]Reconquista R. [d]none. [e]Biblioteca de las Facultades de Filosofia y Teologia S. I. (1931) 105.

La Matanza: [a]Buenos Aires. [b]alt Matanza; for San Justo. [c]Matanza R. [d]none. [e]Biblioteca Juan Bautista Alberdi de la Escuela Normal Mixta Republica de Mexico (?) 4; Biblioteca Bartolome Mitre (?) ?

Lanus: [a]Buenos Aires. [b]none. [c]nr Plata R. [d]none. [e]Biblioteca Jose Maria Cao (1919) 7.

La Plata: [a]*Buenos Aires. [b]alt Plata; for Eva Peron. [c]Plata R. [d]Universidad Nacional de La Plata (1884) 73 – 4216; Universidad Catolica de La Plata (1968) 1 – 293; Universidad Notarial Argentina (1965) 0.5 – 36. [e]Biblioteca Publica de la Universidad Nacional de La Plata (1884) 450.

Lomas de Zamora: [a]Buenos Aires. [b]none. [c](Arroyo) Santa Catalina Creek. [d]Universidad Nacional de Lomas de Zamora (1897?) 5 – 121. [e]Biblioteca del Maestro de la Sociedad Popular de Educacion Antonio Mentruit (1900) 28.

Mar del Plata: [a]Buenos Aires. [b]for General Pueyrredon, Pueyrredon. [c]Atlantic O. [d]Universidad Nacional de Mar del Plata (1961) 7 – 849; Universidad Catolica de Mar del Plata Stella Maris (1964) 1 – 276. [e]Biblioteca Municipal Publica (?) 30.

Mendoza: [a]*Mendoza. [b]none. [c]Tulumaya R. [d]Universidad Nacional de Cuyo (1939) 9 – 949; Universidad de Mendoza (1959) 1 – 162; Universidad Juan Agustin Maza (1960) 1 – 242; Universidad del Aconcagua (1968) 0.8 – 147. [e]Biblioteca Central de la Universidad Nacional de Cuyo (1939) 120; Biblioteca Publica General San Martin (1814) 75.

Merlo: [a]Buenos Aires. [b]none. [c]Reconquista R. [d]none.

Moron: [a]Buenos Aires. [b]for Seis de Septiembre. [c](Arroyo) Moron Creek. [d]Universidad de Moron (1960) 7 – 856. [e]Biblioteca Municipal Domingo Faustino Sarmiento (1912) 20.

Parana: [a]*Entre Rios. [b]for Bajada de Santa Fe. [c]Parana R. [d]Universidad Nacional de Entre Rios (1968?) 1 – 181. [e]Biblioteca Popular (?) 51; Biblioteca del Museo de Entre Rios (1917) 21.

Quilmes: [a]Buenos Aires. [b]none. [c]Plata R. [d]none. [e]Biblioteca Publica Municipal Domingo Faustino Sarmiento (1872) 29.

Rosario: [a]Santa Fe. [b]none. [c]Parana R. [d]Universidad Nacional de Rosario (1889) 30 – 3542. [e]Biblioteca Argentina Doctor Juan Alvarez (1909) 120; Biblioteca de la Universidad . . . (1910) 100.

Salta: [a]*Salta. [b]for Lerma. [c]Arias R. [d]Universidad Nacional de Salta (1967) 5 – 208; Universidad Catolica de Salta (1963) 1 – 160. [e]Biblioteca de la Universidad Catolica de Salta (1963?) 20.

San Isidro: [a]Buenos Aires. [b]none. [c]Plata R. [d]none. [e]Biblioteca del Instituto de Botanica Darwinion (1908) 65; Biblioteca Popular Juan Martin de Pueyrredon (?) 20.

San Juan: [a]*San Juan. [b]none. [c]San Juan R. [d]Universidad Nacional de San Juan (1964) 6 – 518; Universidad Catolica de Cuyo (1953) 0.7 – 265. [e]Biblioteca de la Facultad de Ingenieria y Ciencias Exactas, Fisicas y Naturales de la Universidad Nacional de Cuyo (?) 21.

Santa Fe: [a]*Santa Fe. [b]none. [c]Salado (alt Salado del Norte) R. [d]Universidad Nacional del Litoral (1889) 15 – 1016; Universidad Catolica de Santa Fe (1957) 2 – 297. [e]Biblioteca de la Universidad Nacional del Litoral (1920) 96; Biblioteca Publica Municipal Bernardino Rivadavia (1945) 10.

Tres de Febrero: [a]Buenos Aires. [b]for Caseros. [c]Reconquista R. [d]none. [e]Biblioteca Popular Juan Bautista Alberdi (1914) 32.

Tucuman: [a]*Tucuman. [b]off San Miguel de Tucuman. [c]Dulce R. [d]Universidad Nacional de Tucuman (1912) 10 – 1756; Universidad del Norte Santo Tomas de Aquino (1965) 1 – 252. [e]Biblioteca Central de la Universidad Nacional de Tucuman (1917) 100; Biblioteca Alberdi (1903) 44.

Vicente Lopez: [a]Buenos Aires. [b]inc Olivos. [c]Plata R. [d]none. [e]Biblioteca Popular (1921) 10; Biblioteca Vicente Lopez de la Asociacion de Cultura (?) ?

Bolivia

Cochabamba: [b]for Oropeza. [c]Rocha R. [d]Universidad Boliviana Mayor de San Simon (1826) 4 – 360. [e]Biblioteca Municipal (?) 30; Biblioteca Central de la Universidad . . . (1832) 25.

*La Paz: [b]alt Paz; for Choqueyapu; off La Paz de Ayacucho. [c]Choqueyapo R. [d]Universidad Boliviana Mayor de San Andres (1830) 17 – 900; Universidad Catolica Boliviana (1966) 0.5 – 90. [e]Biblioteca de la Direccion de Cultura (1832) 140; Biblioteca Central de la Universidad Boliviana Mayor de San Andres (1830) 120; Biblioteca Municipal Mariscal Andres de Santa Cruz (1836) 85.

Potosi: [b]none. [c]Potosi R. [d]Universidad Boliviana Tomas Frias (1892) 2 – 207. [e]Biblioteca Municipal Ricardo Jaime Freires (1920) 32; Biblioteca Central de la Universidad . . . (1942) 20.

Santa Cruz: [b]off Santa Cruz de la Sierra. [c]Piray R. [d]Universidad Boliviana Mayor Gabriel Rene Moreno (1880) 2 – 270. [e]Biblioteca Central de la Universidad . . . (1880) 40; Biblioteca Municipal (?) ?

*Sucre: [b]for Charcas, Chuquisaca. [c]Quirpinchaca R. [d]Universidad Boliviana Mayor, Real y Pontificia de San Francisco Xavier de Chuquisaca (1624) 4 – 410. [e]Biblioteca de la Universidad . . . (1624?) 62; Biblioteca y Archivo Nacional de Bolivia (1825) 53; Biblioteca Municipal (?) ?

Brazil

Aracaju: [a]*Sergipe. [b]none. [c]Cotinguiba R. [d]Universidade Federal de Sergipe (1948) 3 – 299. [e]Biblioteca Publica do Estado de Sergipe (1851) 160.

Belem: [a]*Para. [b]for Para. [c]Para R; Guama R; Guajara R. [d]Universidade Federal do Para (1902) 11 – 940. [e]Biblioteca e Arquivo Publico do Para (1871) 55; Biblioteca da Universidade . . . (1962) 42.

Belo Horizonte: [a]*Minas Gerais. [b]for Curral d'El-Rei. [c]Arrudas R. [d]Universidade Federal de Minas Gerais (1892) 16 – 2188; Universidade Catolica de Minas Gerais (1943) 9 – 537; Universidade Mineira de Arte (1957) 0.1 – 56; Universidade de Tecnologia e de Ciencias de Minas Gerais (1970) ? – ? [e]Biblioteca da Universidade Federal de Minas Gerais (1892) 333; Biblioteca Publica de Minas Gerais Professor Luis de Bessa (1954) 74; Biblioteca Publica Municipal (1893) 28.

*Brasilia: [a]*Distrito Federal. [b]none. [c]Torto R; Fundo R. [d]Universidade de Brasilia (1961) 8 – 654. [e]Biblioteca Central da Universidade . . . (1962) 300; Centro de Documentacao e Informacao da Camara dos Deputados (1866 at Rio de Janeiro, relocated 1960?) 235.

Campina Grande: [a]Paraiba. [b]none. [c]nr Mamanguape R. [d]Universidade Regional do Nordeste (1966) 4 – 219. [e]Biblioteca Central da Universidade . . . (1968) 9; Biblioteca Municipal Felix Araujo (1938) 3.

Campinas: [a]Sao Paulo. [b]for Sao Carlos. [c]Atibaia R. [d]Universidade Pontificia Catolica de Campinas (1941) 11 – 682; Universidade Estadual de Campinas (1962) 4 – 845. [e]Biblioteca Publica do Centro de Ciencias, Letras e Artes (1901) 30; Biblioteca da Universidade Pontificia Catolica de Campinas (1942) 20; Biblioteca Publica Municipal (1946) 18.

Campos: [a]Rio de Janeiro. [b]for Sao Salvador. [c]Paraiba R. [d]none. [e]Biblioteca Municipal (1872) 15.

Curitiba: [a]*Parana. [b]for Corityba. [c]nr Iguazu R. [d]Universidade Federal do Parana (1912) 12 – 1387; Universidade Catolica do Parana (1937) 7 – 424. [e]Biblioteca Publica do Parana (1857) 120; Biblioteca da Universidade Federal do Parana (1912) 118.

Duque de Caxias: [a]Rio de Janeiro. [b]for Caxias. [c]Guanabara B of Atlantic O. [d]none. [e]Biblioteca da Refinaria Duque de Caxias da Petrobras (1965) 4.

Feira de Santana: [a]Bahia. [b]for Feira de Sant'Anna. [c]Jacuipe R. [d]none. [e]Biblioteca Publica Municipal Arnold Silva (1890) 10.

Florianopolis: [a]*Santa Catarina. [b]for Desterro. [c]Norte B of Atlantic O; Sul B of Atlantic O. [d]Universidade Federal de Santa Catarina (1932) 7 – 875; Universidade para o Desenvolvimento do Estado de Santa Catarina (1965) 5 – 288. [e]Biblioteca da Universidade Federal de Santa Catarina (1951) 85; Biblioteca Publica do Estado de Santa Catarina (1885) 45.

Fortaleza: [a]*Ceara. [b]for Ceara. [c]Atlantic O. [d]Universidade Federal do Ceara (1903) 9 – 1030; Universidade de Fortaleza (1973) 4 – 157. [e]Biblioteca da Universidade Federal do Ceara (1918) 155; Biblioteca Publica do Ceara (1867) 220.

Goiania: ^{a}*Goias. ^{b}none. ^{c}nr Meia Ponte R. ^{d}Universidade Catolica de Goias (1898 at Goias, relocated 1935?) 8 – 232; Universidade Federal de Goias (1947) 6 – 794. ^{e}Biblioteca da Universidade Federal de Goias (1948) 21; Biblioteca Publica Municipal (?) 8.

Guarulhos: ^{a}Sao Paulo. ^{b}none. ^{c}Tiete R. ^{d}none. ^{e}Biblioteca Publica Municipal (1940) 6.

Jaboatao: ^{a}Pernambuco. ^{b}none. ^{c}Jaboatao R. ^{d}none. ^{e}Biblioteca Publica Municipal (1952) ?

Joao Pessoa: ^{a}*Paraiba. ^{b}for Parahyba. ^{c}Paraiba do Norte (alt Paraiba) R. ^{d}Universidade Federal da Paraiba (1947) 11 – 900. ^{e}Biblioteca da Universidade . . . (1955) 27; Biblioteca Publica do Estado da Paraiba (1859) 12.

Juiz de Fora: ^{a}Minas Gerais. ^{b}for Parahybuna. ^{c}Paraibuna R. ^{d}Universidade Federal de Juiz de Fora (1915) 6 – 613. ^{e}Biblioteca da Universidade . . . (1914) 32; Biblioteca Municipal (1934) 8.

Jundiai: ^{a}Sao Paulo. ^{b}for Jundiahy. ^{c}Jundiai (for Jundiahy) R. ^{d}none. ^{e}Biblioteca do Gabinete de Leitura Rui Barbosa (1908) 11.

Londrina: ^{a}Parana. ^{b}none. ^{c}nr Tibagi R. ^{d}Universidade Estadual de Londrina (1971) 8 – 536. ^{e}Biblioteca da Universidade . . . (1971?) 27; Biblioteca Publica Municipal (1940) 7.

Maceio: ^{a}*Alagoas. ^{b}none. ^{c}Atlantic O; Mundau R. ^{d}Universidade Federal de Alagoas (1931) 4 – 518. ^{e}Biblioteca da Universidade . . . (1950) 55; Biblioteca do Instituto Historico de Alagoas (1869) 10; Biblioteca Publica Estadual (1865) 8.

Manaus: ^{a}*Amazonas. ^{b}alt Manaos. ^{c}Negro R. ^{d}Universidade do Amazonas (1909) 4 – 412. ^{e}Biblioteca Publica do Amazonas (1873) 109.

Natal: ^{a}*Rio Grande do Norte. ^{b}none. ^{c}Atlantic O; Potengi R. ^{d}Universidade Federal do Rio Grande do Norte (1948) 6 – 727. ^{e}Biblioteca da Universidade . . . (1945) 34; Biblioteca Camara Cascudo (1948) 3.

Niteroi: ^{a}*Rio de Janeiro. ^{b}for Nictheroy, Praia Grande. ^{c}Guanabara B of Atlantic O. ^{d}Universidade Federal Fluminense (1912) 17 – 1596. ^{e}Biblioteca Publica do Estado do Rio de Janeiro (1927) 100; Biblioteca da Universidade . . . (1947) 40.

Nova Iguacu: ^{a}Rio de Janeiro. ^{b}for Maxambomba. ^{c}nr Sarapui R. ^{d}none. ^{e}Biblioteca Desembargador Acacio Aragao (1958) 30.

Olinda: ^{a}Pernambuco. ^{b}none. ^{c}Atlantic O. ^{d}none. ^{e}Biblioteca O Luzeiro (?) 8; Biblioteca do Mosteiro de Sao Bento (1917) 7 (in 1939); Biblioteca Osvaldo Guimaraes (1961) 2.

Osasco: ^{a}Sao Paulo. ^{b}none. ^{c}Tiete R. ^{d}none. ^{e}Biblioteca Publica Municipal Monteiro Lobato (?) 2.

Pelotas: ^{a}Rio Grande do Sul. ^{b}none. ^{c}Sao Goncalo Channel. ^{d}Universidade Catolica de Pelotas (1939) 3 – 290; Universidade Federal de Pelotas (1883) 3 – 441. ^{e}Biblioteca Publica Pelotense (1875) 130.

Petropolis: ^{a}Rio de Janeiro. ^{b}none. ^{c}Piabanha R; Quitandinha R; Palatinado R. ^{d}Universidade Catolica de Petropolis (1954) 3 – 250. ^{e}Biblioteca Municipal (1871) 100.

Porto Alegre: ^{a}*Rio Grande do Sul. ^{b}none. ^{c}Guaiba R; Patos Lagoon. ^{d}Pontificia Universidade Catolica do Rio Grande do Sul (1931) 14 – 1133; Universidade Federal do Rio Grande do Sul (1896) 12 – 1626. ^{e}Biblioteca da Universidade . . . (1898) 265; Biblioteca Publica do Estado do Rio Grande do Sul (1871) 98; Biblioteca da . . . Universidade (1942) 80.

Recife: ^{a}*Pernambuco. ^{b}for Pernambuco. ^{c}Atlantic O; Capiberibe R; Beberibe R. ^{d}Universidade Federal de Pernambuco (1827) 12 – 1640; Universidade Catolica de Pernambuco (1912) 8 – 361; Universidade Rural de Pernambuco (1914) 3 – 247. ^{e}Biblioteca da Universidade Federal de Pernambuco (1830) 339; Biblioteca Publica do Estado de Pernambuco (1852) 85.

Ribeirao Preto: ^{a}Sao Paulo. ^{b}none. ^{c}Preto R. ^{d}none. ^{e}Biblioteca Cultural Altino Arantes (1959) 13; Biblioteca Publica da Sociedade Legiao Brasileira (1903) 12; Biblioteca da Faculdade de Medicina de Ribeirao Preto da Universidade de Sao Paulo (1952) 9.

Rio de Janeiro: ^{a}*Guanabara. ^{b}off Sao Sebastiao do Rio de Janeiro. ^{c}Atlantic O; Guanabara B of Atlantic O; Rodrigo de Freitas Lagoon. ^{d}Universidade Federal do Rio de Janeiro (1808) 20 – 2780; Universidade Gama Filho (1972) 18 – 899; Universidade do Estado da Guanabara (1950) 9 – 848; Pontificia Universidade Catolica do Rio de Janeiro (1937) 7 – 675; Associacao Universitaria Santa Ursula (1939) 5 – 423. ^{e}Biblioteca Nacional do Rio de Janeiro (1810) 1800; Biblioteca da Universidade Federal do Rio de Janeiro (1833) 577, inc Biblioteca do Museu

Nacional (1863) 310; Biblioteca do Instituto Historico e Geografico Brasileiro (1839) 300; Biblioteca do Real Gabinete Portugues de Leitura (1837) 121; Biblioteca Estadual (1874) 55.

Salvador: [a]*Bahia. [b]for Bahia, Sao Salvador da Bahia. [c]Todos os Santos B of Atlantic O. [d]Universidade Federal da Bahia (1808) 14 − 1478; Universidade Catolica do Salvador (1961) 7 − 737. [e]Biblioteca Central do Estado da Bahia (1811) 300; Biblioteca da Universidade Federal da Bahia (1909) 224.

Santo Andre: [a]Sao Paulo. [b]none. [c]Tamanduatei R. [d]none. [e]Biblioteca Publica Municipal (1954) 13.

Santos: [a]Sao Paulo. [b]none. [c]Santos R. [d]none. [e]Biblioteca da Faculdade de Filosofia, Ciencias e Letras de Santos (1955) 18; Biblioteca da Sociedade Humanitaria dos Empregados no Comercio (1888) 19; Biblioteca Publica Municipal (1876) 12.

Sao Bernardo do Campo: [a]Sao Paulo. [b]none. [c]nr Grande R. [d]none. [e]Biblioteca Publica Municipal Monteiro Lobato (1952) 6.

Sao Goncalo: [a]Rio de Janeiro. [b]none. [c](Riacho) Imbuacu Creek. [d]none. [e]Biblioteca Municipal (1940) ?

Sao Joao de Meriti: [a]Rio de Janeiro. [b]for Mirity. [c]Sao Joao de Meriti R. [d]none. [e]Biblioteca Sao Boaventura (1940) 2.

Sao Luis: [a]*Maranhao: Sao Luis I. [b]for Maranhao, Sao Luiz. [c]Sao Marcos B of Atlantic O. [d]Universidade do Maranhao (1945) 5 − 584. [e]Biblioteca Publica do Estado do Maranhao Benedito Leite (1829) 80; Biblioteca da Universidade . . . (1914) 20.

Sao Paulo: [a]*Sao Paulo. [b]none. [c]Tiete R. [d]Universidade de Sao Paulo (1827) 28 − 3965 [Sao Paulo campus: (1827) ? − ?]; Universidade Mackenzie (1870) 12 − 703; Pontificia Universidade Catolica de Sao Paulo (1908) 9 − 951. [e]Biblioteca da Universidade de Sao Paulo (1827) 1156 [Sao Paulo campus: (1827) ?]; Biblioteca Municipal Mario de Andrade (1925) 1101.

Sorocaba: [a]Sao Paulo. [b]none. [c]Sorocaba R. [d]none. [e]Biblioteca do Gabinete de Leitura Sorocabano (1867) 11.

Teresina: [a]*Piaui. [b]for Therezina. [c]Parnaiba R; Poti R. [d]Universidade Federal do Piaui (1958) 3 − 284. [e]Biblioteca da Casa Anisio Brito (1937) 13; Biblioteca da Universidade . . . (1968?) 12.

Vitoria: [a]*Espirito Santo; Vitoria I. [b]for Victoria. [c]Vitoria B of Atlantic O; Santa Maria da Vitoria R. [d]Universidade Federal do Espirito Santo (bef 1954) 6 − 568. [e]Biblioteca da Universidade . . . (1933) 50; Biblioteca Estadual (1855) 32; Biblioteca Municipal (1941) 13.

Chile

Antofagasta: [b]none. [c]Moreno B of Pacific O. [d]Universidad del Norte (1956) 6 − 550. [e]Biblioteca de la Universidad . . . (1958) 44.

Concepcion: [b]none. [c]Bio-Bio R. [d]Universidad de Concepcion (1919) 14 − 1100. [e]Biblioteca Central de la Universidad . . . (1920) 220.

*Santiago: [b]alt Santiago de Chile. [c]Mapocho R. [d]Universidad de Chile (1738) 64 − 14,000 [Santiago campus: (1738) ? − ?]; Universidad Catolica de Chile (1888) 16 − 4091; Universidad Tecnica del Estado (1849) 10 − 590. [e]Biblioteca Nacional de Chile (1813) 1500; Biblioteca de la Universidad de Chile (1738?) 1200; Biblioteca de la Universidad Catolica de Chile (1895) 450.

Valparaiso: [b]none. [c]Pacific O. [d]Universidad Catolica de Valparaiso (1928) 7 − 1000; Universidad Tecnica Federico Santa Maria (1926) 4 − 460. [e]Biblioteca de la Universidad Catolica de Valparaiso (1928) 155; Biblioteca Severin (1873) 110.

Vina del Mar: [b]none. [c]Pacific O; Marga Marga R. [d]none. [e]Biblioteca del Estacion de Biologia Marina de Montemar (1941) 11; Biblioteca del Instituto Britanico-Chileno (1942) 9.

Colombia

Barranquilla: [b]none. [c]Magdalena R. [d]Universidad del Atlantico (1941) 3 − 250; Fundacion Universidad del Norte (1966) 1 − 120; Universidad Autonoma del Caribe (1967) 1 − 150; Universidad Libre, Barranquilla campus[1] (1956) 0.4 − ? [e]Biblioteca Publica Departamental (1923) 32.

---

[1]Main campus, at Bogota, founded 1923.

*Bogota: ^{b}for Santa Fe de Bogota, Teusaquillo. ^{c}San Agustin R; San Francisco R. $^{d^1}$ Universidad Nacional de Colombia (1563) 16 – 2208; Pontificia Universidad Javeriana (1623) 9 – 930; Universidad la Gran Colombia (1951) 7 – 650; Fundacion Universidad de Bogota Jorge Tadeo Lozano (1954) 6 – 615. ^{e}Biblioteca Nacional de Colombia (1777) 400; Biblioteca de la Pontificia Universidad Javeriana (1623?) 273.

Bucaramanga: ^{b}none. ^{c}nr Oro R; nr Surata R. ^{d}Universidad Industrial de Santander (1940) 5 – 380; Universidad de Santo Tomas, Bucaramanga campus^2 (?) 0.5 – ? ^{e}Biblioteca Departamental (1898) 28.

Cali: ^{b}none. ^{c}Cali R. ^{d}Universidad del Valle (1945) 7 – 750; Universidad Santiago de Cali (1958) 3 – ? ^{e}Biblioteca de la Universidad del Valle (1946) 94; Biblioteca Municipal del Centenario (1910) 25.

Cartagena: ^{b}none. ^{c}B of Cartagena of Caribbean Sea. ^{d}Universidad de Cartagena (1774) 2 – 197. ^{e}Biblioteca Universitaria Fernandez Madrid (1827) 25.

Cucuta: ^{b}off San Jose de Cucuta. ^{c}Pamplonita R. ^{d}Universidad Francisco de Paula Santander (1962) 1 – 126. ^{e}Biblioteca de la Universidad . . . (1962?) 5; Biblioteca Julio Perez Ferrero (1912) 4; Biblioteca del Centro de Historia (?) 3.

Ibague: ^{b}off San Bonifacio de Ibague. ^{c}nr Chipalo R; nr Combeima R. ^{d}Universidad del Tolima (1945) 3 – 230. ^{e}Biblioteca General de la Universidad . . . (1963) 7; Biblioteca Departamental de Tolima (1954) 2.

Manizales: ^{b}none. ^{c}nr Chinchina R. ^{d}Universidad de Caldas (1937) 3 – 280. ^{e}Biblioteca Central de la Universidad . . . (1958) 25; Biblioteca Departamental de Caldas (1954) 14.

Medellin: ^{b}none. ^{c}Porce R. ^{d}Universidad de Antioquia (1801) 12 – 1234; Pontificia Universidad Catolica Bolivariana (1936) 10 – 560; Universidad de Medellin (1950) 3 – 352; Universidad Nacional de Colombia, Medellin campus (1887) 3 – ?; Universidad Autonoma Latinoamericana (1966) 2 – 170. ^{e}Biblioteca General de la Universidad de Antioquia (1935) 150; Biblioteca Publica Piloto de Medellin para la America Latina (1954) 65.

Pereira: ^{b}none. ^{c}Otun R. ^{d}Universidad Tecnologica de Pereira (1958) 2 – 232. ^{e}Biblioteca de la Universidad . . . (1960) 6; Biblioteca Jesus Maria Ormaza (?) 4.

Ecuador

Guayaquil: ^{b}for Santiago de Guayaquil. ^{c}Guayas R. ^{d}Universidad Estatal de Guayaquil (1867) 4 – 400; Universidad Catolica de Santiago de Guayaquil (1962) 3 – 350; Escuela Superior Politecnica del Litoral (univ) (1958) 2 – 105; Universidad Laica Vicente Rocafuerte (1847) ? – ? ^{e}Museo y Biblioteca Municipal (1862) 85.

*Quito: ^{b}none. ^{c}Machangara R. ^{d}Universidad Central del Ecuador (1586) 50 – 2350; Pontificia Universidad Catolica del Ecuador (1946) 10 – 315; Escuela Politecnica Nacional (univ) (1870) 3 – 127. ^{e}Biblioteca de la Universidad . . . (1586) 110; Biblioteca Ecuatoriana y Archivo (1928) 65; Biblioteca Nacional (1792) 60.

French Guiana

*Cayenne: ^{a}Cayenne I. ^{b}none. ^{c}Atlantic O; Cayenne R. ^{d}none. ^{e}Bibliotheque Franconie (?) 3.

Guyana

*Georgetown: ^{b}none. ^{c}Atlantic O; Demerara R. ^{d}University of Guyana (1963) 2 – 170. ^{e}National Library (1909) 152; University . . . Library (1963?) 88.

Paraguay

*Asuncion: ^{b}for Nuestra Senora de la Asuncion ^{c}Paraguay R. ^{d}Universidad Nacional de Asuncion (1883) 8 – 500; Universidad Catolica Nuestra Senora de la Asuncion (1960) 6 – 410. ^{e}Biblioteca, Museo y Archivo Nacionales (1869) 45.

---

[1] Universities with 5000 or more students.
[2] Main campus, at Bogota, founded 1580.

Peru

Arequipa: [b]none. [c]Chili R. [d]Universidad Nacional de San Agustin (1821) 11 – 481; Universidad Particular Catolica de Santa Maria (1961) 6 – 264. [e]Biblioteca de la Universidad Nacional de San Agustin (1828) 126; Biblioteca Publica Municipal (1821) 28.

Callao: [b]none. [c]Pacific O. [d]Universidad Nacional Tecnica del Callao (1967) 2 – 170. [e]Biblioteca Municipal Piloto (1936) 35; Biblioteca del Instituto del Mar del Peru (1960) 30.

Chiclayo: [b]none. [c]Reque R. [d]Universidad Nacional Pedro Ruiz Gallo (1962) 5 – 263. [e]Biblioteca Municipal (bef 1942) ?

Cuzco: [b]for Cusco. [c]Chunchulmayu R; Huatanay R; Tullamayo R. [d]Universidad Nacional de San Antonio Abad (1598) 8 – 425. [e]Biblioteca Central de la Universidad . . . (1696) 55.

*Lima: [b]for Rimac. [c]Pacific O; Rimac R. [d][1] Universidad Nacional Mayor de San Marcos (1551) 22 – 2394; Universidad Nacional Federico Villarreal (1963) 14 – 686; Universidad Particular San Martin de Porres (1962) 10 – 256; Universidad Nacional de Ingenieria (1876) 9 – 780; Universidad Particular Inca Garcilaso de la Vega (1964) 7 – 240; Universidad Particular Ricardo Palma (1969) 7 – 228; Universidad Catolica del Peru (1917) 6 – 647; Universidad Particular de Lima (1963) 5 – 250. [e]Biblioteca Nacional (1821) 630; Biblioteca Central de la Universidad Nacional Mayor de San Marcos de Lima (1551) 400.

Trujillo: [b]none. [c]Moche R. [d]Universidad Nacional de Trujillo (1824) 6 – 506. [e]Biblioteca Central de la Universidad . . . (1824?) 32.

Surinam

*Paramaribo: [b]none. [c]Suriname R. [d]Universiteit van Suriname (1968) 0.5 – 34. [e]Bibliotheek Cultureel Centrum Suriname (1947) 38.

Uruguay

*Montevideo: [b]none. [c]Plata R. [d]Universidad de la Republica (1833) 40 – 2149; Universidad del Trabajo del Uruguay (college) (1942) 35 – 4016. [e]Biblioteca Nacional del Uruguay (1816) 500; Biblioteca de la Facultad de Medicina de la Universidad de la Republica (1884) 225; Biblioteca del Poder Legislativo (?) 180.

Venezuela

Barquisimeto: [b]none. [c]Cojedes R. [d]Universidad Centro-Occidental (1963) 10 – 506. [e]Biblioteca Publica Pio Tamayo (1911) 22.

*Caracas: [b]for Santiago de Leon de Caracas. [c]Guaire R. [d]Universidad Central de Venezuela (1696) 50 – 2700; Universidad Catolica Andres Bello (1953) 7 – 550; Universidad Simon Bolivar (1967) 7 – 672; Universidad Santa Maria (1953) 4 – 300; Universidad Metropolitana (1970) 0.2 – 20. [e]Biblioteca Nacional (1841) 438; Biblioteca Central de la Universidad Central de Venezuela (1850) 129; Biblioteca de la Universidad Catolica Andres Bello (1953) 112.

El Valle: [b]alt Valle. [c]nr Guaire R. [d]none.

Maracaibo: [b]none. [c]Maracaibo L. [d]Universidad del Zulia (1891) 27 – 1415; Universidad Rafael Urdaneta (1973) ? – ? [e]Biblioteca Baralt (1961) 35; Biblioteca Central de la Universidad . . . (1946) 30.

Maracay: [b]none. [c]nr Valencia L. [d]none. [e]Biblioteca del Centro de Investigaciones Agronomicas (1938) 26.

Petare: [b]none. [c]Guaire R. [d]none.

San Felix de Guayana: [b]alt Ciudad Guayana, Guayana; for San Felix, Santo Tome de Guayana. [c]Orinoco R; Caroni R. [d]none. [e]Biblioteca Publica Caroni (1963) 3 (in 1964).

Valencia: [b]none. [c]Cabriales R. [d]Universidad de Carabobo (1833) 24 – 943. [e]Biblioteca Central de la Universidad . . . (1852) 12; Biblioteca Publica del Estado (1876) 4.

---

[1]Universities with 5000 or more students.

SELECTED BIBLIOGRAPHY

O f the thousands of books and periodical articles consulted in the preparation of *World Facts and Figures*, 260 of the most useful sources of information are listed here. For the convenience of readers who seek more detailed information on a particular subject than can be given in this book, these sources are listed, alphabetically by author (by title if the authorship is anonymous), under the following 28 headings:

Encyclopedias
    Universal
    National and regional
Official statistical yearbooks
Unofficial yearbooks and almanacs
Gazetteers
    Universal
    National and regional
Atlases
Geography
    Universal
    National and regional
Physiography
    Oceans and seas
    Islands
    Rivers
    Mountains
    Lakes
    Waterfalls

Demography
    Universal
    National and regional
Climate
    Universal
    National and regional
Energy
Productivity
Civil engineering
    Bridges
    Tunnels
    Dams
Universities
    Universal
    National and regional
Libraries
    Universal
    National and regional

Encyclopedias: Universal

1. *Aschehougs Konversasjons Leksikon.* Oslo, 1954–1961. 18 v.
2. *Bolshaya Sovetskaya Entsiklopediya.* Moscow, 1970–. v 1–26: A-Ulyanovo.
3. Idem. Moscow, 1949–1958. 53 v.
4. Idem. Moscow, 1926–1931. 65 v.
5. *Brockhaus Enzyklopadie.* Wiesbaden, 1966–1975. 21 v.
6. *Chambers's Encyclopaedia.* Oxford, 1966. 15 v.
7. *Collier's Encyclopedia.* [New York], 1977. 24 v.

8. *(New) Columbia Encyclopedia,* 4th ed. New York, 1975. 3052 pp.

9. *Diccionario Enciclopedico Salvat Universal.* Barcelona, 1969–1974. 20 v.

10. *Diccionario Enciclopedico U.T.E.H.A.* Mexico, 1950–1952. 10 v.

11. Idem; *Apendice.* Mexico, 1964. 2 v.

12. *Dizionario Enciclopedico Italiano.* Rome, 1955–1961. 12 v.

13. *Enciclopedia Italiana di Scienze, Lettere ed Arti.* [Rome], 1929–1939. 36 v.

14. *Enciclopedia Universal.* Sao Paulo, 1969, 10 v.

15. *Enciclopedia Universal Ilustrada Europeo-Americana* (short title: *Espasa*). Barcelona, [1907]–1930. 70 v.

16. Idem; *Apendice.* Madrid, 1930–1933. 10 v.

17. *(New) Encyclopaedia Britannica,* 15th ed. Chicago, 1977. 30 v.

18. *Encylopaedia Britannica,* 11th ed. New York, 1910–1911. 29 v.

19. Idem, 8th ed. Boston, 1860. 22 v.

20. *Encyclopedia Americana.* New York, 1977. 30 v.

21. *Grande Encyclopedie.* Paris [1886–1902]. 31 v.

22. *Grand Larousse Encyclopedique.* Paris, 1960–1964. 10 v.

23. *Gran Enciclopedia del Mundo.* Barcelona, 1961–1964. 20 v.

24. *Grosse Brockhaus.* Wiesbaden, 1952–1957. 12 v.

25. Idem. Leipzig, 1928–1935. 20 v.

26. *Grote Winkler Prins.* Amsterdam, 1966–1975. 20 v.

27. *Meyers Enzyklopadisches Lexikon.* Mannheim, 1971–. v 1–22: A-Sud.

28. *Meyers Grosses Konversations-Lexikon.* Leipzig, 1909. 20 v.

29. *Revai Nagy Lexikona,* Budapest, [1911–1926]. 19 v.

30. *Schweizer Lexikon.* Zurich, 1945–1948. 7 v.

31. *Svensk Upplagsbok.* Malmo, 1947–1955. 32 v.

32. *Uj Magyar Lexikon.* Budapest, 1961–1962. 6 v.

33. *Wielka Encyklopedia Powszechna PWN.* Warsaw, 1962–1969. 12 v.

Encyclopedias: National and Regional

34. *Australian Encyclopaedia.* Sydney, 1958. 10 v.

35. *Diccionario Porrua,* 3rd ed. Mexico, 1970–1971. 2 v (2465 pp).

36. *Enciklopedija Jugoslavije.* Zagreb, 1955–1971. 8 v.

37. *Encyclopaedia of New Zealand.* Wellington, 1966. 3 v.

38. *Encyclopaedie van Nederlandsch-Indie.* Hague, 1917–1939. 8 v.

39. *Encyclopedia Canadiana.* Toronto, 1975. 10 v.

40. *Encyclopedia of Canada.* Toronto, 1935–1937. 6 v.

41. *Gran Enciclopedia Argentina.* Buenos Aires, 1956–1963. 8 v.

42. *Modern Encyclopaedia of Australia and New Zealand.* Sydney. 1964. 1199 pp.

Official Statistical Yearbooks

43. Argentina, Instituto Nacional de Estadistica y Censos. *Anuario Estadistico de la Republica Argentina, 1973.* Buenos Aires, 1974. 423 pp.

44. Australia, Australian Bureau of Statistics. *Official Year Book of Australia . . . 1975 and 1976.* Canberra, 1977. 1148 pp.

45. Austria, Statistisches Zentralamt. *Statistisches Handbuch fur die Republik Osterreich . . . 1976.* Vienna, 1976. 668 pp.

46. Belgium, Institut National de Statistique. *Annuaire Statistique de la Belgique . . . 1976.* Brussels, ND. 810 pp.

47. Brazil. Fundacao Instituto Brasileiro de Geografia e Estatistica. *Anuario Estatistico do Brasil—1976.* Rio de Janeiro, 1977. 813 pp.

48. Bulgaria, Tsentralno Statistichesko Upravlenie. *Statisticheski Godishnik na Narodna Republika Bulgariya, 1977.* Sofia, ND. 635 pp.

49. Canada, Statistics Canada. *Canada Year Book, 1976-77.* Ottawa, 1977. 1142 pp.

50. Cuba, Direccion Central de Estadistica. *Anuario Estadistico de Cuba, 1972.* Havana, 1974. 293 pp.

51. Czechoslovakia, Federalni Statisticky Urad. *Statisticka Rocenka Ceskoslovenske Socialisticke Republiky, 1977.* Prague, 1977. 678 pp.

52. Denmark, Danmarks Statistik. *Statistisk Arbog, 1977.* Copenhagen, 1977. 592 pp.

53. East Germany, Staatliche Zentralverwaltung fur Statistik. *Statistisches Jahrbuch 1977 der Deutschen Demokratischen Republik.* Berlin, 1977. 456+96+16 pp.

54. Ethiopia, Central Statistical Office. *Statistical Abstract, 1976.* Addis Ababa, 1977. 254 pp.

55. Finland, Tilastokeskus. *Suomen Tilastollinen Vuosikirja . . . 1976.* Helsinki, 1977. 529 pp.

56. France, Institut National de la Statistique et des Etudes Economiques. *Annuaire Statistique de la France, 1977.* Paris, 1977. 767+55 pp.

57. Greece, Ethnike Statistike Iperesia. *Statistike Epeteris tes Ellados . . . 1976.* Athens, 1976. 474 pp.

58. Hungary, Kozponti Statisztikai Hivatal. *Statisztikai Evkonyv, 1975.* Budapest, 1976. 493 pp.

59. India, Central Statistical Organisation. *Statistical Abstract, India, 1974.* New Delhi, 1975. 696 pp.

60. Indonesia, Biro Pusat Statistik. *Statistik Indonesia . . . 1976.* Jakarta, 1976. 392 pp.

61. Iran, Statistical Centre of Iran. *Statistical Yearbook of Iran, 1352* [1973-1974]. Tehran, 1976. 517 pp.

62. Israel, Central Bureau of Statistics. *Statistical Abstract of Israel, 1976.* Jerusalem, 1976. 732+115 pp.

63. Italy, Istituto Centrale di Statistica. *Annuario Statistico Italiano, Edizione 1977.* Rome, 1977. 458 pp.

64. Japan, Bureau of Statistics. *Japan Statistical Yearbook, 1977.* Tokyo, 1977. 706 pp.

65. Kenya, Central Bureau of Statistics. *Statistical Abstract, 1975.* Nairobi, ND. 279 pp.

66. League of Nations. *Statistical Year-Book, 1926-1942/1944.* Geneva, 1927-1945. 17 v.

67. Mexico, Direccion General de Estadistica. *Anuario Estadistico de los Estados Unidos Mexicanos, 1970-1971.* Mexico, 1973. 754 pp.

68. Netherlands, Centraal Bureau voor de Statistiek. *Statistisch Zakboek, 1976.* Voorburg, Netherlands, 1976. 365 pp.

69. New Zealand, Department of Statistics. *New Zealand Official Yearbook, 1976.* Wellington, 1976. 1092 pp.

70. Norway, Statistisk Sentralbyra. *Statistisk Arbok, 1977.* Oslo, 1977. 464 pp.

71. Philippines, National Census and Statistics Office. *Philippine Yearbook, 1977.* Manila, 1977. 1095 pp.

72. Poland, Glowny Urzad Statystyczny. *Rocznik Statystyczny, 1975.* Warsaw, 1975. 642 pp.

73. Portugal, Instituto Nacional de Estatistica. *Anuario Estatistico . . . 1976.* Lisbon, 1977. 423 pp.

74. Romania, Directia Centrala de Statistica. *Anuarul Statistic al Republicii Socialiste Romania, 1977.* [Bucharest], ND. 598 pp.

75. South Africa, Departement van Statistiek. *Suid-Afrikaanse Statistieke, 1976.* Pretoria, 1976. unp.

76. South Korea, Bureau of Statistics. *Korea Statistical Yearbook, 1976.* Seoul, 1976. 509 pp.

77. Spain, Instituto Nacional de Estadistica. *Anuario Estadistico de Espana, 1976.* [Madrid], ND. 814 pp.

78. Sweden, Statistiska Centralbyran. *Statistisk Arsbok for Sverige, 1977.* Stockholm, 1977. 586 pp.

79. Switzerland, Eidgenossisches Statistisches Amt. *Statistisches Jahrbuch der Schweiz . . . 1977.* Bern, 1977. 661 pp.

80. Syria, Central Bureau of Statistics. *Statistical Abstract, 1976.* [Damascus], 1976. 999 pp.

81. Taiwan, Directorate-General of Budgets, Accounts and Statistics. *Statistical Abstract of the Republic of China, 1976.* [Taipei], ND. 566 pp.

82. Turkey, Devlet Istatistik Enstitusu. *Turkiye Istatistik Yilligi, 1975.* Ankara, 1976. 463 pp.

83. UK, Central Statistical Office. *Annual Abstract of Statistics, 1976.* London, 1976. 488 pp.

84. United Nations, Statistical Office. *Statistical Yearbook, 1948-1976.* New York, 1949-1977. 28 v.

85. US, Bureau of the Census. *Statistical Abstract of the United States, 1977.* Washington, 1977. 1048 pp.

86. USSR, Tsentralnoye Statisticheskoye Upravleniye. *Narodnoye Khozyaystvo SSSR v 1975 G.* Moscow, 1976. 846 pp.

87. Venezuela, Direccion General de Estadistica y Censos Nacionales. *Anuario Estadistico, 1973.* Caracas, 1975. 2 v.

88. West Germany, Statistisches Bundesamt. *Statistisches Jahrbuch 1977 fur die Bundesrepublik Deutschland.* Wiesbaden, 1977. 736 pp.

89. Yugoslavia, Savezni Zavod za Statistiku. *Statisticki Godisnjak Jugoslavije, 1977.* Belgrade, 1977. 730 pp.

Unofficial Yearbooks and Almanacs

90. *Almanach de Gotha, 1763-1944.* Gotha, 1763-1944. 181 v.

91. *Almanaque Abril, 1977.* Sao Paulo, 1976. 784 pp.

92. *(1977) Almanaque Mundial.* Panama, 1976, 576 pp.

93. *CBS News Almanac, 1978.* Maplewood, New Jersey, USA, 1977. 1040 pp.

94. *China Yearbook, 1976.* Taipei, 1976. 816 pp.

95. *Europa Year Book, 1978.* London, 1978. 2 v.

96. *Hubner's Weltstatistik . . . 1939.* Vienna [1939]. 327 pp.

97. *Information Please Almanac . . . 1978.* New York, 1977. 1008 pp.

98. *Japan Almanac, 1975.* Tokyo, 1975. 486 pp.

99. *Pacific Islands Year Book,* 12th ed. Sydney, 1977. 430 pp.

100. *Pakistan Year Book, 1976.* Karachi, 1976. 519 pp.

101. *Statesman's Year-Book, 1864-1977/78.* London, 1864-1977. 114 v.

102. *(1976-77) West Indies & Caribbean Year Book.* Toronto, ND. 952 pp.

103. Whitaker, Joseph. *Almanack for the Year of Our Lord 1978,* 110th ed. London, 1977. 1220 pp.

104. *World Almanac and Book of Facts, 1867-1978.* New York, 1868-1977. 110 v.

Gazetteers: Universal

105. Brookes, R. *General Gazetteer; or, Compendious Geographical Dictionary.* London, 1815. unp.

106. Chisholm, George G., Ed. *Times* [of London] *Gazetteer of the World.* London, 1899. 1787 pp.

107. Clarke, J. W. *(New) Geographical Dictionary.* London, 1822. 2 v.

108. *Columbia Lippincott Gazetteer of the World.* New York, 1952. 2148 pp.

109. *Edinburgh Gazetteer, or Geographical Dictionary.* Edinburgh, 1822. 6 v.

110. *Gazetteer of the World, or, Dictionary of Geographical Knowledge.* Edinburgh, [1850]-1856. 7 v.

111. *Harper's Statistical Gazetteer of the World.* New York, 1855. 1952 pp.

112. *Imperial Gazetteer.* Glasgow, 1855. 2 v.

113. Johnston, A. Keith. *General Dictionary of Geography.* London, 1877. 1513 pp.

114. *Kratkaya Geograficheskaya Entsiklopediya.* Moscow, 1960-1966. 5 v.

115. *Lippincott's Gazetteer of the World.* Philadelphia, 1902. 2636 pp.

116. *Lippincott's Pronouncing Gazetteer.* Philadelphia, 1855, 2182 pp.

117. Malte-Brun, [Conrad]. *Diccionario Geografico Universal.* Paris, 1828. 2 v.

118. McCulloch, J. R. *Dictionary Geographical, Statistical, and Historical of the Various Countries, Places and Principal Natural Objects in the World.* London, 1866. 4 v.

119. *Ritters Geographisch-Statistisches Lexikon.* Leipzig, 1910. 2 v.

120. *Webster's (New) Geographical Dictionary.* Springfield, Massachusetts, USA, 1977. 1370 pp.

121. Worcester, J. E. *Geographical Dictionary, or Universal Gazetteer, Ancient and Modern.* Boston, 1823. 2 v.

Gazetteers: National and Regional

122. Canada, Board on Geographical Names. *Gazetteer*[s of individual provinces]. Ottawa, 1952–1962. 10 v.

123. *Diccionario Geografico de Espana.* Madrid, 1956–1961. 17 v.

124. *Dicionario Geografico Brasileiro,* 2nd ed. Porto Alegre, 1972. 619 pp.

125. *Imperial Gazetteer of India.* Oxford, 1907–1909. 26 v.

126. Orth, Donald J. *Dictionary of Alaska Place Names.* Washington, 1967. 1084 pp.

127. Papinot, E. *Historical and Geographical Dictionary of Japan.* Tokyo, [1909]. 842 pp.

128. Playfair, G. M. H. *Cities and Towns of China; a Geographical Dictionary,* 2nd ed. Shanghai, 1910. 582+76 pp.

129. US, Office of Geography. *Gazetteer*[s of individual countries and regions].[1] Washington, 1955–. v 1–.

Atlases

130. *Chungkuo Fensheng Titu (Provincial Atlas of China).* Peking, 1963. 142 pp. (US Joint Publications Research Service, number 27,071).

131. *Pergamon World Atlas.* Warsaw, 1968. 525 pp.

132. Rand McNally & Company. *(1977) Commercial Atlas & Marketing Guide,* 109th ed. Chicago, 1978. 669 pp.

133. *Rand McNally International Atlas.* Chicago, 1977. 312+222 pp.

134. *Stielers Hand-Atlas.* Gotha, 1925. 108 leaves.

135. Idem; *Namenverzeichnis.* Gotha, 1925. 315 pp.

136. *Times* [of London] *Atlas of the World,* 5th ed. New York, 1975. 123 plates+223 pp.

137. Touring Club Italiano. *Atlante Internazionale.* Milan, 1968. 173 leaves.

138. Idem; *Indice dei Nomi.* Milan, 1968. 1032 pp.

139. US, Geological Survey. *National Atlas of the United States of America.* Washington, 1970. 417 pp.

Geography: Universal

140. Agostini, Federico de. *Enciclopedia Geografica; Imago Mundi.* Turin, 1970–1975. 13 v.

141. Vivien de Saint-Martin, [Louis], and Louis Rousselet. *(Nouveau) Dictionnaire de Geographie Universelle.* Paris, [1879–1895]. 7 v.

142. *Worldmark Encyclopedia of the Nations,* 5th ed. New York, 1976. 5 v.

[1] By 1977, more than 130 separate gazetteers in this series had been published, and many of these had been revised. From March 1968 to July 1972 they were prepared by the Geographic Names Division of the Army Map Service, which in January 1969 became the Army Topographic Command. Since July 1972, however, the gazetteers have been prepared by the Topographic Center of the Defense Mapping Agency.

Geography: National and Regional

143. *Anuario Geografico Argentino.* Buenos Aires, 1941. 650 pp.

144. Arango Cano, Jesus. *Geografia Fisica y Economica de Colombia.* Bogota, 1956. 338 pp.

145. *Argentina, Suma de Geografia.* Buenos Aires, 1958–1963. 9 v.

146. Brazil, Conselho Nacional de Geografia. *Geografia do Brasil.* Rio de Janeiro, 1959–1960. v 1–2.

147. Canada, Energy, Mines and Resources Canada. *Facts from Canadian Maps.* Ottawa, 1974. 62 pp.

148. Federal Writers' Project (or Writers' Program). *Alabama* [etc] (*American Guide Series*) [guide-books to individual US states]. Various places, 1937–1949. 50 v.

149. India, Ministry of Information and Broadcasting. *India, a Reference Annual, 1976.* New Delhi, 1976. 486 pp.

150. Nugroho. *Indonesia, Facts and Figures.* Jakarta, 1967. 608 pp.

151. (*New*) *Official Guide: Japan.* Tokyo, 1975. 1088 pp.

152. Philippines, Bureau of Coast and Geodetic Survey. *Geographical Data of the Philippines.* Manila, 1962. 22 pp.

153. Sociedade Geographia do Rio de Janeiro. *Geographia do Brasil.* Rio de Janeiro, [1922–1923]. v 1–2, 10.

154. Spate, O. H. K., and A. T. A. Learmonth. *India and Pakistan.* London, 1967. 877 pp.

155. UK, Central Office of Information. *Britain 1977, an Official Handbook.* London, 1977. 509 pp.

156. Vila, Pablo. *Geografia de Venezuela.* Caracas, 1960–. v 1 (454 pp).

157. Zaychikov, V. T., Ed. *Fizicheskaya Geografiya Kitaya* (*Physical Geography of China*). Moscow, 1964. 650 pp. (US Joint Publications Research Service, number 32,119).

Physiography: Oceans and Seas

158. Fairbridge, Rhodes W., Ed. *Encyclopedia of Geomorphology.* New York, 1968. 1295 pp.

159. Fairbridge, Rhodes W., Ed. *Encyclopedia of Oceanography.* New York, 1966. 1021 pp.

160. International Hydrographic Bureau. *Limits of Oceans and Seas* (Special Publication number 23). Monte Carlo, Monaco, [1953]. 35 pp.

161. Menard, H. W., and Stuart M. Smith. "Hypsometry of Ocean Basin Provinces," *Journal of Geophysical Research,* v 71, pp 4305–4325, 15 Sep 1966.

Physiography: Islands

162. Huxley, Anthony, Ed. *Standard Encyclopedia of the World's Oceans and Islands.* New York, 1962. 383 pp.

163. US, National Ocean Survey. *America's Islands.* Rockville, Maryland, USA, 1974. 31 pp.

Physiography: Rivers

164. Akademiya Nauk SSSR, Institut Geografii. *Ocherki po Gidrografii Rek SSSR.* Moscow, 1953. 323 pp.

165. Grande, Jose Carlos P. "O Maior Rio do Mundo," *Boletim Geografico* [Rio de Janeiro], v 13, pp 183–192, Mar–Apr 1955.

166. Gresswell, R. Kay, and Anthony Huxley, Ed. *Standard Encyclopedia of the World's Rivers and Lakes.* New York, 1965. 384 pp.

167. Iseri, Kathleen T., and W. B. Langbein. *Large Rivers of the United States* (US Geological Survey Circular number 686). Washington, 1974. 10 pp.

168. Parde, Maurice, and Roy E. Oltman. "Nouvelles Donnees Experimentales et Evaluations sur les Debits de l'Amazone," *Comptes Rendus . . . de l'Academie des Sciences,* v 264 (series D), pp 1401–1406, 13 Mar 1967.

169. Voskresenskiy, K. P. *Norma i Izmenchivost Godovogo Stoka Rek Sovetskogo Soyuza.* Leningrad, 1962. 546 pp.

Physiography: Mountains

170. *Alpinistes Celebres.* Paris, 1956. 416 pp.

171. Bolinder, Anders, and G. O. Dyhrenfurth. "List of the World's Known Peaks of Over 7,400 Metres (24,280 Feet)," *Mountain World, 1964-1965,* [London, 1966], pp 196-199.

172. Burrard, S. G., et al. *Sketch of the Geography and Geology of the Himalaya Mountains and Tibet.* Delhi, 1933. 359+32 pp.

173. Echevarria C., Evelio. "Survey of Andean Ascents," *American Alpine Journal,* v 13, pp 155-192, 425-452, 1962-1963.

174. Filippi, Filippo de. *Ruwenzori.* London, 1909. 407 pp.

175. Freshfield, Douglas William. *Exploration of the Caucasus,* 2nd ed. London, 1902. 2 v.

176. Frison-Roche, Roger. *Montagnes de la Terre.* Paris, 1964. 2 v.

177. Gansser, Augusto. *Geology of the Himalayas.* London, 1964. 289 pp.

178. Huxley, Anthony, Ed. *Standard Encyclopedia of the World's Mountains.* New York, 1962. 383 pp.

179. International Volcanological Association, Ed. *Catalogue of the Active Volcanoes of the World Including Solfatara Fields.* Naples, 1951-1966. 19 parts.

180. *K Vershinam Sovetskoy Zemli.* Moscow, 1949. 575 pp.

181. Noyce, Wilfrid, and Ian McMorrin. *World Atlas of Mountaineering.* London, 1969. 224 pp.

182. Spencer, Sydney, Ed. *Mountaineering.* London, [1934]. 383 pp.

183. Thomas, Lowell. . . . *Book of the High Mountains.* New York, 1964. 512 pp.

Physiography: Lakes

184. Bue, Conrad D. *Principal Lakes of the United States* (US Geological Survey Circular number 476). Washington, 1963. 22 pp.

185. Canada, Environment Canada. *Canadian Survey on the Water Balance of Lakes.* Ottawa, [1974?]. 92 pp.

186. Halbfass, Wilhelm. *Seen der Erde.* Gotha, 1922. 169 pp.

Physiography: Waterfalls

187. Rashleigh, Edward C. *Among the Waterfalls of the World.* London, 1935. 288 pp.

Demography: Universal

188. *Bevolkerung der Erde* (in *Petermanns Geographische Mitteilungen*). Gotha, 1872-1931. 14 v.

189. Chandler, Tertius, and Gerald Fox. *3000 Years of Urban Growth.* New York, 1974. 431 pp.

190. Hassel, G. *Statistische Uebersichts-Tabellen der Sammtlichen Europaischen und Einiger Ausser-europaischen Staaten.* Gottingen, 1809. 106 pp.

191. *Statistique Internationale des Grandes Villes . . . 1931.* Hague, 1931. 708+26 pp.

192. United Nations, Statistical Office. *Demographic Yearbook, 1948-1976.* New York, 1949-1977. 28 v.

Demography: National and Regional

193. Beloch, Karl Julius. *Bevolkerungsgeschichte Italiens.* Berlin, 1937-1961. 3 v.

194. Sanchez-Albornoz, Nicolas. *Population of Latin America; a History.* Berkeley, 1974. 431 pp.

195. Yazaki, Takeo. *Social Change and the City in Japan.* Tokyo, 1968. 549 pp.

Climate: Universal

196. Sokhrina, R. F., et al. *Davieniye Vezdukha, Temperatura Vezdukha i Atmosfernyye Osadki Severnogo Polushariya.* Leningrad, 1959. 473 pp.

197. UK, Meteorological Office. *Tables of Temperature, Relative Humidity and Precipitation for the World*. London, 1958. 6 v.

198. World Meteorological Organization. *Climatological Normals (Clino) for Climat and Climat Ship Stations for the Period 1931-1960*. Geneva, 1962. unp.

199. *World Weather Records, 1921-1960*. Washington, 1934-1968. 9 v.

Climate: National and Regional

200. Argentina, Servicio Meteorologico Nacional. *Estadisticas Climatologicas, 1901-1950*. Buenos Aires, 1958. 44 pp.

201. Argentina, Servicio Meteorologico Nacional. *Estadisticas Climatologicas, 1941-1950*. Buenos Aires, 1958. 161 pp.

202. Canada, Atmospheric Environment Services. *Temperature and Precipitation, 1941-1970*. Toronto, ND. 6 v.

203. China, Office of Climatological Research. *Chungkuo Chihou Tu (Atlas of Chinese Climatology)*. Peking, 1960. 601 pp. (US Joint Publications Research Service, number 16,321).

204. Czechoslovakia, Vydava Hydrometeorologicky Ustay. *Podnebi Ceskoslovenske Socialisticke Republiky–Tabulky*. Prague, 1961. 379 pp.

205. Gherzi, E. *Meteorology of China*. Macao, 1951. 2 v.

206. India, Meteorological Department, *Climatological Tables of Observatories (1931-1960)*. [New Delhi, 1967]. 470 pp.

207. Japan, Kishocho. *Climatic Table of Japan . . . 1941-1970*. Tokyo, 1972. Parts 2-3.

208. *Klimaticheskiy Spravochnik Afriki*. Leningrad, 1962. 2 v.

209. *Klimaticheskiy Spravochnik Zarubezhnoy Azii*. Leningrad, 1974. 2 v.

210. Leningrad, Glavnaya Geofizicheskaya Observatoriya. *Klimat SSSR*. Leningrad, 1958-1963 8 v.

211. *Spravochnik po Klimatu SSSR*. Leningrad, 1964-. 34 v.

212. Taesler, Roger. *Klimatdata for Sverige*. Stockholm, 1972. 672 pp.

213. UK, Meteorological Office. *Averages of Rainfall for Great Britain and Northern Ireland, 1916-1950*. London, 1958. 36 pp.

214. UK, Meteorological Office, *Averages of Temperature for Great Britain and Northern Ireland, 1921-50*. London, 1953. 36 pp.

215. US, Environmental Data Service. *Monthly Normals of Temperature, Precipitation, and Heating and Cooling Degree Days 1941-70* [by states]. Asheville, 1973. unp.

216. Vivo, Jorge A., and Jose C. Gomez. *Climatologia de Mexico*. Mexico, 1946. 11+73 pp.

Energy

217. United Nations, Statistical Office. *World Energy Supplies, 1971-1975* (Statistical Papers, series J, number 20). New York, 1977. 229 pp.

Productivity

218. World Bank. *World Bank Atlas*, 12th ed. Washington, 1977. 32 pp.

Civil Engineering: Bridges

219. Gies, Joseph. *Bridges and Men*. New York, 1963. 343 pp.

220. Matsuzaki, Yoshimaro. "Honshu-Shikoku Bridge Project," *Civil Engineering in Japan, 1973*, pp 27-45.

221. Smith, H. Shirley. *World's Great Bridges*. New York, 1964. 250 pp.

222. Virola, Juhani. "World's Greatest Bridges," *Civil Engineering* [New York], v 38, pp 52-55, Oct 1968.

223. Wittfoht, Hans. *Triumph der Spannweiten*. Dusseldorf, 1972. 314 pp.

Civil Engineering: Tunnels

224. Pequignot, G. A. *Tunnels and Tunnelling.* London, 1963. 555 pp.

Civil Engineering: Dams

225. International Commission on Large Dams. *World Register of Dams.* Paris, ND. 4 v.

Universities: Universal

226. *International Handbook of Universities,* 6th ed. Paris, 1974. 1326 pp.
227. *Minerva; Jahrbuch der Gelehrten Welt.* Berlin, 1966–1969. 3 v.
228. *World Guide to Universities; Internationales Universitats-Handbuch.* New York, 1971–1972. 4 v.
229. *World of Learning, 1977–78.* London, 1977. 2 v (2036 pp.)

Universities: National and Regional

230. *American Universities and Colleges,* 11th ed. Washington, 1973. 1879 pp.
231. Association of International Education, Japan, Ed. *Colleges and Universities in Japan, 1975.* Tokyo, 1975. 50 pp.
232. *Commonwealth Universities Yearbook, 1976.* London, 1976. 4 v (2467 pp)).
233. Japan Society for the Promotion of Science. *Directory of Colleges and Universities in Japan, 1972.* Tokyo, 1972. 193 pp.
234. *Universities and Colleges of Canada, 1975.* Ottawa, 1975. 583 pp.
235. *Universities Handbook, India, 1975.* New Delhi, 1975. 1035 pp.
236. US, National Center for Education Statistics. *Education Directory, Colleges and Universities, 1976–77.* Washington, 1977. 533 pp.
237. US, National Center for Education Statistics. *Fall Enrollment in Higher Education, 1975.* Washington, 1976. 445 pp.

Libraries: Universal

238. *International Library Directory,* 3rd ed. London, 1968. 1222 pp.
239. *World Guide to Libraries; Internationales Bibliotheks-Handbuch,* 4th ed. New York, 1974. 2 v (1603 pp).

Libraries: National and Regional

240. *Adresar Ustrednich Knihoven Siti, Statnich Vedeckych Knihoven, Lidovych Knihoven a Vysokoskolskych Knihoven v CSR.* Prague, 1973. 307 pp.
241. *All India Educational Directory.* Chandigarh, 1972. 1262 pp.
242. *American Library Directory,* 30th ed. New York, 1976. 1389 pp.
243. Beirens, Gerard. *Bibliotheekgids van Belgie.* Brussels, 1974. 9 v.
244. *Biblioteki SSSR . . . Spravochnik.* Moscow, 1973–1974. 2 v.
245. *Bibliotheek- en Documentatiegids voor Nederland, Suriname en de Nederlandse Antillen.* Hague, 1966. 442 pp.
246. Brazil, Instituto Nacional do Livro. *Guia das Bibliotecas Brasileiras,* 4th ed. Rio de Janeiro, 1969. 502 pp.
247. Dayrit, Marina G., et al. *Directory of Libraries in the Philippines.* Quezon City, 1973. 131 pp.
248. France, Direction des Bibliotheques et de la Lecture Publique. *Repertoire des Bibliotheques et Organismes de Documentation.* Paris, 1971. 733 pp.
249. Giuffra, Carlos Alberto. *Guia de Bibliotecas Argentinas.* NP, 1967. 334 pp.

250. Italy, Direzione Generale delle Academie e Biblioteche. *Annuario delle Biblioteche Italiane.* Rome, [1969]–. v 1–3: A–Rol.

251. Idem. Rome, 1956–[1960]. 3 v.

252. *Jahrbuch der Bibliotheken, Archive und Informationsstellen der Deutschen Demokratischen Republik,* 8th ed. Leipzig, 1975. 563 pp.

253. *Jahrbuch der Deutschen Bibliotheken,* 46th ed. Wiesbaden, 1975. 562 pp.

254. Kelly, Thomas. *History of Public Libraries in Great Britain, 1845–1975.* London, 1977. 582 pp.

255. Klimowiczowa, Irena, and Ewa Suchodolska. *Informator o Bibliotekach i Osrodkach Informacji Naukowej w Polsce.* Warsaw, 1973. 557 pp.

256. *Libraries, Museums and Art Galleries Year Book, 1976.* Cambridge, UK, 1976. 254 pp.

257. Mexico, Departamento de Bibliotecas. *Directorio de Bibliotecas de la Republica Mexicana,* 4th ed. [Mexico], 1970. 285 pp.

258. Moldoveanu, Valeriu, et al. *Ghidul Bibliotecilor din Romania.* Bucharest, 1970. 475 pp.

259. Pan American Union, Columbus Memorial Library. *Guia de Bibliotecas de la America Latina.* Washington, 1963. 165 pp.

260. Pretoria, State Library. *Handbook of Southern African Libraries.* Pretoria, 1970. 939 pp.

INDEX

Alphabetizing is letter-by-letter, in accordance with the English alphabet, without regard for spaces, punctuation marks, or foreign combination letters. For further explanation, see under Alphabetization on p. 15 in the Introduction. Note also that Tables 7a through 8h (pp. 185-253) are not indexed *in detail,* since, for individual countries and cities, all the data presented in these comparative tables are also given in Tables 10a through 11c (pp. 308-678).

Ivai, R: 98
Ivankovo, dam: 290
Ivanovo, city: 433, 544, 635
Ivanovo-Voznesensk, city, see
  Ivanovo
Iviza, city, see Ibiza, city
Iviza, I, see Ibiza, I
Ivory Coast, ctry: 309, 322,
  334, 349
Iwaki, city: 387, 505, 597
Iwo, city: 368, 489, 583
Ixtacihuatl, mt: 134
Iyawadi, R, see Irrawaddy
Izabal, L: 162, 173
Izhevsk, city: 433, 544, 635
Izhevskiy Zavod, city, see
  Izhevsk
Izhma, R: 79
Izmir, city: 396, 512, 605
Iztaccihuatl, mt, see Ixtacihu-
  atl

Jabalpur, city: 381, 499, 591
Jabbul, L: 174
Jaboatao, city: 478, 575, 675
Jackson, city (Michigan, USA):
  456, 561, 656
Jackson, city (Mississippi,
  USA): 456, 561, 656
Jackson, mt: 123
Jackson, R, see James, R
  (Virginia, USA)
Jacksonville, city: 456, 561,
  656
Jacques Cartier, bridge: 270
Jacques-Cartier, city, see
  Longueuil
Jaffa, city, see Tel Aviv-Jaffa
Jahangirnagar, city, see Dacca
Jailolo, city: 28
Jailolo, I, see Halmahera
Jaipur, city: 381, 499, 591
Jakarta, city: 28, 351, 384,
  502, 594
Jalapa, city: 443, 552, 645
Jalapa Enriquez, city, see
  Jalapa
Jamaica, ctry: 317, 329, 343,
  355
Jamaica, I: 42, 57
Jambi, city: 384, 502, 594
Jamdena, I, see Yamdena
James, I, see San Salvador, I
  (Ecuador)
James, R (South Dakota,
  USA): 83
James, R (Virginia, USA): 87
Jamestown, city: 27, 350

Jammu, city: 351, 391, 508,
  600
Jammu and Kashmir, ctry,
  see Kashmir-Jammu
Jamnagar, city: 381, 499, 591
Jamshedpur, city: 381, 499,
  591
Jan Mayen, I: 38
Jan Pieterszoon Coen, mt, see
  Wisnumurti
Japan, ctry: 312, 324, 337,
  351
Japan, sea: 22
Japan, National Diet Library,
  lib: 301
Japen, I, see Yapen
Japura, R, see Caqueta
Jarbah, I, see Djerba
Jari, dam: 288
Jarmaq, mt, see Meron
Jassy, city, see Iasi
Jastrzebie Zdroj, city: 419,
  531, 622
Java, I: 28
Java, sea: 22
Javari, R: 97
Jawa, I, see Java, I
Jawa, sea, see Java, sea
Jaya, mt: 136, 144
Jayawijaya, mts: 136
Jazair, city, see Algiers
Jazair, ctry, see Algeria
Jedda, city, see Jidda
Jefferson, R, see Mississippi
Jefferson City, city: 456, 561,
  656-657
Jena, city: 407, 522, 613
Jeonju, city, see Chonju
Jequitinhonha, R: 97
Jerez, city, see Jerez de la
  Frontera
Jerez de la Frontera, city:
  422, 534, 625
Jersey, ctry, see Channel
  Islands
Jersey, I: 39
Jersey City, city: 456, 561,
  657
Jerusalem, city: 351, 386,
  503, 596
Jesselton, city, see Kota
  Kinabalu
Jesus, I: 46
Jhansi, city: 381, 499, 591
Jhelum, R: 72
Jibuti, city, see Djibouti, city
Jibuti, ctry, see Djibouti, ctry
Jidda, city: 352, 394, 511, 603

Jinja-Njeru, city: 350
Jinsen, city, see Inchon
Jizah, city, see Giza
Joao Pessoa, city: 478, 575,
  675
Jodhpur, city: 381, 499, 592
Jog, wf, see Gersoppa
Jogjakarta, city, see
  Yogyakarta
Johannesburg, city: 350, 370,
  490, 584
John Day, R: 91
John Hancock Center, bldg:
  257
John Hancock Tower, bldg:
  257
John Rylands University
  Library of Manchester, lib:
  304
Johnson, Daniel, dam: 283
Johnston, mt: 120
Johnston and Sand Islands,
  ctry: 318, 331, 345, 356
Johnstone, city, see Renfrew
Johnstown, city: 456, 561,
  657
Joliet, city: 456, 561, 657
Jolla, La, city, see San Diego
Jonkoping, city: 423, 536,
  626
Jonquiere, city: 439, 549, 641
Jordan, ctry: 312, 324, 337,
  351
Jordan, R: 74, 115
Jordanian Jerusalem, city, see
  Jerusalem
Joseph Regenstein Library of
  the University of Chicago,
  lib: 302
Joshin, city, see Kimchaek
Jotunheimen, mts: 132, 133,
  146
Juarez, city, see Ciudad Juarez
Juba, R: 61
Jubbulpore, city, see Jabalpur
Jucar, R: 79
Judaea and Samaria, ctry, see
  West Bank
Juddah, city, see Jidda
Jugoslavia, ctry, see Yugo-
  slavia
Jugoslavija, ctry, see Yugo-
  slavia
Juiz de Fora, city: 479, 575,
  675
Juliana, mt, see Mandala
Juliet, city, see Joliet
Jullundur, city: 381, 500, 592

lib: 301
Leipzig, Universitatsbiblio-
thek, lib: 303
Leith, city, *see* Edinburgh,
city (UK)
Leman, L, *see* Geneva, L
Le Mans, city: 403, 520, 611
Lemberg, city, *see* Lvov
Lemesos, city, *see* Limassol
Lemnos, I: 37
Lempa, R: 93, 114
Lena, R: 66, 101, 105, 109
Lenin, dam: 288
Lenina, mt: 127
Leninakan, city: 435, 545,
636
Leningrad, city: 354, 435, 545,
636
Leningrad A. A. Zhdanov
State University, M. Gorkiy
Scientific Library of the,
lib: 303
Leningrad, M. E. Saltykov-
Shchedrin State Public
Library, lib: 301
Lenin State Library of the
USSR, lib: 301
Lenin State Library of the
White Russian SSR, lib: 303
Lens, city: 403, 520, 611
Leon, city (Mexico): 443, 552,
645
Leon, city (Nicaragua): 355
Leon, city (Spain): 422, 534,
625
Leon de los Aldamas, city, *see*
Leon, city (Mexico)
Leoni, Raul, dam: 285, 288
Leopold II, L, *see* Mai
Ndombe
Leopoldville, city, *see* Kinshasa
Lerida, city: 422, 535, 625
Lerma, city, *see* Salta
Lerma, R, *see* Santiago, R
Lerwick, city: 36
Lesatima, mt: 121
Lesbos, I: 35
Les Escaldes, city: 353
Lesina, city, *see* Hvar, city
Lesina, I, *see* Hvar, I
Lesotho, ctry: 309, 322, 334,
349
Lesser Slave, L: 160
Lesvos, I, *see* Lesbos
Leucas, city: 38
Leucas, I: 38
Leuven, city, *see* Louvain
Leverkusen, bridge: 272

Leverkusen, city: 410, 524,
615
Levkas, city, *see* Leucas, city
Levkas, I, *see* Leucas, I
Levkosia, city, *see* Nicosia
Lewes, R, *see* Yukon
Lewiston, city: 458, 562, 658
Lewiston-Queenston, bridge:
271
Lewis with Harris, I: 35
Lexington, city: 458, 562, 658
Leyden, city, *see* Leiden
Leyland, city, *see* South
Ribble
Leyte, I: 30
Lhasa, city: 375, 495, 587
Lhotse, mt: 123
Lhotse Shar, mt: 123
Liao, R: 68
Liaoyang, city: 375, 495, 587
Liaoyuan, city: 375, 495, 587
Liard, R: 83
Liberia, ctry: 309, 322, 334,
349
Libertador, mt: 139
Liberty, Statue of, bldg: 262
Libiyah, ctry, *see* Libya
Libraries: 10, 298-304, 360,
579-678
Libreville, city: 349, 365, 486,
580
Libya, ctry: 309, 322, 334,
349
Licheng, city, *see* Tsinan
Lichfield, city: 428, 539, 630
Licking, R: 89
Liechtenstein, ctry: 315, 327,
340, 353
Liege, city: 400, 517, 609
Lienyunkang, city, *see* Sin-
hailien
Lierasen, tunnel: 277
Life Expectancies: 6, 205-208,
333-348
Lifou, I: 50
Lifu, I, *see* Lifou
Lihsien, R, *see* Black, R
(Vietnam)
Lille, city: 403, 520, 611
Lillebaelt, bridge, *see* Little
Belt
Lille, Universite de, univ: 297
Lilongwe, city: 350, 367, 488,
581
Lima, city (Peru): 357, 483,
578, 678
Lima, city (USA): 458, 562,
658

Limassol, city: 351
Limbang, R: 113
Limbe, city, *see* Blantyre
Limerick, city: 413, 527, 618
Limin, city, *see* Thasos, city
Limin Vatheos, city, *see*
Samos, city
Limnos, I, *see* Lemnos
Limoges, city: 403, 520, 611
Limpopo, R: 61
Lincoln, bldg: 259
Lincoln, city (UK): 428, 539,
630
Lincoln, city (USA): 458,
563, 658
Lincoln, mt: 135
Lincoln I, tunnel: 279
Lincoln II, tunnel: 279
Lincoln III, tunnel: 279
Linh, mt: 128, 147
Linkoping, city: 423, 536, 626
Linyanti, R, *see* Chobe
Linz, city: 399, 516, 608
Lions Gate, bridge: 268
Lipetsk, city: 435, 545, 636
Lisboa, city, *see* Lisbon
Lisbon, city: 353, 420, 533,
623
Little Belt, bridge: 267
Little Colorado, R: 91
Little Current, city: 43
Little Mecatina, R, *see* Petit
Mecatina
Little Missouri, R: 84
Little Rock, city: 458, 563,
658
Liu, R: 73
Liuchia, dam: 283
Liuchou, city, *see* Liuchow
Liuchow, city: 375, 495, 587
Liverpool, city: 428, 539, 630
Liverpool Public Library, lib:
304
Livonia, city: 458, 563, 658
Livorno, city, *see* Leghorn
Ljubljana, city: 438, 548, 640
Llanquihue, L: 165, 167
Llullaillaco, mt: 139
Loa, mt: 137
Loa, R: 99, 113
Loanda, city, *see* Luanda
Lob, L, *see* Lop
Lobito, city: 349
Localities, *see* Cities
Lodz, city: 353, 419, 532,
622
Lofoi, wf, *see* Kaloba
Logan, mt: 133, 134, 135, 143